Sources and Studies
in the History of Mathematics and
Physical Sciences

Springer
New York
Berlin
Heidelberg
Barcelona
Hong Kong
London
Milan
Paris
Singapore
Tokyo

Sources and Studies in the History of Mathematics and Physical Sciences

K. Andersen
Brook Taylor's Work on Linear Perspective

J. Cannon/S. Dostrovsky
The Evolution of Dynamics: Vibration Theory from 1687 to 1742

B. Chandler/W. Magnus
The History of Combinatorial Group Theory

A.I. Dale
A History of Inverse Probability: From Thomas Bayes to Karl Pearson, Second Edition

A.I. Dale
Pierre-Simon Laplace, Philosophical Essay on Probabilities, Translated from the fifth French edition of 1825, with Notes by the Translator

P.J. Federico
Descartes on Polyhedra: A Study of the *De Solidorum Elementis*

B.R. Goldstein
The Astronomy of Levi ben Gerson (1288–1344)

H.H. Goldstine
A History of Numerical Analysis from the 16th through the 19th Century

H.H. Goldstine
A History of the Calculus of Variations from the 17th through the 19th Century

G. Graßhoff
The History of Ptolemy's Star Catalogue

A.W. Grootendorst
Jan de Witt's *Elementa Curvarum Linearum, Liber Primus*

T. Hawkins
Emergence of the Theory of Lie Groups: An Essay in the History of Mathematics 1869–1926

A. Hermann, K. von Meyenn, V.F. Weisskopf (Eds.)
Wolfgang Pauli: Scientific Correspondence I: 1919–1929

C.C. Heyde/E. Seneta
I.J. Bienaymé: Statistical Theory Anticipated

J.P. Hogendijk
Ibn Al-Haytham's *Completion of the Conics*

A. Jones (Ed.)
Pappus of Alexandria, Book 7 of the *Collection*

Continued after Index

Thomas Hawkins

Emergence of the Theory of Lie Groups

An Essay in the History of Mathematics 1869–1926

With 50 Illustrations

Springer

Thomas Hawkins
Department of Mathematics
Boston University
111 Cummington Street
Boston, MA 02215
USA
twh@math.bu.edu

Sources and Studies Editor:
Gerald J. Toomer
2800 South Ocean Boulevard, 21F
Boca Raton, FL 33432
USA

Library of Congress Cataloging-in-Publication Data
Hawkins, Thomas, 1938–
Emergence of the theory of Lie groups : an essay in the history of mathematics, 1869–1926 / Thomas Hawkins.
p. cm. — (Sources and studies in the history of mathematics and physical sciences)
Includes bibliographical references and index.
ISBN 0-387-98963-3 (hardcover : alk. paper)
1. Lie groups—History. I. Title. II. Series.
QA387 .H39 2000
512′.55—dc21 99-056073

Printed on acid-free paper.

Production managed by Robert Bruni; manufacturing supervised by Erica Bresler.
Photocomposed copy prepared from the author's TEX files.
Printed and bound by Maple-Vail Book Manufacturing Group, York, PA.
Printed in the United States of America.

9 8 7 6 5 4 3 2 1

ISBN 0-387-98963-3 Springer-Verlag New York Berlin Heidelberg SPIN 10750754

Preface

This book is both more and less than a history of the theory of Lie groups during the period 1869–1926. No attempt has been made to provide an exhaustive treatment of all aspects of the theory. Instead, I have focused upon its origins and upon the subsequent development of its structural aspects, particularly the structure and representation of semisimple groups. In dealing with this more limited subject matter, considerable emphasis has been placed upon the motivation behind the mathematics. This has meant paying close attention to the historical context: the mathematical or physical considerations that motivate or inform the work of a particular mathematician as well as the disciplinary ideals of a mathematical school that encourage research in certain directions. As a result, readers will obtain in the ensuing pages glimpses of and, I hope, the flavor of many areas of nineteenth and early twentieth century geometry, algebra, and analysis. They will also encounter many of the mathematicians of the period, including quite a few not directly connected with Lie groups, and will become acquainted with some of the major mathematical schools. In this sense, the book is more than a history of the theory of Lie groups. It provides a different perspective on the history of mathematics between, roughly, 1869 and 1926. Hence the subtitle.

I have divided the book into four parts, each bearing the name of a mathematician. The cast of characters of each part is large, but in each part one mathematician stands out as the central figure. The first part is devoted to the geometrical and analytical origins of the theory of continuous transformation groups of Sophus Lie (1849–1899)—the precursor of the modern theory of Lie groups—and to some of the features of the theory he went on to develop that are important in what is to follow. In Part II the central figure is Wilhelm Killing (1847–1923), whose interest in the foundations of non-Euclidean geometry led him down a path of research that culminated in his discovery of almost all the central concepts and theorems on the structure and classification of semisimple Lie algebras. Part III is primarily concerned with developments that would now be interpreted as bearing on the representation of Lie algebras, particularly simple and semisimple algebras. Élie Cartan (1869–1951), who began his career by sifting through Killing's theorems and providing many of the correct ones with rigorous proofs, stands out as the principal figure. In Part IV that role is played by Hermann Weyl (1885–1955), whose contributions to the structure and representation of Lie groups served to bring the earlier developments into something of a completed whole, while at the same time opening up significant new avenues for research. Readers daunted by the size of the entire book should note that each part constitutes a fairly self-contained monograph that can be read independently of the others.

It is my hope that this book will find an audience among students of mathematics in addition to mathematicians, physicists, and historians of mathematics. My experience is that the understanding of a theory is deepened by familiarity both with the considerations that motivated various developments and with the less formal, more intuitive manner in which they were initially conceived. Hence in addition to stressing the motivational, historical context of the mathematics, I have tried to expound the original mathematical conceptions and reasoning clearly and fairly extensively. There are many fine texts devoted to a systematic presentation of the theory as it stands today—too many to list here. Texts from which I have profited include the books by Humphreys, Olver, Samelson, Serre, and Varadarajarian included in the list of references.

In composing this first book-length study of the history of Lie groups, I have benefited from the historical writings of many others, as is evident from the citations that follow. Features of this study that I believe are original to me include, in Parts I–II: (1) A detailed analysis of the geometrical and analytical considerations that combined to inspire and inform the creation of Lie's theory, including as well an assessment of the role of Felix Klein (Chapters 1–2). (2) The first and only careful analysis and assessment of Killing's papers on Lie algebras and a depiction of the human side of his efforts based upon his extensive correspondence with F. Engel (Chapter 5, Section 6.2). (3) A study of the historical context of Killing's seminal work in foundations of geometry that takes into consideration the combined effect upon Killing of the contemporaneous developments in the foundations of geometry and his educational background as Weierstrass's student (Chapter 4).

In Part III I would mention: (4) A portrayal of Cartan's theory of weights as a gradual but brilliant revamping of Killing's faulty theory of secondary roots (Sections 6.3–6.4, Sections 8.2 and 8.4,). (5) A study of the disciplinary ideals of the Paris community of mathematicians at the turn of the century and a discussion within that context of the directions taken by Cartan's research (Chapters 6 and 8). In addition to providing an important part of the historical context of Cartan's work, it makes understandable the twenty-year hiatus between Cartan's thesis work, where we already find the beginnings of his theory of weights and representations, and the memoirs of 1913–14, where the theory is developed fully. (6) An examination of work within Lie's school by mathematicians other than Cartan on what would now be described as dealing with the representation of Lie algebras (Chapter 7). This work is of interest in its own right as an indication of where, to borrow a Kuhnian phrase, the "normal" mathematics within Lie's school was leading. Among other things we find that during 1906–13 G. Kowalewski had devised his own theory of "weights" to deal with a problem equivalent to the one Cartan completely solved in 1913. The comparison of Cartan's work with the quite respectable normal mathematics of Lie's school provides the basis for a genuinely historical appreciation of the magnitude of his genius.

Finally, in Part IV: (7) By his own admission, Weyl was greatly influenced by Hilbert and the climate he fostered at Göttingen. In Chapter 9, drawing upon the extensive researches of other historians, I have attempted to paint a picture of Hilbert's Göttingen that indicates the extent to which Weyl's work on the representation of Lie groups bears the imprint of his experiences there. (8) A careful analysis of Weyl's preoccupation with the mathematics of general relativity, which for the first time makes complete sense out of his remark that "the wish to understand what really is the mathematical substance behind the formal apparatus of relativity theory led me to the study of representations and invariants of groups" (Chapter 11).

The book includes a substantial index, which I hope will prove useful. In addition to the usual names of individuals and of mathematical terms, the index lists themes that run through the book, e.g., "conceptual thinking," "generic reasoning," and "local vs. global viewpoint." I have also given names to important theorems so as to facilitate referring to them in the text in a way that will help the reader to recognize them, but if, e.g., the exact statement of what I call Lie's exponential map theorem is desired, the page reference will be found in the index. In addition to theorems, I have also given names to mathematical notions that are not part of the present-day vocabulary (e.g., "Hurwitz integral," "Conditions I–III" of Killing). These terms are also included in the index, and in the case of multiple page references, the one containing the definition is underlined.

In referring to publications listed in the References section, I have followed the standard practice among historians of mathematics. Thus a reference such as [Lie, 1876c:12]—or simply [1876c:12] when authorship is clear—refers to p. 12 of the publication listed under "Lie, S., 1876c" in the reference section. The year may be the year it was submitted or the year of publication. In the case of someone like Lie, I have used the dates of submission, since his publications were numerous and the submission dates give a better sense of the chronological progression of his mathematical ideas. Both the date of submission and the date of publication are indicated in the reference. For mathematicians whose collected works have been published, the page references generally refer to the original paginations when these are printed in the collected works. When the original pagination is not included in the collected works, I usually give the pagination of the collected works. A glance at the reference, which includes both original and collected works paginations, will make it clear which is being used. The key to the abbreviations used in the text for manuscript archives is given at the beginning of the References section.

More than twenty-five years have passed since I began the research that has made it possible for me to write this book. During that time many individuals and institutions have provided considerable encouragement and assistance. Among institutions I would mention first of all the National Science Foundation, which through its programs in the history of science provided me with vital financial support on several occasions. In particular, I am grateful to Ron Overman, who as program director was always at hand to facilitate my dealings with the NSF. The story that unfolds on the following pages has been enhanced throughout by the use of manuscript sources, and I am indebted to the many individuals at the archives in Göttingen, Olso, and Zürich (listed at the beginning of the References section) who promptly provided me with copies of requested manuscripts. I am especially indebted to the late Hermann Boerner and to Helga Bertram, of the Mathematisches Institut Giessen, for all they did to make the resources of the Engel archive available to me. Over the years my work on this book has been encouraged in special ways by three departments of mathematics. To my colleagues in the department at Boston University I am grateful for their patient and respectful support of such a long-term project. In particular, I am indebted to Andrew Lyasoff for generously sharing his expertise on TeX with me. I also wish to thank the departments at the University of Chicago and Yale University for the special encouragement they provided by periodically inviting me to talk about my latest findings.

Sizable portions of the research behind the book were done while on sabbatical leaves of absence granted by Boston University. I am grateful to the Department of History of Science at Harvard University for arranging for me to be a visiting scholar during 1980–81 and to A. Borel for arranging membership in the School of Mathematics of the Institute for Advanced Study in Princeton during 1988–89. A first draft of the entire book was written during 1996–97 while I was a resident fellow at the Dibner Institute for History of Science and Technology at M.I.T., and I wish to thank the directors and staff of the institute for providing me with a friendly, supportive environment that proved ideal for book-writing. Without that happy year of duty-free tranquillity I am not sure I ever would have managed to bring all the strands of my research together into a book.

During the many years of researching and writing this book, my attention was focused entirely upon the written word, without any thoughts about a visual component. It was only after the final chapter was written that I realized that it would be especially appropriate in a book such as this to provide the reader with portraits of many of the mathematicians who enter the story. I quickly discovered that locating and obtaining suitable photographs is a difficult, time-consuming task, and I am grateful to the many archives and individuals who assisted me in carrying it out. (The sources of all portraits are given in the References section at the end of the book.) In particular, I want to express my thanks to the following individuals, who far exceeded the expected level professional cooperation and cordiality in their generous efforts to help me: Henri Cartan, Walter Gautschi, Livia Giacardi, Bjorn Jahren, Yvonne Schezt, Arild Stubhaug, and Martha Weyl Thompson.

Finally, I would like to record my gratitude and indebtedness to the following mathematicians and historians who have helped me with their advice and expertise or simply with their encouragement as I labored on this project: Garrett Birkhoff, John Coleman, Leo Corry, Jean Dieudonné, Walter Feit, Jeremy Gray, Bob Hermann, J. E. Humphreys, N. Jacobson, George Mackey, Peter Olver, David Rowe, George Seligman, Jacob Towber, and B. L. van der Waerden. Above all, I am indebted to A. Borel and J.-P. Serre, who throughout the years have encouraged my project and have generously taken the time to read over and comment upon much of what has gone into the book. Except for myself, however, there is no one who has seen and approved everything that is in this final version; the finished product, with whatever faults and idiosyncrasies it may have, is mine alone.

Contents

Part II: Wilhelm Killing

Part III: Élie Cartan

Part IV: Hermann Weyl

Chapter 1
The Geometrical Origins of Lie's Theory

According to Lie, it was during the winter of 1873–74 that his theory of groups was born. The sense in which this is correct will be indicated in the following two chapters.

The discoveries of the winter of 1873–74 and the reason why they did indeed prove to be so decisive as to lead to the creation of the theory of continuous groups are related to Lie's prodigious research activity during the four-year period from the fall of 1869 to the fall of 1873. It was during this period that continuous groups of transformations and infinitesimal transformations entered his work through geometrical considerations and through his related interest in differential equations. And it was during this period that group-theoretic problems were first posed and that the mathematical machinery was developed that not only made the discoveries of the winter of 1873–74 technically possible but also made them assume in Lie's mind such a decisive significance that he henceforth channeled virtually all of his creative energies into the arduous task of building up a theory of continuous transformation groups. Mathematically speaking, one could say that the theory of transformation groups commenced in the winter of 1873–74, but from a historical point of view nothing could be further from the truth. The events of the winter of 1873–74 represented the culminating tip of a historical iceberg that involved Lie's research activity during the period 1869–73. The purpose of Chapters 1–2 is to present an analysis of that iceberg, to relate it to the the events of the winter of 1873–74, and to draw some conclusions about the relative significance of the manifold factors involved in the creation of Lie's theory.

Lie's mathematical activity in 1869–73 can be divided into two periods, which could be described loosely as the periods with and without Felix Klein (1849–1925). During the first three years of the above-mentioned four-year period, namely from the fall of 1869 to the fall of 1872, Lie worked in close contact with Klein. He saw him on a daily basis in Berlin, where they first met, from the end of October 1869 to end of February 1870, in Paris from the end of April to mid-July 1870, and in Göttingen and Erlangen during September and October 1872. During these periods they freely discussed and shared their mathematical ideas.[1] When they were apart, they continued to communicate through letters. It was during this period, especially the first two years, that geometry supplied most of the inspiration for their work. After the fall of 1872, Lie and Klein were separated both physically and also intellectually in the sense that their mathematical interests began to diverge significantly. It was during this period that Lie focused all his attention on the theory of n-variable partial differential equations. This theory was dominated, especially in Germany, by Jacobi's work. As will be seen in Chapter 2, Lie's work on n-variable partial differential equations was greatly influenced by the ideas and analytical notation of Jacobi and his followers. It was through Lie's creative response to Jacobi — a response shaped by his experiences during the "geometrical" years 1869–71 — that the mathematical machinery was developed that made the events of the winter of 1873–74 possible.

1.1. Tetrahedral Line Complexes

In 1865, when Lie completed his studies at the University of Christiania (now Oslo), he had no idea he was destined to become a mathematician. He had done well, but not brilliantly, in all subjects; he was toying with the idea of becoming an observational astronomer. He even gave lectures on the subject in the student union. He had a real talent for explaining the geometry of the heavens. To support himself financially while in this state of career indecision he gave private instruction in mathematics. In this connection, he began to read the geometrical works by Poncelet, Chasles, and, above all, Plücker. Inspired by his reading, he did some original mathematical research on the real representation of imaginary quantities in projective geometry, a portion of which was accepted for publication by one of the leading mathematics journals of the time—Crelle's journal in Berlin.[2] On the basis of this experience, he decided to devote himself to geometrical research, to become a mathematician. He was twenty-six years old.

On the basis of the display of mathematical creativity exhibited by Lie in his essay on the real representation of imaginaries, he was granted a stipend to leave Norway and to travel to various centers of mathematics.

[1] For further biographical background consult [Rowe 1989a].
[2] On this work see [Rowe 1989a:§3].

Figure 1. Sophus Lie circa 1870.

His first stop was Berlin, where the mathematical scene was dominated by Kummer and Kronecker and, above all, by Weierstrass. Berlin was one of the world's foremost centers of mathematics in 1869, and more will be said about it in Chapter 4. The mathematics cultivated in Berlin and the spirit in which it was done, however, did not appeal to Lie. He met another visitor there who felt as he did. He was Felix Klein, and he was just twenty years old—seven years younger than Lie—although he already had his doctorate. Klein had been Plücker's student, and after his mentor's untimely death in 1868, he had edited Plücker's lectures on line geometry for publication. Lie had in fact studied these lectures, and when they met in Berlin they were both actively engaged in line-geometric research. So Klein and Lie enjoyed each other's mathematical company. They were self-styled "synthesists" in the midst of analysts and arithmeticians.

Although Lie and Klein had much in common in terms of mathematical interests, their mathematical backgrounds and personalities were quite different. Klein had a relatively solid mathematical education behind him, having studied at Bonn under Plücker (1801–1868) and then at Göttingen, where Alfred Clebsch (1833–1872) had attracted a circle of students about him. As a mathematician, Klein enjoyed learning about the work of others, because he wanted to understand it from his own point of view and to place it within a more encompassing, conceptually unified picture. By contrast, Lie had a limited mathematical background and no doctorate yet. He tended to focus rather exclusively on developing his own idiosyncratically original ideas and became interested in the work of others only when it was clear that it was relevant to his own interests. This difference in personalities dictated the way they related to each other as mathematicians: Lie developed his ideas, explained them to Klein, who was interested in hearing about them; Klein reacted to Lie's ideas; Lie responded to Klein's reactions; and so on.

Line geometry and its connections with the development of projective geometry have been discussed in detail by Rowe [1989a:§1]. Here it will suffice to explain that line geometry is an approach to the study of geometrical objects based upon considering straight lines, rather than the customary points, as the basic elements of investigation. One then studies configurations comprised of lines, such as what Plücker called *line complexes.* These may be described in the elegant algebraic notation favored by Klein as follows. Let ℓ be a line in three-dimensional complex projective space, $\mathbb{P}^3(\mathbb{C})$. Let $x_1, \ldots, x_4$ and $y_1, \ldots, y_4$ denote the homogeneous coordinates of two points lying on ℓ and hence determining it. For $j \neq k$, let

$$p_{jk} = \begin{vmatrix} x_j & x_k \\ y_j & y_k \end{vmatrix} = x_j y_k - x_k y_j. \tag{1.1}$$

The six quantities $p_{12}, p_{13}, p_{14}, p_{23}, p_{42}, p_{34}$, may be taken as homogeneous coordinates of ℓ. They are related by the equation

$$\Omega = p_{12}p_{34} + p_{13}p_{42} + p_{14}p_{23} = 0.$$

Line complexes are sets of lines whose line coordinates p_{jk} satisfy a homogeneous polynomial equation (in addition to $\Omega = 0$). The degree of the complex is the degree of the equation. The tetrahedral line complexes that interested Lie when he met Klein in Berlin are second-degree complexes, which may be defined geometrically as follows. Let Δ denote a fixed tetrahedron, with faces determined by planes $\pi_1, \ldots, \pi_4$. Each line ℓ in $\mathbb{P}^3(\mathbb{C})$ meets Δ in four points. A tetrahedral line complex $\mathfrak{T}$ is the set of lines ℓ for which the cross ratio of these points is the same. Other mathematicians, such as Theodore von Reye (1838–1919), had considered the geometrical properties of tetrahedral complexes before Lie, but Lie's approach was totally original. The originality began with, and was informed by, his manner of characterizing tetrahedral complexes. He considered the totality $\mathfrak{G}$ of all projective transformations of space that fix the vertices of the tetrahedron Δ. Then for any fixed line ℓ_0, the set $\mathfrak{T}$ of all lines $T[\ell_0]$ such that $T \in \mathfrak{G}$

Figure 2. Klein (top) shared with Lie an interest in the line geometry of Julius Plücker (bottom).

is a tetrahedral complex. In effect, a tetrahedral complex for Lie was the orbit of some fixed line under the transformations of $\mathfrak{G}$. For example, if Δ is the tetrahedron with vertices at the origin and the points at infinity on the coordinate axes, then the $T \in \mathfrak{G}$ are given in Cartesian coordinates by

$$(1.2) \qquad x' = \lambda x, \quad y' = \mu y, \quad z' = \nu z, \quad \lambda\mu\nu \neq 0.$$

In his study of the geometry of tetrahedral complexes, Lie took advantage of the following properties of the transformations of $\mathfrak{G}$, many of which are evident from (1.2): If $T_1, T_2 \in \mathfrak{G}$, then (a) $T_1 \circ T_2 \in \mathfrak{G}$; (b) $T_1 \circ T_2 = T_2 \circ T_1$; (c) for points $p, q \in \mathbb{P}^3(\mathbb{C})$ "of general position," i.e., not lying in the faces of the tetrahedron, a unique $T \in \mathfrak{G}$ exists such that $T(p) = q$; (d) there is a threefold infinity (∞^3) of transformations in $\mathfrak{G}$; (e) if for a point p in general position $T(p) = p$, then $T = I$, the identity transformation.

For us today, the fact that $\mathfrak{G}$ is a *group* stands out. We see Lie using the commutativity and simple transitivity of $\mathfrak{G}$. But in 1869, group theory was still not a part of basic mathematics; it was not something familiar to, and deemed of fundamental importance by, the majority of active mathematicians. In 1869 "group theory" meant the theory of finite permutation groups, that was known to have a nice application to algebraic equations as shown by Galois.[3] But relatively few mathematicians were actively interested in this subject in 1869. Lectures or seminars on the subject were rare at this time. It is remarkable that one place where permutation groups and Galois theory were the subject of lectures was at the University in Oslo (then Christiania), where Ludwig Sylow (1832–1918) presented such lectures in 1862.[4] In fact, Lie had actually attended these lectures! If history were more "rational," this would have been a turning point in Lie's life, for as we shall see, eventually the underlying ideas of Galois theory—construed in a very vague and general sense—were to become a driving force in his creative work involving continuous groups. But apparently history is not so rational. According to Lie, he understood hardly anything that Sylow was saying.[5] This is not surprising, since throughout his life Lie never displayed an interest in algebra or the theory of numbers; he was at heart a geometer. Also, in 1862, he was still five years away from his decision to become a mathematician.

In his work on the geometry of tetrahedral complexes, Lie made fundamental use of the above properties of $\mathfrak{G}$. To illustrate how this occurred something must be said about his notion of a complex curve. According to Lie, any "continuous succession" of lines from $\mathfrak{T}$ in which each intersects its predecessor is a complex curve [1870:69]. The typical example of a complex curve is a space curve whose tangent lines belong to $\mathfrak{T}$; as a point traces out

[3] For an exposition of Galois's own presentation of his theory, see [Edwards 1984].
[4] Birkeland has published a transcription of these lectures [1995], as well as a summary of their content in English [1996].
[5] This fact is documented and discussed further on.

Figure 3. Evariste Galois at age fifteen. He died six years later after sketching the theory that bears his name.

the curve the corresponding tangent lines succeed each other continuously. Thus, for example, any line $\ell \in \mathcal{T}$ is a complex curve. Although Lie's conception of a complex curve was more general than this typical type, since it was not clearly articulated it will be best for the purpose at hand — to convey some idea of the role of $\mathfrak{G}$ in Lie's geometrical deliberations — to restrict attention to complex curves that are space curves with tangents in $\mathcal{T}$.

Lie divided complex curves into *species* (*Gattungen*). Two complex curves, γ_1 and γ_2 are of the same species if $T[\gamma_1] = \gamma_2$ for some $T \in \mathfrak{G}$. In effect he divided complex curves into equivalence classes called species. The species containing a given complex curve γ_0 consists of all curves of the form $T[\gamma_0]$ as T runs through $\mathfrak{G}$. I shall denote this species by $\mathcal{S}(\gamma_0)$. In effect, $\mathcal{S}(\gamma_0)$ is the orbit of γ_0 under $\mathfrak{G}$. If $\gamma_0 = \ell_0 \in \mathcal{T}$, then $\mathcal{S}(\gamma_0) = \mathcal{T}$. Since there are ∞^3 transformations in $\mathfrak{G}$, Lie assumed that each species contains ∞^3 distinct curves. As we shall see in Section 2, this assumption turns out to admit certain mathematically interesting exceptions (W-curves).

Lie's study of complex curves was by means of certain geometrical transformations that took $\mathcal{T}$, suitably conceived, into itself. These transformations, which he called p-, g-, and r-transformations, were constructed by means of a uniform procedure with the aid of the transformations of $\mathfrak{G}$.[6] Let p_0 and q_0 be two points "of general position." Then, by the simple transitivity of $\mathfrak{G}$, if p is any point of general position, then there is a unique transformation $T \in \mathfrak{G}$ such that $T(p_0) = p$. Thus we may define a point transformation P by setting $P(p) = q$, where $q = T^{-1}(q_0)$. In other words, as T runs through $\mathfrak{G}$, $P(T(p_0)) = T^{-1}(q_0)$. P is an example of what Lie called a "reciprocal transformation," by which he meant that $P^2 = I$, I being the identity transformation.[7] If the point q_0 is replaced by a straight line ℓ_0, then $R : T(p_0) \rightarrow T^{-1}[\ell_0]$ defines a point-to-line transformation called by Lie an r-transformation. It is an example of what Lie later called a contact transformation (Section 3). If q_0 is replaced by a plane π_0, then $G : T(p_0) \rightarrow T^{-1}[\pi_0]$ defines a point-to-plane transformation, called a g-transformation, which is also a contact transformation. As Lie explained [1870:69], all these transformations take $\mathfrak{T}$, suitably conceived, into itself, and he used them to study properties of complex curves. Although he was more concerned with the g- and r-transformations, I shall illustrate his group related results and methods for the simpler p-transformations.

As a preliminary to illustrating Lie's group-flavored geometry, some general remarks are in order. These I express on the naive level of eighteenth-century analysis, which is commensurate with Lie's more geometrically articulated reasoning. A species of complex curves $\mathcal{S}(\gamma_0)$ contains (so Lie assumed) ∞^3 curves. Consider all the curves from such a family that pass through a fixed point p_1. Then "in general" they form a surface. For example, if the ∞^3 curves in the family are expressed by equations

$$x(s) = f(\lambda, \mu, \nu, s), \quad y(s) = g(\lambda, \mu, \nu, s), \quad z(s) = h(\lambda, \mu, \nu, s),$$

then the curves passing through $p_1 = (x_1, y_1, z_1)$ satisfy the system of equations

$$f(\lambda, \mu, \nu, s) = x_1, \quad g(\lambda, \mu, \nu, s) = y_1, \quad h(\lambda, \mu, \nu, s) = z_1.$$

This is a system of three equations in four unknowns, so that (taking for granted the requisite implicit function theorem) all the unknowns may be expressed in terms of one, which I will denote by u. Thus the solutions to the above system may be expressed in the form $x(t) = F(u, t)$, $y(t) = G(u, t)$, $z(t) = H(u, t)$ and "in general" define a family of ∞^1 curves—one for

---

6 Here I have relied on Lie's recollections [Lie 1893ms:3ff.], which on this matter appear to me reliable. The same uniform procedure is presented in [Klein and Lie 1871:§4]. In [1870] Lie did not present the above construction of p-, g-, and r-transformations by means of $\mathfrak{G}$. Instead, in each case, he (or Klein, who wrote up the paper for him) translated his definitions into terms more familiar to his readers.

7 To see this, observe that by simple transitivity there is a unique $T_0 \in \mathfrak{G}$ such that $T_0(p_0) = q_0$. Thus, using the commutativity of $\mathfrak{G}$ we have $P^2(p) = P(q) = P(T^{-1}q_0) = P(T^{-1}T_0(p_0)) = [T^{-1}T_0]^{-1}q_0 = TT_0^{-1}q_0 = T(p_0) = p$.

each value of u—that evidently define a surface parametrized by t and u and passing through p_1 with the curves u = const on this surface being the curves of the family passing through p_1. When $\mathcal{S}(\gamma_0) = \mathcal{S}(\ell_0) = \mathcal{T}$, the surface through p_1 consists of the lines from $\mathcal{T}$ that pass through p_1. In general, there are, as noted, ∞^1 such lines, and they form a cone-like object with p_1 as its vertex. It is called the *complex cone at* p_1.

Now let us consider one of Lie's geometrical theorems [1870:71]:

(1.3) **Theorem.** *Let $\mathcal{S}$ be a fixed species of complex curves. Fix a point p_1 and consider the surface $\Sigma(\mathcal{S}, p_1)$ of all curves from $\mathcal{S}$ that pass through p_1. Let $\mathcal{S}^* = P[\mathcal{S}]$, the "p-reciprocal" species. Then the curves of $\mathcal{S}^*$ that pass through p_1 describe the same surface, i.e., $\Sigma(\mathcal{S}, p_1) = \Sigma(\mathcal{S}^*, p_1)$.*

As an application of Theorem 1.3, Lie took $\mathcal{S} = \mathcal{S}(\ell_0) = \mathcal{T}$. Then the surface $\Sigma(\mathcal{S}, p_1)$ is the above mentioned complex cone with vertex p_1. Now, when $\mathcal{S} = \mathcal{T}$, the transformed species $\mathcal{S}^* = P[\mathcal{S}]$ turns out upon calculation to consist of third-degree curves that pass through the vertices of the tetrahedron.[8] The fact that $\Sigma(\mathcal{S}, p_1) = \Sigma(\mathcal{S}^*, p_1)$ thus implies that at each point p_1 the complex cone of $\mathcal{T}$ is covered by the third-order curves of this species that pass through its vertex.

Lie provided no proof of Theorem 1.3, or of any other theorem in [1870]. But it is not difficult to imagine how he might have arrived at it, since it follows readily from the construction of P by means of $\mathfrak{G}$. To suggest some idea of the nature and extent of group-theoretic reasoning in Lie's earliest work, I will present a reconstructed proof. Of course, Lie's own reasoning would have been less formal and more verbal and geometrical. Let $\mathcal{S} = \mathcal{S}(\gamma_0)$, and let $\gamma_0(t)$ denote the points traced out by γ_0 as t varies. Then by simple transitivity, for each t there is a unique $S_t \in \mathfrak{G}$ such that $\gamma_0(t) = S_t(p_0)$. Since $\mathcal{S} = \mathcal{S}(\gamma_0)$ consists of the curves $\gamma = T[\gamma_0]$, we have

$$\gamma(t) = TS_t(p_0). \tag{1.4}$$

To see when a curve of the form (1.4) passes through p_1, use transitivity to write $p_1 = T_1(p_0)$ for some $T_1 \in \mathfrak{G}$. Thus the curve γ in (1.4) passes through p_1 if for some $t = u$ one has $\gamma(u) = p_1$, i.e., $TS_u(p_0) = p_1 = T_1(p_0)$, which in view of property (e) of $\mathfrak{G}$ means that T must be such that $TS_u = T_1$, or

$$T = T_1 S_u^{-1}. \tag{1.5}$$

By virtue of (1.4) and (1.5), the one-parameter family of curves $\gamma_u \in \mathcal{S}$ that pass through p_1 is given by $\gamma_u(t) = T_1 S_u^{-1} S_t(p_0)$, and the surface $\Sigma(\mathcal{S}, p_1)$ consists of all points

$$\sigma(t, u) = T_1 S_u^{-1} S_t(p_0). \tag{1.6}$$

Now consider the surface $\Sigma(\mathcal{S}^*, p_1)$. Since $\gamma^* \in \mathcal{S}^*$ if and only if $\gamma^* = T[\gamma]$, $\gamma \in \mathcal{S}$, then from (1.3) and the definition of P we see that

---

[8] Lie gave the calculations in his lectures [1896:332,340].

$$\gamma^*(t) = P[TS_t(p_0)] = S_t^{-1}T^{-1}(q_0) = S_t^{-1}T^{-1}T_0(p_0), \tag{1.7}$$

where by simple transitivity T_0 is the unique member of $\mathfrak{G}$ such that $T_0(p_0) = q_0$. We can then ask, as above, when is T in (1.7) such that $\gamma^*(u) = p_1$? The answer is that T must satisfy $S_u^{-1}T^{-1}T_0 = T_1$, so $T^{-1} = S_uT_1T_0^{-1}$. Thus the one-parameter family of curves γ_u^* from $\mathcal{S}^*$ through p_1 is given by

$$\gamma_u^*(t) = S_t^{-1}[S_uT_1T_0^{-1}]T_0(p_0) = \sigma^*(t,u). \tag{1.8}$$

Finally, by virtue of the commutativity of $\mathfrak{G}$, it follows that $\sigma(t,u) = \sigma^*(u,t)$, and so the two surfaces are identical.

The commutativity of the transformations, which is essential to the above proof, was used in most of the reasoning involving groups associated to tetrahedral complexes. Thus the system of transformations $\mathfrak{G}$, and specifically the above-mentioned properties, were fundamental to Lie's exploration of the geometry of $\mathcal{T}$ There is, however, no evidence to indicate that before meeting Klein in Berlin, Lie had imagined that the reasoning he based on the properties of the particular system $\mathfrak{G}$ might extend with profit to other families of transformations with the same properties and that these properties defined a type of mathematical object of general significance to geometry in particular and to mathematics in general.

1.2. W-Curves and W-Surfaces

In Berlin Lie and Klein were drawn to each other because of their mutual interest in geometry, especially Plücker's line geometry. In August, just before coming to Berlin, Klein had submitted a paper [1869] on line complexes that are related to a type of surface first studied by Kummer in 1864.[9] If, following Klein, new homogeneous line coordinates $p_{jk} \to x_1, \ldots, x_6$ are introduced so that the defining condition on line coordinates takes the form $\Omega = \sum_{i=1}^6 x_i^2 = 0$, then these complexes are given by a quadratic equation of the form $\Phi = \sum_{i=1}^6 k_i x_i^2 = 0$. The k_i are assumed to be distinct and nonzero, so that the complex $\mathcal{K}$ defined by $\Omega = 0$, $\Phi = 0$ will be referred to as a *generic complex.* The locus of all points p such that the complex cone of $\mathcal{K}$ with vertex p degenerates into a pair of lines is a Kummer surface, and Klein showed that the line-geometric study of $\mathcal{K}$ is tantamount to the study of Kummer surfaces.

Klein's paper also contained a division of lines and planes into "closed systems" that is reminiscent of Lie's division of complex curves into species. Consider, for example, lines. They are divided into "closed systems" of 32 lines each, the line $(a_1, a_2, \ldots, a_6)$ being associated with the 32 lines of

[9] See in this connection the detailed discussion by Rowe [1989a:§2].

Figure 4. A Kummer surface. Such surfaces also played a role in Lie's work on his line-to-sphere mapping (Section 4).

the system $(\pm a_1, \pm a_2, \ldots, \pm a_6)$ [1869:64]. The significance of such a grouping of lines—as well as groupings of other geometrical elements naturally related by the choice of coordinates x_i—was emphasized by Klein in his introductory remarks: Objects so grouped together have identical geometrical relationships to the complex and the associated Kummer surface. Thus, in particular, the lines in a closed system have the same geometrical properties vis à vis $\mathcal{K}$. A simple illustration of this fact is that a line $(a_1, a_2, \ldots, a_6)$ belongs to $\mathcal{K}$ if and only if all the lines $(\pm a_1, \pm a_2, \ldots, \pm a_6)$ of the closed system belong to $\mathcal{K}$. Klein's study also tacitly involved a collection $\mathfrak{G}$ of distinguished projective transformations, namely the $2^6 \div 2 = 32$ projective transformations given in the homogeneous coordinates $x_1, \ldots, x_6$ by

$$x_1' = \pm x_1,\ x_2' = \pm x_2, \ldots,\ x_6' = \pm x_6. \tag{2.1}$$

At least superficially, these transformations had properties similar to those of the system $\mathfrak{G}$ in Lie's study of $\mathcal{T}$. In addition to being closed under composition and commutative, the transformations take $\mathcal{K}$ into itself; and each closed system is the orbit of any one of its members. As with lines of the same closed system, complex curves of the same species possessed identical geometrical properties relative to $\mathcal{T}$ and the associated tetrahedron. For example, if one curve of a species passes through the vertices of the tetrahedron, then they all do by virtue of the fact that the transformations of $\mathfrak{G}$ fix the vertices.

Eventually, Klein and Lie recognized the above-mentioned similarities, but it is unclear when this occurred. Soon after meeting in Berlin, they formulated the common element in their work as deriving from a simple principle, which I state as follows:

(2.2) **Equivalence Principle**. *Suppose that a geometrical configuration $\mathcal{G}_0$ is taken into itself by a transformation T. If another configuration $\mathcal{G}$ has a certain geometrical relation to $\mathcal{G}_0$, denoted by $\mathcal{G} \sim \mathcal{G}_0$, and if $\sim$ is preserved under T, so that $T[\mathcal{G}] \sim T[\mathcal{G}_0]$, then one can conclude that $T[\mathcal{G}] \sim \mathcal{G}_0$* [Klein and Lie 1871:50].

The conclusion is immediate, since $T[\mathcal{G}_0] = \mathcal{G}_0$. As Klein explained to Max Noether, by virtue of their above-described work on line complexes this principle lay quite close at hand for both him and Lie when they met in Berlin.[10] In Lie's work, however, the equivalence principle (2.2) was applied systematically to transformations T with additional properties that in retrospect can be summed up by saying that the totality $\mathfrak{G}$ of T's form a continuous commutative simply transitive group. It is unclear exactly when Klein and Lie realized the fundamental importance in Lie's work of the totality $\mathfrak{G}$ and its properties as a group.

When Klein arrived in Berlin, he may have had in the back of his mind the thought that "groups" (meaning permutation groups) were geometrically relevant. In Göttingen, where Klein spent over seven months before going to Berlin, Clebsch was encouraging a young French mathematician in Paris by the name of Camille Jordan (1838–1921). Jordan was in the process of developing the seminal ideas of Galois into a systematic treatise on the theory of permutation groups and its applications to algebra and geometry, his *Traité des substitutions* [1870]. In connection with this enterprise, he was applying Galois theory to various algebraic equations that determine geometrical objects of interest. For example, in a paper in Crelle's journal [1869e] he considered the polynomial equation of degree 16 that yields the singular points of a Kummer surface. The geometry of the situation enabled him to determine its Galois group, and he then used its structure to conclude that the resolution of the equation of degree 16 reduces to resolving equations of degrees 6 and 2. Clebsch actively encouraged Jordan's application of group theory to geometry. In the preface of his *Traité* Jordan said that the material on geometrical applications (Book III, Ch.III) was made possible by the "libérales communications de M. Clebsch" [1870:viii]; many of the examples considered there were drawn from papers by Clebsch. The inaugural volume of Clebsch's new journal, *Mathematische*

[10] This letter, dated 17 December 1870, is quoted in [Klein 1916ms:13ff]; the original is in NSUB Göttingen (Nachlass Klein XII).

Figure 5. Jordan (top) devoted much of his career to developing and expounding the seminal ideas of Galois. Clebsch (bottom) encouraged Jordan's application of the resultant theory to geometry.

Annalen, also contained Jordan's "Commentaire sur Galois" [1869a]. Thus during Klein's stay in Göttingen when he composed his paper on generic line complexes and Kummer surfaces [1869], he may have overheard some general talk about the theory of Galois and its application to geometry through Clebsch and his students. In any case, Klein referred to Jordan's paper [1869e] in his own, which was written while he was in Göttingen [1869:n.16].

Thus when Klein arrived in Berlin, he may have been more disposed than Lie to perceive groups arising in a geometrical context. As he learned the details of Lie's ongoing investigation of the geometry of tetrahedral complexes, his interest in it increased to the point of active participation. Two episodes in that participation (while he was in Berlin) served to focus attention on the fact that Lie's work involved systems of transformations (later called continuous groups) that were recognized as analogues of groups of permutations: (i) Klein's discovery of W-curves and (ii) Klein's observations about Lie's study of differential equations related to a tetrahedral complex $\mathfrak{T}$.

Episode (i) appears to have occurred first. According to Lie's recollections, the discovery of W-curves happened as follows.[11] Lie had posed to Klein the question, Do curves γ exist with the property that whenever a point $p \in \gamma$ is taken by a $T \in \mathfrak{G}$ into another point $T(p)$ that also lies on γ, then the tangent line to γ at p is taken by T into the tangent line to γ at $T(p)$? Lie realized that if such curves existed, they would possess many interesting geometrical properties. Of course, if a curve γ existed that is taken into itself by all $T \in \mathfrak{G}$, then the tangent property would follow and γ would be a curve of the type Lie sought. However, no such curves seemed to exist. That is, "in general," as T runs through the ∞^3 transformations in $\mathfrak{G}$, $T[\gamma]$ represents ∞^3 different curves. Lie, however, overlooked the possibility that in some cases there might be fewer than ∞^3 curves. By means of the analytical set-up of (1.2), Klein showed that curves exist that are taken into themselves by precisely ∞^1 transformations of $\mathfrak{G}$. It turns out that these curves actually have the property stipulated by Lie in his question. The curves Klein discovered are special examples of what he and Lie called W-curves in German and V-curves in French.[12]

Some of the discoveries Lie and Klein made about these curves, and about the associated W-surfaces, were briefly described in two notes presented in the *Comptes Rendus* during their stay in Paris [Klein and Lie 1870]. Already in these notes and even more so in their detailed study of W-curves in the plane [Klein and Lie 1871], we witness a shift in emphasis from Klein's essay on generic complexes [1869] and Lie's on tetrahedral complexes [1870], a shift that marked a significant step toward the

[11] What follows is based on Lie's recollections [1893ms:6]. Judging by the marginalia in [Klein 1892ms:5], Lie reminded Klein of his role in the discovery of W-curves, which Klein then stated as a fact in his *Ges. Abh.* **1**, 415.

[12] See [Rowe 1989a:§5] for an explanation of this terminology.

emergence, as an independent mathematical entity, of a continuous group of transformations, and a concomitant recognition of an analogy with the theory of permutation groups. The shift begins to be evident from the definition in their *Comptes Rendus* note of a W-curve in space. Whereas in their respective papers on line complexes, groups of transformations had arisen naturally from a given geometrical object, now the starting point is any family $\mathfrak{H}$ of ∞^1 projective transformations of space T_λ that are continuously related in the sense that they are generated by an infinitesimal projective transformation [1870:1222]. Thus for the first time not only is a continuous "group" the starting point for an investigation, but also for the first time in print we have the idea that infinitesimal transformations are a characteristic and useful feature of continuous systems of transformations.

In homogeneous coordinates $x = (x_1, \ldots, x_4)$, such an infinitesimal transformation $x \to x + dx$ is defined by a system of linear differential equations, which we may write with matrix notation —not used by Klein and Lie—as $dx/d\lambda = Ax$, where A is a fixed 4×4 matrix. Matrix notation also enables us to write $T_\lambda = e^{\lambda A}$, although Klein and Lie could not. From the nature of the T_λ it follows that

$$T_\lambda T_\mu = T_{\lambda+\mu}. \tag{2.3}$$

This is an immediate consequence of the exponential form for T_λ, but it was realized by Klein and Lie as well. From (2.3) it follows that the family $\mathfrak{H} = \{T_\lambda\}$ has the following further properties:

$$\text{For all } T, U \in \mathfrak{H}: \text{ (a) } TU \in \mathfrak{H}; \text{ (b) } T^{-1} \in \mathfrak{H}; \text{ (c) } TU = UT. \tag{2.4}$$

Klein and Lie realized that $\mathfrak{H}$ possesses these properties, although without the uniform representation afforded by matrix exponentiation, these properties are not immediately apparent. Later, in their joint paper [1871], they were to say that families with property (2.4a) form *closed systems.* Property (2.4a) together with (2.4b) implies that $\mathfrak{H}$ is a group in the modern sense. For a long time Klein and Lie took it for granted that in general for infinite systems of transformations, property (2.4a), which they regarded as *the* group property, implies (2.4b), as it does for finite permutation groups.[13]

Given such a family $\mathfrak{H}$ of transformations T_λ, the W-curve through point p_0 in space is the locus of points $\gamma_0(\lambda) = T_\lambda(p_0)$ obtained by varying

[13] Jordan made the same tacit assumption in his memoir [1869b] on groups of rigid motions. They all took it for granted that groups as they defined them always possess inverses—that closure under composition entails closure under inversion, as it does for permutation groups. As Lie first began to develop his theory (from 1874 on), he thought he could prove this for r-parameter continuous transformation groups [1876b:14,19]. When he realized that he could not prove it, closure under inversion was explicitly assumed, although he still conjectured that a proof was possible [1880:5]. Finally, he withdrew the general conjecture but showed that it was true for groups comprising certain types of transformations [1888:20–22]. Cf. Klein's addition (1893) to note 8 of his *Erlanger Programm* [1872b:463].

λ. Since $T_\mu(\gamma_0(\lambda)) = T_{\mu+\lambda}(p_0) = \gamma_0(\mu+\lambda)$, every $T_\mu \in \mathfrak{H}$ takes γ_0 into itself. Every such family $\mathfrak{H}$ defines a family of W-curves. Furthermore, associated to every such $\mathfrak{H}$ is a family of ∞^3 transformations $\mathfrak{G}$ containing $\mathfrak{H}$. The transformations in $\mathfrak{G}$ are the projective transformations that fix the vertices of a (possibly degenerate) tetrahedron (determined in effect by a canonical form of the matrix A). Even in the degenerate cases, these families $\mathfrak{G}$ share the properties of the transformations (1.2) in Lie's study of tetrahedral complexes, namely the properties in (2.4) and simple transitivity. The starting point of the theory is thus the family $\mathfrak{H}$ and the associated family $\mathfrak{G}$, which they later (in [1871]) called closed systems of commuting transformations and that are commutative groups in the modern sense.

Prior to 1872, Lie and Klein used the words "closed system", "cycle" and "group" interchangeably with regard to systems of transformations closed under composition—and assumed also to be closed under inversion (as was always the case in the examples they considered).[14] In the following discussion of the pre-1872 work, the word "group" will be employed, but the reader should be mindful that such terminological consistency is to be found with Lie and Klein only after 1871.

Further subgroups of the above generalized "tetrahedral" groups $\mathfrak{G}$ arise from the consideration of W-surfaces, that were introduced by Lie. They were defined as follows. Consider distinct systems of W-curves determined, respectively, by two families $\mathfrak{H} = \{T_\lambda\}$ and $\mathfrak{K} = \{S_\mu\}$, associated with the same "tetrahedron" and therefore included in the same collection $\mathfrak{G}$ of ∞^3 transformations leaving its vertices invariant. Fix a W-curve γ_0 associated with $\{S_\mu\}$. Then through each point of this curve passes exactly one W-curve determined by $\{T_\lambda\}$. The totality of these curves as p varies along γ_0 generates a surface called a W-surface: If γ_0 is given by $\gamma_0(\mu) = S_\mu(p_0)$, then the W-curve through $\gamma_0(\mu)$ is the locus of points $\lambda \to T_\lambda(\gamma_0(\mu))$, so that the W-surface is the locus of points $\sigma(\mu,\lambda) = T_\lambda\gamma_0(\mu) = T_\lambda S_\mu(p_0)$. It follows immediately from the commutativity of all the transformations involved that a W-surface is taken into itself by the ∞^2 transformations $\mathfrak{L} = \{T_\lambda S_\mu\}$ (that form a subgroup of $\mathfrak{G}$).

The study of W-curves and surfaces thus involved the consideration of many continuous families of commuting projective transformations closed under composition, and these families formed the starting point for the theory of W-curves and surfaces. As Klein explained to Max Noether in a letter of 17 December 1870:[15]

> At the moment I conceive of [the work on W-curves and surfaces] ... like this: .... In connection with the above-stated general

[14] Klein (rather than Lie) used "closed system" and "transformation cycle" interchangeably. In the margin of [Klein 1871ms-b:1], he asked Lie whether "cycle" would be preferable to "closed system."

[15] This letter is quoted in [Klein 1916ms]; the quoted passage from pp. 15–16 is the basis for my translation; the original is in NSUB Göttingen (Nachlass Klein XII).

> proposition [the equivalence principle (2.2)], now we do not think of the geometrical configuration as given and ask about the transformations; rather we consider the system of transformations as given and and ask about the geometrical configurations that are left invariant by it. Then among the most varied conceivable types of transformation cycles, the systems of commutative linear transformations that depend upon one ... or more continuously varying parameters lie closest at hand. Our theory of W-curves and surfaces, as well as that of the [tetrahedral] ... and linear complexes, and that of the complex whose lines intersect a fixed conic, are now nothing but an implementation of the general mode of reasoning for the special case of commutative linear transformations.

Thus Klein envisioned the work on tetrahedral complexes and W-configurations as a small aspect of a far reaching geometrical method, according to which the starting point was a group of transformations. Indeed, the above quotation suggests that Klein conceived the possibility of applying their mode of reasoning to other "cycles of transformations" besides the systems of commuting projective transformations that arose readily from the study of tetrahedral complexes.

Since the equivalence principle does not assume the commutativity or linearity of transformations, it seems likely that in the above passage Klein did not envision his and Lie's "general mode of reasoning" as necessarily limited by either of these assumptions. The choice of linear, commuting transformation cycles was just the simplest cycle type with which to begin developing the more general principles. This seems to be what Klein is saying. Nonetheless, in their joint paper on W-curves in the plane [1871] both assumptions play a fundamental role in the reasoning. Indeed, in all the work on the tetrahedral complex, the commutativity of the groups involved is critical to most of the reasoning utilizing them (as, e.g., in Theorem 1.3). It is not at all clear that Klein saw how to obtain interesting geometrical results without these assumptions.

During the first few months of 1871, while separated from Lie, Klein took on the responsibility for composing their joint paper [1871]. Initially he hoped to treat both W-curves in the plane and W-curves and surfaces in space, but the latter subject proved to be so extensive and complex that he did not have the patience to work through the details. Dealing with the situation in the plane he found tedious enough.[16] In March of 1871, after Lie had studied Klein's draft and made suggestions for changes, the paper was finally submitted to *Mathematische Annalen*. In [1871] a system of transformations is defined to form a closed system if it is closed under composition [1871:235]. Furthermore, they realized that their definition is the exact analogue of the customary definition of a group of substitutions (i.e., permutations) [1871:235, n.2]. They were fully aware that what they

[16] See the quotations from Klein's letters to Lie presented by Rowe [1989a:237–8].

were doing was analogous to the theory of substitutions and algebraic equations (Galois theory), for which by now the standard reference was Jordan's *Traité des substitutions* [1870], a work to which they referred [1871:233, n.1], and that seems to have impressed Klein.[17] Thus they wrote: "The following considerations of closed systems of commutative transformations have an intimate relation to investigations that occur in the theory of substitutions and the attendant theory of algebraic equations."

Since the group concept was central to the latter theory, in making the analogy Klein and Lie certainly regarded closed systems of commutative transformations as likewise central to their own theory. But why, then, did they decide to use the term "closed system" rather than "group"? Although they did not provide a reason, the continuation of the above quotation suggests one: "There is, however, a far-reaching difference between the form in which these problems occur in the above-mentioned discipline and here; namely, in that discipline the variables are always discrete, while here it is a matter of continuously variable quantities" [1871:233]. This fundamental difference may have prompted the choice of a different terminology. Klein and Lie boldly proceeded to suggest that since the continuous theory was more tractable than the discrete, it might be helpful to first treat and resolve problems in the continuous case and then apply the results to the discrete case by restricting the values of the variables.

Although Klein gave up the intention of doing for space what he had done for the plane in their joint paper [1871], probably shortly after it was completed he sent off to Lie a draft of an essay describing the highlights of the three dimensional case [Klein 1871ms-b]. The introductory remarks to this essay are of interest. In their joint paper Klein and Lie had begun their investigation of W-curves by first determining all continuous groups of commuting projective transformations in the plane generated by infinitesimal transformations. Then all W-curves were determined. The analogous problem for space, Klein observed, was to investigate all configurations taken into themselves by a continuous family of commuting projective transformations. "If we wished to provide an exhaustive overview of the totality of these configurations, then first of all we would have to carry out for space the investigation that at the time we placed in the forefront in the plane: *What closed systems of commuting linear [= projective] transformations exist in space*?"[18] Here we have an example of what may be referred to as a (vaguely articulated) group classification problem. Klein felt that the solution of this classification problem, posed for the plane, for space, or for a domain of any fixed dimension presented no essential difficulties; all the finite number of possibilities could be checked out one after the other. Even for the case $n = 3$, however, he soon decided that the enormity of the computational task made unfeasible an exhaustive, systematic treatment

---

[17] See [Hawkins 1989:317, n.13].

[18] The German text on which the above translation is based is given in [Hawkins 1989:317, n.15].

analogous to what they had done for $n = 2$.[19] And certainly he did not see how to deal with a space of arbitrary dimension n. The group classification problem for an unspecified n, Klein admitted, was a problem of a different sort and very difficult.[20]

The joint paper on W-curves of 1871 was the last publication of Klein and Lie on W-configurations. To carry the work further in the direction taken there meant dealing with the associated group classification problem. Even for a relatively small value of n it was clear that the number of commutative closed systems involved is considerable. Each such system of ∞^k transformations ($k < n$) defines a class of k-dimensional "W-configurations" (*W-Gebilde*) with potentially interesting geometrical properties by virtue of the same sort of reasoning that had been used to deal exhaustively with $n = 2$ and partially with $n = 3$. Evidently, Klein was not inclined to get bogged down in such considerations, and certainly Lie was similarly disinclined. They felt that their ideas in this direction were important, but importance is a relative notion; they were also attracted by other important mathematical ideas. Another factor that probably encouraged them to suspend their work on W-configurations in favor of other research projects was the fact that their work had not been well received by other mathematicians. Neither the French nor Clebsch understood the *Comptes Rendus* notes, while the word Klein received from Berlin was that the mathematics of the notes was "really too simple."[21]

There is something to the Berliner's criticism, for their group-related geometry was not in any sense profound. In addition, there was the problem that the work on tetrahedral complexes and W-configurations involved many radically new geometrical ideas of Lie's. This was also true, and more so, of the material that they finally decided not to write up for publication. For example, regarding Lie's results on integral surfaces of partial differential equations associated to tetrahedral complexes (a bit of which is discussed below), Klein declared, "No one understands these matters" [1871ms-a:1]. Thus, although Klein and Lie felt that their work on W-configurations was important, they apparently did not wish to give priority to the laborious task of educating mathematicians about it. The joint paper on W-curves in the plane was perhaps partly intended as related to that task, but, despite his pedagogical talents, Klein had clearly had enough of such work. After all, most of the results were Lie's anyway. Being pre-

[19] Letter from Klein to Lie dated 26 January 1872. Klein's letters to Lie during 1870–75 have recently been deposited in the National Library of Oslo. Plans are under way for their publication by Springer-Verlag Berlin. I am grateful to Elin Strøm and Tom Lindstrøm, of the University of Oslo, for allowing me to use their preliminary transcript of these letters.

[20] "Eine ganz andere sehr schwierige Frage scheint die Beantwortung bei einem unbestimmt gelassenen n zu sein. Cf. C. Jordan." [1871ms-b:1] See [Hawkins 1989:317, n.16] for a discussion of the possible meaning of the reference to Jordan.

[21] That is, "eigentlich zu leicht" (Klein to Lie, 4 February 1871; quoted on p. 744 of Lie, *Ges. Abh.* **1**).

occupied with other problems (the sphere mapping and its implications—Section 4) Lie himself had neither the time nor the temperament and talent for systematic exposition. It was not until the end of his life that Lie, with the assistance of his student G. Scheffers, presented some of his results in [1896]. Thus the development of ideas related to the group concept might have ended with the publication of [1871]. That it did not was due to two reasons. The first was that Klein, so to speak, put a bee in Lie's bonnet, that became for Lie an idée fixe that eventually was to play an important role in the events of the winter of 1873–74.

1.3. Lie's Idée Fixe

Klein put the "bee" into Lie's bonnet while they were in Berlin, and it involved another direction in his work on tetrahedral complexes. The study of W-surfaces had led Lie to introduce a logarithmic mapping that played a significant role in encouraging his hopes of a group-theoretic study of partial differential equations. When the tetrahedron is the one corresponding to (1.2), the mapping is given by

$$X = \log x, \quad Y = \log y, \quad Z = \log z. \tag{3.1}$$

By means of (3.1) a correspondence is established between a space $\mathfrak{r}$ in coordinates x, y, z and a space $\mathfrak{R}$ in coordinates X, Y, Z.[22] By virtue of (3.1) the transformations (1.2) of $\mathfrak{r}$ comprising $\mathfrak{G}$ correspond in $\mathfrak{R}$ to translations,

$$X' = X + \alpha, \quad Y' = Y + \beta, \quad Z' = Z + \gamma, \tag{3.2}$$

where $\alpha = \log \lambda$, $\beta = \log \mu$, $\gamma = \log \nu$. (Thus $\mathfrak{G}$ and the group $\mathfrak{G}^*$ of translations (1.6) are *similar*, as Klein and Lie would later say.[23]) The W-curves in $\mathfrak{r}$ associated with $\mathfrak{G}$, being curves of the form $\gamma(t) = e^{tA}p_0$ with A diagonal, thus correspond to straight lines in $\mathfrak{R}$ left invariant by the ∞^1 translations along this line, and the W-surfaces in $\mathfrak{r}$ therefore correspond in $\mathfrak{R}$ to planes. In this way the equations of the W-surfaces associated with the above tetrahedron are obtained. That is, each such surface in $\mathfrak{r}$ corresponds to some plane $AX + BX + CZ = D$ in $\mathfrak{R}$, so by (3.1) its equation is of the form $x^A y^B z^C = D'$. As Klein and Lie pointed out in their *Comptes Rendus* note [Klein and Lie 1870], surfaces with such equations had been studied by Serret, who determined their lines of curvature. The equations

[22] The following presentation draws upon Lie's recollections [1893ms:7ff,11,16ff] and on Ch.8, §5 of Lie's lectures [1896]. According to Lie the developments in §5 "stammen alle von Lie und zwar aus der Zeit vom winter 1869 auf 1870" [1896:356n].

[23] Two groups are similar if the equations defining the one group can be transformed into those defining the other by a change of variables and parameters. Lie and Klein came to regard similar groups as essentially the same.

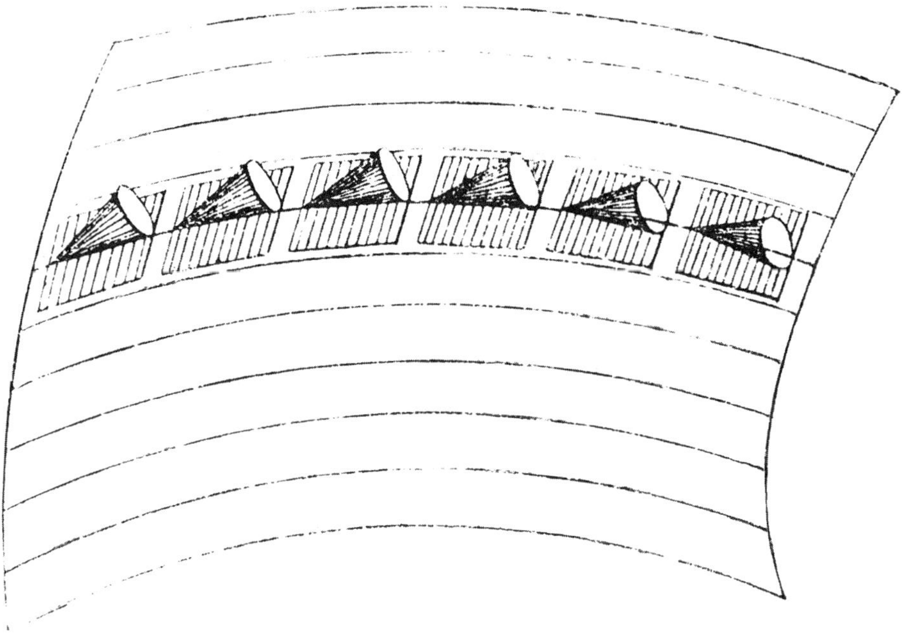

Figure 6. Complex cones touching a surface.

of W-curves for any proper tetrahedron then follow readily, since it may be transformed projectively into the above one.

Lie's approach to the geometry of the tetrahedral complex through the transformations of $\mathfrak{G}$ afforded him many avenues of investigation besides the ones so far related. For example, associated with each point p in space is the complex cone with vertex p, $\mathcal{C}(p)$, which consists of all the lines of $\mathcal{T}$ that pass through p. The complex cone $\mathcal{C}(p)$ was a basic element in Plücker's theory of quadratic line complexes. Among other things, he used this notion to define surfaces, called complex surfaces, that are associated to a given quadratic line complex [1868:159; 1869:335]. In a similar spirit, Lie also studied surfaces associated to the particular quadratic line complex $\mathcal{T}$. Unlike Plücker's approach, Lie's involved infinitesimal rather than algebraic geometry. For example, he considered the problem of determining all surfaces Σ with the property that at each point $p \in \Sigma$, the cone $\mathcal{C}(p)$ "touches" Σ at p. That is, the tangent plane to Σ at p meets the cone $\mathcal{C}(p)$ in exactly one straight line, as in Figure 6.

This geometrical condition on Σ translates readily into a first-order partial differential equation [Lie 1896:260,169], that is, an equation of the form

$$f(x,y,z,p,q)=0, \quad p=\frac{\partial z}{\partial x}, \quad q=\frac{\partial z}{\partial y}.$$

The solutions to this equation are the surfaces Σ satisfying the condition of the problem. It is geometrically evident that any $T \in \mathfrak{G}$ takes a solution surface Σ into another solution surface $\Sigma' = T[\Sigma]$. For if $p' = T(p)$, T takes the tangent plane to σ at p onto the tangent plane of σ' at p' and the cone $\mathcal{C}(p)$ onto the cone $\mathcal{C}(p')$. Thus $\mathcal{C}(p')$ touches Σ' at p', and Σ' also satisfies the conditions of the problem. The transformations of $\mathfrak{G}$ thus take solutions of the partial differential equation into solutions, or, as Lie would later say, *the differential equation admits the transformations of* $\mathfrak{G}$. For the sake of brevity, Lie's subsequent terminology will be used in what follows.

Lie solved this problem by using the logarithmic mapping (3.1).[24] That is, assume without loss of generality that the tetrahedron defining $\mathcal{T}$ is the one corresponding to (1.2). The mapping (3.1) then transforms the partial differential equation $f(x,y,z,p,q)=0$ into another such equation $F(X,Y,Z,P,Q)=0$, where P,Q denote partial differentiation of Z with respect to X and Y, respectively. Solving this new equation is tantamount to solving the original, since the solution can be pulled back by (3.1) to yield the solution to the original equation. Since $F(X,Y,Z,P,Q)=0$ admits $\mathfrak{G}^*$ and hence all translations, Lie concluded that F is a function of P and Q alone (cf. [Lie 1896:584ff.]). The transformed differential equation is therefore of the special form $F(P,Q)=0$ and can be solved directly and the solutions described geometrically. (Cf. [Lie 1896:265].) Lie also solved other partial differential equations arising from the geometry of $\mathcal{T}$ by similar considerations.

When Klein learned (in Berlin) of this aspect of Lie's research on the tetrahedral complex, he made an observation that was to have a great and lasting impression upon Lie. He pointed out that there seemed to be an analogy between Lie's method of integrating differential equations admitting ∞^3 commuting transformations (illustrated here by the above example) and Abel's work on what became known as abelian polynomial equations.[25] Abel had shown that Gauss's method for proving that cyclotomic equations can be solved by radicals could be generalized to any polynomial $f(x)$ with distinct roots $a_1,\ldots,a_n$ that possesses the following two properties: (i) Every root is a rational function of one of them, say $a_i=\theta_i(a_1)$ for all i; (ii) $\theta_i(\theta_j(a_1))=\theta_j(\theta_i(a_1))$ for all i and j. That is, Abel showed that if $f(x)$ satisfies (i) and (ii), then it is solvable by radicals [1829]. The proof involved showing that the resolution of $f(x)=0$ reduces to that of an equation with roots a_i satisfying the following condition: (iii) $a_i=\theta^{i-1}(a_1)$, where $\theta(x)$ is rational and θ^{i-1} denotes its $(i-1)$-fold composite. Polynomials satisfying these conditions came to be known as abelian ([Jordan 1870:286–92;

---

[24] Lie [1893ms:7] described his solution to the above differential equation as a corollary to his discovery that the group $\mathfrak{G}$ is similar to the group $\mathfrak{G}^*$ of all translations. Cf. [Lie 1893ms:17].

[25] For documentary support of this episode see [Hawkins 1989:319n.22].

Kronecker 1853:6; 1877:65–6]).

Property (ii) is a sort of commutativity property, as is (iii). It is possible that this is all Klein had in mind when he called Lie's attention to the analogy, although anyone familiar with the rudiments of Galois theory would have been in a position to see that the Galois group associated to an abelian $f(x)$ consists of commuting permutations. Little is known about Klein's knowledge of Galois theory in 1869. As was noted in Section 2, during his stay in Göttingen prior to coming to Berlin, he may have overheard some general talk about Jordan's work on Galois' theory and its application to geometry through Clebsch and his students. In any case he was familiar with one example of such work: Jordan's paper [1869e] on the singularity equation of a Kummer surface. Jordan's exposition of Galois's memoir [1846], his "Commentaire sur Galois," which appeared in the inaugural volume of Clebsch's *Mathematische Annalen*, makes no reference to Abel's theorem. But anyone who understood Theorem I in Jordan's "Commentaire" (which corresponds to Proposition I of Galois's memoir [1846]) and who asked whether the Galois group of an abelian $f(x)$ consists of commuting permutations would see readily that the answer is affirmative. Thus Klein may have picked up this piece of information in Göttingen. In any case, in Berlin he saw some analogy between Abel's theorem and Lie's method of "resolving" differential equations, and eventually he and Lie perceived this analogy to be that in each case knowledge of a group admitted by an equation yields information about its resolution. (Since the permutations in the Galois group permute the roots of the associated polynomial equation, which are the solutions of the equation, they take solutions into solutions.)

That Klein and, following him, Lie should see an analogy with the Galois theory of commutative permutation groups on the basis of the above sort of example may at first seem surprising, since the success of Lie's method would seem to depend on the special form (1.2) of the transformations comprising $\mathfrak{G}$ rather than on their commutativity per se. However, they showed that Lie's method had a broader scope. For example, both Lie and Klein confirmed that all of the groups of ∞^3 commuting projective transformations associated to some (possibly degenerate) "tetrahedron" can be transformed into groups of translations.[26] Thus, for example, any partial differential equation $f(x, y, z, p, q) = 0$ that admits any such group of transformations could be transformed into the form $F(P, Q) = 0$ and therefore integrated.

One reason Klein's observed analogy was received with such enthusiasm by Lie was that his own deliberations had prepared him to believe

[26] In [1893ms:17] Lie wrote (to Klein): "Wir bemerkten wohl beide, dass (transitive) Gruppen von projektiven vertauschbarer Transformationen der Ebene, resp. Raumes in den Ausartungsfällen in Translationen umge[hen]. . . ." In the nondegenerate case, exemplified by the tetrahedron yielding (1.2), the transformation of the group is given by (3.1)–(3.2).

in such an analogy. Independently of Klein's observation, Lie had noted that special types of differential equations with known methods of solution admit known infinitesimal transformations, and he suspected that the existence of these transformations was the "real" reason why they could be integrated. A simple example is afforded by the first-order ordinary differential equation $dy/dx = f(y/x)$. This equation can be integrated by making the variable change $v = y/x$, which transforms it into a separable equation in x, v. This equation admits the ∞^1 commuting transformations $x' = ax$, $y' = ay$, which Lie saw as the underlying reason why the above technique is successful [1871:§7]. Klein's observation thus supported Lie's opinion that knowledge of transformations, "finite" or infinitesimal, admitted by a differential equation was the true basis for its solution or at least for its reduction to a simpler equation. To Lie it thus seemed plausible that a bona fide theory of differential equations admitting known, commuting infinitesimal transformations could be developed that would serve a function analogous to that served by Galois's theory for polynomial equations. In this connection, he extended his method to yield further results about the integration of first order partial differential equations admitting a group of ∞^2 or ∞^1 commuting projective transformations. For easy reference all these results are summarized below.

(3.3) **Theorem.** *(i) If $f(x, y, z, p, q) = 0$ admits three commuting infinitesimal projective transformations, then it can be transformed into $F(P, Q) = 0$. (ii) If $f(x, y, z, p, q) = 0$ admits two commuting infinitesimal projective transformations, then it can be transformed into $F(Z, P, Q) = 0$. (iii) If $f(x, y, z, p, q) = 0$ admits one infinitesimal projective transformation, then it can be transformed into an equation without the third coordinate $F(X, Y, P, Q) = 0$.*

Lie's statement and proof of Theorem 3.3 were very vague and intuitive. The idea of his proof, which appears to have been unconsciously local in nature, was to show that the given infinitesimal transformations may be used to introduce new coordinates in terms of which they represent translations. Lie's result could be regarded as an adumbration of the theorem that if $X_1, \ldots, X_e$ are commuting C^∞ vector fields that are defined in a neighborhood of a point m of a manifold $\mathcal{M}$ of dimension $d \geq e$ and are linearly independent over $C^\infty(\mathcal{M})$, then in a neighborhood of m they can be represented in the form $X_i = \partial/\partial x_i$.[27] The significance of Theorem 3.3 to Lie was that in each of cases (i)–(iii), information about the solution is provided by knowledge of infinitesimal transformations the equation admitted.

[27] See [Bishop and Chrittenden 1964:20]. Parts (ii) and (iii) were first presented in Lie's papers on the sphere map, since by means of this map Lie had obtained an unlimited number of partial differential equations to which Theorem 3.3 applied [1871b:§16; 1871c:§17]. Apparently, the results of Theorem 3.3 were obtained (possibly with Klein's help) in Paris in 1870 [Lie 1893ms:27]. An analogous result for first-order ordinary differential equations was added, at Lie's insistence, to his joint paper with Klein on W-curves in the plane [1871:§7].

For example, from the conclusion of (ii) Lie concluded from known results that the integration of the transformed equation reduces to a quadrature.

Although the prospect of a meaningful analogy between Galois theory and the type of results on differential equations suggested by Theorem 3.3 was to play an important role in Lie's eventual creation of his theory of transformation groups, his understanding of Galois theory was minimal. The mathematical apparatus of Galois theory was far removed from his interests, which until 1872 were focused on geometrical matters. As noted in Section 1, in 1862 Lie had attended Sylow's lectures on permutation groups while a student at the University of Christiania, but he understood very little of these lectures. It therefore seems likely that he realized what a group of permutations was and had some vague idea of what Galois theory accomplished but that his understanding went no further. Probably, Klein had a better, though limited, grasp of the theory and refreshed Lie's memory on these matters.[28]

What Klein might have known about Galois theory in 1869 was discussed above. Whether Klein told Lie anything about Jordan's paper [1869e] is not known, but Lie's subsequent remarks about the gist of Galois theory suggest that Jordan's application of groups there was the sort of service Lie saw Galois theory as performing: from knowledge of a group to obtain a priori information about the resolution of an equation, such as for example that the resolution of a certain equation of degree 16 reduces to the resolution of equations of degrees 6 and 2. Thus in a letter received by Adolph Mayer on 3 February 1874 Lie wrote,

> Before Galois, in the theory of algebraic equations, one only posed the question: Is the equation solvable by radicals and how is it solved? Since Galois, one also poses ... the question: How is the equation solved *most simply* by radicals? It can be proved e.g., that certain equations of degree six are solvable by means of equations of the second and third degrees and not, say, by equations of the second degree. [1873–4:586]

Likewise, Theorem 3.3 above gave specific a priori information about partial differential equations based upon knowledge of a group of commuting projective transformations admitted by the equation (the continuous group generated by the infinitesimal transformations).

As will be seen in Chapter 2, Lie aspired to more general results of this type; they would constitute for him an analogue of Galois theory. In particular, starting in 1872, he began to consider the case of differential equations admitting groups of noncommuting transformations (possibly stimulated in this regard by the developments described in Section 4). The expectation that some sort of an analogue with the Galois theory of algebraic equations

---

[28] In [1893ms:19] Lie wrote (to Klein): "You probably thought I had learned my little bit of substitution theory from you. That is not the case. Once I attended lectures on it by Sylow, to be sure understanding little." See [Hawkins 1989:319, n.25] for the German text.

could be developed on a scale far greater than that implied by Theorem 3.3 became an idée fixe with Lie and played a role in the events of the winter of 1873–74. In this way the group element in Lie's work was kept alive, albeit in the background, while he pursued his geometrical ideas.

The other immediately more compelling reason why an interest in group related ideas continued and indeed expanded in a conceptual sense, was due to the rash of new research directions that were generated by Lie's discovery of his sphere mapping. Lie's research on the sphere mapping led Klein and Lie, through an interesting web of mutual influences, to appreciate the importance of noncommutative groups of transformations. Without the work motivated by the discovery of this mapping it is difficult to imagine how the events of the winter of 1873–74 could have occurred.

1.4. The Sphere Mapping

It was while Lie and Klein were together in Berlin that Klein had inspired Lie's idée fixe as described in the previous section. After one semester at Berlin, they moved to the next obvious center of mathematics—Paris — where they spent the spring of 1870. Toward the end of their stay in Paris, Lie discovered a remarkable line-to-sphere mapping that switched the focus of his research from tetrahedral complexes to the properties and implications of his new mapping. Before discussing this work, part of which became his doctoral dissertation [1871a], submitted back in Christiania on 25 March 1871, a word is in order regarding Lie's adventures after leaving Paris. Because of the outbreak of the Franco-Prussian war in July 1870, Klein left Paris and returned to Germany. Lie, however, being Norwegian and an avid hiker, opted to return to Norway by first hiking to Italy. According to Klein's recollections, Lie had some interesting hiking techniques. For example, in order to keep his clothing dry while hiking in the rain, he would take it off and put it in his backpack. He must have made an interesting sight!

Perhaps this was one reason why Lie was apprehended by the French authorities. He was suspected of being a German spy and thrown into prison. The letters he had written encouraged the authorities in their suspicions, for when Lie wrote in German about "lines" and "spheres" they thought he was writing about "infantry" and "artillery." When Lie said it was mathematics and began to explain "let x, y and z be rectangular coordinates ...," they decided he was insane! Eventually, he was released through the intervention of Darboux, and he returned to Norway to write up the new mathematical discoveries he had made in Paris—and in prison—as his doctoral dissertation.

Lie and Klein's stay in Paris had proved to be a far more congenial mathematical experience than their stay in Berlin. Camille Jordan was there, and they met him. In fact, Jordan's great treatise on permutation

Figure 7. Gaston Darboux's mathematical affinities were closest to Lie's.

groups and Galois theory [1870] rolled off the press while they were in Paris. Klein, at least, appears to have taken a look at it and was impressed—impressed both by the monumental nature of the theory expounded by Jordan and by how little he understood of it.[29] Since Lie and Klein were interested primarily in geometry, however, the person they spent most of their time talking to in Paris—the French mathematician with whom Lie said he had the greatest intellectual affinity (see Section 6.1)—was the geometer Gaston Darboux (1842–1917).

Darboux, who was Lie's age, was part of a group of French mathematicians who sought to develop an approach to geometry that combined concepts from differential geometry—for which there was a tradition in

[29] See in this connection [Hawkins 1989:317, n.13].

France going back to Monge—and concepts from projective geometry in the tradition of Poncelet. Lie and Klein tended to refer to this methodological amalgam as French "metrical geometry."[30] French metrical geometry was inspired in part by a theorem of Liouville (1846) that showed that, in contrast to the case of the plane, conformal transformations of space are less abundant. They can all be generated by composition of certain projective transformations and transformations closely related to projective transformations, what were called transformations by reciprocal radii. Transformations by reciprocal radii are inversions in spheres—the 3-dimensional analogue of the inversions in circles discussed nowadays in complex analysis texts. By virtue of Liouville's theorem the connections between "conformal geometry" and projective geometry were much closer in the three-dimensional case, and this fact served as the basis of French metrical geometry. One of the tools used by Darboux and his colleagues was a "pentaspherical coordinate" system. Darboux pointed out to them that there was a remarkable formal similarity between some of Klein's line-geometrical results (expressed in 6 homogeneous line coordinates) and Darboux's results (expressed in 5 pentaspherical coordinates). In July 1870, shortly before his departure from Paris, Lie discovered a basis for such analogies in what he called his sphere mapping.

By the time he returned to Norway in December 1870, the discovery and some of its consequences had been announced to the academies of science in Christiania, Paris, and (jointly with Klein) Berlin. The first detailed presentation of Lie's results occurred in his above-mentioned doctoral dissertation, on the basis of which Lie received his doctorate in June 1871, and in a lengthy sequel [1871b], submitted to the scientific society of Christiania in the summer of 1871. At Clebsch's request, in October and November 1871 Lie submitted (through Klein) slightly revised versions of his dissertation and its sequel to *Mathematische Annalen* [1871c]. Unless otherwise noted, the *Annalen* version will serve as the basis for the presentation here.

Lie's sphere mapping derived from a basic notion in projective geometry: reciprocity—a notion that Plücker in particular had emphasized and himself generalized. One of the simplest examples is the duality between points and planes in space. It may be defined by the equation

$$z - xX - yY - Z = 0. \tag{4.1}$$

Thus corresponding to a fixed point whose coordinates are given by x, y, z, and that we may imagine as lying in (complex projective) space $\mathfrak{r}$ is the plane of points X, Y, Z satisfying (4.1) and lying in a corresponding space $\mathfrak{R}$.

Lie's sphere mapping is based upon the reciprocity determined by *two* linear equations:

---

30 See in this connection [Rowe 1989a:§6].

$$(4.2) \qquad \begin{aligned} (X+iY) - zZ - x &= 0, \\ z(X-iY) + Z - y &= 0. \end{aligned}$$

By virtue of these equations it is likewise possible to correlate various geometrical objects in spaces $\mathfrak{r}$ and $\mathfrak{R}$. All of these correlations Lie referred to collectively as his sphere mapping. Some important examples are as follows: (A) For each fixed point $P = (X, Y, Z) \in \mathfrak{R}$, (4.2) represents a system of two linear equations in three unknowns x, y, z, and the solutions $p = (x, y, z)$ form a line $\ell \subset \mathfrak{r}$. Thus (4.2) establishes a correspondence $\Psi : P \to \ell$ between points of $\mathfrak{R}$ and certain lines of $\mathfrak{r}$. (B) The lines $\ell = \Psi(P)$ obtained by varying $P \in \mathfrak{R}$ form in $\mathfrak{r}$ a first-degree, or linear, line complex $\mathfrak{L} = \{\ell : \ell = \Psi(P),\ P \in \mathbf{S}\}$, and the map $\Psi : \mathfrak{R} \to \mathfrak{L}$ is one-to-one and onto. (C) Given any line $\ell \subset \mathfrak{r}$ let $\mathcal{C}(\ell)$ denote the set of lines l^* from the complex $\mathfrak{L}$ that meet the given ℓ. Then the corresponding set of points in $\mathfrak{R}$, $S(\ell) = \Psi^{-1}[\mathcal{C}(\ell)]$, is a sphere. Here by a sphere I mean a set of points in $\mathfrak{S}$ satisfying the usual Cartesian equation of a sphere—but the center and/or the radius may be complex numbers.

In this manner, Lie set up a correspondence between all lines in $\mathfrak{r}$ and spheres in $\mathfrak{R}$—and thus between the line geometry relative to a linear line complex $\mathfrak{L}$ in $\mathfrak{r}$ and a "sphere geometry" in $\mathfrak{R}$. For example, lines in $\mathfrak{r}$ that intersect correspond to spheres in $\mathfrak{R}$ that touch. Lie's sphere geometry (in any dimension, but with emphasis on the real case) has recently become of interest to a number of mathematicians, as can be seen from [Cecil 1992]. As for Lie himself, he pursued his investigations related to the sphere mapping in two directions, both of which are historically important. As will be seen, one direction helped inspire Klein to formulate the key ideas on groups and geometries later set forth in his *Erlanger Programm.* The second direction inspired Lie's theory of contact transformations, which in turn played a fundamental role in the birth of his theory of groups.

The first direction was encouraged by a discovery that Lie made shortly before leaving Paris. By virtue of the sphere-mapping reciprocity, surfaces in the space $\mathfrak{r}$ may be made to correspond to surfaces in the space $\mathfrak{R}$. Under this correspondence, the principal tangent curves on the surface in $\mathfrak{r}$ correspond to the lines of curvature on the associated surface in $\mathfrak{R}$. Principal tangent curves and lines of curvature are two types of curve lying on a surface that are of interest to differential geometers. In Paris Lie and Klein learned that Darboux and Moutard had recently determined the lines of curvature of a certain class of fourth degree surfaces that arise naturally from pentaspherical coordinates and were called generalized cyclides. Lie observed that if one thought of these cyclides as "living" in the space $\mathfrak{R}$, then the corresponding surfaces in $\mathfrak{r}$ were the Kummer surfaces that Klein had studied. In this way, Lie was able to transfer the results of Darboux and Moutard on lines of curvature of generalized cyclides to a description of the principal tangent curves on Kummer surfaces—a new result at the time and one that greatly impressed Klein.

With the encouragement of Klein, Lie set out to explore more fully the relations, established by his sphere mapping, between line geometry in the space $\mathfrak{r}$ and the "metrical geometry" of the Darboux school in the space $\mathfrak{R}$. Now much of the work of the Darboux school involved studying geometrical properties left invariant by conformal transformations. So Lie asked, What do conformal transformations in space $\mathfrak{R}$ correspond to in space $\mathfrak{r}$? He showed that the "tenfold infinity" of conformal transformations of $\mathfrak{R}$, which of course form a group, corresponds to the tenfold infinity of projective line transformations that take the linear line complex $\mathfrak{L}$ into itself. This group of transformations is the projective version of what is now called the symplectic group of structure type C_2. (This will become clear from the discussion of symplectic groups in Chapter 3.) It was in this connection that symplectic groups first arose in Lie's work, and that is why he always referred to them in general as higher-dimensional analogues of "the projective group of a linear line complex." Since the conformal groups have structure type B, one could say (anachronistically) that Lie's correspondence between the projective line geometry of a linear complex in $\mathfrak{r}$ and the "metrical geometry" of $\mathfrak{R}$ is a reflection of the "accidental isomorphism" of types C_2 and B_2.[31]

Lie then went on to show what various special types of conformal transformations—such as rotations, translations, etc.—correspond to as projective transformations in space $\mathfrak{r}$. In this manner he established a correspondence between various types of geometrical operations in the two spaces. For example, he showed that the totality of ∞^7 conformal transformations of $\mathfrak{R}$ generated by rotations, similarity transformations, and translations corresponds in $\mathfrak{r}$ to the ∞^7 projective transformations of $\mathfrak{L}$ that fix a certain line $\ell_\infty \in \mathfrak{L}$ (the intersection of the xy-plane and the plane at infinity). Each type of operation formed a group, although Lie did not make a point of this fact in his publications on the sphere mapping. One application he made of these group correspondences is noteworthy, because it reflects his continuing interest in group classification problems. In [1869b] Jordan had classified all groups (discrete and continuous) of "motions," i.e., subgroups of the group of ∞^6 transformations generated by rotations and translations—the so-called group of rigid motions. From Lie's perspective, these transformations acted on the space $\mathfrak{R}$ and so corresponded to projective transformations of $\mathfrak{r}$ leaving $\mathfrak{L}$ invariant. By virtue of this correspondence, Jordan's classification could be pulled back to yield a classification of subgroups of the ∞^6 projective transformations of $\mathfrak{r}$ corresponding to rigid motions of $\mathfrak{R}$.

Jordan's memoir was the first publication in which continuous groups of transformations had been considered, but as my account in Sections 1–3

[31] By setting up a correspondence between the lines of $\mathfrak{r}$ and the spheres of $\mathfrak{R}$, Lie's sphere mapping also sets up a correspondence between all of line geometry and a sphere geometry. Hermann pointed out that this can be interpreted as reflecting the accidental isomorphism of types A_3 and D_3 [1976:38].

suggests, it does not appear to have been a significant motivating factor behind the initial consideration of continuous groups by Lie and Klein.[32] Geometrical considerations naturally led Klein and Lie to continuous groups. That is not to say that Jordan's memoir had no impact upon them. They— and especially Lie — seem to have viewed it as an example of a successful (albeit limited) group classification project. However, they were primarily interested in continuous groups, and, as we saw in Section 2, were already showing some interest in the vastly more formidable classification problems for transformations acting on a space of arbitrarily high dimension. As we shall see in Chapters 2 and 3, for Lie, in particular, the problem of finding all groups of a particular sort was always considered to be a central task of a prospective theory of groups.

Klein found Lie's mapping between $\mathfrak{r}$ and $\mathfrak{R}$ and the concomitant correspondence between groups of projective transformations in $\mathfrak{r}$ with groups of conformal transformations in $\mathfrak{R}$ very stimulating. It convinced him that he had become too dogmatic in his assumption, encouraged by his association with Clebsch in Göttingen, that projective geometry was more fundamental than other types of geometry. The equivalence established by Lie's sphere mapping put three dimensional metrical geometry on an equal footing with the line geometry relative to the line complex $\mathfrak{L}$. As we shall see in Section 5, Klein then went on to discover that all of line geometry could also be considered, on similar grounds, as equivalent to 4-dimensional metrical geometry. Such considerations, together with his work on the projective representation of non-Euclidean geometries, led him to formulate the radically new conception of "geometrical methods" that ultimately led to his *Erlanger Programm* [1872b].

As for Lie himself, it was another implication of the sphere mapping investigation that had a greater long-term effect on him by suggesting a future direction for research. The reciprocity defined by (4.2) was, he realized, but one example of an infinite family of geometrical reciprocities of 3-dimensional space. That is, each of the following types of systems of equations defines a reciprocity between space $\mathfrak{r}$ with coordinates x, y, z and space $\mathfrak{R}$ with coordinates X, Y, Z:

$$\text{(A)} \quad F(x,y,z,X,Y,Z) = 0, \qquad \text{(B)} \quad \begin{cases} F(x,y,z,X,Y,Z) = 0, \\ G(x,y,z,X,Y,Z) = 0, \end{cases}$$

$$\text{(C)} \quad \begin{cases} F(x,y,z,X,Y,Z) = 0, \\ G(x,y,z,X,Y,Z) = 0, \\ H(x,y,z,X,Y,Z) = 0. \end{cases}$$

Thus the point-plane duality is a type (A) reciprocity, and the sphere mapping is a type (B) reciprocity. A type (C) reciprocity is just an ordinary

[32] See Scholz [1989a; 1989b] for a different opinion.

point-to-point mapping between $\mathfrak{r}$ and $\mathfrak{R}$, since, e.g., for a fixed point x, y, z in space $\mathfrak{r}$, system (C) defines (generally) a point in space $\mathfrak{R}$—the intersection of the three surfaces defined by the equations in (C). What was especially important for Lie was that each such reciprocity defines an associated *contact transformation.*

Analytically, contact transformations are certain transformations of five variables,

$$T : (x, y, z, p, q) \to (X, Y, Z, P, Q). \tag{4.3}$$

For example, for a type (A) reciprocity, the transformation T can be defined analytically in the following manner. Fix X, Y, Z and regard $F = 0$ as defining z as a function of x and y to obtain, by differentiation of $F = 0$ with respect to x and then y, the equations $F_x + F_z p = 0$ and $F_y + F_z q = 0$, where $F_x = \partial F/\partial x$, $p = \partial z/\partial x$, etc. Similarly, with x, y, z held fixed, differentiation of $F = 0$ with respect to X and then Y yields two analogous equations. The resulting five equations

$$F = 0,\ F_x + F_z p = 0,\ F_y + F_z q = 0,\ F_X + F_Z P = 0,\ F_Y + F_Z Q = 0$$

in the ten unknowns $x, y, z, p, q, X, Y, Z, P, Q$ can be solved for X, Y, Z, P, Q (assuming that the requisite implicit function theorem applies) to obtain five equations expressing each of X, Y, Z, P, Q as a function of x, y, z, p, q. These equations define the contact transformation T associated to the reciprocity $F(x, y, z, X, Y, Z) = 0$. In particular, if the type (A) reciprocity is (4.1), which establishes the duality between points and planes in $\mathbb{P}^3(\mathbb{C})$, then the associated contact transformation is given by

$$X = -q, \quad Y = p, \quad Z = z - px - qy, \quad P = y, \quad Q = -x. \tag{4.4}$$

The contact transformation associated to the sphere-mapping reciprocity (4.2) may be described geometrically as follows. For Lie the quintuple (x, y, z, p, q) represented the coordinates of a "surface element"—an infinitesimal portion of a surface at point $a = (x, y, z)$ and with tangent plane at a defined by $z' - z = p(x' - x) + q(y' - y)$. A contact transformation (4.3) was a transformation of surface elements. This was an expression of the fact that contact transformations preserve the contact (or tangency) of surfaces.[33] In the case of the sphere-mapping reciprocity, to see how it determines a contact transformation, consider the tangent plane associated

[33] It should be noted, however, that a contact transformation need not take a surface into a surface. For example, the transformation (4.4) takes the surface $z = 1$ into the point $(0, 0, 1)$. However, it takes the 2-dimensional manifold of its surface elements $ds = (x, y, 1, 0, 0)$ into the 2-dimensional manifold of elements $dS = (0, 0, 1, y, -x)$, which consists of all planes through the point $(0, 0, 1)$. Such sets of elements (along with those that constitute a genuine surface) are included in Lie's concept of an "element union." A characteristic property of contact transformations is that they take element unions into element unions. Regarding surface element unions, see [1890a:44ff.; Lie 1896:522ff.].

to the surface element (x, y, z, p, q) as composed of the straight lines lying in this plane and passing through the point $a = (x, y, z)$. By the properties of the line-to-sphere mapping mentioned above, these lines in $\mathfrak{r}$ correspond to spheres in $\mathfrak{R}$ that touch in a common point $A = (X, Y, Z)$. If the common tangent plane to these spheres at A is $Z' - Z = P(X' - X) + Q(Y' - Y)$, then the contact transformation underlying the sphere mapping reciprocity takes (x, y, z, p, q) into (X, Y, Z, P, Q).

Although contact transformations had originated in Lie's work through geometrical reciprocities, his interest in developing the general theory was due mainly to its potential importance in the study of partial differential equations. In [1864:163–73] the German mathematician Paul Du Bois-Reymond (1831–1889) had focused attention on what Lie realized was a very special type of contact transformation, a type that had been used occasionally in the eighteenth century by Euler, D'Alembert, and Legendre to transform partial differential equations. Typical of the type is the transformation (4.4) above. What these eighteenth-century mathematicians realized was that given a first order partial differential equation $F(X, Y, Z, P, Q) = 0$, where $P = \frac{\partial}{\partial X}$ and $Q = \frac{\partial}{\partial Y}$, if $f(x, y, z, p, q) = 0$ is the equation in x, y, z, p, q formally obtained by the substitutions in (4.4), then *as a partial differential equation*, $f(x, y, z, p, q) = 0$ is equivalent to $F(X, Y, Z, P, Q) = 0$, i.e., the solutions of the one partial differential equation are also transformed into the solutions of the other by means of (4.4). Thus solving the one partial differential equation is tantamount to solving the other. The same is of course true for any "point transformation" $X = \varphi_1(x, y, z)$, $Y = \varphi_2(x, y, z)$, $Z = \varphi_3(x, y, z)$, which from Lie's view point defines a contact transformation associated to a "type C" reciprocity. What Euler and Legendre had discovered was that transformations more general than point transformations existed that possessed the same property of transforming differential equations into equivalent ones—and possibly simplifying their resolution.

Du Bois-Reymond had tried to generalize the transformation (4.4) to get a larger class of transformations with this property, but he did not get very far. Lie, however, realized that his contact transformations were precisely the class of transformations that possessed the requisite property. He envisioned developing a theory of partial differential equations that would draw heavily upon geometrical notions and in which, in particular, the concept of a contact transformation would be fundamental. As Lie had learned while in Paris, Monge had showed how to develop the eighteenth-century theory of partial differential equations in terms of geometrical constructs. Lie dreamed of doing something analogous by means of the geometrical notions associated with contact transformations. Thus, after applying the contact transformation of his sphere mapping to show that certain types of partial differential equations are equivalent in the above sense, he wrote, "To a certain extent, the following developments certainly possess a particular character, in that I am concerned only with special classes of differential equations. Nonetheless, I would stress that the path taken here, namely that of linking the treatment of partial differential equations to broader geometrical concepts, seems to be a method from which general progress in the direction taken by Monge may be expected" [1871c:48].

As we shall see in the next section Lie and Klein learned, much to their mutual surprise and delight, that Lie's program of a Mongean type theory of partial differential equations based on the notion of a contact transformation could be understood within the conceptual framework of geometrical methods that Klein had developed.

1.5. The Erlanger Programm

In Paris Klein had been struck by the analogy between results of French geometers such as Darboux in the realm of differential geometry and results of his own on line geometry [Klein 1892ms:7]. The sphere map revealed the basis of such a connection. For example, Kummer surfaces "live" in $\mathfrak{r}$, since, as Klein had shown in 1869 (see Section 1), they could be studied line-geometrically as the singularity surface of a quadratic line complex, but they correspond to surfaces in $\mathfrak{R}$ that had been studied by the French; and it was by virtue of this correspondence that the algebraic nature of the asymptotic curves to Kummer's surface was revealed. In Lie's work, the space $\mathfrak{r}$ was regarded as the space of projective line geometry in the spirit of Plücker, and $\mathfrak{R}$ the space of the metrical, infinitesimal French geometry in the spirit of Monge, Chasles, and Darboux. A guiding idea behind his geometrical study of the sphere map was to utilize it to bring out connections between Plückerian line geometry and French geometry and to use results known in one space to obtain results in the other. That is, he sought to investigate more generally how the line geometry of $\mathfrak{r}$ corresponds to the "metrical geometry" of $\mathfrak{R}$.

Lie had used the word "group" only at the end of his paper, in the application of Jordan's memoir [1869b] discussed in Section 4. Nonetheless, he certainly realized that the types of transformations discussed earlier in his paper have the group property. This fact, however, seems to have first gained in significance for Lie and Klein as a result of Klein's mathematical reaction to the sphere map work. Impressed by Lie's discovery concerning the asymptotic curves on Kummer surfaces, Klein considered Dupin's theorem, which asserts that three one-parameter families of surfaces that are mutually orthogonal in the sense that surfaces from two distinct families always intersect orthogonally have the further property that the intersections are along lines of curvature. Dupin's theorem lived in $\mathfrak{R}$ and was closely related to the generalized cyclides studied by Darboux and Moutard (which define families to which Dupin's theorem applies). Just as Lie had transferred the results of Darboux and Moutard on lines of curvature of these surfaces to Kummer surfaces in $\mathfrak{r}$, Klein sought the line-geometric analogue of Dupin's theorem.[34] The theorem he ended up with, after dropping unnecessary restrictions, could, he observed, be proved by an argument that paralleled one of the proofs of Dupin's theorem, except that his proof involved four variables rather than the three in the proof of Dupin's theorem [1871c:99]. This circumstance led Klein to pose and answer in [1871d] a question of considerable historical importance. That is, the formulation of the question was important because of the new underlying conceptual viewpoint involved.

The viewpoint consisted in implicitly regarding all the geometries or geometrical modes of treatment relevant to his question as completely determined by specifying a manifold of elements and a distinguished set of transformations acting on the elements of the manifold and defining the permissible geometrical operations. Thus Klein interpreted Lie's above-described correspondence between the projective group $\mathfrak{G}$ of a linear line complex $\mathfrak{L}$ in $\mathfrak{r}$ and the conformal group $\mathfrak{G}^*$ in $\mathfrak{R}$ as establishing an "analogy" between the "geometry of a linear complex" $\mathfrak{L}$ and "ordinary metrical geometry." The former geometry is determined by the manifold of lines $\mathfrak{L}$ and the transformations of $\mathfrak{G}$; the latter is determined by the manifold of points $\mathfrak{R}$ $(= \mathbb{P}^3(\mathbb{C}))$ and the transformations of $\mathfrak{G}^*$. The analogy is established by the mapping $\Psi : \mathfrak{R} \to \mathfrak{L}$, by means of which the transformations of $\mathfrak{G}^*$ go over into those of $\mathfrak{G}$.

Motivated by this analogy, on the one hand, and by the analogy between his line-geometric proof and the proof of Dupin's theorem on the other, Klein asked, Does a similar analogy exist between all of line geometry and the metrical geometry of four-dimensional space? From Klein's viewpoint line geometry is determined by the four-dimensional quadratic hypersurface—$M_4^{(2)}$ in Klein's notation—of elements $(p_{12}, \ldots, p_{42}) \in \mathbb{P}^5(\mathbb{C})$ satisfying $\Omega = p_{12}p_{34} + p_{13}p_{42} + p_{14}p_{23} = 0$ and the group $\mathfrak{G}$ of all projective line transformations, i.e., all $T \in \mathbf{PGL}(6, \mathbb{C})$ that take the "line manifold $M_4^{(2)}$ into itself. Likewise, "the metrical geometry of a space of four dimensions" is defined, by analogy with three-dimensional metrical geometry, as follows. The manifold in question is four-dimensional complex projective space, $\mathbb{P}^4(\mathbb{C})$. Let $\mathfrak{G}_1^*$ denote all projective transformations $T \in \mathbf{PGL}(5, \mathbb{C})$ that take the imaginary sphere into itself. (In homogeneous coordinates $y_1, \ldots, y_5$ this sphere is defined by the equations $y_1^2 + \cdots + y_4^2 = 0$ and $y_5 = 0$.) Then the group $\mathfrak{G}^*$ of metrical geometry is the group generated by $\mathfrak{G}_1^*$ and all four-dimensional transformations by reciprocal radii. Klein then constructed a mapping $\Phi : M_4^{(2)} \to \mathbb{P}^4(\mathbb{C})$ and showed that Φ takes $\mathfrak{G}$ to $\mathfrak{G}^*$, thereby establishing the equivalence of all of projective line geometry with four-dimensional metrical geometry.

More important here than the actual result is the underlying idea of geometries as defined (implicitly) by specifying a manifold and a group

[34] A more detailed discussion of Klein's work on Dupin's theorem is given by Rowe [1989a:§7].

of transformations acting on it. Not only did Klein characterize known geometries in this conceptually new way, he also defined new geometries by specifying a manifold and its distinguished transformations. For example, he generalized his characterization of line geometry to obtain the "geometry of an $M^{(2)}_{n-1}$" in $\mathbb{P}^n(\mathbb{C})$ defined by the manifold of all $(x_1, \ldots, x_{n+1}) \in \mathbb{P}^n(\mathbb{C})$ such that $\Omega = x_1^2 + \cdots + x_{n+1}^2 = 0$ and by all $T \in \mathbf{PGL}(n+1, \mathbb{C})$ taking $M^{(2)}_{n-1}$ into itself. He also showed that there is an analogous correspondence between this geometry and the "metrical geometry" of $\mathbb{P}^{n-1}(\mathbb{C})$, which he also defined.

Thus by the time he submitted his paper "On Line Geometry and Metrical geometry" [1871d], in October of 1871, Klein had arrived at something close to the viewpoint he was later to present in his *Erlanger Programm* about a year later. All that was missing was an explicit statement that the transformations defining a geometry always form a group and the declaration that any specification of a (continuous) group of transformations acting on a manifold defines a geometry. According to Klein's recollections, these points were made in an essay (no longer extant) entitled "Geometrical Methods" ("Methoden der Geometrie"), that he composed in December 1871 [1892ms:20]. Although the work on the sphere mapping was, as indicated, a major source of inspiration for this essay, Klein's work on non-Euclidean geometries [1871a] was certainly another significant source, since his "projective derivation" of these geometries by means of metrics inspired by the work of Cayley fitted nicely into the above-described conceptual framework [1892ms:19]. It was in his second major paper on non-Euclidean geometries, submitted in June 1872, that Klein first wrote up his ideas on geometrical methods for public presentation [1872a:316–325].[35] As we shall see, the ideas of Klein's "Geometrical Methods" made a strong impression on Lie because he perceived a connection with the theory of contact transformations and partial differential equations. To fully appreciate why this was so, it is first necessary to return to March of 1871, when Klein published his line-geometric analogue of Dupin's theorem. Lie's reaction to this paper and the research project it inspired are relevant to his reaction, nine months later, to Klein's essay "Geometrical Methods."

What impressed Lie about Klein's line-geometric analogue of Dupin's theorem was not the underlying analogy between line geometry and four-dimensional metrical geometry that it suggested, and that Klein explicitly developed in his seminal paper on line and metrical geometry some seven months later. Lie was impressed by the link Klein's work implied between Dupin's theorem in dimensions 3 and 4. This led him to pose to himself the problem of extending to n-dimensional space those aspects of differential geometry related to Dupin's theorem, which Lie referred to as "curvature theory," and to establishing a connection between the curvature theory

[35] Due to a printer's strike, Klein's memoir [1872a] did not appear until 1873, so that Klein's second version for publication, his *Erlanger Programm*, actually appeared first.

of dimensions n and $n + 1$. Lie's results are contained in presentations written between April and December of 1871.[36] Although they will not be described here, it is noteworthy that they involve the n-dimensional analogue of a contact transformation and the theory of first-order partial differential equations in n independent variables. As we shall see in the next chapter, at the time, the theory of such equations was dominated, especially in Germany, by the ideas of Jacobi. Even while he was in Paris with Klein working on W-curves and the sphere mapping, Lie had shown a serious interest in Jacobi's theory. According to Klein's recollections, at that time Lie was excitedly studying "to the point of surfeit" the exposition of Jacobi's methods and related material on partial differential equations published by Imschenetsky [1869].[37]

Jacobi's theory was quintessentially analytical, whereas Lie's approach to partial differential equations, in his work on tetrahedral complexes and on the sphere map, was geometrical. Jacobi's analytical innovations were eventually to benefit Lie considerably, but at this time (1871), Lie's approach to the study of partial differential equations continued to be thoroughly geometrical. Now, however, the geometry was conceived on a level of abstraction commensurate with Jacobi's theory. That is, in his work on n-dimensional curvature theory, Lie made the transition from what he later called higher-dimensional geometry in the sense of Plücker to higher-dimensional geometry in the sense of Grassmann's *Ausdehnungslehre*.[38] Following Plücker's lead, Lie had considered higher-dimensional manifolds, but they always had a concrete intuitive interpretation. For example, the manifold of straight lines in space is a four-dimensional manifold, but it corresponds to objects (lines) in ordinary space. Now he saw the value of considering n-dimensional manifolds abstractly, in the sense that its elements $(x_1, \ldots, x_n)$ did not necessarily correspond to the coordinatization of some configuration in ordinary space. For the purposes of the following discussion this sort of n-dimensional geometry will be referred to as "Grassmannian geometry." Lie regarded the formulation of the theory of partial differential equations within the framework of Grassmannian geometry as "a step forming the natural continuation and development of Monge's l L'application de l'analyse à la géométrie."[39]

It was during the same period (April–December 1871) in which Lie's work on curvature theory in n dimensions (done in Norway) was leading to this conclusion that Klein's work (in Germany) was leading him towards

---

[36] Lie, *Ges. Abh.* **1**, 215–28, 267–70, 271–85; **7**, 1–10.

[37] "Lie studierte in Paris den Imschenetski sehr eifrig, bis zum Ueberdruss" [Klein 1916ms:10].

[38] Lie may have picked up the "Plücker–Grassmann" terminology from Klein [Hawkins 1989:320, n.33].

[39] Lie, *Ges. Abh.* **7**,101–102. Lie discussed his transition from n-dimensional Plückerian to n-dimensional Grassmannian geometry in unfinished drafts probably written in late 1872 or 1873. See Lie, *Ges. Abh.* **7**, 57, 59, 101–102,105. Lie's use of Grassmann's work was admittedly superficial (*Ges. Abh.* **7**, 101n.10).

the conclusions expressed in his essay "Geometrical Methods." According to Klein's recollections [1892ms:20], he sent a copy of this essay to Lie, who at first did not understand what was meant by "geometrical methods" but soon wholeheartedly agreed with its contents and provided Klein with "new rich material" drawn from his work on partial differential equations and supporting Klein's viewpoint. When Klein and Lie were together again in the fall of 1872, Lie, with Klein's editorial assistance, presented some of this material in the pages of *Göttinger Nachrichten* [1872b:§II], and Klein, with Lie's encouragement, composed his *Erlanger Programm*, "Comparative Considerations on Recent Geometrical Research" [1872b], with an entire section (Section 9) devoted to Lie's work. Let us consider what these two publications had in common.

Within the context of Grassmannian geometry, Lie reformulated and generalized his notions of surface element and contact transformation. Thus corresponding to first-order partial differential equations

$$f(z, x_1, x_2, \ldots, x_n, p_1, p_2, \ldots, p_n) = 0, \quad p_i = \frac{\partial z}{\partial x_i},$$

is the $(2n+1)$-dimensional space $\mathfrak{R}_{2n+1}$ of (hypersurface) elements

$$(z, x, p) = (z, x_1, \ldots, x_n, p_1, \ldots, p_n),$$

where the p_i are now regarded abstractly as variables rather than partial derivatives of some function z of the x_i. Within this geometrical context, Lie discovered that contact transformations could be conceived as those invertible transformations $T : (z, x, p) \to (Z, X, P)$ of $\mathfrak{R}_{2n+1}$ that leave the Pfaffian equation $dz - (p_1 dx_1 + \cdots + p_n dx_n) = 0$ invariant in the sense that

$$(5.1) \qquad dZ - (P_1 dX_1 + \cdots + P_n dX_n) = \rho\left[dz - (p_1 dx_1 + \cdots + p_n dx_n)\right],$$

where the function $\rho = \rho(x, z, p)$ does not vanish. If $\mathfrak{B}$ denotes the totality of all contact transformations, it is evident that the composition of two contact transformations is a contact transformation (as in the case of the contact transformations of ordinary space introduced in Lie's sphere map papers). Thus $\mathfrak{B}$ has the group property. Lie also realized that the inverse of a contact transformation is a contact transformation, so that $T \in \mathfrak{B}$ implies that $T^{-1} \in \mathfrak{B}$. Of course, strictly speaking, the domain of definition of a typical contact transformation is not all of $\mathfrak{R}_{2n+1}$, so that $\mathfrak{B}$ is really what would now be called a pseudogroup, but such careful distinctions were not made by Lie.

The manifold $\mathfrak{R}_{2n+1}$ together with the group $\mathfrak{B}$ thus determines a geometrical method in the sense of Klein. In this connection, it should be pointed out that Klein's ideas on geometrical method were also formulated in the context of Grassmannian geometry, so that the new geometrical conceptual apparatus that Lie was creating fit in perfectly with Klein's ideas. This instance of intellectual harmony must have been mutually encouraging. To appreciate the full extent of the harmony, it is necessary to keep in mind why Lie was interested in general contact transformations. As can be seen from his introduction of contact transformations in his sphere map work, the reason such transformations were of interest to him was that two equations or, more generally, two systems of equations $F_i(x, z, p) = 0$ and $G_i(X, Z, P) = 0$, $i = 1, \ldots, r$, such that $F_i = G_i \circ T$ for all i, where $T(x, z, p) = (X, Z, P)$ is a contact transformation, are *equivalent* in the sense that knowing how to solve one is tantamount to knowing how to solve the other. Indeed, T takes the solutions of $F_i = 0$ into those of $G_i = 0$. One could thus attempt to classify systems up to equivalence in this sense and for each equivalence class seek the simplest representative equation to solve.

Connected with such an enterprise is a notion of invariance. That is, suppose that a system has a certain property P that is invariant under contact transformations. Then evidently it can be transformed only into another system with property P. This suggests the problem of seeking to determine properties invariant under all contact transformations and the related problem of determining necessary and sufficient conditions that one system be transformable into another by means of a contact transformation. This, in turn, is related to group classification problems. For example, given a system of first-order partial differential equations, consider the contact transformations T that are "admitted" by the system in Lie's sense (as introduced in Section 3) that T takes solutions of the system into solutions. The totality of all such T certainly has the group property. Thus each system has associated to it a group of contact transformations admitted by the system (which might be the trivial group consisting of the identity transformation). If two systems are equivalent in the sense that a contact transformation T exists that takes the equations of the one system into those of the other, then the two groups, say $\mathfrak{G}$ and $\mathfrak{H}$, are related by $\mathfrak{G} = T^{-1}\mathfrak{H}T$. Expressed in modern terms this means $\mathfrak{G}$ and $\mathfrak{H}$ are conjugate within the group $\mathfrak{B}$ of all contact transformations. Expressed in the language of Lie and Klein, it meant that the equations defining the transformations of $\mathfrak{H}$ are transformed into those of $\mathfrak{G}$ by the variable change defined by T. As we have seen (Section 3), Lie and Klein called such groups *similar* and regarded them as essentially the same. The problem of determining all groups of a particular type always meant classifying them up to similarity. Thus the classification of systems of equations is related to the classification of groups of contact transformations up to similarity. Two systems with nonsimilar groups cannot be equivalent in the above sense.

By the time of Lie's paper [1872a], he had discovered that any two first-order partial differential equations could be transformed into one another by contact transformations, so that a single equation has no such invariants. He also realized that systems of more than one equation do possess invariants such as the property of being in involution (see Chapter 2). The problem then is to develop the theory of such invariants. Lie perceived that this was an exemplification of Klein's views on geometrical methods. For example, in the *Erlanger Programm* Klein had characterized the fundamental problem of a geometrical method in the following words: "Given a manifold and a group of transformations of it, determine the theory of invariants relative to the group" [1872b:464]. For the manifold $\mathfrak{R}_{2n+1}$ of hypersurface elements (z, x, p) and the group $\mathfrak{B}$ of contact transformations this was precisely what Lie hoped to do, and for this reason, he referred to the theory he eventually created as his invariant theory of contact transformations. Of course, when Klein spoke of "invariant theory" in the *Erlanger Programm*, he had in mind primarily, although not exclusively, the paradigm of the Clebsch–Gordan theory of algebraic invariants, which generalizes readily to the notion of algebraic invariants associated to any linear group.[40] Lie's invariant theory of contact transformations exemplified the ideas of the *Erlanger Programm* in a manner not anticipated by Klein and indicated that their scope was greater than originally imagined.

Lie's work on the sphere map had thus led, through the above described interplay of ideas between between him and Klein, to a more general connection between groups and geometry, namely the one articulated in the *Erlanger Programm.* Whereas the theory of W-configurations could now be seen as an example of a geometrical method, determined by a commutative group of projective transformations, the *Erlanger Programm* was not at all restricted to commutative groups. On the contrary, the groups that motivated it—the groups acting on $\mathfrak{r}$ and $\mathfrak{R}$ in Lie's sphere-mapping paper, the groups defining geometries in Klein's papers, and the group $\mathfrak{B}$ of contact transformations — were all noncommutative. Now it was clear that all continuous groups, both commutative and noncommutative, were relevant to the geometrical study of manifolds and the concomitant "Mongean" study of partial differential equations.

Perhaps it was this fact, made all the more impressive by the convergence of Lie's ideas on contact transformations and partial differential equations with Klein's ideas on geometrical method, that prompted Klein to call for the development of an autonomous theory of continuous transformation groups. This occurred at the end of the *Erlanger Programm.* There Klein turned once again to the analogy with permutation groups that had been noted already in their paper on W-curves in the plane. Now, however, the analogy was more complete, since noncommutative groups were definitely perceived to be of geometrical interest. He pointed out that in the treatises on the subject such as Jordan's *Traité des substitutions* [1870], the theory of permutation groups formed a field of investigation in its own right; Galois theory followed as an application of that theory. He therefore called for the development of a theory of transformation groups that would have as an application the theory of geometrical methods (understood as the study of invariant relations within a manifold determined by a given transformation group). Undoubtedly Lie was in full sympathy with these sentiments and encouraged Klein to include them.

---

[40] See note 13 (added in 1893) on p. 464 of [Klein 1872b].

In Section 2 we saw that the work of Lie and Klein on W-configurations, despite their realization of the analogy with the theory of permutation groups and Galois theory, did not produce any activity on their part to create an analogous theory of the continuous groups that had formed the basis of their study of these configurations. In particular, the associated group classification problem was not tackled. The situation seems to have been the same after the *Erlanger Programm* was composed. Once again, neither Klein nor Lie turned to the development of the theory of continuous transformation groups. Klein began to work on matters related to complex function theory, and did not further develop the ideas of his program until twenty years later (after Lie had begun to publish his monumental treatise on the theory of continuous transformation groups). Lie himself proceeded to develop his invariant theory of contact transformations. Thus although the work of Klein and Lie in 1869–72 led to a clear recognition of the notion of a continuous group of transformations, to an appreciation of its actual and potential importance in applications to geometry and differential equations, and even to an explicit acknowledgment of the value of a theory of these groups, developing the theory was not a project to which either of them immediately turned. The nature of such a project undoubtedly seemed just as difficult and unappealing as it had in 1871 when their joint paper on W-curves was published. This interpretation is supported by Lie's attitude toward one aspect of such a project: the group classification problem suggested by the *Erlanger Programm*.

Since any specification of a continuous group determines, in effect, a geometrical method, it is of interest to know just how many groups there are that act on an n-dimensional manifold and what they, and the geometry they define, are like. Since similar groups were regarded as determining equivalent geometries, the above considerations suggest the problem of classifying, up to similarity, all continuous groups acting on an n-dimensional manifold, for any n. Although this problem is not explicitly formulated in the *Erlanger Programm*, there is evidence that Lie had perceived this implication.

The evidence is in the form of a letter Lie wrote to Klein in 1879, five years after he had begun to create his theory of transformation groups. At that time Lie was writing up an exposition of his theory for the mathematical world at large, to be published in *Mathematische Annalen* [Lie 1880]. He had decided to conclude the work with a section (§20) indicating all publications before 1874 that were related to his new theory. Although the *Erlanger Programm* was mentioned, it was not emphasized. Klein, to whom Lie sent the draft of the paper for editing, apparently objected in a letter to this slighting of the program, for in a letter to Klein, Lie responded by explaining that of course he esteemed Klein's program and certainly it had proved stimulating to him. "But on the other hand, it must be pointed out that in your essay the problem of determining *all* groups is not posited, probably on the grounds that at the time such a problem seemed to you absurd or impossible, as it did to me. Also ... your essay gave no means for resolving my problem, at least nothing beyond what was known earlier."[41]

Thus it would seem that Lie had perceived the above mentioned group classification problem suggested by the program but had dismissed it as a problem that was not feasible, since the means to resolve it were not at hand. To resolve it would have required a kind of mathematics outside Lie's (and Klein's) mathematical interests in 1872. This classification problem, like the one associated with the theory of W-configurations, therefore would have seemed difficult and unappealing to Lie—even unapproachable. Slightly over a year after the *Erlanger Programm* was written, however, Lie completely changed his mind. By the winter of 1873–74 he had decided that he could make substantial progress on such classification problems, that he could develop a theory of continuous groups, and that the effort to do so was warranted. What had caused him to change his mind? In a word, the answer is Jacobi. The purpose of the next chapter is to explain what this means.

[41] "Dass ich Deine Programmschrift sehr schätze, weiss Du. Ich habe das Wort 'gedankenreich' eingeschaltet. Ganz sicher ist sie mir anregend gewesen. Aber anderseits ist zu bemerken, dass die Aufgabe: alle Gruppen zu bestimmen nicht in Deiner Schrift gestellt wurde; und wohl aus dem Grunde dass eine solche Aufgabe sich damals Dir wie mir als absurd oder unmöglich stellte. Anderseits giebt Deine Schrift auch nicht Mitteln zur Erledigung meines Problems, jedenfalls nicht andere als solche die früher bekannt waren." (Letter to F. Klein, NSUB Göttingen, Nachlass Klein X, Nr. 673. Undated, but probably written in 1879, since it discusses [Lie 1880].)

Chapter 2
Jacobi and the Analytical Origins of Lie's Theory

Lie began to study the general theory of first-order partial differential equations assiduously while in Paris in 1870. His main source was a well-written monographic essay on the subject by V. G. Imschenetsky [1869]. Imschenetsky's essay, which ran to almost two hundred pages, was symptomatic of the interest in the theory of first order equations that had been generated by the contributions of Jacobi. Although Jacobi had developed his ideas in the 1830s (he died in 1851) and lectured on their relation to dynamical systems in the early 1840s, it was not until the 1860s that most of this work was published by Clebsch [Jacobi 1862; 1866]. The appearance of these papers stimulated considerable interest in the theory of first order partial differential equations, particularly in France and Germany. By the time he was in Paris, Lie was already interested in the application of groups to differential equations, for Klein's observations (discussed in Section 1.3) of an analogy between some of his work on tetrahedral complexes and Galois theory had been made in Berlin in 1869.

Given this interest, it is not surprising that Lie began to study Imschenetsky's essay. Indeed, the applications of group-related notions to differential equations and to geometry that he and Klein had so far made were rather limited in scope. Moreover, much of it was tied up with Lie's idiosyncratic study of tetrahedral complexes; it did not have a broad based appeal. By contrast, the general theory of first-order partial differential equations was a theory of considerable current interest both by virtue of its mathematical content and because of its links with the Hamilton–Jacobi approach to dynamics. And it boasted a history with which notable names such as those of Lagrange, Monge, Cauchy, Hamilton, and Jacobi were associated. No doubt as Lie studied Imschenetsky's essay he had in the back of his mind the hope that it would help him develop his idée fixe within this broader and more established context. Let us consider what he found, how he reacted to it, and how those reactions culminated in his decision, in the winter of 1873–74, to devote himself to the monumental task of creating a theory of continuous transformation groups.

In what follows certain liberties have been taken for the sake of brevity and clarity. As Lie began to digest and react to the post-Jacobi theory of first order equations, he continued to use the intuitive, geometrical thought modes that he had developed in the period 1869–71, albeit now extended to spaces of any dimension. He called such reasoning "synthetic" to distinguish it from the analytical line of reasoning that was more typical of the period, especially among followers of Jacobi. Since other mathematicians, such as Clebsch and Mayer, were not convinced by his synthetical reasoning, he gradually translated his ideas into a more analytical form. In what follows, Lie's ideas are presented in the analytical form they had acquired by the fall of 1873. It was in the analytical form that they proved inspirational to him, as I show in my paper [1991]. There the reader will find a more chronologically accurate and fully documented account of the progression of Lie's thinking. The conclusions presented here are based upon the evidence and analysis provided there.[1]

2.1. Jacobi's Two Methods

A first-order partial differential equation with n independent variables $x_1, \ldots, x_n$ and one dependent variable z will be expressed with the notation

$$(1.1)\quad F(x_1, \ldots, x_n, z, p_1, \ldots, p_n) = 0, \quad p_i = \frac{\partial z}{\partial x_i}, \quad \text{or} \quad F(x, z, p) = 0.$$

Ever since Lagrange laid the foundations of the theory in the eighteenth century, the goal had been to reduce the problem of determining solutions of (1.1) to the problem of integrating systems of first-order ordinary differential equations.[2] For an equation (1.1) with two independent variables ($n = 2$) Lagrange showed that the integration of (1) could be reduced to the problem of integrating a *linear* partial differential equation in three independent variables. The fact that the linear equation involved more variables than the original was not a problem, since Lagrange knew how to reduce the integration of a linear partial differential equation in any number of variables to that of a system of ordinary differential equations. Lagrange's

[1] In composing this chapter I have drawn upon my earlier synopsis in [1992].

[2] This section is based on Sections 2–3 of my essay [1991], which should be consulted for further details and a complete bibliography.

result became a fundamental theorem in the ensuing theory. For future reference it is stated here without loss of generality for *homogeneous* linear equations as follows:

(1.2) **Lagrange's Theorem**. *The problem of determining all solutions* $f = f(x_1, \ldots, x_n)$ *to a linear homogeneous equation*

$$\xi_1 \frac{\partial f}{\partial x_1} + \cdots + \xi_n \frac{\partial f}{\partial x_n} = 0, \qquad \xi_i = \xi_i(x_1, \ldots, x_n) \tag{1.3}$$

is equivalent to the problem of completely integrating the system of ordinary differential equations

$$\frac{dx_i}{dt} = \xi_i(x_1, \ldots, x_n), \quad i = 1, ..., n. \tag{1.4}$$

Moreover, every solution to (1.3) is of the form $f = \Theta(f_1, \ldots, f_{n-1})$, *where* $f_1, \ldots, f_{n-1}$ *are functionally independent solutions to (1.3) and* θ *is an arbitrary function of* $n-1$ *variables.*

Here "completely integrating" (1.4) meant obtaining a solution to (1.4) with arbitrary initial conditions. Since in (1.4) the functions ξ_i do not depend on t, the system (1.4) is equivalent in turn to a single ordinary differential equation of order $n-1$. For this reason, the system (1.4) was said to be of order $n-1$, a terminology that will be employed in what follows.

Neither Lagrange nor his immediate successors were able to extend his method to deal with (1.1) when $n > 2$. It was not until 1815 that Johann Pfaff (1765–1825) was able to show by a totally different method (involving the more general problem of integrating what became known as Pfaffian equations) that the integration of (1.1) could be reduced to the complete integration of $n-1$ systems of ordinary differential equations of respective orders $2n-1$, $2n-3$, $2n-5$, $\ldots$,3, 1. For large n the number of systems of ordinary differential equations that must be completely integrated in Pfaff's method is quite large. Jacobi's first major contribution [1837] was to show that only the first system of ordinary differential equations occurring in Pfaff's method needed to be completely integrated in order to integrate (1.1). Jacobi was led to this discovery through his study of Hamilton's papers on dynamical systems [1834; 1835]. Here Jacobi's result will be stated in the special case of equations that do not explicitly involve the dependent variable z:

(1.5) **Theorem (Jacobi's First Method)**. *The integration of the partial differential equation* $F(x_1, \ldots, x_n, p_1, \ldots, p_n) = 0$ *can be reduced to the complete integration of the system of ordinary differential equations of order* $2n-1$

$$\frac{dx_i}{dt} = \frac{\partial F}{\partial p_i}, \quad \frac{dp_i}{dt} = -\frac{\partial F}{\partial x_i}, \quad i = 1, 2, \ldots, n. \tag{1.6}$$

Figure 8. Carl Gustav Jacob Jacobi, the quintessential analyst, provided an analytical framework for Lie's geometrical ideas.

When x_i, p_i represent the generalized position and momentum coordinates of a dynamical system of $n/3$ masses with Hamiltonian $H = F(x, p)$, the equations (1.6) are the canonical equations of motion introduced by Hamilton.

With this result Jacobi had tacitly begun to modify the main objective of the theory of first-order partial differential equations. In the past, the goal had been simply to reduce the integration of such an equation to the integration of auxiliary systems of ordinary differential equations. Now the main goal started to become that of determining a reduction that minimized as much as possible the number of integrations of auxiliary systems required for the reduction. Thus Jacobi's first method, as compared with Pfaff's, cut down tremendously on the number of systems of ordinary differential equations requiring integration. A year later Jacobi discovered a new method of integrating first-order equations that further encouraged the minimizing tendency and gave it a new twist because it did not require finding all solutions to an auxiliary system.

Jacobi's new method—first published posthumously [1862]—was based upon the approach behind Lagrange's original method. Once again it suffices to consider equations (1.1) that do not explicitly involve z. The key to avoiding the impasse that Lagrange and his successors had encountered is contained in the following theorem of Jacobi's:

(1.7) **Jacobi Bracket Theorem.** *A partial differential equation* $F_1(x,p)=0$ *can be integrated if* $n-1$ *functions* $F_2,\ldots,F_n$ *of the* $2n$ *variables* $(x,p)=(x_1,\ldots,x_n,p_1,\ldots,p_n)$ *can be determined such that* $F_1,\ldots,F_n$ *are functionally independent and satisfy the relations* $(F_i,F_j)=0$ *for all* i,j, *where for any functions* G, H *of* (x,p),

$$(G,H)=\sum_{i=1}^{n}\Big(\frac{\partial G}{\partial p_i}\frac{\partial H}{\partial x_i}-\frac{\partial G}{\partial x_i}\frac{\partial H}{\partial p_i}\Big). \tag{1.8}$$

The bracket notation (1.8) had been introduced by Poisson in a paper on problems in Lagrangian mechanics [1809]. It and analogous bracket notations were employed by Lagrange in his treatise on analytical mechanics to simplify the expression of equations. Jacobi had been familiar with the Poisson bracket notation for over a decade, but now it became the key to his new method.

Based upon the foundation provided by his bracket theorem, Jacobi's new method consisted in a method for determining the functions $F_2, \ldots, F_n$. His approach was to determine the F_i successively. Finding F_2 involves determining a solution $f=F_2$ to the equation $(F_1,f)=0$. In view of the definition of the Poisson bracket in (1.8), this equation is a linear homogeneous partial differential equation in the $2n$ variables x_i, p_j. By the reasoning behind Lagrange's theorem, it follows that determining a solution is equivalent to determining a solution to a system of ordinary differential equations. Similarly, to determine next F_3 is to determine a solution $f=F_3$ to the *system* of linear partial differential equations $(F_1,f)=0$, $(F_2,f)=0$. In general, once $F_2,\ldots,F_m$ have been determined, the determination of F_{m+1} involves determining a solution to the system of m linear partial differential equations

$$(F_1,f)=0, \quad (F_2,f)=0, \quad \ldots, \quad (F_m,f)=0. \tag{1.9}$$

Jacobi's new approach thus involved determining one solution each—rather than the complete solution—to successive systems of linear homogeneous partial differential equations. He also discovered that the systems he had to consider, namely the systems (1.9) defined by Poisson brackets, have a special property that facilitates their integration. To express this property and its implications succinctly he introduced some new notation. Given variables $y=(y_1,\ldots,y_s)$, for $f=f(y)$ let

$$A_i(f) = \sum_{k=1}^{s} a_k^{(i)}(y) \frac{\partial f}{\partial y_k}, \quad i = 1, ..., m.$$

Then $A_i(f) = 0$, $i = 1, \ldots, m$, denotes a system of linear homogeneous partial differential equations. Moreover, as Jacobi observed,

$$A_i\big(A_j(f)\big) - A_j\big(A_i(f)\big) = \sum_{k=1}^{s} \big(A_i(a_k^{(j)}) - A_j(a_k^{(i)})\big) \frac{\partial f}{\partial y_k},$$

and so defines another linear differential operator. In operator notation Jacobi could succinctly express his discovery regarding the systems of homogeneous equations (1.9) that occur in his new method: The systems $A_i(f) = (F_i, f) = 0$, $i = 1, ..., m$ satisfy

(1.10) $$A_i\big(A_j(f)\big) - A_j\big(A_i(f)\big) = 0, \quad i, j = 1, \ldots, m.$$

Jacobi's discovery was really an observation of a property of Poisson brackets that had not been observed earlier. He also stated the property in a more symmetric form:

(1.11) **Jacobi Identity**. *If F, G, H are any functions of $2n$ variables x_i, p_j, then*

(1.12) $$((F, G), H) + ((G, H), F) + ((H, F), G) = 0.$$

In sum, Jacobi had discovered that the integration of a nonlinear first-order partial differential equation can be reduced to the problem of finding one solution each of several systems of linear homogeneous partial differential equations that possess the special commutativity property expressed in (1.10). By solving this problem, he arrived at the end result of his new method: *The problem of integrating $F_1(x, p) = 0$ reduces to the problem of determining (in the worst possible case) one solution each of the following systems of ordinary differential equations: 1 system of order $2n - 2$, 2 systems of order $2n - 4$, 3 systems of order $2n - 6 \ldots, n - 1$ systems of order 2*. This result, together with the result of Jacobi's first method (Theorem 1.5), focused the attention of mathematicians on the problem of reducing as much as possible the number of systems of ordinary differential equations that need to be solved in order to achieve the integration of a partial differential equation. For example the work of J. A. Weiler [1863] and Clebsch [1866:§5] showed that the number of systems of ordinary differential equations of a given order in Jacobi's new method does not have to increase as the order decreases: *The integration of $F(x, p) = 0$ can be reduced to obtaining one solution to one system of order $2n - 2$ and two systems each of the following orders: $2n - 4, 2n - 6, 2n - 8, \ldots, 2$*. Clebsch's paper [1866] became the starting point for Adolph Mayer (1839–1908), who, like Clebsch but somewhat later, had studied with Franz Neumann at the University of Königsberg during the post-Jacobi era and was well versed in Jacobi-style

Figure 9. Adolph Mayer, an expert on differential equations, became a sounding board for Lie's evolving ideas on connections between such equations and transformation groups.

analysis as well as mathematical physics.[3] Mayer [1872] devised a method that required finding a solution to a single system of ordinary differential equations for each of the above orders, thereby cutting almost in half the total number of solutions of auxiliary systems required by the Clebsch–Weiler modification of Jacobi's new method: *The integration of* $F(x,p)=0$ *can be reduced to obtaining one solution to one system each of orders* $2n-2, 2n-4, 2n-6, \ldots, 2$.

Shortly after he submitted [1872], Mayer learned that Lie had discovered essentially the same result by a completely different, geometrically inspired, method. Thus began a long-term association between Lie and Mayer, who played a role in the birth of Lie's theory of groups by encouraging Lie to translate his ideas into the language of Jacobian analysis.[4]

---

3 On Mayer's career, see [Tobies and Rowe 1990:8ff.].

4 On Lie's method see [Hawkins 1991:§4]. Portions of Lie's correspondence with

The focus in the theory of first-order partial differential equations epitomized by the theorems of Jacobi, Clebsch–Weiler, and Mayer fit in well with Lie's idée fixe, for it made more specific the manner in which knowledge that a differential equation admits a group of transformations should yield information about its integration: It should translate into information about the number and order of the systems of ordinary differential equations required to integrate the partial differential equation. The systems of ordinary differential equations played somewhat the role of the auxiliary polynomial equations associated to the composition series of the Galois group of a polynomial equation and utilized to analyze its algebraic resolution. If Lie could see how to translate the group-related information into information about the required auxiliary systems, the analogy with Galois theory would be greater than had been the case with his earlier results. Before I can explain how Lie managed to find such a translation, it is necessary to consider another aspect of Jacobi's work, one that helped him see how to proceed.

Jacobi's identity (1.12), which had proved fundamental to his new method, was also seen by him as of great significance in connection with his first method (Theorem 1.5). Combined with Lagrange's theorem, Theorem 1.5 implies that the integration of $F_1(x,p) = 0$ is equivalent to the integration of the linear partial differential equation $(F_1, f) = 0$ and thence to the problem of determining $2n-2$ solutions $f = F_2, \ldots, F_{2n-1}$ to $(F_1, f) = 0$ such that $F_1, \ldots, F_{2n-1}$ are functionally independent. Jacobi believed that the integration of $(F_1, f) = 0$ could be greatly facilitated by the following immediate consequence of his identity:

(1.13) **Poisson–Jacobi Theorem.** *If $f = \Phi$ and $f = \Psi$ are two solutions to $(F, f) = 0$, then $f = (\Phi, \Psi)$ is a third.*

Theorem 1.13 was called the Poisson–Jacobi theorem because, as Jacobi realized after he had discovered it, an equivalent version was known to Poisson and Lagrange [Jacobi 1840]. Jacobi criticized them, however, for failing to realize what he regarded as a fundamentally important implication of the theorem for the integration of $F_1(x,p) = 0$.

Suppose, Jacobi reasoned, that two solutions $f = F_2$ and $f = F_3$ to $(F_1, f) = 0$ are known such that F_1, F_2, F_3 are functionally independent. Then "in general," simply by taking Poisson brackets, the requisite $2n-1$ functionally independent solutions may be determined. That is, by Theorem 1.13, $F_4 = (F_2, F_3)$ is a solution, and therefore, by Theorem 1.13 again, $F_5 = (F_3, F_4)$ is another solution, and so on. Of course, it could happen that, e.g., $F_4 = 0$ or that F_4 is some function of F_1, F_2, F_3. But Jacobi claimed that "in general" this would not happen. Thus in general, given the two solutions F_2 and F_3 the rest of the F_i, and thence the solution to $F_1(x,p) = 0$, could be obtained without further integrations—simply by

---

Mayer were published by Engel in his commentaries in Lie's *Abhandlungen*; see especially [Lie 1873–4].

the differentiations involved in computing Poisson brackets. Jacobi realized that in "particular cases" such as when the solutions F_2 and F_3 are known from general mechanical principles and are not peculiar to the mechanical system defined by $F_1(x,p) = 0$, the above bracketing process can fall short of its goal of generating $2n-1$ independent solutions. He made the further claim that in such cases, knowledge of the deficient number of independent solutions could be used to advance the integration of $F_1(x,p) = 0$.

Perhaps due to the poor health that plagued him during the last decade of his life, Jacobi never published any detailed justification of his claims. Most mathematicians doubted that "in general" $2n-1$ independent solutions to $(F_1, f) = 0$ could be obtained by the bracketing procedure. When the bracketing procedure falls short of its goal, one ends up with $r < 2n-1$ independent functions $F_1, \ldots, F_r$ such that for all i and j the function defined by (F_i, F_j) is functionally dependent on $F_1, \ldots, F_r$. According to Jacobi's claims, knowing the functions $F_1, \ldots, F_r$ should simplify the problem of integrating $F_1(x,p) = 0$. Thus implicit in his remarks, especially in view of the prevailing scepticism about his claims regarding the "general case," was a general problem that will be referred to as Jacobi's problem.

(1.14) **Jacobi's Problem.** *Suppose $f = F_1, \ldots, F_r$ are r functionally independent solutions to $(F_1, f) = 0$ with the property that bracketing produces no more solutions in the sense that for all i, j, (F_i, F_j) is functionally dependent upon $F_1, \ldots, F_r$, so that functions Ω_{ij} of r variables exist for which*

$$(F_i, F_j) = \Omega_{ij}(F_1, \ldots, F_r), \quad i, j = 1, \ldots, r. \tag{1.15}$$

How does knowledge of $F_1, \ldots, F_r$ simplify the problem of solving $F_1(x,p) = 0$?

Although after Jacobi's death several mathematicians had dealt with this problem in the case that the equations in (1.15) have the special form $(F_i, F_j) = 0, \pm 1$,[5] Lie was the first to tackle the general problem, inspired by the fact that for him, it was an expression of his idée fixe. To understand how he could perceive the problem in this manner it is first necessary to consider those aspects of his work during 1869–71 that were conducive to such a perception.

2.2. The Calculus of Infinitesimal Transformations

As we saw in Chapter 1, the notion of a continuous group of transformations that had emerged with Klein's assistance and encouragement from

5 Lützen [1990:670ff.] discusses the work on the problem by Bertrand, Bour, and Liouville.

Lie's work on tetrahedral complexes was rather vague, but two characteristic properties of such groups were emphasized. The first was that a family $\mathfrak{G}$ of invertible transformations $T : y \to y'$, where $y = (y_1, \ldots, y_m)$, has the "group property" and thus forms a group if the composite of two transformations from $\mathfrak{G}$ is also in $\mathfrak{G}$. Lie and Klein took it for granted that, as in the case of permutation groups, it followed that if $T \in \mathfrak{G}$, then $T^{-1} \in \mathfrak{G}$. Closure under composition is what made $\mathfrak{G}$ a group. What made it continuous was the fact that its transformations were generated by *infinitesimal transformations.* Already in their *Comptes Rendus* notes on W-curves and surfaces (Section 1.2) it is clear that for Klein and Lie an infinitesimal transformation

$$\delta T : (y_1, \ldots, y_m) \to (y_1 + dy_1, \ldots, y_m + dy_m), \quad dy_i = \eta_i(y)dt,$$

was regarded as associated to the system of ordinary differential equations

$$\frac{dy_i}{dt} = \eta_i(y), \quad i = 1, \ldots, m. \tag{2.1}$$

If $y_i(t) = f_i(t, y^{(0)})$, $i = 1, \ldots, m$, is the solution to (2.1) satisfying the initial condition $y_i(t_0) = y_i^{(0)}$, then these equations defined the one-parameter family of transformations $T_t : y^{(0)} \to y(t)$ generated by δT. In fact, by 1872, Lie had more or less identified the infinitesimal transformation δT with the system (2.1). In documents from that period he would speak of an infinitesimal transformation but write down a system of ordinary differential equations. Lie thought of the infinitesimal transformations of a group as belonging to it. Indeed, he tended to think of groups in terms of their infinitesimal transformations.

To mathematicians familiar with the post-Jacobi theory of partial differential equations, the system (2.1) could be identified with the linear equation

$$X(f) = \sum_{i=1}^{m} \eta_i(y) \frac{\partial f}{\partial y_i} = 0,$$

thanks to Lagrange's theorem and Jacobi's operator notation. As Lie became steeped in this theory, he eventually identified the infinitesimal transformation δT with the differential operator $X(f)$. This identification will be expressed by the notation

$$\delta T \leftrightarrow X(f).$$

By virtue of the identification, relations involving infinitesimal transformations translate into relations on the Jacobi differential operators $X(f)$. I will now consider the basic properties of infinitesimal transformations discovered by Lie and their expression in terms of Jacobi's differential operators. It would seem that Lie had discovered them by the end of 1872, although their consistent expression in terms of the Jacobi operators $X(f)$

came about a year later.[6] As we shall see, they facilitated his ability to see Jacobi's problem in group-related terms, as a manifestation of his idée fixe, and, combined with his discoveries regarding contact transformations, ultimately convinced him that a general theory of continuous transformation groups was within his reach.

Suppose that $\delta U : y \to y + dy$, $dy_i = \mu_i(y)dt$, $i = 1, \ldots, m$, is a second infinitesimal transformation, so that $\delta U \leftrightarrow Y(f)$, where

$$Y(f) = \sum_{i=1}^{m} \mu_i(y) \frac{\partial f}{\partial y_i}.$$

We wish to consider the result of applying first δT and then δU to y, so as to obtain the composite infinitesimal transformation that I will denote by $\delta U \circ \delta T$. By virtue of the remarks following (2.1), if the infinitesimal transformation δT is applied "continually" to a point y, then it generates the ∞^1 "finite transformations" $y' = T_t(y)$ that represent the solution $y'_i = f_i(y,t)$, $i = 1, \ldots, m$, to the system of ordinary differential equations $dy'_i/dt = \eta_i(y')$ satisfying the initial condition $y' = y$, when $t = 0$.

Consider the Taylor expansion of $y'_i = f_i(y,t)$ with respect to the variable t:

$$y'_i = f_i(y,t) = f_i(y,0) + \frac{\partial f_i}{\partial t}(y,0)t + \frac{\partial^2 f_i}{\partial t^2}(y,0)t^2 + \cdots.$$

Since direct calculation shows that

$$\frac{\partial f_i}{\partial t}(y,0) = \eta_i(y) \quad \text{and} \quad \frac{\partial^2 f_i}{\partial t^2}(y,0) = \sum_{k=1}^{m} \frac{\partial \eta_i}{\partial y_k}(y)\eta_k(y),$$

the series expansion may be written as

$$y'_i = y_i + \eta_i(y)t + \left[\sum_{k=1}^{m} \eta_k(y) \frac{\partial \eta_i}{\partial y_k}\right] \frac{t^2}{2} + \cdots. \tag{2.2}$$

The infinitesimal transformation δT is given by the expansion (2.2) with t infinitely small, say $t = d\tau$, so that, neglecting higher-order infinitesimals, $y'_i = y_i + \eta_i(y)d\tau$ or $dy_i = y'_i - y_i = \eta_i(y)d\tau$, which defines δT.

Consider now the ∞^1 transformations $y'' = U_s(y')$ generated by δU with analogous series expansion

$$y''_i = y'_i + \mu_i(y')s + \left[\sum_{k=1}^{m} \mu_k(y') \frac{\partial \mu_i}{\partial y_k}(y')\right] \frac{s^2}{2} + \cdots. \tag{2.3}$$

Then the expansion for $y'' = U_s(y') = U_s(T_t(y))$ is obtained by substituting (2.2) in (2.3) and expanding terms such as $\mu_k(y') = \mu_k(y + \eta_k(y)t + \cdots)$ in

---

6 See [Hawkins 1991:§7] for the evidence supporting these dates, as well as the evidential basis for attributing to Lie the lines of reasoning given below.

Taylor expansions about y. After simplification, the result is, up to terms of the second order,

$$y_i'' = y_i + \eta_i t + \mu_i s + \Phi_i(s,t) + \cdots, \tag{2.4}$$

where

$$\Phi_i(s,t) = \left[\sum_{k=1}^{m} \eta_k \frac{\partial \eta_i}{\partial y_k}\right]\frac{t^2}{2} + \left[\sum_{k=1}^{m} \mu_k \frac{\partial \mu_i}{\partial y_k}\right]\frac{s^2}{2} + \left[\sum_{k=1}^{m} \frac{\partial \eta_i}{\partial y_k}\mu_k\right] ts,$$

and the functions η_j, μ_j and their derivatives are all evaluated at y.

If the higher-order infinitesimals $t^2, s^2, st, \ldots, t = d\tau$, $s = d\sigma$, are neglected, as was customarily done in considering the composition of infinitesimal motions, then the composite $\delta U \circ \delta T$, sends y_i to $y_i'' = y_i + \eta_i d\tau + \mu_i d\sigma$. Lie accepted this notion of composition as well, which means the following:

(2.5) *If* $\delta T \leftrightarrow X(f)$ *and* $\delta U \leftrightarrow Y(f)$, *then* $\delta U \circ \delta T \leftrightarrow X(f) + Y(f)$.

Of course, (2.5) says that composition is always commutative at the infinitesimal level, but this is not always so for the corresponding finite transformations. In order to characterize, at the infinitesimal level, when the finite transformations T_t and U_s generated respectively by δT and δU commute, Lie focused upon the second-order terms $\Phi_i(s,t)$ in the expansion (2.4). That is, by similar reasoning, the series expansion for $y^{**} = T_t(U_s(y))$ is

$$y_i^{**} = y_i + \eta_i t + \mu_i s + \Psi_i(s,t) + \cdots, \tag{2.6}$$

where

$$\Psi_i(s,t) = \left[\sum_{k=1}^{m} \eta_k \frac{\partial \eta_i}{\partial y_k}\right]\frac{t^2}{2} + \left[\sum_{k=1}^{m} \mu_k \frac{\partial \mu_i}{\partial y_k}\right]\frac{s^2}{2} + \left[\sum_{k=1}^{m} \frac{\partial \mu_i}{\partial y_k}\eta_k\right] ts.$$

Comparison of (2.4) and (2.6) shows that these expressions for y_i'' and y_i^{**} differ in the ts term. If δT and δU commute in the sense that $y'' = U_s(T_t(y)) = T_t(U_s(y)) = y^{**}$ for all y, then the ts terms in $\Phi_i(s,t)$ and $\Psi_i(s,t)$ must agree for all y and for all i, which means that

$$\sum_{k=1}^{m}\left(\eta_k \frac{\partial \mu_i}{\partial y_k} - \mu_k \frac{\partial \eta_i}{\partial y_k}\right) = 0, \quad i = 1, \ldots, m. \tag{2.7}$$

For anyone (such as Lie) familiar with Jacobi's calculus of differential operators, (2.7) states that $X(\mu_i) - Y(\eta_i) = 0$. Furthermore, since Jacobi had shown in [1862:§23] that

$$X(Y(f)) - Y(X(f)) = \sum_{i=1}^{m}\big(X(\mu_i) - Y(\eta_i)\big)\frac{\partial f}{\partial y_i}, \tag{2.8}$$

it follows that (2.7) means that $X(Y(f)) - Y(X(f)) \equiv 0$. On the basis of such considerations Lie concluded that the infinitesimal transformations $\delta T \leftrightarrow X(f)$ and $\delta U \leftrightarrow Y(f)$ generate finite transformations that commute precisely when $X(Y(f)) - Y(X(f)) \equiv 0$. The above reasoning has another consequence that Lie also realized by the end of 1872:

(2.9) *If infinitesimal transformations $\delta T \leftrightarrow X(f)$ and $\delta U \leftrightarrow Y(f)$ belong to a continuous transformation group $\mathfrak{G}$ (in the sense that they generate 1-parameter families T_t and U_s that belong to $\mathfrak{G}$), then the infinitesimal transformation corresponding to $Z(f) = X(Y(f)) - Y(X(f))$ belongs to $\mathfrak{G}$.*

This can be seen as follows. Since $T_t U_s [U_s T_t]^{-1}$—the commutator of T_t and U_s as we would now say— is clearly in $\mathfrak{G}$ and since $y^{**} = T_t U_s [U_s T_t]^{-1}(y'')$, it follows from (2.4) and (2.6) that for $t = d\tau$ and $s = d\sigma$ infinitely small,

$$(2.10)\quad y_i^{**} = y_i'' + \Big[\sum_{k=1}^{m} \eta_k \frac{\partial \mu_i}{\partial y_k} - \mu_k \frac{\partial \eta_i}{\partial y_k}\Big] d\tau d\sigma = y_i'' + \big[X(\mu_i) - Y(\eta_i)\big] d\tau d\sigma.$$

The infinitesimal transformation (2.10), which may be denoted by $dy_i = \big[X(\mu_i) - Y(\eta_i)\big]dt$, corresponds to the differential operator $X(Y(f)) - Y(X(f))$ by virtue of (2.8), and so (2.9) follows. In terms of the composition of infinitesimal transformations, $[\delta T \circ \delta U] \circ [\delta U \circ \delta T]^{-1} \leftrightarrow Z(f)$.

It was also obvious to Lie that if $\delta T \leftrightarrow X(f)$ belongs to $\mathfrak{G}$, then so does the infinitesimal transformation corresponding to $cX(f)$, for any constant c, since this infinitesimal transformation generates the same one-parameter family of transformations T_t of $\mathfrak{G}$.[7] This fact, together with (2.5) and (2.9), indicates that by the end of 1872 Lie had realized that the infinitesimal transformations belonging to a continuous group $\mathfrak{G}$ form what is now called a Lie algebra. As momentous as this realization may seem to us today—by virtue of the important role Lie algebras have come to play in the theory of Lie groups—to Lie in 1872 it evidently did not appear so promising. It was not until he was able to see that the "Lie algebra" of the $X(f)$ could be used to solve group classification problems and to realize the hopes of his idée fixe that he came to appreciate the importance of this construct. As we shall see in the following sections, this had occurred by the winter of 1873–74.

In this connection some additional properties of infinitesimal transformations known to Lie by 1872 need to be mentioned. Recall that Lie's idée fixe involved seeking to imitate the Galois theory of polynomial equations by deducing information about the integration of a differential equation from knowledge that it admits a group of transformations, i.e., transformations that take solutions into solutions. From the viewpoint of Lie's theory of contact transformations (Section 1.5), a partial differential equation is just an

[7] If $y_i = \varphi_i(t)$ is a solution to the original system (2.1), then $y_i = \varphi_i(ct)$ is a solution to the system corresponding to $cX(f)$. This implies that the one-parameter family of transformations T_t^* generated by $cX(f)$ is $T_t^* = T_{ct}$.

equation $F(y_1, \ldots, y_m) = 0$ where $(y_1, \ldots, y_m) = (x_1, \ldots, x_n, z, p_1, \ldots, p_n)$ and the permissible transformations are contact transformations. With this in mind, suppose that an equation $F(y) = F(y_1, \ldots, y_m) = 0$ admits an infinitesimal transformation $\delta T \leftrightarrow X(f)$. Thus if $y^{(0)}$ is a solution to $F(y) = 0$, then $F(\delta T(y^{(0)}) = F(y^{(0)} + dy) = 0$, where $\delta T : y \to y + dy$. Since the Taylor expansion of $F(y^{(0)} + dy)$ is

$$F(y^{(0)}) + \sum_{i=1}^{m} \frac{\partial F}{\partial y_i} \eta_i(y^{(0)}) dt + \cdots = F(y^{(0)}) + X(F) dt + \cdots,$$

where $X(F)$ is evaluated at $y = y^{(0)}$, Lie was led to the following conclusion:

(2.11) *The equation $F(y) = 0$ admits $\delta T \leftrightarrow X(f)$ if and only if $F(y^{(0)}) = 0$ implies $X(F) = 0$ for $y = y^{(0)}$.*

Similar Taylor series considerations led to the analogous result:

(2.12) *The function $F(y)$ is left invariant by $\delta T \leftrightarrow X(f)$, i.e., $F(y + dy) = F(y)$ for all y, if and only if $X(F) = 0$ for all y.*

Let us now consider how these observations about infinitesimal transformations were related by Lie to Jacobi's problem (1.14).

2.3. Function Groups

Lie's observations about infinitesimal transformations apply in particular to infinitesimal contact transformations, i.e., infinitesimal transformations of the $2n + 1$ variables $(x, z, p) = (x_1, \ldots, x_n, z, p_1, \ldots, p_n)$,

$$\delta T : (x, z, p) \to (x + dx, z + dz, p + dp),$$

that leave invariant the Pfaffian equation $dz - (p_1 dx_1 + \cdots + p_n dx_n) = 0$ in the sense of Section 1.5. As we saw at the end of Chapter 1, by 1872 Lie's efforts were focused on the goal of an "invariant theory of contact transformations" in the sense of Klein's *Erlanger Programm.* And of course in the back of his mind was his idée fixe, which, within the context of groups of contact transformations, becomes: Given a partial differential equation (or a system of such) that admits a group of contact transformations, i.e., a group of contact transformations taking solutions into solutions, show how this information translates into information about the integration of the equation (or system).

At about this time (1872) Lie made a discovery that enabled him to carry out his idée fixe. He discovered that corresponding to an infinitesimal contact transformation δT is a partial differential equation $W(x, z, p) = 0$ that completely determines δT and makes it possible to view Jacobi's problem (Section 1) as a manifestation of his idée fixe.[8] The function $W(x, z, p)$

[8] See Section 5 of [Hawkins 1991], especially Theorems 5.1, 5.2, and 5.4.

he later termed the *characteristic function* of the contact transformation δT. For the purposes at hand it will suffice to state Lie's discovery of characteristic functions for the special class of contact transformations he called *transformations of* (x, p). These are contact transformations with equations of the form

$$dx_i = \xi_i(x,p), \quad dz = \zeta(x,p)dt, \quad dp_i = \pi_i(x,p). \tag{3.1}$$

In the notation of Section 2, $\delta T \leftrightarrow X(f)$, where

$$X(f) = \sum_{i=1}^{n} \xi_i \frac{\partial f}{\partial x_i} + \zeta \frac{\partial f}{\partial z} + \sum_{i=1}^{n} \pi_i \frac{\partial f}{\partial p_i}. \tag{3.2}$$

For them Lie's discovery takes the following form:[9]

(3.3) **Lie's Characteristic Function Theorem.** *If δT is an infinitesimal contact transformation of (x, p) with equations (3.1), then a function $W(x,p)$ exists such that $\xi_i = \frac{\partial W}{\partial p_i}$, $\pi_i = -\frac{\partial W}{\partial x_i}$, and $\zeta = \sum_{k=1}^{n} p_k \frac{\partial W}{\partial p_k} - W$. Conversely, if $W(x,p)$ is any function of (x,p), the equations*

$$dx_i = \Big(\frac{\partial W}{\partial p_i}\Big)dt, \quad dz = \Big(\sum_{k=1}^{n} p_k \frac{\partial W}{\partial p_k} - W\Big)dt, \quad dp_i = -\Big(\frac{\partial W}{\partial x_i}\Big)dt \tag{3.4}$$

define an infinitesimal contact transformation of (x, p).

By virtue of this theorem, in addition to the correspondence $\delta T \leftrightarrow X(f)$ we have a correspondence $\delta T \leftrightarrow W$ and so a three-way correspondence

$$\delta T \leftrightarrow X(f) \leftrightarrow W$$

that needs to be kept in mind. If the relations in (3.4) are substituted in (3.2), we have the following:

(3.5) *If $\delta T \leftrightarrow X(f) \leftrightarrow W$ is a contact transformation of (x, p), then $X(f) = (W, f)$ for all functions $f = f(x,p)$.*

Another important fact about characteristic functions that Lie realized is this:

(3.6) *If $\delta T \leftrightarrow X(f) \leftrightarrow W$ is a contact transformation of (x, p), then a partial differential equation $F(x,p) = 0$ admits δT if and only if $F(x_0, p_0) = 0$ implies $(W, F)_{(x_0,p_0)} = 0$.*

In view of (3.5), (3.6) is a direct consequence of Lie's characterization (2.11) of what it means for an infinitesimal transformation to be admitted by an equation. Expressed in the language of the period, (3.6) says that when a

---

[9] See Theorem 7.5 of [Hawkins 1991] and the related discussion.

partial differential equation $F = 0$ admits an infinitesimal contact transformation, then the partial differential equation defined by its characteristic function, namely $W = 0$, is in involution with $F = 0$.[10]

It was by virtue of (3.5) and (3.6) that Lie was able to understand Jacobi's problem in terms of his idée fixe and more generally to interpret the theory of first-order equations in group-related terms. Before considering how (3.5) and (3.6) play this important role, it is instructive to pause for a moment and consider how Lie was led to the connection between an infinitesimal contact transformation and its characteristic function W. His discovery of this connection illustrates the important role played by his geometrical background and orientation in coloring his creative response to the theory of partial differential equations. Technical details will be omitted but can be found in my essay [1991:227–30].

As we saw in Section 2, by virtue of Lagrange's theorem (Section 1) Lie naturally associated an infinitesimal *point transformation* such as $\delta T : (x, z) \to (x + dx, z + dy)$, $(x, z) = (x_1, \ldots, x_n, z)$ with a partial differential equation $X(f) = 0$. Now, point transformations are a special type of contact transformation. This was noted in Section 1.4 for contact transformations of ordinary space, and it is true of contact transformations in general. (In effect, a point transformation $(x, z) \to (x', z')$ can be "prolonged" to define a contact transformation $(x, z, p) \to (x', z', p')$.) As we saw in Section 1, for infinitesimal point transformations δT, by virtue of Lagrange's theorem a linear partial differential equation $W(x, z, p) = 0$ is naturally associated to the system of ordinary differential equations corresponding to δT, and this W turns out to be the characteristic function.[11] To extend this relatively simple observation to a general result, Lie proceeded by a line of reasoning that was basic in projective geometry, as he surely realized: Take a simple observation (e.g., about a circle) and extend it to a general theorem (e.g., about conics) by the application of projective transformations. The only difference was that Lie replaced projective by contact transformations; it was the geometry defined by the group of contact transformations—his invariant theory of contact transformations—that had engaged his interest since his work on the line-to-sphere mapping. By applying noninfinitesimal contact transformations, Lie transformed his simple observation about infinitesimal point transformations into his general characteristic function theorem (of which theorem 3.3 is a special case). In this way a basic proposition of the theory of partial differential equations (Lagrange's theorem) was transformed into the proposition that provided the key to Lie's development of group-related ideas within the context of Jacobi's theory.

---

[10] Two partial differential equations $F(x, p) = 0$ and $W(x, p) = 0$ are in involution when $F(x, p) = 0, W(x, p) = 0$ implies $(F, W)_{(x,p)} = 0$. See [Hawkins 1991:225, 231].

[11] Actually, the inhomogeneous form of Lagrange's theorem is needed; see the more detailed discussion in [Hawkins 1991:201–3].

Let us now consider how Lie applied (3.5)–(3.6) to Jacobi's problem: *Suppose* $f = F_1, \ldots, F_r$ *are* r *functionally independent solutions to* $(F_1, f) = 0$ *with the property that functions* Ω_{ij} *of* r *variables exist for which* $(F_i, F_j) = \Omega_{ij}(F_1, \ldots, F_r), i, j = 1, \ldots, r$. *How does knowledge of* $F_1, \ldots, F_r$ *simplify the problem of solving* $F_1(x, p) = 0$? It follows from the chain rule that any function of $F_1, \ldots, F_r$, $F = \Theta(F_1, \ldots, F_r)$, is also a solution to $(F_1, f) = 0$. Thus in Jacobi's problem all solutions of the form $F = \Theta(F_1, \ldots, F_r)$ are in effect known, and the problem is to use this information to simplify the integration of $F_1(x, p) = 0$. In what is to follow, let $\mathfrak{F}$ denote the set of all functions of the form $F = \Theta(F_1, \ldots, F_r)$ for some Θ, i.e., the set of all "given" solutions $f = F$ to $(F_1, f) = 0$. Equipped with the Poisson bracket, $\mathfrak{F}$ forms an infinite-dimensional Lie algebra.

Now, in Lie's mind $F(x, p) = \Theta(F_1, \ldots, F_r) \in \mathfrak{F}$ could be regarded as a characteristic function and hence identified with an infinitesimal transformation $\delta T \leftrightarrow X(f) \leftrightarrow F$. Moreover, by virtue of (3.5)–(3.6) the fact that $(F_1, F) = 0$ means that δT is admitted by the partial differential equation $F_1(x, p) = 0$. Thus if $\mathfrak{g}_{\mathfrak{F}}$ denotes the set of all infinitesimal contact transformations $\delta T \leftrightarrow X(f) \leftrightarrow W$ with characteristic functions W that belong to $\mathfrak{F}$, they have the group property, since if $\delta T_i \leftrightarrow X_i(f) \leftrightarrow W_i \in \mathfrak{g}_{\mathfrak{F}}$ for $i = 1, 2$, then $\delta T_1 \circ \delta T_2 \leftrightarrow X_1(f) + X_2(f) \leftrightarrow W_1 + W_2$ is in $\mathfrak{g}_{\mathfrak{F}}$, since $W_1 + W_2$ is in $\mathfrak{F}$. Likewise, the Jacobi identity for Poisson brackets shows that the commutator $[\delta T_1 \circ \delta T_2] \circ [\delta T_2 \circ \delta T_1]^{-1} \leftrightarrow X_1(X_2(f)) - X_2(X_1(f))$ has characteristic function (W_1, W_2) and so belongs to $\mathfrak{g}_{\mathfrak{F}}$. Apparently, in Lie's mind the infinitesimal transformations comprising $\mathfrak{g}_{\mathfrak{F}}$ generated by integration a group $\mathfrak{G}_{\mathfrak{F}}$ of noninfinitesimal (or finite) transformations that are admitted by the partial differential equation $F_1(x, p) = 0$. However, he tended to be content to work with the "group" (i.e., Lie algebra) of infinitesimal transformations $\mathfrak{g}_{\mathfrak{F}}$. Adhering to this preference on Lie's part, we may express Jacobi's problem as a manifestation of Lie's idée fixe: *Given that the partial differential equation* $F_1(x, p) = 0$ *admits the group* $\mathfrak{g}_{\mathfrak{F}}$, *how does knowledge of* $\mathfrak{g}_{\mathfrak{F}}$ *simplify the problem of solving* $F_1(x, p) = 0$? Lie provided an answer to this question and called the attendant theory his theory of function groups, a theory that can now be seen as a precursor of the modern theory of Poisson structures.[12]

Lie realized that the theory of function groups transcended its original objective, the solution of Jacobi's problem, and provided the mathematical core of his prospective invariant theory of contact transformations. Thus his final definition was independent of Jacobi's problem and runs as follows. Given functionally independent functions of (x, p), $F_1, \ldots, F_r$ that satisfy relations of the form $(F_i, F_j) = \Omega_{ij}(F_1, \ldots, F_r)$ (but that need not be solutions to $(F_1, f) = 0$), let $\mathfrak{F}$ consist of all functions $F(x, p)$ that are functionally dependent on $F_1, \ldots, F_r$, so that $F = \Theta(F_1, \ldots, F_r)$ for some

[12] See [Olver 1986] for the definition of a Poisson structure (p. 380) and for the historical connections with function groups (pp. 417–18). See also [Hermann 1976:146–7] and [Weinstein 1983].

function Θ. Then $\mathfrak{g}_{\mathfrak{F}}$ consists of all infinitesimal contact transformations of (x, p) with characteristic functions in $\mathfrak{F}$. In his publications, written for nineteenth-century analysts with no understanding of group-related concepts, Lie officially designated $\mathfrak{F}$ as an "r-term function group" and made no reference to $\mathfrak{g}_{\mathfrak{F}}$, except for the telltale word "group," which he usually used without the adjective "function." Of course, equipped with the Poisson bracket, $\mathfrak{F}$ is a Lie algebra in its own right and the mapping $F \to \delta T \leftrightarrow F$ is a homomorphism with kernel consisting of the constant functions, so that the distinction between $\mathfrak{F}$ and $\mathfrak{g}_{\mathfrak{F}}$ is more significant conceptually and psychologically than mathematically.

Lie's solution to Jacobi's problem involved considering those "distinguished" infinitesimal transformations $\delta T \leftrightarrow W$ from $\mathfrak{g}_{\mathfrak{F}}$ that commute with all $\delta U \in \mathfrak{g}_{\mathfrak{F}}$. Since, as noted above, the commutator of δT and $\delta U \leftrightarrow V$ has characteristic function (W, V), $\delta T \leftrightarrow W$ is distinguished when $(W, V) = 0$ for all $V \in \mathfrak{F}$. Suppose $\mathfrak{g}_{\mathfrak{F}}$ contains m such "distinguished" transformations $\delta T_i \leftrightarrow W_i$ with the properties that the characteristic functions W_i are functionally independent and if $\delta T \leftrightarrow W$ is any distinguished transformation of $\mathfrak{g}_{\mathfrak{F}}$, then W is functionally dependent on the W_i, so that $W = \Theta(W_1, \ldots, W_m)$ for some function Θ. The distinguished transformations thus form an m-term "subgroup" (i.e., Lie subalgebra), $\mathfrak{h}_{\mathfrak{F}}$, of $\mathfrak{g}_{\mathfrak{F}}$. Lie's solution to Jacobi's Problem depended upon knowing m:

(3.7) **Theorem.** *If $F_1(x, p) = 0$ admits the r-term group $\mathfrak{g}_{\mathfrak{F}}$ and if the distinguished subgroup $\mathfrak{h}_{\mathfrak{F}}$ is m-term, then the integration of $F_1 = 0$ can be achieved by obtaining one solution each to systems of ordinary differential equations of respective orders $m-1, m-2, \ldots, 3, 2, 1$ and $2(n-q-m)-2, 2(n-q-m)-4, \ldots, 6, 4, 2$, where $q = \frac{1}{2}(r-m)$ is always an integer.*

For example, if $F_1 = 0$ involves $n = 10$ independent variables x_i and if it admits an 8-term group $\mathfrak{g}_{\mathfrak{F}}$ with $m = 4$, then its integration requires one solution each of systems of orders $3, 2, 1$ and $6, 4, 2$.

Lie's result does bear a certain superficial resemblance to results from Galois theory. If the orders of the subgroups occurring in a composition series of the Galois group of a polynomial equation are known, then the degrees of the auxiliary polynomial equations needed to resolve the given equation can be specified. In Lie's theorem if something akin to the order of the subgroup $\mathfrak{h}_{\mathfrak{F}}$ of distinguished transformations is known, namely the number m, then the orders of the auxiliary systems of ordinary differential equations required for the solution of the given partial differential equation can be specified. By viewing Jacobi's problem within the framework of groups of contact transformations—the framework of his invariant theory of contact transformations—Lie had begun to realize the hopes of his idée fixe. Hopeful that he could push the analogy even further in this direction, he wrote to Mayer that "the invariant theory of contact transformations reigns over the integration theory of partial differential equations of the first order in the same sense that the theory of substitutions [permutation groups] rules over the theory of algebraic equations!!!"[13]

By means of the identifications $\delta T \leftrightarrow X(f) \leftrightarrow W$ Lie had become convinced that group-related notions were fundamental to the theory of partial differential equations. The above interpretation and resolution of Jacobi's Problem, of course, supported this conviction. Equally important from a psychological standpoint was his discovery that the Poisson–Jacobi theorem of Section 1 could be understood in group-related terms. Lie had been critical of Jacobi's new method because it was not "transparent" (*durchsichtig*).[14] According to Lie, the problem was that the "inner essence" of the much-acclaimed Poisson–Jacobi theorem had not been investigated. What he meant was that he could not see through the theorem to the underlying true grounds for it. For Lie such grounds were to be found in the synthetic, conceptually oriented mode of reasoning he practiced. Jacobi had derived the theorem as a simple consequence of the Jacobi identity because, as he explained, "It is always useful and not without elegance to reduce all propositions to pure identities" [1862:§28]. The pleasure that Jacobi evidently took in reducing theorems to identities reflected his penchant for elegant formal analysis. For Lie, however, the essence of mathematics did not consist in elegant formalisms. Everything should be understandable by means of group-related geometry. The correspondence $\delta T \leftrightarrow X(f) \leftrightarrow W$ provided Lie with the conceptual means for such an understanding of the Poisson–Jacobi theorem.

Recall that the theorem says that if $f = \Phi$ and $f = \Psi$ are two solutions of $(F, f) = 0$, then so is $f = (\Phi, \Psi)$, and this was, for Jacobi, an immediate consequence of the Jacobi identity for the Poisson brackets, which can be expressed in the form

$$((\Phi, \Psi), f) = (\Phi, (\Psi, f)) - (\Psi, (\Phi, f)). \tag{3.8}$$

Lie's explanation of the theorem proceeded along the following lines.[15] Given Φ and Ψ satisfying the hypothesis of the theorem, consider the corresponding infinitesimal contact transformations $\delta T \leftrightarrow X(f) \leftrightarrow \Phi$ and $\delta U \leftrightarrow Y(f) \leftrightarrow \Psi$. From (3.8) it follows that the commutator $\delta T \circ \delta U \circ [\delta U \circ \delta T]^{-1}$ of $X(f) = (\Phi, f)$ and $Y(f) = (\Psi, f)$ has characteristic function (Φ, Ψ). Now, by hypothesis $X(F) = Y(F) = 0$, which means by (2.12) that δT and δU leave F invariant. The totality of all contact transformations of (x, p) that leave F invariant evidently form a group $\mathfrak{G}$, to which δT and δU belong (in Lie's sense that they generate finite transformations in $\mathfrak{G}$). Thus the commutator $\delta T \circ \delta U \circ [\delta U \circ \delta T]^{-1}$ belongs to $\mathfrak{G}$, which means that it leaves F invariant. Since (Φ, Ψ) is the characteristic function of the commutator, this means $((\Phi, \Psi), F) = 0$, which is the conclusion of the Poisson–Jacobi theorem.

---

[13] Letter of 5 July 1874, quoted by Engel on pp. 582–3 of Lie, *Abhandlungen* **4**.
[14] [Hawkins 1991:224].
[15] Cf. the more detailed and documented discussion in [Hawkins 1991:253].

Jacobi's entire proof of the Poisson–Jacobi theorem consisted in observing that his identity for Poisson brackets can be rewritten as (3.8). The desired conclusion then follows with $f = F$. For Lie, by contrast, (3.8) shows that (Φ, Ψ) is the characteristic function of the commutator. The conceptual layer added by Lie is, from a purely formal point of view, superfluous. But elegant formalism was not relevant to Lie's perception of why something was true. From Lie's point of view the Poisson–Jacobi theorem was true because the totality $\mathfrak{G}$ of contact transformations leaving F invariant forms a *group*. Lie's group-related interpretation of the Poisson–Jacobi theorem epitomized his belief that by means of the correspondence $\delta T \leftrightarrow X(f) \leftrightarrow W$ he could understand the theory of first-order partial differential equations completely in terms of groups of infinitesimal transformations. "I have generally succeeded," he wrote cryptically to Mayer in November 1873, "in conceiving of the theory of differential equations ... as a theory of infinitesimal transformations. Such an interpretation is of value because, among other things, it gives a new viewpoint for research."[16] This realization on his part, based on this correspondence and fully perceived by him by November of 1873 when he wrote the above words to Mayer, was a major psychological factor behind his decision, during the winter of 1873–74, to take on the task of creating a theory of continuous transformation groups. Group theory was now perceived by Lie to be relevant to a theory in the mainstream of mathematics—the general theory of first-order partial differential equations.

2.4. The Invariant Theory of Contact Transformations

The mathematical machinery Lie developed to deal with Jacobi's problem—his theory of function groups—was also seen by him as providing a foundation for the study of matters related to the group of all contact transformations and its subgroups, what he referred to loosely as his invariant theory of contact transformations. Eventually, he came to see it as providing as well the mathematical foundation for the theory of transformation groups, so that task of developing such a theory and the concomitant classification problem no longer seemed "absurd and impossible" as it had circa 1872. In this section I will briefly indicate some of the more important features of the theory of function groups that were to support such a perception on Lie's part.

We saw in Chapter 1 that during the period of his geometrical researches Lie had already exhibited some interest in group classification problems. As noted in Section 1.5, he perceived that Klein's scheme for

[16] Quoted by Engel on pp. 582–3 of Lie, *Abhandlungen* **3**.

classifying geometries involved a group classification problem. Earlier, he and Klein had realized that the problem of determining W-configurations in n-dimensional projective space required resolving a group classification problem, although they succeeded in resolving it only for $n = 2$ (Section 1.2). Lie had also observed in passing that by virtue of the line-to-sphere mapping, Jordan's classification of groups of rigid motions in ordinary space could be transferred to a classification of the corresponding groups of projective transformations in 3 variables (Section 1.4). The results achieved by Lie and Klein involved classifying groups of transformations in 2 or 3 variables, but they balked at the prospect of classifying groups involving transformations in an arbitrary number of variables. These were the classification problems that appeared "absurd or impossible."

In developing his theory of function groups, however, Lie obtained a classification result valid for any number of variables. Expressed in the notation of the last section, what he showed was that if $\mathfrak{g}_{\mathfrak{F}}$ is an r-term "group" of infinitesimal contact transformations with m-term distinguished subgroup $\mathfrak{h}_{\mathfrak{F}}$ and if $\mathfrak{g}^*_{\mathfrak{F}^*}$ is r^*-term with m^*-term distinguished subgroup $\mathfrak{h}^*_{\mathfrak{F}^*}$, then $\mathfrak{g}_{\mathfrak{F}}$ and $\mathfrak{g}^*_{\mathfrak{F}^*}$ are similar if and only if $r = r^*$ and $m = m^*$.[17] He regarded this as the most important result of his papers on function groups [1873b:36,49–50]. This is certainly understandable, since it provides the basis for a complete resolution of the classification problem for function groups. In fact, Lie showed that each equivalence class is represented by a simple canonical form. The classification of function groups certainly represented his most satisfying encounter thus far with group classification problems. Perhaps it also served to encourage him to begin thinking that he might be able to tackle classification problems for more extensive classes of groups of contact transformations—at least if he could bring the machinery of his theory of function groups to bear on them as well. As we shall see, about six months after he had developed the theory of function groups, he saw how to do this for all finite-dimensional groups of contact transformations.

Because Lie suppressed the group theory in his publications on function groups, his characterization of similar function groups was couched in terms of the "official" function groups $\mathfrak{F}$ and $\mathfrak{F}^*$. Let $F_1, \ldots, F_r$ denote the r functionally independent functions of (x, p) that form (so to speak) the basis for $\mathfrak{F}$ and let $F_1^*, \ldots, F_{r^*}^*$ denote the same for $\mathfrak{F}^*$. He had proved that a contact transformation of (x, p) exists that transforms the totality of functions in $\mathfrak{F}$ into those in $\mathfrak{F}^*$ if and only if $r = r^*$ and $m = m^*$. His theorem did not assert that the mapping $\mathfrak{F} \to \mathfrak{F}^*$ was effected by transforming each "basis function" F_i into the corresponding F_i^*. If it had, then it would have provided necessary and sufficient conditions that the

[17] Recall that two groups of transformations are similar if there is a mapping of the underlying manifolds on which the groups act that takes the transformations of the one group into those of the other. Here the mapping must also be a contact transformation. See [Hawkins 1991:239ff.] for details.

two systems of partial differential equations, $F_i(x,p) = 0$ and $F_i^*(x^*,p^*) = 0$, could be transformed into each other by a contact transformation, an equivalence of significance to Lie's prospective invariant theory of partial differential equations as noted Section 1.5.

By the fall of 1873 he had discovered how to use the machinery of the theory of function groups to obtain such conditions, not only for systems defining function groups but for more general systems. The first step, a theorem that establishes the conditions for systems that do define function groups, was instrumental in convincing him that he had the tools at his disposal to classify all finite dimensional transformation groups. It may be stated as follows:

(4.1) **Theorem.** *Let functions $F_i(x,p)$ and $F_i^*(x^*,p^*)$ define two r-term function groups $\mathfrak{F}$ and $\mathfrak{F}^*$. Then a contact transformation of (x,p) exists that takes each F_i into the corresponding F_i^* if and only if for all i and j, (F_i, F_j) is the same function of $F_1, \ldots, F_r$ as (F_i^*, F_j^*) is of $F_1^*, \ldots, F_r^*$.*

In other words, the necessary and sufficient condition is that if $(F_i, F_j) = \Omega_{ij}(F_1, \ldots, F_r)$ and $(F_i^*, F_j^*) = \Omega_{ij}^*(F_1^*, \ldots, F_{r^*}^*)$ (by virtue of the function group property), then $\Omega_{ij} = \Omega_{ij}^*$ for all i, j. Theorem 4.1 enabled Lie to establish necessary and sufficient conditions such that any system of functions of (x,p) be transformable by a canonical transformation T into any other system of functions of (x',p') [Lie 1874a: 62–3 (Thm. XIV)]. He had thus solved a problem that he regarded as "of the utmost importance in my theory of invariants."[18] The solution made it possible to determine when two systems of partial differential equations not containing the dependent variable can be transformed into one another by a canonical transformation of (x,p).

There is one final aspect of Lie's invariant theory of contact transformations that needs to be mentioned: his theory of *homogeneous contact transformations.* This theory offers another example of the manner in which Lie combined ideas drawn from projective geometry with those from the theory of partial differential equations. Lie's theory of function groups was applicable to the problem of integrating differential equations of the form $F(x,p) = 0$, that is equations that do not explicitly involve the dependent variable z. This had been the context of the work of Jacobi and his successors. Jacobi had justified the restriction by showing that the integration of $F(x,z,p) = 0$ can be reduced to that of an equation of the form $F(x,p) = 0$, but his demonstration was shown by Bertrand to be flawed. In Imschenetsky's essay [1869:§14]—Lie's textbook on partial differential equations—the following considerations were presented to justify the reduction. Given $F(x,z,p) = 0$, express any solution in the form $\gamma(x_1, \ldots, x_n, z) = 0$. Differentiation of this equation with respect to x_i yields

[18] Letter to Mayer (received 10 November 1873), quoted by Engel in Lie, *Abhandlungen* **4**, p. 513.

$$\frac{\partial\gamma}{\partial x_i} + \frac{\partial\gamma}{\partial z} p_i = 0 \quad \text{or,} \quad p_i = -\frac{\frac{\partial\gamma}{\partial x_i}}{\frac{\partial\gamma}{\partial z}},$$

and therefore $F(x, z, p)$ can be regarded as a function of the variables x_i, z, $\partial\gamma/\partial x_i$, and $\partial\gamma/\partial z$, i.e.,

$$F(x, z, p) = G\big(x_1, \ldots, x_n, z, \frac{\partial\gamma}{\partial x_1}, \ldots, \frac{\partial\gamma}{\partial x_n}, \frac{\partial\gamma}{\partial z}\big).$$

The equation $G = 0$ is a first-order partial differential equation in the $n+1$ independent variables $x_1, \ldots, x_n, z$ that does not explicitly involve the dependent variable γ. Thus Jacobi's new method applies to $G = 0$ and yields a complete solution, which implicitly defines a complete solution to the original equation $F(x, z, p) = 0$ [Hawkins 1991:243]. Thus, as Jacobi had claimed, it suffices to develop a theory of integration for equations of the form $F(x, p) = 0$.

Lie was familiar with these considerations by Imschenetsky.[19] They implied that he was justified in developing his theory of function groups, which relates exclusively to partial differential equations of the form $F(x, p) = 0$. But to Lie, who approached differential equations with the background and inclinations of a geometer, the above considerations had another implication, which in fact motivated him to develop another theory of function groups. Imschenetsky's reduction procedure involves transforming the original partial differential equation $F(x, z, p) = 0$ by means of the variable change

$$(4.2) \qquad x_i = y_i \ \ (i = 1, \ldots, n), \quad z = y_{n+1}, \quad p_i = \frac{-q_i}{q_{n+1}} \quad (i = 1, \ldots, n),$$

where $q_i = \partial\gamma/\partial x_i$, $q_{n+1} = \partial\gamma/\partial z$. By means of (4.2) $F(x, z, p) = 0$ is transformed into $G(y, q) = 0$. The variable change (4.2) became the starting point for Lie's theory of homogeneous function groups [1873a], created largely in February 1873.[20]

The variable change (4.2) formally resembles the variable change

$$p_i = \frac{q_i}{q_{n+1}}, \ i = 1, \ldots, n,$$

from Cartesian coordinates $p_1, \ldots, p_n$ in n-dimensional complex projective space $\mathbb{P}^n(\mathbb{C})$ to homogeneous coordinates $q_1, \ldots, q_{n+1}$. The change to homogeneous coordinates was, of course, fundamental to projective geometry. It makes any polynomial in the p_i correspond to a homogeneous polynomial in the q_i, and it makes any projective transformation of the p_i, namely a transformation of the form

[19] In presenting his theory of homogeneous function groups, Lie expressly referred to Section 14 of "the valuable work of Imschenetsky" [1873a:64, n.1].

[20] See Engel's remarks on pp. 647–8 of Lie, *Abhandlungen*, **3**.

$$p_i' = \frac{a_{i,1}p_1 + \cdots + a_{i,n}p_n + a_{i,n+1}}{a_{n+1,1}p_1 + \cdots + a_{n+1,n}p_n + a_{n+1,n+1}}, \quad i = 1, \ldots, n,$$

where the $a_{i,j}$ form a nonsingular $(n+1) \times (n+1)$ matrix, correspond to the family of linear homogeneous transformations

$$\rho \cdot q_i' = a_{i,1}q_1 + \cdots + a_{i,n+1}q_{n+1}, \quad i = 1, \ldots, n+1.$$

This, of course, corresponds to identifying the group $\mathbf{PGL}(n+1, \mathbb{C})$ of all projective transformations of $\mathbb{P}^n(\mathbb{C})$ with $\mathbf{GL}(n+1, \mathbb{C})/(\rho I)$, where (ρI) denotes the one-dimensional group of all scalar multiples of the identity transformation I.

By virtue of these correspondences, the algebraic aspects of projective geometry can be developed with greater facility and elegance in terms of homogeneous coordinates. Given Lie's background in projective geometry and his expressed interest in applying the ideas of geometry to advance the theory of differential equations, it is likely that the analogy of (4.2) with the change to homogeneous coordinates encouraged him to see whether (4.2) enabled the analytic aspects of the theory of contact transformations and function groups to be developed with greater facility and elegance. Certainly, this is in effect what he discovered, and the fact that he referred to the resultant theory as his theory of homogeneous contact transformations and function groups suggests that he appreciated the analogy.

Consider, for example, the function G into which F is transformed by (4.2). It turns out that G is a homogeneous function of degree zero in the variables q_i, provided that the notion of a homogeneous function is interpreted in a manner suitable to the consideration of differential rather than polynomial equations. Accordingly, a function $G(y, q)$ is said to be *homogeneous of degree h in the q_i* if h is a nonnegative integer such that G satisfies the Euler condition $\sum_{i=1}^{n+1} q_i \partial G/\partial q_i = hG$. It is easily checked that (4.2) transforms any partial differential equation $F(x, z, p) = 0$ into a partial differential equation $G(y, q) = 0$ for which G is homogeneous in the q_i of degree 0 in this generalized sense. Thus there is an analogy with the change to homogeneous coordinates in projective geometry.

Lie discovered that the analogy goes further. Just as projective transformations are the transformations that define projective geometry in the sense of Klein's *Erlanger Programm*, contact transformations define Lie's invariant-theoretic approach to partial differential equations. Projective transformations correspond in homogeneous coordinates to linear homogeneous transformations, which take homogeneous polynomials of a given degree into the same and are generally more amenable to algebraic treatment. Lie considered what type of transformations correspond, via (4.2), to contact transformations. If

$$x_i' = X_i(x, z, p), \quad z' = Z(x, z, p), \quad p_i' = P_i(x, z, p) \tag{4.3}$$

denotes the general contact transformation of the $2n+1$ variables (x, z, p), then under the variable change (4.2), (4.3) takes the form

$$y_i' = Y_i(y, q), \quad q_i' = Q_i(y, q), \quad i = 1, \ldots, n+1, \tag{4.4}$$

where Y_i and Q_i are of homogeneous of degrees 0 and 1, respectively, in the variables q_i. Moreover, the defining condition that the transformation (4.3) be a contact transformation, namely that it leave the Pfaffian equation $dz - (p_1 dx_1 + \cdots + p_n dx_n) = 0$ invariant in the sense of Section 1.5, becomes the condition that

$$q_1 dy_1 + \cdots + q_{n+1} dy_{n+1} = 0$$

be left invariant, which turns out to say that (4.4) defines a special type of contact transformation of the space $\mathfrak{R}_{2N+1}$ of $2N+1$ variables (y, w, q), $N = n + 1$.[21]

In sum, and in the customary x, z, p notation, homogeneous contact transformations are a special type of the contact transformation of (x, p) of the previous section. Among other things, they keep z fixed and are defined by equations with homogeneity properties. Especially important is the fact that infinitesimal homogeneous contact transformations δT are infinitesimal contact transformations of (x, p) for which $dz = 0$, which means, by Lie's characteristic function theorem and its corollaries, that if $\delta T \leftrightarrow X(f) \leftrightarrow W$, then $X(f) = (W, f)$ for all $f = f(x, z, p)$.[22] Although homogeneous contact transformations are a special type of what is already a special type of contact transformation, since *every* contact transformation of $\mathfrak{R}_{2n+1}$ corresponds to a homogeneous contact transformation of $\mathfrak{R}_{2N+1}$, $N = n + 1$, Lie regarded the theory of homogeneous contact transformations as equivalent to the general theory of contact transformations and consequently as especially important.[23]

In particular, the above considerations imply that the theory of partial differential equations $F(x, p) = 0$, where F is homogeneous of degree 0 in the p_i, and of their transformation under homogeneous contact transformations of (x, p) is worthy of special attention because of its relevance to the integration of first-order partial differential equations involving the dependent variable z. In this connection Lie developed his theory of *homogeneous function groups* [1873a]. These are function groups $\mathfrak{F}$ with the property that there exist independent p-homogeneous functions $H_1, \ldots, H_r \in \mathfrak{F}$ such that $\mathfrak{F}$ consists of all functions $f(x, p)$ that are functionally dependent on $H_1, \ldots, H_r$ so that $f = \Theta(H_1, \ldots, H_r)$ for some function Θ of r variables. (Of course, not every function that belongs to a homogeneous group $\mathfrak{F}$ need be homogeneous.) The theory of homogeneous function groups was

---

[21] See [Hawkins 1991:244–5].

[22] Because $dz = 0$, which turns out to be equivalent to the fact that W is homogeneous of degree 1 in the p_i, so that (in the characteristic function theorem of Section 3) $dz = (\sum_{i=1}^n p_i \frac{\partial W}{\partial p_i} - W)dt = 0$.

[23] This opinion, which is implicit in his publications on homogeneous function groups, is made explicit in his paper on the analytical theory of contact transformations [1873c:116], although he nowhere explained it fully. See also [Lie 1874a:24].

the core of his invariant theory of homogeneous contact transformations. Except for the fact that these function groups possess an additional invariant, their theory is analogous to the general theory of function groups.[24] In particular, the analogue of Theorem 4.1 holds:

(4.5). **Theorem.** *Let p-homogeneous functions $H_i(x,p)$ and $H_i^*(x^*,p^*)$ define two r-term homogeneous function groups $\mathfrak{F}$ and $\mathfrak{F}^*$. Then a homogeneous contact transformation exists that takes each H_i into the corresponding H_i^* if and only if for all i and j, (H_i, H_j) is the same function of $H_1, \ldots, H_r$ as (H_i^*, H_j^*) is of $H_1^*, \ldots, H_r^*$.*

As we will now see, Theorem 4.5 played a special role in the considerations that triggered Lie's decision to devote himself to the creation of a theory of continuous groups of transformations, the precursor of the modern theory of Lie groups.

2.5. The Birth of Lie's Theory of Groups

Lie had been led to the study of function groups through Jacobi's problem. These groups share with the group of all contact transformations the property of being infinite-dimensional. Jacobi's work was also ultimately responsible for leading Lie to the realization that finite-dimensional transformation groups are also of importance to the general theory of first-order partial differential equations.[25] As we have seen, Jacobi's new method had reduced the integration of $F(x,p) = 0$ to the problem of determining solutions to systems of linear homogeneous equations

$$X_i(f) = \sum_{j=1}^{m} \eta_{ij}(y) \frac{\partial f}{\partial y_j} = 0, \quad i = 1, \ldots, q,$$

with the property expressed by equation (1.10) above, viz. $X_i(X_j(f)) - X_j(X_i)) = 0$ for all i and j. Following Clebsch we will refer to such systems as *Jacobian systems*. In the course of extending the ideas of Jacobi's new method to the problem of integrating Pfaffian equations, Clebsch [1866] discovered that the problem reduces to the integration of systems $X_i(f) = 0$ with the property that

$$X_i(X_j(f)) - X_j(X_i(f)) = \sum_{k=1}^{q} a_{ijk}(y) X_k(f).$$

[24] The number m' of independent distinguished functions that are homogeneous in the p_i of degree 0 is also an invariant. Either $m' = m$ or $m' = m - 1$ [1873a:92]. Corresponding to the two possibilities for m' are two canonical forms for a homogeneous function group.

[25] The following account is based upon the more extensive and fully documented study in Section 8 of [Hawkins 1991].

Clebsch called such systems *complete systems*. They evidently include all Jacobian systems. Building upon the ideas Jacobi had introduced to deal with Jacobian systems, Clebsch developed the theory of complete systems. The resultant theory represented a generalization of Lagrange's theorem from Section 1, which deals with a system consisting of a single equation. Clebsch's work on Pfaffian equations, which like that of his predecessors focused primarily on the generic case, prompted the work of Frobenius [1877] on Pfaffian equations that contains the result that has since evolved into the Frobenius Theorem on completely integrable systems of vector fields [Olver 1986:70].

The integration of complete systems was thus fundamental to the general theory of partial differential and Pfaffian equations as it stood in 1870. It is consequently not surprising that Lie sought to develop his idée fixe within this context as well. Indeed, he knew this aspect of Clebsch's work very well, since the theory of complete systems had been one of the primary mathematical tools he employed in developing the analytical theory of function groups.

In his deliberations on Jacobi's problem, Lie had seen that if the partial differential equation $F_1(x,p) = 0$ admits a finite number of infinitesimal contact transformations, then, by repeated application of the Poisson–Jacobi theorem, it followed that it could be assumed that $F_1(x,p) = 0$ admits a function "group" $\mathfrak{g}_{\mathcal{F}}$ of infinitesimal transformations. Within the context of complete systems, somewhat analogous reasoning led him to the discovery that if a complete system admits a finite number of infinitesimal transformations, then it can be assumed that the system admits infinitesimal transformations $X_1(f), \ldots, X_r(f)$ that satisfy relations of the form

$$X_i(X_j(f)) - X_j(X_i(f)) = \sum_{i=1}^{r} c_{ijk} X_k(f), \tag{5.1}$$

where the c_{ijk} are constants. In modern terms this means that the $X_i(f)$ form the basis of a finite-dimensional Lie algebra. To Lie it meant that the complete system admits a group that depends continuously upon a finite number of parameters, the type of group that had arisen in his early work related to tetrahedral complexes. These groups could now be seen to be important to the general theory of partial differential equations. Moreover, Lie discovered a link between these groups and his theory of function groups, a link that encouraged him to believe that he was now in a position to solve, for these groups, the sort of classification problem that he had earlier dismissed as impossible.

On the infinitesimal level on which Lie operated, the problem of classifying all finite-dimensional groups of transformations was to determine up to similarity all finite-dimensional groups (Lie algebras) of infinitesimal transformations in m variables,

$$X(f) = \sum_{i=1}^{m} \eta_i(y) \frac{\partial f}{\partial y_i}, \quad y = (y_1, \ldots, y_m).$$

Today Lie's problem would be formulated as that of determining all Lie algebras of vector fields up to local diffeomorphism. In the case $m = 1$, Lie was able to use (5.1) to conclude by elementary considerations that there are only three distinct possibilities, the projective group of the line and its subgroups, which Lie wrote in the form

$$(1) \quad y' = y + b; \quad (2) \quad y' = ay + b; \quad (3) \quad y' = \frac{ay+b}{y+c}.$$

This success undoubtedly helped him appreciate the value of (5.1), but his confidence that he could achieve comparable success in the case of any number of variables came from another source.

As soon as the number of variables exceeds one, there are contact transformations as well as point transformations to consider. Since (as noted in Section 3) point transformations are special types of contact transformations, Lie's problem was to classify all Lie algebras of infinitesimal contact transformations. Furthermore, in view of the discussion in Section 4, it should be clear that Lie could assume without loss of generality that his contact transformations were homogeneous. Lie's problem reduced, in effect, to classifying up to similarity all Lie algebras of infinitesimal homogeneous contact transformations with basis $X_i(f) = (H_i, f)$, $i = 1, \ldots, r$, within the larger group of all homogeneous contact transformations. Since in this case $X_i(f)$ is given by a Poisson bracket, it follows from Jacobi's identity that $X_i(X_j(f)) - X_j(X_i(f)) = ((H_i, H_j), f)$, and so the Lie algebra property (5.1) implies

$$(H_i, H_j) = \sum_{k=1}^{r} c_{ijk} H_k. \tag{5.2}$$

Equation (5.2) showed Lie that a finite-dimensional group has associated to it a function group. That is, suppose that $\rho \leq r$ of the functions H_i are functionally independent and that the notation is chosen so that $H_1, \ldots, H_\rho$ are independent. Then for $k > \rho$, H_k is some function of $F_1, \ldots, F_\rho$ and (5.2) implies that (F_i, F_j) is expressible as a function of $F_1, \ldots, F_\rho$, which is precisely the condition that $F_1, \ldots, F_\rho$ determine a ρ-term function group $\mathfrak{g}_{\mathcal{F}}$. By means of this connection between the r-dimensional group and the ρ-term function group, Lie believed he could classify the r-dimensional groups into general similarity types by applying the results of the theory of function groups. To give some idea of how this worked—and some of the critical points Lie overlooked—it suffices to consider the case $\rho = r$ in which $H_1, \ldots, H_r$ are functionally independent. Suppose $X_i(f) = (H_i, f)$ and $Y_i(f) = (K_i, f)$ define two groups of this type and that in addition (possibly by replacing the original $Y_i(f)$ by suitable linear combinations) the constants c_{ijk} are the same for the $Y_i(f)$. In other words, the assumption is that the two Lie algebras are isomorphic. In view of (5.2) this means, of course, that (H_i, H_j) is the *same* function of $H_1, \ldots, H_r$ as (K_i, K_j) is of $K_1, \ldots, K_r$. Thus the condition of Lie's Theorem 4.5 of Section 4 is satisfied, so that he could infer the existence of a homogeneous contact transformation that takes each function K_i into the corresponding H_i. This meant that the two r-dimensional groups defined by the $X_i(f)$ and by the $Y_i(f)$ were similar.

Lie was able to conclude, in effect, that groups of the type $\rho = r$ are similar if and only if their Lie algebras are isomorphic, so that the classification problem reduces in this case to that of classifying Lie algebras up to isomorphism. Lie envisioned a "limited number" of nonisomorphic possibilities and evidently felt it would not be difficult to determine them. Perhaps he had in mind the fact that the constants c_{ijk} have many constraints upon them due to the Jacobi identity $((H_i, H_k), H_j) + ((H_k, H_j), H_i) + ((H_j, H_i), H_k) = 0$,[26] and of course since $(H_i, H_j) + (H_j, H_i) = 0$, namely,

$$c_{ijk} + c_{jik} = 0, \quad i, j, k = 1, \ldots, r, \tag{5.3}$$

$$\sum_{s=1}^{r} \{c_{iks}c_{sjt} + c_{kjs}c_{sit} + c_{jis}c_{skt}\} = 0, \quad i, j, k, t = 1, \ldots, r. \tag{5.4}$$

The necessity that the constants c_{ijk} satisfy these relations may have convinced him that the number of distinct possibilities was "limited" (as he put it). In the fall of 1873 Lie thus glossed over an algebraic problem that has never been completely resolved! As we will see in Chapter 5, however, Killing solved it, with some help from E. Cartan, for semisimple Lie algebras.

Lie's classification of groups with $\rho < r$ also led to their division into a finite number of types such that two groups within the same type were similar, provided that the constants c_{ijk} could be chosen to be identical [Hawkins 1991:265–6]. Thus the classification of groups involving a finite number of parameters reduced to the consideration of the constants c_{ijk}. Since Lie glossed over the remaining algebraic classification problem of these constants, he naturally felt he had established the tractable nature of his classification problem—everything reduced to the consideration of the structure constants c_{ijk}. Thus, immediately after sketching his classification scheme to Mayer in a letter of 3 February 1874, Lie wrote, "I suspect that these propositions will form the foundations for what I would almost like to call a new discipline. This transformation theory would occupy the same position within the theory of continuous magnitudes that is occupied within the theory of discrete magnitudes by the theory of substitutions

[26] See [Lie 1873–4:591].

[= permutation groups].... I have posited many general propositions on transformation groups, but they do not have the import of the preceding" [1873–4:594].

There is one further aspect of Lie's classification scheme for groups of homogeneous contact transformations that seems relevant to his perception that his work on partial differential equations had prepared him to successfully launch a new discipline. In 1879 Lie explained the events leading to his research program on transformation groups as follows:[27]

> In the course of investigations on first-order partial differential equations, I observed that the formulas that occur in this discipline become amenable to a remarkable conceptual interpretation by means of the concept of an infinitesimal transformation. In particular, the so-called Poisson–Jacobi theorem is closely connected with the composition of infinitesimal transformations. By following up on this observation I arrived at the surprising result that all transformation groups of a simply extended manifold can be reduced to the linear form by a suitable choice of variables, and also that *the determination of all groups of an n-fold extended manifold can be achieved by the integration of ordinary differential equations.* This discovery ... became the starting point of my many years of research on transformation groups.

From what has been already presented, what Lie meant by these remarks should be clear, except for the italicized sentence. As I have explained in [1991:268–9], it would seem that Lie could already have seen that the results and techniques of his theory of function groups could be used in conjunction with the above-mentioned classification scheme to prove the following: *Suppose c_{ijk} is any system of r^3 constants that satisfy (5.3) and (5.4). Then by the integration of ordinary differential equations it is possible to determine r linearly independent homogeneous functions $H_1(x,p), \ldots, H_r(x,p)$ such that*

$$(H_i, H_j) = \sum_{k=1}^{r} c_{ijk} H_k.$$

In other words, given an *abstract* Lie algebra with structure constants c_{ijk} there exist functions of (x,p) that define the infinitesimal transformations $X_i(f) = (H_i, f)$ of an r-dimensional group of homogeneous contact transformations. Thus the problem of determining all r-parameter groups acting on an n-dimensional manifold had been solved in much the same way that, e.g., Jacobi's new method solved the problem of integrating a first-order partial differential equation.

During the critical period when Lie was deciding to commit himself to the task of creating a theory of finite-dimensional transformation groups,

[27] [1880:93]; the emphasis on the penultimate sentence is Lie's.

the link between finite-dimensional groups and his invariant theory of contact transformations played a vital role in convincing him that he was prepared for, perhaps even destined for, this task. The creation of a theory of continuous groups would build upon and be a continuation of his previous work on contact transformations and function groups—his invariant theory of contact transformations. At the beginning of the above-quoted letter to Mayer (3 February 1874), he expressed such a view in the following words:

> You will be interested ... to see that I have found beautiful interpretations of the symbols $A(f)$, $A_iA_k - A_kA_i$, (H_i, H_k) etc. If I am not mistaken the so called operational calculus gains herewith an unexpected conceptual content. It is worth noting that my investigations on [function] groups, homogeneous [function] groups, contact transformations, as well as my older works, lay ready at hand, so to speak, in order to found the new theory of transformation groups [1873–4:586].

The key to Lie's transition from partial differential equations to transformation groups was his notion of a characteristic function (Theorem 3.3), which enabled him to express infinitesimal transformations in terms of Poisson brackets. On the one hand it enabled him to see Jacobi's problem as a manifestation of his idée fixe, and on the other hand, it enabled him to apply the theory of function groups that resulted from this insight to the problem of classifying all finite-dimensional groups, viewed without loss of generality as consisting of homogeneous contact transformations. He put the matter far more succinctly than I have when he wrote that the theorem that every infinitesimal homogeneous contact transformation is of the form $X(f) = (H, f)$ "has become the starting point of my investigations on transformation groups" [1874a:26].

In analyzing the birth of Lie's theory of groups, I have divided his work during 1869–73 into two periods, a geometrically oriented period (1869–71) discussed in Chapter 1 and the period (1871–73) covered in this chapter. From this chapter it should be clear that the influence of Jacobi on Lie, both directly through his work and indirectly through the work of his successors, was pervasive. Without the work of Jacobi it is difficult to imagine the birth of Lie's theory of groups occurring as it did in the winter of 1873–74. However, the same can be said concerning Lie's geometrically oriented investigations as traced in Chapter 1. Although Jacobi's influence upon Lie was indeed pervasive, it was always subservient to the concepts and research programs that had emerged during his geometrical period and that continued to guide his research thereafter. As we saw, during 1869–71 his work was dominated by two successive research projects: the study of the geometry of tetrahedral complexes (1869–70) and the study of the sphere mapping (1870–71). From the first research project came a fundamental concept and a fundamental idea: the concept of a continuous group of transformations and the idea of a continuous analogue of the Galois theory of algebraic equations—his idée fixe. Likewise, a fundamental concept and a related idea originated in the sphere mapping work: the concept of a contact transformation and the idea of an invariant theory of contact transformations.

During the period 1871–73, these concepts and ideas were developed within the remarkably fertile context of the theory of first-order partial differential equations in the form given to it by Jacobi and his successors. As a result the ideas were realized. Lie was able to create a substantial invariant theory of contact transformations and to work out his idée fixe both within that context (the theory of function groups, Theorem 3.7) and within the context of complete systems. In the process, he discovered his interpretive proof of the Poisson–Jacobi theorem, which epitomizes his realization that group-related concepts are fundamental to the general theory of partial differential equations. In Lie's mind, this realization must have increased the significance of the challenge issued at the end of Klein's *Erlanger Programm*, where Klein (speaking also for Lie) had called for the development of a theory of continuous transformation groups to parallel the theory of permutation groups. As was the case in Jordan's *Traité des substitutions*, the theory was to be developed independently of its applications, which would then follow. In October 1872, when Klein wrote his program, Lie felt that such an enterprise was too far-fetched to be taken seriously. By the end of 1873, his pessimism had been replaced by exuberant optimism. He had discovered in the mathematics that he had created in developing his invariant theory and his idée fixe the approach and requisite mathematical tools for laying the foundations of a theory of finite-dimensional continuous groups. During the winter of 1873–74 he began to take up the challenge of the *Erlanger Programm*, to develop the theory of continuous groups. It became his life work.

Chapter 3

Lie's Theory of Transformation Groups: 1874–1893

At the time that Lie was making the decision to devote himself to working out a theory of continuous transformation groups, he was ideally situated to carry out such a monumental task.[1] In February 1872 the Norwegian National Assembly established by a large majority vote a personal chair of mathematics for him. This gesture on the part of the Norwegian government was prompted by Lie's application for a position that had opened up in Lund, Sweden. The Norwegian professorship put Lie in a position to devote himself with few distractions to his creative work. Teaching duties seem to have been minimal and to have required little or no preparation time, and although he was involved with the publication of Abel's collected works during the period 1873–81, the lion's share of the work was done by Sylow.

3.1. The Group Classification Problem

For Lie the central problem in the theory of transformation groups was the classification problem: the problem of determining, up to similarity, all transformation groups in n variables for any n. As we saw in Section 2.5, he was encouraged to grapple with this far-reaching problem by his tentative classification of groups of homogeneous contact transformations representing all groups of both point and contact transformations acting on an n-dimensional manifold $(x_1, \ldots, x_n)$ and generated by r infinitesimal transformations. It implied that there were only a finite number of types to

---

[1] For biographical details on Lie, I have drawn on [Engel 1899] and [Strøm 1994].

consider and that if somehow one knew a viable set of structure constants c_{ijk}, i.e., a set satisfying the conditions (given in Section 2.5) that would in effect define an abstract Lie algebra, then actual infinitesimal transformations could be constructed for this group by integrating systems of ordinary differential equations.

Although such results were theoretically satisfying, they left open the practical problem of actually determining all transformation groups. That is, can the integrations actually be carried out explicitly? And even if they could, how does one arrive at viable structure constants c_{ijk}? Compounding the problem was the fact that as soon as $n > 1$, there are groups for any value of r and hence an infinite number. Along with his efforts to polish up his theoretical treatment of the general classification problem, Lie expended considerable effort on the actual determination (up to similarity) of all groups for small values of n. Judging by his brief note in the Göttingen *Nachrichten* [1874c], by the end of 1874 he had resolved the problem for $n = 2$ to his own satisfaction, using in part geometrical means. But it was not until 1878 that he managed to translate his results into publishable, analytical terms [1878a:78]. At that time, he also announced that he had solved the problem for $n = 3$ but restricted to groups of point transformations [1878a:78, n.2]. However, the calculations needed to do this remained too extensive to make publication feasible [Lie 1893a:122], and Lie contented himself with partial results [1893a:122–262].[2] As for the problem for n arbitrary, in [1888:598] he expressed the view that it would probably never be resolved.

Although Lie's success in dealing with the group classification problem was not as great as he had initially hoped, in the course of dealing with it he developed the mathematical machinery of his theory of transformations groups. According to Lie, all the principal results of the theory were obtained by 1884 [1893a:xiv]. During the 1870s, except for the abovementioned *Nachrichten* note [1874c], he published his results on transformation groups exclusively in Norwegian journals, the proceedings of the scientific society of Christiania and the mathematics journal *Archiv for Mathematik*. Lie was an editor of this journal and may have been instrumental in its creation. In any case, the first issue of the first volume contains Lie's second paper on transformation groups [1876b] (the first being the *Nachrichten* note). Although he wrote his papers in German, by limiting himself to Norwegian publications he achieved quick publication at the expense of ready accessibility of his work outside Norway.[3]

During the 1880s Lie made a greater effort to make his work better known through publications in *Mathematische Annalen*, of which Klein was

[2] For recent work on the $n = 3$ problem for point transformations, see [González-Lopez et al. 1992a; 1992b; 1996].

[3] Lie did write abstracts of his work for *Jahrbuch über die Fortschritte der Mathematik* and Darboux's *Bulletin des sciences mathématiques*, but these were cryptic.

now editor. Part of this was prompted by his discovery in 1882 that a French mathematician by the name of Georges Halphen (1844–89) had developed a theory of differential invariants for the general projective group that yielded special cases of some of his results about differential equations that he had obtained in pursuing his idée fixe.[4] In the fall of 1882 Lie visited Paris and lectured on his own ideas on groups and differential equations. He was beginning to realize the need to make his work better known and, in particular, the need to develop and publish his ideas on the application of groups to the study of differential equations—an area of his work that had been pushed aside by the need to develop the general theory and to deal with the classification problem. This can be seen in his letters to Mayer and Klein, both of whom were then on the faculty of the University of Leipzig.[5] For example, in the spring of 1883 he wrote to Klein, "With each passing year I am more and more certain that the theory of transformation groups will bring the theory of differential equations into an entirely new state. If only I could collect together and edit all my results."[6]

Realizing that Lie was incapable of the task on his own, Klein and Mayer encouraged a young mathematician by the name of Friedrich Engel (1861–1941) to help Lie out. Engel had received his doctorate in 1883 with a dissertation on differential equations written under Mayer's direction. The plan was that Engel would write a *Habilitationsschrift* on a topic on transformation groups while with Lie and at the same time assist Lie with the chore of editing his results into a systematic, readable form—a book on transformation groups. Lie was delighted with the plan, and so Engel arrived in Christiania in September 1884. There he met twice daily with Lie to work on the book, which Lie envisioned as a systematic development of his theory rather than a popular exposition. According to Engel, Lie reasoned out the contents of each chapter for him in conversation and gave an outline of the facts to be presented, "a sort of skeleton, to be clothed by me with flesh and blood" [1899:l]. Engel was amazed at the magnitude of the mathematical edifice that Lie carried almost entirely in his head, i.e., without having presented most of it somewhere in print. Lie studied what Engel wrote, and this led to revisions and the inclusion of new material. By the end of June 1885, when Engel left Christiania, a sizable manuscript had been created. Neither of them realized that it was just the first of many drafts that would grow in size by a factor of five over the course of eight years.

Engel returned to Leipzig, presented his *Habilitationsschrift* on trans-

---

4 See Lie's communication of 10 July 1882 to the Scientific Society in Christiania, on pp. 236–7 of *Abhandlungen* **5** and Engel's discussion [1899:xlvi–xvviii]. Halphen's theory of differential invariants was developed in [1878] and applied to differential equations in a memoir [1884] that won the *Grand Prix* for 1880 of the Académie des Sciences.

5 See [Engel 1899:xlix].

6 NSUB Göttingen, Nachlass Klein X, Nr. 688. See [Hawkins 1982:145] for a transcription of the German text.

Figure 10. Friedrich Engel devoted his career to the development and promulgation of Lie's ideas. He also encouraged Killing to work out his classification of simple groups (Chapter 5).

formation groups [1885], and became a lecturer (*Privatdozent*). Soon he was joined there by Lie. In 1886 Klein accepted a professorship at Göttingen, where he was instrumental in building up a great school of mathematics (discussed in Chapter 9). Lie was persuaded to leave his beloved homeland to become Klein's replacement as professor of geometry at Leipzig. He did so in the hopes of making his work better known to mathematicians and to have students to help with the development of the theory. In particular, he was able to continue the book project with Engel, and the monumental *Theorie der Transformationsgruppen* finally appeared in three volumes [1888; 1890a; 1893a]. Among Lie's students, many of the best came from France, encouraged by Poincaré, Picard, and Darboux to do so in order to learn first-hand about Lie's theory and return to Paris to write a dissertation. (More will be said about Lie's school in Leipzig and about the reception of his theory in Paris in Part III.)

Lie discovered that teaching at Leipzig involved a great deal of work. As he explained to a friend back in Norway, "While in Norway I hardly spent five minutes a day on preparing the lectures; in Germany I had to spend an average of about 3 hours. The language is always a problem and above all, the competition implied that I had to hold 8–10 lectures a week."[7] The additional burden of teaching, combined with the efforts to publish his work, resulted in a nervous breakdown in 1889 from which he never fully recovered. In 1894 the Norwegian National Assembly established a chair in the theory of transformation groups for him. Lie wanted return to his homeland, but his wife and three children were reluctant to leave Leipzig and so Lie remained there until 1898. By that time, he was suffering from cerebral anemia, and it became clear to all concerned that he should return to Norway. During the fall of 1898 he delivered a series of lectures for two months but then had to give up, although he continued to hold seminars in his home for the two students that had accompanied him from Leipzig. Finally the seminars, too, had to stop. Lie died on 18 February 1899.[8]

It is impossible to do justice within the confines of a single chapter to the vast theory that Lie created. The goal of the rest of this chapter is simply to provide an overview of the salient features of Lie's general theory (Section 2) and to discuss in more detail particular aspects of it that are germane to the developments traced in subsequent chapters, such as the adjoint group (Section 3), the role of simple groups in Lie's application of group concepts to differential equations (Section 4), and the symplectic group (Section 5).

3.2. An Overview of Lie's Theory

Faithful to its prehistorical phase (Chapters 1–2), Lie's groups always consisted of sets $\mathfrak{G}$ of invertible transformations $x' = Tx$ acting on the elements $x = (x_1, \ldots, x_n)$ of some portion of complex n-dimensional space $\mathbb{C}^n$.[9] For Lie such a set constituted a group when it was closed under composition. As we saw in Chapters 1–2, this was always the group property for Lie. It makes $\mathfrak{G}$ into what would now be called a semigroup. Most, if not all, of the groups Lie considered, however, were also closed under inversion and thus contained as well the identity transformation. As indicated in Chapter 1, for a long time Lie believed that, as in the case of permutation groups, the additional properties of his transformation groups could be deduced

[7] Quoted from [Strøm 1994:xxi].

[8] For more details on Lie's last years see [Strøm 1994:xxvi–xxviii], which is the basis for my account.

[9] Of course, in the case of contact transformations $x = (x_1, \ldots, x_n, p_1, \ldots, p_n)$ and the ambient space is $\mathbb{C}^{2n}$. Also, for Lie $(x_1, \ldots, x_n)$ usually represented the homogeneous coordinates of a point in $\mathbb{P}^{n-1}(\mathbb{C})$.

from the semigroup property. By the time he began publishing *Theorie der Transformationsgruppen* he had realized that this could not be done, but he retained the "semigroup" definition of a transformation group and developed as much of the theory as he could within this broader framework. In the following description of Lie's theory, however, I will be concerned exclusively with his results as they apply to transformation groups that have the additional properties making them what is now called a group.

We also saw in Chapters 1–2 that closely tied up with the continuity of a group of transformations $\mathfrak{G}$ for Lie was the existence of infinitesimal transformations that generate the finite transformations of the group. For Lie, intuitively, this is what constituted the continuity of $\mathfrak{G}$. Thus in [1883a:314] he wrote, "A group is called *continuous* when all of its transformations are generated by repeating infinitesimal transformations infinitely often ...." By the time the first volume of *Transformationsgruppen* was written he had modified his characterization of continuity to the following: "A transformation group is called *continuous* if it is possible for any given transformation belonging to the group to specify certain other transformations of the group that differ only infinitely little from the given transformation, [and] if, on the other hand, it is not possible to decompose the totality of transformations in the group into individual discrete families" [1888:3]. In both of these definitions, we see Lie trying to convey as a part of his idea of "continuity" a notion of connectedness. In the first definition the connectedness is achieved through the one-parameter families of transformations T_t generated by an infinitesimal transformation and linking each finite transformation through the "curve" T_t with the identity transformation. In the later definition, written with perhaps semigroups in mind, the connectivity enters in the second stipulation that the group does not decompose into "discrete" families, i.e., families that are separated from one another.

This was before the time when topological notions were a common part of mathematics. Indeed, topological considerations remained outside Lie's theory until the mid-1920s when Weyl and, independently, Schreier began to introduce them, as will be seen in Section 12.1.[10] As for Lie, despite his natural proclivity for geometrically flavored intuitive thinking, he turned to analysis to lend concreteness to his vague words. For groups generated by a finite number of infinitesimal transformations, the type that he had returned to in the winter of 1873–74 and on which he now focused most of his attention, this was easy because the "continuity" of such a group was for him assured when its elements are defined by a single system of equations involving a finite number of parameters:

$$x_i' = f_i(x_1, \dots, x_n, a_1, \dots, a_r), \quad i = 1, 2, \dots, n. \tag{2.1}$$

It will be helpful in what is to follow to let T_a denote the transformation defined by (2.1), so that (2.1) can be expressed by $x' = T_a x$. Lie himself used such notation when seeking to express himself in a less formal,

[10] See also my essay [1999b].

more intuitive manner. In *Transformationsgruppen* it is assumed that the functions in (2.1) are analytic in the $n + r$ variables $x_1, \ldots, x_n, a_1, \ldots, a_r$ [1888:11], and this was seen as guaranteeing that the family defined by (2.1) is continuous [1888:311].

A prime example of a continuous transformation group is the projective group of the line, which is defined by the equation

$$x' = \frac{ax + b}{cx + d}, \quad ad - bc \neq 0. \tag{2.2}$$

As an example of a group that is not continuous, Lie gave the group of all coordinate changes in the plane that take one (right-angled) Cartesian coordinate system into another [1888:7], i.e., all translations, rotations and reflections in planes. The transformations of the group are given by two different sets of equations, defining two separated[11] "continuous" families of transformations that are defined respectively by

$$\begin{aligned} x' &= a + x\cos\alpha - y\sin\alpha, \\ y' &= b + x\sin\alpha + y\cos\alpha, \end{aligned} \quad \text{and} \quad \begin{aligned} x' &= a + x\cos\alpha + y\sin\alpha, \\ y' &= b + x\sin\alpha - y\cos\alpha. \end{aligned}$$

The family on the left involves orthogonal transformations of determinant +1, while that on the left involves those with determinant −1.

A group $\mathfrak{G}$ of transformations T_a defined by a single set of equations of the form (2.1) is thus the principal object of study in *Transformationsgruppen*. These are the "finite continuous transformation groups" of the theory. However, an additional assumption on such equations is necessary before they can provide the theoretical starting point of Lie's theory. In (2.1) all the parameters $a_1, \ldots, a_r$ need not be *essential*. According to Lie they are essential "if it is impossible to introduce as new parameters independent functions of $a_1, \ldots, a_r$ so that as a result the equations $x_i' = f_i(x, a)$ contain fewer than r parameters" [Lie 1888:12]. I have given Lie's definition *verbatim* to illustrate the manner in which he typically masked the "local" nature of what he was discussing. For example, the four parameters a, b, c, d of the projective group of the line are not essential; there are for Lie only three essential parameters involved. In a sufficiently small neighborhood of (a_0, b_0, c_0, d_0) where, e.g., $d_0 \neq 0$, one will have $d \neq 0$, and so the equations (2.2) may be written as

$$x' = \frac{\frac{a}{d}x + \frac{b}{d}}{\frac{c}{d}x + 1} = \frac{\alpha x + \beta}{\gamma x + 1}.$$

In Lie's actual definition of essential parameters, however, there is no mention of the fact that new parameters need only be defined locally as functions of the original parameters.

In his theory of transformation groups Lie studied continuous groups defined by equations of the form (2.1) in which the parameters are assumed

[11] The German word used is *diskret*.

to be essential. This is what he calls an "r-term finite continuous group" and I will call an *r-dimensional* continuous group. As the simple example (2.2) illustrates, Lie's theory is really focused on a small neighborhood of a group about the identity transformation, although this is rarely made explicit (as illustrated by the definition of essential parameters). One could say that Lie's theory of transformation groups is a theory of group germs in the sense of Bourbaki [1975:228], but this would be misleading if taken to mean that Lie explicitly and carefully formulated his theory in such terms, for he did not. Quite the contrary, as in his definition of essential parameters, Lie tended to suppress much mention of the truly local nature of his theory. Further examples will be given below.

From the equations (2.1) defining an r-term group $\mathfrak{G}$ it is easy to obtain the corresponding infinitesimal transformations. Let $a^0 = (a_1^0, \ldots, a_r^0)$ correspond to the identity transformation, so $x_i = f_i(x, a^0)$. Then for a fixed integer k between 1 and r if $da = \delta_{kl}dt$, where δ_{kl} is the usual Kronecker delta, then by Taylor expansion in the a-variables,

$$x_i' = f(x, a^0 + da) = f_i(x, a^0) + \left(\frac{\partial f_i}{\partial a_k}\right)_{a^0} dt + \cdots = x_i + \xi_{ki}(x)dt + \cdots,$$

which, neglecting higher powers of dt, leads immediately to the infinitesimal transformation $\delta T_k : x_i \to x_i + \xi_{ki}(x)dt$ or $X_k(f) = \sum_{i=1}^n \xi_{ki}(x)\frac{\partial f}{\partial x_i}$. More generally, setting $da = (\lambda_1 dt, \ldots, \lambda_r dt)$ leads to the infinitesimal transformation $X(f) = \sum_{k=1}^r \lambda_k X_k(f)$. Although Lie thought of infinitesimal transformations of $\mathfrak{G}$ as deriving in this intuitively simple way from (2.1),[12] he never sought to introduce them "officially" in this manner in developing the theory, since among other considerations, he wished to introduce infinitesimal transformations without taking for granted the existence of an identity element.

It was by means of a less simple, more indirect route, much of it developed under the weaker semigroup hypothesis, that Lie established the existence of r linearly independent infinitesimal transformations corresponding to an r-term continuous group $\mathfrak{G}$ and established the following basic theorem [1888: 75]:

(2.3) **Lie's Exponential Map Theorem.** *Given an r-dimensional continuous transformation group $\mathfrak{G}$, there exist r linearly independent infinitesimal transformations $X_k(f) = \sum_{i=1}^n \xi_{ki}(x)\frac{\partial f}{\partial x_i}$, $k = 1, \ldots, r$, that generate $\mathfrak{G}$ in the following sense: If $e_1, \ldots, e_r$ are arbitrary parameters and $X(f) = \sum_{k=1}^r e_k X_k(f)$, then the equations*

$$x_i' = x_i + X(x_i) + \frac{1}{2!}X^2(x_i) + \frac{1}{3!}X^3(x_i) + \cdots \tag{2.4}$$

[12] See [1876a:48–9; Lie 1876b:33].

define a one-dimensional group of finite transformations such that the transformations of $\mathfrak{G}$ are arranged into ∞^{r-1} families of ∞^1 transformations each. "The totality of all these finite transformations is identical with the totality of all transformations of the group $x'_i = f_i(x, a)$."

In (2.4) I have used the notation $X^2(f) = X(X(f))$, $X^{n+1}(f) = X(X^n(f))$, although Lie did not. Lie called the equations (2.4) the "canonical equations" of the group. I will denote them with the more succinct notation $x' = \exp X \cdot x$, although he never employed such notation.

I have stated Lie's theorem more or less as he did to illustrate his tendency to ignore the local nature of his theorems and indeed, as the verbatim part of the theorem indicates, to state them in a way suggesting that they are global results when they are not.[13] In many cases the local nature of Lie's results is never mentioned: systems of equations with nonvanishing Jacobian will be inverted without mentioning that the inversion may hold only locally; likewise, the solutions of differential equations will be used without mention of the local nature of the solution. Of course, Lie was just following suit, since the analysts such as Jacobi and Clebsch, who were his sources for results on differential equations, also ignored such details. However, in the case of the exponential map theorem, its local nature was made more apparent to the attentive reader, since Lie deduced it from a more general theorem about semigroups $\mathfrak{S}$ of transformations [1888:72] in which, uncharacteristically, he made the local nature of the result explicit: Given any $T_{a^0} \in \mathfrak{S}$, then every transformation T_a "in a certain neighborhood" of a^0 is expressible as a composite $T_a = \exp X \circ T_{a^0}$, where $X(f)$ is a suitable linear combination of $X_1(f), \ldots, X_r(f)$.

In Chapter 9 of *Transformationsgruppen* I [1888:146ff.] Lie presented what he termed in *Transformationsgruppen* III [1893a:590] the main theorem (*Hauptsatz*) of his theory of groups:

(2.5) **Lie's Main Theorem.** *Given an r-dimensional continuous group $\mathfrak{G}$, its r linearly independent infinitesimal transformations $X_1(f), \ldots, X_r(f)$ satisfy relations of the form*

$$X_k(X_l(f)) - X_l(X_k(f)) = \sum_{m=1}^{r} c_{klm} X_m(f), \tag{2.6}$$

where the c_{klm} are constants.[14] *Conversely, given r linearly independent infinitesimal transformations $X_1(f), \ldots, X_r(f)$ which satisfy relations of the form (2.6), the totality of all ∞^{r-1} 1-dimensional groups*

$$x' = \exp(e_1 X_1 + \cdots + e_r X_r) \cdot x$$

form an r-dimensional continuous group $\mathfrak{G}$.[15]

[13] Lie's exponential map theorem would correspond in present-day Lie group theory to the theorem that $X \to \exp X$ defines a diffeomorphism from a neighborhood of $(0, \ldots, 0)$ onto a neighborhood of the identity element of $\mathfrak{G}$.

[14] Theorem 22 [1888:150].

[15] Theorem 24 [1888:158].

Lie spoke of the infinitesimal transformations $X_1(f), \ldots, X_r(f)$ as belonging to and generating the group $\mathfrak{G}$, and in fact he denoted $\mathfrak{G}$ by $X_1(f), \ldots, X_r(f)$. Thus for him there was strictly speaking no separate notion of what would now be called a Lie algebra. However, in what is to follow, it will be helpful to let $\mathfrak{g}$ denote the Lie algebra spanned by $X_1(f), \ldots, X_r(f)$.

Lie also introduced the notation (X_k, X_l) for the left-hand side of (2.6). The inspiration for this notation is clear when one recalls that in his early work the infinitesimal transformations $X(f)$ were homogeneous contact transformations and thus given by $X_k(f) = (H_k, f)$, where the right-hand side is the Poisson bracket with f of the associated characteristic function $H_k = H_k(x,p)$. The Jacobi identity then implies that $X_k(X_l(f)) - X_l(X_k(f))$ has characteristic function (H_k, H_l). In what is to follow, I will adhere to the now customary notation of $[X_k, X_l]$ instead of (X_k, X_l). Also, from now on I will drop the "f" in $X(f)$ and write, e.g., $X = \sum_{i=1}^r \xi(x) \frac{\partial}{\partial x_i}$, which is more in line with the usual notation for a vector field.

We saw in Chapters 1–2 that for Lie two groups $\mathfrak{G}$, $\mathfrak{H}$ were to be considered identical precisely when they were similar. Expressed in terms of the finite equations $T_a : x_i' = f_i(x,a)$ and $U_b : y_i' = g_i(y,b)$, defining $\mathfrak{G}$ and $\mathfrak{H}$ respectively, this means that $m = n$ and $r = s$ and invertible analytic variable changes $y = \Phi(x)$ and parameter changes $b = \Psi(a)$ exist that transform the equations of $\mathfrak{H}$ into those of $\mathfrak{G}$ so that $T_a = \Phi^{-1} \circ U_{\Psi(a)} \circ \Phi$.[14] He realized that in the theory of permutation groups two groups $\mathfrak{G}$, $\mathfrak{H}$, were identified when they are isomorphic, i.e., when a bijective mapping $\mu : \mathfrak{H} \to \mathfrak{G}$ exists such that $\mu(S \circ T) = \mu(S) \circ \mu(T)$. Lie realized that his notion of similar groups was not the same as the notion of a group isomorphism. That is, if two transformation groups $\mathfrak{G}$ and $\mathfrak{H}$ are similar, then they are isomorphic, since in the above notation, we may take $\mu(U_b) = \Phi^{-1} \circ U_{\Psi(a)} \circ \Phi$. On the hand, it is easy to see that two groups can be isomorphic without being similar. For an example close to Lie's interests, let $\mathfrak{G}$ denote the general projective group of the plane $\mathbb{P}^2(\mathbb{C})$. In nonhomogeneous coordinates, its equations involve $n = 2$ variables, x_1, x_2,

$$x_i' = \frac{a_{i1}x_1 + a_{i2}x_2 + a_{i3}}{a_{31}x_1 + a_{32}x_2 + a_{33}}, \quad i = 1, 2, \quad \det(a_{ij}) \neq 0.$$

Now consider the action induced by $\mathfrak{G}$ on the straight lines of $\mathbb{P}^2(\mathbb{C})$. In Cartesian coordinates the lines are parametrized by five coordinates, and so the action of $\mathfrak{G}$ on the lines defines a group $\mathfrak{H}$ of projective transformations

[14] In a passage set in small print, Lie mentioned that "in some circumstances, however, it is purely a matter of investigating the family $x_i' = f_i(x,a)$ in the neighborhood of an individual place $x_1, \ldots, x_n a_1, \ldots, a_r$. This investigation is often facilitated by introducing new variables or parameters that satisfy the above stipulated conditions in the *neighborhood of the place in question*" [1888:25].

in $m = 5$ variables that is clearly isomorphic to $\mathfrak{G}$ but cannot be similar to $\mathfrak{G}$, since $m \neq n$.

Lie realized that the permutation group notion of isomorphism is related to what he called the "composition" (*Zusammensetzung*) of the infinitesimal transformations of a group. Two r-dimensional groups $\mathfrak{G}$ and $\mathfrak{H}$ were said to be "equally composed" (*gleich zusammengesetzt*) if the r independent infinitesimal transformations associated to these two groups could be chosen, by taking suitable linear combinations, such that the constants c_{ijk} in (2.6) may be taken to be the same in both groups. This, of course, means in present-day language that the corresponding Lie algebras $\mathfrak{g}$ and $\mathfrak{h}$ are isomorphic. Incidentally, in France, the first two students to make the trip from Paris to Leipzig (in 1887) to study with Lie, J. de Tannenberg and E. Vessiot, used the French word *structure* for Lie's "Zusammensetzung," and this practice was continued by Cartan (Part III) and thereby, it would seem, into present-day mathematics.[15] Expressed in this terminology, what Lie recognized was that it was the structure of Lie groups that decided whether or not they were (locally!) isomorphic.

To establish this result Lie introduced what he called "the parameter group" associated to $\mathfrak{G}$ [Lie 1888:401ff.], which is the closest he came to the idea of an abstract Lie group, i.e., the idea that a Lie group is a manifold of points $a = (a_1, \ldots, a_r)$ with a multiplication defined. If $T_a, T_b \in \mathfrak{G}$, then $T_a \circ T_b \in \mathfrak{G}$, and so $T_a \circ T_b = T_{a'}$. For Lie the equations $a'_i = \varphi(a_1, \ldots, a_r, b_1, \ldots, b_r)$, $i = 1, \ldots, r$, defined a new r-dimensional group of transformations $P_b : a \to a'$ acting (simply transitively) on r-dimensional space. This group he called the parameter group of $\mathfrak{G}$. If we define a multiplication on the "manifold" of points $a = (a_1, \ldots, a_r)$ by $a * b = a'$, then we obtain a Lie group in the now customary sense, and the transformations P_b of the parameter group correspond to right multiplications $P_b : a \to a * b$.

Lie showed that $\mathfrak{G}$ and its parameter group have the same structure (i.e., have isomorphic Lie algebras) and concluded that this meant they are isomorphic [1888:407]. Given that the isomorphisms occurring in the theory of permutation groups are global, one might have expected him to point out the local nature of his isomorphism, but he did not. On the same level, he showed that two groups $\mathfrak{G}$ and $\mathfrak{H}$ have the same structure if and only if they have (after possibly transforming parameters) the same parameter group [1888:414]. He used this result to show that two groups are isomorphic if and only if they have the same structure [1888:418]. He even emphasized that the isomorphism was a one-to-one and onto mapping.[16]

There is little doubt that Lie realized that his theorems were established only locally. As we saw in connection with the exponential map theo-

[15] Regarding the use of "structure" by de Tannenberg and Vessiot, see the end of Section 5.4.

[16] "[E]s [ist] möglich ..., die Transformationen der einen ... eindeutig umkehrbar auf die Transformationen der andern zu beziehen ...."

rem, he even occasionally stated explicitly the local nature of his results. It may have been a reflection of the "generic reasoning" of the analysts upon whom he drew for analytical results that he tended to ignore the local character of his reasoning. That is, it may be that Lie's global statements were his generic theorems, the theorems that would hold "in general," i.e., with the exception of certain atypical or "degenerate" cases. (More will be said about generic reasoning in Part II.) In such a frame of mind, possible exceptions to the generic theorems are not something of concern and may not even be expected.

Perhaps that is why Engel was somewhat surprised by his discovery that not every transformation of $\mathbf{SL}(2, \mathbb{C})$ is generated by an infinitesimal transformation, so that Lie's exponential map theorem does not hold globally. When he communicated the discovery to his friend, the mathematician Eduard Study, who was working at the time within Lie's sphere on the applications of Lie's theory to the theory of invariants (see Section 7.2), Study replied that this property of $\mathbf{SL}(2, \mathbb{C})$ was new to him, although he had always found Lie's exponential map theorem "amazing" (*wunderbar*).[17] This shows that Study interpreted Lie's statement of his theorem literally, albeit with suspicion. Most of Study's letter was devoted to a geometrical explanation of why $\mathbf{SL}(2, \mathbb{C})$ had this surprising property, and he suggested they write a paper on the matter with the title "A Paradox in the Theory of Groups." In a postcard the next day he added that such a paper was a good idea because most mathematicians "know nothing about such things."

Such a joint paper was never written, but Engel went on to write a two-part paper on the matters raised by his discovery [1892; 1893a]. It did not, however, bear Study's suggested title, because that title was not in keeping with the main thrust of Engel's paper. In this connection it should be recalled that although in Lie's theory "linear" groups play the expected central role, by virtue of Lie's background in projective geometry, he tended to couch everything "linear" in terms of projective space and projective transformations. In his paper Engel showed that for the general projective group $\mathbf{PGL}(n, \mathbb{C})$ and two of its most important subgroups, namely the projective orthogonal group $\mathbf{PSO}(n, \mathbb{C})$ —the identity component of the group of all the $T \in \mathbf{PGL}(n, \mathbb{C})$ that take a nondegenerate quadratic hypersurface into itself— and the projective symplectic group $\mathbf{PSp}(2n, \mathbb{C})$, Lie's exponential map theorem is globally valid. These results formed the bulk of Engel's papers, which thus downplayed the "paradoxical" phenomenon displayed by $\mathbf{SL}(2, \mathbb{C})$.

Nonetheless, in his second paper Engel did present Study's geometrical explanation of the basis for that phenomenon. He also called attention to the implication the examples $\mathbf{SL}(2, \mathbb{C})$ and $\mathbf{PGL}(2, \mathbb{C})$ had for Lie's isomorphism theorem. That is, it was well known that these two groups have isomorphic Lie algebras, but they are not globally isomorphic as Lie's iso-

[17] Letter to Engel dated 9 November 1890 located in Engel Archive Giessen, as are the two postcards Study sent the next day by way of a postscript.

morphism theorem literally asserts. Engel put it this way: Since the two groups have isomorphic Lie algebras, "it follows, as Lie has proved ... that both groups are also isomorphic in the sense of the theory of permutations *so long as one restricts oneself to the transformations in a certain neighborhood of the identical transformation.* This sort of limitation is assumed in all the theories of ... [*Transformationsgruppen* I] ... and only under this assumption is it there proved that groups with the same structure are also isomorphic" [1893a:695]. In particular, Engel pointed out that $\mathbf{SL}(2, \mathbb{C})$ and $\mathbf{PGL}(2, \mathbb{C})$ are not isomorphic "in the strictest sense of the word," i.e., on a global level.

Engel's remarks suggest that Lie was aware of the local nature of his reasoning, although we are left to ponder why he expressed his conclusions so misleadingly. Indeed, a decade later, Engel [1902] had once again to explain that Lie's theorems were to be understood only as local theorems based on local reasoning after an American mathematician, Steven Slocum [1900a; 1900b; 1901], having studied *Transformationsgruppen* but oblivious to this point, claimed to have discovered a flaw in Lie's proof of his main theorem (2.5). My guess is that, as remarked above, in the spirit of the mathematical analysis on which he was raised, Lie felt more at home discoursing on the "generic level" about his transformation groups, and that these are the theorems we see, but cannot necessarily accept in a literal sense, in *Transformationsgruppen.* As we shall see, it was not until the work of Weyl (Part IV) that serious attention was paid to the global aspects of Lie's theory. At that time, the topology that was unavailable to Lie and was later avoided by his students was brought in to deal with the resultant complications. For the remainder of Part II and throughout Part III it is on the local Lie algebra–group germ level that the theory was developed.

3.3. The Adjoint Group

The idea of the adjoint group associated to a transformation group was first introduced by Lie in 1876 and became of fundamental importance in Wilhelm Killing's work on structure (Part II) and remained so in the work of Cartan (Part III) and Weyl (Part IV). Before discussing Lie's conception of this group, let us recall how it is usually introduced today. If $\mathfrak{G}$ is a Lie group, then each $a \in \mathfrak{G}$ determines a conjugation mapping $C_a : g \to aga^{-1}$. The differential dC_a of this diffeomorphism maps the tangent space at the identity element onto itself, and since this tangent space can be identified with $\mathfrak{g}$, the Lie algebra of $\mathfrak{G}$, dC_a may be regarded as an element in $\mathbf{GL}(\mathfrak{g})$. The mapping $\mathrm{Ad} : a \to dC_a$ is easily seen to be a group homomorphism (called the adjoint representation), and the image of $\mathfrak{G}$ under this mapping, $\bar{\mathfrak{G}} = \mathrm{Ad}\,\mathfrak{G}$, is called the adjoint group. It is a subgroup of $\mathbf{GL}(\mathfrak{g})$. The differential of the adjoint representation yields a representation ad : $\mathfrak{g} \to \bar{\mathfrak{g}}$, $\bar{\mathfrak{g}} \subset \mathbf{gl}(\mathfrak{g})$. Each $\bar{X} \in \bar{\mathfrak{g}}$ is given by $\bar{X}(Y) = [X, Y]$ for all $Y \in \mathfrak{g}$.

When Lie first introduced the adjoint group in [1876a:§12], he provided no explanation but merely remarked that any transformation group $\mathfrak{G}$ with infinitesimal transformations[18] $X_1, \ldots, X_r$ satisfying $[X_i, X_j] = \sum_{k=1}^{r} c_{ijk} X_k$ has the same structure as the linear group in the variables $e_1, \ldots, e_r$ with infinitesimal transformations $A_1, \ldots, A_r$ given by

$$A_i = \sum_{j=1}^{r} \left(\sum_{k=1}^{r} c_{kij} e_j \right) \frac{\partial}{\partial e_j}. \tag{3.1}$$

In other words, in 1876 Lie thought he had proved via (3.1) that every $\mathfrak{G}$ is (locally) isomorphic to a linear group, although he eventually realized that the "isomorphism" can have a nontrivial kernel, so that it is only a homomorphism.[19] Before he realized his oversight, he thought he had discovered a quick way to prove what he called in *Transformationsgruppen* III the "third fundamental theorem" of his theory of groups [1893a:597], namely that given constants c_{ijk} satisfying the conditions necessary to define an abstract Lie algebra (given toward the end of Section 2.5) there exists a set of infinitesimal transformations that have the c_{ijk} as their structure constants. As we shall see below and in the next section, after he realized his error, he still found many uses for the adjoint group.

Lie gave various theoretical explanations of the adjoint group that lead to (3.1). The one given in Chapter 16 of *Transformationsgruppen* I [1888] is quite formal and I doubt it conveys the way he originally conceived of the adjoint group. In what follows, I will present the version of his ideas on the adjoint group that is most likely representative of his original conception. On the infinitesimal level, this version can be found in his important paper [1885a:91–5], which is discussed further in the next section. Lie began with the example of a group of r permutations $S_1, \ldots, S_r$ of n objects $x_1, \ldots, x_n$. The S_k permute the x_j, but "as one knows," each permutation S_k induces as well a permutation of $S_1, \ldots, S_r$, namely $S_1, \ldots, S_r \to S_k S_1 S_k^{-1}, \ldots, S_k S_r S_k^{-1}$. That is, if $S_i x_j = x'_j$ and $y_j = S_k x_j$, then $S_i y_j = S_k S_i S_k^{-1} y_j$, so that under the "variable change" $y_j = S_k x_j$, S_i, as a permutation of the y_j's, becomes $S_k S_i S_k^{-1}$.

These considerations, Lie explained, extend directly to continuous groups. Thus let $\mathfrak{G}$ be a continuous transformation group with infinitesimal transformations

[18] Actually, in 1876 Lie was still representing his groups as homogeneous contact transformations, so that $X_i(f) = (H_i, f)$ in the notation of Section 2.5, but the following construction is independent of this representation of the $X_i(f)$.

[19] Lie continued to believe in the truth of the theorem that every group is locally isomorphic to a linear group. In fact, in [1885b:§3, Satz 9] he gave another erroneous proof. In *Transformationsgruppen* III he admitted his error and called for a valid proof [1893a:598, n.2]. As we shall see in Part III, Cartan, too, conjectured the truth of the theorem, but Ado was the first to prove it [1934].

$$X_i = \sum_{k=1}^{n} \xi_{ik}(x)\frac{\partial}{\partial x_k}, \qquad i = 1, \ldots, r.$$

Make the infinitesimal variable change corresponding to X_k, i.e.,

$$y_i = x_i + \xi_{ki}(x)\delta t. \tag{3.2}$$

Under any variable change $y = \Phi(x)$, X_i is transformed (in effect by the differential $d\Phi$) into a function of y: $X_i = \sum_{k=1}^{r} \eta_{ik}(y)\frac{\partial}{\partial y_k}$. For the special transformation (3.2), neglecting higher powers of δt one gets

$$X_i = \sum_{k=1}^{r} \xi_{ik}(y)\frac{\partial}{\partial y_k} + [X_i, X_k]\delta t,$$

which Lie interpreted as meaning that by means of the variable change (3.2) corresponding to X_k, the infinitesimal transformation X_i is changed into the "infinitely close" transformation $X_i + [X_i, X_k]\delta t$.

If $X = \sum_{j=1}^{r} e_j X_j$ denotes an arbitrary infinitesimal transformation of $\mathfrak{G}$, then the variable change induced by X_k sends X to $X + [X, X_k]\delta t = \sum_{i=1}^{r}(e_i + \delta e_i)X_i$, and application of the formula for $[X_j, X_k]$ yields $\delta e_i = (\sum_{j=1}^{r} c_{jki}e_j)\delta t$. Thus we have an infinitesimal linear transformation $\delta T_k : e_i \to e_i + \delta e_i$, where δe_i is as given above. It follows that, in the notation of Chapter 2, $\delta T_k \leftrightarrow A_k$, where A_k is as given in (3.1). Lie also thought about the linear transformation defined by A_k in the following manner. The infinitesimal transformation X may be regarded as a point in a "homogeneous projective space," i.e., with homogeneous coordinates $e_1, \ldots, e_r$. Thus $\delta T_k : X \to X + \delta X$ and $\delta X = [X, X_k]\delta t$.

Since, as we have seen throughout Chapters 1–2, Lie tended to think about groups in terms of their infinitesimal transformations, the above presentation is probably close to how Lie originally conceived of it. By the time *Transformationsgruppen* was published, he had various equivalent characterizations of the noninfinitesimal group $\bar{\mathfrak{G}} = \mathrm{Ad}\,\mathfrak{G}$. The one that is probably closest to his original conception is given in the third volume of *Transformationsgruppen*, where he gave more intuitive presentations of several notions and results presented more formally in the first volume. There [1893a:667–9] the adjoint group (*adjungierte Gruppe*) was introduced as follows. If $x' = T_a x$ are the equations of an r-dimensional continuous transformation group $\mathfrak{G}$, then the variable change $y = T_\alpha x$ transforms the equations of $\mathfrak{G}$ into $y' = T_\alpha T_a T_\alpha^{-1} y$. Thus, if, following Lie, compositions are read from left to right, the transformation T_a is "permuted" to $T_\alpha^{-1} \circ T_a \circ T_\alpha$. In a fashion analogous to what he did in defining the parameter group of $\mathfrak{G}$ (Section 2), Lie considered the group of transformations $R_\alpha : a \to a'$, where $T_\alpha^{-1} \circ T_a \circ T_\alpha = T_{a'}$. Now let the equations $x' = T_a x$ of $\mathfrak{G}$ be written in canonical form, i.e., $x' = \exp(\sum_{i=1}^{r} e_i X_i) \cdot x$, so that the e_i are a new set of "canonical" parameters for $\mathfrak{G}$. Then $R_\alpha : e \to e'$, where $T_\alpha^{-1} \circ \exp(\sum_{i=1}^{r} e_i X_i) \circ T_\alpha = \exp(\sum_{i=1}^{r} e'_i X_i)$. The transformations

R_α are linear transformations of the e_i and define the finite adjoint group; its infinitesimal transformations are spanned by the A_k given in (3.1).

If $(e_1, \ldots, e_r)$ is identified with X (as Lie was wont to do) then Lie's adjoint group may also be conceived as a subgroup of $\mathbf{GL}(\mathfrak{g})$. Using the notation of the presentation of the adjoint group given at the beginning of the section, we have $R_\alpha(\exp X) = C_{\alpha^{-1}}(\exp X) = \exp[dC_{\alpha^{-1}}(X)]$, and so $R_\alpha = dC_{\alpha^{-1}} = dC_\alpha^{-1}$, rather than dC_α as at the beginning of the section, which is why on the infinitesimal level Lie gets $\operatorname{ad} X(Y) = [Y, X] = -[X, Y]$.

Among the applications of the adjoint group made by Lie, in this section I will consider some that involved what are now called solvable Lie algebras, since these algebras will play a role in what is to follow. Recall from Section 2.5 that Lie's decision to develop a theory of transformation groups grew out of his study of complete systems of partial differential equations

$$A_k(f) = \sum_{l=1}^{n} \alpha_l(x_1, \ldots, x_n)\frac{\partial f}{\partial x_l} = 0, \quad k = 1, \ldots, q, \tag{3.3}$$

that admit infinitesimal transformations $X_i = \sum_{j=1}^{n} \xi(x)\frac{\partial}{\partial x_j}$, $i = 1, \ldots, r$, $r = n-q$, forming a group, so that $[X_i, X_j] = \sum_{k=1}^{r} c_{ijk}X_k$.[20] Completeness meant that $[A_i, A_j] = A_i(A_j(f)) - A_j(A_i(f)) = \sum_{k=1}^{q} d_{ijk}(x)A_k(f)$. Lie published some of his results in [1874b], although he suppressed the group-related elements pending his development of his theory of transformation groups. Here the notion of a solvable Lie algebra originated in the following manner.[21] The fact that each X_i is admitted by the system (3.3) means, as Lie showed, that

$$[A_i, X_j] = \sum_{k=1}^{q} \varphi_{ijk}(x)A_k(f). \tag{3.4}$$

Among other things he called attention to a special case in which (3.3) can be solved by $r = n-q$ quadratures, i.e., by the determination of $r = n-q$ integrals. This occurs when one has

$$[X_i, X_k] = b_1X_1 + \cdots + b_{k-1}X_{k-1}, \quad i < k. \tag{3.5}$$

The relation (3.5) implies that the system of partial differential equations

$$A_1(f) = 0, \ \ldots\ , A_q(f) = 0, \ X_1(f) = 0, \ \ldots\ , X_{r-1}(f) = 0 \tag{3.6}$$

is complete and admits the infinitesimal transformation X_r.[22] Completeness follows from the assumed completeness of the original system (3.3) together

[20] In addition, the X_i are assumed not functionally related to the A_k by linear equations.

[21] According to Lie [1893b:679, n.*].

[22] Actually, since the group-related aspects were suppressed, Lie made the more general assumption that $[X_i, X_k]$ is equal to the right-hand side of (3.5) plus $\sum_{k=1}^{q} \gamma_k(x)A_k$, since the same conclusion follows by the same sort of reasoning.

with (3.4) and (3.5) (the latter ensuring that X_r does not occur in the completeness relations). To show that the system (3.6) admits X_r is to show that if f is a solution to (3.6), then $X_r(f)$ is also. That $A_k(X_r(f)) = 0$ follows by writing $A_k(X_r(f)) = X_r(A_k(f)) + (A_k, X_r)$ and using (3.4) and (3.5); and $X_k(X_r(f)) = 0$, $k < r$, follows from (3.5).

Since the complete system (3.6) involves $q + r - 1 = n - 1$ equations, it followed from Clebsch's theory that a solution Π_n is obtained by quadrature, i.e., by computing an integral. Of course, Π_n is also a solution to the original system (3.3). If $x_1, \ldots, x_{n-1}, \Pi_n$ are introduced as new variables, then (in the new variables) $A_1(f) = 0, \ldots, A_q(f) = 0, X_1(f) = 0, \ldots, X_{r-2}(f) = 0$ is a complete system admitting X_{r-1} for the same sort of reasons as before, and so a solution Π_{n-1} is obtained by a quadrature, and so on until solutions $\Pi_n, \ldots, \Pi_{n-r}$ are obtained by quadrature.

A group $\mathfrak{G}$ whose infinitesimal transformations $X_1, \ldots, X_r$ satisfy (3.5) Lie later termed "integrable" [1893a:679] by virtue of the above result. To avoid confusion, I will term $\mathfrak{G}$ and its corresponding Lie algebra solvable, since this is precisely the class of groups picked out by Lie. In fact, in [1893a] he defined a group to be integrable in the now familiar way of characterizing solvability: Let $\mathfrak{g}' = [\mathfrak{g}, \mathfrak{g}]$ denote the Lie algebra spanned by all $[X_i, X_j]$, and set $\mathfrak{g}^{(k+1)} = [\mathfrak{g}^{(k)}]'$ for $k = 1, 2, \ldots$; then the group is integrable if $\mathfrak{g}^{(k)} = \{0\}$ for some k. As Lie realized, integrability is equivalent to finding a basis for $\mathfrak{g}$ such that (3.5) holds. For example, (3.5) implies that $\mathfrak{g}^{(k)} \subset \mathrm{span}\{X_1, \ldots, X_{r-k}\}$.

In [1876a:94] Lie showed that (3.5) implies it is possible to replace $X_1, \ldots, X_r$ by suitable linear combinations $Y_1, \ldots, Y_r$ such that the stronger relation

$$(3.7) \qquad [Y_i, Y_k] - \mu_1 Y_1 + \cdots + \mu_i Y_i, \quad i < k$$

holds. He proved this by using, on the infinitesimal level, the adjoint group. Since $\bar{\mathfrak{g}} = \mathrm{ad}\,\mathfrak{g}$ consists of linear transformations, he first proved in effect a theorem, which now bears his name, about any linear Lie algebra $\bar{\mathfrak{g}}$.[23]

(3.8) **Lie's Theorem.** *If $\bar{\mathfrak{g}} = \mathrm{span}\,\{X_1, \ldots, X_r\} \subset \mathbf{gl}(n, \mathbb{C})$ is solvable, then a basis for $\mathbb{C}^n$ may be chosen so that the matrices representing the $X \in \bar{\mathfrak{g}}$ are all upper triangular.*

Lie expressed and proved this theorem using differential operator notation rather than linear transformations and matrices, and he used solvability in the form of (3.5). He noted explicitly that an immediate consequence of his theorem is that a $v_0 \neq 0$ exists in $\mathbb{C}^n$ such that $X_k(v_0) = a_k v_0$ for $k = 1, \ldots, r$. Of course, this means that for any $X \in \bar{\mathfrak{g}}$, $X(v_0) = a(X)v_0$, where $a(X)$ is a linear homogeneous function of X.

To deduce (3.7) from Theorem 3.8 all Lie had to do was apply the latter to $\bar{\mathfrak{g}} = \mathrm{ad}\,\mathfrak{g}$. Another result he obtained via the adjoint group can

[23] See Theorem III, p. 92, of [1893b].

be stated as follows: *If* $\mathfrak{g}$ *is a solvable subalgebra of* $\mathfrak{h}$ *with* $\dim \mathfrak{h} > \dim \mathfrak{g}$, *then there is a subalgebra* $\mathfrak{k}$ *such that* $\mathfrak{g} \subset \mathfrak{k} \subset \mathfrak{h}$ *and* $\dim \mathfrak{k} = \dim \mathfrak{g} + 1$.[24] According to Lie, in his original determination of all transformation groups acting on the complex plane he had applied this result "almost at each step." To prove it, what he in effect did was consider $\operatorname{ad} \mathfrak{g}$ acting on the vector space $\mathfrak{h}/\mathfrak{g}$. Since $\operatorname{ad} \mathfrak{g}$ is solvable, by the above-mentioned corollary to Lie's theorem there is a K_0 in $\mathfrak{h}$ but not in $\mathfrak{g}$ such that for all $X \in \mathfrak{g}$, $\operatorname{ad} X(K_0) \equiv a(X)K_0 \pmod{\mathfrak{g}}$, which means that the Lie algebra generated by $\mathfrak{g}$ and K_0 is the desired $\mathfrak{k}$.

3.4. Complete Systems and Lie's Idée Fixe

Lie's results on complete systems admitting infinitesimal transformations forming a solvable Lie algebra were a part of his early investigations of complete systems from a group-related point of view. His discovery in 1882 of Halphén's work (mentioned at the beginning of this chapter) spurred him on to stake his claim to priority by further developing his idée fixe within this context and by making the resultant theory known and accessible. And so in July 1884 Lie submitted a paper to *Mathematische Annalen* entitled "General investigations on differential equations that admit a continuous finite group" [1885a]. This paper is of considerable historical importance because it was one of the few papers readily accessible to Wilhelm Killing in the mid-1880s when his work on the foundations of geometry turned his interest to Lie's theory (Chapter 4).

In [1885a] Lie devoted considerable space to those aspects of his general theory of groups that would be needed in the presentation of his theory of complete systems admitting an r-dimensional group. It was here that he introduced the infinitesimal adjoint group along the lines indicated in the previous section. By means of the adjoint group he was also able to characterize on the infinitesimal level when a subgroup $\mathfrak{H}$ of $\mathfrak{G}$ is normal (or "invariant" as was then commonly said). The condition was that $[X, Y] \in \mathfrak{h}$ for all $X \in \mathfrak{g}$ and all $Y \in \mathfrak{h}$ [Lie 1885a:95]. In other words, the subalgebra $\mathfrak{H}$ is what is now called an *ideal*, the term that I will employ in what follows.[25] That this condition implied normality in the usual, permutation group, sense, Lie explained by pointing out that the variable change $x \to y$ induced by X sends Y to $Y' = Y + [Y, X]\delta t$, and since $[Y, X] \in \mathfrak{h}$ for all $Y \in \mathfrak{h}$, $\mathfrak{h}$ is left invariant by this variable change. Recalling for the reader that by "infinite repetition" of X a one-dimensional continuous group $y = T_t \cdot x$

[24] See Theorem V, p. 96, of [1893b].

[25] This terminology was later borrowed from Dedekind, whose work on ideals of algebraic integers was contemporaneous with Lie's work on groups, although there is no evidence that Lie knew it. On the origin of Dedekind's theory see [Edwards 1983], and, for a more detailed analysis of its development, see [Edwards 1980].

is generated (namely $y = (\exp tX) \cdot x$), Lie concluded that the variable change $y = T_t \cdot x$ leaves all of $\mathfrak{H}$ invariant. But as he had explained—see my discussion in Section 3—such a variable change sends the transformations of $\mathfrak{H}$ into those of $T_t^{-1} \circ \mathfrak{H} \circ T_t$ (reading composition from left to right). Thus $T_t^{-1} \circ \mathfrak{H} \circ T_t \subset \mathfrak{H}$ and $\mathfrak{H}$ is normal in the usual sense.[26] Having established this Lie-algebraic characterization of normality, Lie defined a group $\mathfrak{G}$ to be *simple* if $\mathfrak{g}$ contains no (nontrivial) ideals.

Let us now consider how Lie developed his idée fixe within the context of complete systems [1885a:§10]. According to him, the problem of integrating a complete system admitting infinitesimal transformations that form an r-dimensional continuous group can always be reduced to the following type integration problem, which, since it depends upon a positive integer n, will be referred to as Problem n.

Problem n. *Determine the solutions to*

$$A(f) = \frac{\partial f}{\partial x} + \sum_{k=1}^{n} \alpha_k(x, y_1, \ldots, y_n) \frac{\partial f}{\partial y_k} = 0$$

given that this equation admits an n-dimensional Lie algebra $\mathfrak{g}_n$ of infinitesimal transformations $B_k = \sum_{i=1}^{n} \eta_{ki}(x, y_1, \ldots, y_n) \frac{\partial}{\partial y_i}$.

All problems of integrating a complete system admitting a finite-dimensional group reduce to solving Problem n for various positive integers n.

Lie explained that in dealing with Problem n he was assuming that the composition (i.e., structure) of $\mathfrak{g}_n$ had already been fully investigated: "In taking up my old problem again ... I assume that ... the composition of this group has been determined by a preliminary algebraic discussion" [1885a:136].[27] He then showed that if $\mathfrak{g}_n$ contains an $(n-q)$-dimensional ideal, $\mathfrak{g}_{n-q}$, problem n can be reduced to a Problem $n-q$ (with group $\mathfrak{g}_{n-q}$) and a Problem q with group $\mathfrak{g}_n/\mathfrak{g}_{n-q}$. Thus the problem of integrating complete systems admitting a finite group ultimately reduces to the consideration of Problem n for a *simple* Lie algebra $\mathfrak{g}_n$. To this extent, then, Lie's theory was analogous to the Galois theory of algebraic equations.

Problem n with $n = 1$ is easily solved because $\mathfrak{g}_1$ in that case is also solvable (being 1-dimensional), so that Lie's old theory of complete systems admitting solvable groups, discussed in the previous section, applies and shows that the equation can be solved by means of a quadrature. Turning to the analysis of Problem n with $n > 2$ and $\mathfrak{g}_n$ simple, Lie showed that if $\mathfrak{g}_n$ contains a subalgebra $\mathfrak{g}_{n-q}$ of dimension $n-q$ but none of higher dimension, then further simplifications to Problem n were possible. In order to take advantage of these further simplifications, he invoked the following

---

[26] Here Lie was tacitly using the exponential map theorem of Section 2 to conclude that $T^{-1} \circ \mathfrak{H} \circ T \subset \mathfrak{H}$ for all $T \in \mathfrak{G}$ (in a neighborhood of the identity).

[27] Lie also assumed that the finite equations of the group are known. In Section 11 he showed how the case in which these equations are not assumed known can be reduced to this case.

theorems concerning simple groups that he had presented earlier in the paper [1885a:§9]:

> **Theorems V–VII.** *Let $\mathfrak{g}_n$ be simple and let $\mathfrak{g}_{n-q}$ be as above. If $q = 1$, then $n = 3$ and $\mathfrak{g}_n$ is isomorphic to the general projective group of the line. If $q = 2$, then $n = 8$ and $\mathfrak{g}_n$ is isomorphic to the general projective group of the plane. If $q = 3$, then either $n = 15$ or $n = 10$. If $n = 15$, then $\mathfrak{g}_n$ is isomorphic to the general projective group of three-dimensional space. If $n = 10$ then $\mathfrak{g}_n$ is isomorphic to the group of all conformal transformations of three-dimensional space, or, equivalently, to the projective group of a linear line complex.*[28]

The general projective groups to which Lie referred were, of course, the groups $\mathbf{PGL}(n, \mathbb{C})$ with $n = 1, 2, 3$, respectively. As was indicated in Section 2.4, $\mathbf{PGL}(n, \mathbb{C})$ can be defined directly by linear fractional equations or, utilizing homogeneous coordinates, as $\mathbf{GL}(n+1, \mathbb{C})/(\lambda I)$. The Lie algebra of $\mathbf{PGL}(n, \mathbb{C})$ is isomorphic to that of the special linear group $\mathbf{SL}(n+1, \mathbb{C})$, as Lie realized. The conformal group of 3-dimensional space and the projective group of a linear line complex were well known to Lie from his work on the sphere mapping (Section 1.4). As we saw there, the contact transformation defined by the line-to-sphere correspondence transformed the one group into the other. What these last two simple groups correspond to in present-day terms will be indicated further on.

By virtue of Theorems V–VII, Problem n with $\mathfrak{g}_n$ simple and $q \leq 3$ reduced to the consideration of the problem for a specific n and a specific simple group: the general projective groups of one-, two-, and three-dimensional projective space and the conformal group of ordinary space. In each case he was able to use this information to show that the corresponding Problem n could be reduced to a specific canonical form [1885a:Thms.VII–X].

To understand why Lie was able to deal in this way with Problem n for only $q \leq 3$, it is necessary to understand how he had obtained Theorems V–VII. They follow from the following general theorem of his:

> *Suppose $\mathfrak{G}$ is simple and r-dimensional with maximal proper subgroup $\mathfrak{H}$ of dimension $r - q$. Then $\mathfrak{G}$ is locally isomorphic to a transformation group acting on $\mathbb{C}^q$.*

Since for $q \leq 3$ he had determined locally (up to similarity) all groups acting on $\mathbb{C}^q$ (as noted at the beginning of the chapter), he needed only to look through them to pick out the ones that were simple and had maximal proper subalgebras $\mathfrak{h}$ of dimension $r - q$. This led to Theorems V–VII. Analogous theorems would follow for a $q > 3$ once all transformation groups acting on $\mathbb{C}^q$ were likewise determined; but as mentioned, Lie never succeeded in doing this even for $q = 4$. Indeed, by the time he wrote [1885a], he probably realized that such an enterprise was not feasible.

[28] These results were first announced in [1883b:313].

Lie's proof of his general theorem is of interest because it was yet another application of the adjoint group. He realized that he could identify $\mathfrak{G}$ with its adjoint group because he knew that the homomorphism $\mathfrak{G} \to \mathrm{Ad}\,\mathfrak{G}$ was a (local) isomorphism when there are no $Z \neq 0$ in $\mathfrak{g}$ such that $[X, Z] = 0$ for all $X \in \mathfrak{g}$, and that this occurs when $\mathfrak{g}$ is simple, since the set of all such Z, the center of $\mathfrak{g}$ as we would now say, is evidently an ideal of $\mathfrak{g}$. His idea was to show that locally $\mathrm{Ad}\,\mathfrak{G} \subset \mathbf{GL}(\mathfrak{g})$ may be regarded as acting faithfully on $\mathbb{C}^q$. To this end he observed that since $\mathfrak{g}$ is simple, the transformations $\mathrm{Ad}\,g$, $g = \exp X \in \mathfrak{G}$, map $\mathfrak{h}$ onto various $(r-q)$-subalgebras of $\mathfrak{g}$. If $X_1 \equiv X_2 \bmod \mathfrak{h}$, then $\mathrm{ad}\,X_1[\mathfrak{h}] = \mathrm{ad}\,X_2[\mathfrak{h}]$, and so $\mathrm{Ad}(\exp X_1)[\mathfrak{h}] = \mathrm{Ad}(\exp X_2)[\mathfrak{h}]$. Thus the distinct subalgebras $\mathfrak{h}' = \mathrm{Ad}(\exp X)[\mathfrak{h}]$ can be coordinatized by the elements of the vector space $\mathfrak{g}/\mathfrak{h} \cong \mathbb{C}^q$. That is, if $c \in \mathbb{C}^q$ represents the equivalence class of $\mathfrak{g}/\mathfrak{h}$ with representative $X_c \in \mathfrak{g}$, let $\mathfrak{h}_c = \mathrm{Ad}(\exp X_c)[\mathfrak{h}]$.

To show that locally $\mathrm{Ad}\,\mathfrak{G}$ acts faithfully on $\mathbb{C}^q$ is to show that if

$$\mathrm{Ad}(\exp K)[\mathfrak{h}_c] = \mathfrak{h}_c \text{ for all } c, \tag{4.1}$$

then $K = 0$. Now, on the Lie algebra level (4.1) says that

$$[K, \mathfrak{h}_c] \subset \mathfrak{h}_c \text{ for all } c. \tag{4.2}$$

From (4.2) it follows that $K \in \bigcap_c \mathfrak{h}_c$. For $q > 1$ this follows from the maximality of the dimension of $\mathfrak{h}$, since (4.2) shows that if $K \notin \mathfrak{h}_c$, then $\mathfrak{h}_c$ and K would span a subalgebra of dimension $r - q + 1$. For $q = 1$ it follows because if $K \notin \mathfrak{h}_c$, then $\mathfrak{g}$ is spanned by $\mathfrak{h}_c$ and K, but (4.2) says that $\mathfrak{g}' = [\mathfrak{g}, \mathfrak{g}] \subset \mathfrak{h}_c$, so $\mathfrak{g}' \neq \mathfrak{g}$ and $\mathfrak{g}$ cannot be simple. Thus in either case we have $K \in \bigcap_c \mathfrak{h}_c$. But $\bigcap_c \mathfrak{h}_c$ is evidently a proper ideal in $\mathfrak{g}$, and since $\mathfrak{g}$ is simple, we must have $K = 0$. Except for the notation employed, this is the gist of Lie's proof. It is a good example of the geometrically flavored "conceptual" reasoning that he preferred but frequently suppressed in favor of the more orthodox, strictly formal-analytical style characteristic of nineteenth-century analysis.

Lie's "Galois theory" of complete systems made it clear that the problem of determining the structure of all simple groups was of fundamental importance, although he never explicitly formulated this problem in his paper, possibly deeming it too difficult. By way of concluding this section, I consider what Lie knew about the existence of simple groups circa 1885–86, i.e., just before the work of Killing. As Theorems V–VII already suggest, he realized that the general projective groups $\mathbf{PGL}(n, \mathbb{C})$ are simple for all $n \geq 1$ [1885b:§9,Hilfsatz 2].[29] Another type referred to in Theorems V–VII was the conformal group of space. In n dimensions this group is defined[30]

[29] Lie gave no proof, saying simply that it could be proved readily from the elements of his theory. Killing gave a proof in [1888a:271–6].

[30] See, e.g., *Transformationsgruppen* III [1893a:347].

to consist of all transformations $(x_1, \ldots, x_n) \to (X_1, \ldots, X_n)$ of $\mathbb{C}^n$ such that

$$dX_1^2 + \cdots + dX_n^2 = \rho(x)\left(dx_1^2 + \cdots + dx_n^2\right), \quad \rho(x) \neq 0.$$

When $n = 1$ the above condition simply says that $X = f(x)$ is analytic with nonvanishing derivative and hence conformal. But for $n > 1$ a theorem of Liouville's shows that the transformations of this group are more special (see Section 1.4). Lie proved that the conformal group is simple for $n \geq 3$ [1886:550]. He also knew that this group is locally isomorphic to the continuous projective group of a nonsingular quadratic hypersurface in $(n+1)$-dimensional projective space, which is therefore simple for $n \geq 3$ [1886:551]. This isomorphism was a reflection of Klein's discovery in [1871d] of a correspondence between the "metrical geometry" of $\mathbb{P}^n(\mathbb{C})$ and the generalized line geometry of $\mathbb{P}^{n+1}(\mathbb{C})$ (Section 1.5). The continuous projective group of a quadratic hypersurface in $\mathbb{P}^n(\mathbb{C})$ was already mentioned in Section 2 in discussing Engel's work on projective groups generated by infinitesimal transformations. Its Lie algebra is isomorphic to $\mathbf{o}(n+1, \mathbb{C})$. Thus in effect Lie recognized the simplicity of the Lie algebras of types A, B, and D. As we now know, the only other general class of simple groups consists of the symplectic groups (type C). These groups were also introduced by Lie, and the following section is devoted to explaining how they arose in his work and why they were important to him.

3.5. The Symplectic Groups

The projective analogue of the symplectic groups $\mathbf{Sp}(2n, \mathbb{C})$ was introduced by Lie as a generalization of the 10-dimensional projective group of a linear line complex $\mathfrak{L}$ that had figured so prominently in his work on the sphere mapping (Section 1.4). Of course, as a consequence of the sphere mapping, this particular simple group was known to be isomorphic to the conformal group of space as well as to the projective orthogonal group of $\mathbb{P}^4(\mathbb{C})$ and so did not itself represent something new in the way of simple groups. The motivation for the generalization came from Lie's group classification program.

Recall that initially, i.e., for groups acting on two-dimensional space, Lie sought to determine all contact as well as point transformations. Thus Lie's problem was to determine all groups of homogeneous contact transformations $T : (x_1, x_2, p_1, p_2) \to (x_1', x_2', p_1', p_2')$. In his first published announcement of his theory of groups in December 1874 [1874c], he stated that he had succeeded in solving the classification problem for $n = 2$ and summarized his findings by observing that a group of contact transformations of the plane was admitted by either a second- or a third-order ordinary differential equation. Most of the groups arose in connection with the first possibility. The second, he noted, occurred for a 10-dimensional group—here denoted by $\mathfrak{G}_{10}$—concerning which he remarked, "By regarding x_1, x_2, p_1, p_2 as the coordinates of a point in space, one can transform the 10-term group into the group of those collineations of space that transform a linear line complex into itself" [1874c:6].

What Lie meant can be figured out from his later publications.[31] The transformations $T \in \mathfrak{G}_{10}$ can be regarded as point transformations in three-dimensional space with homogeneous coordinates (x_1, x_2, p_1, p_2). Since T is a homogeneous contact transformation, it leaves the Pfaffian equation $p_1 dx_1 + p_2 dx_2 = 0$ invariant. He had observed that a change of variables $(x_1, x_2, p_1, p_2) \to (y_1, \ldots, y_4)$ can be made that transforms the Pfaffian equation into

$$y_1 dy_2 - y_2 dy_1 + y_4 dy_3 - y_4 dy_3 = 0 \tag{5.1}$$

and transforms $\mathfrak{G}_{10}$ into a group $\mathfrak{P}_{10}$ of *projective* transformations that leave (5.1) invariant. For him, (5.1) represented the equation of a linear line complex (as discussed in Section 1.1). Recall that if $(y_1, \ldots, y_4), (z_1, \ldots, z_4)$ are the homogeneous coordinates of two points on a line $\ell \in \mathbb{P}^3(\mathbb{C})$, then the homogeneous Plücker coordinates of ℓ are $p_{12}, p_{13}, p_{14}, p_{23}, p_{42}, p_{34}$, where $p_{ik} = y_i z_k - y_k z_i$. A linear or first-degree line complex is, for example, given by the equation $p_{12} + p_{34} = 0$.[32] In terms of the coordinates $(y_1, \ldots, y_4), (z_1, \ldots, z_4)$ of the points on ℓ this equation becomes

$$B(y, z) = (y_2 z_1 - y_1 z_2) + (y_4 z_3 - z_4 y_3) = 0, \tag{5.2}$$

and, taking $z_i = y_i + dy_i$, (5.2) turns into (5.1). Thus $\mathfrak{P}_{10}$ is the group of all projective transformations that, when expressed in homogeneous coordinates, leave the *equation* (5.2) invariant. The group may be identified with the quotient group $\mathfrak{H}_{10}/(\lambda I)$, where $\mathfrak{H}_{10}$ consists of all $T \in \mathbf{GL}(4, \mathbb{C})$ such that $B(Ty, Tz) = \rho B(y, z)$, where ρ is a nonzero constant. The Lie algebra of $\mathfrak{P}_{10}$ is isomorphic to $\mathbf{sp}(4, \mathbb{C})$.

The connection between $\mathfrak{G}_{10}$ and $\mathfrak{P}_{10}$ was of course of interest to Lie by virtue of the role $\mathfrak{P}_{10}$ had played in the work on the sphere mapping. But that was now incidental to the primary significance of the group $\mathfrak{G}_{10}$ of contact transformations, which became especially important to Lie because, in working out the details of his classification for transformations of the plane ($n = 2$), he also considered the problem of determining which of the groups he had found were genuine groups of contact transformations in the sense that they were not similar to a group of point transformations. These groups he termed *irreducible groups of contact transformations*. Part of his group classification problem became the problem of determining all the irreducible groups of contact transformations in a given number n of

[31] Namely, [Lie 1878b:Abschnitt VI (p. 156ff.); 1879; 1890a:Chs.23–24].

[32] Of course, in addition, we always have the equation $\Omega = p_{12}p_{34} + p_{13}p_{42} + p_{14}p_{23} = 0$.

variables. In this connection for $n = 2$ the group $\mathfrak{G}_{10}$ proved to be fundamental, for he discovered that, up to similarity, there were precisely three irreducible groups, $\mathfrak{G}_{10}$ and two of its subgroups, of dimensions six and seven, respectively [1878b:Abschn.VI]. Shortly after he published this result, he discovered a more satisfying proof [1879:199]. One of the reasons it was more satisfying is that the reasoning extended readily, so he claimed, to transformations in any number of variables and yielded three examples of irreducible groups acting on a space of any dimension n. Although the method no longer implied that these three represented all irreducible groups in n-dimensional space, it provided information he deemed significant for the development of the more difficult n-dimensional theory.

Lie promised to describe the n-variable result in a subsequent paper but, having mentioned it, he evidently did not assign a high priority to publishing a paper with the details. The promised paper was not written until after he had become familiar with Killing's work. As we shall see in Chapter 5, in [1888b] Killing thought he had discovered a possible new infinite family of simple groups—his type C groups—and had labored long and hard but in vain attempting to show that they actually exist. Then in [1889] Lie pointed out that $\mathfrak{G}_{10}$ is simple and that the same is true more generally for the n-variable analogue $\mathfrak{G}_{n(2n+1)}$ and for the corresponding projective group $\mathfrak{P}_{n(2n+1)}$ of all projective transformations that, in homogeneous coordinates, leave the equation $B(y, z) = \sum_{i=1}^{n}(y_{2i-1}z_{2i} - y_{2i}z_{2i-1}) = 0$ invariant.[33] He made no mention whatsoever of Killing's work in [1889], but it was undoubtedly with it in mind that he stressed that the groups $\mathfrak{P}_{n(2n+1)}$, which he also referred to as the projective groups of a complex, together with the general projective groups ($\mathbf{sl}(n + 1, \mathbb{C})$) and the continuous projective groups of a quadratic hypersurface ($\mathbf{o}(n + 1, \mathbb{C})$), represented "three essentially different categories of simple group, each of which supplies infinitely many structures [*Zusammensetzungen*]" [1889:265]. Of course, the groups $\mathfrak{P}_{n(2n+1)}$ have Lie algebras isomorphic to $\mathbf{sp}(2n, \mathbb{C})$ and corresponded to Killing's type C structures. According to the recollections of Engel, by 1884 Lie had dealt with the n-variable case and realized that the groups $\mathfrak{P}_{n(2n+1)}$ are simple.[34] In so far as Engel could recall,[35] he and Lie also realized at that time that for $n > 2$, $\mathfrak{P}_{n(2n+1)}$ possesses a structure different from that of the special linear or orthogonal groups.

It should be noted that much earlier, in [1869c; 1869d], Camille Jordan had already introduced the analogue of the symplectic groups for the coefficient field of the integers modulo a prime p. This was an outcome of his study, inspired by the work of Clebsch [1864] and mentioned briefly

---

33 It was not until Lie published the second volume of his *Transformationsgruppen* [1890a], devoted exclusively to contact transformations and groups of contact transformations, that he worked out the details regarding these irreducible groups for $n = 2$ (Chapters 23–24) and for $n > 2$ (Chapter 25).

34 Letter to Killing dated 16 September 1897 [Hein 1997].

35 "soweit ich mich entsinne ...."

in Section 1.2, of the Galois groups of polynomials that arise in geometry. The group of all $T \in \mathbf{GL}(2n, \mathbb{F}_p)$ that take $B(y,z)$ as defined in (5.2) into a nonzero constant multiple of itself he called "the abelian group." The symplectic groups arise then as the (unnamed) subgroup of all T that leave $B(y,z)$ invariant. The abelian group and its subgroup were also treated by Jordan in his *Traité des substitutions* [1870: Livre II, §VIII].[36] There is no evidence to suggest that Lie was familiar with this aspect of Jordan's work, nor is there any need to imagine any influence from Jordan, since as the above discussion shows, Lie's work on the sphere mapping and his interest in irreducible groups of contact transformations in conjunction with his classification program led naturally to the symplectic groups $\mathfrak{P}_{n(2n\,|\,1)}$.

[36] See [Dieudonné 1961] for a detailed summary of Jordan's work with finite linear groups.

Chapter 4
The Background to Killing's Work on Lie Algebras

In 1868 Wilhelm Killing (1847–1923), a relatively unknown professor of mathematics at the Lyceum Hosianum in Braunsberg, East Prussia, published the first of a series of papers entitled "The composition of continuous finite transformation groups." By virtue of these papers Killing earned for himself a place in the annals of mathematics, for they contained the fundamentals of what would today be described as the structure theory of semisimple Lie algebras. Such key notions as the rank of an algebra, semisimple algebra, Cartan subalgebra, root systems, and Cartan integers originated with Killing, as did the striking theorem enumerating, via root systems, all possible structures for finite-dimensional simple Lie algebras over the field of complex numbers. The extraordinary and unprecedented nature of Killing's work attracted the attention to Élie Cartan (Chapters 6–8), who devoted the early stages of his career to perfecting and extending Killing's methods and results. Both Cartan and Theodore Molien (Section 10.5) also used Killing's results as a paradigm for the development of the structure theory of finite-dimensional linear associative algebras over the complex field, obtaining thereby the theorems on semisimple algebras later extended by Wedderburn to abstract fields and then applied by Emmy Noether to the matrix representation of finite groups.[1] Furthermore, in the 1920s Hermann Weyl (Chapter 12), who, unlike Cartan, was familiar

[1] See [Hawkins 1972]. In particular, it should be noted that the structure of linear associative algebras (hypercomplex numbers) was already a topic of investigation, and by mathematicians employing the tools of *Lie's* theory of transformation groups (e.g., Scheffers and Study), but that the fruitful approach came only after and because of Killing's work.

Figure 11. Wilhelm Killing circa 1890, shortly after his papers on the structure of Lie algebras appeared.

with the theory of group characters and representations of Frobenius and I. Schur (Chapter 10), recognized in the Killing–Cartan structure theory the requisite tools for the determination, by means of a complete reducibility theorem, of all representations of semisimple groups, thereby contributing decisively to the extension of representation theory to continuous groups.

4.1. Non-Euclidean Geometry and Weierstrassian Mathematics

Killing's investigations thus launched a series of developments that have since become an integral part of contemporary mathematics. Because of their greater scope, rigor, and clarity, the later developments have rendered Killing's work obsolete and erased all vestiges of its historical dimension. In the next chapter careful attention will be given to Killing's papers on Lie algebras and the personal and mathematical struggles involved in writing them. The present chapter focuses on how Killing came to have something to write about in the first place. That is, I examine the intellectual climate within which he formulated the research program that eventually produced his celebrated series of papers.[2] Two major influences can be distinguished. The discoveries in non-Euclidean geometry and the concomitant reflections on the foundations of geometry formed the context of his work. His research program was not concerned with algebra as such but with the foundations of geometry in light of the existence of non-Euclidean geometries. In addition, his work bears the imprint of Weierstrass, and more generally, the atmosphere at the University of Berlin. During the decade (1867–77) he spent in Berlin as student and then teacher, Killing acquired both specific mathematical tools and a general attitude toward mathematics that imbued his geometrical work with a distinctive cast that was atypical of the period and that provided the framework for him to articulate the notion of a Lie algebra and the problem of determining all possible structures.

Although Killing conceived his research program as early as 1878–79, he first sketched his ideas publicly in two brief essays, *Fundamental Concepts and Propositions of Geometry* [1880b] and *Extension of the Concept of Space* [1884], each published by the high school (*Gymnasium*) at which he was currently teaching and containing as appendix the program of courses for the forthcoming term.[3] In the second essay, Killing introduced the equivalent of the concept of a finite-dimensional Lie algebra and proposed, as a central problem of his program on the foundations of geometry, the determination of all such algebras, i.e., their classification up to isomorphism.

The year after [1884] appeared, Killing learned through Klein that Lie had been working on related matters since 1873. Lie's priority and the fact that Killing was aware of his work for several years before 1888, however, in no way diminishes the historical significance of Killing's independently conceived research program and its background. This will become clear after reading Chapter 5. Then the following should be evident: Killing's papers on semisimple Lie algebras would never have been written had not Killing been involved with his own research program on foundations of geometry when he learned about Lie. Furthermore, even after he was apprised of the existence of Lie's work and had obtained copies of some of his papers, their effect upon him was more inspirational than educational. From Killing's correspondence with F. Engel we learn that Killing had pursued his algebraic problem more extensively than his essay of 1884 indicates; and although his results were far from completely established or fully understood in terms of their general

2 This chapter is based upon my earlier, more detailed, essay [1980].

3 These essays did not have a large circulation and are now difficult to locate. A copy of Killing's [1884], however, has been included in [Hein 1997]. See [Hawkins 1980:331, n.3] for information on archival locations of [1880b] as well as transcriptions of key portions of the text.

import, he had devised the *methods* and *approach* by means of which he was ultimately to achieve success. They reflect his training at Berlin rather than the influence of Lie.

The correspondence between Killing and Engel shows that although they were both concerned with the structure of Lie algebras, they had difficulty communicating their methods because they were entirely different. When confronted with Lie's work, Killing was reluctant to build upon it as opposed to pursuing his own goals by his own methods. Indeed, Killing never managed to master Lie's methods. As we shall see in Chapter 6, it was in part because Cartan was well versed in the methods of both Lie and Killing that he succeeded in filling in the gaps left by Killing so as to confirm his main results and go beyond them. Lie's work served to inspire Killing and also to focus his attention upon the more manageable problem (of prime importance to *Lie's* research program) of determining the structures of all simple algebras over the complex field, a problem the solution of which Killing was able to fashion from his prior, but tentative, investigations of special cases of the general structure problem.

The remainder of this chapter consists of five sections. In Section 2 I discuss Killing's experiences as a student in Berlin (1867–72) and point out the Weierstrassian tenor of his doctoral dissertation, which bears many resemblances to his essay of 1884, at least in terms of tools employed and general attitude. Section 3 traces Killing's involvement with non-Euclidean geometry and foundations from his participation in Weierstrass's seminar on foundations of geometry in 1872 to his first public declaration of his general research program in his 1880 essay *Fundamental Concepts and Propositions of Geometry*. In Section 4 I indicate how, in the 1884 essay *Extension of the Concept of Space*, he transformed the general program into analytical terms, thereby introducing the equivalent of the notion of a Lie algebra and the problem of determining all possible structures. Having thus indicated the nature of Killing's research program on foundations of geometry and how it leads to the algebraic problem concerning Lie algebras, I return in the remaining sections to the context of, and influences upon, Killing's research program. In Section 5 Killing's work is related to the speculations on the foundations of geometry of Riemann and Helmholtz. Although, as I show, Killing's work bears the unmistakable imprint of their ideas, it stands out as atypical in its response to their work. This point is developed further in Section 6, where a comparison with Klein's *Erlanger Programm* (Section 1.5) serves to educe the underlying Weierstrassian element in Killing's work, which seems linked to precisely that peculiar slant to Killing's work that eventuated in the algebraic problem of the structure of Lie algebras.

4.2. Student Years in Berlin: 1867–1872

Killing arrived at the University of Berlin for the winter semester, intent upon a career as a professor of mathematics in a *Gymnasium* such as he had attended.[4] Although his principal contributions to mathematics were published over twenty years later, the orientation of his entire mathematical output was established during the decade he spent in Berlin. The stunning impression that the university made upon him was partly by way of contrast with his disappointing experiences at Münster, where he had begun his university studies in 1865. At Münster there were no mathematicians. The mathematics courses were taught by an observational astronomer who confessed to a limited training in mathematics. The students were equally disappointing, for, as Killing later recalled, they "showed almost no interest whatsoever in science itself; they wished (with very few exceptions) to study only what was needed for the examinations ..." [Oellers 1925: 22]. Killing experienced the same sort of intellectual isolation at Münster that he had experienced in the earlier stages of his education. Once again he found it necessary to learn mathematics through self study.

After four semesters at Münster, Killing enrolled at the University of Berlin, by then the center of mathematics in Germany. Here mathematics was not taught by observational astronomers but by Kummer, Weierstrass, and Kronecker, who coordinated their lectures so as to provide students with a solid foundation in the principal branches of mathematics.[5] There was also a growing population of talented mathematics students at the pre- and postdoctoral levels, a student-run mathematical union that sponsored lectures, discussions, and problem-solving contests, and the famous mathematics seminar conducted by Weierstrass and Kummer. As we saw in Chapter 1, a few years before Killing's arrival, Klein and Lie had come to Berlin and attended Kummer's seminar, although their reaction to the Berlin style of mathematics was rather negative. Another student at Berlin in Killing's time was Georg Frobenius, who went on to become a leading algebraist and, eventually, in 1892, a Berlin professor. It was as a Berlin professor that Frobenius created his theory of group characters and representations (Section 10.1), which Weyl was ultimately to combine with the structure theory of semisimple Lie algebras initiated by Killing, thereby greatly advancing Lie's theory (Chapter 12).

Like many other students, including Frobenius, Killing was attracted especially by Weierstrass, "who, as no other [teacher], influenced my scientific education" [1897:2]. Part of his appeal may have been due to their common background. Like Killing, Weierstrass was a solid Westphalian, whose speech betrayed his origins [Lampe 1897, 23]; and he was also a Catholic, which may have been significant to Killing, an extremely devout

---

[4] This information and that relating to Killing's educational experiences before coming to Berlin are taken from his "Biographische Notizen," a manuscript in Killing's *Nachlass* at the Universitätsbibliothek Münster. Extensive portions of the manuscript have been transcribed by Oellers [1925].

[5] The following information on the University of Berlin is drawn from [Biermann 1973], especially Chapter 5.

Figure 12. Karl Weierstrass inspired many students besides Killing. Among them was Georg Frobenius, whose work on group representations is discussed in Chapter 10.

Catholic.[6] There were also more universal grounds for his appeal. Killing appreciated his openness with students, his willingness to engage in scientific discussion outside the lecture hall, his concern for the personal welfare of his students, and his generosity with mathematical ideas [1897:16]. Furthermore, although Weierstrass is nowadays thought of primarily as an analyst, his actual mathematical interests were much broader than his published opus might imply. For example, three of the four prize problems posed by Weierstrass between 1866 and 1872 had a geometrical orientation in the sense that geometrical considerations were the motivation for an analytical treatment of the matter [Biermann 1973:251].

Weierstrass's lectures also seem to have struck a responsive chord in Killing. In them he found the challenge and inspiration that had been sorely

---

[6] On the matter of Killing's piety, consult the works by Oellers [1925; 1927; 1929].

lacking in his previous mathematical instruction, for the lectures were aimed at students who "with a genuine enthusiasm for their discipline" were willing to expend the considerable time and energy required "in order to be able to enjoy the noble joys of mathematical knowledge and research" [1897:15]. Through Weierstrass's cycle of lectures on analysis, Killing came in contact with the critical and highly theoretical tendency that characterized all his mathematics but especially the lectures developing the foundations of analysis by arithmetical means that excluded any reliance upon intuition or physical considerations.[7] Killing was impressed by the critical attitude expressed by Weierstrass and by Kronecker.[8] As I shall attempt to show, these Weierstrassian traits are reflected, albeit idiosyncratically, in Killing's geometrical investigations.

Writing a geometrical dissertation under Weierstrass was facilitated for Killing by Weierstrass's theory of elementary divisors [1868]. During Killing's first semester at Berlin, Weierstrass probably presented the theory in his lectures, "Theory and Applications of Determinants," which in all likelihood Killing attended. The theory and, more generally, the transformation of quadratic and bilinear forms were also discussed in Berlin, since Kronecker and students such as Frobenius were also interested in the subject. It is necessary to have a general understanding of the theory of elementary divisors and the background to its creation, since the theory embodied the critical spirit of Weierstrass's mathematics in ways that are particularly significant for Killing's work. I have discussed these matters in detail elsewhere [1977a; 1977b], and will restrict myself here to mentioning some essential points.

Weierstrass's theory is concerned with the necessary and sufficient conditions that one family of bilinear forms $\lambda\Phi - \Psi$, where λ is a complex parameter and $\Phi = \sum_{i,j=1}^{n} a_{ij}x_i y_j$, $\Psi = \sum_{i,j=1}^{n} b_{ij}x_i y_j$, can be transformed into another such family $\lambda\bar{\Phi} - \bar{\Psi}$, by means of (possibly different) linear transformations of the x and y variables. It is assumed that the families are nonsingular in the sense that the determinant of Φ is not zero.[9] That same year Kronecker published a paper dealing with the singular case. Elementary divisors are still a part of linear algebra, but for Weierstrass, who followed in the tradition of Cauchy and Jacobi, the entire theory was formulated in terms of determinants. Set $D_0(\lambda) = \det[\lambda\Phi - \Psi]$ and define $D_1(\lambda)$ to be the greatest common divisor of all the $(n-1) \times (n-1)$ minors of $D_0(\lambda)$, considered as polynomials in λ. Likewise, $D_2(\lambda)$ is the greatest common divisor of all the $(n-2) \times (n-2)$ minors of $D_0(\lambda)$, and so on, so that $D_i(\lambda)$ is defined for $i = 0, 1, \ldots, n-1$. It can then be shown that $D_i(\lambda)$ divides $D_{i-1}(\lambda)$ and that if $E_i(\lambda) = D_i(\lambda)/D_{i-1}(\lambda)$, then E_i divides E_{i-1} as well. It follows from the definition of the E_i and the divisibility

7 On Weierstrass's lectures on the foundations of analysis, consult [Dugac 1973].
8 Regarding Kronecker's role see [Hawkins 1980:332, n.10].
9 Weierstrass began by considering families $pP + qQ$ with $\det[pP + qQ] \not\equiv 0$ but then showed that this family can be expressed in the above form.

properties of the D_i that $D_0(s) = \det[\lambda\Phi - \Psi]$ differs from the product of the E_i by at most a constant factor. Thus if $a_1, \ldots, a_m$ are the distinct roots of $D_0(\lambda)$, then $E_i(\lambda) = \gamma_i \prod_{j=1}^{m}(\lambda - a_j)^{m_{ij}}$, where γ_i is a constant and the m_{ij} are nonnegative integers. Each factor $e_{ij}(\lambda) = (\lambda - a_j)^{m_{ij}}$ with $m_{ij} > 0$ is called by Weierstrass an elementary divisor of the determinant $D_0(\lambda)$ or of the associated family $\lambda\Phi - \Psi$.

Weierstrass's main theorem was that two such families, $\lambda\Phi - \Psi$ in variables x, y and $\lambda\bar{\Phi} - \bar{\Psi}$ in variables $\bar{x}, \bar{y}$, are equivalent in the sense that nonsingular linear transformations $x = H\bar{x}$, $y = K\bar{y}$ exist for which $\lambda\Phi - \Psi$ is transformed into $\lambda\bar{\Phi} - \bar{\Psi}$ if and only if the two families have the same elementary divisors. He also showed that when the bilinear forms are symmetric, then in the above theorem one can take $H = K$. Expressed in terms of matrices (which Weierstrass did not employ[10]), the theory shows when two matrices A and B are similar. That is if in matrix notation

$$\Phi = x^T I y, \quad \Psi = x^T A y, \quad \bar{\Phi} = \bar{x}^T I \bar{y}, \quad \bar{\Psi} = \bar{x}^T B \bar{y},$$

then the fact that $\lambda\Phi - \Psi$ is transformed into $\lambda\bar{\Phi} - \bar{\Psi}$ by virtue of the variable changes $x = H\bar{x}$ and $y = K\bar{y}$ implies that $H^T K = I$ and $H^T A K = B$, and hence that $K^{-1}AK = B$. Likewise, when the forms are symmetric, so that $H = K$, we have $K^T K = I$, which means that the variable change is by means of an orthogonal transformation. Furthermore, in the course of proving his theorem, Weierstrass introduced what amounts to the Jordan canonical form J of the matrix A because the elementary divisors of $\lambda\Phi - \Psi$ with $\Phi = x^T I y$ and $\Psi = x^T J y$ can be read off immediately. Each elementary divisor of $\lambda\Phi - \Psi$ corresponds to a Jordan block; an $N \times N$ block with a's on the diagonal and 1's on the superdiagonal corresponds to the elementary divisor $(\lambda - a)^N$ of $\lambda\Phi - \Psi$.

Independently of Weierstrass, Camille Jordan introduced the same canonical form, albeit in the context of linear transformations, a few years later. Weierstrass's paper, however, seems to have exerted an especially strong influence upon the early development and application of linear algebra. Frobenius, for example, further developed it in many respects and used it as the principal algebraic tool in developing his theory of group characters and representations. Likewise, as we shall see, it was the principal tool used by Killing in dealing with Lie algebras. It was also the form in which Weyl learned linear algebra (see Section 11.2), although, influenced by his Göttingen experiences, in his book *Space, Time, Matter*, published in many editions from 1918 onwards, he began to imbue linear algebra with its now familiar axiomatic, geometrically flavored conceptual format (Section 11.3).

Before discussing Killing's work along these lines, however, I want to point out that in his paper on elementary divisors Weierstrass demonstrated more than theorems. To his colleagues and students he demonstrated the

---

[10] On the role of matrix algebra in the development of linear algebra see my papers [1975b; 1977a].

possibility and desirability of a more rigorous and systematic approach to algebraic analysis than that provided by the type of reasoning that had been prevalent during the eighteenth and early nineteenth centuries. It will be helpful to explain this point a bit further, to place it in a broader historical perspective. The mathematicians of the eighteenth century had inherited from the previous century an entirely new way of doing mathematics, the method of symbolical analysis, which had been introduced by Viète and Descartes and further developed by Newton and Leibniz. The method consisted in reducing mathematical problems to the consideration of "equations" from which the solution was to be deduced by suitable symbolical manipulations. For Viète and Descartes the equations were algebraic, whereas for Newton and Leibniz differential and integral equations and power series were among the types of equations to be considered. Continental mathematicians such as D'Alembert, Euler, Lagrange, and Laplace perfected the analytical methods of their predecessors and demonstrated their power by systematically and successfully applying them to problems in terrestrial and celestial mechanics.

Two works that summed up and symbolized these accomplishments were Lagrange's *Méchanique analitique* (1788) and Laplace's *Traité de méchanique céleste* (1799). They represented unequivocal triumphs for the method of analysis. Thus Lagrange in the preface to *Méchanique analitique* boasted that his book contained neither geometrical figures nor geometrical constructions and that the methods expounded by him involved only "algebraic operations subjected to a uniform progression," so that mechanics becomes a new branch of analysis. And Laplace, in the course of contrasting the older geometrical methods with his own, wrote that although algebraic analysis necessitates consideration of "abstract combinations" far removed from the particulars of the problem at hand, "by abandoning oneself to the operations of Analysis ... one is led, by the generality of this method and by the inestimable advantage of transforming reasoning into into mechanical procedures, to results often inaccessible to synthesis. Such is the fruitfulness of Analysis that it suffices to translate particular truths into this language in order to see emerge from their very expression a multitude of new and unexpected truths" [1835:465].

Being understandably dazzled by the power of the method of analysis, mathematicians such as Lagrange and Laplace tended to slur over certain concomitant difficulties. As the above quotations indicate, the great virtue of analysis was its generality and the resulting uniform, even mindlessly mechanical, nature of its procedures. The generality of the method had been perceived from the outset as its great virtue. Thus Viète stressed that the new method of analysis "does not employ its logic on numbers—which was the tediousness of the ancient analysts—but uses its logic through a logistic that in a new way has to do with species. This logistic is much more successful and powerful than the numerical one ..." [1591:321–2]. Analysis became a method for reasoning with, i.e., manipulating, symbolical expressions involving symbols with "general" values; and a tendency developed to think almost exclusively in terms of the "general" case with little, if any, attention given to potential difficulties that might arise by assigning certain specific values to the symbols. Such reasoning with "general" expressions I shall refer to as *generic reasoning*.

The generic mode of reasoning continued to be practiced by many mathematicians during the first half of the nineteenth century, Jacobi being a prime example. As already mentioned in Section 3.2, during the period when Lie was preoccupied with differential equations (Chapter 2) he had assimilated much work on the generic level including that by Jacobi, and his tendency toward generic theorems and reasoning is probably a reflection of this background. During the early stages of the nineteenth century some mathematicians, concerned with raising the standards of rigor in analysis, rejected the legitimacy of generic reasoning. Thus Cauchy insisted in the preface to his *Cours d'Analyse* (1821) that mathematical reasoning must not be based on arguments "drawn from the generality of algebra," arguments that "tend to attribute an indefinite scope to the algebraic formulas, whereas in reality the majority hold true only under certain conditions and for certain quantities of the variables involved." This can be see in Cauchy's efforts to prove (in 1829) that any quadratic form Ψ can be transformed by an orthogonal transformation into a sum of squares. Using determinants, Cauchy found a formula for the orthogonal transformation, although it was only valid, as he pointed out, when the eigenvalues associated to Ψ are distinct. Using the fact that by varying the coefficients of Ψ infinitely little a form with distinct roots would be attained, Cauchy claimed that a limit argument would push through the theorem in the case of multiple roots, but it is doubtful he could have done so rigorously.[11] A few years later Jacobi gave his own elegant determinant theoretic formula for the orthogonal transformation taking Ψ into a sum of squares and in a posthumously published work showed how to transform any bilinear form into "diagonal" form. Weierstrass's theory of course demonstrated rigorously (and without limits) Cauchy's theorem and gave a precise condition on when Jacobi's generic transformation is possible, namely when all the elementary disvisors are of degree one.

Of course, generic reasoning would be unobjectionable if the conditions of the validity of a proposition were made precise, but this was never done, and the result was completely unsatisfactory. Thus Jacobi "proved" that any quadratic form can be orthogonally diagonalized and that any bilinear form can be diagonalized; the former proposition turns out to be generally true, whereas the latter does not. Nonetheless, even after 1850 generic reasoning continued to be favored by some mathematicians, including Felix Klein. An episode exemplifying this will be recounted further on. Even at the end of his career, Klein was ambivalent about the generic versus the

[11] See in this connection [Hawkins 1977b:126–7].

"Weierstrassian" approach. Thus in his history of mathematics he wrote [1927:16, n.1], after mentioning Jacobi's generic theorem on the transformation of bilinear forms:

> This is connected with his entire mode of presentation, which—in contrast to the precedent set by Gauss—takes no time for an exact treatment of individual cases. Jacobi's approach was influential for a long time among mathematicians .... The recent Berlin school (under the influence of Weierstrass and Kronecker) was the first to return again to the exactness of the Gaussian mode of presentation. Both [modes] have their advantages; one could discourse at length on this. The one [mode] emphasizes formal reasoning (which strives above all for an overview); the other emphasizes concrete reasoning (which has in mind the application to individual cases).

Kronecker articulated the Berlin attitude when he wrote, with Weierstrass's paper [1868] expressly in mind [1874:405]:

> It is common—especially in algebraic questions—to encounter essentially new difficulties when one breaks away from those cases that are customarily designated as general. As soon as one penetrates beneath the surface of the so-called generality, which excludes every particularity, into the true generality, which comprises all singularities, the real difficulties of the investigation are usually first encountered, but at the same time, also the wealth of new viewpoints and phenomena that are contained in its depths.

The link between the theory of elementary divisors and geometry was provided for Killing by the theory of quadric surfaces. In homogeneous coordinates, a quadric surface is given by an equation $P = 0$, where P is a quadratic form in the four homogeneous coordinates of space. A pencil of such surfaces is determined by $P + \lambda Q = 0$ or $pP + qQ = 0$, where Q is likewise a quadratic form. The leading texts on analytic geometry did not go beyond the generic level in their discussions of these pencils. An attempt to rectify the situation was made by Jacob Lüroth [1868]. He recognized that the geometrical nature of the pencil $P + \lambda Q = 0$ depends upon the characteristic equation $|P + \lambda Q| = 0$ and upon whether one or more of its roots are also roots of all minor determinants, but his analysis of the possibilities, done without benefit of the theory of elementary divisors, was incomplete. Killing sought to treat the matter thoroughly in his doctoral dissertation, *Bundles of Second-Order Surfaces* [1872].

In the introduction, Killing wrote that his intention was "to give the geometrical interpretation of a work that my esteemed teacher Herr Weierstrass read before the Academy on 18 May 1868." The reference was, of course, to Weierstrass's paper on elementary divisors; and in presenting its geometrical interpretation, Killing tacitly transferred the Weierstrassian spirit to the realm of geometry: In geometrical investigations also it is necessary to explore all the possibilities disclosed by the analytical framework and not just those of primary geometrical interest. The general tendency among geometers at this time was not in accordance with this viewpoint. (See in this connection the discussion of Klein's doctoral dissertation in Section 5 below.) Adhering to it, Killing began by classifying the pencils according to the elementary divisors of the associated determinant—there are thirteen cases—and then proceeded to make further distinctions according to whether the characteristic roots are real or complex. As Weierstrass explained in his official evaluation of the dissertation, the problem dealt with is not particularly difficult but it requires "great care and circumspection," precisely the sort of characteristics cultivated by Weierstrass. Killing's dissertation displayed the systematic thoroughness that Weierstrass appreciated, and he cited it as the "sole exhaustive work" on the subject in lectures delivered as late as 1886–87 [Hawkins 1980:332, n.15].

I have stressed these Weierstrassian elements in Killing's dissertation because they can be seen as well in his essay [1884] on the foundations of geometry. Both works have the same commitment to an exhaustive consideration of all the geometrical possibilities, however interesting they may be, that are deducible by means of the analytical framework. One notable difference, however, is the level of difficulty involved in carrying out the respective programs. Killing never achieved the level of rigor and completeness in his work on the foundations of geometry that he did in his dissertation, but instead opened up a vast new world for further exploration.

4.3. Non-Euclidean Geometry and General Space Forms

Killing received his doctorate in March of 1872. He had interrupted his studies at Berlin for two semesters during 1870–71 in order to teach in a small *Rektoratsschule* in Rüthen, Westphalia, where his father was then *Bürgermeister*. The school was threatened by extinction through lack of faculty, and Killing taught in all subjects, sometimes as much as thirty-six hours per week [Oellers 1925:27; 1927:270]. After 1872 he continued his studies at Berlin through the summer semester of 1872 as a participant in the mathematics seminar. The decision to attend the seminar was fateful because Weierstrass lectured on the foundations of geometry with particular reference to non-Euclidean geometry.[12] The choice of this subject is not as surprising as it may at first appear, since despite the narrow range of his own publications, Weierstrass kept abreast of all the latest mathematical developments [Lampe 1897:10]. Certainly, non-Euclidean geometry would fall within that category, for by 1870 it had begun to attract the attention it deserved, owing to a combination of events.

---

[12] Killing is apparently the sole source of information concerning these lectures; see [1878:74n; 1880a:273; 1884:4].

The publication of Gauss's extensive correspondence with Schumacher (1860–65), with its approving reference to the work of Lobachevsky, served to rescue it from oblivion. And Beltrami's paper [1868] showing that plane Lobachevskian geometry could be interpreted as the geometry of a surface of constant, negative curvature convinced many mathematicians of its logical consistency. Equally significant was the appearance of the philosophically oriented essays of Riemann [1868] and Helmholtz [1868a; 1868b] on the foundations of geometry, which attracted considerable attention among mathematicians and philosophers alike. Probably much of the motivation for Weierstrass's lectures was supplied by these essays. Riemann had been the highly respected rival of Weierstrass, and his work was studied with great care and interest in Berlin [Hawkins 1980:333, n.19]. Furthermore, Helmholtz was, since 1870, Weierstrass's colleague at the university, and we know that Weierstrass held him in high esteem [Biermann 1973:113, n.149].

Weierstrass's lectures thus exposed Killing to the subject that was to occupy him for the rest of his life. In order to fully appreciate the direction taken by his reflections on the foundations of geometry, it will be helpful to briefly recount the sequence of discoveries regarding non-Euclidean geometries with the perspective of Killing's work in mind.[13] Non-Euclidean geometry began with the discovery of Lobachevskian geometry, which showed that Euclidean geometry is not the only geometry that is logically consistent and, as far as could be determined, compatible with experience. Lobachevskian geometry involves a parameter k that may be regarded as the radius of curvature and that for $k = \infty$ yields Euclidean geometry as a limiting case. The value of k, however, cannot be determined a priori so that geometry becomes an empirical science. The discoverers of Lobachevskian geometry, however, tended to regard it much as Euclidean geometry had been regarded in the time of Kant: Lobachevskian geometry, including thereby the limiting case of Euclidean geometry, was tacitly regarded as embracing all the geometrical possibilities.

Riemann's simple observation [1868:284] that the unboundedness of space has greater empirical certainty than its infinitude, so that a geometry corresponding to a manifold of constant positive curvature is also possible, was consequently sensational in its impact: It destroyed the absolute nature of Lobachevskian geometry. Although Riemann's discussion of manifolds of constant positive curvature is vague, it would seem that he identified the geometry of such a manifold with spherical geometry—at least that is how his essay was interpreted at the time. In spherical geometry, two points do not necessarily determine a unique "straight line," since they may be diametrically opposite. Spherical geometry thus violates what Beltrami termed the postulate of the straight line [1868:377], and with Riemann's essay in

[13] The standard works on the history of non-Euclidean geometry are [Bonola 1912; Stäckel and Engel 1895; 1899–1913]. More recent studies include [Gray 1979; Rosenfeld 1988].

mind, he implied that this was a characteristic property of geometries of constant positive curvature. This misconception, which was held by many mathematicians, including Weierstrass, was corrected by Klein [1871a] and Simon Newcomb [1877], who showed, independently, that a geometry of positive curvature that does satisfy the postulate of the straight line is possible. Klein gave it the name "elliptic geometry."

Each new discovery revealed the unjustified limitations of the current perception of the geometrical possibilities. The situation was somewhat analogous to the contemporaneous one in analysis, where the discovery of new types of functions, such as continuous functions which oscillate infinitely often in any interval and fail to possess a derivative at most points, revealed to Weierstrass and his circle the untenable limitations of current perceptions of the nature of functions and their properties. Not everyone reacted to the analytical discoveries as they did in Berlin. Some sought to prove that continuous functions are "generally" differentiable [Hawkins 1975a:42–54], whereas Klein [1873] proposed replacing the function concept with that of a "function strip" so as to make differentiability follow from continuity, thereby restoring the validity of our intuition of functions and their applicability to physical phenomena. Likewise, not everyone reacted to the discoveries in non-Euclidean geometry as did Killing, and I believe that the Weierstrassian response to the existence of pathological functions colored Killing's attitude toward the foundations of geometry. This point is developed further in Section 5.

Although the seeds were sown in Weierstrass's seminar lectures, it was some time before they bore fruit. After obtaining his doctorate, Killing prepared for and passed the examinations qualifying him to teach at the high school (*Gymnasium*) level. Until 1878 he taught at that level in Berlin. Preoccupied with his teaching duties as well as the rites of courtship and marriage, he published nothing for several years. The first sign of life we have is a review [1876] of Johann Frischauf's book *Elements of Absolute Geometry* [1876]. Frischauf, a mathematics professor at the University of Graz, Austria, had earlier (1872) published a paraphrase of Bolyai's essay (1832) on absolute geometry. The purpose of his second book was to update it by including a discussion of the more recent work of Beltrami, Riemann, and, especially, Helmholtz. At the conclusion of his generally favorable review, Killing expressed his gratitude to the author for writing such an informative and stimulating book. But Frischauf's book was in fact not as informative as it might have been. He had clearly failed to digest Klein's series of papers on non-Euclidean geometry, which are given only passing references in a few footnotes. As a consequence he followed in the footsteps of Beltrami and identified the geometry of a finite space with spherical geometry [1876:101 ff.]. Killing was also at this time unaware of the contents of Klein's papers. Shortly after he reviewed Frischauf's book, Newcomb's paper [1877] appeared in Crelle's journal and initiated Killing's lifelong research activity on non-Euclidean geometry and its foundations.

Figure 13. The essays of Riemann (top) and Helmholtz (bottom) and the "space problem" they inspired influenced both Killing and Weyl (Chapter 11) in ways of decisive importance to our story.

Convinced of the mathematical validity of Newcomb's geometry, Killing discussed his views with Weierstrass in the fall of 1877. Many years later, he recounted those discussions in a letter to Mittag-Leffler:

> At the time Weierstrass was entirely of Beltrami's opinion [regarding the geometry of a manifold of constant positive curvature] which was undoubtedly also shared by Riemann. At first I had some trouble bringing him to my view; but he then immediately made some investigations into the matter, although he did not wish to communicate anything of their content to me. Thereafter he always urged me to bring my investigations to a conclusion. He repeatedly expressed his appreciation of my work and indicated that he was completely familiar with its content ... [Hawkins 1980:33, n.4].

Thus encouraged, Killing published two papers [1878; 1880a] aimed at showing, with the aid of the coordinates Weierstrass had introduced in his seminar lectures of 1872, that there were three distinct, equally legitimate non-Euclidean geometries of constant curvature: Lobachevskian geometry, the spherical geometry of Riemann, and the geometry of Newcomb and Klein. These geometries, all of which Killing accepted as compatible with experience, he termed non-Euclidean space forms to distinguish them from the more general space forms that had come to occupy his attention and to which we must now turn.

In 1878 Killing left Berlin for a position as *dritter Oberlehrer* in the *Gymnasium Petrinum* in Brilon, Westphalia, where he had once been a student. Then in 1880 he was appointed to a professorship at the *Lyceum Hosianum*, an academic training facility for prospective Catholic clergymen located in Braunsberg, East Prussia (now Braniewo, Poland). Although far removed from Berlin and burdened with the duties of his office, Killing maintained his active interest in non-Euclidean geometry. Following the advice of Weierstrass, he studied mechanics in non-Euclidean space forms and the purely geometrical theorems thereby required—including theorems about quadric surfaces in non-Euclidean geometries that involved the theory of elementary divisors [1883a; 1883b]. Killing's work with non-Euclidean space forms culminated in a comprehensive book, *Non-Euclidean Space Forms Treated Analytically* [1885], that Frischauf had encouraged him to write. The contents of the book, however, represented only a subsidiary part of his interests and deliberations relating to non-Euclidean geometry. As he explained in the foreword, "My own investigations in this domain [non-Euclidean space forms] were primarily intended as preliminary studies for more general space forms."

It was his dissatisfaction with the present state of the foundations of geometry that led Killing to introduce these more general space forms. Already in his review of Frischauf's book there is a hint of this dissatisfaction, for the one aspect of the book he criticized was the treatment of foundational matters [1876:465]: "Here it is a question of the development of the concepts: space, surface, line, point; congruent; bounded and unbounded; finite and infinite. As valuable as the given exposition is, it cannot be regarded as exhaustive; for it is not based on general concepts but rather upon representations of a very limited character." Killing outlined what he felt was the correct way to proceed in his essay *Fundamental Concepts and Propositions of Geometry* [1880b]. In order to provide the basis for an exhaustive exposition of the foundations of geometry, he began with the problem of determining the fundamental concepts and propositions. According to him, concepts are formed by means of definitions that combine several previously given concepts to form a new one. Some concepts must therefore remain undefined and form the basis of all others; they are the fundamental concepts. Likewise, some propositions must be accepted without proof, and they are called the fundamental propositions. In order to obtain them it is necessary to examine the representations associated with the fundamental concepts and derive from them propositions about these concepts that "make the construction of geometry possible" [1880b:3].

Killing pointed out that the science of geometry that results from this procedure is necessarily abstract. It would be tempting, he explained, to believe that we could specify so many properties of the fundamental concepts from our representation of space that the resulting geometric system holds only for those representations. With the work of Beltrami and Klein in mind (he studied Klein's papers carefully in 1878), Killing informed the reader that such beliefs are unfounded: The same system of propositions can have several interpretations; thus the theorems of plane Euclidean geometry apply to a surface of zero curvature and to Klein's projective model (parabolic geometry). Not only is it impossible to uniquely determine the interpretation of a system of geometrical propositions, Killing continued, it is also impossible to uniquely determine the system itself: For example, if from the assumptions that Euclid posits implicitly in his definitions and explicitly in his axioms only the parallel postulate and that regarding the infinitude of the straight line are dropped, four possibilities arise, all of which, "as far as we can presently judge," agree with experience [1880b:4].

On the basis of these observations, which in turn reflect his assimilation of the developments in non-Euclidean geometry, Killing came to the following conclusions, which set the tone for his own work:

> It is therefore appropriate to posit as fundamental propositions only such truths as afford a consistent development and lead, through their investigation, to the various possibilities. We regard all investigations of this type as branches of the same science and designate every individual possibility, with its further consequences, as a space form. Many may directly contradict experience; others may be very unlikely; but they nonetheless all rest upon the same foundation and exhibit in their procedures of proof an unmistakable similarity. The fundamental propositions do not suffice for the unique determination of a space form; for

this further assumptions are required that may be designated as conditional propositions. [1880b:4]

The discoveries in non-Euclidean geometry thus convinced Killing that it is necessary to conceive and investigate the foundations of geometry on such an abstract, general level as to admit space forms that run counter to our intuitions and experiences of space, just, I suggest, as, e.g., Weierstrass's example (1872) of a continuous, nowhere differentiable function contradicted the usual intuition of a continuous curve.

There is a further connection between Weierstrass and Killing's exceedingly abstract and general conception of geometry that should be mentioned at this point, even though Killing does not refer to it in *Fundamental Concepts and Propositions.* In a later discussion [1892:123–124] of his conviction that geometry in the narrow sense of the non-Euclidean space forms is but a small part of a broad scientific domain, the theory of general space forms, Killing remarked that the following considerations speak in favor of his view: If in a proper space form a straight line or plane or, more generally, some configuration left invariant by all transformations of a subgroup is regarded as the element, one is led to a system that exhibits all the characteristics of a general space form. The underlying idea of regarding lines or planes as the elements, rather than points, goes back, of course, to Plücker. But it was Weierstrass who made Killing aware of the above implications of Plücker's idea, for in his first paper [1878] on non-Euclidean geometry Killing indicated an awareness of these implications and stated that he arrived at such conclusions on the basis of some remarks made by Weierstrass.

Although Killing had already begun to systematically explore the possibilities for space forms, in [1880b] he focused his attention on an exposition of the fundamental concepts and propositions. In formulating them he followed the lead of his former teacher, Helmholtz, whose work on foundations Weierstrass had discussed with him in Berlin [Hawkins 1980:333, n.28]. Helmholtz perceived the facts at the basis of geometry in terms of the properties of mobile rigid bodies. Accordingly, Killing's fundamental concepts are rigid body, part of a body, to fill (or cover) a space, time, rest, and motion. "That these concepts cannot be dispensed with in geometry scarcely requires mention. Indeed, the ancients often attempted to ban motion from geometry, but their fruitless attempts, which led only to a restriction and had many disadvantages in method of proof as a result, demonstrate that the concept is, at least for the time being, indispensable" [1880b:5]. Such sentiments were already implicit in many discussions relating to the parallel postulate, including the work of Lobachevsky, in the sense that some critical demonstrations depended upon the assumption that geometrical figures could be moved without altering their form [Hawkins 1980:333, n.29]. Thus J. Hoüel, Lobachevsky's popularizer in France, declared explicitly in his *Critical essay on the fundamental principles of geometry* [1867:47], "We require that a figure of invariable form can be transported in any manner in its plane or in space. —All geometry is founded on the idea of the invariability of forms." Killing's work, consequently, was also in harmony with the spirit of the purely mathematical literature on foundations of geometry.

4.4. From Space Forms to Lie Algebras

As the title of his book [1885] indicates, Killing's work on non-Euclidean space forms focused upon their analytical treatment. It is thus not surprising to find that by the time *Basic Concepts and Propositions of Geometry* was published in 1880, he had already begun exploring the possibilities for space forms by the methods of analysis. A hint of these early analytical investigations is provided by Killing's correspondence with Felix Klein in 1880.[14] In a letter of October 5, Killing informed Klein that he had proved that space forms as conceived in *Basic Concepts and Propositions* can be represented by continuous manifolds in the sense of Riemann, that is, by a system of n-tuples $(x_1, \ldots, x_n)$ of real numbers where the x_i vary continuously. An analytical treatment of general space forms was thus possible, and Killing implied he had done extensive work in this area, for he claimed that he was able to specify all space forms for which the number n of dimensions and the degree of mobility m are given and satisfy $n \leq m < (n+1)^2$. (What Killing meant by "degree of mobility" is indicated below.) He promised to communicate further results of his investigations to Klein as soon as he found a more satisfying form for the equations involved, but Klein did not hear from him again on this matter until July 1884, when he received a copy of Killing's essay *The Concept of Space Extended* [1884] in the mail.

The opening paragraph of the essay effectively summarizes the viewpoint from which it was written:

> The aim of the present work is to extend the concept of space in a manner that appears to me scarcely less necessary than that which previously led from Euclidean to non-Euclidean space. In this respect, I am following a work that I published in the 1880 Easter Program of the Brilon Gymnasium .... There I listed those concepts and judgments that are indispensable to geometry and that cannot be reduced to a smaller number .... Every system of further concepts and judgments that is compatible with the basic system I call a space form. I consider the most important space forms to be those that satisfy the assumptions implicitly posited by the primary definitions of Euclid. Aside from the system treated by him, the Euclidean space form, these [assumptions] are satisfied by three other systems. In order to signify, on the one hand, their close relation to the primary definitions of Euclid and, on the other hand, their distinctness from his geometry, they may

[14] See [Hawkins 1980:333, n. 31] for information about this correspondence.

> be designated as non-Euclidean space forms. They can be carried over to an arbitrary number of dimensions. Herewith the concept of a space form is, however, in no way exhausted.

It is by the methods of analysis that Killing proposed to systematically explore the possibilities inherent in the general concept of a space form.

The analytical starting point is an n-dimensional continuous manifold of points $(x_1, \ldots, x_n)$ endowed with m degrees of mobility. Following Helmholtz,[15] Killing focused upon the behavior of infinitely small motions. Such a motion sends $x = (x_1, \ldots, x_n)$ into $x+dx = (x_1+dx_1, \ldots, x_n+dx_n)$, where

$$dx_\rho = u^{(\rho)}(x_1, \ldots, x_n)dt, \quad \rho = 1, \ldots, n. \tag{3.1}$$

Integration of the system of differential equations (3.1) yields the corresponding finite motion, but it played a secondary role in Killing's analysis. The problem of determining all space forms was conceived as commencing with the problem of classifying the associated infinitesimal motions. In order to make the classification problem tractable, he began by prescribing some general properties of the infinitesimal motions of a space form, properties that, in effect, endow them with the structure of a Lie algebra.

The first property derives from the observation that if (3.1) defines a motion in an n-dimensional space form, then if p denotes a real constant, then $dx_\rho = pu^{(\rho)}dt$, $\rho = 1, \ldots, n$, represents the same motion "if we ignore the velocity" (p. 10). Next Killing considered the traditional conception of the composition of infinitely small motions, according to which if $dx_\rho = u^{(\rho)}dt$ and $dx_\rho = v^{(\rho)}dt$ are applied successively, the resultant motion is $dx_\rho = (u^{(\rho)} + v^{(\rho)})dt$. The composition of infinitesimal motions is therefore analogous to the parallelogram law of the composition of forces. (The earliest justifications of the latter presume the former.) Several geometers, including Möbius (1838), Klein (1871), and Klein's student F. Lindemann (1874), had used this analogy to reduce the study of the composition of rigid motions in space to an application of statics. Killing had in fact studied Lindemann's paper, which extends the earlier work to the context of the rigid motions in Klein's parabolic, elliptic, and hyperbolic geometries.[16] As we saw in Section 2.2 Lie interpreted the composition of infinitesimal transformations in the same manner.

With this traditional conception in mind, Killing asserted that if a space form admits m infinitely small motions

$$dx_\rho = u_\chi^{(\rho)}(x_1, \ldots, x_n)dt, \quad \chi = 1, \ldots, m, \tag{3.2}$$

then the motion

---

[15] See [Helmholtz 1868b:625–636] or the clearer (albeit no more rigorous) exposition of Helmholtz's proof by Frischauf [1876:123–129].

[16] For details on early discussions of the composition of forces and infinitesimal motions, see [Hawkins 1980:304].

$$dx^{(\rho)} = \left(\sum_{\chi=1}^{m} p_\chi u_\chi^{(\rho)}\right)dt, \quad \rho = 1,\ldots,n$$

(where the p_χ denote real numbers), is also possible. The conditions so far described imply that the infinitesimal motions form, in modern terms, a real vector space. The number m is called the degree of mobility of the space form if (in modern terminology) the m motions defined in (3.2) are linearly independent and form a basis for all possible motions. Killing restricted his attention to the case of finite m "just as we assumed that the number n of dimensions is finite" (p. 10).

Thus far Killing had not ventured beyond the customary conventions for dealing with infinitely small motions in mechanics and geometry. These had proved adequate in dealing with the rigid motions of ordinary space or, as with Lindemann, the additional two non-Euclidean spaces introduced by Klein. But Killing was not dealing with known space forms. He was seeking to determine all possible space forms. Evidently, sensing the need for additional analytical tools, he imposed a further, unconventional, condition on his infinitesimal motions that implies that they form a finite-dimensional real Lie algebra. The presentation and explication of the condition are extremely obscure, and what follows is my attempt to explain the basis of Killing's reasoning with greater clarity and detail than he did. The relevant portion of the text is reproduced in Appendix 4 of my paper [1980] along with a brief commentary, and the entire text is contained in [Hein 1997].

Killing began by declaring that "every motion of a body must still be possible for it after it has been moved in any way" and proceeded to examine the effect on a point $x = (x_1,\ldots,x_n)$ of the successive application of two infinitesimal motions $dx_\rho = u_\iota^{(\rho)} d\sigma$ and $dx_\rho = u_\chi^{(\rho)} dt$. Strict adherence to the previously described conventions would imply that regardless of the order of application, x is moved to $x + dx$, where $dx_\rho = \big(u_\iota^{(\rho)} + u_\chi^{(\rho)}\big)dt$. Infinitesimal motions commute. If, however, M_ι and M_χ denote the corresponding finite motions, they need not commute; and since $M_\iota M_\chi x$ can differ from $M_\chi M_\iota x$, $(M_\chi M_\iota)(M_\iota M_\chi)^{-1}$ need not leave x fixed. Killing's vague deliberations seem aimed at determining the infinitesimal motion corresponding to the "commutator" motion $(M_\chi M_\iota)(M_\iota M_\chi)^{-1}$, which must be a permissible infinitesimal motion and consequently expressible as a linear combination of the given "basis" motions (3.2). Part of the obscurity probably stems from the need to consider second-order infinitesimals while giving the appearance of adhering to the traditional convention that ignores them. Of course, what Killing was doing was quite similar to what, unbeknownst to him, Lie had already done (Section 2.2).

The motion u_ι sends x into $y = x + dx$, where

$$y_\rho = x_\rho + u_\iota^{(\rho)}(x)d\sigma, \quad \rho = 1,\ldots,n.$$

The motion u_χ then sends y into $z = y + dy$, where $dy_\rho = u_\chi^{(\rho)}(y)dt$. If $u_\chi^{(\rho)}(y) = u_\chi^{(\rho)}(x + u_\iota d\sigma)$ is expanded about x and higher-order infinitesimals are neglected, we obtain

$$dy_\rho = \left(u_\chi^{(\rho)}(x) + \sum_{\nu=1}^{n} \frac{\partial u_\chi^{(\rho)}}{\partial x_\nu} u_\iota^{(\nu)} d\sigma \right) dt.$$

Thus the effect of following the motion u_ι by u_χ is to send x to z, where

$$z_\rho = x_\rho + u_\iota^{(\rho)}(dx)d\sigma + u_\chi^{(\rho)}(x)dt + d\sigma \left(\sum_{\nu=1}^{n} \frac{\partial u_\chi^{(\rho)}}{\partial x_\nu} u_\iota^{(\nu)} \right) dt. \tag{3.3}$$

By ignoring the second-order infinitesimal term, we would arrive at the conventional conception of composition. If the order of application of u_ι and u_χ is reversed, similar considerations show that x is sent into w, where

$$w_\rho = x_\rho + u_\iota^{(\rho)}(x)d\sigma + u_\chi^{(\rho)}(x)dt + d\sigma \left(\sum_{nu=1}^{n} \frac{\partial u_\iota^{(\rho)}}{\partial u_\nu} u_\chi^{(\nu)} \right) dt. \tag{3.4}$$

Now, by the traditional convention, the inverse of $x \to x+dx$ is $x \to x-dx$. By subtracting (3.4) from (3.3) an infinitesimal motion corresponding to $(M_\chi M_\iota)(M_\iota M_\chi)^{-1}$ is thus obtained (by conveniently omitting the dt):

$$dx_\rho = U_{\iota\chi}^{(\rho)} d\sigma, \quad U_{\iota\chi}^{(\rho)} = \sum_{\nu=1}^{n} \left(u_\iota^{(\nu)} \frac{\partial u_\chi^{(\rho)}}{\partial x_\nu} - u_\chi^{(\nu)} \frac{\partial u_\iota^{(\rho)}}{\partial x_\nu} \right). \tag{3.5}$$

Since the motion (3.5) must be a linear combination of the "basis" motions u_χ, Killing obtained the condition

$$U_{\iota\chi}^{(\rho)} = \sum_\mu a_{\mu,\iota\chi} u_\mu^{(\rho)}, \quad \rho - 1, \ldots, n, \tag{3.6}$$

where the $a_{\mu,\iota\chi}$ denote real constants. This should appear quite similar to Lie's line of thought using Jacobi's calculus of differential operators (Section 2.2). To express Killing's condition (3.6) in operator form, let

$$X_\iota = \sum_{\rho=1}^{n} u_\iota^{(\rho)} \frac{\partial}{\partial x_\rho} \quad \text{and} \quad X_\chi = \sum_{\rho=1}^{n} u_\chi^{(\rho)} \frac{\partial}{\partial x_\rho}.$$

Then

$$[X_\iota, X_\chi] = \sum_\rho U_\iota^{(\rho)} \frac{\partial}{\partial x_\rho},$$

and (3.6) asserts that

$$[X_\iota, X_\chi] = \sum_\mu a_{\mu,\iota\chi} X_\mu,$$

which is the condition that the operators $X_1, \ldots, X_m$ form a Lie algebra.

By virtue of (3.6), at most m of the m^2 motions $U_{\iota\chi}$ can be linearly independent, and Killing sought relations expressing dependency. Two are immediately evident:

$$U_{\iota\chi}^{(\rho)} + U_{\chi\iota}^{(\rho)} = 0, \quad U_{\iota\iota}^{(\rho)} = 0, \quad \rho = 1, \ldots, n. \tag{3.7}$$

In addition to these, he obtained (pp. 13–14) the less obvious

$$\sum_{\nu} \left(a_{\nu,\chi\lambda} U_{\iota\nu}^{(\rho)} + a_{\nu,\lambda\iota} U_{\chi\nu}^{(\sigma)} + a_{\nu,\iota\chi} U_{\lambda\nu}^{(\sigma)} \right) = 0. \tag{3.8}$$

In the operator notation, (3.7) and (3.8) become, respectively,

$$[X_\iota, X_\chi] + [X_\chi, X_\iota] = 0, \quad [X_\iota, X_\iota] = 0$$

and the Jacobi identity,

$$[X_\iota, [X_\chi, X_\lambda]] + [X_\chi, [X_\lambda, X_\chi]] + [X_\lambda, [X_\iota, X_\chi]] = 0.$$

The infinitely small motions of Killing's space form satisfy precisely the conditions defining a finite-dimensional Lie algebra. According to Killing, he discovered the Jacobi identity (3.8) no later than 1879 and originally proved it as presented in 1884 [Hawkins 1980:333, n.35].

There are three stages to the resolution of the problem of determining all possible space forms, which may be appropriately referred to as the algebraic, analytical, and geometrical stages. In the algebraic stage the symbols $u_\chi^{(\rho)}$ forming the basis for the infinitesimal motions are viewed as abstract symbols satisfying the relations given in (3.7) and (3.8). The problem here is to choose the basis motions $u_\chi^{(\rho)}$ so that the associated matrix of coefficients $a_{\nu,\iota\chi}$ is as simple as possible, to facilitate determination of the actual functions of $x, u_\chi^{(\rho)} = u_\chi^{(\rho)}(x)$, from the differential equations (3.6). The solution of these equations constitutes the analytical stage. At the analytical stage the infinitesimal motions are thus determined, and the final stage involves using them to describe the geometrical nature of the space form. It is, of course, at the algebraic stage that Killing was confronted with what amounts to the problem of determining the structures of all real Lie algebras.

Within the confines of a *Programmschrift* Killing naturally could not discuss the mathematics of any of the stages in detail, for the problems involved are extremely general and difficult. Furthermore, although he had already worked on these problems, his results were far from definitive. In the essay he therefore chose to limit himself to what is little more than a summary of results for the problem of determining all n-dimensional space forms with m degrees of mobility when neither n nor m exceeds 3. Although the treatment of the algebraic stage is limited in scope and specialized accordingly, traces of the characteristic elements of his later publications on Lie algebras are evident, especially in the discussion of the least trivial of the cases, $m = 3$.

Killing began consideration of this case with the following remarks (p. 15):

> For three degrees of mobility, equation (10) [an immediate consequence of the Jacobi identity (3.8)] takes the form
>
> $$(a_{2,21} + a_{3,31})U_{23} + (a_{3,32} + a_{1,12})U_{31} + (a_{1,13} + a_{2,23})U_{12} = 0.$$
>
> If the coefficients of the quantities U in this equation do not vanish, then there exist in the system motions that are permutable with one another. Consequently, there are three distinct cases:
>
> I. $a_{2,21} + a_{3,31} = 0, \; a_{3,32} + a_{1,12} = 0, \; a_{1,13} + a_{2,23} = 0.$
>
> II. $U_{12} = 0, \; a_{3,31} = a_{3,32} = 0.$
>
> III. $U_{12} = U_{13} = 0.$
>
> For the first condition, three real motions can always be chosen so that the corresponding equations are
>
> $$U_{23} = \alpha u_1, \quad U_{31} = \beta u_2, \quad U_{12} = \gamma u_3.$$

Cases I–III are determined by the number p of linearly independent U and hence, in modern terms, by the dimension of the derived algebra $\mathfrak{g}' = [\mathfrak{g}, \mathfrak{g}]$.[17] Case I corresponds to $p = 3$, case II to $p = 2$, and case III to $p = 0$ or 1. The conditions stated in I are an immediate consequence of the assumption $p = 3$. Those stated in II also follow from the hypothesis $p = 2$. The condition $a_{3,31} = a_{3,32} = 0$ indicates that Killing has chosen a new basis u_1, u_2, u_3 so that u_1 and u_2 span the $\overline{U}_{\iota\chi}$ of the original basis $\overline{u}_1, \overline{u}_2, \overline{u}_3$, and that he recognized the property we would now describe by stating that the motions spanned by u_1 and u_2 form an ideal (he could, of course, have added $a_{3,12} = 0$ in II). That, in addition u_1 and u_2 can be chosen so that $U_{12} = 0$ follows from the Jacobi identity as expressed in (3.8) above.

The resolution of the algebraic problem in case II is especially noteworthy. Killing wrote (pp. 15–16):

> If the simplest values of the coefficients $[a_{\nu,\iota\chi}]$ are desired, a quadratic equation must be solved; it has either two real and distinct, two imaginary, or two equal roots, and therefrom the three special cases result:
>
> (a) $U_{12} = 0, \; U_{13} = u_1, \; U_{23} = u_2;$
>
> (b) $U_{12} = 0, \; U_{13} = \alpha u_1 + u_2, \; U_{23} = -u_1 + \alpha u_2;$

[17] Killing formally introduced p in print later [1886:7], although it is clear from that work and the above determination of cases that he used this number as a principle of classification from the outset of his investigations.

(c) $U_{12} = 0,\ U_{13} = u_1,\ U_{23} = u_2 + \alpha u_1$.

The quadratic equation referred to is the characteristic equation of the linear transformation $u \to [u, u_3]$, where u denotes a linear combination of u_1 and u_2. The three special cases correspond to the three possibilities for the elementary divisors, and the form of the $U_{\iota\chi}$ is designed to yield these possibilities; they are simple canonical forms for the linear transformation $\mathrm{ad}\, u_3 : u \to [u, u_3]$.

Admittedly, the algebraic problem on this level ($m \leq 3$) is very special and manageable without profound methods, but even here Killing's work exhibits some of the features that characterize his subsequent treatment of the general problem: classification of systems by means of the value of $p = \dim \mathfrak{g}'$; extensive reliance on the Jacobi identity; application of the theory of elementary divisors to the characteristic equation of an adjoint mapping to obtain canonical forms for the "structure constants" $a_{\nu,\iota\chi}$ and to distinguish possibilities. There are, however, no complex Lie algebras in *The Space Concept Extended.* Unlike Klein and Lie, Killing was reluctant to admit space forms on manifolds of complex numbers into his treatment of the foundations of geometry, a reluctance that continued after his success in studying the structure of complex Lie algebras in his papers of 1888–90. (See [Killing 1892:124].) There is some evidence, however, from Killing's correspondence with Engel, disussed in the next chapter, to indicate that even before his acquaintance with Lie's transformation groups (which act on complex manifolds), Killing carried his investigations over to the complex case, since it forms a more convenient setting for the analysis of the elementary divisors of the characteristic equation. But his intention undoubtedly was to use the results of such an investigation to deal with the enumeration of real space forms. As will be seen in the next chapter, it was the contact with Lie and, particularly, his colleague and assistant Friedrich Engel, that encouraged Killing to pursue his analysis of the complex case for its own sake and to the point where it bore fruit.

4.5. Riemann and Helmholtz

To gain a greater historical perspective on Killing's work on the foundations of geometry as discussed in the previous sections, it is enlightening to consider more carefully its relation to the essays of Riemann and Helmholtz. As we shall see more clearly, Killing drew heavily upon their work for inspiration. It is doubtful that the *The Space Concept Extended* would ever have materialized without the essays of Riemann and Helmholtz to serve as a guide. Nonetheless, Killing's research program diverges in significant respects from those of his eminent predecessors, respects that are crucial in the sense that they involve an emphasis congenial to the algebraic problem of determining the structure of all Lie algebras. Furthermore, it is in terms of these contrasts with the work of Riemann and Helmholtz that further manifestations of the influence of Weierstrass and Killing's training at Berlin become evident.

A distinguishing feature of Killing's approach to the foundations of geometry is its extreme generality. In insisting upon the need to proceed from general concepts, Killing was following the precedent set by Riemann in his essay "On the hypotheses lying at the basis of geometry" [1868]. Riemann approached the foundations of geometry through the general concept of an n-fold extended manifold of real numbers $(x_1, \ldots, x_n)$, in which, he suggested tentatively, metric relations are determined by a quadratic differential form $ds^2 = \Sigma_{i,j=1}^{n} g_{ij}(x_1, \ldots, x_n) dx_i dx_j$. Although Riemann's approach was exceedingly general, it was motivated by physical concerns. As he explained at the conclusion of the essay, the purpose of the general approach was to avoid any unnecessary restrictions that might serve to hinder the progress of scientific knowledge [1868:286]. Within the general framework he proposed, the various hypotheses of geometry could be objectively examined and the extent of their basis in experience judged. Despite the obscurity of the concluding section of the essay it is fairly clear that Riemann did not at all dismiss the possibility that the needs of physical science might even require conceiving of space as a manifold of variable curvature (with zero average curvature). The essay was written at a time when he was preoccupied with the study of the "forces" of nature—gravity, light, heat, electricity, and magnetism—and their interrelations, and there is some evidence to indicate that he believed that the mathematical analysis of such interrelations might require differential forms corresponding to manifolds of variable curvature.[18]

These speculations are only hinted at toward the end of the essay, whereas much more attention is focused upon manifolds of constant curvature. Such manifolds, Riemann stressed, possess properties in agreement with experience and our geometrical intuition: Figures can be moved about freely without distortion, and the sum of the interior angles of a triangle is independent of the triangle. Modern readers of Riemann's essay are immediately struck by the remarks concerning manifolds of variable curvature in which Riemann seems to adumbrate general relativity. In the nineteenth century, however, the essay was regarded in a different light. Except for W. K. Clifford's brief, sympathetic exposition [1870], Riemann's speculations about a variably curved space were either ignored altogether or peremptorily dismissed [Hawkins 1980:334, n. 39].

Riemann's discussion of manifolds of constant curvature, on the other hand, impressed his nineteenth-century readers. Felix Klein expressed the prevailing sentiment, when, in his lectures on non-Euclidean geometry (1889) he declared that "this concept of an n-fold extended manifold of con-

[18] See pp. 285–6 (top pagination) of Riemann's *Werke* [1990] as well as pp. 507–508, 521–538, 547–549; and see also [Speiser 1927].

stant curvature is the most essential result of the Riemannian approach."[19] The homogeneity of space and the free mobility of bodies seemed essential to geometry and incontrovertible facts of experience. Just as previous generations of mathematicians had offered proofs of the parallel postulate, those of the second half of the nineteenth century provided many proofs that there are only three or, in some cases, four possible geometries compatible with experience: Euclidean and Lobachevskian geometry, the spherical geometry of Riemann, and elliptic geometry. All these geometries correspond analytically to manifolds of constant curvature. The proper context for geometry was thus a manifold of constant curvature. The advantage of Riemann's abstract approach was that it brought out the significance of spatial homogeneity. Nor was the restriction to manifolds of constant curvature deemed highly restrictive. The emphasis was rather on the great generality and intellectual daring brought to bear upon geometry by conceiving of it in terms of manifolds of constant curvature. For example, it admitted the possibility of an unbounded yet finite space, which to J. Rosanes represented "most remarkably the height of abstraction in Riemann's train of thought" [1871:18].

The bounds of non-Euclidean geometry were thus prescribed by the concept of a manifold of constant curvature. Because local and global properties of a geometry were not clearly distinguished in the 1870s and 1880s, the geometrical possibilities in n dimensions thus seemed limited to (at most) the three non-Euclidean space forms (to use Killing's terminology) and Euclidean geometry. These space forms of course provided a rich source of diverse and interesting research problems and gave rise to an extensive literature, as can be seen from Killing's own book on the subject [1885] as well as from [Loria 1888]. In choosing to go beyond the realm of non-Euclidean space forms, Killing was not proceeding in the spirit of Riemann. Unlike Riemann, Killing did not envision any physical significance to the exploration of more general space forms. He shared the view of his contemporaries that the non-Euclidean space forms were the ones that agreed with experience; and in arguments with his skeptical colleague, R. von Lilienthal, he was even willing to admit that the geometry of Euclid is the only "true" geometry [Hawkins 1980:Appendix 5]. It was rather his critical attitude toward previous treatments of the foundations of geometry that led him to pursue geometry beyond the bounds imposed by experience.

Killing was especially critical of the arguments given by Riemann and Helmholtz to establish the metrical properties of space. In *The Concept of Space Extended* he expressed his dissatisfaction with Riemann's presentation in the following passage:

> Riemann assumes that the expression for the line element could be any function of the differentials that is always positive and

[19] [Klein 1893a:212]. Klein also said essentially the same thing much earlier [1872a:313].

> that increases in a certain ratio when all differentials are multiplied by this constant. The assumption would be permissible only if in measuring a line its position in space did not at all come into consideration. Now, the stipulation that the length should be independent of its position does in no way state that the line exists independently of the surrounding space and can be measured independently of the same. Consequently, in my opinion, a special investigation is required to explore which functions are suitable as expressions for the line element, and I deem it a mistake that Riemann did not undertake such an investigation [1884:4].

This critique of Riemann's treatment of metric relations in space seems to echo Weierstrass's dissatisfaction with the present state of affairs because, according to Killing, Weierstrass conveyed the following viewpoint on foundations of geometry in his seminar lectures of 1872:

> Above all, one must strive for a purely geometrical foundation. If, however, one wishes to proceed from the concept of an n-fold extended manifold, the concept of a distance function can be adjoined. For this concept the characteristic properties would be specified, and it would then be required to seek the general analytical form of such a function [1884:4].

These remarks suggest that Weierstrass agreed with Riemann's emphasis upon the primacy of metrical relations but felt that no one had as yet analyzed the possibilities in a sufficiently rigorous manner.

In his essay [1868b] Helmholtz had already addressed himself to this point of Riemann's essay and attempted to show that Riemann's hypothesis that metric relations are given by a quadratic differential form could be derived from certain facts of experience concerning the existence and mobility of rigid bodies that presume that spatial measurements are possible by means of congruence. Because these facts formed the starting point of his exploration of the foundations of geometry, Helmholtz explained [1868b:621], his own work lacked "the great generality" characteristic of Riemann's essay prior to the introduction of similar assumptions (in the discussion of manifolds of constant curvature). But once Riemann made these assumptions, "my results are in complete agreement with his." It was to support his claim of complete validity of Riemann's assumption concerning the nature of ds [1868b:625–639]. Helmholtz's proof, with its emphasis upon the consideration of infinitesimal motions, and his approach to the foundations of geometry clearly influenced Killing; but Killing's work involved a degree of generality akin to that of Riemann's essay prior to the assumptions leading to manifolds of constant curvature. Helmholtz avoided such generality by introducing metrical assumptions in his hypotheses. In postulating the existence of rigid bodies (Hypothesis II [1868b:622]), he assumed the existence of an "equation" $J(x_1, \ldots, x_n; y_1, \ldots, y_n)$ between any two points x and y of a rigid body that is unaffected by the motion of the body. From the existence of this equation he then concluded that his third hypothesis, the "completely free mobility of rigid bodies," implied that $n(n+1)/2$ quantities are necessary to determine the position of a rigid system.

In Killing's terminology, free mobility in Helmholtz's sense implies degree of mobility $m = n(n+1)/2$. In fact, in a preliminary announcement of his results, Helmholtz used this terminology to explain that:

> The stipulation that Riemann first introduces at the conclusion of his investigation, namely that spatial configurations should have, without change of form, that degree of mobility that is assumed in geometry, I had introduced from the outset; and this stipulation then limits the possibilities of the hypotheses that can be made regarding the expression for the line element [ds] to such an extent that only the form accepted by Riemann remains, all others being excluded [1868a:613–614].

Helmholtz had in effect assumed the existence and properties of a distance function in order to limit the geometrical possibilities.

It was probably the general lack of clarity and rigor of Helmholtz's demonstration that convinced Killing that it would be best to begin without any assumptions about distance functions. As he put it in the foreword of his book [1885:iv]:

> Attempts to create a natural foundation for geometry have not hitherto been accompanied with the desired success. The reason lies, in my opinion, in this: Just as geometry had to abandon the concept of direction in the sense stipulated by the parallel axiom, so also that of distance cannot be maintained as a fundamental concept, and consequently geometry must go beyond the non-Euclidean space forms in the narrower sense ....

After a thorough investigation of general space forms, in which the degree of mobility need not satisfy $m = n(n+1)/2$, an adequate and appropriate conceptual framework would be at hand to determine conditions under which a metric in Riemann's sense exists in a space form and to rigorously characterize the special Euclidean and non-Euclidean space forms to which the analysis of Helmholtz leads.[20] In his attempt to make Helmholtz's approach rigorous, Killing thus found it necessary to imbue it with a degree of generality commensurate with that found in Riemann's essay.

[20] After a detour—of unexpectedly long duration—into the realm of Lie algebras and their structure, Killing returned to his theory of space forms. Using his and Lie's results on transformation groups, he studied the special class of "proper space forms" [1892:159] that satisfy an additional, eighth, axiom [1892:130]. Killing argued that proper space forms have degree of mobility $n(n+1)/2$ and admit a metric ds in Riemann's sense. From the fact that an infinitesimal motion must leave ds invariant, Killing deduced the equations defining what is now called a Killing vector field [1892:167]. Using these equations, he concluded that the proper space forms have constant curvature. In this way he sought to clarify Helmholtz's deliberations. For more details and references on these matters see [Hawkins 1980:334, n.43].

In Section 3, I suggested that the discovery of non-Euclidean geometries and Killing's reaction to those discoveries were analogous to the discoveries of pathological functions and the attempts of Weierstrass to rigorously establish the theory of functions without recourse to physical considerations or geometrical intuition. I also suggested that the analogy was historically significant, that it colored the manner in which Killing dealt with the foundations of geometry. I shall now adduce some evidence in support of these claims. Although I have found nothing as convincing as an explicit affirmation by Killing himself, I believe that the evidence is sufficient to make them at least plausible. In this connection and as first evidence of the validity of my claims, I cite the opinion of R. von Lilienthal, Killing's colleague for many years at Münster: "Weierstrass's strongly theoretical bent found in Killing, who, so to speak, grew up on scholastic philosophy ... a splendid sounding board. Non-Euclidean geometry, on which Weierstrass had given some lectures in his seminar, became Killing's favorite sphere of activity."[21] Although Lilienthal offered no proof, his opinion is not lacking in authority, because not only did he know Killing well, he had also attended the University of Berlin and had written his dissertation under the direction of Weierstrass, as well an essay on him [1931].

A more direct and specific link between the Weierstrassian attitude toward the theory of functions and non-Euclidean geometry is provided by Killing himself in *The Concept of Space Extended* when he considered an essay on the foundations of geometry written by the Belgian engineer J.-M. de Tilly [1880]. De Tilly had attempted to prove that the only possible geometries are those of Euclid and Lobachevsky and the spherical geometry of Riemann. In refuting de Tilly, Killing pointed to the existence of continuous non-differentiable functions that revealed the untenability of de Tilly's assumption that his distance function varies monotonically in a sufficiently small neighborhood. Since Weierstrass presented his celebrated example of a continuous nowhere-differentiable function to the Berlin Academy in July of 1872 [Weierstrass 1895:71–74], it is likely that he also presented it in the mathematics seminar, so that Killing may have had this example specifically in mind. In his later book on the foundations of geometry Killing invoked other Weierstrassian counterexamples, such as his counterexample to the so-called Dirichlet principle, to support certain geometrical arguments [1893:172–73; 1898:49–54, 186–190, 209, 212–214].

Further evidence that Weierstrass's critical approach to analysis could be congenial to work on the foundations of geometry is afforded by the example of Moritz Pasch, whose book *Lectures on Modern Geometry* [1882b] was just as atypical, although in a different way, of the work being done on non-Euclidean geometry and foundations as were Killing's *Programmschriften* of 1880 and 1884. After obtaining his doctorate from the Univer-

---

[21] "Nachruf für Killing," typescript dated 24.2.23 (Engel Archive Giessen). A transcript of some passages is given in Appendix 5 of my [1980].

sity of Breslau, Pasch spent the academic year 1865–66 at the University of Berlin, where he attended the lectures of Weierstrass and Kronecker. As he explained later in an autobiographical essay [1930:9], their lectures exerted a "great influence" upon the development of his approach to the foundations of geometry. His unprecedented book achieved a degree of clarity and rigor that was in line with the emphasis at Berlin, as was his avoidance of any appeal to intuition in his proofs. There is also evidence that Pasch perceived a connection between the latest developments in geometry and those in the theory of functions. In his autobiographical notes [1930:8] he pointed out that Euclid, who had resolved the problem of the foundations of mathematics in his time, "was unable to go as far as we now, after a long development that has led to the theory of functions and non-Euclidean geometry, recognize as necessary." Although Pasch articulated this connection between non-Euclidean geometry and the theory of functions toward the end of his life, it seems fairly certain that it was at the back of his mind when he composed *Lectures on Modern Geometry*, since the same year he also published a book on analysis in which the foundations of the theory of functions are developed in the spirit of, and with many direct references to, the work of Weierstrass [1882a:47n, 55n, 173n, 179].

I have attempted to show that the "strongly theoretical bent" of Weierstrass' mathematics did indeed strike a responsive chord in Killing just as it did in Pasch. In order to bring out more fully the nature of Weierstrass's influence on work on non-Euclidean geometry and its foundations, it is instructive to consider the work in this area of a mathematician who was not influenced by Weierstrass, who, in fact, was consciously opposed to the spirit in which Weierstrass approached mathematics. Felix Klein was such a mathematician. The next section is devoted to an examination of his work on non-Euclidean geometry. Klein's work is doubly significant because despite the entirely different spirit in which it was undertaken, Killing was encouraged by it, for he viewed it as leading necessarily to general space forms.

4.6. Killing and Klein on the Scope of Geometry

By disavowing the primacy of metric relations in the foundations of geometry, Killing was going against the judgment of Weierstrass himself, as the above-quoted summary of Weierstrass's 1872 lectures indicates. Indeed, in 1870 Klein had presented a lecture in the Berlin seminar on Cayley's "Sixth Memoir upon Quantics" [1859] in which he suggested in conclusion that Cayley's general metric, and hence projective geometry, might serve as the basis for non-Euclidean geometry just as it did, as Cayley had showed, for Euclidean geometry. According to Klein [1921:50–51], Weierstrass disagreed on the grounds that the distance between two points must form the starting point for the founding of geometry. Weierstrass may not have been as adamant in his opinion as Klein's recollection suggests, for in his unpublished lectures of 1886 the same viewpoint is expressed less dogmatically [Hawkins 1980:335, n.45], but it is certain that he felt that the distance between two points afforded the most promising starting point for the development of the foundations of geometry.

Killing was absent from Berlin when Klein gave his seminar talk, but when he finally studied Klein's papers "On so-called Non-Euclidean Geometry" [1871a; 1872a] between 1878 and 1879 he found encouragement for his own developing approach to the foundations of geometry. In his brief correspondence with Klein in 1880 Killing spoke of "the close connection between my investigations and your interesting results on projective space forms,"[22] and in the postcard announcing the posting of a copy of *The Concept of Space Extended* to Klein he expressed his confidence that Klein would recognize the close connection of their work [Hawkins 1980:Appendix 3C]. In *The Concept of Space Extended* itself, Killing went on to declare that "as soon as the step taken by Mr. Klein is followed through with complete consistency, only a small generalization is required in order to arrive at our space forms" [1884:3].

These remarks warrant a closer look at Klein's publications on non-Euclidean geometry and their relation to Killing's theory of space forms. We shall see that Klein's work does indeed contain elements that must have proved encouraging, and possibly inspirational, to Killing as he developed his views on the foundations of geometry. On the other hand, the works of Killing and Klein are lacking in any genuine affinity. When the attitudes and concomitant objectives behind their respective works are considered, essential differences emerge, differences that in particular show that Klein's research program would never have convinced him of the need to consider the problem of determining the structures of all possible finite-dimensional Lie algebras. Klein was not interested in consistently carrying out the approach he had introduced so as to achieve an exhaustive analysis of all the logical possibilities in the spirit of Weierstrassian thoroughness. Nor was it, from his perspective, a matter of a "little generalization" to arrive at the general space forms Killing wished to study. By focusing upon these differences between Killing and Klein we shall be in a position to bring out more clearly and fully the link between Killing's work and his training at Berlin. Indeed, just as Killing's work reflects his educational background, so too does Klein's reflect his contact with Plücker at Bonn and Clebsch at Göttingen.

The kinship that Killing sensed upon reading Klein's papers was naturally due in large part to the fact that they broke with the precedent of taking metric relations as fundamental to geometry. But the sense of kinship has further grounds as well. Under the influence of Riemann's essay [1868], Klein stressed the importance of a more general understanding of the

[22] Letter to Klein dated 5 October 1880. See Appendix 3B of my paper [1980] for the German text and location of the original.

domain of non-Euclidean geometry: Just as the parallel postulate had been the object of careful scrutiny by the founders of non-Euclidean geometry (Gauss, Lobachevsky, Bolyai), so too should all the other assumptions that lay at the basis of our geometrical representations; Lobachevskian geometry "is the first step in a direction whose general possibility is indicated by Riemann's work" [1872a:312]. These remarks are similar in tone to Killing's pronouncements quoted in Sections 4–5 and reveal the common influence of Riemann's essay.

Killing undoubtedly also appreciated the Helmholtzian touch in Klein's discussion of hyperbolic and elliptic geometry: The linear transformations that leave invariant the variety—called the "absolute"—which determines Cayley's metric are regarded as the motions of the geometry. For Klein these motions formed a transformation group, and he perceived the important role groups of transformations could play in bringing unity to the ostensibly disparate fields of contemporary research in geometry—non-Euclidean geometry being but one such field. Klein's definitive expression of his views was presented in his *Erlanger Programm* of 1872 (Section 1.5). There is no evidence to indicate that Killing ever read this work, but Klein included a preliminary sketch of his ideas in his paper [1872a] on non-Euclidean geometry, which Killing did read. In the paper we find assertions such as, "It is possible to speak of a definite [geometrical] treatment of a manifold of n dimensions only after a transformation group is given ...." (p. 318). Killing would certainly have regarded such statements as in agreement with the generalized Helmholtzian approach to the foundations of geometry expressed in his own essays of 1880 and 1884.

Killing's declarations of a close relationship between his theory of space forms are consequently understandable, but the above similarities belie a more fundamental divergence of attitude and objectives. In the first place, Klein was always concerned about the physical relevance of non-Euclidean geometry [1872a:312–13]. He later explained that although his own work dealt with the purely mathematical side of the subject, he never shared the complete indifference to physical compatibility "that is often prized by modern axiomatizers" [1921:253]. Indeed, when Klein first announced his results in a paper submitted to the Göttingen Academy of Science, he included a proof that the sixfold motions of physical space must correspond to the linear transformations leaving a surface of second degree (the infinitely distant points) invariant and hence to one of the three geometries (parabolic or Euclidean, hyperbolic, and elliptic) that he had derived by specializing the general Cayley metric [1871a:251–252]. Klein's interest in physical problems and his conviction that pure mathematics is relevant to physical science goes back to his student days at the University of Bonn, where he came under the influence of Julius Plücker, who simultaneously held professorships in mathematics and physical science and had, in fact, made original contributions in both disciplines. Klein's intention while at Bonn was to pursue a career in physical science, and probably he would have done so had not Plücker decided shortly before Klein's arrival in Bonn in 1865 to halt, after twenty years, his researches on experimental physics in order to develop his ideas on line geometry, a geometry based upon the line as element (Section 1.1). By turning to line geometry, Plücker was not turning his back on physics in any sense; the physical relevance of line geometry to mechanics and optics was foremost in his mind.[23]

It is in the light of his concern with physical relevance that Klein's call for an examination of all the assumptions of geometry must be interpreted, for although Klein acknowledged and paraphrased Riemann's objective of proceeding from general concepts so as to avoid any unnecessary limitations in the study of nature [1872a:313], he never seems to have shared Riemann's vision of the physical possibility of a manifold of variable curvature. As noted above, Klein regarded Riemann's chief contribution to non-Euclidean geometry to be the concept of a manifold of *constant* curvature. Like Killing and the vast majority of mathematicians sympathetic to non-Euclidean geometry, Klein felt that experience implied a manifold of constant curvature, which, as already noted, seemed to greatly limit the geometrical possibilities. Unlike Killing, Klein felt no compulsion to pursue the science of geometry beyond the bounds imposed by experience in order to rigorously clarify and establish its foundations.

Klein also preferred to impose limits upon the scope of non-Euclidean geometry for another reason. Under the influence of Clebsch and his students at Göttingen, Klein had been inculcated with "the consistently projective way of thinking" of the modern geometry of the period as expounded by Clebsch [1921:52,412]. The objective of his papers on non-Euclidean geometry was to complete what Cayley had begun in his "Sixth Memoir on Quantics" when he derived Euclidean from projective geometry to show that "descriptive [i.e., projective] geometry is *all* geometry" [1859:592]. Klein sought to show that Cayley's dictum applied to non-Euclidean geometries as well. He would consequently have had little sympathy for Killing's program of going beyond projective space forms. After all, the idea was to encompass all geometries within the framework of projective geometry. For this reason Klein was not particularly enthusiastic about spherical geometry,[24] which Killing accepted as equally legitimate with elliptic geometry. Spherical geometry violated the basic tenet of projective geometry that two points uniquely determine a straight line. As Klein later explained [1921:241], the fact that Beltrami identified the geometry of a manifold of constant positive

---

[23] See the documents quoted by Ernst [1933:38] and Plücker's papers [1866; 1965:Sects. II, IV].

[24] Klein's discussion of spherical geometry [1872a:324] is limited to the remark that "die Kugel ist... nicht das einfachste Bild für eine Mannigfaltigkeit von positivem konstantem Krümmungsmasse," and that the simplest is his own model [1871b:Sect. 10], according to which points of the geometry are straight lines through the center of a sphere and a line consists of all such points lying in the plane of a great circle. In this Plücker-type geometry, two points uniquely determine a straight line.

curvature with spherical geometry showed how far removed Beltrami was from the "consistently projective way of thinking" that characterized his own work.

Thus despite the appearance of great generality, Klein's attitude toward the scope of non-Euclidean geometry was considerably different from, and more restrictive than, Killing's. His concern with the need for compatibility with experience and his "projective bias" effectively limited his conception of non-Euclidean geometry to the parabolic, hyperbolic, and elliptic geometries, with spherical geometry relegated to a mathematical limbo. Later, especially because of his acquaintance with the discoveries of W. K. Clifford while on a visit to England in 1873 to attend the meetings of the British Association for the Advancement of Science, Klein recognized the need to distinguish between the local and global properties of a geometry [Klein 1921:241]. Stimulated by Clifford's discovery of a two-dimensional manifold of zero curvature that is finite in extent, Klein went on to uncover the existence of infinitely many geometries of constant curvature with global distinctions. He therefore revised his opinions on the possibilities for non-Euclidean geometries accordingly. Exactly when this occurred is unclear, for Klein's mathematical interests had turned elsewhere, and it was not until he lectured on non-Euclidean geometry at the University of Göttingen during the academic year 1889–90 that he began to make his revised views public. They are of interest because although more open-minded, they continue to reflect the characteristics that distinguish Klein's viewpoint from Killing's.

In these later pronouncements, Klein expressed with greater fervor his distaste for conceiving geometry so broadly as to transcend experience. The new non-Euclidean geometries that he had discovered were all locally identifiable with parabolic, hyperbolic, or elliptic geometry. Since the differences occur on the global level and our experience of the external world is effectively limited to a relatively small portion of space, all the new geometries were deemed in agreement with experience. Killing enthusiastically accepted and studied the new "Clifford–Klein space forms," as he termed them [1891]. In typical fashion, however, Killing pushed the geometrical analysis beyond what Klein felt were the bounds imposed by experience [1928:256]. Klein stressed that *he* used the term "non-Euclidean geometry" to refer to "the real discipline" of geometries compatible with experience "and not merely the abstract mathematical [discipline] that it has occasioned" [1898:386]. Further on in the same publication he contrasted "the actual geometry of our empirically given space" with "generalizations that have been formed and that may be mathematically valuable in other respects" (p. 386). Klein may well have had Killing's theory of space forms in mind when he wrote the above passage, by that time, and thanks to Klein's observation, Killing's work was recognized as a part of Lie's theory of transformation groups, which, of course, was deemed of considerable value for its applications to differential equations. It should also be noted that the new geometries did not cause Klein to relinquish his "projective bias": The new distinctions, he emphasized, arise naturally "with the projective view taken as basis [1890:380].

The research program underlying Killing's "Braunsberger Programm" was thus far removed in spirit from the *Erlanger Programm.* Certainly, the differences reflect divergent personalities, but it is also the case that the divergent tendencies were reinforced by the different educational experiences; Klein's attitude was colored by his experiences at Bonn and Göttingen, whereas Killing's was colored by his experiences at Berlin. The Weierstrassian spirit underlying Killing's work also comes out through comparison of *The Concept of Space Extended* [1884] with his doctoral dissertation [1872] on pencils of quadratic surfaces. In Section 2 I indicated the manner in which Killing, by providing a geometrical interpretation of the theory of elementary divisors, transferred the Weierstrassian penchant for systematic thoroughness to the realm of geometry. The research program outlined in [1884] is motivated by similar concerns: Just as the objective of the dissertation was to determine analytically all the geometrical possibilities for a pencil of quadric surfaces, so, too, the objective of [1884] was to analytically determine all possible space forms. Furthermore, the theory of elementary divisors, which had been the principal tool in the dissertation, was also recognized by Killing as basic to his program of determining all space forms. Indeed, it was by perceiving the problem in terms of the framework of the theory of elementary divisors that Killing gave to the study of Lie algebras a direction completely foreign to Lie's work, as will be seen in the next chapter.

In order to add a further dimension to the discussion of Weierstrass's influence upon Killing along the above lines, it is instructive to return to the comparison with Klein. After the untimely death of Plücker in 1868, Klein decided, with the encouragement of Clebsch, to write his doctoral dissertation on a topic in Plücker's line geometry: line complexes of the second degree. As we saw in Section 1.1, in line geometry, a line is determined by six coordinates $x_1, \ldots, x_6$, which are related by an equation $\Omega = 0$, where Ω is a real, nonsingular quadratic form. A line complex of the second degree consists of all lines in three-dimensional space with coordinates satisfying an additional equation $\Phi = 0$, where Φ is also a real quadratic form in $x_1, \ldots, x_6$. Klein decided to approach the geometric study of these complexes by first simultaneously transforming Ω and Φ into sums of square terms by means of a linear coordinate transformation. Such a transformation, however, is not always possible; it depends upon the nature of the elementary divisors of $\det[\lambda\Omega + \Phi]$.[25] Weierstrass's theory was first published in 1868 and was not known to Klein, who, knowing that the desired transformation was possible "in general," since the roots of $\det[\lambda\Omega+\Phi] = 0$

[25] Even though Ω and Φ are symmetric, Ω is not definite; consequently, the roots of $\det[\lambda P + \Omega] = 0$ need not be real and Ω, Φ need not be simultaneously diagonalizable.

are "in general" distinct, considered in effect only this case "as was customary at the time in geometrical investigations" [1921:4]. In the terminology of Section 2, Klein was reasoning generically.

Had Plücker been living, he undoubtedly would have approved Klein's study of the generic case, which involves various possibilities depending upon the number of real roots λ. But now Klein had to submit it to the other professor at Bonn, Rudolph Lipschitz, who had been appointed in 1864 even though Plücker had favored Clebsch for the position. Official documents indicate that the "mathematico-physical direction" of Plücker's interests were so far removed from the "analytical direction" of Lipschitz's work that it was deemed infeasible to have the mathematical seminar run jointly by them [Ernst 1933:34–5]. Plücker's coolness toward Lipschitz was undoubtedly reinforced by Lipschitz's ties with Berlin, for Plücker felt he had been unfairly treated by the Berlin mathematicians [Ernst 1933:22–23]. Lipschitz had received his doctorate from Berlin in 1853 and regarded himself as a student of Dirichlet. From 1857 to 1862 he was again in Berlin as instructor (*Privatdozent*). At that time he undoubtedly became aware of Weierstrass's researches on the transformation of quadratic forms [1858] which presaged his theory of elementary divisors and would have been of special interest to Lipschitz since Weierstrass's work was encouraged by Dirichlet's critical reworking of a theorem in Lagrange's *Mécanique analytique* [Hawkins 1977b:129–30]. That would explain why when Klein presented his results to Lipschitz, the latter had in his possession a copy of the page proofs of Weierstrass's 1868 paper on elementary divisors.

Upon learning of Klein's results, Lipschitz informed him of Weierstrass's paper, provided him with his copy, and "required... that I take into consideration all other, more special, cases ..." [1921:4]. Lipschitz thus required of Klein the same sort of systematic thoroughness that was later to characterize Killing's dissertation. Neither his training under Plücker nor the literature on the "new geometry" to which Clebsch introduced him after Plücker's death, however, had prepared Klein to sympathize with the Weierstrassian viewpoint. As he later explained, he always regarded the "general" case as, *geometrically*, the most interesting, while the remaining cases seemed to constitute "degeneracies" [1921:4]. Indeed, despite Lipschitz's intervention, Klein still did not pursue the geometrical analysis with the Weierstrassian thoroughness exemplified by Killing's dissertation. In deference to Lipschitz, Klein expounded the theory of elementary divisors and obtained some *algebraic* results relating to the general theory. He also specified the eleven cases of the elementary divisors of $\det[\lambda\Omega+\Phi]$ corresponding to a line complex of second degree. But the *geometrical* study of these complexes is restricted to the case of distinct characteristic roots—a subcase of one of the above eleven cases. Furthermore, after Klein left Bonn for postdoctoral study at Göttingen, he completely ignored the theory of elementary divisors in his further papers on line complexes, published in Clebsch and Neumann's new journal, *Mathematische Annalen*, such as his paper [1869] on Kummer surfaces discussed in Section 1.2.[26]

Klein's distaste for an exhaustive analysis of all possibilities regardless of their intrinsic geometrical interest was much more pronounced when it was a question of possibilities that were contrary to experience, as can happen when the subject is non-Euclidean geometry. Klein's approach to this subject was through the Cayley metric associated with a fundamental surface $\Omega = 0$, where Ω is a quadratic form in the homogeneous coordinates of the manifold (not to be confused with the Ω of line coordinates). There are many possibilities for the metric, and hence the geometry, depending upon the nature of Ω.[27] In his papers on non-Euclidean geometry Klein considered only the three possibilities that yielded geometries he deemed compatible with experience. In his lectures on non-Euclidean geometry of 1889–90, other possibilities were mentioned as follows: "If we wished to be completely systematic, a sizable number of further special cases would have to be enumerated, depending upon the type of associated fundamental surface .... Nevertheless, we proceed no further, since our interest lies essentially with the three mentioned cases" [1893b:228]. Concerning one such further case (when Ω is ring-shaped) Klein revealed the basis for his attitude: "This case is without interest for us if we wish to maintain the analogy with ordinary geometry" [1893c:227].

Elsewhere in his lectures, Klein placed his attitude in a broader context that explicitly brings out the contrast with the prevailing climate of opinion at Berlin [1893b:361]:

> With what should the mathematician concern himself? Some say, Certainly "intuition" is of no value whatsoever; I therefore restrict myself to the pure forms generated within myself, unhampered by reality. That is the password in certain places in Berlin. By contrast, in Göttingen the connection of pure mathematics with spatial intuition and applied problems was always maintained and the true foundations of mathematical research recognized in a suitable union of theory and practice.

In describing the viewpoint found in Berlin, Klein clearly had Weierstrass principally in mind. Although Weierstrass no doubt would have denied such an exaggerated description of his views, the spirit that Klein attempted so unsympathetically to describe was a part of the Berlin ambience, as Klein sensed, and indeed it is, I believe, reflected in Killing's work on foundations of geometry. The contrasts that Klein drew between Berlin and Göttingen apply in particular to the work of Killing and Klein on non-Euclidean geometry. In the next chapter we will see where Killing's approach to geometry led him when he discovered its relevance to Lie's theory of groups.

---

[26] See p. 53 of [1869]. The geometrical investigation of the nongeneric cases became a task for one of Klein's doctoral students at Erlangen, A. Weiler.

[27] All the possibilities seem to have been studied first by Sommerville [1910].

Chapter 5
Killing and the Structure of Lie Algebras

As will be seen in Section 2 below, shortly after Killing published his essay *The Concept of Space Extended* [1884], he learned that Lie had earlier introduced similar ideas. Before long, Killing decided to temporarily postpone his research on his theory of space forms and to apply the underlying ideas he had developed to the problem of classifying Lie algebras, with simple algebras particularly in mind. Thanks to his extensive correspondence with Engel, we are able to glimpse the excitement, frustration, and uncertainty that attended what we can now see—thanks largely to the subsequent work of Élie Cartan (Chapter 6)— as the birth of the theory of the structure of semisimple Lie algebras.[1]

Killing's correspondence with Engel indicates that even before his contact with Lie's theory, Killing had already developed his ideas to a point where he could see their relevance to problems within Lie's theory. Unfortunately, Killing does not seem to have left behind any documents that enable us to know with certainty exactly what he had achieved prior to learning of Lie's theory. In Section 1, based upon the hints in Killing's letters and upon the contents of his eventual papers, I present my guess as to what he had done prior to, or shortly after, his contact with Lie and Engel. Indirect support for my reconstruction comes from the fact that, as

[1] An annotated transcription of the Killing–Engel correspondence has recently been published by Hein [1997]; transcriptions of the letters cited here by date can be found there. All English translations of quotations from this correspondence are mine and are based upon copies of the original letters. In writing this chapter, I have drawn heavily upon my paper [1982], where I first analyzed the correspondence.

I point out in the following sections, it fits in well with, and makes sense out of, what Killing writes in his letters to Lie and Engel. In Section 2 I discuss Killing's contact with Lie and his work, and in Section 3, making extensive use of the Killing–Engel correspondence, I present the manner in which Killing, with Engel's encouragement, began to develop his fruitful but tentative and unfocused mathematical ideas and to relate them specifically to Lie's theory of transformation groups and its problems, notably the problem of determining simple groups. The main technical difficulties revealed in Killing's theory through the efforts of Engel and his students form the subject of Section 4. Here again the Killing–Engel correspondence is enlightening. In the final section I briefly discuss the fate of both Killing and his theory after the publication of the latter in 1888–90.

5.1. Space Forms and Characteristic Equations

In discussing Killing's theory of space forms I will use the notation of Lie. Thus, as indicated in Section 4.4, the motion $dx_i = u_i(x)dt$, $i = 1, \ldots, n$, will be identified with $X = \sum_{i=1}^{n} u_i(x)\frac{\partial}{\partial x_i}$. Recall from Section 4.4 that for a space form with r degrees of mobility there exist r linearly independent motions $X_1, \ldots, X_r$ such that

$$(1.1) \qquad [X_i, X_j] = \sum_{k=1}^{r} c_{ijk} X_k.$$

Killing also recognized, albeit in his own notation, that $[X_j, X_i] = -[X_i, X_j]$ and that the Jacobi identity

$$(1.2) \qquad [X_i X_j X_k] \stackrel{\text{def}}{=} [X_i, [X_j, X_k]] + [X_j, [X_k, X_i]] + [X_k, [X_i, X_j]] = 0$$

obtains. In order to determine all n-dimensional space forms with r degrees of mobility, his first task was therefore to determine all the inequivalent possibilities for the c_{ijk}, which, since only real numbers were used, amounts to the problem of determining all r-dimensional real Lie algebras. The next step would be to integrate the differential equations implied by (1.1) so as to obtain the actual functions $u_i = u_i(x)$, hitherto treated as formal symbols. Killing realized that the integrations are greatly simplified when most of the c_{ijk} are zero, which is the case when $[X_i, X_j] = \omega X_h$. This observation led him to the problem of determining, for a given X, solutions Y to the equation

$$(1.3) \qquad [X, Y] = \omega Y.$$

Except for the operator notation, all of the above mathematical considerations are contained in Killing's essay [1884]. This publication appeared before Killing was aware of Lie's work and actually represents only a fragment of his investigations at the time. On the basis of the hints scattered throughout the essay itself and Killing's remarks in his subsequent correspondence with Engel, it is fairly certain that the approach and methods underlying his papers of 1888–90 on the structure of Lie algebras were already in existence in 1884. The purpose of this section is to indicate them and the probable extent to which they had been developed prior to the stimulation afforded by Lie's work and by his contact with Engel. Although the exposition, insofar as it estimates the extent of development, is necessarily hypothetical, it is based upon the information and clues contained in the Killing–Engel correspondence, which will be presented in greater detail in Section 3. The actual extent of development is of secondary importance, however, because, as will be shown in Section 2, Killing's entire approach to the problem of structure was quite different from anything that Lie had imagined; the effect of Lie's work upon Killing was more inspirational than educational.

Killing's starting point is equation (1.3). If $X = \sum_{i=1}^{r} e_i X_i$ and $Y = \sum_{j=1}^{r} y_j X_j$, then in view of (1.1), (1.3) represents a system of linear equations:

$$(1.4) \qquad \sum_{j=1}^{r} \gamma_{kj} y_j = \omega y_k, \quad k = 1, \ldots, r, \quad \text{where} \quad \gamma_{kj} = \sum_{i=1}^{r} e_i c_{ijk}.$$

Thus solutions are possible if and only if ω is a root of the characteristic equation

$$(1.5) \quad \Delta(\omega) = |\gamma_{kj} - \omega\delta_{kj}| = (-1)^r[\omega^r - \psi_1(e)\omega^{r-1} + \psi_2(e)\omega^{r-2} - \cdots] = 0.$$

In more familiar notation, $\Delta(\omega) = \det[\operatorname{ad} X - \omega I]$. The characteristic equation, of course, was a familiar sight to Killing due to his training at Berlin and his dissertation applying Weierstrass's theory of elementary divisors (Section 4.2).

Killing naturally focused his attention upon (1.3) and the corresponding characteristic equation (1.5). In order to have the c_{ijk} as simple as possible, one would naturally seek to choose X such that the corresponding Weierstrass–Jordan canonical form is as simple as possible. For a mathematician not trained at Berlin, the natural tendency would most likely have been to reason on the basis of a "general" X with no multiple characteristic roots, and to overlook the singularities that can actually occur. Indeed, Lie did just that, as we shall see in the following section. By contrast, Killing was more sensitive to the potential weaknesses of the generic mode of reasoning. The fact that $[X, X] = 0$, for example, implies that $\omega = 0$ is always a characteristic root, contrary to the spirit of the "general" approach. Furthermore, in his essay [1884], Killing had listed all possible space forms in 3 dimensions with at most 3 degrees of mobility, and this list contained examples in which, no matter how X is chosen, $\omega = 0$ is a multiple root of the characteristic equation. Even earlier, he had written down the infinitesimal transformations of a non-Euclidean space form of constant curvature [1880a:280]. They constitute the Lie algebra of the orthogonal group. In 1880 he did not regard them as a Lie algebra, i.e., in terms of the bracket operation $[X, Y]$, but by 1884 Killing perhaps realized that also in this important case zero is a multiple characteristic root for every choice of X as soon as the dimension of the underlying space is sufficiently large. Killing's sensitivity to the need to consider the possibility of multiple roots was partly responsible for leading him to fundamental considerations for the structure of Lie algebras.

An important parameter in Killing's deliberations is thus the minimal multiplicity k of $\omega = 0$ as a characteristic root. The characteristic equation corresponding to any $X = \sum_{i=1}^{r} e_i X_i$ is thus of the form

$$\Delta(\omega) = (-1)^r [\omega^r - \psi_1(e)\omega^{r-1} + \cdots \pm \psi_{r-k}(e)\omega^k] = 0,$$

where $\psi_{r-k}(e) \not\equiv 0$. In order to simplify the multiplication constants as much as possible Killing chose $X = X_1'$ such that $\psi_{r-k}(e) \neq 0$, i.e. such that k is the multiplicity of $\omega = 0$ in the characteristic equation for X_1'. According to Weierstrass's theory, $X_2', \ldots, X_k'$ may then be chosen so that $[X_1', X_i'] = \epsilon_i X_{i=1}'$ where ϵ_i is 0 or 1. The $X_i', i = 1, \ldots, k$, thus represent, in more modern terminology, a basis consisting of ordinary or generalized eigenvectors for $\omega = 0$, of the generalized null space $\mathfrak{g}_0(X_1')$ of $\operatorname{ad} X_1'$.

Since X_1' has been chosen such that the dimension of its null space is minimal—one says that X_1' is regular —it turns out that $\mathfrak{g}_0(X_1')$ is what is now called a Cartan subalgebra, namely a nilpotent subalgebra equal to its own normalizer.[2] The parameter k figures prominently in Killing's work. Although today k is defined to be the rank of the Lie algebra $\mathfrak{g}$ spanned by $X_1, \ldots, X_r$, Killing reserved that term and the letter l for another parameter associated with $\mathfrak{g}$ that he probably introduced somewhat later. (See Section 3.) In similar fashion the remaining X_i', $i > k$, can be chosen to span the other "root spaces" $\mathfrak{g}_\alpha(X_1'), g_\beta(X_1')$, *etc.*, corresponding to nonzero roots $\omega = \alpha, \beta, \ldots$. With such a choice for $X_1', \ldots, X_r'$, the multiplications $[X_1', X_j']$ are determined, but it still remains to determine the expressions $[X_i', X_j']$ for $i, j \neq 1$. In this connection Killing appreciated the value of the Jacobi identity (1.2), but he also probably realized that the problem was too formidable to be considered in its completely general form and introduced further classifying principles to break the problem into more manageable pieces.

One such principle, already in evidence in the essay [1884], was to classify space forms by the number p of linearly independent expressions $[X_i, X_j]$, which represents, in modern terminology, the dimension of the

---

[2] See [Samelson 1969:36] for a proof. See Section 5 for a discussion of how the nilpotency of $\mathfrak{g}_0(X_1')$ was established.

derived algebra $\mathfrak{g}' = [\mathfrak{g}, \mathfrak{g}]$ of $\mathfrak{g}$. Killing apparently began by concentrating most of his attention upon the case $p = r$. Even this restriction leaves too many possibilities to consider simultaneously, and he seems to have concentrated upon those $\mathfrak{g}$ satisfying two further conditions. The three conditions are what I shall refer to as Killing's conditions.

Killing's Conditions:

I. $p = r$ ($\mathfrak{g}' = \mathfrak{g}$).

II. $[X_i, X_j] = 0,\ i, j = 1, \ldots, k$ ($\mathfrak{g}_0$ is abelian).

III. All nonzero roots are distinct ($\dim \mathfrak{g}_\alpha = 1$ for all $\alpha \neq 0$).

Today these conditions are recognizable as characteristic properties of semisimple Lie algebras, although they do not by themselves imply that $\mathfrak{g}$ is semisimple. For Killing, of course, they held no comparable significance circa 1884.

Killing's correspondence with Engel, as well as his publications, suggests that his general practice was to work through special cases when confronted with a difficult problem. Condition I probably seemed of some importance because it was satisfied by the infinitely small motions of non-Euclidean space forms of constant curvature and also by the corresponding motions of projective geometry, i.e., the Lie algebras of the orthogonal and special linear groups. Condition II appeared to Killing to be always true when $p = r$. A proof is easy to give when $k = 2$, and Killing was unaware of any examples with $p = r$ and $\mathfrak{g}_0$ not abelian. (They do exist; see the end of Section 6.2.) In fact, Killing later believed that he had succeeded in proving that I implies II. Condition III was probably entirely a matter of convenience, at least before contact with Lie. Killing had every intention of treating all the possibilities revealed by Weierstrass's theory of elementary divisors when multiple nonzero roots are permitted. Even after the work of Lie had focused his attention upon simple groups, which seemed on the basis of known examples to satisfy Condition III, Killing continued his efforts to achieve an exhaustive classification.

By virtue of his approach to classification through the characteristic equation and Weierstrass's theory of elementary divisors, Killing was in a position to make a simple but consequential observation that Lie apparently never made, or at least never appreciated. If X_α, X_β satisfy $[X_1, X_\alpha] = \alpha X_\alpha$, $[X_1, X_\beta] = \beta X_\beta$, where α and β are nonzero, then the Jacobi identity $[X_1 X_\alpha X_\beta] = 0$ shows that

$$[X_1, [X_\alpha, X_\beta]] = (\alpha + \beta)[X_\alpha, X_\beta]. \tag{1.6}$$

Hence if $[X_\alpha, X_\beta] \neq 0$, either $\beta = -\alpha$ and $[X_\alpha, X_\beta]$ is a linear combination of $X_1, \ldots, X_k$, or $\gamma = \alpha + \beta$ is another root of the characteristic equation. This is, of course, fundamental to the structural analysis of Lie algebras. By making use of this insight, Killing readily determined the possible structures for all $\mathfrak{g}$ satisfying Conditions I–III and $k = 1$ [1888a:§7]. These conditions on $\mathfrak{g}$ correspond to the "general" case in which all the characteristic roots are distinct. Probably, Killing explored this case toward the beginning of his investigations. If so, he undoubtedly found the results encouraging for the task of exploring the situation when $k > 1$. But Killing, as a mathematician trained at Berlin, did not presume further that he had established, by analyzing the "general" case, genuinely general results that obviated the need to study carefully the case $k > 1$. Indeed, as a result of his investigation of the case $k > 1$, Killing was led to the discovery of the "Cartan" integers and concomitant relations among the roots (indicated below) that he came to regard as demonstrating the impossibility of making a generic approach rigorous by passage to a limit. Thus in [1888a:58] he wrote, "However, while in many investigations where it is a question of the roots of an equation only simple roots need be considered with the case of multiple roots then being handled by continuous passages, that is not possible for the characteristic equation of the group, since entirely fixed relations always hold among the roots."

Before turning to Killing's analysis of the case $k > 1$, one further general remark is in order. As noted, the space forms of Killing correspond to real Lie algebras, whereas Weierstrass's theory applies most conveniently when complex numbers are permitted. Probably, Killing dealt with what we would now call complex Lie algebras for the sake of convenience, to survey the possibilities with an eye toward ultimately bringing his results to bear upon the "real" case. Later, when he learned of Lie's theory of transformation groups, the study of the complex case was seen by Killing to be of interest in itself, since Lie's work was formulated entirely within the context of complex-valued coefficients. Killing nonetheless remained reluctant about admitting complex Lie algebras directly into his research program on space forms, despite the fact that complex manifolds were admitted in geometrical investigations by other geometers such as Klein and Lie [1892:124]. But he never attempted to apply his results to the study of real algebras. That formidable task was left to Cartan, as will be seen in Section 8.5.

We now sketch the principal discoveries Killing made under Conditions I–III. Judging by his correspondence with Engel, Killing made them prior to his acquaintance with the work of Lie, although the only extant exposition of them is embodied in Killing's paper [1888b], which I have followed here. As will be seen, Conditions I–III lead very quickly to the fundamental relations that were eventually to provide Killing with the basis for his classification of simple Lie algebras. Let $X_1, \ldots, X_r$ denote the specially chosen basis described above (with the primes now deleted). Then the relation $[X_1, X_\alpha] = \alpha X_\alpha$, when combined with the Jacobi identity and the abelian nature of $\mathfrak{g}_0$ (Condition II), implies that

$$[X_i, X_\alpha] = \alpha^{(i)} X_\alpha, \quad i = 1, \ldots, k. \tag{1.7}$$

Thus if $X = \sum_{i=1}^{k} e_i X_i \in \mathfrak{g}_0$, then

$$[X, X_\alpha] = \alpha(X)X_\alpha, \text{ where } \alpha(X) = \sum_{i=1}^{k} e_i\alpha^{(i)}. \tag{1.8}$$

The characteristic roots are therefore linear homogeneous functions of the coordinates of X. Killing [1888b:§12] did not use the abbreviated notation $\alpha(X)$.

By the time Killing published his results, he had used the Jordan–Weierstrass canonical form to prove that in any Lie algebra $\mathfrak{g}$, $[\mathfrak{g}_\alpha, \mathfrak{g}_\beta] \subset \mathfrak{g}_{\alpha+\beta}$ for all α and β (including α or $\beta = 0$) [1888a:§8]; but for Lie algebras satisfying I–III this relation is an immediate consequence of (1.6) and (1.7). From this relation, it follows that when $\mathfrak{g}' = \mathfrak{g}$ (Condition I) $\mathfrak{g}_0$ must be spanned by elements in the sets $[\mathfrak{g}_\alpha, \mathfrak{g}_{\alpha'}]$, where $\alpha' = -\alpha$ and, since $[\mathfrak{g}_0, \mathfrak{g}_0] = 0$ (Condition II), $\alpha \neq 0$, i.e., $\alpha(X) \not\equiv 0$. Hence k pairs of indices α, α' can be chosen such that if

$$H_\alpha = [X_\alpha, X_{\alpha'}], \tag{1.9}$$

then the H_α form a basis for $\mathfrak{g}_0$. (Killing did not use the H_α notation.) By virtue of (1.8),

$$[H_\alpha, X_\beta] = \beta(H_\alpha)X_\beta, \tag{1.10}$$

and since $[H_\alpha, X_\beta] = [[X_\alpha, X_{\alpha'}], X_\beta]$, it is natural to apply the Jacobi identity and to compare the result with (1.10). In this manner Killing obtained, with the assistance of (1.6), the formula

$$\beta(H_\alpha) = -\sum_\epsilon (c_{\alpha\beta\epsilon}c_{\alpha'\epsilon\beta} - c_{\alpha'\beta\epsilon}c_{\alpha\epsilon\beta}). \tag{1.11}$$

In this formula, by virtue of (1.6), $c_{\alpha\beta\epsilon} = 0$ either for all ϵ or for all ϵ except $\epsilon = \beta + \alpha$ (in the sense that $[X_\alpha, X_\beta] = c_{\alpha\beta\epsilon}X_\epsilon$). The latter case corresponds to the situation $[X_\alpha, X_\beta] \neq 0$, so that $\omega_\beta + \omega_\alpha$ is a root. Similarly, $c_{\alpha'\beta\epsilon}c_{\alpha\epsilon\beta} = 0$ either for all ϵ or for all ϵ except $\epsilon = \beta - \alpha$, in which case $\omega_\beta - \omega_\alpha$ is a root. Thus from (1.11) Killing was able to conclude that whenever $\beta(H_\alpha) \neq 0$, either $\beta + \alpha$ or $\beta - \alpha$ is a root of the characteristic equation [1888b:15].

Starting with β, further roots are therefore obtained by adding or subtracting α whenever $\beta(H_\alpha) \neq 0$. Suppose that in this manner roots $\beta + t\alpha$, $-t' \leq t \leq t''$ are obtained, where t' and t'' denote nonnegative integers such that $\beta + t\alpha$ is *not* a root for $t = -(t'+1)$ or $t = t''+1$. (Killing left it unclear at this point whether $\beta + t\alpha$ could be a root for some integer outside the interval $[-t', t'']$.) Let $\beta_0 = \beta - t'\alpha$, $\beta_1 = \beta_0 + \alpha$, $\beta_2 = \beta_1 + \alpha$, *etc.* Killing observed that if (1.11) is written down for $\beta = \beta_0, \beta_1, \ldots, \beta_s (s = t' + t'')$, and if the results are added then cancellations occur on the right-hand sides, so that

$$\sum_{j=0}^{s} \beta_j(H_\alpha) = 0. \tag{1.12}$$

Since $\beta_j = \beta_0 + j\alpha$, (1.12) can be written as $2\beta_0(H_\alpha) + s\alpha(H_\alpha) = 0$, or

$$2\beta(H_\alpha) + a_{\beta\alpha}\alpha(H_\alpha) = 0, \tag{1.13}$$

where $a_{\beta\alpha} = t'' - t'$ [1888b:16, eqn. (7)]. Killing also defined $a_{\alpha\alpha} = -2$, so that (1.13) remains true for $\beta = \alpha$ as well. From the definition of the integer $a_{\beta\alpha}$, Killing could conclude that

(1.14) *$\beta + a_{\beta\alpha}\alpha$ is a root, as well as $\beta + t\alpha$ for all t between 0 and $a_{\beta\alpha}$.*

The integers $a_{\beta\alpha}$ are the negatives of what are now usually called the Cartan integers: $a_{\beta\alpha} = -\langle\beta, \alpha\rangle$. (Cartan also worked with the $a_{\beta\alpha}$ rather than the $\langle\beta, \alpha\rangle$.) The fundamental identity $\langle\beta, \alpha\rangle = 2(\beta, \alpha)/(\alpha, \alpha)$ is missing from Killing's theory, since it involves the so-called Killing form, which he did not employ. (As we shall see in Section 6.2, what amounts to the Killing form was introduced into the theory by Cartan, whereas Cartan integers and Cartan subalgebras are already to be found in Killing's work.) From the modern theory, however, it follows that the H_α may be chosen such that $(\beta, \alpha) = \beta(H_\alpha)$. Equation (1.13) thus follows immediately from the fundamental identity. In Killing's theory (1.13) plays a role analogous to that now played by the fundamental identity. For example, it follows immediately from (1.13) that if $H_1, \ldots, H_k$ denote the H_α's and $\alpha_1, \ldots, \alpha_k$ the corresponding roots, then if $\det[a_{ij}] \neq 0$, every characteristic root α_μ is, as a linear homogeneous function, a rational linear combination of $\alpha_1, \ldots, \alpha_k$ and likewise that $a_{\mu\alpha}$, $1 \leq \alpha \leq k$, is a rational linear combination of $a_{1\alpha}, \ldots, a_{k\alpha}$ [1888b:17].

On the basis of the above remarkable properties of the characteristic roots, Killing eventually discovered that the integers a_{ij}, $i, j = 1, \ldots, k$, satisfy relations that greatly limit the numerical values that they can assume [1888b:§13]. Exactly when he made this discovery, i.e., before or after contact with Lie and Engel, is unclear, although it is plausible that he had discovered these relationships for small values of k prior to that contact. In any case, it is convenient at this point to indicate how Killing obtained these relations in the simplest case ($k = 2$).

Let us suppose (with $k = 2$) that $\det(a_{ij}) \neq 0$, so that every root β is a rational linear combination of α_1 and α_2: $\beta = m_1\alpha_1 + m_2\alpha_2$. Then it follows from (1.14) that $\mu = \beta + a_{\beta 1}\alpha_1$ is also a root. Likewise, $\nu = \mu + a_{\mu 2}\alpha_2$ is a root. From the way ν is obtained, it is clear that $\nu = m_1'\omega_1 + m_2'\omega_2$. By means of (1.13) the coefficients m_i' can be expressed in terms of the m_i:

$$\begin{aligned} m_1' &= -\, m_1 + a_{21}m_2, \\ m_2' &= -\, a_{12}m_1 + (a_{12}a_{21} - 1)m_2. \end{aligned}$$

Killing considered the linear transformation B defined by these equations. (In effect, B corresponds to the Coxeter transformation $\nu = \sigma_2\sigma_1(\beta)$, where σ_i denotes the reflection determined by $\alpha_i, i = 1, 2$.) By virtue of the definition of B it follows that starting with any root $m_1\omega_1 + m_2\omega_2$, application of

B yields another root $m'_1\omega_1 + m'_2\omega_2$. Application of B to this root yields a further root $m''_1\omega_1 + m''_2\omega_2$, and so on. Since there are only a finite number of roots, not all roots generated in this fashion by B can be distinct. Killing was thus able to conclude that $B^n = I$ for some integer n. Hence the roots of

$$D(s) = \det[B - sI] = s^2 + (2 - a_{12}a_{21})s + 1$$

must be roots of unity, and the elementary divisors must be of the first degree.[3] From these properties of $D(s)$ it follows easily that $a_{12}a_{21} = 0, 1, 2, 3$ and that $a_{12}a_{21} = 0$ if and only if $a_{12} = a_{21} = 0$. For $k > 2$ precisely the same considerations with α_1 and α_2 replaced by α_i and α_j yield the same results about $a_{ij}a_{ji}$. In his paper, Killing used the above properties of $D(s)$ when $k > 2$ to obtain further relations among the a_{ij}.

Even assuming that Killing had developed his theory of space forms to the extent suggested here, his discoveries would have appeared far less auspicious to him than they do to us today. The discovery that the values of the a_{ij} are very limited would be encouraging only if it were assured that *all* roots can be generated from the k roots $\alpha_1, \ldots, \alpha_k$ by means of the coefficients a_{ij}. Even if that were the case, what would it signify? That is, the above discoveries were made under a number of ad hoc hypotheses to which Killing at that time could not have attached any great importance. Furthermore, he had permitted complex numbers into the calculations to facilitate the analysis, but eventually, for his classification of space forms, he must deal with the "real" case. It is no wonder that Killing did not publish these investigations. They were far too inconclusive for public exposure, even in the form of a *Programmschrift*. It would have been more reasonable for him to have abandoned his attempt to classify space forms, since he had at least pursued the problem far enough to realize just how formidable it was. Had it not been for the contact with Lie and Engel, Killing's preliminary investigations probably would never have borne fruit.

5.2. Encounter with Lie's Theory

Shortly after Killing posted a copy of his essay *The Concept of Space Extended* [1884] to Felix Klein in July of 1884, Klein informed him that its contents seemed to be closely related to the theory of transformation groups

[3] Killing [1888b:19] referred to a recent paper by Lipschitz [1887] for the theorem that $B^n = I$ (B a matrix of any size) if and only if these conditions hold. But there is no reason why Killing could not have realized these properties of a B satisfying $B^n = I$ before 1887, since they follow easily from the Weierstrass–Jordan canonical form, especially in the two-by-two case considered here. Killing himself suggested that Lipschitz's theorem was not a revelation because in referring to it he added that Lipschitz's paper was based upon "works of other mathematicians." Frobenius had, in fact, proved a more general theorem characterizing those B with only a finite number of distinct powers much earlier [1878:§3, VIII], although Lipschitz did not cite his paper.

of his friend Sophus Lie. As we saw in Chapter 3, Lie, who was then professor of mathematics at the university in Christiania, Norway (now Oslo), had been publishing papers on his theory since 1874. His work, however, was not well known in Germany. For example, when Killing had previously discussed his theory of space forms with other German mathematicians more aware than he of the latest developments, he was assured that no one had considered such matters.[4] As we saw, Lie was at least partly to blame for the neglect of his work because much of it was published in the Norwegian journal *Archiv for Mathematik*, which was not widely read or distributed in Germany.

Upon learning about Lie, Killing sent him a copy of his essay on 4 August 1884,[5] but he made no attempt to obtain or read Lie's papers, being preoccupied with the final stages of preparation of his book *Non-Euclidean Space Forms Treated Analytically* [1885]. Lie apparently did not respond to Killing's overture and definitely did not reciprocate and send Killing some copies of his own work. It is clear that Lie read Killing's essay, however, for his annotated copy is preserved in the University Library of Oslo. Also, Lie quickly published a note in the *Archiv* [1884b] in which he showed how a theorem stated without proof by Killing in his essay could be derived from some of his previously published results. When Killing finally completed the manuscript of his book, he began to study the one paper on the theory of transformation groups that Lie had published in *Mathematische Annalen* [Lie 1880]. He also wrote again to Lie on 27 October 1885 and requested copies of the *Archiv* papers, assuring him that his own concern with the theory was limited to its geometrical applications: "Whereas the transformations are for you an end in themselves, for me they are only significant by virtue of their geometrical applications. Thus, in addition to wishing to be able to specify all those points at which you have arrived before me, I would like in general to be able to rely on your works."[6]

Lie sent Killing copies of his *Archiv* papers, although Killing had access to them for only a limited period of time. For whatever reason, Killing regarded them as on loan to him, and a few months later, on 16 February 1886,[7] he wrote to Lie requesting permission to keep them a bit longer. Apparently Lie did not then tell Killing he could keep the papers permanently, for they were soon returned. Killing later explained to Engel that the papers were in his possession for such a short time that he was unable to digest fully their contents.[8] The publications of Lie most accessible to Killing were therefore those appearing in *Mathematische Annalen* [1880; 1885a]. The former contains Lie's determination of all transformation groups in

---

4 Letter of Killing to Engel, dated 11 November 1885.
5 According to a letter from Killing to Lie bearing that date (UB Oslo, Brevsamling Nr. 289).
6 UB Oslo, Brevsamling Nr. 289; transcription in [Hawkins 1982:137].
7 See letter bearing that date (UB Oslo, Brevsamling Nr. 289).
8 Letter dated 9 February 1888.

two variables, catalogued by their associated infinitesimal transformations (or Lie algebra). Further, it provided Killing with a vast supply of examples on which to test his theory. It must have also encouraged Killing to pursue his own, quite different, approach to the problem of determining all transformation groups. The 1885 paper was discussed at length in Section 3.3. Recall that it was there that Lie gave a clear discussion of the adjoint group and stressed the importance to the study of differential equations of simple groups and, in particular, the classification of their Lie algebras up to isomorphism. Killing had already observed the fact, now expressible in Lie's terminology, that the transformations $[X_i, X_j]$ corresponding to a basis for a Lie algebra $\mathfrak{g}$ always determine an "invariant subgroup" (i.e., ideal) of $\mathfrak{g}$, namely $\mathfrak{g}'$ [Killing 1886:8]. Hence in order for $\mathfrak{g}$ to be simple, $\mathfrak{g}' = \mathfrak{g}$: Simple $\mathfrak{g}$ must satisfy Killing's Condition I ($p = r$). The known examples of simple groups indicated by Lie further satisfy Conditions II and III. The problem of determining all simple groups was tailor-made for Killing's approach and provided a needed focus. By contrast, as we saw in Section 3.3, Lie's approach to the problem was severely limited in what it could achieve because it depended upon his dimension-by-dimension approach to the classification of *all* transformation groups. As we shall see, the idea arose in Killing's mind that he could possibly accomplish something here beyond what Lie had done.

Killing's methods built upon this knowledge of "linear algebra" as he had learned it at Berlin. Lie's knowledge of the subject appears to have been limited to what was generally known circa 1850 and used in the study of differential equations. He was familiar with the concept of the characteristic equation and knew that a linear transformation could be placed, by a change of variables, in a triangular form. In his *Archiv* papers he used this knowledge of linear algebra to obtain some significant results, which relate to the adjoint group such as Lie's theorem on solvable Lie algebras (Section 3.2). Because of the limited access Killing had to these papers and because of his reluctance to borrow ideas from Lie that he had not in some form conceived independently, he was not as familiar with these aspects of Lie's theory and did not use them. Engel was the first to apply them to Killing's theory (Section 4).

It would seem that the closest Lie ever came to Killing's approach to the theory of transformation groups occurred in the early stages of development of the theory. In [1876a:§7] he made his first attempt in print to prove that an r-dimensional Lie algebra of vector fields determines an r-dimensional transformation group. By way of preparation for the proof, he assumed that $X_1, \ldots, X_r$ could be chosen such that

$$(2.1) \qquad [X_r, X_i] = \sum_{j=1}^{r-1} c_{ij} X_j, \quad |c_{ij}| \neq 0.$$

Then, like Killing several years later, he sought to choose $Y = \sum_{i=1}^{r-1} \nu_i X_i$

such that $[X_r, Y] = \omega Y$. In view of (2.1), ω must be a root of an equation of degree $r-1$ (the characteristic equation of $\operatorname{ad} X_r$ acting on the span of $X_1, \ldots, X_{r-1}$). Lie assumed that this equation had $r-1$ distinct roots $\omega_1, \ldots, \omega_{r-1}$, so that $X_1, \ldots, X_{r-1}$ could be replaced by $Y_1, \ldots, Y_{r-1}$, where $[X_r, Y_i] = \omega_i Y_i, i = 1, \ldots, r-1$. Here Lie considered the "general case" described in Section 1 and in Section 4.2.

Using the simplified expressions $[X_r, Y_i]$ Lie established a proposition (Satz 11) required for his proof of the above-mentioned existence theorem. Regarding its validity he commented as follows [1876a:177]:

> Of course, our proof for this proposition is valid only under the two limiting assumptions that the determinant [of the c_{ij}] ... differs from zero [the assumption leading to (2.1)] and that, in addition, the above-mentioned characteristic equation of degree $r-1$ possesses $r-1$ distinct roots. That the proposition remains valid in all remaining cases could be seen by a passage to the limit. Or the proof could also be directly carried out for the various cases, which affords no major difficulties.

Since the lemma itself was to play no role in his theory (except to assist in the sufficiency proof), Lie decided that he could postpone carrying out the complete proof of his lemma until "another occasion." Although he was more explicit than most regarding the assumptions limiting the generality of his proof, his attitude that the generic proof is indicative of what is generally true, as *could* be shown without encountering major difficulties, was not uncommon among practitioners of generic reasoning. It is doubtful that had Killing never entered the picture, Lie would have eventually pursued the route taken by Killing, who, as noted in Section 1, came to regard his discoveries on structure as revealing the untenability of taking a limit from the "general" case as Lie claimed could be done.

Killing actually read Lie's proof when he had the *Archiv* papers on loan and commended Lie for abandoning it, while at the same time hinting obscurely at the nongeneric nature of $\mathfrak{g}_0$: "You were right to completely abandon your first proof [of existence]. Probably you have yourself noted the following, on which I base the general proof of the proposition, that for $r > 3$ the group necessarily contains commuting transformations."[9] Killing thus congratulated Lie for abandoning his generic proof. The reference to commuting transformations seems to be an oblique reference to the fact that $\omega = 0$ is generally a root of multiplicity $k > 1$. Killing had some general theorems about commuting elements in the root space $\mathfrak{g}_0$ that will be discussed in Section 4 below. (See also the quotation given in Section 3 from Killing's letter to Engel dated 12 April 1886.)

---

[9] Letter to Lie dated 16 February 1886 (UB Oslo, Brevsamling Nr. 289); see [Hawkins 1982:144] for the German text.

5.3. Correspondence with Engel

We saw that when in 1884 Killing sent Lie a copy of his essay *The Concept of Space Extended*, Lie did not reply and, in particular, did not reciprocate by sending copies of some of his own papers. Killing therefore wrote to Klein on 21 October 1885 requesting his assistance in obtaining copies of Lie's papers.[10] Klein simply advised Killing to request them directly from Lie, but he also mentioned that a young mathematician at Leipzig by the name of Engel was now working on transformation groups. Indeed, as we saw in Chapter 3, Engel was now a lecturer at Leipzig, after having spent nine months working with Lie in Christiania. Killing therefore decided that in addition to writing to Lie with a request for copies of his papers he would write to Engel and request a copy of his *Habilitationsschrift* in exchange for a copy of his essay. This letter, dated 6 November 1885, initiated what was to be a substantial scientific correspondence that played a considerable role in providing Killing, isolated as he was in Braunsberg, the stimulation and encouragement to develop his ideas into a publishable form.

After he had examined Killing's essay, Engel wrote back on 9 November assuring Killing that he was justified in his claim that he had "discovered transformation groups in Lie's sense." Killing returned the compliment by declaring to Engel that "your work is of the greatest significance to me." He then repeated for Engel's sake the noncompetitive nature of his interest in transformation groups (application to space forms) while at the same time managing to suggest that he had obtained results on groups not yet published by Lie:[11]

> The fact that I must also be concerned with these groups will, I hope, induce you and Lie to publish more rapidly. I, of course, do not want to enter into a competition over this theory; but all the more so is it desirable to me to be able to use the results for my theory of space. In this sense you must include me as the third member of the alliance. Herr Lie was so kind as to send me a large number of his relevant investigations. I see that with my work, in one important point, I have been anticipated by him but that I have been led to results that at least until now have not been published.

Killing's first revelation of some of his unanticipated results came in the form of another *Programmschrift*, entitled as a contribution "Toward the Lie Theory of Transformation Groups" [1886]. This work did not contain the investigations discussed in Section 1. They were still too fragmentary to present even in a *Programmschrift*. Instead, Killing published some less fragmentary results of lesser significance. They will not be discussed here

[10] NSUB Göttingen, Nachlass Klein X.
[11] Letter of 11 November 1885.

except to note that they also involved extensive reliance upon the Jacobi identity and tended to emphasize the case in which $p = r$ (Condition I).

Killing (like Lie) was extremely sensitive to criticism, and it is fortunate that Engel praised Killing's first conscious contribution to Lie's theory, thereby ensuring the continuance of the correspondence. Engel also pointed out that recently he, too, had become interested in the structure of groups (*Zusammensetzung der Gruppen*). Although he did not realize it in 1886, Killing was destined to enrich greatly the meaning and content of "structure" in Lie's theory. In his letter, Engel did not say much about his own research on structure, since, he confessed, it "mostly still sits in the ink pot."[12] But on March 15 he was pleased to announce that his work had finally left the ink pot:

> Eight hours ago today Herr Mayer presented to the Leipzig Soc. of Science a note "On the structure of finite continuous transformation groups." I hope that the printing will not be delayed too long, so that I can soon send you a copy. Today I would like to communicate only one proposition. I prove that every group with more than four parameters contains four-term subgroups. The proof is completely analogous to a proof, not yet published, by Lie that every group contains three-term subgroups. It follows, e.g., from the above theorem that there is no 7-term simple group.

Here, of course, Lie's terms "n parameter" and "n-term' mean, in present terminology, "n-dimensional."

Engel's activity regarding the structure of Lie algebras and especially his theorem implying that a 7-dimensional simple Lie algebra could not exist turned Killing's thoughts to his old, tentative investigations. In so far as they were valid, they implied the existence of three-dimensional subalgebras, namely those spanned by the elements $H_\alpha, X_\alpha, X_{\alpha'}$ discussed in Section 1. And as the above quotation suggests, the existence of three-dimensional subalgebras was clearly the key to Engel's own theorem.

It was apparently with the investigations in mind that are described in Section 1 above that Killing, in reply, asked whether any simple groups were known besides the special linear and orthogonal groups. For on 29 March 1886 he wrote:

> Your work is a great delight to me. Lie's theorem that every group contains three-term subgroups I posited already many years ago, without being able to prove it, as the foundation for my own investigations; for four-term subgroups it is entirely new to me. Your proposition that there is no 7-term simple group suggests the thought: How large must r be in order that r-term simple groups generally exist. I am familiar with only the two classes of simple groups for which $r = n(n+2)$ or $r = n(n-1)/2$ (for $n > 4$), and indeed the former class is the general projective transformation

[12] Letter of 18 February 1886.

> of an n-dimensional space, the latter the group of rigid motions of a Riemannian space (respectively, what is the same thing, the group of all the projective transformations of a space that take a general configuration of the second order into itself). Lie recently proved that the latter group is simple ....

On the Lie algebra level, the two classes of simple type to which Killing was referring can be identified with the Lie algebras of the special linear and orthogonal groups $\mathbf{sl}(n, \mathbb{C})$ and $\mathbf{o}(n, \mathbb{C})$. The former were to exemplify type A in the classification system Killing eventually created and the latter types B and D. Notice that Killing was not aware of the existence of a third general class, namely the symplectic Lie algebras $\mathbf{sp}(2n, \mathbb{C})$ (Killing's type C). As we saw in Section 3.4, this class arises by generalization of the projective group of a linear line complex, and although Lie had apparently already made this generalization and probably realized that it constituted a new class of simple groups, he had not yet presented anything in print to this effect when Killing wrote the above words.

Engel responded to Killing's inquiries by suggesting the possibility of a third class of simple groups, namely the above-mentioned generalization of the projective group of a linear line complex:[13]

> The question as to which parameter numbers r simple groups belong has also interested me for some time, and I believe that it must be possible to resolve it. There seem to be relatively few simple groups. Thus first of all, the two types mentioned by you. If I am not mistaken, the group of a linear complex in space of $2n + 1$ ... dimensions ($n > 1$) with $(2n+1)2n/1 \cdot 2$ parameters is distinct from these. In threefold space this group is isomorphic to that of surface of 2nd degree in fourfold space. I do not know whether a similar proposition holds in fivefold space. The projective group of fourfold space that leaves invariant a trilinear expression of the form
>
> $$\sum_{ikj}^{1\cdots 5} a_{ijk} \begin{vmatrix} x_i & y_i & z_i \\ x_k & y_k & z_k \\ x_j & y_j & z_j \end{vmatrix} = 0$$
>
> will probably also be simple. This group contains 15 parameters, the corresponding [group] in 5-fold space contains 16, in 6-fold [space] 14, in 8-fold [space] 8 parameters. In 9-fold space there is no such group. These numbers are already interesting. Are the corresponding groups simple? Probably this is worth investigating. But also Lie, who long ago thought about these things, has not yet done so.

Among other things, in this passage Engel clearly alerted Killing to the possibility of a third general type of simple group. Of course, what "the

[13] Letter of 8 April 1886.

group of a linear complex" signified would not have been clear to him. Later, he apparently forgot about Engel's remark and, as a result, expended considerable effort trying unsuccessfully to justify the existence of a "new" type C of simple group.

With the above reply Engel thus confirmed the significance of Killing's question, suggesting that the determination of all simple groups was a worthwhile, manageable problem, although he and Lie had made very little advance in this direction. Killing was further encouraged by the receipt of Engel's paper [1886], in which his theorem on the existence of four-dimensional subalgebras is made to depend, in Engel's words [1886:91], on "the important theorem" that every Lie algebra of more than three dimensions contains commuting elements, which "Killing first posited, without communicating the proof," in his essay of 1884. Killing did not communicate his proof because it was inextricably intertwined with his whole (still tentative) approach to the problem of determining all Lie algebras (as sketched in Section 1). The contact with Engel did more than recall his old investigations; it stimulated him to renewed efforts to bring them to a definitive conclusion. On 12 April 1886 Killing informed Engel of this effect of their contact and even boldly revealed the still conjectural implication of those investigations, viz., that the special linear and the orthogonal groups provided the only examples of simple groups:

> Your work was the occasion for me to bring an older investigation of mine, on a related theme, to publishable form. It turned out that the results are much more generally valid than I first conjectured. I, of course, do not know whether I am being deceived by my conjecture or whether I am on a completely secure path. The proofs are completed up to a certain point; other parts turn out to be likely. I also cannot tell whether the little that is still lacking will come easily or with difficulty. The propositions for which correctness only seems likely are the following: "The only simple groups are of the two known kinds." If the proposition is not correct, I hope to be able to determine the missing types; if it is correct, then the following proposition arises: If $r > m(2m+1)$, then relative to the structure of an r-term group two cases are possible: (1) either to each one-term group $p_1X_1f+\cdots+p_rX_rf = Y_1f$ m infinitesimal transformations can be found that together with Y_1 are independent so that ... $(Y_1Y_2) = \ldots = (Y_1Y_{m+1}) = 0, \ldots, (Y_mY_{m+1}) = 0$ ...; or (2) the group contains an $[r - m(2m+1)]$-term *invariant* subgroup $Z_1f, \ldots, Z_\nu f$ $(\nu = r - m(2m+1))$, where each two of the latter commute.
>
> These are conjectures, proved in the simplest cases. Next time I will go into the certain results of my investigations.

Killing's conjectured theorem that the special linear and the orthogonal groups are the only examples of simple groups was based upon his calculations for small values of k.[14] Once he became acquainted with the concept of a simple group, Killing came to realize, or at least to believe, that the type of considerations sketched in Section 1 apply to simple groups and, in fact, apply under the most favorable conditions. That is, for simple groups Killing realized that the determinant of the $k \times k$ system a_{ij} does not vanish and that all roots are obtainable from the k roots $\alpha_1, \ldots, \alpha_k$ by means of the integers a_{ij}. Furthermore, if α is a root, then so is $\alpha' = -\alpha$, and $[X_\alpha, X_{\alpha'}] \neq 0$. Thus for $k = 2$, there are at most four possibilities for the roots and hence for a simple $\mathfrak{g}$, depending upon whether $a_{12}a_{21}$ has the value 0, 1, 2, or 3. It would have been easy for Killing to see that when $a_{12}a_{21} = 0$, $\mathfrak{g}$ will contain a nontrivial ideal. That would leave the three possibilities $a_{12}a_{21} = 1, 2, 3$. The first two correspond to the types A_2 and B_2 of the special linear and orthogonal groups, while the third is the exceptional type G_2.

Killing's conjecture indicates that he had ruled out the third possibility. Later he blamed this oversight upon "errors of calculation" (see below). Probably, he had correctly obtained the root system corresponding to $a_{12}a_{21} = 3$ but had then made some calculational errors in the course of attempting to construct a multiplication table—the c_{ijk}—for this case. Perhaps the Jacobi identity did not seem to be satisfied. In any case, Killing was initially not mentally prepared for the discovery of new simple groups. It was more reasonable for him to doubt the existence of a group corresponding to $a_{12}a_{21} = 3$ than to doubt his calculations that implied that no such group existed. As Killing later confided to Engel in a letter of 22 October 1888, the existence of exceptional simple groups was so unexpected that "an error in reasoning should almost be conjectured." If initially Killing also considered the case $k = 3$, his conjecture indicates that he had overlooked the simple type C_3 as well, possibly confounding it with the type B_3 of the orthogonal group. It must be kept in mind that Killing was either unfamiliar with the general symplectic group $\mathbf{sp}(2n, \mathbb{C})$ or thought it was always locally isomorphic to the orthogonal group. In either case, the type C_3 would also have represented something new and hence unanticipated, although, as we saw, Engel had warned him of this possibility.

Killing's second conjectured theorem supported his claim that if $r > 3$, then any r-dimensional group contains commuting elements. Exactly how he imagined the second theorem following from the first is unclear. Perhaps the first case corresponds to $\mathfrak{g}$ being simple or what he later termed "semisimple" (or at this stage to a $\mathfrak{g}$ satisfying Conditions I–III of Section 1). If, e.g., $\mathfrak{g}$ is simple and k is the multiplicity of $\omega = 0$ in the characteristic equation for Y_1, then $Y_1, \ldots, Y_{m+1}$ could correspond to all or part of the basis for $\mathfrak{g}_0(Y_1)$. Since by the first conjectured theorem $r = k(2k+1)$ or $r = k(k+2)$, the condition that $r > m(2m+1)$ would imply in either case that $k = \dim \mathfrak{g}_0 \geq m+1$. Likewise, the second case would then correspond

[14] Killing to Engel, 27 April 1888.

to $\mathfrak{g}$ not being semisimple. As we shall see, Killing posited some theorems about the existence of nilpotent ideals (as we would now describe them) in this case, although it is unclear when he arrived at them. There is a possibility, at least, that he was here thinking of those results. (They are discussed below.)

It is not difficult to imagine the amazement with which Engel read Killing's letter with its bold but informed conjectures. Here was an obscure professor at a Lyceum dedicated to the training of clergymen in the far-away reaches of East Prussia, discoursing with authority and conjecturing profound theorems on Lie's theory of transformation groups, a theory that had seemed an area of mathematics known to relatively few mathematicians and mastered by even fewer. In his reply dated 19 April 1888 Engel stressed the great value the determination of all simple groups would have for the theory of transformation groups itself and its application to differential equations:

> If all structures of simple groups could be directly specified, it would be a tremendous advance. In the theory of structure generally it would be of incalculable utility, and no less so for the theory of integration [of partial differential equations]. That is why, as you can imagine, I am waiting in suspense for your definitive communication on these most important questions.

Engel also wrote to his friend David Hilbert, who was visiting Paris, and told him of the "very nice" results of Killing and of the "mighty advance" the determination of all simple groups would represent.[15]

While Engel's enthusiasm over Killing's conjectured theorem is understandable, it is surprising that he did not temper it by once again bringing up the matter of the projective group of a linear complex, which has its Lie algebra isomorphic to $\mathbf{sp}(2n, \mathbb{C})$. In his letter of April 8, Engel had suggested that $\mathbf{sp}(2n, \mathbb{C})$ represented a third type of simple structure for $n > 2$, and this agrees with Engel's recollections (noted above in Section 3.4) to the effect that he and Lie knew that, for $n > 2$, $\mathbf{sp}(2n, \mathbb{C})$ is a distinct structure type from those of $\mathbf{sl}(n, \mathbb{C})$ and $\mathbf{o}(n, \mathbb{C})$. How, then, could Engel accept Killing's conjectured theorem as "very plausible"[16]? Furthermore, at this point in time Lie had assumed a professorship at Leipzig, where Engel was instructor; and Killing himself even briefly visited Lie and Engel there in the summer of 1886 (see below). Given all this contact, it is difficult to imagine that Lie was not aware of Killing's conjectured theorem. Yet there is no evidence that Lie pointed out that the theorem could not be correct because of the projective group of a linear complex. Indeed, after Lie declared in print [1893a:682n,776] that he had "long ago" called attention to the four general classes (types A–D), Killing expressed his surprise

[15] Letter of 24 April 1886 (NSUB Göttingen, Nachlass Hilbert 94:1).
[16] Letter to Killing dated 19 April 1886.

to Engel.[17] As will be seen below, Killing's discovery of the simple type G_2, rather than any criticism Engel and Lie perhaps could have made of his conjectured theorem, revealed to him its untenability and made him receptive to the possibility of further new simple types, which he then proceeded to discover, including the "new" general type C.

In his next letter (24 April 1886), Killing made it clear that the promised communication developing the ideas behind his conjectured theorems was a long way off. Other duties prevented him from concentrating upon his research; and with the summer semester fast approaching, he anxiously wondered when he would find time to conclude what he had begun:

> In reply to your letter, I regret that I cannot yet talk about the theory of structure along the lines recently indicated. Unfortunately, other business has many times kept me from by own work, and when I cannot throw myself into my own project with full force I achieve nothing. But, apart from that, much "hard work" is still required before I achieve the desired goal. Now the summer semester is beginning with its diverse problems of a different sort, and I ask myself anxiously, When will I have finished that work on the theory of groups, which I regard as necessary for the further development of my theory of space?

Killing's theory of space forms called for nothing less than the determination of all possible finite-dimensional Lie algebras, not just simple algebras. The approach to this problem devised by Killing and now applied tentatively to simple groups was, as suggested in Section 1, definitely in need of "hard work" before it could be regarded as definitive, even as applied to the problem of determining all simple groups. Given the mathematical challenge facing him and the demands of his teaching and administrative duties at the Lyceum Hosianum, the sense of frustration conveyed in Killing's letter is certainly understandable.

5.4. Killing's Theory of Structure

Engel was also preoccupied with his teaching obligations at the University of Leipzig, and Killing's letter of April 24 was the last that passed between them until, on 16 July 1886, Killing sent Engel a postcard, inquiring half seriously whether Engel's silence meant that he was working on proofs of the conjectured theorems: "So have you completely forgotten me! ... Or do you wish to surprise me with the proof of the theorem we finally set down as likely?" Killing had nothing to fear in the way of competition from

[17] Killing to Engel, 15 September 1897. In his reply the following day, Engel could only say that in 1884, when he visited Lie in Norway, Lie had already considered the projective version of $\mathbf{sp}(2n, \mathbb{C})$ for $n > 2$. As we saw in Section 3.4, Lie did not publish anything on the case $n > 2$ until 1889.

Engel, who still had no idea of his approach to structure. The purpose of the postcard was to arrange a meeting with Engel in Leipzig. Killing was *Rektor* of the Lyceum Hosianum and in that capacity had been asked to visit the sister institution in Heidelberg. The plan was to visit Leipzig en route. There was a further reason for visiting Leipzig. In April Lie had assumed the professorship vacated by Klein, who had moved to Göttingen. Killing's tentative plan was to arrive on July 31 and leave the following day. When he learned that Lie and Engel, as well as Friedrich Schur and Eduard Study, would be in Leipzig to greet him, the plan became definite.

Shortly before the day of departure, Killing forwarded a manuscript to Engel:[18]

> So as not to appear with entirely empty hands, I enclose here the beginning of a work with which I am presently occupied. Unfortunately, during the summer I have seldom been able to work regularly; official duties have again severely cut into my time, other contingencies interfere in manifold ways. Now I would like most of all to devote myself to work undisturbed; but I could not rightly go against the request of my colleagues to visit the Lyceum in Heidelberg, although I tried to for a long time.

Although the manuscript has not been preserved, the correspondence indicates that it was a draft of the first part [1888a] of Killing's four-part essay, "The Structure of finite continuous transformation groups" [Hawkins 1982:152, n.17]. Here Killing introduced the concept of the characteristic equation of $\mathfrak{g}$ and proved that its coefficients, the functions $\psi_i(e)$ in (1.5), are invariants of the adjoint group. Engel had been encouraging Killing to study the adjoint group and its invariants [Hawkins 1982:152, n.18], and this result was the fruit of that encouragement. Although this result stands out as an interesting but somewhat isolated fact in Killing's theory, it will be seen in Section 6.2 that this property of the coefficients provided the key to Cartan's vindication of Killing's classification of simple groups.

Another concept introduced by Killing in part one and presumably in the manuscript was that of the rank l of $\mathfrak{g}$. This he defined as the smallest number l of functions $P_1(e), \ldots, P_l(e)$ in terms of which all coefficients $\psi_i(e)$ of the characteristic equation are expressible: $\psi_i(e) = F_i(P_1(e), \ldots, P_l(e))$, $i = 1, 2, \ldots$ [1888a:§4]. Probably, Killing was attempting to express the idea of the number l of functionally independent characteristic roots $\alpha_1(e), \alpha_2(e), \ldots, \alpha_r(e)$, since each $\psi_i(e)$ is expressible as an elementary symmetric function of these roots and $\partial\alpha_i(e)/\partial e_j$ has rank l.

Since at most k of the functions $\psi_i(e)$ do not vanish identically, it is clear from Killing's definition of l that $l \leq k$. In general, however, the Killing rank l and the modern rank k of $\mathfrak{g}$ are not equal. It will be helpful in what is to follow to explain how they are related. When $\mathfrak{g}$ is semisimple, it is true that $k = l$ (as Killing realized), so that both definitions of rank coincide.

[18] Letter of 28 July 1886.

More generally, if $\mathfrak{g} = \mathfrak{s} + \mathfrak{r}$, where $\mathfrak{r}$ is the radical of $\mathfrak{g}$ and $\mathfrak{s}$ a semisimple subalgebra, the Killing rank l of $\mathfrak{g}$ satisfies the equation $l = l_1 + \lambda$, where l_1 is the rank (in either sense) of $\mathfrak{s}$ and λ is the Killing rank of $\mathfrak{r}$. (This result is due to Cartan. See Section 6.2.) Since Killing spent most of his time and effort studying Lie algebras $\mathfrak{g}$ satisfying $\mathfrak{g}' = \mathfrak{g}$, it should be noted that in this case $\lambda = 0$, so that the Killing rank of $\mathfrak{g}$ coincides with the rank of its semisimple component $\mathfrak{s}$. Furthermore, since $\mathfrak{r}$ can be regarded as an $\mathfrak{s}$-module, $k - l$ represents the multiplicity of 0 as a weight of $\mathfrak{r}$. Killing deemed it a defect of his theory that k, rather than l, played the dominant role [1888a:59].

Although Killing's manuscript contained many new and important ideas, his theory of simple groups was not included. It was for this reason that the above-quoted letter has an apologetic tone. He had still not found time to work out the possibilities for the integers $a_{\alpha\beta}$ and the corresponding root systems for any value of k. And there was also the nagging question, Do simple groups necessarily satisfy Conditions I–III? The answer was certainly yes with regard to Condition I, but Killing did not have a proof that II and III hold for any simple group. Thus in the manuscript Killing demonstrated his approach by applying it to the one case he could work out without considerable difficulty: the determination of all $\mathfrak{g}$ satisfying I–III and with $k = 1$ as well.[19] This is the "general case." Since $l \leq k$, in this case $l = 1$ also. Thus, in the notation employed above, $\mathfrak{s}$ is isomorphic to $\mathbf{sl}(2, \mathbb{C})$. From a modern standpoint, Killing's structural analysis of the "general case" is equivalent to the determination of all irreducible $\mathfrak{s}$ modules $\mathfrak{r}$ without 0 as a weight and thus of odd highest weight.

Even for this restricted class of groups, the novelty and potential power of Killing's methods would have been evident to Lie and his colleagues in Leipzig. Lie's reaction to Killing's visit is recorded in a letter to Klein. It shows that even though Lie respected Killing's new mathematical ideas, other things about Killing did not impress him favorably: "Killing was just here. He has some really nice ideas. In many other respects, however, he does not make a solid impression."[20] The negative note in these remarks, perhaps due in part to a clash of personalities, was a harbinger of the great animosity that developed between them, spurred on by the breakdown of Lie's health.

The remainder of 1886 seems to have been a period of minimal mathematical activity by Killing. He had spent August at the Lyceum in Heidelberg. Upon his return he was distracted by the serious, long-term illness of his daughter and preoccupied with his manifold duties as *Rektor*. "It would

[19] The published version of Part I also contains an analysis of all $\mathfrak{g}$ with $k = 1$ which satisfy Conditions I and II (so that multiple nonzero roots are admitted) [1888a§8]; but Killing's letter to Engel dated 27 April 1887 shows that this was not in the original communication to Leipzig. The incompleteness of Killing's analysis of this case was revealed by Cartan [1894b:125–6].

[20] NSUB Göttingen, Nachlass Klein X, Nr. 722; see [Hawkins 1982:153].

be useless under these circumstances to squander my energy on transformation groups," Killing explained to Engel, adding sadly that he hoped soon to resume his work, since it was becoming unfamiliar to him: "Even those results at which I had previously arrived seem foreign to me .... I hope that soon I can take up my work regularly again and also be able to communicate something to you."[21] Killing was finally able to fulfill his wish in a letter of 27 April 1887. The letter indicates that he had focused on the general problem of the nature of groups satisfying Condition I ($p = r$) rather than upon the determination of simple groups. He still believed that the latter problem was essentially resolved, that only the two known classes of simple groups existed, even though he had limited his calculations to small values of k. There was a further reason for emphasizing the study of groups satisfying $p = r$: Killing hoped thereby to establish the validity of Conditions II and III for simple groups.

Killing's letter of April 27 indicates that he had made significant progress in his study of groups satisfying $\mathfrak{g}' = \mathfrak{g}$. At their meeting in Leipzig, Killing explained,

> it was not possible for me to directly write down all groups with $p = r$, $l > 0$. That is now possible, but I lack a suitable designation for this generalized case, which hopefully will come to me. Those groups with $p = r$ that are composite [i.e., not simple] fall into two essentially different categories, depending on whether or not there is an invariant subgroup with strictly commuting transformations. There is only an entirely restricted number of groups of the latter kind for each l, and all others can be built up from them by adding an invariant subgroup with strictly commuting elements. These groups (I would almost like to call them semisimple, were the name not too strange) share many of the properties of simple [groups].

Killing had thus arrived at the notion of a semisimple Lie algebra, which he characterized as one with no abelian ideal, as is still done today. But it would also appear that he was alluding to an incorrect radical splitting theorem, according to which $\mathfrak{g}$ (with $\mathfrak{g}' = \mathfrak{g}$) decomposes into a simple or semisimple subalgebra and an *abelian* ideal. (Since $\mathfrak{g}' = \mathfrak{g}$, the radical of $\mathfrak{g}$ is always nilpotent but need not be abelian.) The final form of the research reported on in the above letter is contained in the third part of Killing's "Zusammensetzung" papers [1889a:§§21–22]. If we use it as a guide, it is possible to give a plausible reconstruction of Killing's line of thought as reported in his letter of 27 April 1887, approximately a year and a half before he submitted part three for publication.

Consider the $k \times k$ system of "Cartan" integers $a_{\alpha\beta}$ introduced in Section 1 above. Killing considered the roots generated by the $a_{\alpha\beta}$ and the Lie algebra they determine. When the $a_{\alpha\beta}$ are "simple" (connected

[21] Letter of 29 November 1886.

Coxeter graph), the corresponding $\mathfrak{g}$ will be simple.[22] When the system $a_{\alpha\beta}$ is "composite," $\mathfrak{g}$ decomposes into what would now be termed a direct sum of simple ideals. These are the $\mathfrak{g}$ Killing tentatively called semisimple. Killing next considered the possibility, already in evidence from his analysis of the "general case," that $\mathfrak{g}$ is such that its characteristic equation admits more roots than are generated—"required" in Killing's terminology—by the $a_{\alpha\beta}$. These additional roots he called secondary and the necessary ones principal roots. Corresponding to the secondary roots are elements $X \in \mathfrak{g}$ that span an invariant subgroup (ideal) $\mathfrak{r}$, which in his letter Killing apparently thought was abelian. We note in passing that, viewed in modern terms, $\mathfrak{g} = \mathfrak{s} + \mathfrak{r}$. Killing's theory of secondary roots represents the origin of the modern theory of weights, as will be seen in Chapters 6 and 8.

We are now in a position to understand the likely grounds upon which Killing declared that $\mathfrak{r}$ is always abelian. In his paper [1889a] Killing gave special attention to the case of $\mathfrak{g} = \mathfrak{g}'$ satisfying $k = l$, and presumably in his original research this case was also accorded special consideration. This case appears to have misled Killing into incorrect assumptions about the more general situation in which k and l may differ. For example, when $k = l$, $\mathfrak{g}_0$ is abelian (Condition II). (See Section 6.2.) Perhaps the abelian nature of $\mathfrak{g}_0$ when $k = l$ led Killing to think that $\mathfrak{g}_0$ is always abelian, that Condition I implies II (the erroneous Theorem C of the next section). Killing's belief that $\mathfrak{r}$ is abelian may have been based upon a similar case of deception. That is, Killing apparently thought that when $\mathfrak{g} = \mathfrak{g}'$ satisfies $k = l$, then $\mathfrak{r}$ is abelian, although this is not correct. When Killing extended his analysis of the "general case" to include all $\mathfrak{g} = \mathfrak{g}'$ for which $k = l = 1$ (published in [1888a:§8]), he discovered, in effect, that $\mathfrak{g} = \mathfrak{s} + \mathfrak{r}$ in the sense of the (correct) radical splitting theorem (Section 6.2), where $\mathfrak{s} \cong \mathbf{sl}(2, \mathbb{C})$ and $\mathfrak{r}$ is abelian. Although in this case $\mathfrak{r}$ is abelian, it need not be when $k = l > 1$.[23] Not realizing this, Killing, in the letter of April 27, probably concluded too rashly that the apparent abelian nature of $\mathfrak{r}$ when $k = l$ carries over to the general situation.

By the time he composed his paper [1889a], Killing realized that $\mathfrak{r}$ could be nonabelian if $k > 1$, although it would always be of "rank zero." A group $\mathfrak{G}$ was said by Killing to be of rank zero when the corresponding Lie algebra $\mathfrak{g}$ has Killing rank $l = 0$. This means that the characteristic equation of $\mathfrak{g}$ reduces to $\Delta(\omega) = \pm\omega^r$, so that all characteristic roots are zero [1888a:285]. In modern terms, this means that $\operatorname{ad} X$ is nilpotent for every X in $\mathfrak{g}$ and therefore by Engel's theorem $\mathfrak{g}$ is nilpotent. Thus Killing's groups of rank zero are precisely the nilpotent Lie algebras. This characterization of rank-

---

22 Killing's definition of a simple system [1888b:23] is too cumbersome to repeat here. Cartan [1894b:62] expressed Killing's notion much more succinctly and perspicuously as follows: (a_{ij}) is simple if for every i and j indices $m_1, \ldots, m_s$ exist such that $a_{im_1} a_{m_1 m_2} \cdots a_{m_s j} \neq 0$. In geometrical terms, this means that the corresponding Coxeter graph is connected. There is no evidence that either Killing or Cartan thought in terms of graphs

23 I am indebted to George Seligman for an example illustrating this fact.

zero groups emerged from the efforts of Engel and his students to fill in the gaps in Killing's proofs (Section 5). Killing's theorem [1889a:107] that when $\mathfrak{g} = \mathfrak{g}'$, $\mathfrak{g}$ can be decomposed into a simple or semisimple subalgebra and an invariant subgroup (ideal) of rank zero was thus a correct version of the radical splitting theorem. His proof, however, was not at all correct. Among other faults, it depended upon the false theorem that $\mathfrak{g}_0$ is always abelian when $\mathfrak{g}' = \mathfrak{g}$.

Engel and Lie were quite pleased with the new notion of a semisimple group: "Your division of groups," Engel wrote, "into semisimple (why not?) and others is by all appearances a very felicitous one and also meets with Lie's full approbation.[24] Killing was himself more cautious about the designation "semisimple": "Before I can recommend a name, I must first understand more exactly the essence of these groups ...."[25]

The satisfying discoveries regarding groups satisfying $\mathfrak{g}' = \mathfrak{g}$ apparently encouraged Killing to resume his classification of simple groups in terms of simple systems $a_{\alpha\beta}$ and the roots they generate. In fact, in his paper [1889a:102] the type of considerations indicated above are used by Killing to prove that if $\mathfrak{g}$ is simple or semisimple, then $k = l$ and Conditions I–III are satisfied. Perhaps in the spring of 1887 Killing was near enough to establishing this result to feel convinced that his method of classifying simple groups was indeed valid and consequently worthy of being pursued more diligently and thoroughly. In a postcard dated 23 May 1887, Killing had announced the discovery of errors in his earlier computations and the possibility that other simple groups might exist: "The remark in my last letter that for small values of l only the two stated types of simple group exist is based upon calculational errors. If I am not mistaken, there are still more simple groups."

In his next letter, dated 23 May 1887, Killing wrote down the root system and the multiplication table for a 14-dimensional Lie algebra $\mathfrak{g}$ of the type now known as G_2. Although he did not attempt to obtain a concrete realization of this group, he did remark that as a group of point (rather than contact) transformations such a realization would require at least five variables and that the group "must 'live' already for this number." Engel was fascinated by this "extremely important" example and eventually obtained some concrete representations in five variables as Killing had predicted.[26] According to Cartan [1893b:785], the group of all contact transformations of ordinary 3-dimensional space that (by prolongation) leave the system of second-order partial differential equations $r = \frac{4}{3}t^3$, $s = t^2$ invariant is simple of type G_2, where r, s, t denote, respectively, the second order derivatives

[24] Letter of 5 May 1887.
[25] Letter of 23 May 1887.
[26] See Engel's letters to Killing dated 4 June 1887, 14 June 1888, 18 February 1890, 8 March 1890, 27 August 1890, as well as Killing's replies of 18 June 1888 and 24 April 1890. Engel finally published a note [1893c] on the matter at the time when Cartan announced [1893b] his independent discovery of a transformation group corresponding to each of the exceptional types.

$\partial^2 z/\partial x^2$, $\partial^2 z/\partial x \partial y$, $\partial^2 z/\partial y^2$.

The discovery that type G_2 actually exists seems to have transformed Killing's attitude toward the possible existence of further new simple groups. As the above quotation suggests, he was now prepared to encounter more new simple groups. Indeed, whereas initially Killing's errors of calculation had convinced him that no new simple groups exist, after this discovery, when he obtained the a_{ij} and associated root systems for the exceptional types E and F and the "new" general type C_l, Killing maintained that simple groups for these types exist even though he never managed to carry his calculations far enough to be able to write down multiplication tables for them.

Killing's latest revelations prompted Engel to urge him "at once to publish, at least provisionally, a part of your results."[27] Killing's answer is noteworthy. Given the constraints upon his time for scientific research, his desire for a complete, rigorously worked out theory was overridden by the need to force his ideas, by means of a commitment to publish, onto paper before they were lost forever:[28]

> Your wish that I might publish at least a part of my results will be fulfilled as soon as is possible. It is indeed awkward to publish things before they have been brought to a conclusion. On the other hand, it is too important for me myself to focus on a part of my things, to check the proofs thus far supplied for rigor and the like, to prevent the loss of the main proof idea, which is often only in my head but not written down ....

Engel expressed his delight at Killing's decision to commence publication and relayed Klein's invitation that he use the pages of *Mathematische Annalen* to that end. All of Killing's previous journal publications had appeared in *Journal für die reine und angewandte Mathematik*, the journal founded by Crelle and controlled by the mathematicians at Berlin, where Killing had received his training. *Mathematische Annalen* was the journal of the rival Göttingen school of Clebsch and Klein. Klein, who knew of the significance of Killing's research through Lie and, especially, Engel, wanted to claim his work for his own journal, particularly because it involved the theory of groups.

Spurred on by the decision to publish, Killing spent the summer and early fall attempting to work out the details of his theory. On 18 October 1887 he announced to Engel that, among other things, he had now succeeded in determining the structure (*Bildung*) of all simple groups. Although he declined in the letter to take the time to describe his results in detail, he did list the dimensions r for all simple groups of rank $l \leq 4$. The list shows that Killing was now looking for possible root systems with much greater care and a willingness to encounter systems that did not cor-

[27] Letter of 15 May 1887.
[28] Letter of 23 May 1887.

respond to the special linear or orthogonal groups. In addition to these groups and the 14-dimensional group of rank 2 (G_2), Killing's list included a second group of rank 3 and dimension 21 (C_3) and a second group of rank 4 and dimension 36 (C_4), as well as *two* 52-dimensional groups of rank 4. The latter both correspond to type F_4, but even in his paper [1888b:30–31] Killing overlooked their equivalence, leaving it for Cartan to point out. It would thus seem that this time Killing had really determined all simple groups, that he had essentially achieved the classification he set forth three months later in his paper [1888b:§§13–15], which includes the exceptional types E_l, $l = 6, 7, 8$.[29] There Killing determined all simple systems $a_{\alpha\beta}$ and corresponding root systems by exploiting the properties of the "Coxeter transformation" B (defined for $l = 2$ in Section 1) and using the fact that a simple system $a_{\alpha\beta}$ of rank $l + 1$ contains one of rank l, $l \geq 2$.

Despite the impressive discoveries he had made regarding simple groups, the tone of Killing's letter of October 18 was despondent. The brilliance of his discoveries regarding simple groups seemed to pale in the light of what was still left to accomplish:

> Probably I will publish my investigations in many sections, since *one* work would be far too voluminous. One part is ready for printing. Incidentally, no work has been so unpleasant for me as the one under consideration. You must realize that I have always worked without delight and that time after time I was tempted to leave the investigation unfinished. Only the hope of finally achieving a satisfying result caused me to again take up the work afresh. Even now I am not yet finished. It is quite easy for me to write down all simple groups for values of l that are not too large; but it still remains to characterize these groups in general. But more on these matters some other time! Likewise, I have not yet completely carried out the study of groups with $p = r$ with a "distinguished" transformation [i.e., an X belonging to the center of $\mathfrak{g}$]. I have already come upon many results, but they are insufficient to finish the systematic development. Similarly for groups with $p < r$. Here it is above all the multitude of different cases which have to be distinguished that will hinder me.

This remarkable passage indicates that in all areas of his theory—simple groups, groups with $p = r(\mathfrak{g}' = \mathfrak{g})$, and groups with $p < r$—Killing was painfully aware that he had not achieved the desired completeness and conclusiveness. His remarks about simple groups, for example, hint at the

[29] The modern notation A_l, B_l, etc. for simple complex structure types is a variant of Killing's original notation [1888b:§15]. Killing denoted the types A_2, B_2, G_2 by IIA, IIB, IIC, but for rank $l \geq 4$ he used the letters A to D for the four general simple types. Thus for $l = 4$ types A_4–D_4 are respectively IVA, IVB, IVC, IVD. In addition, Killing found (he thought) two exceptional types, IVE and IVF. Cartan showed they are equivalent. For $l = 6, 7, 8$ Killing discovered the exceptional types VIE, VIIE, and VIIIE. Cartan [1894b:94] used the letter G for Killing's IIC and the letter F for IVE=IVF.

problems he was facing establishing the existence and nature of the new simple types. Writing down the structure constants c_{ijk} for the new general type C and the "six" exceptional types was a formidable problem that Killing obviously did not at all relish. In these calculations heavy use is naturally made of the Jacobi identity, but that hardly guarantees that once the c_{ijk} are determined, they actually define a Lie algebra. It would still be necessary to confirm that the c_{ijk} actually satisfy the conditions that guarantee, for example, that the Jacobi identity holds.

Obviously, it would also be desirable to obtain a concrete representation of each simple type, but this task was apparently never attempted by Killing. When Engel presented him with a concrete representation of G_2, he wrote back, "I myself would probably not have come so quickly to treat the question of the representation of groups whose structure is given."[30] Killing more or less concentrated upon the problem of determining the structure constants c_{ijk}, but this problem, which he considered with such distaste in October 1887, was never fully resolved by him. He seems to have concentrated most of his efforts on the "new" general type C, still regrettably oblivious of the projective group of a linear complex. For type C Killing managed to determine the c_{ijk} to within certain arbitrary constants [1888b:39–43], but he failed to establish consistency with the Jacobi identity. The determination of the c_{ijk} for the types E and F was carried out to a lesser extent [1888b:46–48], in fact not at all for the E types. With typical understatement, Killing explained to Engel that he had left "an entirely insignificant calculation" for the reader.[31] To sum up, the problems regarding simple groups that confronted Killing in October 1887 were indeed formidable, as he sensed; and he never fully resolved them.

Killing's reference to "distinguished transformations" $X \in \mathfrak{g}$ in conjunction with his study of groups with $p = r$ ($\mathfrak{g}' = \mathfrak{g}$) is also enlightening with regard to the difficulties he was facing in this area of his theory. Recall that X is distinguished if $[X, Y] = 0$ for all $Y \in \mathfrak{g}$. In modern terms X belongs to the center of $\mathfrak{g}$. Since such an X is a member of the ideal $\mathfrak{r}$ of rank zero of Killing's radical splitting theorem (discussed above), the existence of distinguished elements means that $k > l$ (as an $\mathfrak{s}$-module, $\mathfrak{r}$ admits the weight 0). Killing's remarks to Engel thus reflect the difficulty he was having in attempting to go beyond the case $k = l$, difficulties not entirely overcome in his publication [1889a] on groups with $p = r$. The "multitude of different cases" that arise for groups with $p < r$, which Killing deemed the major obstacle to their study, finally induced him to abandon altogether his objective of an exhaustive classification. Indeed, included among the groups with $p < r$ were, as Killing noted in the above passage, the groups of rank zero (nilpotent Lie algebras). Although the passage is characteristically obscure, it seems to refer to an approach for writing down the structure constants for groups of rank zero that Killing mentioned briefly

30 Letter of 18 June 1888.
31 Letter of 4 February 1888.

in his first paper [1888a:290] with the comment that it appeared insufficient for $r > 6$. The classification of groups of rank zero was, Killing realized, far from being resolved.

Given the enormity of the problems still facing him and the amount of time and effort already expended in a life with many other demands on his time, Killing's depressed and frustrated frame of mind is thus understandable, and even more so when it is kept in mind that his first love was geometry. His research program on the foundations of geometry had now led him into a complex web of purely algebraic problems of considerable difficulty. That he persevered to the extent that he did is a tribute to his determination. Nor should the role of Engel be ignored. Without his line of communication with Engel, eager to learn of Killing's latest revelation on "Zusammensetzung" and capable of judging its worth, Killing might have lacked the stimulation to persevere as he did.

On 5 November 1887 Killing finally posted the first part of his work to Klein, apologizing for its incomplete nature: "It is with fear and trembling that I send you herewith the manuscript of a work that I respectfully request to appear in the *Annalen*. It is not pleasant for me to publish an unfinished work, and on the whole I prefer to publish only finished ones. This time, however, I believe I should make an exception."[32] Klein had no misgivings about the form of publication: "I was especially delighted to receive your communication. As a matter of fact I welcome wholeheartedly your appearance as a contributor to the *Annalen* and especially the fact that our journal will be put to good use for the development of the theory of groups."[33] Regardless of any shortcomings the form of Killing's papers might present, Klein, who declared himself unable to judge their merit, trusted in the highly favorable opinions of Lie and Engel. Certainly, Klein was also pleased by the prospect of a Berlin-trained mathematician publishing in the *Annalen* on the theory of groups, a field not stressed at Berlin but one that Klein esteemed highly.

Three months later, the second installment containing Killing's remarkable results on simple Lie algebras was forwarded to Klein for publication (as [Killing 1888b]). Killing was painfully aware of the imperfect nature of the theoretical analysis. His dissatisfaction, accompanied by lingering doubts about the veracity of his results, was conveyed in a letter to Engel written at the time part two was sent to Klein. Regarding his determination of all simple types he wrote:[34]

> If I said I was satisfied with the stated results I would be lying. I have attempted on many occasions to find an error in the proof, but so far without success. I have produced a second proof that I have not published, since it is based on the same principles and

---

[32] Transcription and source in [Hawkins 1982:159].
[33] Letter of 17 November 1887; transcription and source in [Hawkins 1982:159].
[34] Letter of 6 February 1888.

> so guarantees no greater degree of certainty. Also other (likewise unpublished) investigations speak for the high likelihood of the stated results. Therefore, I thought it best to publish the results with the proof as quickly as possible, since only then can a serious, thorough examination take place; and it is of the greatest importance to me that complete certainty about these questions (on the structure of simple groups) be achieved as soon as possible.

The "thorough examination" of his work that Killing hoped its publication would stimulate was begun by Engel and his students (Section 5) and brought to a successful completion by Élie Cartan (Section 6.2).

On 18 September 1888 Killing forwarded part three to Klein with the explanation that, seven and a half months ago when part two was submitted, he had thought that the third part would follow much more rapidly.[35] Part three [1889a] contains Killing's study, discussed partially above, of algebras satisfying $\mathfrak{g}' = \mathfrak{g}$ and is almost as significant as part two. Here he tried to prove his assumption (in part two) that simple Lie algebras satisfy Conditions I–III, introduced the concept of a semisimple Lie algebra (defined as one expressible as a direct sum of simple ideals) and posited his radical splitting theorem, now correctly formulated to the effect that if $\mathfrak{g}' = \mathfrak{g}$, then $\mathfrak{g} = \mathfrak{s} + \mathfrak{r}$, where $\mathfrak{s}$ is a semisimple subalgebra and $\mathfrak{r}$ is an ideal of rank zero. The proofs were, however, even more sketchy and the overall presentation even more obscure than previous parts. Particularly burdensome was the need to consider the case in which the characteristic equation admits elementary divisors of higher order (i.e., $\operatorname{ad} X$ is not diagonalizable for all $X \in \mathfrak{g}_0$). With this aspect of his work in mind, Killing wrote as follows to Engel, who was then acting as referee for part three and in this capacity receiving communications from Killing on the need for a more extensive consideration of higher-order elementary divisors: "Despite the aversion I have to the entire work, I believe I must subject myself to this task."[36] This passage captures the admixture of distaste and determination with which Killing approached his work once he realized the magnitude of the difficulties he faced.

Since in the modern theory of Lie algebras "everything diagonalizes," Killing's need to deal with higher-order elementary divisors perhaps requires some clarification. Let $\mathfrak{g} = \mathfrak{s} + \mathfrak{r}$ in accordance with the modern radical splitting theorem, so that $\mathfrak{s}$ is a semisimple subalgebra and $\mathfrak{r}$ is the radical of $\mathfrak{g}$. Then relative to the Cartan subalgebra $\mathfrak{s}_0 \subset \mathfrak{s}$ the characteristic polynomial of $\mathfrak{g}$ admits only linear elementary divisors: $\operatorname{ad} X$ can be diagonalized simultaneously for all X in $\mathfrak{s}_0$. But for Killing, $\mathfrak{g}_0$ rather than $\mathfrak{s}_0$ was the focal point of the theory. The radical splitting theorem, which brings with it the possibility of focusing upon $\mathfrak{s}$ and $\mathfrak{s}_0$, was the ultimate objective of Killing's deliberations and not the starting point. Of course,

[35] NSUB Göttingen, Nachlass Klein X, Nr. 186.
[36] 10 November 1888.

$\mathfrak{g}_0 = \mathfrak{s}_0$ when $k = l$,[37] and corresponding to this fact is Killing's theorem [1889a:81] that when $k = l$ the elementary divisors of the characteristic equation are all linear. But when $k > l$, so that $\mathfrak{g}_0 \neq \mathfrak{s}_0$, the characteristic equation relative to $\mathfrak{g}_0$ can have nonlinear elementary divisors. In the modern theory "everything diagonalizes" because the general theory of the structure of Lie algebras has been replaced by the theory of the structure and representation of semisimple Lie algebras. The first steps in that direction were taken by Cartan in the final chapter of his doctoral thesis of 1894 (Section 6.4).

Even with his radical splitting theorem at hand, Killing realized that his goal of an exhaustive determination of all Lie algebras satisfying $\mathfrak{g} = \mathfrak{g}'$ had not been attained. That is, if $\mathfrak{g} = \mathfrak{s} + \mathfrak{r}$, it still remains to determine the multiplications $[\mathfrak{s}, \mathfrak{r}]$ and the structure of $\mathfrak{r}$. Although Killing admitted that he had not been able to classify all rank-zero algebras $\mathfrak{r}$ [1889a:107], he claimed that this was not necessary, because if one could determine all $\mathfrak{g} = \mathfrak{g}'$ such that $\mathfrak{s}$ is simple and $\mathfrak{r}$ is the only abelian ideal in $\mathfrak{g}$, then the determination of *all* $\mathfrak{g} = \mathfrak{g}'$ posed only "minor difficulties" [1889a:116]. Thus Killing restricted his attention to this case. From a modern perspective, Killing's problem is to determine all irreducible $\mathfrak{s}$-modules $\mathfrak{r}$ corresponding to a simple $\mathfrak{s}$. From a historical perspective, Killing's work on this problem [1889a:§§25–26], though incomplete and partly erroneous, originated the representation theory of semisimple Lie algebras and, in particular, the theory of weights. This aspect of Killing's work will be discussed more fully in Section 6.3.

The effort required to bring part three into a publishable form convinced Killing that he could no longer hope to achieve his original objective of an exhaustive classification of all Lie algebras by proceeding next to deal exhaustively with the case $p < r$ ($\mathfrak{g}' \neq \mathfrak{g}$). Thus when he sent the manuscript of part three to Klein, he wrote, "An exhaustive treatment of all cases in which ... [$p < r$] ... can scarcely be begun due to its great diversity. I would be happy to communicate a number of propositions that in part have been in my possession for a long time, but I do not know whether I will bring it to a conclusion."[38] When the fourth and final part finally appeared [1890b], it conformed to Killing's prediction. He concentrated his efforts on attempting to extend his radical splitting theorem to arbitrary algebras. His main results were the following (stated in modern notation). Given any Lie algebra $\mathfrak{g}$, then $\mathfrak{g}' = \mathfrak{s} + \mathfrak{r}$, where $\mathfrak{s}$ is simple or semisimple and $\mathfrak{r}$ is a nilpotent ideal of $\mathfrak{g}$ [1890b:187]. If $\mathfrak{g}^{(i+1)} = [\mathfrak{g}^{(i)}]'$ for $i = 1, 2, \ldots$ and if $\mathfrak{g}^{(i+1)} = \mathfrak{g}^{(i)} \neq 0$ for some i, then $\mathfrak{g}^{(i)} = \mathfrak{s} + \mathfrak{r}$, where $\mathfrak{s}$ is simple or semisimple and $\mathfrak{r}$ is a nilpotent ideal of $\mathfrak{g}$. Furthermore, it is possible to choose a vector space complement to $\mathfrak{g}^{(i)}$, viz., $\mathfrak{g} = \mathfrak{g}^{(i)} + \mathfrak{h}$, such that

[37] This follows from Cartan's rank formula (presented earlier in this section and again in Section 6.2 and the fact (presumed throughout this discussion) that $\mathfrak{g}' = \mathfrak{g}$, so that $\mathfrak{r}$ has rank zero.

[38] 18 September 1888.

$[\mathfrak{h}, \mathfrak{s}] = 0$ [1890b:189]. In the case $i = 1$ ($[\mathfrak{g}']' = \mathfrak{g}'$), Killing asserted further that $\mathfrak{h}$ is an abelian subalgebra [1890b:184]. As these statements indicate, Killing failed to articulate a completely satisfactory general radical splitting theorem. (Also keep in mind that where I have spoken of nilpotent ideals, Killing spoke of ideals of rank zero; the nilpotency characterization of algebras of rank zero had not yet been discovered.)

In the fourth part Killing also suggested that perhaps Lie's designation "composition of groups" (*Zusammensetzung der Gruppen*) was not the best choice to describe the theory he had now so greatly, albeit tentatively, advanced [1890b:163]. He pointed out that according to Lie, a group was either simple or composite, and yet one also spoke of the composition of simple groups. On the basis of this inconsistency, he advocated speaking instead of the "shape" (*Gestaltung*) of a group rather than its composition.[39] Unbeknownst to Killing, E. Vessiot and W. de Tannenberg, the first two graduates of the École Normale Supérieure to study with Lie in Leipzig (Section 6.2), had already in a sense met Killing's objection. In their lengthy review of the first volume of Lie's *Theorie der Transformationsgruppen* they expressed Lie's idea of *Zusammensetzung* by the French word *structure* [Vessiot and Tannenberg 1889:137]. In his own publications, Cartan always referred to *la structure des groupes*, thereby establishing this expression in the vocabulary of twentieth-century mathematics.

Let us now consider the technical problems posed by Killing's work and the initial efforts to resolve them.

5.5. Groups of Rank Zero

The emphasis in Sections 3–4 was upon Killing's discoveries. I now consider certain aspects of his attempt to demonstrate them. Killing had worked out his classification of simple algebras under Conditions I–III, and it was therefore necessary to establish that they are satisfied by any simple algebra. Since Killing realized that for any Lie algebra the derived algebra $\mathfrak{g}'$ is an ideal of $\mathfrak{g}$, Condition I ($\mathfrak{g}' = \mathfrak{g}$) was not problematic. Conditions II–III did not follow with comparable ease from the hypothesis of simplicity. Killing concentrated on the more fundamental Condition II ($\mathfrak{g}_0$ is abelian). He was able to show that for any $\mathfrak{g}$ (simple or not) $\mathfrak{g}_0$ is a subalgebra [1888b:7] and that as a Lie algebra in its own right, $\mathfrak{g}_0$ has a characteristic equation of the form $\Delta(\omega) = \pm\omega^k$ [1888b:8]. This property of $\Delta(\omega)$ makes $\mathfrak{g}_0$, in Killing's terminology, a group of rank zero. That is, a Lie algebra $\mathfrak{g}$ is of Killing rank $l = 0$ when its characteristic polynomial reduces to $\Delta(\omega) = \omega^r$, which is tantamount to saying that $\operatorname{ad} X$ is nilpotent for all $X \in \mathfrak{g}$. By Engel's theorem groups of rank zero are therefore identifiable with nilpotent Lie

[39] Later, in [1904], he switched to *Bau*.

algebras, but it must be remembered that Engel's theorem, and such an identification, emerged only later, as will be seen.

Even though, as Killing himself emphasized, groups of rank $l = 0$ were not amenable to treatment by his methods, geared as they were to deal with the case $l > 0$, he was forced by the exigencies of his mathematical theory to probe more deeply into their nature [1888b:§9]. His starting point was the observation that any two-dimensional subgroup of a group of rank zero must be abelian. That is, if $\mathfrak{g}$ has rank zero and $X_1, X_2 \in \mathfrak{g}$, then $[X_1, X_2] = aX_1 + bX_2$ implies that $a = b = 0$, since otherwise $\operatorname{ad} X_i$ for $i = 1$ or 2 would possess a nonzero characteristic root. This incidentally shows that when $k = \dim \mathfrak{g}_0(X_1) = 2$, then $\mathfrak{g}_0$ must be abelian and may have been the inspiration behind Killing's Condition II.

Killing also showed that in the Jordan–Weierstrass form for $\operatorname{ad} X$, at least two Jordan blocks must exist. That is, some $Y \neq cX$ exists such that $[X, Y] = 0$. He considered a "completely general" $X_1 \in \mathfrak{g}$, by which he apparently meant an X_1 such that $\operatorname{ad} X_1$ has maximal rank or, equivalently, minimal nullity n, where by the nullity of $\operatorname{ad} X_1$ I mean the dimension of its null space. He then extended X_1 to a basis $X_1, X_2, \ldots, X_k$ for $\mathfrak{g}$ that put $\operatorname{ad} X_1$ in its Jordan–Weierstrass form with Jordan blocks decreasing in size. That is, as Killing expressed it,

$$(5.1)\quad \begin{aligned} &[X_1, X_2] = a_1 X_3,\ [X_1, X_3] = a_2 X_4,\ \ldots,\ [X_1, X_{k-1}] = a_{k-2} X_k, \\ &[X_1, X_k] = 0, \end{aligned}$$

where each a_i is either 0 or 1 and the distance between successive a_i's which are 0 decreases as i increases. If $a_{m-2}, a_{m'-2}, \ldots$ are the vanishing coefficients, then $X_m, X_{m'}, \ldots$ together with X_1 and X_k are the linearly independent eigenvectors for $\operatorname{ad} X_1$, so that the nullity n is always at least 2. Killing's fundamental proposition regarding groups of rank zero is the following incorrect assertion:

> **Theorem A.** *Let $\mathfrak{g}$ have rank zero and let a basis $X_1, \ldots, X_k$ be chosen as described above, and let n denote the nullity of $\operatorname{ad} X_1$. Then the $n - 1 \geq 1$ elements $X_m, X_{m'}, \ldots, X_k$ commute with all of $\mathfrak{g}$.*

This theorem is correct when n, which we shall term the nullity of $\mathfrak{g}$, satisfies $n = 2$. Otherwise it is false, as we shall see. Killing gave very few details of his proof. For example, it is not even clear how the hypothesis that X_1 is "completely general" was to be used. Perhaps, as was customary with him, he first discovered the theorem in the case $n = 2$ and then convinced himself, a bit too hastily, that it was valid for $n > 2$ as well. In any case, Killing felt compelled to abbreviate his proofs in order to keep the length of papers within bounds. Later, he confided to Engel, with the proof of Theorem A specifically in mind, that he regretted his decision to abbreviate [Hawkins 1982:163]. The letter to Engel was written shortly before he learned that a complete proof of Theorem A was an impossibility, since it was false.

As Killing emphasized [1888a:288], Theorem A implied that groups of rank zero always possess "distinguished" elements, that is, elements that commute with the entire group. In more familiar language, Killing's observation is that the center of a rank-zero group is never trivial. This is, of course, correct, although Killing went on to assert mistakenly that in effect the dimension of the center is at least $n-1$, where n is the nullity of $\mathfrak{g}$. From the vantage point of the present, the correct part of Killing's assertion is the essence of Engel's theorem: If $\mathfrak{g}$ has rank zero, then $\mathfrak{g}$ is nilpotent.[40] But Killing did not introduce or discover the nilpotency property of a Lie algebra. That was done later by Engel in the course of his attempts to check the veracity of Theorem A. Thus Theorem A was the source of Engel's theorem. Killing himself inferred some consequences of Theorem A that were headed in the direction of Engel's theorem. He considered the case $n = 2$ in Theorem A and deduced that $\mathfrak{g}'$ is contained in span$\{X_3, X_4, \ldots, X_k\}$, the linear span of $X_3, \ldots, X_k$. Without claiming that he had worked out a proof, Killing also suggested that this result is true for any n. As will be seen momentarily, this correct consequence of Theorem A was assumed by Killing in his efforts to vindicate his assumption that simple groups always satisfy Conditions II and III.

Theorem A was, in fact, fundamental in Killing's efforts in this direction. In the second part [1888b] of his structure papers—the part containing the classification of simple types—Killing presented his first, obviously temporary, argument for the validity of Condition II when $\mathfrak{g}$ is simple. It was temporary, since it assumed for convenience that Condition III held, i.e., that all roots $\alpha \neq 0$ are simple roots. The proof proceeded as follows [1888b:8–9]. Let $X_1, \ldots, X_k, X_\alpha, X_\beta \ldots$ be a basis for $\mathfrak{g}$ in accordance with the setup of Section 1. That is, X_1 has $\omega = 0$ as characteristic root with minimal multiplicity k (X_1 is regular in modern terminology), and $\mathfrak{g}_0(X_1)$ has $X_1, \ldots, X_k$ as a basis. Killing also tacitly assumed that X_1 is "completely general" with respect to $\mathfrak{g}_0(X_1)$ (which can be done without loss of generality), and he chose $X_1, \ldots, X_k$ in accordance with Theorem A so that $X_m, X_{m'}, \ldots, X_k$ commute with all of $g_0(X_1)$. Using the Jacobi identity and the fact that dim $g_\alpha = 1$ for $\alpha \neq 0$, he then showed (sketchily but correctly) that if some of the $a_i = 1$ in (5.1), then $[X_k, X_\alpha] = 0$ for all $\alpha \neq 0$. But by Theorem A, X_k also commutes with all of $\mathfrak{g}_0$. Thus X_k commutes with all of $\mathfrak{g}$. In other words:

Theorem B. *If $\mathfrak{g}$ satisfies Condition III and $\mathfrak{g}$ contains no distinguished elements, then all $a_i = 0$ and therefore (by Theorem A) $\mathfrak{g}_0(X_1)$ is abelian.*

Since a simple $\mathfrak{g}$ cannot contain distinguished elements, Theorem B can be applied—provided, that is, that Condition III need not be assumed in the proof.

[40] That is, the proof of Engel's theorem then follows by induction on the dimension of $\mathfrak{g}$, using the fact that $\mathfrak{g}$ modulo its center is also of rank zero.

If we can judge by his remarks in presenting the proof of Theorem B, Killing was confident that with enough time he could extend it to the case of multiple nonzero roots. In fact, the proof of Theorem B together with the implications of Theorem A implies a stronger result: Not only does $\mathfrak{g}$ possess distinguished elements when $\mathfrak{g}_0(X_1)$ is not abelian (i.e., when some a_i's equal 1 in (5.1)), but $\mathfrak{g}' \neq \mathfrak{g}$. This is because $\mathfrak{g}'$ is spanned by elements from $[\mathfrak{g}_\alpha, \mathfrak{g}_{\alpha'}]$, where $\alpha' = -\alpha$. But Killing realized that $[X_\alpha, X_{\alpha'}] \in \operatorname{span}\{X_m, X_{m'}, \ldots, X_k\}$, which does not include X_1 or X_2 (assuming some $a_i = 1$). Since Killing also believed that X_1 and X_2 are not in $[\mathfrak{g}_0, \mathfrak{g}_0]$, it followed that $\mathfrak{g}' \neq \mathfrak{g}$. In Section 19 of the third part [1889a] Killing attempted to establish these conclusions also in the case of multiple nonzero roots:

> **Theorem C**. *If the rank of* ad X *is* $r-k+1$ *for some* X, *then* $\mathfrak{g}' \neq \mathfrak{g}$ *and* $\mathfrak{g}$ *contains distinguished elements. Hence if* $\mathfrak{g}' = \mathfrak{g}$, $\mathfrak{g}_0(X_1)$ *is abelian for* X_1 *regular.*

The hypothesis on the rank of ad X in Theorem C simply means that the nullity n of a regular X_1 is less than k, so that some a_i's in (5.1) are equal to 1.

Theorem C thus asserts, incorrectly, that Condition I always implies Condition II. In typical fashion Killing wrote down something resembling a complete proof only for a special case, "in order to give a good overview of the development and to discuss that case in which all formulas can be completely written down" [1889a:60]. (Cf. the remarks on this passage in Section 6.2.) Thus he assumed that in the basis $X_1, \ldots, X_k$, chosen in accordance with Theorem A, all $a_i = 1$ (so the nullity of ad X_1 is $n = 2$). He also assumed that each root space $\mathfrak{g}_\alpha$ possessed only a single eigenvector for α. After extensive calculations based upon this type of Jordan–Weierstrass form (which I have not attempted to check), Killing concluded that X_k is distinguished and that not both X_1 and X_2 can belong to $\mathfrak{g}'$. He then argued in a much more vague fashion that the same conclusions follow in the completely general case.

Although Killing's proof of Theorem C was consequently sketchy and vague, one thing was certain: It relied critically upon Theorem A, as did the earlier proof of Theorem B. Using Theorem C, Killing proved that when $\mathfrak{g}$ is simple, then Condition III is satisfied [1889a:102]. The erroneous Theorem A was thus the keystone in Killing's attempt to vindicate his presumption that all simple groups satisfy Conditions I–III. It was the discovery that Theorem A is false that first revealed that Killing's mode of demonstration contained fundamental flaws, so that the confirmation of his discoveries would involve more than simply filling in the details omitted by Killing. The untenability of Theorem A, of course, does not in itself disprove Theorem C; it only invalidates Killing's proof (using Theorem A) of Theorem C. Actually, Theorem C is also false, although this fact was not at all certain until Cartan obtained a counterexample [1894b:124] through his analysis of all algebras of Killing rank $l = 1$. (See Section 6.2.) It was the earlier discovery of counterexamples to Theorem A by Engel and his students, however, that first established the need to prove Condition II by an alternative route. We therefore now consider in some detail how counter examples to Theorem A were discovered.

The first two parts of Killing's paper appeared in 1888 and the third in 1889. Engel was not alone in quickly recognizing the importance of these publications. Lie wrote to Ernest Vessiot that although Killing's papers were "most disagreeably edited," they "contain results of the greatest interest, if only everything is correct ...."[41] About the same time Lie wrote to Klein that as a result of paging through parts I–III of Killing's paper, he had arrived at an essentially more favorable assessment of them: "I have in no way checked everything; but certainly these works are noteworthy. I hope that in the fall I find time to actually read them."[42] Lie expressed this more favorable opinion of Killing during the summer of 1889, but he apparently never found the time to read through the papers carefully as intended. Undoubtedly, the sheer algebraic nature of Killing's work was a factor, as was Lie's ill health [Hawkins 1982:165, n.25]. Nonetheless, Lie remained confident of their importance and essential validity. Thus in May of 1890 he wrote to Klein of the great triumph for the theory of transformation groups that Killing's "beautiful investigations" implied: "Killing has done beautiful research. If, as I believe, the results are correct, he has performed an outstanding service. Generally speaking, now the theory of transformation groups and differential invariants will reign over vast domains of mathematics."[43]

The other mathematicians who had met Killing in Leipzig, F. Schur and Study, were also impressed by the actual papers. Schur wrote to Klein in 1890: "I have even come to terms with [Killing's] ... works on the composition of transformation groups. I count them among the most significant and most difficult investigations that have been carried out during the past ten years."[44] And when Lie ranked Study, as a mathematician, on a level with Killing, Study humbly protested: "I do not at all wish to be compared with Killing, for whose works I have the greatest admiration."[45] Within the small circle of mathematicians involved with Lie's theory of transformation groups, Killing's work was consequently regarded highly, but because of the intrinsic complexity of the theory, compounded as it was by the obscurity of Killing's presentation, it was admired from a distance. Engel was the only one who persevered in his attempt to confirm the validity of Killing's

[41] Undated letter, circa 1889. For transcription and source see [Hawkins 1982:165].
[42] NSUB Göttingen, Nachlass Klein X, Nr. 751; transcription in [Hawkins 1982: 165].
[43] NSUB Göttingen, Nachlass Klein X, Nr. 752); [Hawkins 1982:165].
[44] NSUB Göttingen, Nachlass Klein XI, Nr. 893; [Hawkins 1982:165].
[45] Letter of 31 December 1890 (UB Oslo, Brevsamling Nr. 289; see [Hawkins 1982:165].

theorems by assiduously studying his demonstrations.

From Engel's correspondence with Killing we learn that Engel carefully read most of the first two parts of the structure papers and that he was unable to follow the proofs of Theorems A and B. While checking out the page proofs of part two, Engel apparently sent Killing a postcard regarding Theorem B and the question of its validity without the unjustified hypothesis of distinct nonzero roots. Killing referred Engel to Theorem C of Section 19 of Part III while also admitting that the need always to consider the possibility of higher order elementary divisors was a defect of his approach:[46]

> The proof on p. 8 [i.c., that of Theorem B] is not entirely simple for multiple roots. It is a defect of my starting point that elementary divisors of higher order always require separate consideration. I have, however, given the proof now in Section 19 (part three).

Engel wrote back that even though he had not yet digested the material in Section 19, the publication of Part II should not be delayed any further.

Whether or not Engel ever attempted to digest Section 19, it would have been clear to him that the consequential part of Theorem C required Theorem A. Engel was especially interested in Theorem A because it was directly related to his own investigations on structure. His approach to structure was dominated by Lie's distinction between integrable and nonintegrable groups, i.e., solvable and nonsolvable Lie algebras (Section 3.2). In [1887] Engel presented a new characterization of this distinction. He defined a group $\mathfrak{g}$ to be without conic section groups if it contained no three-dimensional subalgebra with the structure of the "projective group of a conic section" in the plane. This group and its extensions to higher dimensions will be discussed in Section 7.1. Here it suffices to know that the Lie algebra of this group is isomorphic to $\mathbf{sl}(2,\mathbb{C})$. Thus $\mathfrak{g}$ is without conic section groups if it contains no subalgebra isomorphic to $\mathbf{sl}(2,\mathbb{C})$. Clearly, every Lie algebra $\mathfrak{g}$ is either with or without conic section groups. Engel proved that this distinction is tantamount to distinguishing between Lie algebras that are nonintegrable and integrable in Lie's sense. Stated in modern terms, his theorem states that *a Lie algebra $\mathfrak{g}$ is solvable if and only if it is without conic section groups.* It follows immediately from this theorem that any $\mathfrak{g}$ of rank zero must be solvable, since if $\mathfrak{g}$ contains conic section groups, its characteristic equation will admit nonzero roots and $\mathfrak{g}$ will not have rank zero.

When Engel communicated his discoveries to Killing in a letter of 3 February 1887, Killing pointed out that the results were "not entirely new" to him.[47] At least for those nonsolvable $\mathfrak{g}$ satisfying $\mathfrak{g}' = \mathfrak{g}$, the existence of elements $H_\alpha, X_\alpha, X_{\alpha'}$, defining a conic section group was an immediate consequence of his theory. (See (1.9).) On the other hand, Engel's the-

---

[46] Letter of 22 October 1888.
[47] Letter of 7 February 1888.

orem as it applied to rank-zero groups had not, Killing admitted, emerged from his theory, and he ascribed full priority to Engel. Killing could not refrain from remarking that his own approach could add something even here: When $\mathfrak{g}$ has rank zero, then, "if I remember correctly," at least one $X \in \mathfrak{g}$ exists that commutes with all elements in $\mathfrak{g}$. As it stands, Killing's assertion is correct, since a rank zero $\mathfrak{g}$, being nilpotent, has a nontrivial center; but Killing was undoubtedly invoking his Theorem A, which asserts that $n-1 \geq 1$ linearly independent elements of a rank-zero algebra $\mathfrak{g}$ commute with all of $\mathfrak{g}$. By the time Killing submitted Part I nine months later, he had used the false part of Theorem A to prove a theorem "that Mr. Engel first posited and proved under a somewhat more general assumption" [1888a:289]. The theorem may be stated as follows:

> **Theorem D.** *Every k-dimensional Lie algebra $\mathfrak{g}$ of rank zero contains a $(k-1)$-dimensional ideal $\mathfrak{g}_{k-1}$, which in turn contains a $(k-2)$-dimensional ideal $\mathfrak{g}_{r-2}$, and so on.*

Theorem D thus asserts that every group of rank zero is solvable, in accordance with Engel's theorem characterizing groups without conic section groups.

When responding to Engel's letter of February 3, Killing had expressed his regret that he was still not at home with Lie's methods. Indeed, Killing's remark about commuting elements in a rank-zero group follows immediately from Lie's theorem (Section 3.2) on the existence of a universal eigenvector (corresponding to $\alpha = 0$ in the rank-zero case). Although Engel drew this fact to Killing's attention in his reply to Killing's letter, it did not increase Killing's resolve to master Lie's methods. As Killing later confessed to Engel, in reading through Lie's papers he was less concerned with learning Lie's methods than with determining which of Lie's theorems he had discovered independently.[48] Killing thus stubbornly refused to pursue any but his own ideas and methods. Of course, the task of learning Lie's methods in 1886–87 was a difficult one. His papers were difficult to read, and Killing did not have most of them at his disposal for an extended period of time. By the time the first volume of Lie's book *Theorie der Transformationsgruppen* appeared in 1888, Killing was ready to quit his algebraic investigations in favor of more geometrical problems. The task of combining the methods of Lie with those of Killing so as to establish Killing's result rigorously was left to Engel and his students and, above all, to Cartan.

Because of Engel's interest in solvable Lie algebras, Killing's Theorem D, and hence Theorem A, were of special concern to him. The concern was perhaps increased by Engel's discovery that his own proof of his theorem characterizing solvable Lie algebras by means of "conic section groups" contained a gap [Hawkins 1982:167, n.27]. In 1890 Engel organized a seminar devoted to a careful study of Killing's "Zusammensetzung" papers.[49]

---

[48] Letter of 4 February 1889.

[49] This is documented in Engel's letter to Killing of 12 June 1890.

Special attention was accorded to Section 9 of Part I, which deals with Lie algebras of rank zero. One of the three students in the seminar, Karl Umlauf,[50] was assigned the task of studying them. For example, given the gap in the proof of Engel's theorem characterizing solvable groups, it was not even absolutely established that algebras of rank zero must be solvable. Killing's Theorem D asserted this, of course, and the proof was based upon Theorem A. In a letter of 8 May 1890, Engel sought Killing's blessing on Umlauf's task: "In the meantime have you found any results about groups of rank zero? If not and if you have nothing against it, I would like to set one of my students on the investigation of these groups; I almost believe that the general form of these groups will be attained."

Killing naturally had no students to assign such tasks. He was teaching at the Lyceum Hosianum, a school dedicated to the training of future clerics, not future mathematicians. There is consequently a slightly wistful tone to his reply (14 May 1890):

> If I had students in mathematics, I would allow many excursions into structure. The simple groups of ranks 4–8 provide easy material for seminar works; and I believe that the structure of all groups that are their own principal subgroup [i.e., $\mathfrak{g} = \mathfrak{g}'$] would provide a suitable subject for a dissertation. Then all that is lacking are the groups of rank zero, the treatment of which is urgently needed and involves no great difficulty.

As this quotation makes clear, Killing certainly underestimated the difficulty of the tasks he would assign to students. Take, for example, the problem of determining the possibilities for the structure of Lie algebras satisfying $\mathfrak{g}' = \mathfrak{g}$. In Part III [1889a:107ff.] Killing had indicated how he thought this problem could be resolved. His approach is intimately related (mathematically and historically) to the theory of the representation of semisimple Lie algebras and, as such, will be discussed in Section 6.3. Here it suffices to point out that Killing's approach involved, as a subproblem, the equivalent of determining all irreducible modules for all simple Lie algebras. A very challenging problem indeed for a doctoral dissertation!

Section 9 of Killing's Part II on algebras of rank zero raised many questions in the minds of Engel and his students, and Engel conveyed some of them in a postcard to Killing. From Killing's reply of 21 June 1890 we learn that F. Schur had also had difficulty with the same material and had entered into a correspondence with him regarding it. Although that correspondence appears lost, there is a letter from Schur to Engel (17/29 January 1890) relating to it: "Hidden away in Killing's works are a goodly number of proofs that are regrettably false in part. I have been in correspondence

[50] Karl Arthur Umlauf (b. 1866) studied mathematics and science at the University of Leipzig from 1887 to 1891. Information on his career after 1891 is difficult to obtain. He apparently published no mathematical papers beyond his dissertation.

with him about this, but one does not learn anything from his letters."[51] Killing believed that Schur and Engel were having difficulty with the material because they were unfamiliar with Weierstrass's theory of elementary divisors and the associated canonical form, and so he expounded it to Engel. But in reply, Engel had to confess that he still could not follow Killing's proof of Theorem A or, more precisely, the part of the proof showing that $n-1$ independent elements commute with the entire algebra. Engel felt that the problems with the proof could be overcome by means of "determinant considerations" and promised to keep working on the matter.[52]

In his reply, Killing admitted that his proofs were too sketchy but claimed that a letter was not the appropriate vehicle for anything more than additional proof sketches. Killing began by considering the case in which the nullity n equals 2. In this case the theorem is correct, and Killing indicated the line of proof in more detail than was usual for him. He admitted that the proof did not extend to the case $n > 2$, and to deal with $n > 2$ he suggested two approaches, one involving "very hideous calculations," while the other involved a theorem that Killing declined to prove in the letter. This letter suggests that probably Killing originally discovered Theorem A in the case $n = 2$ and concluded a bit too hastily that the proof could be pushed through for higher values of n as well.

Engel's reply (20 July 1890) is especially interesting because it involved the ideas that led to what is usually referred to as Engel's theorem: If $\mathfrak{g}$ has rank zero, then (to use a more recent expression) $\mathfrak{g}$ is nilpotent. In his letter Engel informed Killing that he had also (independently) proved Theorem A for the case $n = 2$ and then he sketched the proof because he felt Killing might find it of interest. Recall from the discussion of Theorem A that when the nullity n equals 2, a basis $X_1, \ldots, X_k$ for $\mathfrak{g}$ can be chosen so that $[X_1, X_2] = X_3$, $[X_1, X_3] = X_4$, $\ldots$, $[X_1, X_{k-1}] = X_k$, and $[X_1, X_k] = 0$. Engel sketched a proof that X_k commutes with all of $\mathfrak{g}$, so that X_k is what they called a "distinguished element" of $\mathfrak{g}$. Engel then observed that he could obtain further information on the multiplications $[X_i, X_j]$. That is, he could prove the following:

(5.2) *For all* $i \le j$, $[X_i, X_j] \in \operatorname{span}\{X_{j+1}, \ldots, X_k\}$.

This shows, in modern terms, that the lower central series $\mathfrak{g}, \mathfrak{g}', \mathfrak{g}^2 = [\mathfrak{g}, \mathfrak{g}'], \mathfrak{g}^3 = [\mathfrak{g}, \mathfrak{g}^2] \ldots$ is such that $\mathfrak{g}^j \subset \operatorname{span}\{X_{j+1}, \ldots, X_k\}$, so that $\mathfrak{g}^k = 0$ and $\mathfrak{g}$ is nilpotent.

Engel based his proof on the simple but useful theorem, apparently first observed by him, that the homomorphic image of an algebra of rank zero is itself of rank zero. This was used to proceed inductively as follows. The algebra $\overline{\mathfrak{g}} = \mathfrak{g}/[\operatorname{span}\{X_k\}]$ is the homomorphic image of $\mathfrak{g}$ and so of rank zero. If $X \to \overline{X}$ denotes this homomorphism, then the basis $\overline{X}_1, \ldots, \overline{X}_{k-1}$

[51] Engel Archive Giessen (Paket "Briefe von F. Schur 1888–1932"); transcription in [Hawkins 1982:168].

[52] Letter of 25 June 1890.

for $\overline{\mathfrak{g}}$ is analogous to the basis $X_1, \ldots, X_k$ for $\mathfrak{g}$ (in particular the nullity is still $n = 2$) and so Engel's reasoning for $\mathfrak{g}$ applies to $\overline{\mathfrak{g}}$ and it follows that $\overline{X}_{k-1}$ is distinguished in $\overline{\mathfrak{g}}$, so that $[X_i, X_{k-1}] \in \text{span}\{X_k\}$ for all $i \leq k$. Likewise $\mathfrak{g}/[\text{span}\{X_{k-1}, X_k\}]$ is of rank zero with basis setup as for $\mathfrak{g}$, so that $[X_i, X_{k-2}] \in \text{span}\{X_{k-1}, X_k\}$ for all $i \leq k-2$, and so on. Thus (5.2) follows.

Although the above proof is limited to the case $n = 2$, it is easy to see, assuming that a valid proof has been given for Killing's Theorem D, that every algebra of rank zero possesses a distinguished element by Lie's theorem. Engel had pointed this out to Killing, as noted above. With the existence of a distinguished element the above reasoning carries over to any group of rank zero, the only difference being that the basis in (5.2) is created progressively from the successive distinguished elements. Thus Engel's argument establishes Engel's theorem modulo a valid proof of Theorem D. In his doctoral dissertation [1891:§5], Umlauf presented rigorous proofs that Theorem D holds, that every rank-zero algebra consequently has a distinguished element, and that a basis may be chosen such that (5.2) holds. Although these results are not specifically ascribed to Engel, Engel's above-described letter to Killing shows that the seminal ideas stem from him, so that the designation "Engel's theorem" is not without its justification even if the proof of Theorem D is actually due to Umlauf.[53] Umlauf's entire analysis of groups of rank zero is based upon Engel's theorem in the form (5.2).

Killing's letter to Engel dated 21 July 1890 indicates that he was impressed by Engel's new viewpoint and encouraged to work further on the proof of Theorem A. His latest assault on the proof involved, once again, lengthy computations, and once again, he did no more than sketch the line of reasoning. Because of "the burdensomeness" of his latest approach, he stressed the urgent need for a simpler proof of Theorem A. Engel assured him that Umlauf was investigating the problematic case of nullity $n > 2$: "I do not know whether he can resolve the general case, but in any case I can say that he will be in a position to determine, say, all types up to 10-term groups" (27 August 1890). Umlauf's procedure for determining all algebras of rank zero of small dimension was based on Engel's observation that the homomorphic image of a rank-zero algebra is also of rank zero. That is, suppose all rank-zero algebras of dimension k are known. Then if $\mathfrak{g}$ has dimension $k+1$, it possesses a distinguished element X, and

---

[53] In the preface of his dissertation Umlauf wrote, "The propositions due to Prof. Engel are indicated as such in the text." In the text Umlauf attributed to Engel the theorem that the homomorphic image of a group of rank zero is of rank zero [1891:§5:Satz 1] and suggested [1891:36, n.2] that his proof of the following lemma draws upon similar ideas used by Engel in [1887]: If a k-dimensional algebra of rank zero contains a $(k-1)$-dimensional subalgebra, then it contains a $(k-1)$-dimensional ideal. The lemma is used to prove that if $\mathfrak{g}$ has rank zero, then $\mathfrak{g}$ is solvable (Killing's Theorem D). Cf. Cartan [1893c:401; 1894b:50], who is probably responsible for the introduction of the term "Engel's theorem."

$\overline{\mathfrak{g}} = \mathfrak{g}/[\text{span}\,\{X\}]$ is also of rank zero. Since $\overline{\mathfrak{g}}$ is of dimension k, it is known; and it is not difficult for small values of k to enumerate all the possible ways that the multiplication table of $\mathfrak{g}$ may be obtained from that of $\overline{\mathfrak{g}}$ [Umlauf 1891:46–47].

In the course of carrying out the determination of all groups of rank zero of small dimension Umlauf (or Engel) discovered the existence of counterexamples to Killing's Theorem A. Engel sent off a postcard to Killing with the bad news. The postcard has not been preserved, so that it is not certain which counterexample was first discovered. Umlauf's calculations revealed many. (Umlauf himself singled out only one counterexample to Killing's Theorem A [1891:45, n.1].) The lowest dimension for which they occur is $k = 5$. Killing responded by indicating a general class of such counter examples (24 February 1891). Let $\mathfrak{g}$ be spanned by $X_1, \ldots, X_k$, where k is odd and at least 5. Set $[X_1, X_2] = [X_4, X_5] = [X_6, X_7] = \cdots = [X_{k-1}, X_k] = X_3$ and $[X_i, X_j] = 0$ otherwise. For any $X = \sum_{i=1}^{k} e_i X_i$, $\operatorname{ad} X$ has nullity $n = k - 1$; but $[X, X_i] = 0$ for all i implies that $X = e_3 X_3$, so that there is only one distinguished element and not $n - 1 = k - 2 \geq 3$ as asserted by Theorem A. Killing's theorem could not be more incorrect. He himself put it slightly less dramatically: "The theorem to which you refer [Theorem A] cannot hold in full generality and requires a fundamental restriction. I had entirely forgotten this and first had to reproduce my work to convince myself that I had stated it." Killing's forgetfulness is puzzling in view of the attention he had given to a better proof of Theorem A just six months earlier.

There is no mention in the correspondence of the implications of the discovery of the untenability of Theorem A for Theorems B and C. Perhaps again Killing's memory failed him and he forgot the role of Theorem A in the proofs of Theorems B and C; or perhaps in his weary state he did not wish to face up to the fact that all of his analysis rested upon a foundation of sand. In any case, we do know that he had no intention of reworking the contents of his papers, for in the letter containing the above class of counterexamples he wrote, "Schur has been repeatedly urging me to thoroughly work through the relevant investigations on structure for publication as a book. So far I have not been completely convinced by this thought, since I prefer first to finish other works." Killing had in mind his unfinished research program on the foundations of geometry, his theory of space forms, which had prompted his somewhat tangential foray into the structure of Lie algebras. Perhaps in view of these sentiments, Engel kindly refrained from mentioning the disastrous consequences of the counterexamples for the proofs of Theorems B and C, although the possibility cannot entirely be ruled out that he failed to see these implications. In any case, Cartan was the first to seriously study the problems posed by the questionable status of these theorems and to establish their untenability, while at the same time verifying Killing's results on the classification of simple groups and establishing the viability of Killing's overall approach to structure.

5.6. The Lobachevsky Prize

As we shall see in Section 6.2, in the pages of the thesis of Élie Cartan, Killing's work received the "thorough examination" he had wished for it. In Section 1 and throughout Chapter 4 stress has been laid upon the relevance of the disciplinary ideals fostered by Weierstrass to the shaping of Killing's research program. In view of the strong emphasis upon rigor within the Weierstrass camp, it may seem contradictory that Killing's work was presented in such a nonrigorous form, that the Weierstrassian ideal of rigor was realized far better in Cartan's thesis than in Killing's papers. The contradiction is, however, more apparent than real. Killing did not reject, consciously or unconsciously, the ideal of rigor taught at Berlin. But he found himself unable to live up to it. Although there are analogies between his thesis problem of 1872, handled with complete rigor, and the problem posed to him by his theory of space forms, the latter problem proved to involve difficulties of a much higher order of magnitude.

Guided by Weierstrass's critique of generic reasoning, Killing had studied the case in which k, the dimension of $\mathfrak{g}_0$, is greater than one and consequently encountered the integers $a_{\alpha\beta}$ and their marvelous properties. Especially after his contact with Lie and Engel, Killing sensed the profundity and importance of the theory being brought to light by his approach. But he soon found himself incapable of securing the envisioned theory. Undoubtedly, there were many reasons for this inability. He may have lacked sufficient mathematical powers to deal successfully with such a formidable problem. Certainly, he lacked sufficient time and peace of mind because of his heavy teaching and administrative duties and an unsettling illness in his family. Notwithstanding his impetuosity, Killing was deeply affected by his experiences at Berlin. They instilled in him the *conscience*, if not the temperament, of a Weierstrassian. He was candid about the weaknesses of his work. It grieved him to find it necessary to publish his work in the tentative form it eventually assumed. The quotations above, particularly those in Section 4, testify to this point abundantly. Killing emphasized the necessity of a "thorough examination," but with the encouragement of Klein, Lie, and Engel, he wisely decided to publish his imperfect work so that others might provide it.

With his paper [1892] Killing returned to the foundations of geometry. The same year he became professor of mathematics at the University of Münster, where he remained for the rest of his career, working primarily on foundations of geometry and on pedagogical matters. The structure of Lie algebras had ceased to be at the center of his scholarly interests.[54] In

[54] In 1890 Killing published two papers [1890a; 1890c] that were related to his investigations on structure. After 1890, he published two further papers on the structure of Lie algebras: a *Programmschrift* [1895] on groups of rank zero which develops his objections, made in correspondence with Engel, to Umlauf's mode of classifying these groups, and a brief paper [1904] proposing a classification

effect, the doctoral thesis of Élie Cartan [1894b] became the book that Killing would not, and probably could not, write.

In 1897 Sophus Lie became the first recipient of the Lobachevsky prize of the Physico-Mathematical Society of Kazan. The prize was awarded for the best recent book on a geometrical subject and especially for books pertaining to non-Euclidean geometry. Upon the strong recommendation of Felix Klein [1898], the prize went to Lie for the third volume of *Theorie der Transformationsgruppen* [1893a], which contains Lie's treatment of the Helmholtz space problem, the problem that had helped inspire Killing's own theory of space forms. By virtue of the second volume of his book, *Introduction to the Foundations of Geometry* [1898], Killing was one of four candidates under consideration for the second awarding of the prize, in 1900.[55] His book was the culmination of his efforts to carry out his research program on the foundations of geometry, the program that had led him into the realm of the structure of Lie algebras. Another candidate was A. N. Whitehead by virtue of his book *Universal Algebra* [1898b] and an essay "The geodesic geometry of surfaces in non-Euclidean space" [1898a].

Whitehead's work was recommended exuberantly by R. S. Ball, while Engel recommended Killing's book for the prize.[56] Engel actually shared Lie's low opinion of Killing's contributions to the foundations of geometry. For example, when Killing's paper "On the Foundations of Geometry" [1892] appeared, Engel wrote to his friend Study: "Killing's latest opus on the foundations of geometry contains real nonsense."[57] Engel realized nonetheless that, especially in Killing's own mind, his work on the structure of Lie algebras was inextricably intertwined with his research program on the foundations of geometry; and Engel had the highest regard for Killing's contributions to the theory of structure. For this reason, Engel wrote a strong recommendation of Killing's book, even though it does not contain a development of his theory of structure. And when the panel leaned toward Whitehead, Engel disputed the accuracy of Ball's report. The panel finally decided that both authors were equally deserving of the prize and drew lots to determine the recipient. Fate then awarded the prize to Killing, who wrote to Engel (14 June 1901):

> Such recognition of my work, whose defects I myself know only too well, had till now seemed impossible. The wish of my youth, that my life should not be entirely unfruitful for mathematics, can now indeed be considered as fulfilled.

---

for a special type of group of rank zero, namely those with a basis $X_1, \ldots, X_k$ satisfying $[X_1, X_i] = X_{i+1}, [X_1, X_k] = 0$.

[55] For references and some additional details, see [Sommerville 1911:180].

[56] The following account is based upon Engel's letter to Killing dated 4 August 1901.

[57] 1 March 1892 (Engel Archive Giessen, Paket "Engel an Study"); see [Hawkins 1982:186].

Figure 14. Killing as Rector magnificus (1897–1898).

Chapter 6

The Doctoral Thesis of Élie Cartan

Élie Cartan (1869–1951) was born in Dolomieu, a small village in southwestern France.[1] His ancestors were peasants. His father, Joseph, was the village blacksmith. According to Cartan, the course of his life was changed by a visit to the village by the representative of the Département Isère, to which Dolomieu belongs. During that visit he recommended that Cartan, whose intellectual abilities had already impressed his teachers, try out for a scholarship to a secondary school (*lycée*). Tutored by one of his teachers, Cartan did well on the scholarship examinations and received a full scholarship to the Collège de Vienne. He was ten years old at the time. The next ten years were spent in schools far from his home. These included the Lycée de Grenoble, where he spent two years and received first prize in mathematics in a nationwide competition among secondary schools. By 1887, when he entered the distinguished Lycée Janson-de-Sailly, in Paris, his prospects for a successful career in mathematics were thus already good. In November 1888 he entered the École Normale Supérieure, in Paris.

Although initially established as a school for the training of teachers and overshadowed by the École Polytechnique as the school for bright aspiring mathematicians, the École Normale had become by 1888 the elite school for the training of mathematicians, as well as scientists and intellectuals generally. Symptomatic of the changing fortunes of the two schools was the highly publicized decision in 1861 by Gaston Darboux (1842–1917), at the time a promising potential mathematician who had placed first on

[1] For biographical information about Cartan I have drawn upon [Akivis and Rosenfeld 1993], in particular the biography of Cartan given in Chapter 1 and Cartan's own recollections, which are reprinted in Appendix C.

Figure 15. Élie Cartan in a photo taken in 1903.

the entrance examinations for both schools, to attend the École Normale. In choosing the École Normale, Cartan had made the best choice for a career in mathematical research. He was graduated first in his class.

6.1. Lie and the Mathematicians of Paris

According to Cartan, in Paris (at the École Normale or at the Sorbonne) he attended lectures of, and was impressed by, mathematicians such as Charles Hermite (1822–1901), Jules Tannery (1848–1910), Darboux, Émile Picard (1856–1941), Henri Poincaré (1854–1912), and Édouard Goursat (1858–1936). His highest esteem was reserved for Poincaré, "a mathematical giant ... whose lectures were flying over our heads."[2] Before proceeding to discuss

[2] I have followed the translation of [Akivis and Rosenfeld 1993:5]; the original text is reproduced by them on p. 276.

the contributions of Cartan, it is enlightening to consider the reception of Lie's theory of groups in Paris and some related attitudes of the leaders of the Paris mathematical community (Poincaré, Picard, Darboux). The latter shed light not only on the reception accorded Lie's theory but also on the course of Cartan's own work as discussed here and in Chapter 8.

The symbiotic relationship that developed between Lie and the mathematicians of Paris can be traced back to two fateful journeys Lie made to Paris in the years 1870 and 1882. His first visit, in the company of Klein, has been recounted in Section 1.4. It was also during that 1870 visit that Lie made the acquaintance of Gaston Darboux. Darboux was roughly Lie's age and both were just commencing their careers as mathematicians. Lie perceived in Darboux a kindred geometrical spirit. This sense of intellectual affinity on Lie's part is conveyed in a letter to Klein written from Paris (before Klein's arrival), where he wrote, "I must speak with Darboux because he understands me best of all."[3] After 1870 they continued to maintain contact through a correspondence that covered matters of common geometrical interest.[4]

In 1881 Darboux succeeded Chasles as professor of higher geometry at the Sorbonne. Chasles had used his professorship to expound projective geometry in a manner that required no mathematical analysis, whereas Darboux, in the words of his student Paul Appell, "developed instruction along a new path, where Bonnet had already been active, the path of general geometry considered as an application of analysis ...; it is in this chair that he founded this brilliant school of geometry, the disciples of which are now dispersed in all countries ...."[5] Darboux's lectures on geometry were regularly attended by the students at the École Normale (where Darboux had been *maître de conférences* the decade preceding 1881), and it was from this source that he drew his disciples.

In developing his lectures at the Sorbonne, Darboux had in mind especially the relation between differential geometry and partial differential equations [1887–96: V. I, p. v]. Initially, he had expected to wrap up the project within a year, but the interest he generated was so great that the project took on far greater proportions and eventually resulted in the publication of his four-volume *Leçons sur la théorie générale des surfaces et les applications géométriques* [Darboux 1887–96]. Among those involved in the project were Goursat, Gabriel Koenigs, and Louis Raffy, all of whom were *maîtres* at the Ecole Normale for many years, including those while Cartan was in attendance. Darboux's lectures, together with his own publications on partial differential equations, also generated a renewed interest in this

[3] "Ich muss mit Darboux sprechen, denn er versteht mich am besten" (NSUB Göttingen, Nachlass Klein X, Nr. 685).

[4] Darboux's letters to Lie and copies of Lie's letters to Darboux are located in UB Oslo, Brevsamling Nr. 289 and 52, respectively. The originals of Lie's letters are in the Bibliothèque de l'Institut de France.

[5] Quoted on p. 457 of [Darboux 1912].

Figure 16. By Cartan's time, Gaston Darboux had become one of the leaders of the Paris community of mathematicians.

area in Paris [Picard 1897b:959]. Within the Darboux school, Lie was highly regarded. Although Darboux never digested the theory of transformation groups, he studied Lie's publications on geometry and partial differential equations and referred to them extensively and favorably in his lectures.[6] Goursat did the same in his lectures on partial differential equations [1891; 1896–98] and even devoted space to a brief account of transformation groups [1891:Ch. 12].

Perhaps Darboux considered himself too old or too established in his own research program to attempt to utilize Lie's theory of groups, but the same was not true of the most promising mathematicians of the next gen-

6 According to the index to all four volumes of Darboux's *Leçons* [1887–96:v.4, 535], Lie's work is referred to in a total of 56 sections or appendixes. Many references are accompanied by complimentary remarks. See, e.g., pp. 232, 340, 348, 376, 405 of Volume 1 (1887), pp. 61 and 218 of Volume 2 (1889) and pp. 358, 381, and 423 of Volume 3 (1894).

eration, Poincaré and Picard. Shortly after completing his doctoral thesis in 1878, Poincaré's involvement with the creation of his theory of Fuchsian functions led him to an appreciation of group-theoretic notions (discontinuous groups of linear fractional transformations).[7] Picard, two years younger than Poincaré, quickly became interested in similar group-theoretic matters [1883a]. Thus when Lie returned to Paris in the fall of 1882 he found in Poincaré and Picard two young, gifted mathematicians who already appreciated group-theoretic ideas. Indeed, Poincaré informed Lie offhandedly that all mathematics was a tale about groups.[8] According to Lie, it was Poincaré and Picard who were quick to grasp the ideas of his theory of groups as he presented them in Paris.

The seeds sown by Lie's 1882 visit to Paris quickly bore fruit. The following year Picard wrote to Lie, expressing his admiration of Lie's *Annalen* paper [1880] and his hope of taking it as a starting point for a memoir on linear differential equations.[9] Picard had in mind a Galois type theory for linear ordinary differential equations—the initial version of what is now called the Picard–Vessiot theory—which he announced in [1883a] and later developed more fully in [1887b]. The following year, stimulated by some remarks by Sylvester on nonions, Poincaré suggested how Lie's theory of groups could be used to classify hypercomplex number systems.[10] Poincaré also indicated in [1886] how group-theoretic ideas in the sense of Lie could be used to re-prove a theorem on the birational transformation of algebraic surfaces obtained by Picard. Picard then went on to write a prize-winning essay on birational transformations of algebraic surfaces that employed Lie's ideas in a substantial way [1889:Ch.3]. "I hope," Picard wrote to Lie concerning his prize-winning essay, "that you will soon see with pleasure the use I have made of your work .... The theory of functions will again certainly profit from your researches in many other ways in the future."[11] In the meantime, Poincaré [1887] had applied Lie-algebraic considerations to obtain a solution to Helmholtz's space problem for two dimensions. Then Picard showed in [1892b] how Lie's theory could be used to explore the question of whether for functions of more than two real variables there existed equations that would play the role of the Cauchy–Riemann equations. As Picard explained to Lie, "It would be a great triumph for your theory if it would lead to equations by means of which one could extend, more or

---

[7] See [Gray 1986] for details.

[8] Writing to Klein from Paris in October 1882, Lie said: "Poincaré sagte gelegentlich, dass alle Mathematik eine Gruppengeschichte war. Ich erzählte ihm dann über dein Programm, das er nicht kannte" (NSUB Göttingen, Nachlass Klein X, Nr. 685).

[9] Letter dated 13 March 1883 (UB Oslo, Brevsamling Nr. 289).

[10] This work and its influence are studied in [Hawkins 1972].

[11] "J'espère que vous verrez bientôt avec plaisir de quelle utilité m'ont été vos travaux .... La théorie des fonctions aura certainment bien d'autres profits à tirer encore ultérieurement de vos recherches." Letter of 6 January 1888 (UB Oslo, Brevsamling Nr. 289).

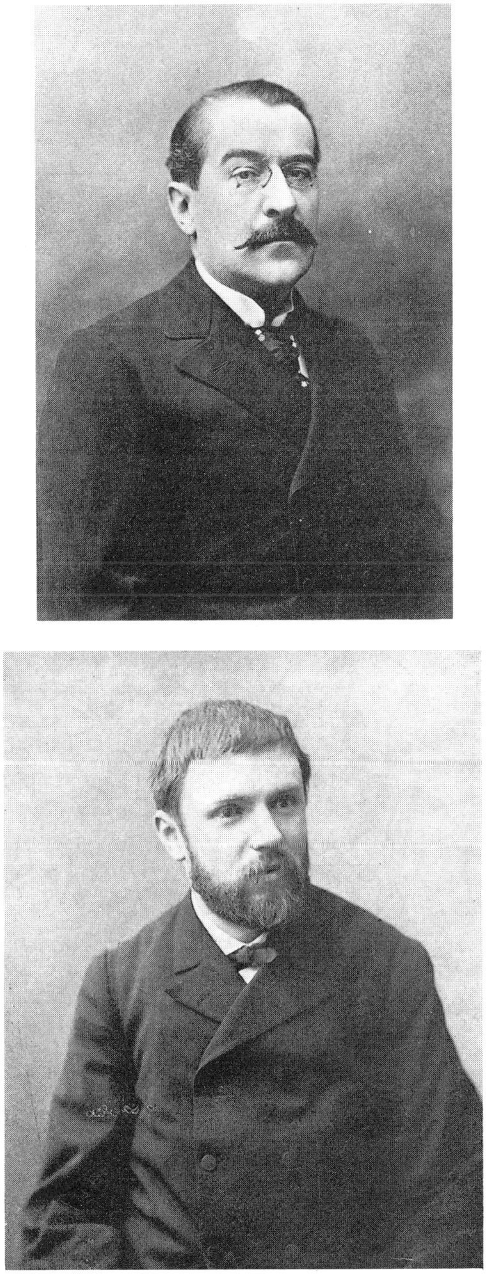

Figure 17. Two other leaders of the Paris community, Picard (top) and Poincaré (bottom), shared Darboux's high regard for Lie's theory of groups, but, unlike Darboux, they themselves sought to apply it.

less, the principles of the theory of functions."[12]

The diverse applications Poincaré and Picard sought for Lie's theory of groups was a token of their regard for it. Thus writing to Lie in 1888, Picard wrote, "You have created a theory of major importance that will be counted among the most remarkable mathematical works of the second half of this century."[13] By 1889 Picard had joined Darboux and Poincaré as members of the Académie des Sciences, and before long they had an opportunity to display their esteem for Lie in a more official manner. The death of Kronecker in 1892 opened up a position as corresponding member in the Académie, and they threw their support behind Lie as the candidate to replace him.[14] Their efforts were successful, as Lie won easily, garnering thirty votes to two for Cremona and one for H.A. Schwarz.[15] By far the greatest show of respect for Lie came earlier, however, when in 1888 Darboux, Picard, and Poincaré supported the plan to encourage graduates of the École Normale to spend a year seriously studying Lie's theory of groups, preferably in Leipzig, in order to further cultivate its applications. By 1893 sufficiently many *normaliens* had taken the advice of the leaders of the Paris community of mathematicians that Picard could write to Lie, "Paris is becoming a center for groups; it is all fermenting in young minds, and one will have an excellent wine after the liquors have settled a bit."[16]

The extraordinary manner in which Lie's theory of groups was welcomed, praised, and developed in Paris was not matched elsewhere. In Berlin, for example, Lie and his theory were not highly regarded. As a document written in 1902 by Frobenius indicates, Weierstrass felt—with a good deal of justification—that Lie's theory was so unrigorously developed that it would have to be redone from the ground up [Biermann 1973:215]. Among Berliners Lie's reputation was also tarnished by remarks made in communication with Weierstrass over the publication of Abel's collected works, which revealed Lie's lack of understanding of the theory of Abelian integrals, a subject cultivated with great care by Weierstrass. In the later Berlin school of Weierstrass's students Schwarz and Frobenius, where the theory of finite groups was cultivated by Frobenius as a branch of the theory

[12] "Ce serait un bien grand triomphe pour votre théorie si on pouvait être ainsi conduit à des équations auxquelles on pourrait étendre plus ou moins les principes de la théorie des fonctions." Letter of 20 June 1892 (UB Oslo, Brevsamling Nr. 289).

[13] "Vous avez créé là une théorie d'une importance capitale, et qui comptera comme une des oeuvres mathématiques les plus remarquables de la seconde moitié de ce siècle ...." Letter dated 25 June 1888 (UB Oslo, Brevsamling Nr. 289).

[14] Poincaré wrote the report recommending Lie, which, according to Picard, "fait admirablement ressortir l'espirit qui a guidé vos recherches et la place énorme que la théorie des groupes et des invariantes tient dans la science mathématique actuelle." Letter to Lie of 20 June 1892 (UB Oslo, Brevsamling Nr. 289).

[15] See *Comptes Rendus* **114** (1892):1329.

[16] "Violà Paris devenir un centre de groupes; tout cela fermente dans ces jeunes cerveaux, et on aura un excellent vin quand les liqueurs seront un peu reposées." Letter dated May 1893 (UB Oslo, Brevsamling, Nr. 289).

of numbers, Lie's theory of groups was actually denigrated. To Frobenius it resembled a wasteland, a useless "theory of methods" (*Methodenlehre*) that did nothing more than show how problems (differential equations) solved by Euler and Lagrange in natural ways could be solved much more circuitously by means of Lie's methods. Thus Lie's theory of groups was criticized on two grounds: In the first place, the foundations of the theory were inadequately developed, and secondly, the theory, even if made rigorous, still did not provide simple or natural solutions to problems.

The same sort of dual criticism is manifest in a paper by Johannes Knoblauch [1906], an assistant professor at the University of Berlin. In [1884a] Lie had briefly indicated how to use his theory of groups to determine some well-known invariants associated with a surface $ds^2 = Edx^2 + 2Fdxdy + Gdy^2$, such as its Gaussian curvature [1884a:§23]. Later, at Leipzig, he encouraged one of his students, Kasimir Zorawski, to use his method to determine the number of functionally independent invariants of any order n for a surface. This Zorawski did, and the results were published in [1892]. Knoblauch solved the same problem after first offering some scathing criticisms of the approach via Lie's theory of groups [1906:247]. First Knoblauch criticized the foundations of the theory by questioning the legitimacy of using a method based on consideration of the group of infinitesimal transformations, presumably since in determining the equations of the infinitesimal transformation of x, y, E, F, G induced by the infinitesimal transformation $\delta x = f(x, y)\delta t$, $\delta y = g(x, y)\delta t$, the higher order infinitesimals δt^2, $\delta t^3, \ldots$ are ignored: "It is doubtful that the method of infinitesimal transformations ... is suitable, since its application, at the first step, destroys the system of equations, which is the basis for the true subject of the investigation."

To this criticism Knoblauch immediately added a second, which, as with Frobenius, questioned the simplicity of the solution afforded by the Lie group approach: "At least, in view of this circumstance, it should be considered as justified only if the objective is achieved in an especially simple way. But the opposite is the case; from the highly stratified formal apparatus that Lie's theory brings with it, one can merely obtain the motivation to solve the interesting and important problem in another way." As one of Lie's students, Gerhard Kowalewski, observed in a letter to Engel, with the publication of Knoblauch's paper "the Berliners" had made public their distaste for Lie's theory.[17] Although the Berlin attitude towards Lie

[17] Letter dated 25 October 1906 (Engel Archive Giessen). The relevant passage reads as follows: "Haben Sie den Knoblauchschen Aufsatz in Crelles Journal gelesen? Nun haben die Berliner öffentlich ihrer Abneigung gegen die Lieschen Theorien Ausdruck gegeben." Lie had an equally great distaste for Weierstrassian style mathematics. In his autobiography Kowalewski related the following anecdote, which captures Lie's attitude very well [1950:52]. Toward the end of his life someone asked Lie why the foundational chapters of *Theorie der Transformationsgruppen* were written in such a "function-theoretic" manner. Lie answered laughingly: "I wanted to show the Berliners that I, too, could

was an extreme one, nowhere was his work acclaimed and developed on such a large scale as it was in Paris. The extraordinary reception there is related to certain attitudes shared by the leaders of the mathematical community, especially Picard and Poincaré. These attitudes are worth describing, because they illuminate the special sense in which Lie's theory of groups was appreciated and cultivated in Paris and thereby provide an instructive perspective from which to view Cartan's work and its tendencies.

The popular writings of Picard and Poincaré reveal a common and recurrent concern that mathematics in its continuing development should maintain a proper balance between the "artistic" tendency within mathematics, i.e., to develop mathematics for its own sake, and the "utilitarian" tendency, i.e., to develop a mathematical theory in relation to its applications to science and engineering, or simply to other branches of mathematics.[18] Both tendencies were defended as legitimate, but each was regarded as dangerous when emphasized to the exclusion of the other. For Picard, Fourier epitomized the utilitarian tendency in mathematics, whereas Galois's highly abstract ideas (to use Picard's characterization) epitomized the artistic tendency [1905:28]. Picard clearly preferred the example of Cauchy, whom he saw as combining the two tendencies [1917:4]. In view of the need to keep the utilitarian role of mathematics in sight, mathematical analysis, as opposed to arithmetic and algebra, was obviously a safer investment of research time. Differential equations in particular afforded one of the safest areas of analysis. As Picard wrote, "There is no more interesting question for applications than the study of differential equations" [1890: 705], and "the whole history of science demonstrates the relations that unite pure analysis and natural phenomena—this solidarity is expressed mathematically when the study of phenomena reduces to the study of a differential equation" [1905:79].[19]

The link between analysis and the external word was stated somewhat differently and more eloquently by Poincaré at the first International Congress of Mathematicians (Zürich, 1897):

> The sole natural object of mathematical thought is the whole number. It is the external world that has imposed the continuous upon us .... Without it there would be no infinitesimal analysis; all mathematical science would reduce itself to arithmetic or the theory of substitutions [in the sense of Jordan's *Traité*].
>
> On the contrary, we have devoted almost all our time and all

write tediously." Kowalewski went on to point out that the function-theoretic touch was due to Engel, who actually wrote the book under Lie's general supervision. As a result, the book remained foreign matter to Lie, who always referred to his papers, rather than the book, in discussing his work with others.

[18] See, e.g., Picard [1897a; 1905:28–9, 82–3; 1915; 1917:3–4] and Poincaré [1889:130–1; 1898; 1908a:36].

[19] See also Picard [1905:195; 1917:24]. Poincaré expressed a similar view, e.g., in [1908a].

> our strength to the study of the continuous. Who will regret it; who will think that this time and this strength have been wasted? Analysis unfolds before us infinite perspectives that arithmetic never suspects; it shows us at a glance a majestic assemblage whose array is simple and symmetric; on the contrary, in the theory of numbers, where reigns the unforeseen, the view is, so to speak, blocked at every step [1898:87].

This passage reads as a paean to analysis—at the expense of arithmetic (the theory of numbers and finite groups). As it implies, according to Poincaré analysis enjoyed a twofold advantage over arithmetic. In the first place, it enjoyed the benefits that derived from forming the link with the external world: a rich source of mathematical ideas and directions for research and a means of fulfilling the utilitarian role of mathematics. In the second place, analysis provided a greater abundance of satisfying, readily obtainable general results, where in arithmetic "every whole number has its own individuality; each of them is a sort of exception, and that is why general theorems will always be less common in the theory of numbers and also why those that do exist will be more hidden and will longer escape detection" [1908a:40].

The emphasis placed upon analysis vis-à-vis arithmetic by Picard and Poincaré does not contradict the fact that they both did some research on arithmetical problems. Such research was not a major concern for either of them, and the same was true of the research activity in Paris during the last decades of the nineteenth century. Mathematical physics, complex function theory, differential geometry, and, uniting them all, differential equations, were the dominant areas of research. For example, of the 127 doctoral dissertations done at the Sorbonne during 1871–1900, 27 (or 21.3%) were on mathematical physics (including mathematical astronomy), 24 (18.9%) were in complex function theory, 26 (20.5%) were in geometry (mostly differential geometry), and 32 (25.2%) were on differential equations. By contrast, only two dissertations (1.6%) were written on the theory of finite groups, with another two on the theory of numbers.[20]

In Paris Galois and his interpreter, Jordan, were treated with respect. Their work had exerted a great influence outside of France. By the close of the nineteenth century the theory of finite groups and Galois theory were being vigorously investigated in Germany, Great Britain, and the United States. This generated a certain degree of national pride, but nonetheless there was a tendency in Paris to see Jordan's work on finite group theory as too abstract, too much an expression of the artistic tendency, and not an attractive source of further research. Hermite, the revered patriarch of the Paris community in the closing decades of the nineteenth century, never studied Jordan's group-theoretic work, despite the fact that he had done some research related to Galois theory as a young man. Hermite de-

[20] Data taken from [Estanave 1901; 1903].

clared Jordan's work "too difficult" and "too abstract" [Lebesgue 1923:li]. Picard (Hermite's son-in-law) expressed similar opinions about the theory of numbers and finite groups [Picard 1897b:960]. Incidentally, Hermite was professor of higher algebra at the Sorbonne until his retirement in 1897, but he chose to lecture on complex function theory, and his lectures were a great inspiration to students.

Although Jordan had indicated the relevance of the theory of finite groups to the solution of certain linear ordinary differential equations, this apparently did not appear of sufficient interest to warrant extensive cultivation of the theory of finite groups per se. Thus when Fuchs called attention to those differential equations with the property that all solutions are algebraic, Jordan pointed out that this occurs precisely when the group is finite.[21] Jordan's subsequent work on this matter concerned the purely group-theoretic problem of determining the finite subgroups of $\mathbf{GL}(n, \mathbb{C})$.[22] Poincaré, by contrast, considered the more analytical side of the matter. Given a finite subgroup of $\mathbf{GL}(n, \mathbb{C})$, "it is then necessary to form the corresponding differential equations. M. Jordan does not stress this point very much"[1881b:97]. (See also [Poincaré 1883b].) Poincaré returned to these matters two decades later and became more deeply involved with its group-theoretic aspects when he discovered the relevance of Frobenius's new theory of group characters and representations. (See in this connection the remarks in Section 8.1.) In his *Traité des substitutions* Jordan had also indicated how finite groups occur in geometrical matters, but probably finite groups did not seem congenial to the infinitesimal geometry cultivated by Darboux and his students. (On the relevance of Lie groups to Darboux-type geometry, see the discussion of Emile Cotton in Section 8.3)

The concern Picard and Poincaré felt to maintain a balance between the artistic and utilitarian tendencies in mathematics relates to their attitude towards general mathematical theories. In his introduction to the collected works of Galois, Picard wrote, "General theories especially need to be illustrated by particular applications. It is easy to specify many domains in modern mathematics where it would, I regret to say, be difficult to do this. Many theories are confined in their too great generality. From the artistic point of view ... nothing is more satisfying than to develop a great theory. Nevertheless ... many regret this tendency, which has perhaps its dangers" [1897a:vii–viii]. The danger was the absence of applications [1917:19].[23] Lie's theory of groups was a general theory, but it was embraced in Paris because of its manifold applications, many of which had

---

[21] Jordan, *Oeuvres* **2**, 1–3.
[22] Jordan, *Oeuvres* **2**, 1–139, 177–217.
[23] In [1915], a hysterical polemic written during the trauma of the First World War, Picard saw all the tendencies he disliked personified by German mathematicians: "une tendance ... de poser *a priori* des notions et des concepts et de suivre indéfinement les consequences, sans se soucier de leur accord avec le réel aucun souci de distinguer ce qui pourra utile au développement ultérieure de la science mathématique" (pp. 167–8).

been discovered by Poincaré and Picard. But it was always the concrete side of Lie's theory that was appreciated and emphasized. For example, when Picard received the doctoral dissertation on transformation groups of Arthur Tresse (1869–1958), Cartan's good friend and one of the *normaliens* who studied with Lie in Leipzig, he wrote to Lie, "I have in my hands Tresse's thesis. I must tell you that reading it alarmed me. These generalities, *far too general*, and with little in the way of applications, are a bit tiring ...."[24]

Although Poincaré was probably more open-minded on the matter of general theories in the sense that he was more far-sighted in his appreciation of their eventual potential for applications, he shared Picard's general sentiments. This is seen clearly in the famous essay [1895] in which he laid the foundations for a new general theory—*analysis situs* as he called it, or algebraic topology as it would now be called. In the introduction to this essay, an essay that surely served as a paradigm to aspiring young French mathematicians, Poincaré took great pains to justify creating this new theory: "It is not enough that a science be legitimate: its utility must be incontestable. So many diverse objects solicit our attention that the most important alone have the right to obtain it" (p. 2). As the quotation makes clear, the importance of a theory is proportional to its utility. Poincaré went on to list diverse applications that could be made of *analysis situs*—applications to complex function theory and the classification of surfaces, to the study of the differential equations of celestial mechanics and to questions in finite group theory related to linear differential equations with the property that all integrals are algebraic. Incidentally, Poincaré also esteemed Darboux's research program in differential geometry because of its relevance to rational mechanics, a subject that Darboux had also cultivated at the Sorbonne.[25]

Lie's theory of groups was thus welcomed in Paris because it was a theory with diverse applications to geometry and mathematical analysis. Whereas Frobenius dismissed Lie's theory as an empty "doctrine of methods" (*Methodenlehre*), a misguided attempt to take the beautiful theory of finite groups—Frobenius grew up in the Berlin school and felt no concern over dangers of artistic tendencies—and extend it mindlessly to continuous ones, Picard shied away from finite group theory as too abstract while lauding Lie's theory for providing for differential equations "an incomparable element of classification" making possible "a vast synthesis by giving a common origin to scattered notions seemingly without connections" [1900:232]. This idea of group theory as a powerful synthesizing agent was developed on a broader philosophical scale by Poincaré in his lecture on the future of mathematics given at the International Congress of Mathematicians in

[24] "J'ai entre les mains la thèse de Tresse. Je dois vous avouer que sa lecture m'effraie. Ces généralités, *par trop générales*, et avec peu d'applications me fatiguent un peu ...." Letter of 2 June 1893 (UB Oslo, Brevsamling Nr. 289).

[25] See Poincaré's remarks in [Darboux 1912:453]. On the matter of Darboux's contribution to the development of a course in rational mechanics, see the remarks of Appell, [Darboux 1912:456].

Rome in 1908.

The central theme of Poincaré's lecture was developed in response to the question, How shall we proceed in the future development of mathematics? In all the conceivable directions that the artistic tendency in mathematics may suggest? Poincaré's answer was firmly negative. If the artistic tendency were given free reign, mathematics, Poincaré opined, would become too vast, too diffuse to comprehend. Mathematicians must follow the lead of physicists and seek to introduce unexpected order into the chaos of mathematical facts, to link apparently diverse mathematical objects. Mathematics, according to Poincaré's vision, was "the art of giving the same name to different things. It is enough that these things, though differing in matter, should be similar in form"[1908a:34]. Poincaré went on to stress in this connection the importance of the concept of a group. The underlying "form" is frequently a group structure. Group theory, not in itself but through its manifold applications, would, in Poincaré's view, play an important role in the future progress of mathematics. Lie's theory of continuous groups was particularly important in this respect because it worked its magic upon the domain of continuous magnitude, where the overwhelming majority of research activity would, and in Poincaré's opinion should, occur.

The emphasis upon applications of transformation groups in Paris also had its effect upon Lie. Although from the outset his work on groups was done in the conviction that it was a fundamental tool for the study of geometry and, especially, differential equations, prior to his second visit to Paris, in 1882, he had channeled most of his energy into building up the pure theory. The trip to Paris helped to convince him to shift the emphasis more toward applications. Thus not long after the visit to Paris we find Lie writing to A. Mayer in Leipzig, "It seems to me that I must first show that the theory of transformation groups is good for something before I dare to continue my publications on the extensive domains of this theory."[26] By the time of his move to Leipzig in 1886 Lie was according more attention to applications, especially in the realm of differential equations as exemplified by his long memoir [1885b] (discussed in Section 3.4), which he had already sketched out by the spring of 1883. Sounding very much like Picard, Lie complained to Vessiot in 1891 that, "Here in Germany a number of mathematicians are working on transformation groups. Killing, Schur, Engel, Study, Scheffers, Maurer. Unfortunately, for the most part they pursue entirely abstract investigations and not any applications ...."[27]

[26] "Ich muss, scheint es mir, zuerst zeigen, dass die Theorie der Transformationsgruppen etwas leistet, ehe ich wagen kann, meine Publicationen auf diesem Theorie ausgedehnten Gebiete weiter fortzusetzen." Letter written ca. 1882–3 (NSUB Göttingen, Nachlass Klein X, Nr. 686).

[27] "Hier in Deutschland arbeiten ja ein Anzahl Mathematiker in Tr[ansformationsgruppen]. Killing, Schur, Engel, Study, Scheffers, Maurer. Leider fahren sie meistentheils fort ganz abstrakte Untersuchungen und keine Anwendungen ...." Letter of 8 October 1891 (UB Oslo, Brevsamling Nr. 52).

In the case of Killing, Lie hastened to add that Killing's were nevertheless "of great interest." The reason was that although they did indeed deal with highly abstract questions of structure and although they were undertaken in the spirit of the "artistic tendency" in mathematics that Killing had absorbed in Berlin, they had a direct bearing upon Lie's envisioned Galois theory of differential equations because they provided a complete classification of all simple (finite-dimensional) structures. As we shall see in the following sections, Cartan undertook as his thesis problem to perfect the abstract theory of structure adumbrated by Killing. From the viewpoint of Paris this was a worthwhile undertaking in view of the perceived relevance to the application of Lie's theory to differential equations. Cartan's dissertation was consequently far more abstract and general than the typical memoir involving transformation groups that came out of Paris. The analogy with Jordan is striking. Jordan had clarified and developed the abstract ideas of Galois, and Cartan did precisely the same relative to Killing's ideas. Cartan, like Jordan, clearly had a penchant for abstract, structural reasoning. It is no wonder that when the Bourbaki school began to emerge in Paris, the elderly Cartan was hailed as a kindred spirit [Chern and Chevalley 1952:217–8].

However, as we shall see here and in Chapter 8, Cartan was guided to a far greater degree in his work on Lie's theory by applications, and, more generally, by the need for a balance between the two tendencies within mathematics to which Picard and Poincaré referred. Indeed, an echo of their sentiments can be seen in the concluding remarks of a talk Cartan gave in Belgrade, Yugoslavia, on "The Influence of France in the Development of Mathematics." The talk was given in February 1940 and hence in the midst of the Second World War, which may account for its nationalistic emphasis.[28] Nonetheless, there is no reason to doubt the heartfelt sincerity of Cartan's concluding remarks, which conjure up those of Picard and Poincaré:

> More than any other science, mathematics develops through a sequence of consecutive abstractions. A desire to avoid mistakes forces mathematicians to find and isolate the essence of problems and the entities considered. Carried to an extreme, this procedure justifies the well-known joke that a mathematician is a scientist who neither knows what he is talking about [n]or whether whatever he is indeed talking about exists or not. French mathematicians, however, never enjoyed distancing themselves from reality; they do know that, although needed, logic is by no means crucial. In mathematical activity, like any other human activity, one should find a balance of values: There is no doubt that it is important to think correctly, but it is even more important

[28] The talk was translated from French into Serbian and published in 1940 and again in 1941. An English translation of the Serbian version, which I have used, is given as Appendix D of [Akivis and Rosenfeld 1993].

to formulate the right problems. In that respect, one can freely say that French mathematicians not only always knew what they were talking about, but also had the right intuition to select the most fundamental problems, those whose solutions produced the strongest influence on the overall development of science.

6.2. Cartan's Theory of Semisimple Algebras

On 14 April 1888 Gaston Darboux wrote a letter to Sophus Lie introducing two graduates of the École Normale Supérieure, Wladimir de Tannenberg and Ernest Vessiot, who were coming to Leipzig for the summer semester.[29] They thus became the first students from the École Normale to pursue graduate study on transformation groups with Lie in Leipzig. A year after Darboux composed his letter of introduction, Jules Tannery, the assistant director of scientific studies at the École Normale, wrote to Lie thanking him for his kind treatment of Tannenberg and Vessiot. The "lively emotion" with which they spoke of their experiences in Leipzig and of their respect for Lie prompted Tannery to hint that other *normaliens* would follow their example. Much to the delight of Lie, Tannery's prediction proved correct. As a consequence, Lie dedicated the third volume of his *Theorie der Transformationsgruppen* [1893a] to the École Normale, with special thanks to Darboux, Picard, and Tannery.

One of the successors of de Tannenberg and Vessiot was Arthur Tresse, who was introduced to Lie by Tannery in a letter of 17 August 1891. Tresse was a good friend of Élie Cartan. They had been graduated together from the École Normale in 1891, and while Cartan spent a year doing military service, Tresse went to Leipzig to study transformation groups. Tresse thus arrived in Leipzig not long after Engel and his students had begun working through Killing's papers and had made the unsettling discovery regarding the falsity of Theorem A and, presumably, about the consequent untenability of the proofs of Theorems B and C as well (Section 5.5). While in Leipzig, Tresse developed close ties with Engel, with whom he spent much time.[30] From Engel Tresse probably learned of Killing's work and the main problem standing in the way of its justification (as indicated in Section 5.5): in view of the untenable nature of Theorem A, is Theorem C correct and, if not, how can Killing's Conditions II ($\mathfrak{g}_0$ is abelian) and III(the roots $\alpha \neq 0$ are simple) be demonstrated for simple groups? However, it seems that Lie also indicated to Tresse and others what needed to be done. When Tresse

[29] The letter is in UB Oslo, Brevsamling Nr. 289, as are the two letters from Tannery mentioned below.

[30] Tresse wrote to Engel of "la dette que j'ai contractée envers vous pour les longues heures que m'avez consacrées. . .lors de mon séjour à Leipzig" (13 December 1893, Engel Archive Giessen, Paket "Briefe von Cartan, Tresse, . . .").

Figure 18. According to Cartan [Akivis and Rosenfeld 1993:276], at the École Normale it was Jules Tannery (pictured) who exerted the most profound influence on students.

returned to Paris in 1892 to work on a thesis problem suggested by Lie, he took up lodgings with Cartan and recounted all he had learned in Leipzig. In particular Tresse informed Cartan of Killing's work and the investigations required in order to vindicate its principal conclusions.[31] According to Tresse's later recollections,[32] Cartan became intensely interested in the challenges afforded by Killing's work and proceeded to spend most of his time in the library, presumably absorbed in his study of Killing's papers

[31] In a letter to Engel, after complaining about the brief and superficial thesis examination given him by Lie, Tresse wrote, "La veille, M. Lie s'etait occupé plus long temps de mon camarade Cartan, à qui je me flatte cependant d'avoir indiqué, sur les indications de M. Lie, d'ailleurs, les recherches à faire sur les travaux de Killing (18 December 1893, Engel Archive Giessen, Paket "Briefe von Cartan, Tresse, ...").

[32] Made to Cartan's son, Professor H. Cartan. I am grateful to Professor Cartan for communicating this information to me.

and Umlauf's dissertation.

Lie learned of Cartan's existence when one of his students, Georg Scheffers, who was visiting Paris, sent him a postcard with the message, "A friend of Tresse by the name of Cartan is occupied with a thesis on Killing."[33] In a letter to Lie written three days later and after a mathematical conversation with Cartan, Scheffers added that "Cartan makes a very good impression."[34] Shortly thereafter, Lie himself arrived in Paris and met Cartan on several occasions. Much later, Cartan wrote to Engel, "I can never forget all that I owe to the great Norwegian scientist whom I had the honor of seeing often in Paris in 1893 when, having left the École normale supérieure, I prepared my thesis on the structure of continuous groups."[35] But by the time Lie arrived in Paris (late March at the earliest), Cartan appears to have already made many of his fundamental discoveries; the first announcement of his results appeared in the *Comptes Rendus* for 17 April 1893 [1893b]. Although Lie's presence in Paris was undoubtedly inspiring and helpful, Lie's direct influence upon Cartan's thesis seems to have been marginal.

The key to Cartan's successful and brilliant reworking of Killing's theory was provided by the function $\psi_2(e)$, the coefficient of ω^{r-2} in the characteristic equation $\Delta(\omega) = \det[\operatorname{ad} X - \omega I] = 0$, $X = \sum_{i=1}^{r} e_i X_i$ of Section 5.1. In order to appreciate Cartan's originality as well as the manner in which it involved developing ideas of Killing, it is necessary to touch upon an aspect of Killing's work that has thus far been neglected. In the course of his study of the structure of algebras satisfying $p = r$ ($\mathfrak{g}' = \mathfrak{g}$), Killing discovered that for any $\mathfrak{g}$, if X is expressible in the form $X = [Y, Z]$, then necessarily $\psi_1(e) = 0$ [1888a:263–4]. Hence when $p = r$, it follows that $\psi_1(e) = 0$ for all $X \in g$. (The reader should note that $\psi_1(e) = \operatorname{tr} \operatorname{ad} X$.) Since

$$\psi_1(e) = \sum_{i,k=1}^{r} e_i c_{ikk},$$

it follows that when $\mathfrak{g}' = \mathfrak{g}$,

$$\sum_{k=1}^{r} c_{ikk} = 0, \quad i = 1, 2, \ldots, r. \tag{2.1}$$

[33] "Ein Freund von Tresse, Namens Cartan, beschäftigt sich mit einer Thèse über Killing." Letter of 20 March 1893 (UB Oslo, Brevsamling Nr. 289).

[34] "Cartan macht einen sehr angenehmen Eindruck." (UB Oslo, Brevsamling Nr. 289).

[35] "Je ne saurais oublier tout que je dois au grand savant norvégien, qui j'ai eu l'honneur de voir souvent à Paris en 1893, alors que, sortant de l'École normale supérieur je préparais ma thèse sur la structure des groupes continues." The letter, written ca 1922, is missing, but the above quotation from it was copied by Engel (Engel Archive Giessen, Paket "Engels Nachforschungen an Lie, Lobachweskij"), apparently for use in supporting publication of Lie's collected works. See also Cartan's reminiscences on Lie's visit to Paris in 1893 [1971], but note that Cartan mistakenly puts the visit in 1892.

Killing made these discoveries before he was acquainted with Lie's publications.[36] When he read them he discovered that Lie had just recently published some similar results [1884b:§3*]. Lie observed that when the expressions $\sum_{k=1}^{r} c_{ikk}$ do not vanish for all i, the elements $X = \sum_{i=1}^{r} e_i X$ such that $\sum_{i,k=1}^{r} e_i c_{ikk} = 0$ define an $(r-1)$-dimensional linear subspace which is also an ideal in $\mathfrak{g}$. Thus if $\mathfrak{g}$ has no ideals, (2.1) must hold.

Although Killing and Lie had discovered essentially the same result, they regarded and justified it in essentially different ways. These differences are of considerable historical significance. From Killing's viewpoint, his discovery involved the coefficient $\psi_1(e)$ of the characteristic equation. It revealed that $\psi_1(e)$ vanishes identically on $\mathfrak{g}'$. Furthermore, the *basis* for this property of $\psi_1(e)$ was, for Killing, as we shall see, that it is an invariant of the adjoint group. It was this viewpoint that proved fruitful to Cartan. Lie, by contrast, did not characterize his discovery in terms of $\psi_1(e)$. In fact, although in a sense the idea of the characteristic equation of a group was known to him, he did not develop it or consider it of great theoretical value. That is, Lie realized that to determine the two-dimensional subgroups containing some $X \in \mathfrak{g}$ one must solve $[X, Y] = \omega Y$ and that the permissible values of ω are the roots of the characteristic equation of the system of linear equations implied by $[X, Y] = \omega Y$. The characteristic equation was for him a notion from the theory of determinants—or linear algebra as we would now describe it—that was occasionally useful. (See, for example, the presentation at the end of Section 5.2.) But the characteristic equation viewed as a function of $X = \sum_{i=1}^{r} e_i X_i \in \mathfrak{g}$ with variable coefficients, viz.,

$$(2.2) \qquad \Delta(\omega) = \pm[\omega^r - \psi_1(e)\omega^{r-1} + \psi_2(e)\omega^{r-2} - \cdots] = 0,$$

was not something to which Lie attached any theoretical significance for his theory of transformation groups. In particular, neither the characteristic equation (2.2) nor the adjoint group is to be found in Lie's proof.[37] Encouraged by Engel to study the invariants of the adjoint group, Killing first pointed out their significance to the proof that $\psi_1(e) = 0$ defines an ideal of $\mathfrak{g}$.

A function $F(e)$ that is invariant under the "finite" transformations of the adjoint group has the characteristic property that $dF = 0$ for every infinitesimal transformation $\operatorname{ad} X_k$ of the adjoint group, where

$$\operatorname{ad} X_k : X \to X + [X, X_k]\delta t = \sum_{i=1}^{r} (e_i + \delta e_i) X_i, \quad X = \sum_{i=1}^{r} e_i X_i.$$

[36] Killing to Lie, 27 October 1885 (UB Oslo, Brevsamling Nr. 289).

[37] Later, Lie did give a proof that employs the adjoint group [1893a:692–693]. He also claimed (p. 692n) that this proof was already indicated by him in his own review of his paper [1884b] for *Fortschritte der Mathematik* (Berlin, 1887, pp. 325–326); but it is difficult to see any trace of it there.

Expressed in Lie's operator notation, the condition that $F(e)$ be invariant becomes $E_k F(e) \equiv 0, k = 1, \ldots, r$, where the differential operator E_k corresponds to $\operatorname{ad} X_k$ (see Section 3.3). In their publications both Killing and Cartan used this characterization of invariance. Killing proved [1888a:§2] that the coefficients $\psi_i(e)$ of the characteristic equation are invariants of the adjoint group. It is enlightening to translate Killing's theorem into more familiar terms. The function $\psi_i(e)$ is a homogeneous polynomial of degree i, so that in particular, $\psi_1(e)$ and $\psi_2(e)$ are, respectively, a linear functional and quadratic form. Expressed in coordinate-free terms, the condition that a linear functional F be an invariant is that $F(D_Y X) = 0$ for all X and Y in $\mathfrak{g}$, where $D_Y X = [X, Y]$. In this case invariance means that F must vanish on $\mathfrak{g}'$. For a quadratic form G the condition of invariance stipulates that G, regarded as a symmetric bilinear form $G(X, Y)$, satisfy $G(X, D_Y X) = 0$ for all $X, Y \in \mathfrak{g}$ or, equivalently, that $G(X, D_Y Z) + G(D_Y X, Z) = 0$ for all $X, Y, Z \in \mathfrak{g}$. In other words, G must be associative: $G(X[Y, Z]) = G([X, Y], Z)$ for all $X, Y, Z \in \mathfrak{g}$. Thus Killing's theorem contains, upon translation, the theorem that ψ_2 is associative. Neither Killing nor Cartan expressed their ideas in these coordinate-free terms, but as we shall see, the important role played by the associativity of the Killing form in the modern theory is a direct consequence of their work.

With the function ψ_1 in mind, Killing proved the following generalization of the theorem proved by Lie in [1884b]:

(2.3) *If $F(e)$ is a nontrivial linear homogeneous function that is also an invariant of the adjoint group, then the set $\mathfrak{h}$ of $X = \sum_{i=1}^{r} e_i X_i \in \mathfrak{g}$ such that $F(e) = 0$ is an $(r-1)$-dimensional ideal in $\mathfrak{g}$ containing $\mathfrak{g}'$. If $F_1, F_2, \ldots, F_s$ are linear homogeneous invariants, then the set $\mathfrak{h}$ of all X such that $F_i(e) = 0$ for all i is an ideal in $\mathfrak{g}$ containing $\mathfrak{g}'$.*

Killing's theorem follows immediately from the coordinate-free characterization of invariance: By hypothesis $F([X, Y]) = 0$ for all X and Y in $\mathfrak{g}$, and this means that $[X, Y] \in \mathfrak{h}$ or, in other words, that $[\mathfrak{g}, \mathfrak{g}] \subset \mathfrak{h}$. Killing's own proof, which used the invariance of F in the form $E_k F(e) \equiv 0$ for $k = 1, \ldots, r$, is essentially an unenlightening version of the coordinate-free proof. In particular, it showed that $[\mathfrak{g}, \mathfrak{g}] \subset \mathfrak{h}$. Killing's theorem is historically significant because it revealed a connection between ideals and invariants of the adjoint group. But for Killing this connection was essentially restricted to linear homogeneous invariants. As a consequence, his insight into the connection between the coefficients ψ_i of degree $i > 1$ and ideals in $\mathfrak{g}$ was limited to the observation [1888a:268] that if some such ψ_i is of the form $[F(e)]^i$ with F linear, then $\mathfrak{h} = \{e : \psi_i(e) = 0\}$ is an ideal containing $\mathfrak{g}'$. This is an immediate consequence of Killing's theorem above, since $\mathfrak{h} = \{e : F(e) = 0\}$ and F is also an invariant $(0 \equiv E_k(F^i) \equiv i F^{i-1} E_k(F))$.

Cartan saw how to extend the connection between invariants and ideals to any homogeneous function of degree i and not just those that are the ith power of a linear function. For $i = 2$ the extension takes the following form:

(2.4) *If $G(e)$ is a homogeneous function of the second degree and also an invariant of the adjoint group, then the collection $\mathfrak{h}$ of all X in $\mathfrak{g}$ such that $\partial G/\partial e_i = 0$ for $i = 1, \ldots, r$ is an ideal of $\mathfrak{g}$.*

If we write $G(e) = \sum_{i,j=1}^{r} a_{ij}e_ie_j$, where $A = (a_{ij})$ is symmetric, then $\mathfrak{h}$ is the nullspace of A. If we think of G as defining a (possibly degenerate) inner product on $\mathfrak{g}$ by $\langle X, Y\rangle = e^T Af$, $X = \sum e_iX_i$, $Y = \sum f_iX_i$, then $\mathfrak{h}$ consists of all $X \perp \mathfrak{g}$, i.e., $\mathfrak{h} = \mathfrak{g}^{\perp}$. Of course, there is no evidence that Cartan thought in these terms. The key for him, as we shall see, was that $\frac{\partial G}{\partial e_i} = \sum_{j=1}^{r} a_{ij}e_j$ is a *linear* homogeneous from.

When Cartan first presented this theorem, he declared the proof to be so simple that it could be omitted [1893c:398]. He did include it in his thesis, however, where it was further generalized for use in establishing his criteria for solvability and semisimplicity [1894b:20–22]. Although the proof is relatively simple, it is not trivial. Killing never perceived the avenue of extension followed by Cartan, even though it exploits the idea behind Killing's own proof of his theorem. The proof is worth sketching as an indication of the creative manner in which Cartan assimilated and transformed Killing's ideas.

Let $G(e)$ be a homogeneous function of degree two that is an invariant of the adjoint group. Each partial derivative $\partial G/\partial e_i$ is a linear homogeneous function but need not be an invariant, even though G is. Thus Killing's above theorem with $F_i = \partial G/\partial e_i$ does not apply. Killing's proof, however, had established more than $[\mathfrak{g}, \mathfrak{h}] \subset \mathfrak{h}$, since it showed that $[\mathfrak{g}, \mathfrak{g}] \subset \mathfrak{h}$. Cartan realized that the $\mathfrak{h}$ of Killing's theorem would still satisfy $[\mathfrak{g}, \mathfrak{h}] \subset \mathfrak{h}$ if the F_i satisfied the weaker condition

$$(2.5) \qquad E_kF_i(e) = 0 \text{ for all } i, k \text{ and all } X = \sum e_iX_k \in \mathfrak{h}.$$

When expressed in coordinate-free terms, (2.5) states that for every i, $F_i([X, Y]) = 0$ for all $X \in \mathfrak{h}$ and all $Y \in \mathfrak{g}$; and that is precisely the condition that $[X, Y] \in \mathfrak{h}$. The vector subspace $\mathfrak{h}$ of Cartan's theorem, which is now called the *radical* of G, is therefore an ideal of $\mathfrak{g}$, provided that (2.5) holds for $F_i = \partial G/\partial e_i$. Cartan showed this was the case by a straightforward calculation. In this way he established his generalization of Killing's generalization of Lie's theorem.

Cartan's theorem with $G = \psi_2$ enabled him to obtain further information about ψ_2 when $\mathfrak{g}$ is simple. Killing had showed that when $\mathfrak{g}$ is simple or, more generally, when $\mathfrak{g}' = \mathfrak{g}$, ψ_1 vanishes identically by virtue of its invariance. By virtue of the invariance of ψ_2, Cartan's theorem showed that $\mathfrak{h} = \{X = \sum e_iX_i : \partial\psi_2/\partial e_i = 0, i = 1, \ldots, r\}$ is an ideal of $\mathfrak{g}$. Thus when $\mathfrak{g}$ is simple, either $\mathfrak{h} = 0$ or $\mathfrak{h} = \mathfrak{g}$. That is, either ψ_2 is nondegenerate or ψ_2 vanishes identically on $\mathfrak{g}$. But Cartan could see that if Killing's theory were essentially correct, then ψ_2 could not possibly vanish identically. This insight was provided to Cartan by consideration of the sum of the squares of the characteristic roots $\alpha_1, \ldots, \alpha_r$ corresponding to any $X = \sum_{i=1}^{r} e_i X_i \in \mathfrak{g}$. From the well-known relations among the roots and coefficients of a polynomial, it followed that if $\mathfrak{g}' = \mathfrak{g}$, then

$$(2.6) \qquad \alpha_1^2 + \cdots + \alpha_r^2 \equiv [\psi_1(e)]^2 - 2\psi_2(e) \equiv -2\psi_2(e).$$

Thus if $\psi_2(e) \equiv 0$, the sum of the squares of the roots would vanish identically. But a fundamental relation in Killing's theory of simple groups was

$$(2.7) \qquad 2\beta(H_\alpha) + a_{\beta\alpha}\alpha(H_\alpha) = 0,$$

where $H_\alpha = [X_\alpha, X_{\alpha'}]$, $\alpha' = -\alpha$, $\alpha(H_\alpha) \neq 0$. (See Section 5.1.) According to (2.7) every root $\beta(H_\alpha)$ is a *rational* multiple, viz., $-a_{\beta\alpha}/2$, of $\alpha(H_\alpha)$. Thus according to (2.7) the sum of the squares of the roots could not vanish for $X = H_\alpha$.

In his attempt to vindicate his analysis of simple groups, Killing argued that (2.7) remains valid for any $\mathfrak{g}$ satisfying $\mathfrak{g}' = \mathfrak{g}$ [1889a:70]. But Killing relied upon Theorem C ($\mathfrak{g}_0$ is abelian when $\mathfrak{g}' = \mathfrak{g}$) and other invalid presumptions, as Cartan observed [1894b:43]. Cartan perceived, however, that Killing's line of reasoning leading to (2.7) (sketched in Section 5.1) could be carried out for *any* $\mathfrak{g}$ and consequently without the aid of any special assumptions about the nature of the Jordan–Weierstrass canonical form or the nature of $\mathfrak{g}_0$ [Cartan 1893c:399–400; 1894b:41–42]. That is, if α and β are any two roots (including the possibility of the root 0) such that $\alpha' = -\alpha$ is also a root, then if $X_\alpha \in \mathfrak{g}_\alpha$, $X_{\alpha'} \in \mathfrak{g}_{\alpha'}$, $H_\alpha = [X_\alpha, X_{\alpha'}]$, the analogue of Killing's relation $\sum_{j=0}^{s} \beta_j(H_\alpha) = 0$ from Section 5.1 holds. Thus $\beta(H_\alpha)$ is a rational multiple of $\alpha(H_\alpha)$. For future reference we single out a special case of this result corresponding to taking $\alpha = 0$:

$$(2.8) \qquad \text{If } X \in [g_0, g_0], \text{ then } \omega_\beta(X) = 0 \text{ for all roots } \omega_\beta.$$

To show that $\psi_2 \not\equiv 0$ from the above result, Cartan had to prove that when $\mathfrak{g}$ is simple, $\alpha(H_\alpha) \neq 0$ for some α and H_α. Even under Conditions I–III Killing never really justified his assumption that $\alpha(H_\alpha) \neq 0$ in (2.7). By making what was probably the first application of Engel's theorem, Cartan showed that when $\mathfrak{g}' = \mathfrak{g}$ (Condition I) an α and H_α must exist for which $\alpha(H_\alpha) \neq 0$.[38]

In this manner Cartan obtained a proposition of fundamental importance to his renovation of Killing's theory:

[38] The proof goes as follows: Since $\mathfrak{g}' = \mathfrak{g}$, $\mathfrak{g}_0$ must be spanned by elements from the various $[\mathfrak{g}_\alpha, \mathfrak{g}_{\alpha'}]$, including $\alpha = 0$. Each $[\mathfrak{g}_\alpha, \mathfrak{g}_{\alpha'}]$ consists of H_α's such that if $\alpha(H_\alpha) = 0$ for all H_α, then $\beta(H_\alpha) = 0$ for all β and H_α. But then $\beta(X) = 0$ for all β and all $X \in \mathfrak{g}_0$. Thus $\mathfrak{g} = \mathfrak{g}_0$ and thus $\mathfrak{g}$ has rank zero. Referring to Umlauf's thesis, Cartan observed that Engel had showed that if $\mathfrak{g}$ has rank zero, then $\dim \mathfrak{g}' \leq r - 2$, so that $\mathfrak{g}' \neq \mathfrak{g}$. Cartan realized that (2.8) can be derived more directly [1893c:398], although in his thesis he obtained it in the manner indicated.

(2.9) *If* $\mathfrak{g}' = \mathfrak{g}$, *then* $\psi_2 \not\equiv 0$.

Since Cartan's proposition (2.4) shows that for $\mathfrak{g}$ simple either $\psi_2 \equiv 0$ or ψ_2 is nondegenerate, an immediate corollary to (2.9) is the proposition that ψ_2 must be nondegenerate when $\mathfrak{g}$ is simple. By exploiting the nondegeneracy of ψ_2, Cartan was able to confirm the properties of simple algebras assumed by Killing, such as Conditions II and III. I will sketch his proof of Condition II ($\mathfrak{g}_0$ is abelian), since it typifies the manner in which he exploited the nondegeneracy of ψ_2 and also because it serves to clarify further the role of the so-called Killing form in the respective theories of Killing and Cartan.

The formulas expressing the coefficients $\psi_i(e)$ of the characteristic polynomial $\Delta(\omega) = \det[\operatorname{ad} X - \omega I] = \det[\gamma_{ij} - \omega\delta_{ij}]$, $\gamma_{ij} = \sum_{k=1}^{r} e_k c_{kji}$, in terms of the e_i and the structure constants c_{ijk} become increasingly complicated as i increases because they involve sums of $i \times i$ determinants from the matrix (γ_{ij}) representing $\operatorname{ad} X$. Killing had observed [1889b; 1890b:164] that simpler expressions are possible for the sums of the mth powers of the characteristic roots, viz., $K_m(e) = \sum_{i=1}^{r}(\alpha_i)^m$, $m = 1, 2, \ldots$. Although the notion of the trace of a matrix had not yet received such recognition that it was given a special name, Killing realized in effect that $K_m(e) = \operatorname{tr}\,[(\gamma_{ij})^m]$. Thus $K_1(e) = \psi_1(e)$ and

$$(2.10) \qquad K_2(e) = \sum_{h,k=1}^{r} \gamma_{kh}\gamma_{hk} = \sum_{i,j=1}^{r} K_{ij} e_i e_j, \quad K_{ij} = \sum_{h,k=1}^{r} c_{ikh} c_{jhk}.$$

$K_2(e)$ is the quadratic Killing form usually written as $\operatorname{tr}\,[\operatorname{ad} X \circ \operatorname{ad} X]$. Because Killing first called attention to the functions $K_m(e)$ it could be said that he was the first to introduce the Killing form. But the appellation "Killing form" is historically misleading if it is taken to imply that Killing grasped the theoretical significance of this form for the theory of the structure of Lie algebras. He simply realized that by virtue of the known relations among the $\psi_i(e)$ and the $K_m(e)$, the $K_m(e)$ are also invariants of the adjoint group and can thus be used in place of the $\psi_i(e)$ in various considerations such as the determination of the rank l of $\mathfrak{g}$. The advantage of employing the $K_m(e)$ was that the expressions for them are simpler. This advantage, however, was never effectively exploited by Killing, who did not introduce the $K_m(e)$ until the fourth part of his "Zusammensetzung" papers and never singled out the "Killing form" $K_2(e)$ for special consideration. For these insights we must turn to the work of Cartan. In this connection, it should be kept in mind that for all algebras satisfying $\mathfrak{g}' = \mathfrak{g}$, $K_1(e) = \psi_1(e) \equiv 0$, and so (as already indicated in (2.6)) $K_2(e) = [\psi_1(e)]^2 - 2\psi_2(e) = -2\psi_2(e)$, so that the Killing form and Cartan's form ψ_2 differ only by a constant factor.

Cartan's proof that $\mathfrak{g}_0$ is abelian when $\mathfrak{g}$ is simple proceeds as follows [1893c:401ff.; 1894b:Ch.IV,§4]. Let a basis $X_1, \ldots, X_r$ be chosen such that $X_1, \ldots, X_k$ span $\mathfrak{g}_0$; $X_{k+1} \ldots, X_{k+k'}$ span the root space g_α; $X_{k+k'+1}, \ldots,$ $X_{k+k'+k''}$ span g_β; and so on. Also, let $i \in (\alpha)$ mean that $X_i \in g_\alpha$. From

(2.10) it follows that if $K_{ij} \neq 0$, then $c_{ikh}c_{jhk} \neq 0$ for some h and k. Suppose that $i \in (\alpha)$, $j \in (\beta), h \in (\gamma), k \in (\delta)$. Then by the relation $[\mathfrak{g}_\mu, \mathfrak{g}_\nu] \subset \mathfrak{g}_{\mu+\nu}$, the nonvanishing of $c_{ikh}c_{jhk}$ requires that $\alpha + \delta = \gamma$ and that $\beta + \gamma = \delta$. From this it follows that for $i \in (\alpha)$ and $\alpha \neq 0$, K_{ij} vanishes except when $j \in (\alpha')$. Consequently, K_2 decomposes into the sum of two forms,

$$K_2(e) = \Sigma' K_{ij} e_i e_j + \Phi(e_1, \ldots, e_k),$$

where Σ' denotes summation over all i, j such that $i \in (\alpha), j \in (\alpha')$, and $\alpha \neq 0$. The nondegeneracy of K_2 implies that these two forms are likewise nondegenerate as forms in $r - k$ and k variables, respectively. From the nondegeneracy of the first it follows that for all $\alpha \neq 0$ a β exists such that $\alpha + \beta = 0$, since otherwise for $i \in (\alpha)$, $K_{ij} = 0$ for all j. The nondegeneracy of Φ—the restriction of the "Killing form" $K_2(e)$ to the "Cartan subalgebra" $\mathfrak{g}_0$—is used to prove that $\mathfrak{g}_0$ is abelian. Here Cartan used the corollary (2.8), which shows that all roots of the characteristic equation vanish on $\mathfrak{g}_0'$. Since Φ represents the sum of the squares of those roots, Φ must vanish identically on $\mathfrak{g}_0'$. This will contradict the nondegeneracy of Φ unless $\mathfrak{g}_0' = 0$. Hence $\mathfrak{g}_0$ is abelian.

Having now sketched some of Cartan's proofs, we may compare his proof techniques with those of Killing. Killing relatively successful analyzing the structure of simple groups under the presumed validity of Conditions II and III, but his approach to justifying these assumptions was primitive from a methodological standpoint. It consisted in choosing a basis $X_1, \ldots, X_r$ for $\mathfrak{g}$ that put ad X_1 in its Jordan–Weierstrass canonical form, where X_1 is the regular element generating $\mathfrak{g}_0$. i.e., $\mathfrak{g}_0 = \mathfrak{g}_0(X_1)$ in the notation of Section 5.1. Thus the form of the products $[X_1, X_i]$ was known, and the object was to determine sufficiently the form of the products $[X_i, X_j]$, $i, j > 1$, so that general structure theorems such as Theorem C (Section 5.5) would emerge. The form of the products $[X_i, X_j]$ was to be determined by sheer calculation involving repeated use of the Jacobi identity. For a completely general canonical form these calculations proved unmanageable to Killing, who was thus forced to carry them out under rather ad hoc restrictive assumptions such as the assumption that the canonical form was relatively simple. As he put it at one typical juncture, "In order to give a good overview of the development and to to discuss first of all those cases where the formulas can be completely written down, we provisionally make the following two assumptions ..." [1889a:60]. Using the calculations in special cases as a guide, Killing would then infer what occurred in general without providing any detailed calculations but claiming that they would follow essentially the same pattern and differ only in complexity.

Killing was acutely aware of the shortcomings of his approach and of the desirability of an alternative approach that would be independent of the diverse possibilities for the Jordan–Weierstrass canonical form. In the introduction of Part III [1889a:58], where the shortcomings of his approach are most evident, he wrote:

> It cannot be denied that the consideration of multiple roots is burdensome, since it becomes necessary to distinguish various cases that cannot easily be reduced to a uniform type. The need always to distinguish many cases brings with it the apprehension that I could have overlooked a possibility. Because of this I cannot deny that it would be desirable to confirm the results obtained by me in a more direct, or better put, in a more unified, manner.

Killing went on to suggest a possible unified approach. It consisted in the direct study of $|\gamma_{ij}| = \det[\operatorname{ad} X]$ and its subdeterminants, much in the spirit of Jacobi and Weierstrass, whose work pertaining to what would today be described as linear algebra was dominated by the consideration of determinants and related identities. We know this because Killing had already described this approach to Engel in a letter of 31 January 1887. At that time he explained to Engel, "These investigations are not yet concluded. Such determinants are not at all my specialty. But I have recognized how important these investigations are for the theory of the structure of groups." When he suggested the same approach in Part III, the situation had not changed. Killing admitted that he had not yet brought the approach to a fruitful conclusion.

As the above-sketched proofs suggest, Cartan did not have to resort to the determinant-theoretic arguments that Killing explored with reluctance and trepidation. He was able to work under the hypothesis of a completely general Jordan–Weierstrass canonical form—as he did, e.g., in the reasoning leading to (2.9) ($\psi_2 \not\equiv 0$ when $\mathfrak{g}' = \mathfrak{g}$)—because he did not need to become embroiled in the details required to work out explicitly simple forms for $[X_i, X_j]$. Of course, both he and Killing relied upon the relation $[\mathfrak{g}_\alpha, \mathfrak{g}_\beta] \subset \mathfrak{g}_{\alpha+\beta}$, but the function ψ_2 provided Cartan with the means of circumventing such details. It facilitated the "more unified manner" that had eluded Killing. Whereas Killing's powers of calculation had been dissipated by his crude approach, the nondegeneracy of ψ_2 enabled Cartan to concentrate his own considerable calculational skills for more effective use. It provided a focus that Killing lacked. Despite Killing's observations regarding the functions $K_m(e)$, he never perceived the advantage to be gained by considering the sum of the squares of the characteristic roots, as Cartan did, for example, in his generalization of Killing's (2.7) and in the proof that $\mathfrak{g}_0$ is abelian. Neither the Killing form nor its nondegeneracy when $\mathfrak{g}$ is simple figured in Killing's attempt to vindicate his results.

Cartan's results on simple groups were incorporated into his thesis [1894b], where they were presented from a broader perspective, as an application of his theory of semisimple groups. Killing had defined $\mathfrak{g}$ to be semisimple if it is decomposable into a direct sum of simple algebras. From this definition it follows immediately that many properties of simple algebras are shared by semisimple ones. In particular, ψ_2 is nondegenerate. On the other hand, Cartan could see that many of his proofs involving simple algebras, such as the above proof that $\mathfrak{g}_0$ is abelian, do not rely specifically on the simplicity of $\mathfrak{g}$ but rather on the nondegeneracy of ψ_2. Killing had arrived at his conception of semisimplicity in the course of the analysis that culminated in his radical splitting theorem. As we have noted, that analysis was irreparably flawed, but Cartan salvaged from it the perception that semisimplicity could be characterized in a different manner. An immediate corollary to Killing's radical splitting theorem was the proposition that $\mathfrak{g}$ is semisimple if and only if $\mathfrak{g}$ contains no invariant abelian subgroup or, equivalently, no invariant solvable subgroup. The starting point of Cartan's theory of semisimple groups was the alternative definition of semisimplicity suggested by the above corollary. Whereas Killing had vainly attempted to deduce the properties of simple algebras from the theory of algebras satisfying $\mathfrak{g}' = \mathfrak{g}$, Cartan showed in his thesis that they could be rigorously obtained from the theory of semisimple groups in the sense of the new definition of semisimplicity—the sense in which the term will be used hereafter.

The keystone of Cartan's theory of semisimple groups was provided, as it still is, by the following criterion.

> **Cartan's Criterion for Semisimplicity.** *The group associated to $\mathfrak{g}$ is semisimple if and only if ψ_2 is nondegenerate.*

By means of this criterion he demonstrated that the two definitions of semisimplicity, his and Killing's, are equivalent. His proof is, in fact, essentially the one still given. The criterion itself was based upon the observation that the ideal $\mathfrak{h} = \{e : \partial\psi_2/\partial e_i = 0, i = 1, \ldots, r\}$ is solvable. This property of $\mathfrak{h}$ followed from the following criterion.

> **Cartan's Criterion for Solvability.** *The group associated to $\mathfrak{g}$ is solvable if and only if ψ_2 vanishes identically on $\mathfrak{g}'$.*

This criterion, in turn, was closely related to two criteria for solvability already observed by Killing.

Killing had learned of the concept of solvability through Engel and his characterization of solvability in terms of the absence of "conic section subgroups" (Section 5.4). Although solvability was a concept from Lie's theory rather than his own, Killing was somewhat interested in it because he recognized its close connection with his own concept of an algebra of rank zero. As he explained to Engel, "The case $l = 0$ is not essentially different from the case in which the coefficients $[\psi_i]$ ... can be represented as *linear* functions of the coordinates, or as you put it, in which no conic section group is contained in the group." Implicit in this remark is a criterion for solvability: $\mathfrak{g}$ *is solvable if and only if the coefficients $\psi_i(e)$ of the characteristic equation are all expressible as homogeneous functions of linear homogeneous invariants.*

The proof of Killing's criterion rests upon his observation [1888a:268] that if $\psi_i = H_i(P_1, \ldots, P_s)$, where H_i is homogeneous and the P_j are linear homogeneous invariants, then by Killing's theorem (2.3), $\mathfrak{g}' \subset \mathfrak{h} = \{e :$

$P_j(e) = 0$, $j = 1, \ldots, s\}$. Thus on $\mathfrak{g}'$, $\psi_i(e) = H_i(0, \ldots, 0) = 0$ for all i, so that $\mathfrak{g}'$ has rank zero. Thus (by Killing's Theorem D of Section 5.5) $\mathfrak{g}'$, and consequently $\mathfrak{g}$, is "without conic section subgroups." The proof proceeding in the reverse direction follows from Lie's theorem, which Killing also knew from Engel's paper on conic section subgroups. Implicit in the above proof is another criterion for solvability, which Killing also recognized: $\mathfrak{g}$ *is solvable if and only if* $\mathfrak{g}'$ *has rank zero* [1890b:180]. It was this criterion that Killing had in mind when he explained to Engel that solvable algebras were "not essentially different" from those of rank zero. And it was this criterion that Cartan used to obtain his own. Indeed, if $\mathfrak{g}$ is solvable, so that $\mathfrak{g}'$ has rank zero, then all the coefficients ψ_i vanish on $\mathfrak{g}'$. To prove the converse, Cartan needed the crucial lemma that had enabled him to show that ψ_2 is nondegenerate when $\mathfrak{g}$ is simple: If $\mathfrak{g}' = \mathfrak{g}$, then $\psi_2 \not\equiv 0$. If ψ_2 vanished identically on $\mathfrak{g}'$ and $\mathfrak{g}$ were not solvable, then for some integer n, $\mathfrak{h} = \mathfrak{g}^{(n)}$ would satisfy $\mathfrak{h}' = \mathfrak{h} \neq 0$. Since ψ_2 vanishes on $\mathfrak{h}$, the lemma is contradicted [Cartan 1894b:47].[39]

Cartan's theory of semisimple groups was thus based on several observations already made by Killing: the alternative characterization of semisimplicity, the invariance of the ψ_i and thus the K_i, the advantage of working with the K_i due to their simpler expression in terms of the structure constants c_{ijk}, the connection between invariants of the adjoint group and solvable ideals and the two criteria for solvability. But for Killing all these observations were interesting albeit isolated pieces of information that did not figure in his theory of structure. By brilliantly developing, and then exploiting, Killing's ideas, Cartan first demonstrated their fruitfulness. While many essential elements of the modern theory of semisimple Lie algebras, such as Cartan subalgebras and Cartan integers, originated with Killing, the fundamental role played by the Killing form is the result of Cartan's insights.[40]

---

[39] Killing stated his criteria explicitly in Part IV [1890b:161, 180–181]. Cartan referred only to Killing's observation (in Part II) that $\mathfrak{g}'$ has rank zero when all ψ_i are expressible by means of linear invariants. It should be noted that in Part IV, although Killing mentioned the intimate relation between his two criteria, he deduced the second one as a corollary to a totally different, more involved, line of reasoning [1890b:173–180]. Perhaps his strained relationship with Lie and his own stubborn pride kept him from making any explicit use of Lie's theorem on solvable algebras.

[40] Independently of Cartan, Theodore Molien [1893] introduced the analogue of the bilinear Killing form in the theory of linear associative algebras. Molien and his work will be discussed in Section 10.1 and Section 10.5. Here I simply note that he introduced the characteristic equation $\pm[\omega^r - \phi_1(u)\omega^{r-1} + \cdots] = 0$ corresponding to the linear transformation $u_R : x \to xu$ and sought to relate the ideals of the associative algebra to the coefficient ϕ_1. In this case, $\{u : \phi_1(u) = 0\}$ is not an ideal, but Molien observed that $\mathfrak{J} = \{u : \phi_1(uv) = 0 \text{ for all } v\}$ is an ideal. Here $\phi_1(uv) = \text{tr }(v_r \circ u_r)$ is the analogue of the Killing form. Since the "associativity" of $\phi_1(uv)$, viz., $\phi_1((uv)w) = \phi_1(u(vw))$, is an immediate consequence of the associativity of the algebra, the proof that $\mathfrak{J}$ is an ideal is more straightforward. Cartan, in 1898, also used his insights into the structure

In the final three chapters of his thesis Cartan turned to the study of algebras that are neither solvable nor semisimple. Such an algebra necessarily contains solvable ideals, and Cartan observed that any two such ideals generate a third that contains both. Thus every such algebra $\mathfrak{g}$ contains a greatest solvable ideal $\mathfrak{r}$ containing all others [1894b:97]. In this manner the concept of the radical, which had lurked in the shadows of Killing's theory of groups satisfying $\mathfrak{g}' = \mathfrak{g}$, was finally brought to light. In agreement with Killing's radical splitting theorem, Cartan showed that when $\mathfrak{g}' = \mathfrak{g}$, $\mathfrak{r}$ has rank zero [1894b:108].

Cartan also observed that by means of the homomorphism of $\mathfrak{g}$ induced by $\mathfrak{r}$ (viz., $\mathfrak{g} \to \mathfrak{g}/\mathfrak{r}$), a familiar concept to readers of Lie, $\mathfrak{g}$ is mapped onto a semisimple algebra $(\mathfrak{g}/\mathfrak{r})$. Initially, Cartan believed that he could prove that this semisimple algebra could be realized as a subalgebra of $\mathfrak{g}$, so that any $\mathfrak{g}$ would decompose into a semisimple subgroup and a solvable ideal. This is, of course, the radical splitting theorem and generalizes Killing's version. Cartan announced it in the *Comptes Rendus* [1893a:963–964], but by the time he composed his thesis he had evidently discovered a flaw in his proof, for there he stated the general radical splitting theorem, attributed the statement to Killing, and indicated that the considerations he had developed in Chapter VI of his thesis showed that it would suffice to demonstrate the theorem for algebras that admit only a single solvable ideal. "But this is a question that would take us too far afield" [1894b:128]. Although Cartan here generously ascribed the general theorem to Killing, the latter never stated it without the restriction $\mathfrak{g}' = \mathfrak{g}$. E. E. Levi published the first proof of the general theorem in [1905].

Cartan did manage to establish a theorem that can be regarded as a special case of the radical splitting theorem. That is, he showed that the three-dimensional simple subgroups $H_\alpha, X_\alpha, X_{\alpha'}$ of $\mathfrak{s} = \mathfrak{g}/\mathfrak{r}$ could be realized as subgroups of $\mathfrak{g}$ [1894b:100–103]. This result is tantamount to Engel's theorem that $\mathfrak{g}$ is solvable if and only if $\mathfrak{g}$ contains no simple three-dimensional subalgebra.[41] Viewed in the light of other theorems established by Cartan, it can also be regarded as the radical splitting theorem for $\mathfrak{g}$ of Killing rank $l = 1$. That is, Cartan proved that for any $\mathfrak{g}$, its Killing rank l equals $l_1 + \lambda$, where l_1 is the Killing rank of $\mathfrak{g}/\mathfrak{r}$ and λ the Killing rank of $\mathfrak{r}$ [1894b:112]. Since for the semisimple algebra $\mathfrak{s} = \mathfrak{g}/\mathfrak{r}$, $l_1 = k_1 = \dim \mathfrak{s}_0$, when $l = 1$ it follows that $l_1 = 1$ and $\lambda = 0$: $\mathfrak{s}$ is the simple three-dimensional

of Lie algebras to develop the analogous theory of linear associative algebras, but he naturally employed the quadratic form $\phi_2(u)$ (the coefficient of ω^{r-2}) rather than the bilinear form $\phi_1(uv)$ of Molien, whose work was unknown to him. This work by Molien and Cartan is discussed in detail in [Hawkins 1972], especially pp. 258–59, 268. Weyl, whose contributions are discussed at length in Chapter 12, apparently initiated the custom of making K_2 and the associated bilinear form tr (ad $X \circ$ ad Y) central to the theory of Lie algebras [Weyl 1925:362].

41 About the same time that Cartan [1893a] announced his proof of Engel's theorem, Engel [1893b] filled in the gap in his original proof.

algebra and $\mathfrak{r}$ has rank zero. Thus by Cartan's proof of Engel's theorem $\mathfrak{g} = \mathfrak{s} + \mathfrak{r}$.

Using the representation $\mathfrak{g} = \mathfrak{s} + \mathfrak{r}$, Cartan set out, in the penultimate chapter of his thesis, to construct systematically algebras of Killing rank $l = 1$. Killing had already done this for the special case in which $k = 1$ also. In general $k \geq 1$. Cartan's interest in generating examples for which $k > 1$ may have been motivated by this desire to test the validity of Killing's Theorem C, for only the proof had been vitiated by the discoveries of Engel and his students (Section 5.5). From Cartan's formula for the Killing rank of $\mathfrak{g}$ it follows that for any $\mathfrak{g}$ satisfying the hypothesis of Theorem C ($\mathfrak{g}' = \mathfrak{g}$), $\lambda = 0$, so that whenever $k = l$ the rank formula yields $\dim \mathfrak{g}_0 = k = l_1 = \dim \mathfrak{s}_0$ since $\mathfrak{s}$ is semisimple. This means that $\mathfrak{g}_0$ may be identified with $\mathfrak{s}_0$, and consequently $\mathfrak{g}_0$ is abelian. In other words, any $\mathfrak{g}$ that satisfies the hypothesis of Theorem C and $k = l$ will satisfy its conclusion: Counterexamples to Theorem C, if they exist, occur when $k > l$, the simplest case being $k > l = 1$. Even in this case, the generation of examples en masse is not easy, because $\mathfrak{r}$ need not be abelian (as it is when $k = l = 1$), and verification of the Jacobi identity is problematic as a result. By some clever devices involving hypergeometric series and commutative hypercomplex number systems, Cartan overcame the difficulties. Among the examples he generated were counterexamples to Theorem C, including one of dimension 18 with $k = 4$ [1894b:124–125]. Killing's entire theory of structure was thus built upon a foundation of sand, but Cartan had already provided a new and firm foundation through the creation of his theory of semisimple groups.

Another matter left unresolved by Killing was the question of the existence of simple groups corresponding to the new simple types E and F. (Cartan, of course, realized that Killing's type C was exemplified by the symplectic groups of Lie.) From the root systems he had obtained for these types, Killing derived by means of the Jacobi identity various relations that the structure constants c_{ijk} would have to satisfy, thereby limiting somewhat the numerical possibilities [1888b:46–48]. But he went no further. He failed to produce specific numerical values and to verify that they satisfy the requisite conditions, notably the Jacobi identity. Cartan completely specified numerical values — integers and simple fractions—for the structure constants corresponding to each exceptional type [1894b:87–93]. But he provided no proof that the resulting multiplication tables satisfy the Jacobi identity. In the case of E_8, $r = 248$, so that there are $\binom{248}{3} = 2{,}511{,}496$ triples X, Y, Z for which the Jacobi identity must be verified. Many triples can be considered together, but the number of cases to be considered is nonetheless formidable. A skilled and intrepid calculator, Cartan presumably carried out the extensive and tedious calculations necessary to confirm that his tables satisfy the Jacobi identity. The readers of his thesis either had to accept in faith that he had done so or themselves perform the necessary calculations. Especially from the vantage point of the present, Cartan's treatment of existence thus leaves something to be desired, although none of his contemporaries appears to have doubted that he had confirmed Killing's belief in the existence of the exceptional types.[42]

So far, we have discussed the work of Killing and Cartan as it pertains to the structure of Lie algebras. However, the seeds for Cartan's theory of what would nowadays be described as the representation of Lie algebras, namely his theory of weights, were also contained in the pages of Killing's papers, in the guise of his theory of "secondary roots," which is discussed in the next section. Then in Section 4 we shall see what Cartan did with Killing's theory in his thesis, and why.

6.3. Killing's Secondary Roots

In pursuing his goal of a classification of $\mathfrak{g}$ satisfying $\mathfrak{g}' = \mathfrak{g}$ in [1889a], Killing distinguished three kinds of characteristic root of $\mathfrak{g}$. First there was the root 0, which Killing regarded as in a class by itself. Then there were the principal roots (*Hauptwurzeln*) and the secondary roots (*Nebenwurzeln*). These distinctions were made en route to Killing's attempt to prove his radical splitting theorem: Any $\mathfrak{g}$ satisfying $\mathfrak{g}' = \mathfrak{g}$ has a vector space decomposition $\mathfrak{g} = \mathfrak{s} + \mathfrak{r}$, where $\mathfrak{s}$ is a semisimple subalgebra and $\mathfrak{r}$ is an ideal of $\mathfrak{g}$ of rank zero, which in view of Engel's theorem (Section 5.5) means that $\mathfrak{r}$ is nilpotent. From the modern viewpoint, given the above radical splitting theorem as a starting point, we may regard $\mathfrak{r}$ as an $\mathfrak{s}$-module in the obvious way. Then $\mathfrak{r}$ decomposes into weight spaces, $\mathfrak{r} = \mathfrak{r}_\mu + \mathfrak{r}_\nu + \cdots$ with weights $\mu, \nu, \ldots$. Thus $\mu, \nu, \ldots$ are roots of the characteristic equation $p(w) = \det[\operatorname{ad} X - \omega I]$ of $\mathfrak{g}$ relative to a Cartan subalgebra $\mathfrak{s}_0$ of $\mathfrak{s}$, which we may take such that $\mathfrak{s}_0 \subset \mathfrak{g}_0$. As will be seen, the nonzero roots $\mu, \nu, \ldots$ are essentially what Killing termed the secondary roots of $\mathfrak{g}$. Any root $\alpha \neq 0$ of $\mathfrak{s}$ is likewise a root of $p(\omega)$, and they are essentially what Killing termed principal roots of $\mathfrak{g}$. Killing's own presentation of principal and secondary roots, however, is much more clouded in obscurity because it was given in

[42] At least one reader of Cartan's thesis, Claude Chevalley, performed the calculations required to confirm that Cartan's multiplication tables do indeed define Lie algebras (personal communication via J.-P. Serre). Chevalley also indicated [1948] how to provide an a priori proof that a Lie algebra exists for every simple type in the Killing–Cartan classification. As Chevalley realized, Ernst Witt [1941] had taken a first step in this direction by giving an a priori proof that if all simple types of rank $l \leq 4$ exist, then all simple types exist. (Witt also constructed models of Lie algebras of types G_2 and F_4 so that this paper contained a complete existence proof.) Chevalley's proof idea inspired and influenced the a priori existence proof given by Harish-Chandra [1951]. Serre's existence and uniqueness theorem was first published in [1966]. An existence proof along the lines pursued by Killing and Cartan was given by Tits [1966], who used the Chevalley basis theorem [1955] to demonstrate that a set of structure constants satisfying the Jacobi identity could be determined for any semisimple structure (Cf. [Frenkel and Kac 1980:40–41].)

the midst of his unsuccessful attempt to prove his radical splitting theorem and because all roots were regarded by him as linear functions on $\mathfrak{g}_0$ rather than on the possibly smaller $\mathfrak{s}_0$. With this in mind, we now proceed to Killing's own presentation of his theory of secondary roots.

Let l be the rank of $\mathfrak{g}$ in the sense of Killing. Recall from Section 5.4 that this means that $l = \dim \mathfrak{s}_0$, i.e., that l is the rank of the semisimple $\mathfrak{s}$ in the modern sense. Killing claimed that it is always possible to determine l root pairs $\pm\alpha_i$ and corresponding elements $X_i, X_{i'}$ in $\mathfrak{g}$ such that the elements $H_i = [X_i, X_{-i}]$, $i = 1, \ldots, l$, are linearly independent. Unless k, the dimension of $\mathfrak{g}_0$, coincides with l, the dimension of $\mathfrak{s}_0$, the H_i will not span $\mathfrak{g}_0$. For these roots and elements the "Cartan integers" a_{ik} may be introduced. When among the roots of the characteristic equation of $\mathfrak{g}$ "only those occur that are required by the a_{ik}," then $\mathfrak{g}$ is simple or semisimple, depending upon whether the system a_{ik} is "simple" or "composite" [1889a:72–3]. (Recall from Section 5.4 that this means that the Coxeter graph associated with a_{ik} is connected or disconnected.) When the characteristic equation admits more roots than are required by the system a_{ik} $(i, k = 1, \ldots, l)$, Killing termed the required ones principal roots while "all further roots of the characteristic equation are designated as secondary roots" [1889a:76].

From Killing's definition it may at first glance appear that a root cannot be simultaneously principal and secondary: Either it is or it is not "required" by the a_{ik}. That he did not intend his definition to be so understood is evident from the remarks he made immediately following the above definitions. "It will be seen that no secondary root can be a principal root when the group satisfies the two conditions posited at the end of Section 20" [1889a:76]. The implication is that generally a secondary root may also be a principal root. If one follows the tortuous path along which Killing leads the reader toward the radical splitting theorem, it becomes clearer what he means by a secondary root. If $\mathfrak{g} = \mathfrak{s} + \mathfrak{r}$, then the root $\mu \neq 0$ is secondary if the corresponding root space $\mathfrak{g}_\mu$ contains nonzero elements from $\mathfrak{r}$. The restriction to nonzero roots reflects Killing's tendency to regard the root 0 in a class by itself, as neither principal nor secondary, even though he realized that 0 would arise in the manner of a secondary root when $k = \dim \mathfrak{g}_0$ is larger than l. Indeed, the two conditions referred to by Killing in the above quotation are that $\mathfrak{g}' = \mathfrak{g}$ and that $k = l$. The latter condition, expressed in modern language, means that we may take $\mathfrak{s}_0 = \mathfrak{g}_0$ and that consequently 0 is not a weight of $\mathfrak{r}$. Since when 0 is not a weight no root of $\mathfrak{s}$ can be a weight, Killing's assertion that when $k = l$ no secondary root can be a principal root is correct. As we shall see, he also recognized the connection between the phenomenon of secondary roots being equal to principal roots and the occurrence of 0 in the manner of a secondary root.

A major source of complexity and attendant difficulty and incompleteness in Killing's reasoning stemmed from the need to consider cases in which the characteristic equation relative to $\mathfrak{g}_0$ has nonlinear elementary divisors—that is, $\operatorname{ad} X$ cannot be put in diagonal form for all X in $\mathfrak{g}_0$. Under the hypothesis that $k = l$ Killing discovered that all the elementary divisors are linear. Since when k, the dimension of $\mathfrak{g}_0$, equals $l, \mathfrak{g}_0 = \mathfrak{s}_0$, Killing's discovery corresponds to the modern theorem that X acts diagonally on the $\mathfrak{s}$-module $\mathfrak{r}$ for each X in $\mathfrak{s}_0$. When $l < k$, so that $\mathfrak{g}_0$ includes not only $\mathfrak{s}_0$ but the weight space for the weight 0 of $\mathfrak{r}$, it can happen that not every X in $\mathfrak{g}_0 \sim \mathfrak{s}_0$ acts diagonally on $\mathfrak{r}$, so that the characteristic equation relative to $\mathfrak{g}_0$ does possess nonlinear elementary divisors. Because of his approach—the study of the structure of $\mathfrak{g}$ with the radical splitting theorem as a goal not a starting point—$\mathfrak{g}_0$ rather than $\mathfrak{s}_0$ was the frame of reference. Thus beyond the case $k = 1$ Killing was confronted with the unpleasant task of considering possibilities corresponding to nonlinear elementary divisors in addition to the problems posed by multiple roots. Probably for this reason he tended to concentrate upon the more tractable case $k = l$.

After finally presenting his proof of the radical splitting theorem, Killing admitted that since the ideal $\mathfrak{r}$ has rank zero and since algebras of rank zero had not been classified, his theorem did not immediately provide the classification of algebras satisfying $\mathfrak{g}' = \mathfrak{g}$ to which he aspired [1889a:107]. But Killing claimed that it was not necessary to await the classification of algebras of rank zero, since the general classification problem could be readily resolved without major difficulties once the following problem had been resolved: Determine the structure of all $\mathfrak{g} = \mathfrak{g}'$ that satisfy the following conditions [1889a:116]:

(3.1) $\mathfrak{g}$ does not decompose in the sense that it cannot be written as a vector space direct sum $\mathfrak{g} = \mathfrak{g}_1 + \mathfrak{g}_2$ of two subalgebras such that $[\mathfrak{g}_1, \mathfrak{g}_2] = 0$ (or, equivalently, $\mathfrak{g}$ cannot be expressed as the direct sum of two *ideals* of $\mathfrak{g}$);

(3.2) $\mathfrak{g}$ possesses only one solvable ideal $\mathfrak{r}$, which is then necessarily abelian;

(3.3) $\mathfrak{s} = \mathfrak{g}/\mathfrak{r}$ is one of the simple types A–D.

If condition (3.1) is applied to $\mathfrak{g} = \mathfrak{s} + \mathfrak{r}$, it says that $[\mathfrak{s}, \mathfrak{r}] \neq 0$, i.e., considered as an $\mathfrak{s}$-module, $\mathfrak{r}$ is not the trivial representation. Condition (3.2) asserts that $\mathfrak{r}$ must be an *irreducible* $\mathfrak{s}$-module, since a submodule $\mathfrak{r}_1$ of $\mathfrak{r}$ would constitute a second solvable ideal. As for condition (3.3), Killing thought incorrectly (as Cartan later showed in his thesis) that when $\mathfrak{g}$ satisfies (3.1), then the semisimple $\mathfrak{s}$ of his radical splitting theorem, $\mathfrak{g} = \mathfrak{s} + \mathfrak{r}$, must, in fact, be simple. Thus Killing's intention in (3.3) was simply to rule out consideration of the exceptional simple types. As we shall see in the next section, Cartan considered Killing's problem with condition (3.3) omitted. With that condition included Killing's problem is tantamount to that of determining all irreducible representations of a simple Lie algebra of the classical types A–D.

Although Killing regarded types A–D as "the most important simple groups" [1889a:83], the decision to exclude the exceptional types in formulating his problem was based upon time constraints more than anything else. He was reaching the limit of his endurance to do battle with the difficult algebraic problems his research program on foundations of geometry had revealed, and the exceptional types were still rather elusive entities, since Killing had no concrete representations for them (although Engel obtained one for type G) and had not even fully worked out the structure constants except for type G. The following passage makes it clear that the restriction imposed by (3.3) was purely a matter of convenience and that he believed that if it were dropped and the resulting more general problem solved, then the classification of all $\mathfrak{g} = \mathfrak{g}'$ would follow easily: "This problem [of determining all $\mathfrak{g}' = \mathfrak{g}$ satisfying (3.1)–(3.3)] is not just of intrinsic interest because the groups satisfying the stated conditions are probably of outstanding importance but rather also because its solution is absolutely necessary for [the determination of] all further groups that are equal to their own principal subgroup [i.e., to $\mathfrak{g}'$] (insofar as other simple groups do not come into consideration). If, however, we can assume the above problem solved for all simple groups, the solution of the general problem of specifying all groups that are their own principal subgroup involves only insignificant difficulties" [1889a:116].

In other words, the larger problem of interest to Killing, but beyond his present capabilities, was that of determining all $\mathfrak{g}$ satisfying conditions (3.1) and (3.2). The problem of determining all $\mathfrak{g} = \mathfrak{g}'$ satisfying (3.1) and (3.2) I shall therefore refer to as *Killing's problem*. Of course, Killing thought that $\mathfrak{g}/\mathfrak{r}$ would be simple. Once this was seen to be incorrect, Killing's problem could be seen to be tantamount to determining all irreducible representations of semisimple Lie algebras. Indeed, as we shall see in the next section, Cartan realized this.

It was in the context of Killing's problem that Killing sought to apply the properties of secondary roots. Let $\mathfrak{g}$ satisfy $\mathfrak{g}' = \mathfrak{g}$ and let α_i, $i = 1, \ldots, l, \ldots, L$, denote the principal roots and $\mu, \nu, \ldots$ the secondary roots. From his general discussion of roots (as expounded in Section 5.1) Killing observed that corresponding to each secondary root μ are integers $a_{\mu i} = t'' - t'$, $i = 1, \ldots, L$, such that $w_\mu + tw_i$ is a root for $-t' \leq t \leq t''$. Also, the fundamental relation $2\mu(H_i) + a_{\mu i}\alpha_i(H_i) = 0$, $i = 1, \ldots, l$, obtains, and every root, principal or secondary, is some rational linear combination of $\alpha_1, \ldots, \alpha_l$ [1889a:§20]. Killing realized that when $\mathfrak{g} = \mathfrak{s} + \mathfrak{r}$ satisfies the conditions (3.1) and (3.2) of his problem, then all secondary roots of $\mathfrak{g}$ are obtainable from a single one, say μ, by forming all roots $\nu = \mu + t\alpha_i$ for t between 0 and $a_{\mu i}$ and then forming all roots $\lambda = \nu + t\alpha_i$ for t between 0 and $a_{\nu i}$, and so on [1889a:83]. He did not introduce any notation for the system of roots thus generated from μ, but we shall denote it by $S(\mu)$. Thus $S(\mu)$ includes the root 0 if it is generated from μ. Expressed in modern language, $S(\mu)$ is the smallest saturated system of weights containing μ. From the modern theory we know that a unique dominant weight λ exists that is conjugate to μ. Thus $S(\mu)$ is the saturated system of weights associated with λ, that is, the system of weights associated with the irreducible $\mathfrak{s}$-

module $\mathfrak{r}$. Killing, however, never characterized the system $S(\mu)$ in terms of something akin to dominant weights. In fact, he was unable to describe the nature of an arbitrary system $S(\mu)$, although he could easily enough calculate the members of the system corresponding to a specific μ.

For a specific secondary root μ Killing computed the members of $S(\mu)$ on the basis of the following considerations [1889a:§25]. Let

$$\mu = m_1\alpha_1 + \cdots + m_l\alpha_l, \tag{3.4}$$

where the m_i denote rational numbers. Then if μ is a secondary root, the integers $a_{\mu i}$, $i = 1, \ldots, l$, exist, and from (3.4) it follows that $a_{\mu k} = \sum_{i=1}^{l} m_i a_{ik}$. Hence the expression (3.4) determines a secondary root if and only if integers $g_1, \ldots, g_l$ exist that satisfy

$$g_k = m_1 a_{1k} + \cdots + m_l a_{lk}, \quad k = 1, \ldots, l. \tag{3.5}$$

Since $\det(a_{ik}) \neq 0$, (3.5) can be solved for the m_i. Equation (3.5) thus completely determines μ, and Killing frequently denoted it by the l-tuple $(g_1, \ldots, g_l)$. In modern terms, (3.5) is precisely the condition that the linear functional μ defined by (3.4) be a weight. In fact, if λ_i denotes the linear functional corresponding to the choice $g_k = \delta_{ik}$, then the λ_i are analogous to the fundamental dominant weights of the modern theory: If $\mu = (g_1, \ldots, g_l)$, then $\mu = \sum_{i=1}^{l} g_i\lambda_i$.[43] Since each α_i, $1 \leq i \leq l$, corresponds to the l-tuple $(a_{i1}, \ldots, a_{il})$ comprising the ith row of the $l \times l$ system a_{ik}, for every integer t between 0 and g_k,

$$\mu + tw_k = (g_1 + ta_{k1}, \ldots, g_l + ta_{kl})$$

defines a member of $S(\mu)$; and in this manner Killing computed them.

It is, of course, one thing to determine the members of $S(\mu)$ and quite another thing to determine their multiplicities. For Killing this distinction did not assume major proportions because he seemed to have been led to think of the case in which $\mathfrak{s} \cong \mathbf{sl}(2, \mathbb{C})$ as paradigmatic. That is, in his determination of all $\mathfrak{g}$ satisfying $k = l = 1$ [1888a:§§7–8], Killing obtained, in effect, the decomposition $\mathfrak{g} = \mathfrak{s} + \mathfrak{r}$, with $\mathfrak{s} \cong \mathbf{sl}(2, \mathbb{C})$ and $\mathfrak{r}$ abelian. (The condition $k = l$ makes all weights of $\mathfrak{r}$ odd integers, and from this fact and the Jacobi identity the conclusion $\mathfrak{r}' = 0$ follows readily.) From this decomposition Killing could see that multiple secondary roots arise when $\mathfrak{r}$ decomposes into several ideals (viz., submodules), and he apparently thought this situation was typical of the manner in which multiple secondary roots arise [1889a:78]. Thus he seems to have taken it for granted that when $\mathfrak{r}$ satisfies the conditions of his problem, notably (3.2), which makes $\mathfrak{r}$ irreducible as an $\mathfrak{s}$-module, multiple secondary roots cannot occur. Cartan also

[43] In relating Killing's notation to present-day notation, it must be kept in mind that Killing's roots $\alpha_1, \ldots, \alpha_l$ do not necessarily form a base, a notion neither he nor Cartan employed.

inferred from Killing's muddled and ambiguous presentation that Killing had presumed this and gave a counterexample in his thesis [1894b:128–9].

In the above discussion I have used the term "secondary root" in the sense of Killing so as to exclude the root 0. Killing realized in his own terms that the irreducibility of $\mathfrak{r}$ does not mean that 0, as a weight of $\mathfrak{r}$, must have multiplicity one when it occurs, although he had an incorrect understanding of what the multiplicity actually is. He realized that 0 will be generated from μ (so that $0 \in S(\mu)$ in our notation) precisely when μ is a sum of principal roots, that is, if and only if all the m_i in (3.4) are integers, in which case some secondary roots would be principal [1889a:112]. In view of (3.5), which may be solved for the m_i, Killing could see that this case always arises when $\mathfrak{s}$ is such that $\det(a_{ik}) = 1$ and hence for types G_2, F_4, and E_8. For $\mathfrak{g} = \mathfrak{s}+\mathfrak{r}$, with $\mathfrak{s}$ one of these types, he asserted that $k \geq 2l$ [1889a:112], or in other words, that the multiplicity $k - l$ of 0 as a weight of $\mathfrak{r}$ is at least as large as its multiplicity as a root of $\mathfrak{s}$. Thus, "In order to compose an r-term group [$\mathfrak{g} = \mathfrak{s} + \mathfrak{r}$] in which ... [$\mathfrak{g}' = \mathfrak{g}$] out of the 14-term simple group [$\mathfrak{s}$] of rank two, at least six secondary roots that are principal roots must be adjoined, and two vanishing roots, so that the number of terms [i.e., $\dim \mathfrak{g}$] amounts to at least 22" [1889a:112]. This statement is equivalent to the assertion that the minimal dimension of an irreducible $\mathfrak{s}$-module of type G_2 is 8. Killing had evidently determined the correct minimal system $S(\mu)$ but had erred regarding the multiplicities of its members, assigning multiplicity 2 to 0 as a weight when it is actually 1, as Cartan was later to observe [1894b:153]. Killing's inability to assign multiplicities correctly was the fundamental flaw that vitiated his attempted solution to his problem.

Killing restricted his attention, as indicated above, to his problem with the additional condition (3.3), limiting the simple types considered to A–D. The problem divides naturally into three parts: (i) Given an expression $\mu = (g_1, \ldots, g_l)$, determine the secondary roots derivable from it, i.e., determine $S(\mu)$; (ii) write down the structure constants c_{ijk} to define the "composite group" $\mathfrak{g} = \mathfrak{s} + \mathfrak{r}$; (iii) find a concrete representation for $\mathfrak{g}$. In formulating (ii), Killing required that the secondary root μ not generate 0 [1889a:83]. Hence it would seem that he wished to construct only $\mathfrak{g}$'s satisfying $k = l$.

Killing treated part (i) in Section 25 of [1889a]. He distinguished two types of root in $S(\mu)$. There are the roots ν such that $S(\mu) = S(\nu)$, which he termed "mutually determining roots" [1889a:114]. They were the roots obtained from μ by operations of the type $\mu + g_i\alpha_i$—in other words, the conjugates of μ as we now express the matter. Further roots in $S(\mu)$ were termed "subordinate." Killing did not characterize the make-up of $S(\mu)$ as is done nowadays by means of the concept of a dominant weight. His general remarks about the nature of $S(\mu)$ were limited to specifying the number of "mutually determining roots" contained in $S(\mu)$, which depends upon the nature of μ. Actually, Killing's inability to achieve a more definitive characterization of $S(\mu)$ probably did not strike him as being as unsatisfactory as it does to us today, because in part (ii) the underlying belief that all secondary roots have multiplicity 1 severely limited the possibilities for the systems $S(\mu)$ that needed to be considered.

Section 26 was devoted to the more difficult part (ii). Given a simple $\mathfrak{s}$ of one of the types A–D, Killing's idea was to consider symbols $Y_1, \ldots, Y_l, \ldots, X_i, \ldots$ representing a canonical basis for $\mathfrak{s}$. Here the Y_i span $\mathfrak{s}_0$ (the Cartan subalgebra), and X_i corresponds to the principal root α_i. Then to each ν in $S(\mu)$ a symbol Z_ν is assigned, where of course one defines $[Y, Z_\nu] = \nu(Y)Z_\nu$. The problem is to determine constants c such that $[X_i, Z_\nu] = cZ_{\nu+i}$, where $Z_{\nu+i}$ corresponds to $\nu + \alpha_i$ (assuming that it is a root). Tacit in the formulation is the assumption that each secondary root corresponds to one such basis element, that it has multiplicity 1. The consequence of this assumption becomes evident in Killing's application of the Jacobi identity to triples such as X_i, Y_j, Z_ν. By such calculations he concluded that when $\mathfrak{s}$ is of type A_l the only systems $S(\mu)$ that permit choosing the c's consist of those with $\mu = (g, g, \ldots, g)$, where g is any nonzero integer. For $l = 2$ these systems correspond, after translation into modern notation, to the saturated systems determined by the dominant weight $\lambda = g\lambda_1$ if $g > 0$ and by $\lambda = -g\lambda_2$ if $g < 0$, where λ_1 and λ_2 denote the fundamental dominant weights.

These conclusions on Killing's part were in conflict with the facts presented by Study in his book on ternary forms, that appeared the same year as Killing's paper (1889). As we shall see in Section 7.2, Study showed that for every nonnegative integer pair $(m, n) \neq (0, 0)$, the "normal form of type (m, n)" determines a projective group with the structure of A_2 that leaves nothing planar invariant. Translated into modern terms, Study's projective group is the irreducible representation of $\mathfrak{sl}(3, \mathbb{C})$ with dominant weight $\lambda = m\lambda_1 + n\lambda_2$. Killing's work on his problem was done before Study's book was published. There is no evidence to suggest that Killing ever read Study's book or that Study ever read Killing's paper, the third in his series on structure. Both book and paper were extremely difficult reading. Even if they had read each other's work, however, it is not clear whether the conflicting results would have been apparent, that, e.g., Killing would have seen the implications of Study's results for his problem of determining "composite" algebras $\mathfrak{g} = \mathfrak{s} + \mathfrak{r}$. The connection between Killing's problem and Study's results on projective groups was first made evident through Cartan's reinterpretation of Killing's problem, as we shall see in the following section.

Similar use of the Jacobi identity when $\mathfrak{s}$ is of type B_l led Killing to conclude that only a very limited number of $S(\mu)$ were permissible: μ can be presumed chosen such that all g_i are 0, 1, or 2. Since the g_i must also be all odd or all even, for $l = 2$, by way of example, this leaves only the two possibilities $\mu = (2, 0)$ and $\mu = (1, 1)$. In modern notation, the systems generated by these choices of μ are those given, respectively, by the fundamental dominant weights λ_1 and λ_2 —the only two saturated systems of type B_2 with all multiplicities equal to 1.[44] Killing thus concluded that: "By contrast with the simple groups of the form (A), there are hence only very few types of groups that can be obtained from the form (B) by composition; to completely specify these forms involves no difficulty" (p. 120). This statement, which is equivalent to the modern assertion that there exists a very limited supply of irreducible modules of type B_l, illustrates, by the magnitude of its inexactitude, how Killing's misconceptions regarding multiplicities of secondary roots led him to see his problem erroneously as far less extensive than it actually was—so much so that he felt it unnecessary to proceed further in this case.[45]

Once the Jacobi identity and the tacit hypothesis that all secondary roots have multiplicity one had been used to limit the feasible $S(\mu)$, Killing next sought to represent the basis symbols Y X, Z concretely by means of differential operators $\Sigma_j \xi_j(x)\partial/\partial x_j$, so that the Jacobi identity is automatically satisfied. This was the standard mode of representation used by Lie and his school. When the ξ_j are linear and homogeneous, the corresponding operators determine a linear Lie algebra. Killing did not restrict himself to linear homogeneous ξ_j, and so his representations are not linear. For this reason I will not discuss this extremely sketchy and incomplete portion of the paper [1889a:119–122].

Killing's monumental program, to find all the Lie algebra structures, had ultimately focused upon two problems particularly congenial to the methods he had devised. The first problem was that of determining all simple structures. The second was what we have termed above Killing's problem; expressed in modern terms it is tantamount to determining all inequivalent irreducible modules for simple structures. Killing's efforts on behalf of the first problem were by far the more successful, for although his demonstrations were irreparably flawed, he had created an essentially viable conceptual and technical apparatus for dealing with the problem and had correctly articulated the nature of its solution. Cartan in effect showed this in his doctoral dissertation of 1894. The same cannot be said regarding Killing's work on the second problem. His misconceptions regarding the multiplicity of secondary roots prevented him from a correct vision of the nature of its solution. Nevertheless, Killing did initiate a theory of secondary roots that, although imperfect, was sufficiently seminal to prove a potent tool in the capable hands of Cartan.

---

44 Killing apparently overlooked the fact that 0 is included in the system generated by $(2,0)$, contrary to his (apparent) stipulation that he would exclude from consideration these cases ($k > l$), as indicated above.

45 As suggested in the last note, in working out the solution to part (ii) Killing may have dropped his apparent stipulation that 0 not be a weight. Even with this stipulation in force, however, his conclusion remains extremely inaccurate, since the number of irreducible B_l modules that do not have 0 as a weight is far from meager. For example, the irreducible B_2 modules that do not admit 0 as a weight are all those with dominant weight $\lambda = m\lambda_1 + n\lambda_2$ such that either m or n is even [Antoine and Speiser 1964:1569].

6.4. Cartan's Application of Secondary Roots

The classification of simple structures, although highly abstract and algebraic, was deemed important in Paris because of its relevance to Lie's envisioned Galois theory of differential equations. Given the disconcerting combination of abstractness and obscurity presented by Killing's papers, however, it is doubtful that the formidable problem of trying to make sense of it all would have seemed appealing as the subject of a dissertation to a typical aspiring doctoral candidate in Paris. But the challenge had appealed to Cartan, who, as we saw in Section 2, brilliantly succeeded in reworking the theoretical foundations of Killing's theory in his dissertation. Although Cartan understandably accorded special attention in his dissertation to the matter of providing a viable theoretical basis for Killing's classification, he did not stop there but took on the even greater task of attempting to make sense out of all aspects of Killing's claims. Killing's problem was one such aspect.

In his thesis Cartan showed by an example [1894b:130] that, contrary to the view implicit in condition (3.3) of that problem, $\mathfrak{g}$ can be indecomposable without $\mathfrak{s}$ being simple, where, in the notation of Section 3, $\mathfrak{g} = \mathfrak{s} + \mathfrak{r}$. Because Cartan was unable to prove the radical splitting theorem, however, he characterized $\mathfrak{s}$ as $\mathfrak{g}/\mathfrak{r}$, where $\mathfrak{r}$ is the maximal solvable ideal of $\mathfrak{g}$, i.e., what is now called the radical of $\mathfrak{g}$. Thus a proper statement of Killing's problem should admit the possibility that $\mathfrak{s}$ is semisimple rather than simple. Cartan therefore articulated Killing's problem as follows: *Determine all indecomposable* $\mathfrak{g}$ *such that* $\mathfrak{r}$ *(the radical of* $\mathfrak{g}$*) is its sole solvable ideal.* This means that $\mathfrak{r}' = 0$, so that $\mathfrak{r}$ is abelian. Thus $\mathfrak{r}$ may be considered as an $\mathfrak{s}$-module, since $\mathfrak{r}' = 0$ means that for $X + \mathfrak{r} \in \mathfrak{s}$ and $R \in \mathfrak{r}$, $[X + \mathfrak{r}, R] \stackrel{\text{def}}{=} [X, R]$ is well-defined. Also, the fact that $\mathfrak{r}$ is the sole solvable ideal in $\mathfrak{g}$ means that $\mathfrak{s}$ acts irreducibly on $\mathfrak{r}$. Finally, Cartan showed that the hypothesis that $\mathfrak{g}$ be indecomposable means that $\mathfrak{r}$ is a faithful $\mathfrak{s}$-module [1894b:Ch.VIII,§§3–4].

Cartan passed over in silence Killing's claim that by solving his problem, the classification of all $\mathfrak{g}$ with $\mathfrak{g}' = \mathfrak{g}$ would be achieved except for some relatively easy "details." He does not appear to have been seriously interested in Killing's Weierstrassian research program of obtaining an exhaustive classification of all Lie algebras, or even all satisfying $\mathfrak{g}' = \mathfrak{g}$. To carry out that program would be extraordinarily difficult, and from the viewpoint of Paris would be of limited significance. Cartan's interest in Killing's problem derived from his perception that it was equivalent, as he proved in considerable detail, to another problem which he regarded as potentially more important for applications [1894b: 134]:

> The study of semisimple linear, homogeneous groups of [dimension] ... r in n variables that leave no planar manifold invariant and the study of nonintegrable groups of [dimension] ... $r + n$

> that admit a sole invariant integrable subgroup of [dimension] ... n and *that do not decompose* ... are reducible to one another.

Cartan established this equivalence in the customary language of bases of infinitesimal transformations represented by differential operators. Expressed in more conceptual terms, his reasoning translates into the following: Given an indecomposable, nonsolvable $(r+n)$-dimensional algebra $\mathfrak{g}$ with n-dimensional radical $\mathfrak{r}$ and $\mathfrak{s} = \mathfrak{g}/\mathfrak{r}$, associate with each $X+\mathfrak{r}$ in $\mathfrak{s}$ the linear transformation $T = \phi(X+\mathfrak{r})$ acting on $\mathfrak{r}$ as vector space and defined by $T(R) = [X, R]$ for all R in $\mathfrak{r}$. The totality $\mathfrak{l}$ of the T's is the desired semisimple linear Lie algebra isomorphic to $\mathfrak{s}$, i.e., $\phi : \mathfrak{s} \to \mathfrak{l} \subset \mathbf{gl}(n, \mathbb{C})$ is a faithful representation of $\mathfrak{s}$. Conversely, given a semisimple $\mathfrak{l} \subset \mathbf{gl}(\mathcal{V})$ that acts irreducibly on $\mathcal{V}$, make the vector space direct sum $\mathfrak{g} = \mathfrak{l} \oplus \mathcal{V}$, into a Lie algebra by setting $[L_1 \oplus v_1, L_2 \oplus v_2] = L_3 \oplus v_3$, where $L_3 = [L_1, L_2]$ and $v_3 = L_1 v_2 + L_2 v_1$. In this sense, the study of the above-described $\mathfrak{g}$'s is equivalent to the study of semisimple linear Lie algebras $\mathfrak{l}$.

This observation on Cartan's part is historically significant because it revealed how Killing's theory of structure, including his theory of secondary roots, was relevant to the study of linear Lie algebras with semisimple structure that act irreducibly on the underlying vector space and how, conversely, the properties of these linear Lie algebras could be used to illuminate aspects of the theory of secondary roots, such as multiplicities. In particular, it enabled Cartan's to translate Killing's problem into what I will refer to as *Cartan's version of Killing's problem: Determine all semisimple Lie algebras of linear transformations* $\mathfrak{l} \subset \mathbf{gl}(\mathcal{V})$ *that leave nothing planar invariant, i.e., act irreducibly on* $\mathcal{V}$.

In his thesis Cartan did not attempt to solve his version of Killing's problem, but he did solve the problem in a special case that, like the classification of simple Lie algebras was perceived to be relevant to the Galois type study of differential equations. Thus he wrote, undoubtedly with Lie's theory of complete systems (Section 3.4) in mind:

> As an application we will find the smallest possible number of variables that can be exchanged among themselves by a *simple* linear and homogeneous group, which obviously leaves no planar manifold invariant. This question has an extreme importance in the theory of the integration of partial differential equations that admit a finite and continuous group, since its solution gives all the simplest possible types of linear differential equations to which the integration of the proposed system reduces [1894b:135].

Thus, in effect, Cartan proposed to show how Killing's theory of secondary roots could be used to solve what from the viewpoint of Paris was a useful, indeed extremely important, problem, the problem (in the language of today) of determining for each simple structure the nontrivial irreducible representation of minimal degree.

There is no doubt that Cartan's interest in this matter was also stimulated by a conjecture of Lie's that relates to it. Recall from Section 3.4 that Lie had hoped to determine simple groups based upon the following theorem: If $\mathfrak{g}$ is a simple, r-dimensional Lie algebra with maximal subalgebra of dimension $r - q$, then there is a Lie algebra $\mathfrak{h}$ in q variables with the structure of $\mathfrak{g}$. In the first place, it is obviously desirable that the simple $\mathfrak{h}$ in q variables be projective, since they are more tractable. (Keep in mind that by virtue of his geometrical background Lie preferred projective as opposed to linear homogeneous groups, although of course the former could always be expressed in terms of linear homogeneous transformations of homogeneous coordinates.) Lie observed that for small values of q this was always possible. He therefore posed the following question in the form of a conjecture [1885a:135]: "Among the projective groups of a q-fold manifold, does there exist ..., I ask, a group with the same structure as [a given simple group in q variables] ...? I conjecture that this question has yes as an answer." Here it is a question of a projective group in q inhomogeneous or $q + 1$ homogeneous variables. For example, if $\mathfrak{g}$ is the Lie algebra of the group of conformal transformations in three-dimensional space, $q = 3$. Since Lie realized that $\mathfrak{g} \cong \mathbf{o}(5, \mathbb{C})$, Lie's conjecture in this case is that $\mathbf{o}(5, \mathbb{C})$ has a linear representation in $q + 1 = 4$ variables. Lie knew that the answer was yes in this case because he realized that $\mathbf{o}(5, \mathbb{C}) \cong \mathbf{sp}(4, \mathbb{C})$.

Lie observed that his conjecture might be false for $\mathbf{o}(n + 1, \mathbb{C})$ with $n > 5$, and he set one of his better students, H. Werner, the task of dealing with $\mathbf{o}(n + 1, \mathbb{C})$ with this matter in mind as his doctoral dissertation [1889]. Werner's approach was to determine the maximal dimension $r - q$ of a subalgebra of $\mathbf{o}(n+1, \mathbb{C})$, where $r = \dim \mathbf{o}(n+1, \mathbb{C})$. By Lie's theorem this would mean that there is a group in q variables with the structure of $\mathbf{o}(n + 1, \mathbb{C})$. Werner showed that $q = n - 1$. Thus Lie's conjecture would imply that there is a projective group in n homogeneous variables with the structure of $\mathbf{o}(n + 1, \mathbb{C})$. Lie apparently realized that this was impossible, for he concluded on the basis of Werner's results that his conjecture was incorrect [1893a:809]. But it is a moot point whether he had a proof. Although Lie stressed the importance of his question–conjecture for applications of his theory to differential equations, neither Werner nor other of Lie's students at Leipzig dealt any further with it, perhaps discouraged by Werner's negative results.

The problem posed by Cartan is clearly relevant to Lie's conjecture, as he realized. In this connection he observed, after having solved it, that by constructing the minimal projective (viz., linear) representation $\mathfrak{l}$ for each simple type he could also conclude that except for types A and C, every simple type can be represented in fewer variables in nonprojective form. This was because he had observed that the minimal $\mathfrak{l}$ for each of the other types leaves invariant a proper nonlinear submanifold of the vector space on which $\mathfrak{l}$ acts. For example, the $\mathfrak{l}$ of type F_4 leaves a quadratic form invariant, as well as a system of quadratic equations. By regarding the action of $\mathfrak{l}$ on the nonlinear manifolds defined by such expressions, a nonprojective representation of $\mathfrak{l}$ is obtained that involves fewer variables [1894b:145–8].

With the tools of Killing's theory at his disposal, especially Killing's secondary roots, Cartan was thus able to go considerably further than Lie and his students. Let us now consider a few of the main ideas behind his solution to the problem. To this end, let $\mathfrak{s}$ be a simple Lie algebra. Cartan proceeded from Killing's point of view. That is, $\mathfrak{s}$ was regarded as associated with an indecomposable Lie algebra $\mathfrak{g}$ with sole solvable ideal $\mathfrak{r}$, where $\mathfrak{s} = \mathfrak{g}/\mathfrak{r}$. Accordingly, we let $\alpha_1, \ldots, \alpha_l, \ldots, \alpha_L$ denote the principal roots of $\mathfrak{g}$ (viz., the roots of $\mathfrak{s}$) and $\mu, \nu, \ldots$ the secondary roots (viz., the weights of the irreducible $\mathfrak{s}$-module $\mathfrak{r}$). As in Section 3 we shall let $S(\mu)$ denote the system of secondary roots, so that μ denotes a secondary root capable of generating all the rest. This property of secondary roots observed by Killing (and discussed in Section 3) was derived more rigorously by Cartan in his dissertation. Furthermore, Cartan did not make Killing's fundamental mistake regarding the multiplicities of secondary roots, although it is regrettable that nowhere in his dissertation did he indicate his method for deducing multiplicities. It seems likely, however, that his multiplicity calculations were facilitated by the new point of view he brought to bear on Killing's problem, according to which secondary roots may be considered as the eigenvalues $\mu(H)$ with respect to the transformations H comprising a Cartan subalgebra $\mathfrak{l}_0$ of some simple linear Lie algebra $\mathfrak{l}$ that leaves nothing planar invariant. This viewpoint allows one to generate eigenvectors by applying transformations of the enveloping algebra of $\mathfrak{l}$ to a maximal vector in the now familiar manner. We shall return to this matter when we discuss Cartan's paper [1913] in Section 8.4. Although he nowhere stated it explicitly, it would also seem that Cartan realized that if (in his own notation) S_i denotes the "substitution" $S_i\nu = \nu + a_{\nu i}\alpha_i$, where α_i, $1 \leq i \leq L$, is some principal root, then ν and $S_i\nu$ have the same multiplicity. He also apparently realized that μ has multiplicity 1.[46]

[46] The closest Cartan came to stating that conjugate roots have the same multiplicity is the following proposition [1894b:104,Thm.III]: Let α_1 be principal and μ principal or secondary. If $\mu + t\alpha_i$ is a root for $-h' \leq t \leq h$, then $\mu + h\alpha_i$ and $\mu - h'\alpha_i$ have the same multiplicity n, while all roots $\mu + t\alpha_i$ have multiplicity at least n. If μ has the property that either $\mu + \alpha_i$ or $\mu - \alpha_i$ is not a root for every $i = 1, \ldots, l$, then Theorem III implies immediately that μ and $\mu + a_{\mu i}\alpha_i$ have the same multiplicity, since $a_{\mu i} = h - h'$. (Such a μ is called a frontier weight in [1913].) Theorem III followed from the considerations involved in Cartan's lengthy proof of Engel's theorem that $\mathfrak{g}$ is not solvable if and only if $\mathfrak{g}$ contains a 3 dimensional simple subalgebra. Actually, all the systems $S(\mu)$ that Cartan needed to consider in his dissertation—those satisfying the conditions of his lemma (stated below)—consist entirely of frontier weights, except that 0 might also be a member. Nonetheless, it would seem that Cartan knew the result to be generally valid as a consequence of the work he had done in his dissertation. Thus in Section 3 of [1913] he referred to his dissertation (although not the part containing Theorem III) and then proceeded to state without proof basic properties of weights, including the property that conjugate weights have equal multiplicity [1913:58]. In [1913] he did prove that the dominant weight (and

Cartan's solution to his problem is based upon the following lemma.

Lemma. *Among the secondary roots in $S(\mu)$ there is at least one root ν with the property that the integers $a_{\nu i}$ satisfy $|a_{\nu i}| \leq 1$ for $i = 1, \ldots, L$, except for $i = \nu$ and $i = -\nu$ if ν (and therefore $\nu' = -\nu$) is a principal root* [1894b:135–6].

If ν is the root described in the lemma, clearly $S(\nu) \subset S(\mu)$, so that the systems described by the lemma include the system(s) of minimal cardinality. Now suppose an algebra $\mathfrak{g}_1$ corresponding to this minimal system $S(\nu)$ actually exists and that $\mathfrak{r}_1$ is its radical. Since the dimension of $\mathfrak{r}_1$ is just the sum of the multiplicities of the roots $\lambda \in S(\nu)$ and a similar relation is true for $\mathfrak{r}$ with respect to $S(\mu)$, it follows that $\dim \mathfrak{r}_1 \leq \dim \mathfrak{r}$, provided that the multiplicity of each λ in $S(\nu)$ with respect to $\mathfrak{r}_1$ is less than or equal to its multiplicity with respect to $\mathfrak{r}$.

Cartan made this proviso [1894b:137], which is certainly necessary, in the sense that the above inequality is not true for all saturated systems. But the existence of such examples, which may or may not have been known to him, did not pose a serious difficulty within the context of his problem because among the roots in $S(\nu)$, where ν is described in the lemma, most have multiplicity 1. For if ν is not a principal root, so that, in accordance with the hypothesis of the lemma, $|a_{\nu i}| \leq 1$ for all $i = 1, \ldots, L$, then $S(\nu)$ cannot contain any "subordinate roots" (to use Killing's term). That is, if $\lambda = S_i \nu$, then, as Cartan observed, the integers $a_{\lambda i}$ are just a permutation of the integers $a_{\nu i}$. (From the modern viewpoint this follows from the orthogonality of all S in the Weyl group with respect to the Killing form; Cartan's proof was different.) Hence for all λ in $S(\nu)$ generated by the substitutions S_i, $|a_{\lambda i}| \leq 1$ for all $i = 1, \ldots, L$; and $S(\nu)$ contains no subordinate roots $\lambda + t a_{\lambda i}$ with t strictly between 0 and $a_{\lambda i}$. The system $S(\nu)$ consequently consists exclusively of the conjugates of ν, and as noted, Cartan realized that this meant that all roots in $S(\nu)$ have multiplicity 1. Thus when ν is not principal, $\text{card}[S(\nu)] = \dim \mathfrak{r}$. When ν is principal, it follows, as Cartan realized, that 0 belongs to $S(\nu)$ as a subordinate root. Nonetheless, 0 is the sole subordinate root, and $S(\nu)$ consists of the conjugates of ν together with 0. Thus 0 is the only root with a multiplicity that might exceed 1. Since Cartan was able to determine this multiplicity, he was able to confirm a posteriori on a case-by-case basis that the "minimal" systems do indeed occur among those described in his lemma.

In order to solve his problem Cartan first computed, for each simple structure type, all systems $S(\nu)$ such that the $a_{\nu i}$ have the property in the lemma. (All computations were omitted, however, from the dissertation.) Except for types G_2, F_4, and E_8 it turned out that the $S(\nu)$ of minimal cardinality arise when ν is not a principal root [1894b:§8]. Thus if a simple linear group could be constructed corresponding to $S(\nu)$, it would be the

hence any frontier weight) has multiplicity 1. This was proved more or less as it is still proved nowadays. See Section 8.4 for a sketch of Cartan's proof.

desired group in the minimal number of variables. In the case of types G_2, F_4, and E_8, the $l \times l$ matrix (a_{ij}) corresponding to a fundamental system of roots has determinant 1, and Cartan realized that in this case the ν of the lemma must be principal roots, which without loss of generality may be assumed to be one of the l "fundamental roots" that generate all the principal roots.[47] Thus it was necessary to consider only these l roots. For G_2 Cartan found in this manner one system of minimal cardinality 7. Of course, 0 is included, but he (somehow) realized that 0 has multiplicity 1. Thus the desired linear group would involve 7 variables. For F_4 he found two systems of minimal cardinality 25 and claimed that 0 "must belong to at least two independent infinitesimal transformations," i.e., 0 has multiplicity at least 2 and the number of variables is at least 26. Since the cardinality of the other $S(\nu)$ was 49, the fact that 0 had multiplicity 2 (as Cartan showed by example) did not affect the "minimal" status of the systems with cardinality 25. In the case of E_8, every choice of ν led to a system containing all principal roots, and Cartan realized that the representation of minimal degree is given by the adjoint representation.

Having determined for each simple structure type the "minimal" system(s) $S(\nu)$, Cartan proceeded to describe for each the corresponding linear Lie algebra $\mathfrak{l}$ [1894b:§10]. For types A through D the $\mathfrak{l}$'s turned out to be the classical types, viz., $\mathbf{sl}(l+1, \mathbb{C})$, $\mathbf{o}(2l+1, \mathbb{C})$, etc., but the $\mathfrak{l}$'s for types E through G were completely new. Unfortunately, Cartan suppressed all his calculations and gave no idea how he had arrived at the groups $\mathfrak{l}$. For each structure he simply specified a basis of linear homogeneous infinitesimal transformations $X = \Sigma a_{ij} x_j \partial/\partial x_i$, leaving it to the reader to verify that they do indeed define a Lie algebra of the appropriate structure and yield the desired linear Lie algebra. In the case of E_7 there are 133 basis elements, each of which corresponds to a 56×56 matrix!

Cartan had thus solved that part of Killing's problem of relevance to the application of Lie's theory to differential equations. Killing's problem in its full generality—the problem that is tantamount to that of determining all irreducible representations of all semisimple Lie algebras—was not solved by Cartan in his dissertation. This is not surprising for a number of reasons. First of all, Killing's problem was itself just a special case of a more general problem aimed at classifying all Lie algebras. Killing had claimed without any proof that solving the special case was the major hurdle, but it is not at all clear that such a conclusion would have been clear to others. The general classification problem requires solving three problems: (1) proving a radical splitting theorem $\mathfrak{g} = \mathfrak{s} + \mathfrak{r}$ for arbitrary $\mathfrak{g}$; (2) determining all possible multiplications $[\mathfrak{s}, \mathfrak{r}]$, which is tantamount to determining all $\mathfrak{s}$-modules $\mathfrak{r}$

[47] The fact that the $l \times l$ matrix (a_{ij}) corresponding to a fundamental system $\alpha_1, \ldots, \alpha_l$ has determinant 1 implies (as Killing had also realized) that ν is an integral linear combination of the fundamental roots α_i, $i = 1, \ldots, l$. Both Killing and Cartan realized that consequently $S(\nu)$ will contain a principal root. If ν were not itself a principal root, then, since as noted above, $S(\nu)$ would consist exclusively of conjugates of ν. But then ν would also be principal.

for all semisimple $\mathfrak{s}$, where now $\mathfrak{r}$ need not be irreducible; (3) classifying all solvable Lie algebras $\mathfrak{r}$. These problems are all formidable. The easiest was (1), which was solved in 1905 by Levi, as noted in Section 6.2. To solve (2) requires, in modern terms, (2a) proving a complete reducibility theorem for all semisimple Lie algebras $\mathfrak{s}$; (2b) determining all irreducible $\mathfrak{s}$-modules. As for (3), a satisfactory solution has yet to be given.

Solving Killing's problem thus only had "artistic value" (in the language of Picard and Poincaré) as a partial solution to the above threefold problem, which in addition to being very challenging, had itself only artistic value, since it would yield the goal of Killing's research program—an exhaustive classification of all Lie algebras. Cartan had picked out just that piece of Killing's problem with a significant application to differential equations, just that piece that seemed of interest in Paris. Cartan undoubtedly realized that the tool he had used—a perfected version of Killing's theory of secondary roots—had the potential to deal with Killing's problem in general. Still, a general solution involved difficulties that Cartan had been able to avoid in the special case of the problem he had considered. He had showed that for a certain small number of special systems $S(\nu)$ there are corresponding linear algebras $\mathfrak{l}$, but the general problem required facing the question as to whether *every* possible system $S(\mu)$, the number of which is infinite, has associated with it a projective group that leaves nothing planar invariant. Killing had believed that this question, regarded from his viewpoint of "composite groups" $\mathfrak{g} = \mathfrak{s} + \mathfrak{r}$, had a negative answer, but he had been deceived by his misconception about multiplicities of secondary roots. Let us assume that Cartan believed that the answer was affirmative. Presumably, then, a general solution would also involve being able to describe the composition of an arbitrary system $S(\mu)$. Cartan had indeed done this for the special systems $S(\nu)$ of relevance to his problem, namely those satisfying the condition of his lemma, but, as we saw, their composition is straightforward. Not only is the structure of general $S(\nu)$ more difficult, but there is an infinite number of them to consider and a linear algebra $\mathfrak{l}$ to specify for each!

Nonetheless, Cartan overcame all these difficulties brilliantly in [1913]. Given the manner in which he accomplished this, it seems doubtful that it required twenty years of constant thought to overcome the obstacles. What seems more likely is that Cartan returned to Killing's problem when its solution was perceived to have what were considered to be significant applications to matters of interest in Paris. Indeed, such an interpretation is born out by the course of Cartan's research after 1894, as we shall see in Chapter 8.

Chapter 7
Lie's School and Linear Representations

We saw that in reformulating and then solving a special case of Killing's problem, Cartan had been led to investigate an aspect of what would now be regarded as the representation theory of semisimple Lie groups. It is important to keep in mind, however, that Cartan's investigations form part of the prehistory of the theory in the sense that in 1894, and in fact throughout the period 1886–1915 considered in Part III, there was no such thing as an established, broadly recognized theory of the representation of Lie groups and algebras to which mathematicians were consciously contributing. As we shall see in Chapter 10, Frobenius created his beautiful theory of the representation of finite groups, with its complete reducibility theorem and theory of characters, in 1896–1903. With the exception of Poincaré, however, none of the mathematicians whose work is described in Part III seem to have been interested in, or even familiar with, Frobenius's theory. And even Poincaré does not seem to have envisioned the feasibility of an analogous theory for continuous groups (See the discussion in Section 8.1.) The feasibility of a continuous analogue does not seem to have been seriously entertained until Hermann Weyl published his celebrated series of papers [1925] on the representation of semisimple groups. These impressive papers, together with the publication of the Peter–Weyl theorem [1926], drew the attention of many mathematicians and served to convince them that the representation of Lie groups and algebras was a viable, fertile field for mathematical research. Indeed, research activity in this area continues unabated to this day.

Because the paradigm of an established theory of the representation of continuous groups was lacking as a source of motivation in the period

1886–1915, in seeking to understand the developments in this period historically, it is especially important to appreciate the considerations that motivated research. Accordingly, throughout this chapter and the next, emphasis will be placed upon the problems and, more generally, upon the attitudes that prompted a mathematician to pursue a particular line of investigation as well as upon the manner in which he imagined his work to relate to the mathematics of the period. In these respects the monumental research program on transformation groups posed by Sophus Lie is of central importance. All of the mathematical investigations considered in this chapter relate, in one way or another, to the programmatic objectives of the school Lie established at the University of Leipzig. The first section, on Lie and his school, thus forms an introduction to, and overview of, the entire chapter.

7.1. Representations in Lie's Research Program

As we saw in Chapter 3, in the spring of 1886 Lie left his homeland of Norway to fill the professorship at the University of Leipzig vacated by his friend Klein. Lie's expressed purpose in coming to Leipzig was to create about himself a school of mathematicians dedicated to the investigation and dissemination of his theory of continuous transformation groups, a theory on which he had been working in isolation and with varying degrees of intensity since 1873. As he wrote to Darboux in 1893, "It is my dream, before I return to Christiania, to build up in Leipzig a healthy mathematical school."[1] Lie remained in Leipzig until 1898, when ill health forced his return to Norway. He died there the following year at the age of 57.

Given the relative brevity of Lie's tenure at Leipzig, the ill health that plagued him throughout the period, and his mercurical, often abrasive personality, it is remarkable how successful he was in achieving his objective. Three *Privatdozenten* at Leipzig, Friedrich Engel, Friedrich Schur, and Eduard Study, turned their research interests to aspects of Lie's theory. Engel, in fact, became Lie's right-hand man and devoted his entire career to the development and exposition of Lie's ideas. It was with Engel's invaluable assistance that Lie was able to express his theory in a systematic, literate form as *Theorie der Transformationsgruppen* (1888–1893), the contents of which—particularly that of the first volume—were indicated in Chapter 3. Lie also managed to attract a number of doctoral students at Leipzig, among whom I mention Georg Scheffers, who edited three volumes of Lie's lectures for publication, and Gerhard Kowalewski, whom Lie described as one of his

[1] Letter dated simply 1893. The original is located in the Bibliothèque de l'Institut de France. A microfilm copy and a typed transcription are located in UB Oslo, Brevsamling Nr. 52, letter 14.

Figure 19. Lie (top) as he looked during his years as professor in Leipzig. Engel (below), Lie's right-hand man at Leipzig, eventually became a professor at Giessen, where he continued to work for the promulgation of Lie's theory.

best "German students."[2] The qualification was necessary because Lie attracted students from several other countries as well. Most significant were the students from the prestigious École Normale supérieure.

In creating his theory of transformation groups, Lie was guided by the paradigm of the theory of finite groups of permutations, especially as epitomized by Camille Jordan's *Traité des substitutions*. Whereas the theory of finite groups was largely developed after its applicability to the study of algebraic equations had been revealed by Galois and his expositors, Lie hoped to develop his theory and its applications more or less simultaneously. Thus the two intertwined and fundamental goals Lie set for the development of his theory were (1) to cultivate those aspects of the pure theory that appeared essential for the (somewhat vaguely) envisioned applications; (2) to demonstrate the fruitfulness of the pure theory by successfully applying it in other branches of mathematics. The application accorded the highest priority by Lie was to differential equations. As we have seen in Chapters 1–3, he devised various theories of differential equations analogous to the Galois theory of polynomial equations.

In connection with (1), Lie stressed the importance of the problem of determining, up to similarity, all transformation groups in n variables. I will refer to this as Lie's classification problem. A solution, however partial, to the classification problem was deemed essential for applications. As one of Lie's students, the American James Page, explained in his doctoral thesis, "In order to derive practical knowledge from the theory of Groups for the theory of Differential Equations and its kindred branches of mathematics, it is necessary to know some, and if possible *all* groups of transformations" [1888:239]. The qualification "if possible" had to be added because this problem did not seem tractable enough to permit anything approximating a general solution. Indeed, Lie had devoted considerable time in the early stages of developing his theory to the classification problem for $n = 1, 2, 3$. The computations involved already in the case $n = 3$ were sufficiently extensive that he never published them. In *Theorie der Transformationsgruppen* [1888:598] Lie expressed the opinion that the solution to the classification problem for any n was a long way off and would probably never be achieved. For this reason, it seemed a wise expedient to concentrate on solving the classification problem for specific types of groups that were especially important.

As we have already seen, one such class consisted of the projective groups, i.e., the continuous subgroups of the general projective group $\mathbf{PGL}(n, \mathbb{C})$. The general projective group was the group of projective geometry in the sense of Klein's *Erlanger Programm* [1872b], and from this standpoint projective groups were important for geometrical reasons. Lie had been in close contact with Klein during the period in which he con-

[2] In a letter of introduction by Lie on behalf of Kowalewski (1 August 1898). The letter is given in its entirety in Kowalewski's autobiography [1950:85–6].

ceived the *Programm*, and in fact, Lie always stressed that the ideas it contained were shared by him at the time as well. For example, in his *Programm* Klein wrote, "When treating a manifold with a group as the starting point we first ask which configurations are left unchanged by all transformations of the group. But there are also configurations that admit some but not all transformations of the group, and these are especially interesting in the sense of the group-based treatment" [1872b:489]. Quoting from this passage, Lie claimed that the problem it implied was one that had been propounded by both of them [1884a:541–2]. In fact, as Lie pointed out, Klein's actual work involving groups and geometry was limited to the consideration of discontinuous groups, whereas continuous groups were evidently of primary concern in the *Erlanger Programm*. It was Lie who had developed a theory of continuous groups that made it possible to implement the ideas of the *Erlanger Programm*. Lie illustrated this himself by determining the curves and surfaces of three-dimensional space that are left invariant by some projective group [1882; 1893a:Ch. 9–10; 1895a]. These investigations were extended to higher dimensions by G. Fano (Section 3), who became an exponent of the application of the theory of continuous groups to geometry. Fano's *Enzyklopädie* article on this subject [1907b] was to prove stimulating to Cartan, as will be seen in Chapter 8.

Projective and linear groups also played an important role in the classification problem for transitive groups, that are defined by analogy with the concept of a transitive permutation group [Lie 1888:212]. According to Lie [1888:Ch.28], each r dimensional transitive group in n variables $\mathfrak{G}$ has associated with it a linear Lie algebra $\mathfrak{l}$ of infinitesimal motions which may be described roughly as follows. If x_0 is a point of "general position," consider all X in $\mathfrak{g}$, the infinitesimal group (Lie algebra) of $\mathfrak{G}$, that leave x_0 fixed. Since X represents the infinitesimal transformation sending any x into $x + X(x)\delta t$, the X that leave x_0 fixed satisfy $X(x_0) = 0$ and define a subalgebra $\mathfrak{h}$ of dimension $r - q$. Now, $\mathfrak{h}$ fixes x_0 but transforms the "line elements" through x_0. That is, if x is near x_0, then an X in $\mathfrak{h}$ sends x into some y and hence the line determined by x_0 and x into the line determined by x_0 and y. By means of suitable approximations, Lie argued that $\mathfrak{h}$ is isomorphic to a linear homogeneous group $\mathfrak{l}$ of infinitesimal transformations $Y = \Sigma a_{ij} x'_j \partial/\partial x'_i$, where $x' = x - x_0$. Furthermore, $\mathfrak{l}$ does not depend upon the choice of x_0 in the sense that if x'_0 is another general point and $\mathfrak{m}$ the corresponding linear algebra, then $S \in \mathbf{GL}(n, \mathbb{C})$ exists such that $\mathfrak{m} = S^{-1}\mathfrak{l}S$. Also, if $\mathfrak{G}'$ is similar to $\mathfrak{G}$, the corresponding linear groups $\mathfrak{l}$ and $\mathfrak{l}'$ satisfy the above relation.

These considerations led Lie to describe a procedure for determining all transitive groups in n variables by constructing them on the basis of knowledge of $\mathfrak{l}$ [1888:615]. It is doubtful that he believed that his procedure could actually be carried out in general, but he nonetheless indicated the steps that should be followed. The first step is to determine all "types" of linear Lie algebras $\mathfrak{l}$. Two such algebras $\mathfrak{l}$ and $\mathfrak{m}$ are of the same type if they are similar in the sense indicated above. In other words, $\mathfrak{l}$ and $\mathfrak{m}$ are equivalent n-dimensional representations of some abstract Lie algebra. The very first step in carrying out Lie's procedure is thus tantamount to determining all inequivalent finite-dimensional representations of all possible structures!

Lie indicated how his procedure could be carried out in a few special cases. He determined all transitive groups that can arise when $\mathfrak{l} \cong \mathbf{sl}(n, \mathbb{C})$ or $\mathfrak{l} \cong \mathbf{gl}(n, \mathbb{C})$ [1888:631]. He also applied the procedure to determine all primitive transformation groups in three variables [1893a:Ch.7]. Primitive groups are special types of transitive groups defined by analogy with the case of permutations. (See Section 5 for the definition.) Lie's procedure is more practicable for them [Kowalewski 1899:113]. Lie felt that it was of more value to know the primitive groups than the imprimitive ones [1893a:122]. One reason for this opinion was probably that such knowledge was necessary for determining simple groups by the method he had devised in his *Mathematische Annalen* paper of 1885 (discussed in Section 3.4). Recall that Lie had hoped to determine simple groups based upon the following theorem: If $\mathfrak{g}$ is a simple, r-dimensional Lie algebra with maximal subalgebra of dimension $r-q$, then there is a Lie algebra in q variables with the structure of $\mathfrak{g}$. Thus if all Lie algebras in q variables were known — or just the *primitive* ones—one could inspect them and pick out those that are simple. Lie had found all Lie algebras (up to similarity) for $q = 1, 2, 3$; and for the above reason he assigned Page the task of writing a dissertation determining all primitive Lie algebras in $q = 4$ variables.

Page's work underscored the computational difficulties that attended Lie's procedure as the number of variables increased. A considerable part of the difficulty arose from the fact that although a primitive group in n variables leads (in the manner indicated earlier) to an infinitesimal group of linear homogeneous transformations in n variables, not every such group corresponds to a primitive group, although there was no known way to determine this unsuitability in advance of extensive computations. Page was therefore forced to carry out pages and pages of the calculations involved in Lie's procedure only to discover, in many cases, that they do not lead to a primitive group. (Most of these negative calculations were too extensive to present in the published version [1888] of his dissertation.)

Shortly after Page completed his dissertation, Engel published some results on Pfaffian equations [1889–90] that had significant implications for the implementation of Lie's procedure in the case of primitive groups. Engel's work suggested the value of separating all linear homogeneous infinitesimal groups $\mathfrak{l}$ into two distinct categories. In the first went all those $\mathfrak{l}$ that, considered as a projective group in $n-1$ dimensions leave no points, lines, planes, etc. invariant. Since $\mathfrak{l}$ acts on the line elements through some x_0, it was natural to regard $\mathfrak{l}$ as a projective group. The $\mathfrak{l}$ that as projective groups "leave nothing planar invariant" (as it was also sometimes expressed) *always* lead via Lie's procedure to primitive groups. The remaining $\mathfrak{l}$, which constitute the second category, leave some system of Pfaffian equations invariant and may be handled using Engel's results, which also make it possible to eliminate many possibilities in advance, thereby avoiding some of the calculations that Page was forced to perform [Kowalewski 1899:115–16; 1950:91–2].

Engel's work thus enhanced the prospects for further implementing Lie's classification procedure for primitive groups. With the encouragement of Engel, one of Lie's doctoral students, Gerhard Kowalewski, dealt with this problem for $n = 5$ in his *Habilitationsschrift* [1899]. A subproblem of Lie's classification problem for primitive groups was thus to determine all projective groups in $n - 1$ dimensions that leave nothing planar invariant. As will be seen in Section 5, this problem in particular was to occupy the attention of Kowalewski, although he was never able to completely solve it. By a theorem first published by Cartan in 1909 (see Section 8.1), we now know that these groups must be semisimple and hence coincide with the faithful irreducible representations of semisimple structures. Thus Kowalewski's problem (as we shall henceforth call it) belongs in retrospect to the representation theory of semisimple Lie algebras, although he himself in no sense so conceived it.

At the end of Chapter 6 I suggested that Cartan did not work on Killing's problem or the more general problem of which it was a special case because these problems, at least in 1894, were not seemed sufficiently important from the viewpoint of Paris. There were, in fact, more important difficult problems to work on. One such problem was to develop the theory of infinite-dimensional transformation groups. Lie repeatedly stressed the value such a theory would have for applications to differential equations, and Lie's sentiments were echoed by Picard in Paris.[3] After a visit to Paris in 1893, Lie stressed the point that before a complete group-theoretic treatment of differential equations could exist, it would be necessary to specify the "structures" of all simple infinite-dimensional transformation groups [1895b:291–2]. Despite the fact that the meaning of "structure" for infinite-dimensional groups had not even been worked out, Lie optimistically conjectured that it was "not unlikely" that the classification of infinite-dimensional simple groups would turn out to be as manageable as the finite-dimensional case had turned out to be—thanks to the work of Killing and Cartan. As we shall see in Section 8.1, Cartan and Vessiot both worked on the problem of developing an adequate theory of infinite-dimensional groups, and in the case of Cartan, some of his discoveries probably played a role in his reassessment of the importance of Killing's problem.

In addition to applications of his theory to differential equations and geometry, Lie also emphasized potential applications to the theory of invariants, both differential and algebraic [Lie 1884a:545–6; 1893a:vii–viii]. The theory of algebraic invariants had originated in England with Boole, Cayley and Sylvester and had been developed extensively in Germany since the

[3] See [Lie 1883a:56; 1884a:538; 1891a:316–18; 1891b:§14; Picard 1892a:727–8].

1860's under the leadership of Clebsch and Gordan. (Some of Cayley's work is discussed in detail in Section 4.) In the minds of many at the time, invariant theory was synonymous with "modern algebra." The starting point of the theory of invariants is the transformation of a form, that is a homogeneous polynomial in variables $x_1, \ldots, x_n$. For simplicity I consider a binary form ($n = 2$) of degree m:

$$\phi(a, x) = \sum_{i=0}^{m} \binom{m}{i} a_i x_1^{m-i} x_2^i. \tag{1.1}$$

Let $x = Sx'$ denote a nonsingular linear transformation of the variables x. (For convenience I employ matrix notation, but the practitioners of invariant theory did not.) The transformation S induces a transformation of ϕ into a new form ϕ' defined by

$$\phi(a, x) = \phi(a, Sx') \stackrel{\text{def}}{=} \phi'(a', x').$$

The relation between the coefficients a' of ϕ' and the a of ϕ is given by a linear transformation $a = \mu(S)a'$. A polynomial $I(a)$ in the a_i is an invariant of the base form *(Grundform)* $\phi(a, x)$ if $I(a') = (\det S)^w I(a)$ for all S in $\mathbf{GL}(2, \mathbb{C})$. The number w is called the weight of the invariant. A basic example is the discriminant $I(a) = a_1 a_3 - a_2^2$ of the binary quadratic form

$$\phi(a, x) = a_1 x_1^2 + 2a_2 x_1 x_2 + a_3 x_2^2.$$

In this case $I(a')$, the discriminant of ϕ', equals $(\det S)^2 I(a)$. A polynomial $J(a; x)$ in the a_i and the x_j such that $J(a', x') = (\det S)^w J(a; x)$ for all S is called a covariant of ϕ. Clearly, $\phi(a, x)$ is a covariant of itself. All these definitions extend to n-ary forms. The objective of the theory of invariants was, stated loosely, to study the properties of invariants and covariants of a form (or system of forms) and, in some satisfactory manner, to describe or represent them all. More will be said about this objective in Section 2.

To anyone inclined to view mathematics in terms of groups, it is not difficult to realize that the transformations $a = \mu(S)a'$ that correspond to $x = Sx'$ in the above manner form a group since $\mu(S)\mu(T) = \mu(ST)$. Invariant theorists such as Clebsch and Gordan never called attention to this property. For them, such a viewpoint was without significance. Lie, however, stressed the group-theoretic aspects of invariant theory. In nonspecific terms, Klein had already alluded to this viewpoint in his *Erlanger Programm* of 1872, and it seems likely Lie recognized it much earlier than 1884 as well. Of course, in modern terms the invariants of the base form ϕ would be called invariants of the group defined by the representation $S \to \mu(S)$ of $\mathbf{GL}(m + 1, \mathbb{C})$ induced by the action of $\mathbf{GL}(2, \mathbb{C})$ on ϕ. The group of $\mu(S)$ will be denoted by $\mathfrak{M}_m$ in what is to follow.

Lie claimed [1884a:546] that the infinitesimal group of $\mathbf{GL}(n, \mathbb{C})$ could be used to easily compute the in- and covariants of ϕ. Some such considerations were presented in his lectures at Leipzig and later published by Scheffers [Lie 1893b:Ch.23]. Although we shall not discuss these matters, Lie's understanding of the infinitesimal group of $\mathfrak{N}_m$ is of interest to what follows. This group may be computed by applying the infinitesimal transformations of $\mathbf{gl}(2, \mathbb{C})$ to the variables x of $\phi(a, x)$, thereby inducing infinitesimal transformations of the parameters. A frequently used basis for $\mathbf{sl}(2, \mathbb{C})$ consisted of

$$P = x_1 \frac{\partial}{\partial x_2}, \quad N = x_2 \frac{\partial}{\partial x_1} \quad \text{and} \quad H = x_1 \frac{\partial}{\partial x_1} - x_2 \frac{\partial}{\partial x_2}$$

[Lie 1888:557]. Thus

$$[P, N] = H, \quad [H, P] = 2P, \quad \text{and} \quad [H, N] = -2N.$$

If to these operators we add $U = x_1 \partial/\partial x_1 + x_2 \partial/\partial x_2$, we have a basis for $\mathbf{gl}(2, \mathbb{C})$.

If we let

$$\phi_i = \binom{m}{i} x^{m-i} x^i,$$

so that $\phi(a, x) = \Sigma a_i \phi_i$, then it is readily seen that

$$H\phi_i = (m - 2i)\phi_i, \quad P\phi_i = (m - i + 1)\phi_{i-1}, \quad N\phi_i = (i + 1)\phi_{i+1}.$$

Expressed in modern terms, the linear space of binary forms of degree m is the irreducible module of $\mathbf{sl}(2, \mathbb{C})$ of highest weight m. Expressed in the notation of Lie and his school, the elements of the infinitesimal group of $\mathfrak{N}_m$ corresponding to H, P, and N are

$$(1.2) \quad \mathcal{H} = \sum_{i=0}^{m} (m - 2i) a_i q_i, \quad \mathcal{P} = \sum_{i=0}^{m} (m - i) a_{i+1} q_i, \quad \mathcal{N} = \sum_{i=0}^{m} i a_{i-1} q_i,$$

where $q_i = \partial/\partial a_i$. That is, $H\phi = \mathcal{H}\phi$, $P\phi = \mathcal{P}\phi$, $N\phi = \mathcal{N}\phi$ for any form $\phi = \Sigma a_i \phi_i$. As will be seen in Section 4, these differential operators had already been introduced into the theory of invariants by Cayley in the 1850s. As we now know, the Lie algebras (1.2) yield all the irreducible representations of $\mathbf{sl}(2, \mathbb{C})$. Lie recognized the equivalent of this fact in a form I will now describe.

Let λ_1, λ_2 denote homogeneous coordinates of points on the complex projective line $\mathbb{P}^1(\mathbb{C})$, and consider the form $\phi(\lambda) = \sum_{i=0}^{m} \binom{m}{i} a_i \lambda_1^{m-i} \lambda_2^i$. If $x = (x_0, \ldots, x_m)$ denotes the homogeneous coordinates a point in complex m-dimensional projective space $\mathbb{P}^m(\mathbb{C})$, then the correspondence $\lambda \to x$, where $x_i = \lambda_1^{m-i} \lambda_2^i$, maps the line onto a curve Γ_m of order m in $\mathbb{P}^m(\mathbb{C})$. (If we set $t = \lambda_2/\lambda_1$, then $x_i = t^i x_0$, so that in inhomogeneous coordinates $X_i = x_i/x_0$, $i = 1, \ldots, m$, $X_i = t^i$.) The curve Γ_m is called a normal curve (*Normkurve*). The m points (counting multiplicities) (λ_1, λ_2) on $\mathbb{P}^1(\mathbb{C})$ such that $\phi(\lambda) = 0$—equivalently $\sum_{i=0}^{m} \binom{m}{i} a_i t^i = 0$—for a specific form ϕ (i.e.

for specific values of the a_i) therefore correspond, under the above mapping, to the points of intersection of Γ_m with the hyperplane

$$a_0x_0 + \binom{m}{1}a_1x_1 + \binom{m}{2}a_2x_2 + \cdots + a_mx_m = 0.$$

By virtue of the mapping $\lambda \to x$, the action of $\mathbf{PGL}(2,\mathbb{C})$ on the line induces a corresponding action on $\mathbb{P}^m(\mathbb{C})$, which in turn defines a projective group in the $m+1$ variables x with the structure of $\mathbf{sl}(2,\mathbb{C})$. This group evidently transforms the normal curve Γ_m into itself. When $m = 2$ the normal curve is the conic section $X_2 = (X_1)^2$. With the representation for $m = 2$ in mind, Lie always referred to $\mathbf{PGL}(2,\mathbb{C})$ as the projective group of a conic section. He also named the projective groups in the $m+1$ variables x the projective groups of a normal curve. These are the groups $\mathfrak{N}_m$ discussed above.

The above presentation of normal curves is essentially that presented by Klein in his lectures on higher geometry at Göttingen in 1892–93 [1893a:388–91]. For Klein the significance of the mapping $\lambda \to x$ was that it showed that the projective study of the points of $\mathbb{P}^1(\mathbb{C})$ satisfying $\phi(\lambda) = 0$, or, equivalently, the invariant theory of binary forms $\phi(\lambda)$, was equivalent to the projective study of the relations between a normal curve and its intersecting hyperplanes—the geometry defined by the projective group associated with the normal curve. In words, and with attention focused primarily upon the case $m = 2$, Klein had already communicated these ideas in his *Erlanger Programm* [1872b:§§4–5]. The case $m = 2$ had been first studied by Hesse [1866], who referred to the underlying idea as a principle of transfer *(Uebertragungsprinzip)*, and Klein adopted this term. Klein's generalization of Hesse's principle of transfer was generalized even further by Cartan, as will be seen in Chapter 8.

A detailed study of Hesse's paper and the metamorphosis of his principle of transfer up to Cartan is given in my paper [1988]. Here it should be emphasized that neither Hesse nor the other mathematicians who proceeded to study normal curves did so from a group-theoretic perspective. By contrast, Lie was primarily interested in the projective groups of normal curves—in effect, the irreducible representations of $\mathbf{sl}(2,\mathbb{C})$ defined by (1.2). Undoubtedly, Lie shared Klein's view [1893a:390] that these groups are particularly important. In the third volume of *Theorie der Transformationsgruppen* he stated that these groups leave nothing planar invariant and that any projective group with the structure of $\mathbf{sl}(2,\mathbb{C})$ leaving nothing planar invariant is the projective group of a normal curve [1893a:758n]. In effect, Lie had characterized all irreducible representations of $\mathbf{sl}(2,\mathbb{C})$. Precisely when he first made these observations, to what extent he proved them and how much significance he attached to them at the time of discovery is regrettably unclear. By 1893 they had assumed an importance by virtue of the work of Eduard (1862–1930), as Lie fully realized.

Lie saw in Study the ideal person to realize his hopes of a fruitful union between the theory of invariants and his theory of transformation groups. As will be seen in the next section, Study held strong opinions on the proper way to conceive of and express mathematics, opinions that were in many respects in harmony with the attitudes fostered by Lie and his school. Study's own outlook was broadened by his contact with Lie and his students, and it was precisely his broadened outlook that led from his work on the theory of invariants to the articulation of complete reducibility theorems. Such theorems, of course, call attention to the components in the reduction—projective groups that leave nothing planar invariant—and the problem of characterizing them. Study was thus led to consider this problem as well.

7.2. Eduard Study

By the time Study came into contact with Lie and his school, he had already formulated a general research program for his career as a mathematician. Under the influence of Grassmann's *Ausdehnungslehre*, which he had digested as a university undergraduate, Study set himself the task of carrying out research aimed at providing geometrical knowledge with a suitable conceptual and algebraic framework.[4] He was less concerned with personally advancing the frontiers of mathematical knowledge than with the problem of providing an appropriate framework for what was already known. The sixth thesis propounded along with his doctoral dissertation (Munich, 1884) expressed his attitude: "Less effort should be expended on extending the boundaries of the mathematical sciences and far more on regarding what is already at hand from more comprehensive viewpoints" [Engel 1931:140]. Initially, Study was of the opinion that Grassmann's *Ausdehnungslehre* provided an adequate basis for geometry, but by the time he penned the above words he had become convinced that such a basis was to be found rather in the theory of invariants and particularly in the symbolical method employed by Clebsch and Gordan and their followers. The theory of invariants thus became, and remained, at the center of Study's master plan for mathematical research.

In 1885 Study completed his *Habilitationsschrift* (1886) under the supervision of Klein at the University of Leipzig. But neither Klein nor the other mathematicians at Jena, Strassburg, Munich, and Leipzig that Study had previously come into contact with during his predoctoral studies appear to have exerted any significant influence upon him and his master plan for mathematical research [Engel 1931:137]. Study's first exposure to the ideas of Lie's theory of groups probably came through contact with Engel. As explained at the beginning of Chapter 3, Engel had returned to

---

[4] See [Engel 1931:140] and [Weiss 1933:109].

Leipzig from Norway in July 1885 after spending ten months there with Lie. Study quickly became a close personal friend. During the four semesters in 1885–1888 that Study spent as instructor (*Privatdozent*) at Leipzig, he was constantly in Engel's company [Engel 1931:137]. Thus when Lie arrived in Leipzig in April of 1886 as Klein's replacement, Study was undoubtedly already favorably disposed toward Lie's theory. Lie, for his part, took a personal liking to Study.[5] Their relationship was much more amicable than the one Study had with Klein. One reason was that Lie did not attempt to do what Klein had done. Seeing that Study was a geometer with considerable knowledge of—and commitment to—the theory of invariants, a subject far too algebraic for Lie's own tastes, Lie encouraged Study to continue his study of invariant theory and to develop the fruits of its connections with the theory of transformation groups.[6] As a further inducement, Lie asked Study to lecture on the theory of invariants with Lie's general viewpoint in mind during the summer semester 1887 for the benefit of his better students.[7] Perhaps this was one reason why Study decided to spend the month of January 1887 in Erlangen, intensively studying the theory of invariants under the personal tutelage of Gordan, who, since the untimely death of Clebsch in 1872, had become the leading authority on the subject on the continent.

By the time of his visit to Erlangen, Study had clearly become a believer in Lie's theory of groups. In his letters to Engel during this period we find him recounting his proselytizing efforts to interest Gordan in transformation groups, a difficult task, since Gordan was completely ignorant of the subject.[8] Lie was delighted to have Study numbered among the members of his school. "I have some hope for Study," he wrote to Klein. "He has become interested in the connection between transformation groups and the theory of invariants. I cannot be certain, but I have the *distinct impression* that he will contribute something in this direction .... I am naturally very pleased to have gained capable mathematicians such as Study ...."[9] The fullest and most forceful expression of Study's belief in the importance of Lie's theory is contained in his twenty-page review [1889b], done at Lie's urging, of the first volume of *Theorie der Transformationsgruppen* [1888]. Here Study endorsed Lie's view that the theory of continuous trans-

5 This is reflected in Lie's letters to Klein (NSUB Göttingen, Nachlass Klein X, Nr.735, 736, 741, 750, 751, 762).

6 See Lie's letters to Klein written ca November 1886 and 8 December 1886 (NSUB Göttingen, Nachlass Klein X, Nr. 728, 730).

7 Letter of Lie to Klein ca November 1886 (NUUB Göttingen, Nachlass Klein X, Nr. 728).

8 Letters of Study to Engel dated 22 January 1887 and 11 April 1887 (Engel Archive Giessen). That Study failed in his efforts to convert Gordan is suggested by a letter from Engel to Study (26 February 1893) describing an evening Engel spent with Gordan, who complained that Lie and Engel bragged about group theory but could not tell him what he wanted to know.

9 Letters of Lie to Klein (NSUB Göttingen Nachlass Klein XI, Nr. 741 (ca 1888) and Nr. 751 (ca 1889)).

Figure 20. The above oil painting of Study by Wilhelm Fassbinder, of Bonn, was done in 1928, many years after the events traced in this section. Just a few years before the painting, however, Study reenters our story, in the self-styled role of a villain (Chapter 11, Section 4).

formation groups is justified in the first place by virtue of the envisioned Galois-type theory of differential equations. The application to differential equations therefore required a "penetrating study" of transformation groups themselves (pp. 171–2). In particular, Study declared Lie's group classification problem to be a fundamental problem not just for Lie's theory but for mathematics in general (p. 184). This was because Study, like Lie, believed that "there are few areas of mathematical science that could not experience an essential advance by adopting the basic ideas of the new discipline" (p. 189). Study's conviction is reflected in his work on the theory of invariants.

A fundamental tool in the theory of invariants was the method of series expansions (*Reihenentwicklungen*), which had been introduced by Clebsch and Gordan. By viewing these expansions within the conceptual framework of Lie and his school, Study discovered that they become what would now be termed complete reducibility theorems. Such a conceptual framework was, however, totally foreign to Clebsch and Gordan. Before indicating Study's reinterpretation of series expansions it is historically enlightening to present series expansions in the spirit of the original formulation. The presentation will employ the notation of the symbolical method esteemed so highly by Study. Besides facilitating a presentation more in the spirit of the period, the notation makes many definitions and properties of forms more evident and should therefore aid in the comprehension of the material.

Consider for simplicity a binary form

$$\phi(a,x) = \sum_{i=0}^{n} \binom{n}{i} a_i x_1^{n-i} x_2^{i}.$$

The symbolical method is based upon the analogy with the binomial expansion

$$(p_1x_1 + p_2x_2)^n = \sum_{i=0}^{n} \binom{n}{i} p_1^{n-i} p_2^{i} x_1^{n-i} x_2^{i}.$$

The form $\phi(a,x)$ may be represented by $\phi = (p_1x_1{+}p_2x_2)^n$, or, more briefly, by $\phi = (p,x)^n$, where $(p,x) = p_1x_1{+}p_2x_2$, provided that $p_1^{m-i}p_2^i$ is identified with a_i.[10] In symbolical notation the transformation of ϕ into ϕ' induced by $x = Sx'$ (in the notation of Section 1) becomes $\phi(a,x) = (p,x)^n = (p,Sx')^n = (S^Tp,x')^n = (p',x')^n = \phi'(a',x')$, and the transformation $a' = \mu(S^{-1})a$ may be computed by expanding

$$a_i' = (p_1')^{n-i}(p_2')^i = (s_{11}p_1 + s_{21}p_2)^{n-i}(s_{12}p_1 + s_{22}p_2)^i,$$

where the s_{ij} are the coefficients of S. From this expression it is also evident that a_i' is a linear homogeneous function of the $a_i = p_1^{n-i}p_2^i$. Many computations and formulas are simplified by use of the symbolical notation. For example, if $\phi = (p,x)^n = (q,x)^n = (r,x)^n = \cdots$, any expressions of the form $[p,q]^k, [p,r]^k, [q,r]^k, \ldots$, where $[p,q] = p_1q_2 - p_2q_1$, etc., that can be interpreted nonsymbolically (with $a_i = p_1^{n-i}p_2^i = q_1^{n-i}q_2^i =$, etc.) will represent an invariant of ϕ, since $[p',q'] = (\det S)[p,q]$. Likewise, any expression involving a product of terms like $[p,q]^k$ and $(p,x)^k$, $(q,x)^k$ etc., that can be interpreted nonsymbolically will represent a covariant of ϕ, since $(p',x') = (p,x)$. In fact, a fundamental theorem of the theory of binary

[10] The notation (p,x) is that of Study [1889a]. Clebsch and Gordan wrote p_x. For a clear, elementary exposition of the polarization process and its role in the symbolical method see Weyl's book on the classical groups [1946:5–6, 243–5]. See [Rota 1982:6–7] or [Kung and Rota 1984:§2] for a modern interpretation. On the origins of the symbolical method, as well as other aspects of the early history of invariant theory, see [Parshall 1989].

forms states that all in- and covariants of f can be represented symbolically as sums of products of the two types $[p, q]$ and (p, x).[11]

Symbolical in- and covariants are examples of what are sometimes called vector invariants. They are homogeneous polynomials in d-dimensional vectors (with $d = 2$ in the binary case considered here) $p, q, x, \ldots$ that are invariants, in the modern sense, of the group $\mathbf{GL}(d, \mathbb{C})$. As such they are easier to compute and deal with then the nonsymbolical invariants, which are homogeneous polynomials in the $N \gg d$ coefficients a_i of the base form $\phi(a, x)$ and are invariants in the modern sense with respect to the group of transformations $\mu(S) \in \mathbf{GL}(N, \mathbb{C})$. That is the great virtue of the symbolical method. It depends upon the fundamental proposition that every symbolical in- or covariant corresponds to an actual, nonsymbolical in- or covariant. See Section 11.4 for further discussion of this proposition in the context of Weyl's response to Study's criticism of his book on general relativity.

The method of series expansions for binary forms was developed independently by Clebsch [1872a:§7] and Gordan [1871: §2]. Consider a binary form in two series x and y of variables $\phi(a, x, y) = (p, x)^m (q, y)^n$. As in the case of one variable series x, the linear transformation $x = Sx'$, $y = Sy'$ induces a transformation of ϕ into $\phi'(a', x', y')$ and a corresponding linear transformation of parameters $a = \mu(S)a'$. Thus the in- and covariants of ϕ may be defined and studied. The method of series expansions showed how the invariant theory of ϕ can be reduced to the invariant theory of a system of forms in the single-variable series x. It accomplished this by means of the following expansion of $\phi(a, x, y)$ in terms of powers of $[x, y] = x_1 y_2 - x_2 y_1$:

$$\text{(2.1)} \qquad \phi = (p, x)^m (q, y)^n = \sum_{i=0}^{l} c_i \phi_i(x, y) \cdot [x, y]^i, \quad l = \min\{m, n\}.$$

In (2.1) the c_i are rational numbers given by explicit formulas. The $\phi_i(x, y)$ are forms in x and y that are derived by a process called polarization from forms

$$e_i = [p, q]^i (p, x)^{m-i} (q, x)^{n-i}$$

in x alone.[12] From their form it is clear that the e_i are covariants of ϕ. Gordan called them the *elementary covariants* of ϕ. It should be noted that the e_i are linear covariants in the sense that they are linear and homogeneous functions of the coefficients a_i of ϕ. This can be seen from the symbolical mode of representing the e_i. Since the polarization process does not alter

---

[11] See [Clebsch 1872a:32] and [Gordan 1887:110ff.]. Clebsch and Gordan denoted $[a, b]$ by (a, b) (which conflicts with Study's notation).

[12] The polarization process that leads from e_i to $f_i(x, y)$ may be defined as follows. If ϕ is any form in x of degree k, set $\mathcal{P}\phi = (1/k)\mathcal{D}\phi$, where $\mathcal{D}$ denotes the differential operator $\mathcal{D} = y_1(\partial/\partial x_1) + y_2(\partial/\partial x_2)$. Hence $\mathcal{P}^2\phi = \{1/k(k-1)\}\mathcal{D}^2\phi$, etc. Then $\phi_i(x, y) = \mathcal{P}^{l-i} e_i$.

this linearity property, the same is true of the $\phi_i(x,y)$ and hence of the terms $c_i\phi_i(x,y)\cdot[x,y]^i$.

I will illustrate these generalities with an example that will also be instructive further on. Let $\phi=(p,x)^2(q,y)$. In nonsymbolical form ϕ has the representation

$$\phi(a,x,y)=a_1x_1^2y_1+a_2x_1^2y_2+2a_3x_1x_2y_1+2a_4x_1x_2y_2+a_5x_2^2y_1+a_6x_2^2y_2.$$

The elementary covariants $e_0=(p,x)^2(q,x)$ and $e_1=[p,q](p,x)$ have the nonsymbolical representations

$$\begin{aligned} e_0 =& b_1x_1^3+3b_2x_1^2x_2+3b_3x_1x_2^2+b_4x_2^3,\\ e_1 =& b_5x_1+b_6x_2, \end{aligned}$$

where

$$\begin{aligned} b_1 =& a_1, & \qquad & b_4 = a_6,\\ b_2 =& \tfrac{1}{3}a_2+\tfrac{2}{3}a_3, & \text{and} \qquad & b_5 = a_2-a_3,\\ b_3 =& \tfrac{2}{3}a_4+\tfrac{1}{3}a_5, & & b_6 = a_4-a_5. \end{aligned} \tag{2.2}$$

The series development is

$$\phi=\phi_0+\frac{2}{3}\phi_1(x,y)\cdot[x,y],$$

where

$$\phi_0=b_1x_1^2y_1+b_2(x_1^2y_2+2x_1x_2y_1)+b_3(2x_1x_2y_2+x_2^2y_1)+b_4x_2^2y_2,$$

and

$$\phi_1(x,y)\cdot[x,y]=b_5\left(x_1^2y_2-x_1x_2y_1\right)+b_6\left(x_1x_2y_2-x_2^2y_1\right).$$

Let us now consider some consequences of the series expansion (2.1). For the sake of concreteness, I will focus upon the above example, but the conclusions are generally valid. In general, the coefficients b_i of the e_i will be linear homogeneous functions of the a_i, as in (2.2), because the e_i are linear covariants of ϕ. Corresponding to this manner of expressing the e_i we have also

$$\begin{aligned} \phi_0 =& \; b_1\psi_1+b_2\psi_2+b_3\psi_3+b_4\psi_4,\\ \phi_1(x,y)\cdot[x,y] =& \; b_5\psi_5+b_6\psi_6, \end{aligned} \tag{2.3}$$

where $\psi_1=x_1y_1$, $\psi_2=x_1^2y_2+2x_1x_2y_1$, etc. The correspondence between the coefficients of the e_i and the $\phi_i(x,y)$ is not a coincidence but a consequence of the polarization process when applied to sums of forms of the same degree in x. In our example the number of b_i equals the number of a_i. This will be generally true, since each e_i, being of degree $m+n-2i$, will have $m+n-2i+1$ coefficients, and $\sum_{i=0}^{l}(m+n-2i+1)=(m+1)(n+1)$, the number of a_i. Now (2.1) asserts that every form $(p,x)^m(q,y)^n$ can be expressed as a linear combination of the $\phi_i(x,y)\cdot[x,y]^i$. Thus the b_i must be linearly independent functions of the a_i. In matrix form, $b = Ma$ with $\det(M) \neq 0$. The independence of the b_i also shows that each e_i represents all forms in x of degree $m+n-2i$ as the a_i run through all possible values.

The transformation $x = Sx'$ induces a transformation of $e_i(b,x)$ into $e_i'(b',x')$ and thus a transformation $b = \nu(S)b'$. Since $b = Ma$ and since (2.1) is a formal identity in a, x, and y, it remains valid with these replaced by a', x', and y', so that $b' = Ma'$ also and therefore $\nu(S) = M\mu(S)M^{-1}$, where $a = \mu(S)a'$. This means that $\phi(a,x,y)$ is a simultaneous covariant of e_0 and e_1. That is, if

$$\phi(a,x,y) = \phi(M^{-1}b,x,y) = \phi^*(b,x,y),$$

then $\phi^*(b',x',y') = \phi^*(b,x,y)$. Consequently, the invariants of ϕ and the simultaneous invariants of e_0 and e_1 amount to the same thing: all polynomials $I(b)$ satisfying $I(b') = (\det S)^w I(b)$. Such an invariant may be expressed as an invariant $J(a)$ for $\phi(a,x,y)$ by setting $J(a) = I(Ma) = I(b)$.

As this example illustrates, by virtue of the series expansion (2.1), ϕ is a simultaneous covariant of its elementary covariants e_i, and the problem of determining all in- and covariants of ϕ reduces to the determination of all in- and covariants of the e_i, a system of forms in x alone. There were other applications of (2.1) as well. As Gordan's student Georg Kerschensteiner explained in his edition of Gordan's lectures, "In the first place, series developments serve to transform complicated symbolical products into simpler ones or to represent unknown forms by known ones. They are thus the most important tool for the symbolical calculus, especially for the specification of the linearly independent co- and invariants of a form of the nth degree" [Gordan 1887:90–1]. Series expansions were used, for example, to prove the fundamental theorem, mentioned earlier, on the symbolical representation of the in- and covariants of a binary form in terms of expressions of the form $[p,q]$ and (p,x).

During the 1860s, 70s, and 80s the theory of invariants had received its most extensive and systematic development in the case of binary forms. Nonetheless, in this period Clebsch and Gordan did concern themselves with certain aspects of the theory of n-ary forms for $n > 2$. For example, Clebsch pointed out in [1872b] that geometrical applications suggested the value of considering, in the n-ary case, $n > 2$, several different types of variable and forms and invariants in these new variables. Clebsch's observations were motivated by Plücker's introduction of point, line, and plane coordinates in ordinary three-dimensional space. Here it will suffice to consider, as did Study, point and line coordinates in the plane. The more general situation is discussed in the next section. Let x_1, x_2, x_3 denote homogeneous point coordinates in the plane. The corresponding line coordinates u_1, u_2, u_3 of the line through points x and y are defined by the determinants

$$(2.4) \qquad u_1 = \begin{vmatrix} x_2 & x_3 \\ y_2 & y_3 \end{vmatrix}, \quad u_2 = \begin{vmatrix} x_1 & x_3 \\ y_1 & y_3 \end{vmatrix}, \quad u_3 = \begin{vmatrix} x_1 & x_2 \\ y_1 & y_2 \end{vmatrix}.$$

Line coordinates transform contragrediently in the sense that if $x = Sx'$, then $u = [S^T]^{-1}u'$, so that $(x', u') = (x, u)$, where $(x, u) = x_1u_1 + x_2u_2 + x_3u_3$. Thus the needs of geometry suggested the value of studying the invariant theory of ternary forms in several series of point coordinates $x, y, z, \ldots$ and line coordinates $u, v, w, \ldots$. The simplest case would be a form in one of each type of variable, viz., $\phi(a, x, u) = (p, x)^m(q, u)^n$. As before, $x = Sx', u = [S^T]^{-1}u'$ transforms ϕ into

$$\phi'(a', x', u') = (p, Sx')^m(q, [S^T]^{-1}u')^n = (S^Tp, x')^m(S^{-1}q, u')^n$$

and defines a linear transformation $a = \mu(S)a'$, which is a representation of $\mathbf{GL}(3, \mathbb{C})$.

As in the case of binary forms, by means of series expansions the study of forms in several series of point and line coordinates can be reduced to the study of forms $\phi(a, x, u)$ [Clebsch 1872b:21]. Gordan showed [1872:§4] that the reduction could be carried one step further by means of another series expansion of the form

$$\phi(a, x, u) = \sum_{i=0}^{l} c_i\phi_i(x, u) \cdot (x, u)^i, \quad l = \min\{m, n\}. \tag{2.5}$$

The coefficients c_i are rational numbers and given by explicit formulas. The form $\phi_i(x, u)$ is of degree $m-i$ in x and $n-i$ in u and satisfies the condition $\mathcal{D}\phi_i = 0$, where $\mathcal{D}$ is the differential operator

$$\mathcal{D} = \frac{\partial^2}{\partial x_1 \partial u_1} + \frac{\partial^2}{\partial x_2 \partial u_2} + \frac{\partial^2}{\partial x_3 \partial u_3}.$$

As in the binary case, the $\phi_i(x, u)$ are linear covariants of ϕ. Study called forms satisfying $\mathcal{D}f = 0$ *normal forms*. For example, if $\phi = (p, x)^2(q, u)$ is the general form of degrees $(2, 1)$ with nonsymbolical representation

$$\phi = a_{111}x_1^2u_1 + \cdots + a_{113}x_1^2u_3 + 2a_{121}x_1x_2u_1 + \cdots + 2a_{233}x_2x_3u_3,$$

then the condition $\mathcal{D}\phi = 0$ implies that

$$\begin{aligned} a_{111} + a_{122} + a_{133} &= 0, \\ a_{121} + a_{222} + a_{233} &= 0, \\ a_{131} + a_{232} + a_{333} &= 0. \end{aligned} \tag{2.6}$$

Thus, whereas the general form of type $(2, 1)$ has 18 independent coefficients a_{ijk}, the normal form of type $(2, 1)$ has only 15 independent coefficients by virtue of (2.6). The condition $\mathcal{D}\phi = 0$ imposed upon a general form of type (m, n) likewise leads to a system of linear homogeneous equations in the coefficients a.

The above presentation of series expansions has been more or less in the spirit in which they were treated by Clebsch and Gordan, namely as theorems and constructs of a purely algebraic, computational nature. This formal machinery was, of course, envisioned as applying to geometry, especially by Clebsch, but the machinery itself was not conceptualized in a geometrical fashion. Study, by contrast, was a firm believer in a geometrical–conceptual mode of thinking, a mode of thinking perhaps encouraged initially through his familiarity with Grassmann's *Ausdehnungslehre*. Such a way of regarding mathematics is already evident in his *Habilitationsschrift* of 1886 on the characteristic problem for conics.[13]

In homogeneous coordinates a conic is represented by the vanishing of a quadratic ternary form $\phi = (p, x)^2$. Study conceived of the six nonsymbolical coefficients a_i of ϕ as the homogeneous coordinates of a point in five-dimensional projective space. Such an interpretation, explained Study, "does not simply serve to provide a much clearer grasp of known things by taking into account, in the form of its theorems, the existing analogies with the relations of ordinary geometry. It also leads to new truths that would be very difficult to discover in another way simply because their very articulation without utilization of this analogy would become extraordinarily complicated" (p. 71). Study pointed out that he was not the first to employ such a viewpoint and referred to a paper [1868] by Cayley, who employed a "quasigeometrical representation of conditions by means of loci in hyperspace." That is, if $\phi(x, y, z)$ is the general homogeneous polynomial in x, y, and z of degree n, its N coefficients a_i can be regarded as the homogeneous coordinates of an arbitrary point in $(N-1)$-dimensional projective space $\mathbb{P}^{N-1}(\mathbb{C})$. Special curves of degree N are given by $\phi(x, y, z) = 0$ by specifying conditions (equations) that the coefficients a_i must satisfy. These conditions then define a locus in $\mathbb{P}^{N-1}(\mathbb{C})$ that may be described geometrically. Cayley, too, had in mind principally conics. For example, in the five-dimensional space of all conics he considered those that pass through two given points and touch a given conic (Annex 3). He described the properties of these conics geometrically by describing the properties of the sextic surface in five-dimensional space determined by their coefficients. According to Cayley, he got the idea of a "quasigeometrical representation" from Salmon.[14] Prior to Study, Veronese [1884] and Segre [1885] had inde-

---

[13] On the problem and Study's contribution to its solution, consult [Coolidge 1945:130–7], and, for a more modern perspective, [Kleiman 1980].

[14] Cayley referred to Salmon's papers [1866; 1867]. In these papers Salmon was concerned with the study of p simultaneous equations in p unknowns for $p > 3$. For $p = 3$, the problem can be interpreted geometrically as the intersection of three surfaces in space. "The question now before us may be stated as the corresponding problem in a space of p dimensions. But we consider it as a purely algebraic question, apart from any geometrical considerations .... We shall however retain the geometrical *language*, both because we can thus avoid circumlocutions, and also because we can the more readily see how to apply to a system of p equations, processes analogous to those which are employed in a system of three" [1866:327–8].

pendently studied conics from Cayley's viewpoint. Study also appreciated the fact that this viewpoint is implicit in the n-dimensional formulation of Hesse's principle of transfer, as explained in Section 1 above.

According to Study's geometrical–conceptual viewpoint it was advantageous to transform algebraic problems into a geometrically flavored conceptual setting derived by analogy with geometrical notions from ordinary space. In this manner algebraic objects such as forms became geometrical objects such as points or linear spaces. For example, a binary form $\phi = \sum_{i=0}^{n} \binom{n}{i} a_i x_1^{n-i} x_2^i$ represented a point with homogeneous coordinates $a_0, a_1, \ldots, a_n$, and the totality of all such forms represented a linear variety of projective dimension n and *Stufenzahl* $n+1$. The *Stufenzahl* is thus the vector space dimension and will be referred to accordingly in what follows. In effect, the "general" binary form of degree n corresponded to projective n-dimensional space. This point of view was closely tied to the tendency in nineteenth-century mathematics to generalize the analytic geometry of the plane and space to "space" of n dimensions, but more was involved. It is one thing to extend to n dimensions or variables propositions known for $n = 2$ and 3. It is another thing to take algebraic or analytical problems and consciously translate them into a geometrical mode so as to deal with them in a conceptually and intuitively satisfying manner by means of n-dimensional geometry. Nowadays, such a practice is commonplace, but relatively few nineteenth-century mathematicians explicitly followed it in their publications, although some may have utilized it as a heuristic device and then suppressed it or clothed it with analytical formalities in their publications. In the case of Study, however, there is no doubt.

Klein and Lie were also practitioners of geometrical–conceptual reasoning. For example, in his lectures Lie would frequently declare, "Let us reason with concepts!" and would then use a picture to replace an analytical proof [Kowalewski 1950:51–2]. Lie's (and Klein's) geometrical–conceptual outlook was, however, special in that it had a strong group-theoretic orientation. In the case of Lie in particular, not only was multivariate analysis translated into—or even initially conceived by means of—the geometrical imagery of higher-dimensional manifolds, but emphasis was placed upon the consideration of how points or other objects are transformed by the action of some group. When Study came into contact with Lie's school he was already a practitioner of geometrical–conceptual reasoning, and this undoubtedly was one reason why Lie took such a liking to him. But through his contact with Lie, Study's own conceptual perspective acquired an additional, group-theoretic dimension. It was from this enriched conceptual perspective that Study regarded the series expansions of Clebsch and Gordan.

Consider the series expansion (2.1) for the binary form $\phi(a, x, y) = (p, x)^m (q, y)^n$. Let $\mathfrak{R}$ denote the vector space determined by all such forms. Each term $c_i \phi_i \cdot [x, y]^i$ of the expansion represents, by allowing the coefficients a of ϕ to take on all possible values, a linear submanifold $\mathfrak{R}_i$ of $\mathfrak{R}$. Thus in the series expansion of ϕ, corresponding to (2.3) in Study's view, the term ϕ_0 corresponds to the vector space $\mathfrak{R}_0$ consisting of all linear combinations of the four forms $\psi_1 = x_1^2 y_1$, $\psi_2 = x_1^2 y_2 + 2x_1x_2y_1$, $\psi_3 = 2x_1x_2y_2 + x_2^2 y_1$, $\psi_4 = x_2^2 y_2$. Likewise, the term $\phi_1(x, y) \cdot [x, y]$ corresponds to a vector space $\mathfrak{R}_1$ of dimension 2 spanned by $\psi_5 = x_1^2 y_2 - x_1x_2y_1$ and $\psi_6 = x_1x_2y_2 - x_2^2 y_1$. Now, $\mathfrak{R}$ is acted on by the group $\mathfrak{G}$ of transformations $a = \mu(S)a'$. In his work on the theory of invariants, Study generally required that $\det S = 1$, so that $\mathfrak{G}$ has the structure of "the projective group of a conic section," the Lie algebra of $\mathfrak{G}$ being isomorphic to $\mathbf{sl}(2, \mathbb{C})$. Since the terms $c_i\phi_i(x, y) \cdot [x, y]^i$ are covariants of ϕ, each $\mathfrak{R}_i$ is left invariant by the group $\mathfrak{G}$. Furthermore, the action of $\mathfrak{G}$ on $\mathfrak{R}_i$ is identical to the action of the group $\mathfrak{G}_i$ induced by applying S in $\mathbf{SL}(2, \mathbb{C})$ to the elementary covariant e_i, or, equivalently, to the general binary form of degree $m + n - 2i$ in x. This is evident in the above example of $\phi = (p, x)^2(q, y)$, and its general validity follows from invariant-theoretic principles.

The group $\mathfrak{G}_i$ is the projective group of a normal curve, as discussed in Section 1, namely the group denoted there by $\mathfrak{N}_{m+n-2i}$. Since $\mathfrak{G}_i$ leaves no linear submanifold invariant, the same is true of the action of $\mathfrak{G}$ on $\mathfrak{R}_i$. Incidentally, it should be clear now why the isomorphism

$$(2.7) \qquad \mathcal{V}(m) \otimes \mathcal{V}(n) \cong \bigoplus_{i=0}^{l} \mathcal{V}(m + n - 2i), \quad l = \min\{m, n\},$$

where $\mathcal{V}(k)$ denotes the irreducible $\mathbf{SL}(2, \mathbb{C})$ module of highest weight k, is called the Clebsch–Gordan series. There is, however, a great conceptual gap between (2.7) and the actual series expansion (2.1) of Clebsch and Gordan, a gap that Study first began to bridge.

Study never presented the above interpretation of the series expansion (2.1) in print because he never had an occasion to expound the theory of binary forms, but he clearly realized it.[15] His publications on the theory of invariants concentrated more on ternary forms, which were of considerable interest for the geometry of the plane. Study was also the first to publish a systematic treatise on the theory of ternary forms in his book *Methoden zur Theorie der ternären Formen* [1889a]. Section 12 of Part II is entitled "geometrical significance of series expansions" and contains the analogous interpretation of the ternary case. Conscious that such material was not customary in a treatise on the theory of invariants, Study began by justifying its inclusion on the grounds that it made it possible to perceive conceptually (*begrifflich einzusehen*) the theorems on series expansions, or at least to elucidate them by means of parallel conceptual considerations. Moreover, "We do not in the least wish to suppress such a supplement to our treatise, since I believe we obtain thereby a deeper sense of the essence of the series expansions of the theory of invariants" (p. 114).

---

[15] Study's preoccupation with the group-theoretic aspects of binary forms, however, is evident from his letters to Engel written from Erlangen, e.g., those of 22 January and 27 February 1887 (Engel Archive Giessen).

Consider the series expansion (2.5) of the ternary form $\phi(a,x,u) = (p,x)^m(q,u)^n$ and the vector space $\mathfrak{N}$ to which it corresponds. $\mathfrak{N}$ is transformed by the linear group $\mathfrak{L}$ of transformations $a = \mu(S)a'$, $S \in \mathbf{SL}(3,\mathbf{C})$, described above. As in the binary case, each term $\phi_i(x,u) \cdot (x,u)^i$ in the expansion corresponds to a vector subspace $\mathfrak{N}_i$ of $\mathfrak{N}$ that is transformed into itself by $\mathfrak{L}$, since $\phi_i(x,u) \cdot (x,u)^i$ is a linear covariant of ϕ. Study called the forms $\phi_i(x,u)$, which are of degrees $m-i$ and $n-i$ in x and u respectively, the elementary covariants of ϕ. If, as in the binary case above, we consider the coefficients b_j of these ϕ_i, they are likewise linear homogeneous functions of the coefficients a_j, but they are not all linearly independent, due to the fact that the b_j corresponding to a ϕ_i must satisfy the linear homogeneous equations imposed by the condition of normality $\mathcal{D}\phi_i = 0$, such as those exemplified by (2.6). Study, however, had proved that Gordan's expansion (2.5) is unique in the sense that if $\phi = \sum_{i=0}^{l} c_i'\phi_i'(x,u) \cdot (x,u)^i$, where the $\phi_i'(x,u)$ are normal and of degrees $m-i$ and $n-i$ in x and u respectively, then $\phi_i' = \phi_i$ and $c_i' = c_i$ for all i (p. 56). From this property he could show that the number of independent b_j is equal to the number of a_j (p. 57).

To translate this algebraic property succinctly into geometrical terms, Study defined two vector subspaces $\mathfrak{N}_1$ and $\mathfrak{N}_2$ of $\mathfrak{N}$ to be completely skew (*vollständig schief*) if the sum of their dimensions equals the dimension of the smallest linear manifold containing both. The series expansion (2.5) then became the geometrical assertion that $\mathfrak{N}$ contains vector subspaces $\mathfrak{N}_0, \ldots, \mathfrak{N}_l$ that are left invariant by $\mathfrak{L}$, are pairwise completely skew, and have dimensions adding up to that of $\mathfrak{N}$. Study also showed that $\mathfrak{L}$ leaves no proper subspace of any $\mathfrak{N}_i$ invariant. (We shall discuss this matter further below.) In effect he recognized what would now be called the complete reducibility of $\mathfrak{N}$. This is what his geometrical interpretation amounted to. In fact, Study also pointed out that $\mathfrak{N}$ is characterized by the property, later embodied in Maschke's theorem (Section 10.4), that to every vector subspace there belongs a second that is completely skew to it and such that the sum of their dimensions is that of $\mathfrak{N}$ (p. 117).

The vector space $\mathfrak{N}_i$ that corresponds to the term $\phi_i(x,u) \cdot (x,u)^i$ can, as an $\mathbf{SL}(3,\mathbf{C})$-module, be identified with the module corresponding to the normal form $\phi = (p,x)^m(q,u)^n$, $\mathcal{D}\phi = 0$. Study recognized this in his own terms and, as a prelude to his geometrical interpretation of series expansions, devoted a section to "The manifold of all normal forms (m,n)" (Part II, Section 11). As Study observed, this is a subspace, $\mathfrak{N}_1$, of the space $\mathfrak{N}$ of all forms $f = (p,x)^m(q,u)^n$ and has dimension

$$\frac{(m+1)(n+1)(m+n+2)}{2}.$$

The fact that no proper subspace of $\mathfrak{N}_1$ is left invariant, however, did not follow from known results on ternary forms, which concentrated on the general form of type (m,n). But it did follow from Study's extension of known theorems on the symbolical representation of the in- and covariants of general ternary forms to normal ones (p. 70). From its dimension it can be seen that $\mathfrak{N}_1$ is the $\mathbf{SL}(3,\mathbb{C})$-module of highest weight $\lambda = m\lambda_1 + n\lambda_2$, where the λ_i are the fundamental dominant weights.

Study's geometrical interpretation of series expansions raised two purely group-theoretic questions, which he discussed with Lie and Engel in 1887. Suppose that $\mathfrak{L}$ is any linear group in the N homogeneous coordinates of a linear manifold $\mathfrak{N}$ that defines a projective group, i.e., a subgroup of $\mathbf{PGL}(N,\mathbb{C})$. Since the Lie algebra of $\mathbf{PGL}(N,\mathbb{C})$ is isomorphic to $\mathbf{sl}(N,\mathbb{C})$, let us assume that the Lie algebra $\mathfrak{l}$ of $\mathfrak{L}$ is a subalgebra of $\mathbf{sl}(N,\mathbb{C})$. The type of questions we are about to discuss were in practice frequently (but tacitly) regarded in terms of $\mathfrak{l}$, and I shall present them explicitly in these terms. The first question is: does $\mathfrak{N}$ admit a decomposition into linear submanifolds $\mathfrak{N}_i$ with the above described properties? The series expansions for binary and ternary forms indicated that for any $\mathfrak{l}$ that can arise from such forms the answer was affirmative. A theorem positing an affirmative answer for all $\mathfrak{l}$ with a given structure will be referred to as a *complete reducibility theorem* for that structure. The second question arises naturally from the first: Given a particular Lie algebra structure, what is the nature of the $\mathfrak{l} \subset \mathbf{sl}(N,\mathbb{C})$ with that structure that leave nothing planar invariant? A theorem providing an answer to this question will be referred to as a *characterization theorem* for that structure. To prove both a complete reducibility theorem and a characterization theorem for a given structure is tantamount to solving the problem of determining all linear representations $\mathfrak{l}$ of that structure, and Study regarded the significance of these potential theorems in this light. That is, he regarded the establishment of such theorems as a contribution to Lie's group classification problem—goal (1) as described in Section 1.

With the assistance of Engel, Study was able to give a proof of the complete reducibility theorem for $\mathfrak{l} = \mathbf{sl}(2,\mathbb{C})$, although it was never published [Lie 1893a: 785]. A corresponding characterization theorem was in effect already known to Lie: The projective groups with the structure of $\mathbf{sl}(2,\mathbb{C})$ that leave nothing planar invariant are the projective groups of normal curves (as indicated in Section 1). Study also posited a complete reducibility theorem for $\mathbf{sl}(3,\mathbb{C})$ and worked out a proof, which, however, was recognized as containing a gap.[16] He had proved to his own satisfaction a corresponding (correct) characterization theorem: Any projective group with the structure of $\mathbf{sl}(3,\mathbb{C})$ that leaves nothing planar invariant arises as the group of a normal form of some type (m, n).[17] Study also examined the

[16] Letter from Study to Engel dated 31 July 1892 (Engel Archive Giessen). Study probably devised his proof in 1887 or thereabouts.

[17] Study stated this theorem in his book on ternary forms (p. 113) and reaffirmed to Engel that he had proved it in a letter of 31 July 1892 (Engel Archive Giessen). There is no documentary evidence indicating the nature of the proof or whether it was without flaws.

geometrical properties of these groups and determined the nonlinear manifold of lowest degree that is left invariant by the group and defines it (as the totality of all projective transformations leaving the manifold invariant).[18] In other words, he imitated what Lie had done when he characterized the irreducible representations of $\mathbf{sl}(2, \mathbb{C})$ as the projective groups of normal curves.

The theory of n-ary forms, despite its highly incomplete state circa 1887–89, naturally suggested the possibility of similar theorems for $\mathbf{sl}(n, \mathbb{C})$, but Study went even further in the realm of conjecture. Evidently under the influence of his contact with Lie and Engel, he had arrived at a much broader conception of geometry and the theory of invariants, according to which $\mathbf{GL}(n, \mathbb{C})$ or $\mathbf{SL}(n, \mathbb{C})$ was replaced by other transformation groups [1889b:202, n.7]. Study's broadened, group-theoretic conception of geometry thus brought with it a correspondingly broadened conception of the theory of invariants, no longer restricted to the special and general linear groups. By 1887 he had begun to explore the broader possibilities. In a letter of 13 March 1887 to Engel, Study wrote, "I have ... nothing less than a theory of invariants for the group of reciprocal radii (in [n-dimensional projective space] R_n) according to which I can write down a priori all the invariants of given surfaces, etc., just as in the symbolical method of the projective group."[19] The group of reciprocal radii has the structure of the orthogonal algebra $\mathbf{o}(n+2, \mathbb{C})$ [Lie 1893a:357]. Study published some of his results on the determination of orthogonal invariants in [1897]. The prospect of a theory of invariants corresponding to other groups perhaps encouraged him to wonder whether something akin to series expansions exist and therefore whether complete reducibility theorems also exist for groups other than those corresponding to $\mathbf{sl}(n, \mathbb{C})$. In any case, he did speculate on the possibility of such theorems in a letter written to Lie in 1890.

Study had communicated verbally some of his discoveries and conjectures regarding projective groups to Lie and Engel in 1887, but he did not write them up for publication, becoming instead preoccupied with other group-theoretic questions of current interest in Leipzig. In 1888 he left Leipzig to become *Privatdozent* at the University of Marburg. For intellectual reasons he would have preferred to remain in Leipzig, but he hoped that the move to Marburg would serve to accelerate his academic career within the Prussian university system.[20] At Marburg he completed and published his book [1889a] on ternary forms and indicated that "on a later occasion" he would treat thoroughly the problem of determining all projective groups with the structure of $\mathbf{sl}(3, \mathbb{C})$ (p. 114).

---

[18] These manifolds are described in Study's book (p. 786ff).

[19] "Ich habe ... Nichts geringer als eine Invariantentheorie der gruppe der reziproken Radii (in $\mathfrak{R}_n$), in der ich alle Invarianten gegebener Flächen usw. a priori hinschreiben kann, gerade so wie in der symbol[ischen] Methode der proj[ektiven] Gruppe" (Engel Archive Giessen).

[20] Study became a full professor at the University of Greifswald in 1897.

But by the end of 1890 nothing along those lines had materialized. At that time Lie wrote to Study urging him to publish his results on projective groups and requesting information on the specifics of what was proved and what was conjecture. Study obliged even though "the matter is far from my thoughts and I myself do not know exactly how far the methods applied go and what I have proved with certainty."[21] After stating his complete reducibility theorem for $\mathbf{sl}(2, \mathbb{C})$, his characterization theorem for $\mathbf{sl}(3, \mathbb{C})$ and his conjectured complete reducibility theorem for $\mathbf{sl}(3, \mathbb{C})$, Study continued: "Etc. I had a bunch of similar theorems, on the G_{15} of R_3 [$\mathbf{PGL}(4, \mathbb{C})$], on the group of reciprocal radii and the like; I believe such theorems can be generally posited first of all for simple and semisimple projective groups."[22]

Study thus believed it would be possible to establish complete reducibility theorems and characterization theorems (in terms of the minimal nonlinear invariant manifold) for a large class of structures, including all that are semisimple. One of the groups mentioned by him, the group of reciprocal radii in R_2, has the structure of $\mathbf{o}(4, \mathbb{C})$, so that it is semisimple but not simple, a fact that had been noted by Killing [1889a:74–5] when he introduced the concept of semisimplicity for the first time. There is no evidence, however, to suggest that Study's belief in the truth of such general complete reducibility theorems was anything more than that. Only in the case of $\mathbf{sl}(2, \mathbb{C})$ had a proof been achieved.

By the summer of 1892 Study still had not published his results on projective groups, and through the intermediary of Engel, Lie renewed his pressure on Study, even going so far as to attempt (without success) to prove the characterization theorem for $\mathbf{sl}(3, \mathbb{C})$ himself.[23] In response Study explained to Lie that he had wished "to let nothing be published on these matters until I have found the time to thoroughly think through all these problems as they relate to one another and to prove as many as possible of the general, conjectured theorems."[24] Study did not feel that he wished to embark on such a major research project at the moment, preferring instead to complete other less formidable projects. Referring to his insecure position as *Privatdozent*, he complained, "thus I must always work under constraint and can only work on matters that I am entirely certain will produce something. That is an unfortunate state of affairs ...."[25]

Study never did return to work on these general problems. As the above

---

[21] Study to Lie, 31 December 1890 (UB Oslo, Brevsamling Nr. 289). The text reads, "die Sache liegt mir schon fern, und ich weiss selbst nicht genau, wie weit die angewandten Methoden tragen."

[22] "U.s.w. Ich hätte noch eine Menge ähnlicher Sätze, über die G_{15} der R_3, über die Gruppe der reciproken Radien und dergleichen; ich glaube, dass man solche Sätze überhaupt zunächst für einfache und halbeinfache projektive Gruppen aufstellen kann."

[23] Engel to Study, 28 July 1892 (Engel Archive Giessen).

[24] Study to Lie, August 1892 (UB Oslo, Brevsamling Nr. 289).

[25] "so muss ich immer haftig arbeiten, und kann nur solche Sachen machen, bei denen ich ganz sicher bin, dass auch gleich etwas herauskommt. Das ist ein unglücklicher Zustand ...."

quotations indicate, he realized their difficulty. Moreover, in 1893 he made a consequential self-discovery. In January of that year he had written to Lie that by Easter he would be free at last to devote himself almost entirely to study and research on the theory of transformation groups. But by July, after attempting to digest *Theorie der Transformationsgruppen*, Study wrote frankly to Lie, "Unfortunately, I fear that I will never succeed in really mastering your theory; I can only work with intuition and have no capacity for abstraction. I wish, however, if it is acceptable, to contribute at least to the spread of your theory through reviews and through teaching."[26] Probably the same sentiments that made him abandon his hopes of mastering the generalities of Lie's theory convinced him of the futility and undesirability of returning to his research program on projective groups.

Despite Study's decision not to publish his fragmentary results on projective groups, his theorems and conjectures were not entirely lost from sight because Lie decided to take matters into his own hands and discuss them in the third volume of *Theorie der Transformationsgruppen* (pp. 785–8). Using a six-page manuscript provided by Study, Lie began by stating (without proof) the "beautiful" complete reducibility theorem for $\mathbf{sl}(2,\mathbb{C})$. Turning then to projective groups with the structure of $\mathbf{sl}(n,\mathbb{C})$, he suggested that, "A theorem similar to the one stated above undoubtedly holds for these groups; hence what will matter above all is to specify those among them that transform their space in such a way that no planar manifold remains at rest" (p. 786). Lie thus conjectured that a complete reducibility theorem was likely for $\mathbf{sl}(n,\mathbb{C})$, so that the corresponding characterization theorem becomes important, but he ventured no further—unlike Study in his 1890 letter to Lie. Whether this precaution was taken by Study in the manuscript he sent to Lie or whether it was due to Lie (and Engel) is regrettably unclear.

The remainder of Lie's presentation was devoted to Study's characterization theorems and contained further information. For example, for projective groups with the structure of $\mathbf{sl}(3,\mathbb{C})$ that leave nothing planar invariant—the groups that arise by consideration of normal ternary forms as noted above—Study obtained the nonlinear invariant manifold of minimal dimension, the analogue of the normal curves for $\mathbf{sl}(2,\mathbb{C})$. (These results had already been described in somewhat greater detail on p. 113 of Study's book on ternary forms [1889a].[27]) For $\mathbf{sl}(4,\mathbb{C})$ Study obtained similar results on the possibilities for the minimal invariant manifold, although (judging by Lie's account) perhaps without as much detail, since the possibilities increase in number. For $\mathbf{o}(4,\mathbb{C})$ Study specified correctly the dimensions $N = (m+1)(n+1)$ of its irreducible representations and

[26] "Leider fürchte ich, es wird mir niemals gelingen, Ihrer Theorien ordentlich zu beherrschen, ich kann nur mit Anschauung arbeiten und habe keine Fähigkeit zur Abstraction. Ich will aber, wenn es angeht, wenigstens durch Referate und durch Lehrtätigkeit zur Verbreitung Ihrer Theorien beitragen." Letter of 12 July 1893 (UB Oslo, Brevsamling Nr. 289).

[27] See in this connection Borel's discussion of Study's results in [2000:Ch. 5].

described the minimal (two-dimensional) invariant manifold. "It is clear," Lie wrote in conclusion, "that all these considerations can be generalized in many directions. It is to be hoped that Study will soon publish them in detail" (p. 788). Study never fulfilled Lie's hopes. When he finally became a professor at Greifswald, however, he arranged for one of Lie's students, Gerhard Kowalewski, to join him as *Privatdozent*. Kowalewski's habilitation research, done under Engel's direction, confronted him with Study's characterization problem, albeit in a somewhat different guise (Section 5). Although Study himself never became personally involved, he encouraged his younger colleague to attempt, by his quite distinct methods, to succeed where he could not.

7.3. Gino Fano

Study's complete reducibility theorem for $\mathbf{sl}(2, \mathbb{C})$ found its first application at the hands of Gino Fano (1871–1952), a student of Corrado Segre (1863–1924) at the University of Turin. During the 1870s and 80s while Lie was preoccupied with the inner workings of his theory of groups and its applications to differential equations, Klein began to correspond with Segre, who although just 20 years old in 1883, had already written his doctoral dissertation. His dissertation was inspired by Klein's approach to Plücker's line geometry, according to which line geometry is regarded as the geometry of a quadratic hypersurface in 5-dimensional projective space and thereby as an aspect of the theory of higher-dimensional projective spaces. Veronese had published an influential paper [1881] that Segre saw as inaugurating a genuinely *geometrical* research program for the study of higher-dimensional projective geometry, a program to which Segre also hoped to contribute. Klein's characterization of line geometry as an aspect of the projective geometry of 5-dimensional projective space was thus especially appealing to Segre, and so he initiated a correspondence with Klein. Klein's papers on line geometry that appeared in the early volumes of Clebsch's new journal, *Mathematische Annalen*, contained many of the ideas later developed more systematically in his *Erlanger Programm*, which Segre also read. "I will never forget," Segre wrote later to Klein, "the effect produced in me by my first reading of your works in the initial volumes of Math. Ann. and the [Erlanger] program of 1872 ...."[28] Indeed in [1885] Segre presented what to my knowledge is the earliest study of geometry in the spirit of the *Erlanger Programm*.

In [1885] Segre considered conics as points in 5-dimensional projective space—the point determined by the 6 coefficients of its equation in homogeneous coordinates. In this connection he cited the paper by Cayley [1868]

[28] Quoted in the original French by Terracini [1926:216].

described in the previous section. From Cayley's quasigeometrical point of view the conic defined in homogeneous coordinates x, y, z by the equation

$$ax^2 + by^2 + cz^2 + 2fyz + 2gxz + 2hxy = 0$$

is to be interpreted as the point in 5-dimensional projective space with homogeneous coordinates (a, b, c, f, g, h). Thus the totality of all conics in the plane is identified with $\mathbb{P}^5(\mathbb{C})$. Within this 5-dimensional space Cayley had considered the locus of all degenerate conics, i.e. those satisfying the equation

$$abc - af^2 - bg^2 - ch^2 + 2fgh = 0,$$

obtained by setting the discriminant equal to 0 [1868:198]. This "discriminant locus" is an algebraic variety of dimension 4 defined by an equation of the third order. Following Segre I shall denote it with Klein's notation by M_4^3. In general, degenerate conics consist of two lines in the plane. Among the conics in M_4^3 are those for which the two lines coincide. In this case the equation of the conic takes the form $(\alpha x + \beta y + \gamma z)^2 = 0$, and the corresponding points in $\mathbb{P}^5(\mathbb{C})$ are given by

$$(a,\ b,\ c,\ f,\ g,\ h) = (\alpha^2,\ \beta^2,\ \gamma^2,\ \beta\gamma,\ \gamma\alpha,\ \alpha\beta).$$

These points lie on a 2-dimensional algebraic variety of the fourth order, which, again following Segre, I shall denote by F_2^4. This is the surface of Veronese, who (unbeknownst to Segre) had made the most thorough study of its properties [1884]. As Cayley realized, these varieties of degenerate conics come into play in the quasigeometrical description of the conics satisfying given conditions.

Cayley's quasigeometrical approach was very congenial to the young Segre, who proposed to combine it with the ideas of the *Erlanger Programm*. Since every projective transformation of the plane, i.e., every $T \in \mathbf{PGL}(3, \mathbb{C})$, takes conics into conics, it determines a transformation T^* of $\mathbb{P}^5(\mathbb{C})$, which takes F_2^4 into itself and M_4^3 into itself (since T takes lines into lines and line pairs into line pairs) [1885:11ff.]. Moreover, the transformations T^* are projective, so that in this manner a subgroup $\mathfrak{G} \subset \mathbf{PGL}(5, \mathbb{C})$ is defined. The linearity of T^* as a transformation of the homogeneous coordinates $a, b, \ldots, h$ was a well-known fact from the theory of invariants. The manifold $\mathbb{P}^5(\mathbb{C})$ together with the group $\mathfrak{G}$ defines a geometry in the sense of the *Erlanger Programm*, and Segre proposed to show that "the projective geometry of the conics of the plane coincides exactly" with this geometry, in the sense of Kleinian equivalence [1885:1].

Although Segre's subsequent work was not as closely linked to the *Erlanger Programm*, he continued to esteem it highly, and shortly after he assumed the chair of geometry at Turin in 1888, he decided it would be worthwhile to have an Italian translation of the *Erlanger Programm* because he felt that its contents were not well enough known among young Italian geometers. In that connection it should be noted that the *Erlanger*

Figure 21. The above photo of Segre is from 1889. Some of Segre's geometrical work that is not group related is discussed in Section 8.5 because Cartan applied it to the theory of groups.

Programm was not widely known *anywhere* at this time. It had a limited circulation because it was simply printed as part of the formalities of assuming a professorship at the University of Erlangen, and Klein, who had become preoccupied with other matters, had not done anything himself to promote or promulgate its ideas. Klein, however, agreed wholeheartedly with Segre's proposal of a translation, especially since, as he noted, the first volume of Lie's *Theorie der Transformationsgruppen* had just appeared in 1888. Now the theory was available to develop the ideas of the *Programm.* Segre convinced one of the students at Turin, Gino Fano, to do the translation, which was published in *Annali di matematiche* in 1890. Fano's translation thus became the first of many translations of the *Erlanger Programm.*

The first young Italian geometer to develop some of the ideas in the *Erlanger Programm* was Federigo Enriques (1871–1946), who had spent some time in 1891 with Segre. He published a paper [1893] on a problem touched on in the *Erlanger Programm.* Given a geometry as determined by a group acting on a manifold, the main object is to study the configurations left invariant by the group. However, it is also of interest to study those configurations left invariant by *subgroups* of this group. Even before the *Erlanger Programm* was composed Klein and Lie had done just this when they studied curves in the plane left invariant by a one-dimensional group of commuting projective transformations—their theory of W-curves (Section 1.2). As we saw, they had also studied curves and surfaces in 3-dimensional space left invariant by commuting projective groups, but they never published a detailed account of this work because it proved too time-consuming to work out the details. In his paper Enriques proposed to determine surfaces in 3-dimensional space left invariant by any continuous group of projective transformations.

Unbeknownst to Enriques, Lie had already published a paper on this subject in a Norwegian journal [1882]. Lie presented these results again in the third volume of his *Theorie der Transformationsgruppen* in 1893 (pp. 190–98), and, in response to criticism, he presented a more detailed account in another paper [1895a]. Between them, Enriques and Lie had more or less resolved the problem of invariant surfaces in 3-dimensional space. The next contribution along these lines was made by Fano, who considered hypersurfaces in 4-dimensional space. Fano's work was not at all just a routine extension of the work of Lie and Enriques to the next dimension. It was rather remarkable for that time period because it involved what appears to have been the first—and only—serious application of the theory of the structure and representation of Lie algebras to geometry before the work of Cartan some fourteen years later.

With Segre's encouragement Fano became interested in geometry in the spirit of the *Erlanger Programm.*[29] His first effort in this direction was the above-mentioned 1890 translation of the *Programm* into Italian. After spending the academic year 1893–94 in Göttingen with Klein, Fano composed a number of papers, some jointly with Enriques, on group-theoretic aspects of geometry. One line of research was aimed at extending the work of Lie and Enriques that dealt with the determination of surfaces left invariant by some continuous projective group. Fano posed the problem of determining $(n-1)$-dimensional algebraic varieties in n-dimensional space that admit a continuous group of projective transformations, with special attention accorded to the case $n = 4$.

Fano rejected the methods that had been employed by Lie and Enriques in the case $n = 3$ as unsuitable for the extended problem because they involved extensive calculations lacking any conceptual coherence [1896:188]. At least for nonsolvable Lie algebras he was able to devise a conceptually simple yet general method based on Study's complete reducibility theorem for $\mathbf{sl}(2,\mathbb{C})$ as follows. Let $\mathfrak{l} \subset \mathbf{sl}(n+1,\mathbb{C})$ be a nonsolvable subalgebra.

[29] For further details and information, see [Hawkins 1984:452ff.].

Then by Engel's theorem characterizing solvable algebras [1893b] (discussed in Section 5.5) $\mathfrak{l}$ necessarily contains a subalgebra $\mathfrak{s}$ with the structure of $\mathbf{sl}(2,\mathbb{C})$. Since an algebraic variety left invariant by $\mathfrak{l}$ is a fortiori left invariant by $\mathfrak{s}$ as well, the sought-for varieties are contained among those left invariant by an $\mathfrak{s} \subset \mathbf{sl}(n+1,\mathbb{C})$ isomorphic to $\mathbf{sl}(2,\mathbb{C})$. Fano therefore sought to determine this class of varieties.

Study's complete reducibility theorem for $\mathbf{sl}(2,\mathbb{C})$ showed that $\mathbb{C}^{n+1} = \bigoplus_i \mathcal{V}_i$, where each (h_i+1)-dimensional vector space $\mathcal{V}_i$ is left invariant by $\mathfrak{s}$ and the restriction $\mathfrak{s}_i$ of $\mathfrak{s}$ to $\mathcal{V}_i$ leaves nothing planar invariant. Fano realized that each $\mathfrak{s}_i$ with $h_i \neq 0$ corresponds to the group that indicates how the coefficients of the general binary form of degree h_i are transformed when its variables are subjected to a linear transformation. (These therefore correspond to the projective groups of normal curves as discussed in Section 1.) Fano's problem reduced to the problem of determining the invariants of these forms. In the case $n = 4$, which he discussed in detail, this problem is manageable, since $\Sigma_i(h_i+1) = 5$ and the possibilities for the h_i are therefore limited. Fano stressed the noteworthiness of this connection between his purely geometrical problem and the fundamental problem of the theory of invariants.

Fano's paper is of some historical interest as a first attempt to apply Study's complete reducibility theorem for $\mathbf{sl}(2,\mathbb{C})$. In addition, since Study's proof had never been published, Fano provided a geometrical demonstration [1896:192ff.]. He admitted somewhat apologetically that it was rather long and difficult to read. The proof had two parts. One part (Section 8) consisted of a valid proof, couched in nineteenth-century geometrical terminology readily translatable into modern idiom, that it suffices to prove the theorem for any $\mathbf{sl}(2,\mathbb{C})$-module $\mathcal{V}$ with a proper submodule $\mathcal{W}$ such that $\mathcal{V}/\mathcal{W}$ is irreducible. The proof of the theorem in this special case (Section 7) was presented in somewhat vague geometrical terms, although, as Borel has shown [1998:§13], the underlying ideas are viable and can be translated into a modern proof. Actually, the first fully rigorous proof of the complete reducibility theorem for $\mathbf{sl}(2,\mathbb{C})$ was already implicit in Cartan's thesis [1894b], although it is embedded in the proof of another theorem.[30]

Fano also applied Study's complete reducibility theorem to differential equations in [1900]. If an ordinary differential equation (as studied in the Picard–Vessiot theory) has a fundamental set of solutions y_i such that all 2×2 determinants of the matrix

$$\begin{pmatrix} y_1 & y_2 & \cdots & y_{n-1} \\ \\ y_2 & y_3 & \cdots & y_n \end{pmatrix}$$

vanish, then the rationality group of the differential equation is the projective group of the normal curve in $\mathbb{P}^{n-1}(\mathbb{C})$, i.e., the irreducible representation of $\mathbf{sl}(2,\mathbb{C})$ of highest weight n. In his doctoral thesis (dedicated to

[30] For full details, see the accounts by Borel [1986:74; 1998:§12].

Picard) F. Marotte had, among other things, made an analytical study of the integration of this type of equation when $n = 3$ or 4 [1898]. Fano did the same for $n = 5$ in [1899]. In his expository article [1900] for *Mathematische Annalen* he treated the case of arbitrary n and then, invoking Study's complete reducibility theorem, the case in which the rationality group is any projective group with the structure of $\mathbf{sl}(2, \mathbb{C})$.

Given Fano's achievements in working out the ideas of the *Erlanger Programm*, it is not surprising that he was given the responsibility of writing the articles in Klein's *Enzyklopädie der mathematischen Wissenschaften* devoted to work in the spirit of the *Erlanger Programm*. These articles were published in 1907 and totaled 168 pages in length [1907a; 1907b]. In them Fano not only brought together all the work related to groups and geometry, he also synthesized some of it in ways that, as will be seen in Chapter 8, proved suggestive to Cartan, who took on the task of preparing Fano's article "Continuous Geometrical Groups" [1907a] for an updated, French version of the *Enzyklopädie*. Here I will indicate some of the matters treated by Fano in [1907a] that Cartan utilized in his work on what would now be described as the classification of the irreducible representations of complex simple Lie algebras.

We saw in Section 1 that the projective groups of normal curves—the irreducible representations of $\mathbf{sl}(2, \mathbb{C})$—were a byproduct of the Hesse–Klein principle of transfer. In his article [1907a] Fano presented this principle and two generalizations of it of his own devising. Although not in any sense profound, these generalizations, I believe, proved suggestive to Cartan for reasons that will be given in Chapter 8.

The Hesse–Klein principle may be expressed in the following manner. Let $\Psi : \mathbb{P}^1(\mathbb{C}) \to \mathbb{P}^n(\mathbb{C})$ be defined by

$$(3.1) \qquad \Psi(x_0, x_1) = (y_0, y_1, \ldots, y_n), \quad \text{where} \quad y_i = \binom{n}{i} x_0^{n-i} x_1^i.$$

Then Ψ maps $\mathbb{P}^1(\mathbb{C})$ one-to-one onto the curve $\mathcal{M}_1$ defined by the equations $y_i = x_0^{n-i} x_1^i$, $i = 0, \ldots, n$. The projective line $\mathbb{P}^1(\mathbb{C})$ is thus mapped onto a curve $\mathcal{M}_1$ in $\mathbb{P}^n(\mathbb{C})$ that for $n = 2$ is a conic and for any n is a rational normal curve. By virtue of the mapping Ψ the projective geometry of the line is "transferred" to the geometry on the curve $\mathcal{M}_1$ defined by the group $\mathfrak{H}_n = \Psi \circ \mathbf{PGL}(2, \mathbb{C}) \circ \Psi^{-1}$.

If $x = Sx'$, $S \in \mathbf{GL}(2, \mathbb{C})$, represents the projective transformation of $\mathbb{P}^1(\mathbb{C})$ taking point $p' = (x'_0, x'_1)$ into $p = (x_0, x_1)$, where

$$x_0 = ax'_0 + bx'_1, \quad x_1 = cx'_0 + dx'_1,$$

then the corresponding projective transformation in $\mathfrak{H}_n$ takes point y' into y, where

$$y_i = \binom{n}{i} (x_0)^{n-i} (x_1)^i = \binom{n}{i} (ax'_0 + bx'_1)^{n-i} (cx'_0 + dx'_1)^i.$$

Expansion of the right-hand side of this equation using the relations

$$\binom{n}{i}(x_0')^{n-i}(x_1')^i = y_i'$$

shows that y_i is a linear homogeneous function of $y_0', y_1', \ldots, y_n'$, which may be denoted by $y = \sigma(S)y'$. This shows that the transformations of $\mathfrak{H}_n$ may be lifted to define projective transformations on all of $\mathbb{P}^n(\mathbb{C})$. In this way the group $\mathfrak{H}_n$ lifts to the group $\mathfrak{K}_n$ of projective transformations of $\mathbb{P}^n(\mathbb{C})$ that take the curve $\mathcal{M}_1$ into itself. Klein was the first to introduce these projective groups $\mathfrak{K}_n$ of a normal curve.

The primary significance of the group $\mathfrak{K}_n$ was that it forged a link between geometry and the theory of invariants. That is, if, as in Section 1, $\phi(a,x) = \sum_{i=0}^n \binom{n}{i} x_0^{n-i} x_1^i$, then $\phi(a,x) = \phi(a, Sx') = \phi'(a', x')$, where, in the notation of Section 1, $a = \mu(S)a'$ and the transformations $\mu(S)$ form the group $\mathfrak{N}_n$. If we think of $a_0, a_1, \ldots, a_n$ as dual or hyperplane coordinates of $y_0, y_1, \ldots, y_n$, where $\sum_{i=0}^n a_i y_i = 0$ is the hyperplane, then the transformation $y = \sigma(S)y'$ of point coordinates induces the transformation $a = \mu(S)a'$ of hyperplane coordinates, which means in algebraic terms that $\mu(S) = [\sigma(S)^T]^{-1}$. The invariant theory of the binary form ϕ could thus be given a geometrical interpretation in terms of the Kleinian geometry of $\mathcal{M}_1 = \Gamma_n$, the normal curve of Section 1, as was done by Meyer in [1883].

After describing the Hesse–Klein principle along the lines indicated above, Fano proceeded to make the first of two suggestive generalizations of it. He considered points $p = (x_0, x_1, \ldots, x_k)$ in $\mathbb{P}^k(\mathbb{C})$ and let $f_0(x), \ldots, f_n(x)$, $n > k$, denote $n+1$ linearly independent homogeneous polynomials of fixed degree d in the x_i that form a "linear system" in the sense that if we define $\phi(a,x)$ by $\phi(a,x) = \sum_{i=0}^n a_i f_i(x)$, then under the variable change $x = Sx'$, $S \in \mathbf{GL}(k+1, \mathbb{C})$, $\phi(a, Sx')$ is expressible in the form $\phi'(a', x') = \sum_{i=0}^n a_i' f_i(x')$. In other words, the polynomials $f_i(x)$ have the property that $f_i(Sx') = \sum_{j=0}^n \mu_{j,i} f_j(x')$, i.e., they span a representation module for $\mathbf{GL}(k+1, \mathbb{C})$. Fano introduced the mapping $\Psi : \mathbb{P}^k(\mathbb{C}) \to \mathcal{M}_k \subset \mathbb{P}^n(\mathbb{C})$ defined by

$$\Psi(x) = (y_0, \ldots, y_n) = (f_0(x), \ldots, f_n(x)). \tag{3.2}$$

This mapping "transfers" the geometry of $\mathbb{P}^k(\mathbb{C})$ to the geometry on $\mathcal{M}_k$ defined by the "transferred" group, and once again the transferred group lifts to a group $\mathfrak{L}_n^{(d,k)}$ of projective transformations of $\mathbb{P}^n(\mathbb{C})$ that are defined by linear homogeneous transformations $y = \sigma(S)y'$.

On the basis of his generalization, Fano could conclude that "the invariant theory of the linear system coincides with the projective geometry of [$\mathcal{M}_k$]" [1907b:362]. What motivated Fano to present his generalization? I believe it was the work of Veronese and Segre, for the only example he gave to illustrate the generalization was to consider the case in which $\phi(a,x)$ is the general homogeneous polynomial of degree 2 in three variables—so

Figure 22. Fano as he appeared later in life. His 1907 article for Klein's encyclopedia on groups and geometry influenced Cartan, who published a revised and expanded version in 1915 (Section 8.3).

$d = 2$, $k = 2$, and $n = 5$ in the above. Then $\mathcal{M}_k$ is the surface of Veronese. The polynomial $\phi(a, x)$ set equal to zero defines a conic in $\mathbb{P}^2(\mathbb{C})$, and the group $\mathfrak{L}_5^{(2,2)}$ is the projective group that Segre had introduced in his 1885 paper on the projective geometry of conics as the geometry on a variety in $\mathbb{P}^5(\mathbb{C})$. More generally, if $\phi(a, x)$ is the general homogeneous polynomial of degree d, then $\mathcal{M}_k$ is what is now called a Veronese variety. From the vantage point of the present, we can see that Fano was pointing out how to generate representations of $\mathbf{GL}(n, \mathbb{C})$ or $\mathbf{PGL}(n, \mathbb{C})$—and thus of $\mathbf{sl}(n, \mathbb{C})$— by means of "linear systems" of homogeneous polynomials.

Fano's second generalization proceeded as follows. Consider $p = (x_0, x_1, \ldots, x_k) \in \mathbb{P}^k(\mathbb{C})$ and $q = (y_0, y_1, \ldots, y_l) \in \mathbb{P}^l(\mathbb{C})$ and let $f_i(x, y)$, $i = 1, \ldots, n$, denote linearly independent homogeneous polynomials of degree d in x and degree e in y that form a "linear system," so that if

$$\phi(a,x,y) = \sum_{i=0}^{n} a_i f_i(x,y),$$

then under the linear variable changes $x = Sx'$, $y = Ty'$, where $S \in \mathbf{GL}(k+1,\mathbb{C})$ and $T \in \mathbf{GL}(l+1,\mathbb{C})$, we have $\phi(a,x,y) = \phi'(a',x',y')$ in the above explained sense, so that the $f_i(x,y)$ span a representation module for $\mathbf{GL}(k+1,\mathbb{C}) \times \mathbf{GL}(l+1,\mathbb{C})$. Then define the mapping

$$\Psi : \mathbb{P}^k(\mathbb{C}) \times \mathbb{P}^l(\mathbb{C}) \to \mathcal{M}_{k+l} \subset \mathbb{P}^n(\mathbb{C}), \tag{3.3}$$

where $\Psi(x,y) = (z_0,\dots,z_n) = (f_0(x,y),\dots,f_n(x,y))$.

Fano's second generalization was motivated by work of Segre, who in [1891] had introduced what are now called Segre maps and varieties. As Fano observed, the variety $\mathcal{M}_{k+l}$ is a Segre variety if $n = (k+1)(l+1)-1$ and the polynomials f_i are the $n+1$ products $x_i y_j$. In this manner a geometry in the sense of the *Erlanger Programm* is defined on Segre's varieties and, more generally, on the varieties $\mathcal{M}_{k+l}$. The groups of projective transformations of $\mathbb{P}^n(\mathbb{C})$ defining these geometries correspond to a matrix equation $z = \mu(T,U)z'$ defining a representation of $\mathbf{GL}(k+1,\mathbb{C}) \times \mathbf{GL}(l+1,\mathbb{C})$. In the case $k = l$ one could also take $T = U$ and so obtain a representation of $\mathbf{GL}(k+1,\mathbb{C})$. Fano himself focused on the mapping Ψ, however, and did not pause to discuss these group-theoretic implications.

Although Fano himself did not make the connection, his article contained other important examples of what he called "linear systems." Let me explain. A basic tenet underlying the *Erlanger Programm* and the early geometrical work of Lie was that the "elements" that are coordinatized in describing a geometry need not be literal points. As we have seen in Chapter 1, this view was inspired by Plücker's introduction of line coordinates in space in 1846. Plücker's ideas were modified to suit the needs of projective geometry by Cayley in 1862, and in that form they were applied extensively in geometrical research by Klein, who had been Plücker's student. In his *Enzyklopädie* article Fano went into considerable detail on the development and application of Plücker coordinates and their relation to Grassmann's ideas [1907b:310, n.62]. As I mentioned in passing in Section 2, in [1872b] Clebsch had extended the idea of line coordinates to $\mathbb{P}^n(\mathbb{C})$ and further generalized it as follows. If $x_0, x_1, \dots, x_n$ are homogeneous coordinates in $\mathbb{P}^n(\mathbb{C})$, then one may consider the $\binom{n+1}{2}$ line coordinates $X_{ij} = \begin{vmatrix} x_i & x_j \\ y_i & y_j \end{vmatrix}$ associated with the line determined by x and y. Likewise, one may consider the $\binom{n+1}{3}$ plane coordinates

$$X_{ijk} = \begin{vmatrix} x_i & x_j & x_k \\ y_i & y_j & y_k \\ z_i & z_j & z_k \end{vmatrix}$$

associated with the plane through x, y, and z, and so on up to the $\binom{n+1}{n} = n+1$ coordinates of a hyperplane. With geometrical applications in mind,

Clebsch stressed the need to study the invariant theory of forms in all these types of variables.

In $\mathbb{P}^l(\mathbb{C})$, for any fixed integer m between 1 and l, the $\binom{l+1}{m}$ mth-order Plücker–Clebsch coordinates $X_{ij\ldots}$ are homogeneous polynomials of degree m in the coordinates of each of the m points $x, y, \ldots$. Since $X_{ij\ldots}$ defines a homogeneous polynomial in the variables $x_0, x_1, \ldots, x_l$, $y_0, y_1, \ldots, y_l, \ldots$, we may denote it by $X_{ij\ldots} = f^{(m)}_{ij\ldots}(x, y, \ldots)$. Then

$$\phi^{(m)}(x, y, \ldots; a) = \sum_{i,j,\ldots} a_{ij,\ldots} f^{(m)}_{ij\ldots}(x, y, \ldots)$$

defines a linear system in Fano's sense. Consequently, the variable changes $x = Sx'$, $y = Sy'$,..., $S \in \mathbf{GL}(l+1, \mathbb{C})$ induce a representation of the groups $\mathbf{GL}(l+1, \mathbb{C})$ and $\mathbf{PGL}(l+1, \mathbb{C})$, or, on the Lie algebra level, a representation of $\mathbf{sl}(l+1, \mathbb{C})$. This representation shows how the general projective group transforms the mth order Plücker-Clebsch coordinates. As we shall see, this representation proved important in the work of Cartan (Chapter 8) as well as that of Schur (Chapter 10).

7.4. Cayley's Counting Problem

So far in this chapter we have been considering work motivated by Lie's theory that would now be regarded as involving the representation of Lie groups or algebras. In this section, however, it is necessary to go back in time to the 1850s, the early days of the theory of invariants. At that time Cayley did some work on a problem in the theory of invariants that can now be interpreted as involving the representations of $\mathbf{sl}(2, \mathbb{C})$. More important from the viewpoint of the present chapter is the fact that some of Cayley's ideas were taken up by Lie's student Gerhard Kowalewski, who devised a theory of weights to solve a general problem we can now see as that of determining all irreducible representations of semisimple Lie algebras—the problem that Cartan succeeded in solving in 1913 with his own theory of weights. Kowalewski's work is discussed in the next section and Cartan's in Section 8.2. As we shall see in Section 10.5, Cayley's problem also figures in a later, vitally important, episode in our story.

The theory of invariants grew up around the general problem of describing, in one way or another, the in- and covariants associated with an n-ary form, i.e., a homogeneous polynomial in n variables. This general problem was first posed by Cayley who was inspired to do so through some papers written by Boole in the 1840s.[31] One way in which Cayley made this problem mathematically specific was in terms of the following counting

[31] [Crilly 1986]. On the early history of invariants see also [Parshall 1989].

Figure 23. Cayley was one of the founders of the theory of invariants. The counting problem considered here played a vital role in the events leading up to Weyl's work, as will be seen in Chapters 10 and 12.

problem. Recall from Section 1 that a polynomial $C(a,x)$ that is homogeneous of degree p in $a_0, \ldots, a_d$ and of degree q in x_1, x_2 is a *covariant* of the binary form $\phi(a,x) = \sum_{i=0}^{d} \binom{d}{i} a_i x_1^{d-i} x_2^i$ if with $x = Sx'$ and $a = \mu(S)a'$ one has $C(a,x) = (\det S)^w C(a', x')$, where $S \in \mathbf{GL}(2, \mathbb{C})$ $a = \mu(S)a'$ is the corresponding change in the coefficients a_i of ϕ. An invariant is a covariant in the "limiting case" in which $q = 0$, so $C(a,x) = C(a)$. Thus the case of invariants is also covered in the following problem, which I will refer to as the counting problem of Cayley.

Cayley's Counting Problem: *For fixed values of p and q, determine the number of linearly independent covariants $C(a,x)$ of degrees p and q in a and x, respectively.*

Cayley posed his counting problem in a paper written in 1852 for Crelle's journal [1854]. Before it appeared in print, he had solved it. The solution was presented in the second of Cayley's famous memoirs on "quantics" [1856].

Cayley's solution was based upon his discovery that the in- and covariants of the binary form $\phi(a,x)$ can be characterized by means of differential equations. That is, if differential operators $\mathcal{X}$ and $\mathcal{Y}$ are defined by

$$\mathcal{X} = \sum_{j=1}^{d} j a_{j-1} \frac{\partial}{\partial a_j}, \quad \mathcal{Y} = \sum_{j=0}^{d-1} (d-j) a_{j+1} \frac{\partial}{\partial a_j},$$

then we have the following theorem.

(4.1) **Theorem.** *$C(a,x)$ is a covariant if and only if $\mathcal{X}C = x_2 \partial C/\partial x_1$ and $\mathcal{Y}C = x_1 \partial C/\partial x_2$. In particular, $C(a)$ is an invariant if and only if $\mathcal{X}C = \mathcal{Y}C = 0$.*

An immediate consequence of Theorem 4.1 is that if $C(a,x)$ is a covariant, then $\mathcal{Z}C = 0$, where $\mathcal{Z}$ denotes the commutator

$$\mathcal{Z} = \left(\mathcal{X} - x_2 \frac{\partial}{\partial x_1}\right)\left(\mathcal{Y} - x_1 \frac{\partial}{\partial x_2}\right) - \left(\mathcal{Y} - x_1 \frac{\partial}{\partial x_2}\right)\left(\mathcal{X} - x_2 \frac{\partial}{\partial x_1}\right).$$

By writing $\mathcal{Z}$ in the simplified form

$$\mathcal{Z} = \mathcal{X}\mathcal{Y} - \mathcal{Y}\mathcal{X} + \left(-x_1 \frac{\partial}{\partial x_2} + x_2 \frac{\partial}{\partial x_1}\right),$$

where

$$\mathcal{X}\mathcal{Y} - \mathcal{Y}\mathcal{X} = \sum_{j=0}^{d} (d-2j) a_j \frac{\partial}{\partial a_j},$$

Cayley discovered that $\mathcal{Z}C = 0$ imposes a condition on the terms of $C(a,x)$.

That is, if

$$C(a,x) = \sum_{j=0}^{q} \binom{q}{j} B_j(a) x_1^{q-j} x_2^j, \tag{4.2}$$

where $B_j(a)$ is a homogeneous polynomial of degree p, viz.,

$$B_j(a) = \sum_{m_0+\cdots+m_d=p} \alpha_{m_0\ldots m_d} a_0^{m_0} a_1^{m_1} \cdots a_d^{m_d},$$

then each term of B_j has the same "weight" $w_j = (pd-q)/2 + j$, where the weight of a term of $B_j(a)$ is defined to be $w_j = \sum_{i=0}^{d} i m_i$. Later a homogeneous polynomial such that each term has the same weight was termed *isobaric*, a term I shall use. If x_1 and x_2 are assigned weights 1 and 0, respectively, so $x_1^{q-j} x_2^j$ is assigned weight $q-j$, then each term of $C(a,x)$ as given in (4.2) has the same weight, so that $C(a,x)$ is also isobaric. That is, we have the following result.

(4.3) **Theorem.** *If $C(a, x)$ is a covariant, then it must be isobaric of total weight $(pd + q)/2$.*

A mathematician reading Cayley's papers today will see immediately that their contents can be interpreted readily in terms of Lie algebras, although that concept was not introduced until twenty years later. Viewed in modern terms, the operators $\mathcal{X}, \mathcal{Y}$, and $[\mathcal{X}, \mathcal{Y}] = \mathcal{X}\mathcal{Y} - \mathcal{Y}\mathcal{X}$ act on the vector space of all homogeneous polynomials $B(a)$ of degree p, thereby making it into a module for the Lie algebra $\mathbf{sl}(2, \mathbb{C})$. Indeed, as Cayley realized, if $B(a)$ has degree p and weight w, $(\mathcal{X}\mathcal{Y} - \mathcal{Y}\mathcal{X})B = (pd - 2w)B$. Thus $B(a)$ is a weight vector of weight $\pi = pd - 2w$ in the modern sense (due to Cartan as will be seen in the next chapter). Furthermore, Cayley's solution to his counting problem for covariants $C(a, x)$ of degrees p and q in a and x respectively, involved, in effect, the irreducible representation of $\mathbf{sl}(2, \mathbb{C})$ of highest weight q.

This follows from the fact that, if one writes out the two differential equations of Theorem 4.1 characterizing a covariant $C(a, x)$, they are equivalent, respectively, to

$$\mathcal{X}B_0 = 0, \ldots, \mathcal{X}B_j = jB_{j-1}, \ldots, \mathcal{X}B_q = qB_{q-1} \tag{4.4}$$

and

$$\mathcal{Y}B_q = 0, \ldots, \mathcal{Y}B_{q-j} = jB_{q-j+1}, \ldots, \mathcal{Y}B_0 = qB_1. \tag{4.5}$$

Since by Theorem 4.3 $B_0(a)$ has Cayley weight $w = (pd - q)/2$, equations (4.4) and (4.5) imply to us that B_0 is a highest-weight vector ($\mathcal{X}B_0 = 0$) of Cartan weight $q = pd - 2w$ and that the B_j generated by B_0 under application of $\mathcal{Y}$ as in (4.5) define a basis for the corresponding irreducible $\mathbf{sl}(2, \mathbb{C})$-module. For Cayley the significance of equations (4.4) and (4.5) was that they showed that a covariant is determined by its leading coefficient B_0. Eventually, he obtained the following converse result.

(4.6) **Theorem.** *If $B_0(a)$ is any homogeneous polynomial of degree p and weight $w = (pd - q)/2$ that satisfies $\mathcal{X}B_0 = 0$, then if $B_1, \ldots, B_q$ are defined by (4.5), the resulting polynomial $C(a, x)$ defined by (4.2) is a covariant.*

To prove Theorem 4.6, Cayley showed (4.4) and (4.5) are satisfied. This meant showing that $\mathcal{Y}B_q = 0$ (or equivalently that $\mathcal{Y}^{q+1}B_0 = 0$) and that $\mathcal{X}B_j = jB_{j-1}$ for $j > 0$.

A homogeneous isobaric polynomial $B(a)$ such that $\mathcal{X}B = 0$ was later called a *semi-invariant* or a *seminvariant*, since it satisfies half the conditions for an invariant given in Theorem 4.1. Adopting this terminology, we may say that Theorem 4.6, by setting up a correspondence between covariants and their leading coefficients, provided Cayley with the following result.

(4.7) **Corollary.** *If $M(p,d,w)$ denotes the number of linearly independent seminvariants of degree p and weight $w = (pd - q)/2$, then $M(p,d,w)$ is the solution to the counting problem.*

Cayley discovered a simple characterization for the number $M(p,d,w)$, which constitutes his main result. I will refer to it as his law.

(4.8) **Cayley's Law.** *If $N(p,d,w)$ denotes the number of nonnegative integral solutions $m_0, \ldots, m_d$ to the two equations*

$$\sum_{j=0}^{d} m_j = p, \qquad \sum_{j=0}^{d} j m_j = w,$$

then $M(p,d,w) = N(p,d,w) - N(p,d,w-1)$.

The idea behind Cayley's proof is quite simple. Each simultaneous solution to the above Diophantine equations determines a monomial $a_0^{m_0} a_1^{m_1} \cdots a_d^{m_d}$ of degree p and weight w. Thus $N(p,d,w) = \dim \mathcal{V}(p,d,w)$, where $\mathcal{V}(p,d,w)$ denotes the vector space of all homogeneous isobaric polynomials in a of degree p and weight w. Since application of $\mathcal{X}$ lowers weight by 1, $\mathcal{X}$ is a linear transformation from $\mathcal{V}(p,d,w)$ into $\mathcal{V}(p,d,w-1)$. The kernel of $\mathcal{X}$ consists precisely of the seminvariants of degree p and weight w. Thus if $\mathcal{X}$ is *surjective*, $\dim[\ker(\mathcal{X})] = \dim \mathcal{V}(p,d,w) - \dim \mathcal{V}(p,d,w-1)$, and Cayley's law follows. Cayley's actual proof was along these lines, albeit expressed in nineteenth-century terminology. He felt that it was evident that $\mathcal{X}$ was surjective. As he expressed it, if $B(a)$ is the "general" polynomial of degree p and weight w, then $\mathcal{X}B$ will represent the general polynomial of degree p and weight $w-1$ [1856:256]. Cayley was correct, but the proof turned out to be rather nontrivial. Valid proofs were given by Sylvester [1878] and by Hilbert [1887].[32]

Cayley's law afforded him a solution to the counting problem congenial to computations. Although there is no simple formula for the number $N(p,d,w)$, it can be represented, as Cayley showed, as the coefficient of $x^w z^p$ in the series expansion of the generating function

$$\phi(x,z) = \frac{1}{(1-z)(1-xz)(1-x^2 z)\cdots(1-x^d z)}.$$

From this representation Cayley obtained a generating function representation of the numbers $M(p,d,w)$ as well and worked out their values in extensive tables.

Shortly after Cayley obtained his solution to the counting problem, his friend J. J. Sylvester rederived the results using the notion of infinitesimal transformations [1852], which, as we have seen, was the way the elements in a Lie algebra were conceived by Lie and his students. Indeed, Sylvester also briefly considered the n-ary analogue of Theorem 1, which involves

[32] For further details see [Hawkins 1987:1647–48].

sl $(n, \mathbb{C})$, and even suggested studying the invariants of an n-ary form defined by considering only orthogonal transformations. In fact, in a series of papers beginning in 1887 and culminating in the monographic essay [1892], the Belgian mathematician Jacques Deruyts (1862–1945) did extend Cayley's theory to the context of the in- and covariants associated with an underlying n-ary form $\phi(a, x)$. He even posed and solved Cayley's problem within this context, although the solution is complicated. Lie algebras had been introduced by the time Deruyts did his work, but he made no reference to them, despite Lie's attempts in the 1880s to call attention to their relevance to invariant theory. This is unfortunate, because Deruyts's papers, although extremely difficult to penetrate, involve, in effect, the irreducible modules for **sl** $(n, \mathbb{C})$, just as Cayley's had for $n = 2$.[33]

Although the above-mentioned work of Cayley, Sylvester, and Deruyts had no direct effect upon the developments leading up to the Killing–Cartan theory of the structure or Cartan's classification of the irreducible representations of semisimple Lie algebras (Section 8.4), Cayley's work did play a role in the subsequent advance of that theory made by Weyl through his proof of the complete reducibility theorem (Chapter 12). That is, as will be seen in Section 10.5, I. Schur's interest in Cayley's problem led him to pose and solve the analogous problem for the invariants of the rotation group, and the ideas involved in his solution provided Weyl with the key to his proof of the complete reducibility theorem. Nowadays, Cayley's solution to the counting problem, namely Cayley's law, is resolved using the complete reducibility theorem,[34] so Cayley's own solution indirectly inspired the theorem behind the modern approach to his work.

7.5. Kowalewski's Theory of Weights

Cayley's work on the counting problem also may have indirectly influenced Cartan, for, as we shall see in the present section Gerhardt Kowalewski (1876–1950) attempted to use a variation of the notion of a weight as found in Cayley's paper to solve the problem of determining all projective groups which leave nothing planar invariant. His efforts may have inspired in part Cartan's own solution. In any case, as we shall see, Kowalewski's work provides an instructive context within which to appreciate fully the historical dimensions of Cartan's achievement.

---

[33] For biographical information on Deruyts see [Godeaux 1949]. A lucid account, in modern mathematical terms, of the mathematical ideas underlying Deruyts's work has been given by Green [1991], who writes there (p. 248): "We can now look back on Deruyts's work and find a wealth of methods which, to our eyes, are pure representation theory; some of these methods are still unfamiliar today. One can speculate that the representation theory of linear groups could have taken a different course, or at least matured more rapidly, if Schur or Weyl had taken up the ideas which are lying just below the surface of Deruyts's memoir!"

[34] For a modern treatment of Cayley's law see [Springer 1977:§3.3].

Kowalewski obtained his doctorate from the University of Leipzig in 1898 with a dissertation written under the supervision of Lie. Lie regarded him as one of his best German students, along with Engel and Scheffers.[35] He was also one of Lie's last students, for Lie returned to Norway in 1898 and died the following year. Kowalewski remained at Leipzig until 1901. Consequently, Engel was an influential force upon him during his post-doctoral development as a mathematician. With the encouragement of Engel, Kowalewski chose as his *Habilitationsschrift* [1899] to work on the problem of determining all *primitive* transformation groups in 5 variables. The distinction between primitive and imprimitive transformation groups was introduced by Lie by analogy with the theory of finite permutation groups. Recall that a finite group $\mathfrak{G}$ of permutations of $S = \{1, 2, \ldots, n\}$ is imprimitive if S can be partitioned into subsets S_i of equal cardinality such that $\mathfrak{G}$ permutes the S_i among themselves. If no such partition exists, $\mathfrak{G}$ is primitive. Lie extended these notions to continuous groups as follows [1888:220–21]. Let $\mathfrak{G}$ denote a continuous group of transformations acting on a region Ω of $\mathbb{C}^n$. Suppose for $0 < q < n$ there exists a family of ∞^{n-q} q-dimensional manifolds in $\mathbb{C}^n$ such that every point of Ω belongs to one and only one such manifold. Then if all the transformations of $\mathfrak{G}$ take each manifold of this family into another manifold of this family, $\mathfrak{G}$ is said to be imprimitive. If no such family exists, $\mathfrak{G}$ is said to be primitive. As with permutation groups, primitive transformation groups are always transitive.

As noted in Section 1, thanks to some papers by Engel in 1889–90, an important subproblem of the problem of determining all primitive groups in n variables was to determine, up to similarity by linear transformation, all $\mathfrak{l} \subset \mathbf{gl}(n, \mathbb{C})$ that regarded as projective groups acting on $\mathbb{P}^{n-1}(\mathbb{C})$ leave nothing "planar" invariant. By a theorem first stated by Cartan in 1909, these projective groups must be semisimple, so that the above-mentioned subproblem can now be seen as tantamount to determining all irreducible representations of semisimple algebras. However, it must be kept in mind that Kowalewski and his colleagues did not realize this connection with semisimplicity. In keeping with the convention of Section 1, I will refer to this subproblem as *Kowalewski's problem*, since it was he who focused his attention upon it as an aspect of the classification of primitive groups.

Kowalewski's approach to his problem was based upon the following considerations. Let $\mathfrak{l}$ be the Lie algebra of a group of projective transformations on $\mathbb{P}^n(\mathbb{C})$ that leaves nothing planar invariant. Without loss of generality we may therefore assume that $\mathfrak{l} \subset \mathbf{sl}(n+1, \mathbb{C})$ and that $\mathfrak{l}$ leaves no proper, nontrivial subspace of $\mathbb{C}^{n+1}$ invariant. Then $\mathfrak{l}$ cannot be solvable (integrable in Lie's terminology) because if it were, Lie's theorem would imply the existence of a flag of subspaces left invariant by $\mathfrak{l}$. (Lie and his students did not speak of flags, but this notion and Lie's theorem (Section 2.2) were very familiar to them.) Now, Engel had proved [1893b] that if $\mathfrak{l}$

[35] See the letter of reference written by Lie on behalf of Kowalewski and quoted on pp. 85–6 of [Kowalewski 1950].

Figure 24. The above portrait of Kowalewski is from his autobiographical book, Bestand und Wandel, *published in the year of his death, 1950.*

is not solvable, then $\mathfrak{l}$ necessarily contains a three-dimensional subalgebra $\mathfrak{m}$ isomorphic to $\mathbf{sl}(2, \mathbb{C})$. Then by Study's complete reducibility theorem for $\mathbf{sl}(2, \mathbb{C})$ all the possibilities for $\mathfrak{m}$ could be specified. That is, Study's theorem implies

$$\mathbb{C}_{n+1} = \mathcal{V}_0 \oplus \mathcal{V}_1 \oplus \cdots \oplus \mathcal{V}_k, \tag{5.1}$$

where $\mathfrak{m}$ leaves each vector subspace $\mathcal{V}_i$ invariant but leaves no nontrivial subspace of $\mathcal{V}_i$ invariant for $i > 0$ and acts trivially on $\mathcal{V}_0$. Thus the restriction $\mathfrak{m}_i$ of $\mathfrak{m}$ to $\mathcal{V}_i$, $i > 0$, is, to use Kowalewski's language, the group of a normal curve, or, simply, a normal group (*Normgruppe*).

The above considerations suggested two approaches to the solution of Kowalewski's problem, both of which were taken by him and his students. One approach was to proceed by the dimension n. That is, for a fixed value of n, write down all the possibilities for $\mathfrak{m}$ given by Study's complete reducibility theorem. Then for each such possibility for $\mathfrak{m}$, use the fact that

$\mathfrak{m} \subset \mathfrak{l}$ to build up all $\mathfrak{l}$'s that could possibly contain $\mathfrak{m}$. This was done for $n = 5$ by Kowalewski in [1899]. There he had actually determined the possible $\mathfrak{l}$'s by a different method and then, belatedly, noted that the above considerations could be used (p. 95). According to Kowalewski, Engel had called his attention to Fano's paper [1896], discussed in Section 3, where Study's complete reducibility theorem for $\mathbf{sl}(2, \mathbb{C})$ was already employed to solve a different problem. Later, Kowalewski's students at the University of Bonn did the same, as doctoral dissertations, for $n = 6$ [Apfelstedt 1906] and $n = 7$ [Kummer 1908]. Already with the cases $n = 6, 7$ the computations involved became extensive: Together Apfelstedt and Kummer produced 140 pages consisting mainly of calculations done within the framework of Kowalewski's method of weights (to be discussed below).

A second approach was as follows. There may be many $\mathfrak{m} \subset \mathfrak{l}$ with the structure of $\mathbf{sl}(2, \mathbb{C})$, yielding different decompositions (5.1). Let k_0 denote the smallest value of k for which a decomposition (5.1) holds. One can then seek to find all $\mathfrak{l}$ by proceeding according to k_0, starting with $k_0 = 1$. Kowalewski's correspondence with Engel (considered below) reveals that he was optimistically pursuing this approach as early as 1902, although he never publicly stated his overall objective. His caution proved prudent, because by 1913 he had managed to deal completely only with the case $k_0 = 1$ and partially with the case $k_0 = 2$, and none of his students pursued the dimension-by-dimension approach for $n > 7$. Nevertheless, his efforts and those of his students are of historical interest as another attempt, following those by Study and also by Killing (Killing's problem, Section 6.3), to solve what can now be seen as the problem of determining all irreducible representations of semisimple Lie algebras. They are also of historical interest for the light they shed on the operational status of the Killing–Cartan structure theory within Lie's school. And finally, they are of interest because they led Kowalewski to a theory of weights based loosely on the ideas Cayley had introduced in his theory of seminvariants a half-century earlier, as indicated in Section 4.

Kowalewski remained in Leipzig as an instructor (*Privatdozent*) until 1901, when he obtained a position as an assistant professor (*ausserordentlicher Professor*) at the University at Greifswald, where Study was professor. "I will always maintain a lively interest in the theory of groups," Kowalewski wrote from Greifswald to Engel, "and I am delighted also to find in Greifswald, in the person of Professor Study, an outstanding mathematician who is sympathetic to this theory."[36] After paying a visit to Study, Kowalewski was delighted to find that Study's appreciation of Lie's theory was greater than anticipated: "I am surprised by how highly he esteems Lie's theory. He goes so far in this regard as to consider the determination

[36] "Für die Gruppentheorie werde ich immer das lebhafteste Interesse bewahren und freue mich sehr, dass ich auch in Greifswald in Herrn Professor Study einen hervorragenden Mathematiker finde, der dieser Theorie sympathisch gegenübersteht." Letter dated 11 October 1901 (Engel Archive Giessen).

of primitive groups as valuable, whereas other mathematicians consider it an activity barely worthy of human beings."[37] At Greifswald, with Study's encouragement, Kowalewski took the first steps along the path of the second approach to his problem—classification according to k_0.

Kowalewski first considered the special case of the case $k_0 = 1$ in which $\mathcal{V}_0 = 0$. In other words, the assumption is that $\mathfrak{l}$ contains a subalgebra $\mathfrak{m}$ with the structure of $\mathbf{sl}(2, \mathbb{C})$ that leaves no nontrivial subspaces invariant. Thus without loss of generality $\mathfrak{l}$ can be assumed to contain the three infinitesimal transformations that form the standard basis for $\mathfrak{m}$ given in Section 1, namely, $\mathcal{P}$, $\mathcal{N}$, $\mathcal{H}$ of (1.2). To determine the possibilities for $\mathfrak{l}$, Kowalewski developed his method of weights which will be discussed below. He found that with the exception of $n = 6$, the only possibilities for $\mathfrak{l}$ were the well-known groups $\mathfrak{l} = \mathbf{sl}(n+1, \mathbb{C})$, $\mathfrak{l} = \mathbf{sp}(n+1, \mathbb{C})$ (n odd), and $\mathfrak{l} = \mathbf{o}(n+1, \mathbb{C})$ (n even) [1902]. But for $n = 6$ an exception occurred. As soon as he discovered it, he wrote excitedly to Engel, "I have now arrived at your 14-term projective group in R_6 in a new way ...."[38] Kowalewski was referring to the irreducible representation of the simple type G_2 in 7 homogeneous coordinates. Such a representation had first been given by Cartan in his thesis [1894b:Ch.VIII]. Engel had earlier obtained nonlinear representations of G_2 in 5 variables and in [1900] showed that Cartan's representation could be derived in another fashion. Kowalewski was familiar with Engel's papers but apparently unfamiliar with those of Cartan, so that to him this group was Engel's. Indeed, during 1899–1913 Kowalewski never once cited Cartan's publications. That this was done largely out of ignorance rather than personal animosity is suggested by the high esteem Kowalewski later expressed in his autobiography [1950] for Cartan, where he declared him to be the greatest living mathematician (p. 73) and specifically acknowledged his successful solution to what I have termed Kowalewski's problem (p. 93).

In the last quoted letter Kowalewski expressed his confidence in his ability to deal with the other possibilities for $\mathfrak{m}$ besides that corresponding to $k_0 = 1$, $\mathcal{V}_0 = 0$: "I can also treat the analogous problem for the other three-term simple projective groups .... There is no limit to how far I can go, since the method (classification of the infinitesimal transformations by weight) always remains the same and makes the calculations very simple once the infinitesimal transformation of maximal weight is determined ...."[39] Nevertheless, Kowalewski does not appear to have dealt

[37] "Ich bin überrascht, wie hoch er die Lie'sche Gruppentheorie schätzt. Er geht sogar so weit darin, dass er auch die Bestimmung primitiver Gruppen für werthvoll hält, während doch andere Mathematiker das als kaum menschenswürdige Beschäftigung betrachten." Letter dated 26 October 1901 (Engel Archive Giessen).

[38] "Ich bin jetzt auf einen neuen Weg zu Ihrer 14-gliedrige proj[ektive] Gruppe im R_6 gelangt ...." Letter dated 15 July 1902 (Engel Archive Giessen).

[39] "Ich kann auch für die andern dreigliedrigen einfachen proj Gruppen die analoge Aufgabe behandeln .... Ich kann ... beliebig weit gehen, da die Methode (Classification der inf Tr nach Gewichten) immer dieselbe bleibt, und die

immediately with the continuation of his problem along the above lines. Study, who had been encouraging him on the problem solved in [1902], suggested the value of the analogous problem for ternary forms.[40] That is, the linear Lie algebra $\mathfrak{m}$ considered by Kowalewski in [1902] arises in the usual invariant-theoretic way by applying the transformations of $\mathbf{sl}(2,\mathbb{C})$ to the general binary form of degree n (as explained at the end of Section 1); one could thus consider the linear Lie algebra that arises by applying the transformations of $\mathbf{sl}(3,\mathbb{C})$ to the general ternary form of degree n. This is a "normal form" of type $(n,0)$, and by Study's formula (Section 2) the $\mathfrak{m}$ it generates involves $n^* = (n+1)(n+2)/2$ homogeneous coordinates and leaves nothing planar invariant. Study therefore suggested to Kowalewski finding all $\mathfrak{l} \subset \mathbf{sl}(n^*,\mathbb{C})$ that contain $\mathfrak{m}$. Kowalewski discovered that the only possibility was $\mathfrak{l} = \mathbf{sl}(n^*,\mathbb{C})$, and since this result was so negative, he never published it.

In 1904 Study moved to a professorship at Bonn, and Kowalewski followed him there in 1905. (Engel, who had never held a tenured professorship at Leipzig, became Study's replacement at Greifswald.) At Bonn Kowalewski returned to his problem. Besides setting his students Apfelstedt and Kummer to pursue the approach by dimension n, he continued to pursue the approach by k_0 and now studied the situation $k_0 = 1$ with $\mathcal{V}_0 \neq 0$. In addition to the groups he had already found, he obtained the well-known eight-dimensional Lie algebra in 6 homogeneous coordinates that expresses "how the ∞^5 conic sections of a plane are permuted when the general projective group is applied to the plane" [1906:407]. In Study's terms this is just the $\mathfrak{l}$ corresponding to the normal form of type $(2,0)$. The other $\mathfrak{l}$ he found was a 21-dimensional group in 8 homogeneous variables with the structure of $\mathbf{o}(7,\mathbb{C})$. (The spin representation of degree 8 for B_3.) This group had already been introduced by Engel in [1900] en route to his new derivation of Cartan's representation of G_2 in 7 homogeneous coordinates. Thus Kowalewski's investigations had not yet revealed any new groups.

Kowalewski was not discouraged by this fact, however, because he felt himself on the brink of a general solution to his problem. Thus he wrote to Engel on 6 July 1906, "By the way, I am, I believe, very close to a completely general resolution of the problem of determining all projective groups which leave nothing planar invariant. The following theorem brings me a lot closer: In a space of the nth degree a projective group that leaves nothing planar invariant always contains a three-term subgroup which leaves invariant a skewed system of planar spaces with sum total degree n, *but leaves nothing planar invariant within these spaces*. This theorem provides a way to determine all projective groups that leave nothing planar invariant. Hopefully, I

Rechnungen sich sehr einfach machen, wenn man die inf Transformationen von Maximalgewicht zuerst bestimmt."

[40] Letter of Kowalewski to Engel dated 16 February 1903 (Engel Archive Giessen).

will achieve this goal before long."[41] As this passage indicates, Kowalewski felt he could solve his problem by the approach described above in connection with (5.1). He expressed the same optimism in a letter of 25 October 1906 to Engel, but it was not until 1913—the year Cartan published his solution to Kowalewski's problem—that he was able to publish anything more on his problem, and what he did publish [1913; 1914] was still far from the anticipated general solution.

Kowalewski's paper [1913] and its continuation [1914] were at the same time his most ambitious and most disappointing attempt to resolve his problem. Their objective was to solve the problem for the case $k_0 = 2$. Let $\mathfrak{l}$ be a subalgebra of $\mathbf{sl}(n+1, \mathbb{C})$ that falls under this case. Then (5.1) becomes $\mathbb{C}_{n+1} = \mathcal{V}_0 \oplus \mathcal{V}_1 \oplus \mathcal{V}_2$, where $\mathfrak{m}$ denotes the simple three-dimensional subalgebra of $\mathfrak{l}$ that yields the above decomposition. Let $\mathfrak{m}^*$ consist of all $X \in \mathfrak{l}$ such that $X[\mathcal{V}_i] \subset \mathcal{V}_i$ for $i > 0$. Clearly, $\mathfrak{m}^*$ contains $\mathfrak{m}$ and is a subalgebra. The restriction of $\mathfrak{m}^*$ to $\mathcal{V}_i$, $i > 0$, is of type $k_0 = 1$ with $\mathcal{V}_0 = 0$ in the notation of (5.1) above, the type Kowalewski had studied in [1902]. Thus the possibilities for these restrictions were known, and this information about $\mathfrak{m}^*$ was used to determine $\mathfrak{l}$. Two cases were distinguished depending upon whether $\mathfrak{m} \neq \mathfrak{m}^*$ (Case I) or $\mathfrak{m} = \mathfrak{m}^*$ (Case II) [1913:1954]. Although Kowalewski seemed (p. 1954) to indicate that he would deal with both cases, he actually never considered the second. (More on this below.)

The analysis of Case I led Kowalewski to 5 groups, all of which were new and consequently required a detailed characterization. We may express them by the following notation: $\mathfrak{l}_{14}(A_1 \oplus G_2)$, $\mathfrak{l}_{10}(A_4)$, $\mathfrak{l}_{16}(A_1 \oplus B_3)$, $\mathfrak{l}_{16}(B_4)$, $\mathfrak{l}_{16}(D_5)$, where the first linear Lie algebra is in 14 homogeneous coordinates and has the structure of the direct sum of simple algebras of types A_1 and G_2, and so on. Kowalewski himself neither employed this notation nor the Killing–Cartan classification for simple and semisimple structures it implies. Consider, for example, his treatment of the group $\mathfrak{l}_{14}(A_1 \oplus G_2)$. Kowalewski wrote the basis for this group as a direct sum (in effect) of the known standard bases for the 14-dimensional and three-dimensional simple Lie algebras, but the term "semisimple" was not part of his vocabulary [1913:1996]. He was obviously completely unfamiliar with the Killing–Cartan structure theory. In order to characterize this group (and the other one corresponding to a semisimple structure), Kowalewski made the following observation: If $\mathfrak{l} \subset \mathbf{sl}(N, \mathbb{C})$ is r-dimensional and leaves no nontrivial subspaces invariant, then one can immediately derive from it an $(r+3)$-dimensional $\mathfrak{l}^*$ in $2N$

[41] "Übrigens bin ich, wie ich glaube, ganz nahe daran, die Bestimmung aller projektiven Gruppen, die nichts Ebenes invariant lassen, ganz allgemein zu erledigen. Der folgende Satz ... brachte mich ein gutes Stück vorwärts: In einem Raume n-ter Stufe enthält eine projektive Gruppe, die nichts Ebenes invariant lässt, stets eine solche dreigliedrige Untergruppe, bei der ein windschiefes System von ebenen Räumen mit der Stufensumme n invariant bleibt, aber in keinem dieser Räume etwas Ebenes. Von diesem Satze aus ergibt sich ein Weg zur Bestimmung aller projektiven Gruppen, die nichts Ebenes invariant lassen. Hoffentlich gelange ich bald zum Ziele" (Engel Archive Giessen).

coordinates with the same property. Kowalewski's construction was equivalent to a now familiar tensor product construction, but tensor products were still not at this time a standard part of mathematics. Kowalewski simply specified a basis for $\mathfrak{l}^*$, without mentioning the idea underlying the construction; but presumably he was thinking along the same lines made explicit by Cartan in his memoir [1913], which is discussed in Section 8.4. That is, if $\mathfrak{l}$ transforms the variables $x_1, \ldots, x_N$ and $\mathbf{sl}(2, \mathbb{C})$ transforms variables y_1, y_2, then $\mathfrak{l}^* = \mathfrak{l} \oplus \mathbf{sl}(2, \mathbb{C})$ transforms the $2N$ variables $x_i y_j$ by $(X \oplus Y)(x_i y_j) = [X(x_j)][Y(y_j)]$, which, when expanded, will be a linear combination of the $2N$ variables $x_k y_l$. In this manner Kowalewski was able to explain that the two $\mathfrak{l}$'s that (in our terms) are semisimple are derived, by this procedure, from, respectively, Engel's simple 14-dimensional group (take $\mathfrak{l} = \mathfrak{l}_7(G_2)$) and Engel's simple 21-dimensional group in 8 homogeneous variables ($\mathfrak{l} = \mathfrak{l}_8(B_3)$).

A total of 60 pages was devoted by Kowalewski to the task of obtaining the above 5 groups and another 17 pages to their characterization. But he never considered Case II. It is easy to see by Kowalewski's own procedures and results that an infinite number of groups arise in Case II. If $\mathfrak{l}$ is any irreducible representation of type $k_0 = 1$ acting on a vector space $\mathcal{W}$ and if $\mathcal{U}$ is the 2-dimensional vector space on which $\mathbf{sl}(2, \mathbb{C})$ acts, then $\mathfrak{l}^* = \mathfrak{l} \oplus \mathbf{sl}(2, \mathbb{C})$ acts irreducibly on $\mathcal{W} \otimes \mathcal{U}$, and it is not difficult to see that $\mathfrak{l}^*$ must be of type $k_0 = 2$.[42] According to Kowalewski's calculations, it is only when $\mathfrak{l} = \mathfrak{l}_7(G_2)$ or $\mathfrak{l} = \mathfrak{l}_6(B_3)$ that the resultant $\mathfrak{l}^*$ falls under Case I. Thus, for example, the choice of $\mathfrak{l} = \mathbf{sl}(n+1, \mathbb{C})$, $n = 1, 2, 3, \ldots$, leads to $\mathfrak{l}^*$'s falling under Case II, which consequently involves an infinite number of possibilities. Whether Kowalewski recognized this way to produce examples of Case II is not clear, although all the facts and constructions used above to produce $\mathfrak{l}^*$'s of Class II were known to him in some form or other. In any case, Case II would have appeared more formidable to Kowalewski simply because when $\mathfrak{m}^*$ reduced to $\mathfrak{m}$ there was very little grist for the mill of his method—his method of weights (*Gewichte*), to which I now turn.

Kowalewski defined weights as follows [1902:371–2]. Let $\mathfrak{l}$ be a linear Lie algebra in n homogeneous coordinates x_i, i.e., $\mathfrak{l} \subset \mathbf{gl}(n, \mathbb{C})$. Weights are defined with respect to an element $D \in \mathfrak{l}$ of the form $D = \sum_{i=1}^{n} a_i x_i \partial / \partial x_i$ such that not all the a_i are equal. Hence D, which is a linear trans-

---

[42] By hypothesis $\mathcal{W} = \mathcal{V}_0 \oplus \mathcal{V}_1$ with respect to some 3-dimensional simple $\mathfrak{m} \subset \mathfrak{l}$, where $\mathfrak{m}$ acts trivially on $\mathcal{V}_0$ and irreducibly on $\mathcal{V}_1$. Thus $\mathcal{W} \otimes \mathcal{U} = \mathcal{V}_0 \otimes \mathcal{U} + \mathcal{V}_1 \otimes \mathcal{U}$, and $\mathfrak{m}^* = \mathfrak{m} \oplus \{0\} \subset \mathfrak{l}^*$ acts trivially on $\mathcal{V}_0 \otimes \mathcal{U}$. If e_1, e_2 denotes a basis for $\mathcal{U}$ and $\mathbb{C}e_i$ the 1-dimensional subspace spanned by e_i, then

$$\mathcal{V}_1 \otimes \mathcal{U} = \mathcal{V}_1 \otimes \mathbb{C}e_1 + \mathcal{V}_1 \otimes \mathbb{C}e_2,$$

and each $\mathcal{V}_1 \otimes \mathbb{C}e_i$ is an irreducible $\mathfrak{m}^*$-module. Thus $\mathfrak{l}^*$ admits the decomposition $\mathcal{W} \otimes \mathcal{U} = \mathcal{V}_0 \otimes \mathcal{U} \oplus \mathcal{V}_1 \otimes \mathbb{C}e_1 \oplus \mathcal{V}_1 \otimes \mathbb{C}e_2$ of type (5.1) with respect to $\mathfrak{m}^*$, and so $k_0 \leq 2$. But $k_0 = 1$ is ruled out by Kowalewski's analysis of this case, which showed that only simple projective groups arise, whereas $\mathfrak{l}^*$ is not simple.

formation on $\mathbb{C}^n$, corresponds to a diagonal nonscalar matrix. (For any $x = (x_1, \ldots, x_n) \in \mathbb{C}^n$, $Dx = (Dx_1, \ldots, Dx_n) = (a_1x_1, \ldots, a_nx_n)$.) With respect to D, the variable x_i is assigned weight $+a_i$ and the operator $\partial/\partial x_i$ weight $-a_i$. Thus the weight of $x_j\partial/\partial x_k$ is $a_j - a_k$. An element $X = \Sigma c_{ij}x_i\partial/\partial x_j$ is *isobaric* if each nonzero term in X has the same weight g. Thus if X is isobaric, $a_i - a_j = g$ for all i and j such that $c_{ij} \neq 0$. The key property of an isobaric X of weight g is that $[X, D] = gX$. Thus a weight g is simply an eigenvalue of ad D and the isobaric X the corresponding eigenvector.

Conceived as such, these notions are fundamental to the Killing–Cartan theory of structure. The fact that Kowalewski presented them with no reference whatsoever to that theory reflects his lack of reliance upon—or even familiarity with—it. According to Kowalewski his theory of weights "is based on a generalization of an idea of Lie that I owe to the encouragement of Herr Engel" [1902:371–2]. As we saw, terms such as those of "weight" and "isobaric" originated with Cayley's theory of seminvariants (Section 4), although Kowalewski gave a different meaning to these terms. There is a mathematical connection between Kowalewski's problem for the structure type of $\mathbf{sl}(n, \mathbb{C})$ and the notions of weights and isobaric functions as they occur in theory of seminvariants, suitably generalized to n-ary forms (as was done by Deruyts), but this was not realized by Kowalewski, Engel and Lie. Nor, it would seem, did Engel stress the connections with the Killing–Cartan theory. Of course, Kowalewski's weights are not roots in the sense of Killing and Cartan because they are not linear homogeneous functions of a Cartan subalgebra, but the lack of any reference whatsoever to the Killing–Cartan theory is nonetheless remarkable.

It follows immediately from Kowalewski's definition that every X in $\mathfrak{l}$ can be expressed as a sum of isobaric elements X_i of distinct weights g_i: $X = \sum_{i=1}^{s} X_i$. By considering $[D, X]$, $[D[D, X]]$, etc., Kowalewski was able to conclude that $\sum_{i=1}^{s} g_i^k X_i$ is in $\mathfrak{l}$ for $k = 0, 1, 2, \ldots$ and thus:

(5.2) $$\textit{If } X = \sum_{i=1}^{s} X_i \in \mathfrak{l}, \textit{ then } X_i \in \mathfrak{l} \textit{ for } i = 1, \ldots, s.$$

He also pointed out that:

(5.3) *If $X, Y \in \mathfrak{l}$ are isobaric with respect to D and of weights g, g', respectively, then $[X, Y]$ is isobaric with respect to D of weight $g + g'$, unless $[X, Y] = 0$.*

Properties (5.2) and (5.3) of weights are the principal means by which Kowalewski and his students solved Kowalewski's problem for a specific dimension n or value of k_0. If, for example, $\mathfrak{l}$ is assumed to contain the group of a normal curve—meaning $k_0 = 1$ and $\mathcal{V}_0 = 0$ in (5.1)—then, working for convenience with $\mathfrak{l}^* = \mathfrak{l} \oplus \{\lambda I\}$ within $\mathbf{gl}(n, \mathbb{C})$, it could be assumed without loss of generality that $\mathfrak{l}^*$ contains the four elements

$$X_{-2} = \sum_{i=0}^{n-1}(i+1)x_i\frac{\partial}{\partial x_{i+1}}, \qquad V = \sum_{i=1}^{n} ix_i\frac{\partial}{\partial x_i},$$

and

$$X_{+2} = \sum_{i=1}^{n}(n-i)x_i\frac{\partial}{\partial x_{i-1}}, \qquad \overline{V} = \sum_{i=0}^{n-1}(n-i)x_i\frac{\partial}{\partial x_i}.$$

(In the notation of (1.2), $X_{-2} = \mathcal{N}$, $X_{+2} = \mathcal{P}$, and $\overline{V} - V = \mathcal{H}$.)

Using (5.2) and (5.3) further elements of $\mathfrak{l}^*$ can be generated. For example, if it is known that some diagonal D is in $\mathfrak{l}$, then if any X in $\mathfrak{l}^*$ is expressed as a sum of isobaric elements with respect to D, it follows by (5.2) that these elements themselves are in $\mathfrak{l}^*$. Likewise, (5.3) can be used to generate elements in $\mathfrak{l}^*$ of specific weight with respect to a given D. (For example, we could take $D = V - \overline{V}$, in which case X_{+2} is isobaric with weight $+2$ and X_{-2} is isobaric with weight -2.) Relative to any D there exists also a highest weight k. The isobaric elements possessing the highest weight can frequently be determined, thereby adding further known elements to $\mathfrak{l}^*$ and thus to $\mathfrak{l}$.

Kowalewski's method of weights was essentially a computational device rather than the conceptual basis for a general theory of classifying and characterizing projective groups that leave nothing planar invariant. In his paper [1913] on $\mathfrak{l}$ of type $k_0 = 2$, however, he did introduce an interesting twist to his method of weights. Recall that he distinguished two cases. In Case I, $\mathfrak{m}^*$ contains $\mathfrak{m}$ as a proper subalgebra, and Kowalewski showed that in this case either $\mathfrak{l}$ contains a copy of $\mathbf{o}(5, \mathbb{C})$ or "Engel's" simple group $\mathfrak{l}_7(G_2)$. Thus in either case $\mathfrak{l}$ contains two "diagonal" elements D_1 and D_2. Since any $X \in \mathfrak{l}$ can be expressed as a sum of elements that are isobaric with respect to *both* D_1 and D_2, to each such X corresponds a pair of weights g_1 and g_2, which Kowalewski considered as the Cartesian coordinates of a "weight point" (g_1, g_2) in the plane. For both cases $\mathbf{o}(5, \mathbb{C}) \subset \mathfrak{l}$ and $\mathfrak{l}_7(G_2) \subset \mathfrak{l}$ he wrote down a "doubly isobaric" basis for these subalgebras of $\mathfrak{l}$ and drew the diagram of the weight points. These diagrams are similar to modern root diagrams. In fact, for $\mathbf{o}(5, \mathbb{C})$ he drew in effect the root diagram for type B_2 [1913:1964], but the coincidence is somewhat fortuitous, depending as it does upon the agreement of the customary inner product in the plane and the Killing form. Thus his diagram for $\mathfrak{l}_7(G_2)$ is a distorted root diagram for type G_2 [1913:1979].

By considering weight points (g_1, g_2) rather than just weights, Kowalewski was now in effect regarding his weights as linear functions on the two dimensional subspace spanned by the weight point diagrams, although he was manifestly unaware of any connection with the transformations $S_w(w_\beta) = w_\beta + a_{\beta\alpha}w_\alpha$ of Cartan's thesis (discussed in Section 6.4). For example, in the case $\mathbf{o}(5, \mathbb{C})$, Kowalewski showed that a certain variable change can be made that has the effect of rearranging the basis elements of $\mathbf{o}(5, \mathbb{C})$ and, in particular, sends D_1 and D_2 into $D'_1 = -D_1$ and $D'_2 = D_2$, respectively. Thus if some X in $\mathfrak{l}$ had weight point (g_1, g_2) with respect to the D_i, it has weight point $(-g_1, g_2)$ with respect to the D'_i. This was interpreted geometrically as a reflection of the weight figure in the y-axis. Kowalewski also showed that the weight points of $\mathfrak{l}$ must be distributed symmetrically with respect to these reflections and, in fact, if there exist m independent isobaric X's in $\mathfrak{l}$ with weight point (g_1, g_2) and if one of the reflections defined by the $\mathbf{o}(5, \mathbb{C})$ diagram takes it into (g'_1, g'_2), then there exist m independent isobaric X's in $\mathfrak{l}$ with respect to (g'_1, g'_2). Such considerations of course greatly facilitated determining the possibilities for $\mathfrak{l}$ in Case I.

Throughout the period 1899–1913, during which Kowalewski and his students dealt, in many pages, with little pieces of the apparently immense problem of determining all projective groups that leave nothing planar invariant, they remained ignorant of, or at least not well versed in, the Killing–Cartan theory of structure. Kowalewski had been encouraged to develop his theory of weights by Engel, whose greater familiarity with the Killing–Cartan theory may have caused him to give such encouragement, since properties (5.2) and (5.3) used in Kowalewski's theory of weights were derived from the same considerations as lay at the basis of the Killing–Cartan theory. Although Engel apparently never mastered the Killing–Cartan theory, he esteemed it highly and may have for this reason been encouraging to Kowalewski.

To this extent Kowalewski may have been indirectly and unconsciously influenced by the work of Killing and Cartan. But in developing his theory of weights Kowalewski proceeded on his own, oblivious of points of contact with the Killing–Cartan theory. As a consequence his theory of weights bears only superficial resemblance to that theory—or to Cayley's theory of seminvariants. Even in the later papers [1913; 1914], Kowalewski's weights are not linear functionals on a Cartan subalgebra of the Lie algebra $\mathfrak{l}$ under consideration. Moreover, Kowalewski's weights correspond to roots, not weights, in the modern theory. His theory of weights was nothing more than a computational device; unlike Cartan's theory of weights, it did not provide a means of establishing general theorems about the nature of *all* projective groups that leave nothing planar invariant.

The work of Kowalewski and his students was far more typical than Cartan's of the work done by Lie's students and even by Lie himself. For example, Lie's approach to the problem of determining all projective groups in 3 homogeneous coordinates that leave some 2-dimensional manifold invariant was not informed by sufficiently general and powerful principles to be capable of dealing with the analogous problem in four or more homogeneous coordinates, as Fano had observed (Section 3). But Fano's method, which also utilized Study's complete reducibility theorem for $\mathbf{sl}(2, \mathbb{C})$, was itself no more general or powerful than Kowalewski's. It is only in comparison with Cartan's brilliant, masterful utilization of structural considerations, particularly those based on the Killing–Cartan theory of semisimple Lie algebras, that the work of Kowalewski and Fano pales in significance. As we shall see in the next chapter, in a trilogy of memoirs published in 1913–14, Cartan solved Kowalewski's problem in complete generality using his theory of Killing's secondary roots (now called weights), realized its equivalence to Killing's problem (Section 6.3), and did the same as well for real Lie algebras.

Chapter 8
Cartan's Trilogy: 1913–14

Two decades after his doctoral dissertation, Cartan published a "trilogy" of memoirs [1913; 1914b; 1914c] of fundamental importance to the theory of Lie groups and algebras and especially to the study of their representations. In [1913] he further developed Killing's theory of secondary roots and applied it to the problem of determining all linear Lie algebras of semisimple structure that leave nothing planar invariant. That is, he solved the representation problem that he had already articulated twenty years earlier in his dissertation. In [1914b] he solved the same problem for real linear Lie algebras. This required the results of [1913] together with the classification of all real simple structures that he developed in [1914c].

The publication of Cartan's trilogy thus marks a major event in the prehistory of the representation of semisimple Lie algebras. However, since these papers are so much a direct continuation and elaboration of the ideas in his doctoral dissertation, the historian may well ask why there was a hiatus of two decades before he published these now classic memoirs. There is no evidence or reason to suspect that Cartan had made a serious, sustained effort to deal with these matters earlier and had run into an impasse. Indeed when he retrospectively described the memoir [1913]—the keystone of the trilogy—he stated simply, "In 1913 I determined all irreducible linear representations of simple and semisimple groups by a method the principle of which was already contained in my thesis ..." [1931:19].

Cartan's neglect of the problems resolved in the trilogy becomes completely understandable when viewed from the perspective of the activities and priorities of the Paris mathematics community in general and of Cartan in particular. From this perspective, the problems appeared relatively unimportant at the turn of the century. In order to clarify this point I will consider (1) the problems suggested by Cartan's thesis work and their relative significance in view of the priorities emphasized by the Paris community of mathematicians (as depicted in Section 6.1) and, in this light, (2) Cartan's mathematical activity in the years following his thesis. We shall see that Cartan chose to work on problems of greater interest and significance to that community—and, of course, in particular to himself as a member. Many of these problems were also more challenging in that they could not be solved by falling back on the machinery developed in his thesis but required profound mathematical innovation, a challenge Cartan evidently relished. Indeed, although I give attention to the manner in which Cartan's mathematical interests were stimulated and shaped in general ways by his background and the intellectual environment within which he worked—that is, by the fact that he was a product of the institutions of Paris and a member of its community of mathematicians—his great originality and intellectual independence should never be in doubt.

By relating Cartan's research activity to the interests and emphasis in Paris, the remarkably creative and unique manner in which he responded to these stimuli is in no way diminished. But when viewed in the light of the priorities and interests prevalent in Paris, the twenty-year hiatus is no more anomalous than is the fact that Cartan belatedly developed the ideas in his thesis and composed the trilogy. Both the hiatus and the eventual return to the representation problem become more understandable when considered with regard to these priorities and interests. That is, we shall see that Cartan's post-thesis research was definitely guided by these priorities and interests, rather than by other possible research directions suggested by his thesis, but that the path his research thus took helped encourage a deeper appreciation of the representation problem and the value of the theory of secondary roots as a means for its resolution. With this in mind, I begin by considering some of the problems suggested by Cartan's thesis work and their fate in his hands.

8.1. Research Priorities 1893–1909

The most obvious problem involved the overriding objective of Killing's research: to effect an exhaustive classification of all Lie algebra structures. This objective involved three problems: proof of the radical splitting theorem; classification of all solvable Lie algebras; resolution of Killing's problem, assuming that it could also be proved (as Killing claimed) that its resolution would essentially yield the classification of all nonsolvable Lie algebras. From Killing's viewpoint, colored as it was by his Weierstrassian research program on foundations of geometry, these problems were of the utmost significance. They were not so regarded in Paris. Killing's work was significant in Paris because of its relevance to the nascent Galois-type theory of differential equations. It was from this viewpoint that Cartan undertook the task of reworking Killing's theory of structure. And from this viewpoint the above problems did not seem to be urgent matters. As was seen in Section 6.4, Cartan had used Killing's theory of secondary roots to resolve precisely that special case of Killing's problem that was perceived to be of significance to the Galois theory of differential equations. Furthermore, when he discovered a flaw in his proof of the radical splitting theorem (announced and sketched in [1893a:104]), he dismissed the matter with the remark, "Considerations analogous to those developed in Chapter VI (Section 3) show that it would suffice to prove this theorem for groups that admit only one invariant integrable subgroup. But that is a question that would take us too far afield" [1894b:128]. For his purposes the homomorphism $\mathfrak{g} \to \mathfrak{g}/\mathfrak{r}$, $\mathfrak{r} = \text{rad}(\mathfrak{g})$, would suffice. It was Levi [1905] who first published a proof of the radical splitting theorem. As for the problem of classifying all solvable algebras, there is no evidence that Cartan ever gave it any serious thought.

Although Cartan did not share Killing's sense of priorities, Killing's work had made a profound and lasting impression upon the young French mathematician. It opened to him a new mathematical world by introducing him to a substantial theory of structure. Consequently, Cartan's applications of group theory possessed a dimension not typical of his colleagues in Paris. Whereas in Paris Lie's theory of groups was deemed important by virtue of its applications, Cartan was unique in his emphasis upon the importance of the new structure theory for applications. Due perhaps to its relative abstractness and technical complexity, the new structure theory of Lie algebras remained his private domain, much as the structure theory of finite groups had remained (among French mathematicians) Jordan's private domain. Nonetheless, Cartan worked in a more encouraging environment than had Jordan, because, as we have seen, in Paris the application of Lie's theory was encouraged and a number of *normaliens* were working in this area. For example, Ernest Vessiot, who had developed Picard's ideas on a Galois theory of linear differential equations [1892], and Jules Drach were especially interested in the problem of developing a Galois theory for differential equations; and although it is doubtful they had digested the Killing–Cartan structure theory, they appreciated the importance of Cartan's thesis for applications to differential equations. Thus Drach in his attempt to articulate the general principles for a Galois theory of differential equations referred to Cartan's thesis results as "classic" within four years of their publication [1898:339].

During the four years following publication of his thesis, most of Cartan's output was directed toward developing the applications of its ideas. In one note [1895] he wrote, probably with the work of Poincaré and Picard in mind (Section 6.1), of the well-known interest in relating complex function theory to Lie's theory of groups. After noting that F. Schur and Maurer had done work on the nature of the complex-valued functions that arise in the equations of a transformations group, Cartan continued, "The goal of this note is to show the importance of *structural* considerations in this type of research and to indicate some very general theorems that follow from recently discovered results on the structure of finite continuous groups." Cartan also showed the importance of structural considerations for the application of his results on simple groups to differential equations. That is, given a differential equation admitting a finite dimensional transformation group in the sense of Lie, it would be necessary to determine the composition series of the group and a canonical form for the simple groups that arise thereby in order to carry out a Galois-type analysis. Cartan applied the new structure theory to indicate the nature of the operations involved [1894a; 1896b].

Another avenue of research suggested by Cartan's thesis was to show how the new structure theory could be utilized to throw light upon the problem of classifying hypercomplex number systems (linear associative algebras of finite dimension over the complex field). The motivation for such a line of research came from Poincaré, who had suggested it (Section 6.1). Poincaré's interest caught the attention of Lie, who set one of his students, Georg Scheffers, the task of carrying out Poincaré's suggestion. Scheffers's results, which were not very satisfying, were based upon the less profound theory of structure of Lie and Engel and raised the obvious question in Cartan's mind—as well as in the mind of F. Schur's student T. Molien—as to what could be accomplished using ideas analogous to those in the new theory of structure inaugurated by Killing.[1] Cartan provided the answer in several publications [1897a; 1897b; 1898]. Poincaré was impressed by this work of Cartan. He realized that it was intimately related to Frobenius's new theory of group characters and that "the theories of the two learned mathematicians mutually clarify one another" [1903:106].

In [1903] Poincaré demonstrated his mastery of the theories of Frobenius and Cartan by applying them to carry much further his earlier work [1883a] on linear differential equations with exclusively algebraic solutions. Poincaré, incidentally, was probably the sole French mathematician (with the possible exception of Jordan) who was familiar with Frobenius's theory of group characters. In fact, in his evaluation of Cartan's work [1912:141] Poincaré acknowledged Frobenius's theory as the most important advance in the theory of finite groups in many years. Apparently, however, Poincaré's interest and enthusiasm for Frobenius's theory of group representations never stimulated Cartan to seriously ponder the possibility of an analogous theory for continuous groups, for he never regarded his trilogy in this light until Weyl had made this fruitful connection.

During 1894–98, in addition to the above mentioned work, Cartan published two papers that indicate the breadth of his mathematical interests. One was a note relating to the structure of finite groups. It concerned

[1] Molien's work is briefly discussed in Chapter 10; for a more detailed treatment of the work of Scheffers, Molien, and Cartan see [Hawkins 1972].

Bertrand's theorem that if a rational function in n variables takes at least two distinct values under the $n!$ permutations of its variables, then, except when $n = 4$, it must take at least n distinct values. Cartan pointed out [1894c] that this theorem follows readily from the fact that, except for $n = 4$, the alternating group is the only normal subgroup of the symmetric group. Cartan's other paper [1896a] was a piece of Darboux-type geometry that is historically significant because it was in this connection that he began to develop his calculus of differential forms, which eventually figured prominently in his later work on the structure of groups.

The problem that concerned Cartan was the following. In the plane it is well known that the integral $\iint_D dx\, dy$, representing as it does the area of the region D, is invariant under displacements. Likewise for the volume integral $\iiint_D dx\, dy\, dz$ in space. Thus:

> In view of the development in this century of the *principle of duality* it is entirely natural to regard the plane as generated by *lines* as well as by *points*, and space as generated by *planes* as well as by *points*, and to ask whether certain multiple integrals extended over *sets of lines* in the plane or *sets of planes* in place would satisfy properties analogous to those of *area* and *volume*, i.e., would be invariant when all lines or all planes are subjected to the same displacement [1896a:140].

For example, in the case of the plane, if u, v denote coordinates of the line composed of the points (x, y) satisfying $ux + vy + 1 = 0$, then, as Cartan showed, the integral

$$\iint_D \frac{du\, dv}{(u^2 + v^2)^{\frac{3}{2}}},$$

is invariant under displacements. If D is the set of lines that meet a closed convex curve, then the integral represents the perimeter of the curve.

Although Cartan's motivation for considering such integrals was apparently geometrical, he pointed out at the close of his paper that his integrals were related to the "invariant integrals" introduced by Poincaré in his prize-winning essay on the 3-body problem [1890:Ch II]. Poincaré considered a system of equations $dx_i/dt = X_i(x_1, \ldots, x_n)$, $i = 1, \ldots, n$, as representing a "law of motion" of points in n-dimensional space. (A special case would be the canonical equations of motion of Hamilton and Jacobi.) Thus, for example, an integral $\int_C \sum Y_i(x_1, \ldots, x_n) dx_i$ is an invariant integral if $\int_{C_0} \sum Y_i dx_i = \int_{C_1} \sum Y_i dx_i$, where C_1 represents the "displacement" of the curve C_0 under the "motion" defined by the differential equations. As Poincaré noted, one can consider as well invariant integrals that are double, triple, ..., n-fold integrals. From the viewpoint of Lie's theory (which Poincaré did not mention) these integrals are invariant over the curves, surfaces, etc. that are transformed by the one-parameter group generated by the infinitesimal transformation associated to the equations $dx_i/dt = X_i(x_1, \ldots, x_n)$ (as indicated in Section 2.2). Cartan realized, and pointed out, that the theory of invariant integrals could be developed as an application of Lie's theory of groups. In fact, his opening remarks suggest that he had first obtained the invariant integrals discussed in [1896a] by group-theoretic calculations.

In order to change to homogeneous coordinates and in order to see how the integral is changed by a displacement, Cartan had to make variable changes in his double integral. In footnotes he indicated how the variable changes could be made by a symbolical calculus based on defining, in effect, the product $\mu \wedge \nu$ of 1-forms and 2-forms. It was in connection with Pfaff's problem that Cartan developed the calculus of forms more fully and generally and, in particular, introduced the other basic operation on forms—differentiation. Pfaff's problem was to transform

$$\omega = a_1(x)dx_1 + \cdots + a_n(x)dx_n$$

by a variable change $x_i = \phi_i(y_1, \ldots, y_n)$ into

$$\omega = b_1(y)dy_1 + \cdots + b_p(y)dy_p$$

with p as small as possible, the motivation being that the transformation resolved the problem of integrating the "Pfaffian equation" $\omega = 0$.[2] The most recent major advance on the study of Pfaff's problem in Cartan's time had been made independently by Frobenius [1877] and Darboux [1882], who in order to study the problem beyond the generic case introduced what Frobenius called the bilinear covariant of ω. Cartan recognized that this bilinear covariant was simply the derivative $d\omega$, where

$$d\omega = \sum_{i=1}^{n} da_i \wedge dx_i \quad \text{and} \quad da_i = \sum_{i=1}^{n} \frac{\partial a_i}{\partial x_i} dx_i.$$

It should be noted that Poincaré had developed many aspects of the theory of n-variable analogues of line and surface integrals in publications spanning the years 1886–1895,[3] and that Gabriel Koenigs, one of Cartan's teachers at the École Normale, had already noted [1895] the connection between Poincaré's results and Darboux's memoir [1882]. But it was Cartan who showed how the calculus of differential forms could be used with great profit to rework the problem of Pfaff. This he did in a lengthy memoir published in the *Annales* of the École Normale [1899] in which he used his calculus of differential forms to generate higher-dimensional analogues of the bilinear covariant. During 1899–1901 Cartan devoted himself to the continuation of the line of research initiated in [1899]. He used his calculus of forms to study systems of Pfaffian equations [1901b] and special types

---

2 On Pffaf's paper and its role in the theory of partial differential equations, see [Hawkins 1991:203ff.].

3 For a detailed historical account of the theory of differential forms, that includes the connections with generalized Stokes theorems and related considerations, see [Katz 1981; 1985].

of Pfaffian equations [1901c]. This aspect of his mathematical output appears to have been the most accessible in Paris, relating as it did to partial differential equations and Poincaré's theory of invariant integrals.

Cartan himself related his work on Pfaffian equations to the work on partial differential equations by former *normaliens* Jules Beudon (b. 1869), Etienne Delassus (b. 1868), Edouard Goursat (b. 1858), and Charles Riquier (b. 1853).[4] A substantial portion of Goursat's monograph [1922] on Pfaffian equations was devoted to an exposition of Cartan's results.[5]

In carrying out his research on Pfaffian equations, Cartan may have had in his mind the possibility of an eventual application to the theory of infinite-dimensional groups of transformations. In any case, Lie had defined the transformations of such groups as solutions to certain partial differential equations [1888], so that the idea of thinking of infinite groups in terms of Pfaffian equations was close at hand. Lie repeatedly stressed the importance of the study of infinite-dimensional groups for the envisioned Galois theory of differential equations [1883b:56; 1884a:538; 1891a:316–8; 1891b:14], and Picard echoed his emphasis [1892a:727–8]. After a visit to Paris in 1893, Lie stressed the point that before a complete group-theoretic treatment of differential equations could exist it would be necessary to specify the "structures" of all simple infinite-dimensional transformation groups [1895b:291–2]. Despite the fact that the meaning of "structure" for infinite-dimensional groups had not even been worked out, Lie optimistically conjectured that it was "not unlikely" that the classification of infinite-dimensional simple groups would turn out to be as manageable as in the finite-dimensional case.

The development of the theory of infinite groups in conjunction with its application to differential equations may have been one of the reasons that the leaders of the Paris community in 1900 announced as the topic for the *Grand prix des sciences mathématiques* for 1902 the problem of perfecting the application of the theory of continuous groups to the study of partial differential equations.[6] The prize was won by Vessiot for a memoir "Sur l'intégration des systèmes différentiels qui admettent des groupes continus de transformations," that was eventually published in *Acta Mathematica*

---

[4] See Cartan's papers [1901a; 1901b] for references to their work.

[5] The final three chapters of Goursat's monograph were devoted to an exposition of recent progress in the theory. "Les plus importants de ces progrès," wrote Goursat, "sont dus à M. Cartan, dont on trouvera le nom presque à chacque page de ces trois Chapitres. Je souhaite que cette exposé rapide puisse engager les jeunes mathématiciens à étudier plus à fond ces élégantes méthodes, dont l'auteur a déjà fait si belles applications" [1922:vii].

[6] For the announcement of the prize, see *Acad. Sci. Paris, C. R.* **131** (1900):1151. The prize committee consisted of Jordan, Picard, Poincaré, Appell, and Painlevé, with the last four making up the *rapporteurs*. Another motivation for the prize may have come from Vessiot's discovery of flaws in Drach's attempt [1898] to develop a Galois theory for partial differential equations. It was common knowledge in Paris. See [Dugac and Taton 1981].

[1904].[7]

In this memoir, Vessiot posed the question as to the nature of the integrations required by his Galois-type theory of differential equations. Since the case in which the group of the differential equation is finite-dimensional was readily treated by known results, he concentrated on infinite-dimensional groups. He assumed that in this case, too, the question reduced to the case of primitive simple groups, an assumption, as we shall see, that Cartan later showed to be incorrect. Lie had discovered four classes of infinite-dimensional primitive simple groups, and Kowalewski's results on primitive groups [1899], Vessiot noted, showed that for five or fewer variables no other such groups exist. These four classes are (I) The group of all transformations in n variables; (II) the group of all volume-preserving transformations in n variables; (III) the group of all transformations in $2n \geq 4$ variables that leave invariant the integral

$$\iint_D x_1 dx_2 + \cdots + x_{2n-1} dx_{2n};$$

(IV) the group of all contact transformations of n-dimensional space, regarded as point transformations in the $2n - 1$ variables (x, z, p) (in the notation of Section 2.3). Vessiot therefore discussed the nature of the integrations involved for these four types, but he stressed the incompleteness of his results: "What, then, is the answer that can be given to the question that was posed to us? One sees that it will be possible to resolve it completely only when all the types of simple groups are known" [1904:309]. As we shall see, by 1907 Cartan was able to provide a solution to the problem posed implicitly by Vessiot's remark.

The year that Vessiot won the prize, Cartan announced that he had succeeded in developing a new approach to the structure of continuous groups that applied equally to finite- and infinite-dimensional groups [1902]. The new approach (discussed below) derived from his work on Pfaffian equations. The details were set out in two memoirs totaling 144 pages [1904; 1905]. Then, in a note [1907] Cartan announced that he had succeeded in classifying all simple infinite-dimensional groups. For transitive groups, his results showed that Lie's four general classes exhausted the possibilities, but he also showed that nontransitive—and therefore nonprimitive—simple groups exist. Indeed, he showed that it could not be assumed (as Vessiot had done) that infinite-dimensional groups necessarily have a composition series of finite length with simple factor groups. Cartan introduced a class of nonsimple groups, called improper simple groups, in the hope that then

[7] For the prize committee's evaluation of the paper see *Acad. Sci. Paris, C. R.* **135** (1900): 1158–61. Next to Cartan, Vessiot was perhaps the best of the normaliens working on applications of Lie's theory. For an overview of his mathematical work, see [Cartan 1947]; the obituary essay by Montel [1953] contains biographical information on Vessiot, including his teaching and administrative activities.

Figure 25. This unidentified photo from Olso bears a striking resemblance to the photo of Vessiot at a more advanced age found in [Cartan 1947]. Vessiot, one of the first normaliens to study with Lie in Leipzig (Section 6.2), later corresponded with him and may have given him this photo.

every group would have a composition series of finite length with proper and improper simple factor groups. He determined all improper simple groups as well. The details of his classification were published in [1909], his sole publication for that year. By that time it had been fifteen years since the publication of his doctoral thesis.

During the first fifteen years of his career as a mathematician, Cartan had spent the first four years (1895–98) developing applications of the ideas in his thesis. This was followed by three years (1899–1901) developing the theory of Pfaffian equations and then nine years (1901–1909) working out his structure theory of infinite-dimensional groups, which he expounded in lengthy memoirs totaling 370 pages.[8] Cartan's nine-year effort was appre-

[8] It was only relatively recently that an effort was initiated, in [Singer and Stern-

ciated in Paris, especially by Poincaré. Thus in his report recommending Cartan for a professorship at the Sorbonne in 1912, Poincaré affirmed that "one of the most important applications of Lie groups is the integration of differential equations" [1912: 139] and then singled out Cartan's work on infinite dimensional groups as the most original and ingenious part of his mathematical output [1912:141,145].

It was, in fact, fitting that Poincaré wrote the report on Cartan, for although it is doubtful that he ever took the time to completely master the intricacies of the Killing–Cartan structure theory or Cartan's subsequent work in this vein, he, far more than Picard or Darboux, had the breadth of vision and facility with abstract mathematical ideas to fully appreciate the importance and applicability of Cartan's relatively abstract type of mathematics. Not only did Poincaré stress the importance of structural considerations in his lecture at the 1908 International Congress of Mathematicians (Section 6.1), after Lie's death he even ventured to publish some papers on Lie groups that dealt with their theory rather than their application [1899a; 1899b; 1901; 1908b].[9] Poincaré's use of Cartan's results on the structure of hypercomplex number systems (mentioned earlier) further exemplifies his ability to perceive the value of abstract structural considerations for applications. Picard never did anything similar. Picard always seemed to respect Cartan from a great distance. He recognized the importance of Cartan's work, but its abstractness (by the standards of the period) made him slightly uncomfortable. Even in his elocution at Cartan's scientific jubilee in 1939, Picard stressed the abstractness of Cartan's work: "The most abstract parts of analysis are indebted to you for highly important advances ..." [Cartan 1939:163]. Cartan's work on Lie groups in general was categorized by Picard as exemplifying the artistic tendency within mathematics (as discussed in Section 6.1): "In the preceding works you show yourself to be a pure mathematician, that is to say, an artist or a poet in the world of numbers and forms" [1939:164]. Picard evidently felt more comfortable discussing Cartan's later work on geometry and spinors because of its relevance to physics via Einstein's theory of general relativity and quantum mechanics, respectively [1939:164–6]. As for Darboux, he made no effort to understand Lie's theory of groups, let alone Cartan's more abstract rendition of it. Until his death in 1912, Poincaré was Cartan's most knowledgeable supporter in Paris.

Given the priorities he shared with his colleagues in Paris, Cartan's mathematical career had certainly taken an appropriate and highly successful course. It had also brought him once again to the representation problem implicit in his doctoral thesis, for as we shall now see, in order to classify the transitive simple groups, Cartan found himself solving another special case of that problem and once again relying on the properties of Killing's secondary roots.

---

berg 1965], to come to grips with Cartan's theory of infinite groups.

9 See [Schmid 1982] for a discussion of these papers.

8.2. Another Application of Secondary Roots

The starting point of Cartan's new approach to structure was the following definition of isomorphic groups (introduced independently by Vessiot [1903:§IX]): A group $\mathfrak{G}'$ of transformations T' in $m+n$ variables $x_1, \ldots, x_m, y_1, \ldots, y_n$ is said to be a prolongment of a group $\mathfrak{G}$ of transformations T in $x_1, \ldots, x_m$ if every T' in $\mathfrak{G}'$ transforms the variables x_i among themselves and in the manner of some T in G. Thus the equations for T' are of the form

$$\text{(7.2)} \qquad \begin{aligned} x_i' &= f_i(x_1, \ldots, x_m), && i = 1, \ldots, m, \\ y_j' &= g_j(x_1, \ldots, x_m, y_1, \ldots, y_n), && j = 1, \ldots, n, \end{aligned}$$

and the first m equations correspond to some T in $\mathfrak{G}$. The correspondence $T' \to T$ clearly defines a homomorphism from $\mathfrak{G}'$ to $\mathfrak{G}$. The prolongment is said to be *holoédrique* if the homomorphism is an isomorphism. Two groups $\mathfrak{G}$ and $\mathfrak{H}$ are defined to be isomorphic if there exist *holoédrique* prolongments $\mathfrak{G}'$ and $\mathfrak{H}'$ of $\mathfrak{G}$ and $\mathfrak{H}$, respectively, in the same total number of variables such that $\mathfrak{G}'$ and $\mathfrak{H}'$ are similar in the sense of Lie: By change of the variables the equations defining the transformations in both groups can be made identical.

All these definitions apply to finite-dimensional transformation groups as well, and Cartan showed that when $\mathfrak{G}$ is r-dimensional it is always possible to prolong it to a group $\mathfrak{G}'$ that leaves r Pfaffian expressions (or 1-forms) $\omega_1, \ldots, \omega_r$ (in the augmented set of variables) invariant. Furthermore, this system of 1-forms is closed, viz., $d\omega_k = \Sigma c_{ijk}\omega_i \wedge \omega_j$, and the c_{ijk} turn out to be the familiar structure constants of the group $\mathfrak{G}$. Suppose now that $\mathfrak{G}$ is not finite-dimensional but, for the sake of simplicity, is transitive. Then it can still be prolonged to a group $\mathfrak{G}'$ consisting of all transformations leaving 1-forms $\omega_1, \ldots, \omega_m$ invariant, the integrals of $\omega_1 = \cdots = \omega_m = 0$ defining the transformations of the group $\mathfrak{G}'$. But the ω_i no longer form a closed system. Rather, constants c_{ijk} and a_{ijk} exist such that

$$d\omega_k = \sum_{i,j} c_{ijk}\omega_i \wedge \omega_j + \sum_{i,r} a_{irk}\omega_i \wedge \pi_r, \quad k = 1, \ldots, m,$$

where the π_r, $r = 1, \ldots, p$, are p additional 1-forms. Here the c_{ijk} and the a_{irk} together play the role of structure constants. In particular, the a_{irk} define a finite-dimensional linear group $\mathfrak{L}$ whose Lie algebra $\mathfrak{l}$ has basis elements

$$U_r = \sum_{i,k} a_{irk} u_i \frac{\partial}{\partial u_k}, \quad r = 1, \ldots, p.$$

The linear group $\mathfrak{L}$ provided the key to Cartan's classification of primitive infinite-dimensional simple groups, and for this reason he called it the adjoint group [1909:97]. He showed that the classification problem reduced to determining all $\mathfrak{l} \subset \mathbf{gl}(n, \mathbb{C})$ that "leave nothing planar invariant" (i.e., leave no vector subspaces invariant) and that have the additional property (discussed below) of being "semi-involutive." In his doctoral thesis Cartan had considered *semisimple* $\mathfrak{l} \subset \mathbf{gl}(n, \mathbb{C})$ that leave nothing planar invariant. He showed that the problem of determining all such $\mathfrak{l}$—his representation problem— was equivalent to Killing's problem, and he solved a special case of this problem that had applications to the Galois theory of differential equations (Section 6.4). The obvious question now was, What is the structure of *any* $\mathfrak{l} \subset \mathbf{gl}(n, \mathbb{C})$ that leaves nothing planar invariant? The example $\mathfrak{l} = \mathbf{gl}(n, \mathbb{C})$ made it clear that such an $\mathfrak{l}$ need not be semisimple, but Cartan perceived that in general, a nonsemisimple $\mathfrak{l}$ is as close to being semisimple as $\mathbf{gl}(n, \mathbb{C})$ is. This is the gist of the following theorem from his paper.

> **Theorem XVIII.** *If a linear homogeneous group leaves no planar multiplicity invariant, then it is either semisimple or it decomposes into a semisimple group and a one-parameter group of homothetic transformations* $T = \lambda I$ [1909:148].

Thus the problem of determining all $\mathfrak{l} \subset \mathbf{gl}(n, \mathbb{C})$ that leave nothing planar invariant was essentially equivalent to the representation problem of Cartan's thesis, and Cartan could therefore use the theory of secondary roots to help determine all semi-involutive $\mathfrak{l}$ that leave nothing planar invariant. Of course, Theorem XVIII also shows that, in the case of *projective* groups that leave nothing planar invariant they are always semisimple, since we can then take $\mathfrak{l} \subset \mathbf{sl}(n, \mathbb{C})$. Thus Theorem XVIII shows that Kowalewski's problem is identical to the representation problem.

Cartan's description of the theory of secondary roots in [1909] does not reveal any noteworthy innovations beyond what was already contained in his dissertation, although he now made explicit what was more or less implicit in his dissertation: Conjugate secondary roots have the same multiplicity [1909:149]. There is, however, some new terminology. He now used the term "weight" *(poid)* to refer to both the principal roots $\alpha, \beta, \ldots$ (i.e., roots in the modern sense) and the secondary roots $\lambda, \mu, \ldots$ (i.e., weights in the modern sense) associated with a semisimple $\mathfrak{l}$. The new terminology was probably prompted by his familiarity with Kowalewski's paper [1899] on primitive transformation groups in five variables. He examined this paper because Kowalewski had there determined all infinite-dimensional primitive groups in five variables as well as those that are finite-dimensional. It was in this paper that Kowalewski first introduced his concept of a weight. As we saw in Section 7.5, Kowalewski's weights relate to the principal roots of $\mathfrak{l}$ rather than to the secondary roots or weights in the modern sense, so that Cartan's usage of "weight" was broader, embracing as well the all-important secondary roots.[10]

[10] Later, in [Cartan 1913], the term "weight" was reserved for the secondary roots alone.

Cartan utilized the theory of secondary roots as follows. Suppose $\mathfrak{l}$ is semi-involutive and leaves nothing planar invariant. Then the associated system of secondary roots $S(\lambda)$ necessarily possesses the following property [1909:152], which I will call Property C.

> **Property C.** *Whenever $\mu, \nu \in S(\lambda)$ are such that $\mu + \alpha$ and $\nu + \alpha$ also belong to $S(\lambda)$ for some root $\alpha \neq 0$, it must be that $\mu - \nu = \beta - \gamma$, where the root 0 is allowed among the roots β and γ.*

Using modern terms we would say that the saturated system $S(\lambda)$ has Property C whenever the existence of a nonzero root α such that μ, ν, $\mu + \alpha$, and $\nu + \alpha$ are all in $S(\lambda)$ implies that the difference of the weights μ and ν is equal to a difference of roots.

Thus, as a first step toward determining all semi-involutive $\mathfrak{l}$ leaving nothing planar invariant, one could determine all systems $S(\lambda)$ possessing Property C. Cartan did not reveal how he went about this determination. All calculations were omitted from his paper on the grounds that they were "rather long and tedious" [1909:150]. It follows from the modern theory of weights that there is only a finite number of cases that need to be considered as serious candidates for satisfying Property C, and this number is relatively small when compared with the infinitude of all possible systems.[11] Nonetheless, the computational task of actually carrying out the determination must have been very time-consuming, as Cartan's remark suggests. Also, once all systems with Property C were known, the computational task was far from completed, since Property C is only a necessary condition for the associated $\mathfrak{l}$ to be semi-involutive. To determine whether such an $\mathfrak{l}$ is actually semi-involutive much more computation was required, including for each $S(\lambda)$ with Property C the determination of the corresponding representation $\mathfrak{l}$. In the end, Cartan was left with four semi-involutive $\mathfrak{l}$'s: the special linear and symplectic groups and the two nonsemisimple groups generated from them by adding the scalar multiples of the identity transformation. Using these results he determined all primitive infinite-dimensional

---

[11] Given a simple structure of rank l, let $\alpha = \sum_{i=1}^{l} k_i\alpha_i$ denote the maximal root with respect to some base $\alpha_1, \ldots, \alpha_l$. Let $k = \max\{k_1, \ldots, k_l\}$ and $m = \max\{3, k+1\}$. Then the saturated system $S(\lambda)$ with highest weight $\lambda = \sum_i m_i\lambda_i$, where the λ_i are the fundamental dominant weights, cannot satisfy Property C if $m_i > m$ for some i. The value of m for types A–D is 3; for E_6 and G_2, $m = 4$; for E_7 and F_4, $m = 5$; for E_8, $m = 7$. Of course, many systems $S(\lambda)$ with all $m_i \leq m$ do not actually have Property C. Thus, for example, the sole system of type G_2 possessing Property C is $S(\lambda_1)$. A saturated system of weights corresponding to nonsimple semisimple structures can have Property C only if (1) there are only two simple summands; (2) if the saturated system corresponds to the module $\mathcal{V}(\lambda) \otimes \mathcal{W}(\pi)$, the saturated systems $S(\lambda)$ and $S(\pi)$ must satisfy the stronger condition that the difference of *any* two weights is a root. Since this condition will not be satisfied, e.g., by $S(\lambda)$ if $\langle \lambda, \alpha \rangle \geq 2$ for some root α, consideration of maximal roots shows immediately that the only candidates to satisfy this stronger condition are as follows: Type A_l: any λ_i, $i = 1, \ldots, l$; type B_l: λ_1; type C_l: λ_1; type D_l: λ_1; type E_6: λ_1, λ_6; type E_7: λ_7.

groups and confirmed that all transitive simple infinite-dimensional groups are represented by the types I–IV given above [1909:101–129].

The classification of infinite-dimensional simple groups, a problem whose importance for applications to differential equations Cartan stressed [1907:1094; 1909:93], had thus led him, via the adjoint group $\mathfrak{l}$ and Theorem XVIII, back to the context of the representation problem, although, as in his doctoral dissertation, only a special case of the problem was relevant to the problem at hand. But Cartan could see from his familiarity with Kowalewski's paper [1899:70,115] that a subproblem of the problem of determining all finite-dimensional primitive transformation groups was that of determining all $\mathfrak{l} \subset \mathbf{sl}(n, \mathbb{C})$ that leave nothing planar invariant—Kowalewski's problem as it was called in Section 7.5.[12] By virtue of Theorem XVIII Cartan could see that Kowalewski's problem was essentially equivalent to the representation problem—as well as to the problem implied by Study's characterization theorems (Section 7.2).

8.3. Continuous Groups and Geometry

With the publication of Cartan's classification of simple infinite dimensional groups in [1909], the major outstanding structure-related problems regarding the application of group theory to differential equations had been resolved. During the following years he began to focus his attention upon the application group theory to geometry. Here, too, Poincaré's views may have been a factor, since he had been stating since 1880 that geometry was the study of groups. The earliest expression of a group-theoretic interpretation of geometry by Poincaré seems to occur in an unpublished supplement to the memoir on Fuchsian functions he submitted in competition for a prize of the Paris Academy of Sciences. The supplement was submitted on 28 June 1880.[13] Poincaré also expressed similar views in conversations with Lie in Paris in 1882, at which time Poincaré learned from Lie about Klein's *Erlanger Programm* (Section 1.5). The first published statement of such views was apparently in his paper on the Helmholtz space problem [1887:90]. They are also present in his important memoir on *analysis situs* [1895:193] and in his review [1902:112] of Hilbert's *Grundlagen der Geometrie*, as well as throughout his popular scientific lectures. Thus it is likely that Cartan had Poincaré's views in mind as he turned his attention to geometrical applications of continuous groups.[14]

---

[12] Cartan gave no sign until after publication of his trilogy [1913; 1914b; 1914c] that he was aware of the work done on Kowalewski's problem after 1899 by Kowalewski and his students. The first reference by Cartan to this subsequent work was in his *Encyclopédie* article on continuous groups and geometry, which went to press in 1915.

[13] See [1981; Gray 1982] for further information.

[14] Rosenfeld states that Poincaré "attracted Cartan's attention to geometric ap-

In this connection he realized that the solution of Kowalewski's problem also possessed a geometrical importance. All of these experiences apparently combined to bolster his resolve to develop the ideas in his thesis and solve the representation problem, and hence Kowalewski's problem, in complete generality. How he accomplished this will be discussed in Section 4. The present section is devoted to a depiction of the background to Cartan's interest in geometrical applications of structure. Two publications seem to have played important roles in stimulating his interest in applying the theory of groups to geometry: a paper by Émile Cotton [1905], which combined Darboux's infinitesimal geometry with Lie's theory, and the article on continuous groups and geometry that Fano wrote for Klein's *Enzyklopädie* [1907b].

We saw in Section 7.3 that Klein's *Erlanger Programm* of 1872 was republished, in Italian translation by Fano, in 1890. The following year a French translation was published.[15] Klein expressed his delight at the translation into French, since, as he explained in a footnote, "now more than ever the theory of groups seems to be of interest in France and so the content of my program will perhaps be of some interest." Klein's hopes were not immediately realized. Although Lie's theory of transformation groups was seriously studied in Paris, especially by recent graduates of the École Normale, the application of Lie's theory to the study of differential equations, rather than to geometry, held center stage in 1891 and for some time thereafter. In this respect Cartan's interests during 1894–1909 were typical. Furthermore, geometrical interests at the time in Paris were dominated by Darboux and his students and thus focused on differential geometry and related differential equations.

Probably for these reasons Klein's *Erlanger Programm* failed to inspire any immediate research activity in Paris. For the same reasons, however, it was perhaps inevitable that some enterprising *normalien* would eventually attempt to bring Lie's theory of transformation groups to bear upon the type of geometry cultivated in Paris. Certainly such an application would have Darboux's blessing, since he had played a role in directing *normaliens* to Leipzig to study transformation groups with Lie (Section 6.1), although he himself, unlike Picard and Poincaré, remained unfamiliar with the group-theoretic aspect of Lie's work. The first *normalien* to attempt such an application was Émile Cotton (b. 1872). In his thesis [1900], which was dedicated to Darboux and Goursat, Cotton studied three-dimensional manifolds that are left invariant by some transformation group in the sense of Lie and obtained canonical forms for the line element ds^2. A few years later he made a more far-reaching observation on the relation of Lie's theory to infinitesimal geometry.

Darboux had introduced the method of moving frames of reference

plications of group theory" [Akivis and Rosenfeld 1993:5], but no supporting documentation is given.

[15] See [Klein 1872b] in the list of references for bibliographic details.

to study the infinitesimal properties of curves and surfaces in Euclidean space. It was a basic part of his lectures on geometry [1887–96:v.1, Livre 1]. In a note in the *Comptes Rendus* of the Paris Academy [1905] the Belgian mathematician A. Demoulin pointed out that a similar method could be introduced into "Géométrie cayleyenne"— the metric geometries introduced in projective space by means of a quadratic form (the absolute)—thereby making possible the extension of Darboux's method to the study of these geometries, such as hyperbolic or elliptic geometry. Demoulin's point of view was not group-theoretic, but his remarks were viewed in this light by Cotton, who observed in [1905] that the method of moving frames could be generalized by replacing the group of Euclidean displacements by any transitive finite-dimensional transformation group. Cotton's paper thus made Lie's theory of groups geometrically appealing by relating it to the strong interest in infinitesimal geometry and kinematics within the Darboux school. In addition, Cotton utilized group-theoretic work on differential equations by Tresse and Vessiot and provided what he regarded as a more intuitive presentation of some of Vessiot's results on differential equations.

"I hope," Cotton wrote, "that the present work will facilitate the utilization of groups (other than the group of motions) that arise naturally in classical questions ..." [1905:43]. His work did far more than he anticipated, for it provided the fillip for Cartan's grand synthesis of his newly developed approach to the structure of groups with the geometrical and kinematical perspective of the Darboux school.[16] In a paper in Darboux's *Bulletin* [1910a] Cartan observed that his new approach to the structure of groups could be articulated in such a way that when applied to finite-dimensional transitive groups it represented a generalization of Darboux's theory of moving frames. That is, if $\mathfrak{G}$ is an r-dimensional transitive group in $n \leq r$ variables, it may be prolonged to an isomorphic group $\mathfrak{G}^*$ in r variables that consists of all transformations that leave invariant the r 1-forms $\omega_1, \ldots, \omega_r$ that yield the equations of structure $d\omega_k = \sum c_{ijk}\omega_i \wedge \omega_j$. In particular, if $\mathfrak{G}$ is the 6-dimensional group of displacements (rigid motions) of ordinary space, then the 1-forms $\omega_1, \ldots, \omega_6$ turn out to be the components with respect to a moving frame of the instantaneous dispacements of the frame, and the equations of structure are simply the first-order partial differential equations of Darboux's theory.

Cartan evidently took great satisfaction in this opportunity to relate the theory of structure to the mathematics he had learned as a student: "By presenting matters from the kinematical point of view the generalization of a classical theory, that of the moving trihedral due to M. Darboux, is achieved. In the particular case in which the group is that of the displacements of space ... the structure formulas coincide straightaway with the first-order partial differential equations that M. Darboux has rendered classical in the theory of displacements with many parameters" [1910a:251].

[16] Cf. Cartan's remarks about Cotton in his *Notice* [1931:62].

In a note [1910b] Cartan pointed out that his approach to Darboux's theory, when applied (in the spirit of Darboux) to geometrical objects such as curves and surfaces, yielded some new and interesting results.

Cartan's papers [1910a; 1910b] inaugurated a new phase in his career as a mathematician. From then on geometrical considerations were to play a far more dominant role in his research. As in the papers of 1910, the interests of the Darboux school, and in particular differential geometry, were central in this phase of his development, although his own originality and penchant for structural considerations continued to set his work apart from that of his colleagues. From his work on moving frames he went on to develop his theory of generalized spaces, starting in 1922, in response to Einstein's general theory of relativity and then, starting in 1926, his theory of symmetric spaces.[17] Although Cartan's research in the geometrical phase of his career was thus dominated by differential geometry, his geometrical interests and knowledge were far broader and included the type of projective and algebraic geometry that was associated with Klein's *Erlanger Programm*, as can be seen from his masterly revision [1915] of Fano's article [1907b] for the French version of the *Enzyklopädie der mathematischen Naturwissenschaften*. As we shall see, it was his preoccupation with this sort of geometry that provided the fillip for his trilogy of memoirs on the representation problem [1913; 1914b; 1914c]. Although in what follows the focus will be upon these memoirs and their algebraic-geometrical background, it should be noted in passing that the content of these memoirs later became fundamental in the theory of symmetric spaces.

Precisely when Cartan began working on the encyclopedia project is unclear. Jules Molk, the editor-in-chief of the French version, was professor of mathematics at the University of Nancy, where Cartan held a position from 1903 until 1909, when he moved to a position as associate professor (*maître de conférences*), and then, in 1912, professor, at the Sorbonne. It therefore seems likely that it was during this period that Cartan committed himself to compose the article, as well as the revision of Study's 1898 article on complex and hypercomplex numbers [1908]. Certainly, it is evident from passages in Cartan's trilogy, especially [1914b; 1914c], that geometrical applications along the lines discussed by Fano were foremost in his mind. This point will be elaborated in the following sections. The geometrical implications of the results in his trilogy were thus accorded prominence in his *Encyclopédie* article [1915], the publication of which was hampered by the vicissitudes of the First World War.

It is evident that Cartan put considerable time and thought into the revision of Fano's article. Whereas Fano's article was 97 pages long, Cartan's ran to 135 pages; and much of the new material was original to Cartan. The emphasis upon Lie groups was also even greater. Whereas Fano had explained to his readers that Lie's theory would be continually used in his

[17] For an account of Cartan's work on generalized and symmetric spaces see [Akivis and Rosenfeld 1993] and [Borel 2000].

article [1907b:294], Cartan declared, "the present article is simply an exposition of the geometrical applications of the theory of continuous groups of S. Lie" [1915:4]. Among new applications that Cartan included were those to infinitesimal geometry facilitated by the group-theoretic generalization of Darboux's moving frames and those involving infinite-dimensional groups. He also greatly expanded Fano's treatment of projective groups. For example, he showed how Fano's generalizations of the Hesse–Klein principle of transfer (Section 7.3) could be used to construct new projective groups with a given structure from known ones, thereby generating geometries equivalent to the ones defined by the original groups (pp. 99–105). Since these ideas are used in his resolution of Kowalewski's problem [1913], they will be discussed in the next section. His discussion of geometries equivalent to the hermitian geometry of Fubini and Study will be considered in Section 5, since it relates to his solution of the real analogue of Kowalewski's problem [1914b]. Here it will suffice to discuss Cartan's interpretation of Klein's notion of equivalent geometries and its implications.

In the *Erlanger Programm*, having introduced the viewpoint that a geometry or geometrical method is determined by a group $\mathfrak{G}$ acting on a manifold $\mathcal{M}$ of "elements," Klein devoted considerable space to showing that many ostensibly different geometries were in fact essentially the same from his point of view. Following Cartan's terminology in his *Encyclopédie* article [1915], I will refer to geometries that are the same for Klein as *equivalent* geometries. Klein introduced two ways in which equivalent geometries are obtained from given ones. In Section 4 of the *Erlanger Programm* he said that if one geometry $(\mathfrak{G}, \mathcal{M})$ is given, then an equivalent one is formed by a one-to-one mapping $\Phi : \mathcal{M} \to \mathcal{N}$ onto another manifold $\mathcal{N}$. If $\mathfrak{H}$ denotes the group of all transformations $\Phi \circ T \circ \Phi^{-1}$ such that $T \in \mathfrak{G}$, then the geometries $(\mathfrak{G}, \mathcal{M})$ and $(\mathfrak{H}, \mathcal{N})$ are equivalent. This principle of transfer was a generalization of the Hesse–Klein principle of transfer. In terms of the groups involved, two transformation groups define geometries equivalent in the above sense if and only if they are similar in the sense of Lie.

In Section 5 of the *Erlanger Programm* Klein took into account the viewpoint, exemplified by Plücker's line geometry, that in geometry points need not be taken as the fundamental elements and coordinatized; some other type of configuration such as lines or spheres could equally play the role of fundamental elements. The choice of another type of configuration as fundamental did not alter the content of the geometry, even though the dimension of the manifold on which the group acted could change. For example, in projective geometry the group of all projective transformations $\mathfrak{G} = \mathbf{PGL}(4, \mathbb{C})$ acts on the 3-dimensional manifold of points $\mathcal{M} = \mathbb{P}^3(\mathbb{C})$, but it thereby induces an action on all lines. Now all lines may be coordinatized by six homogeneous line coordinates p_{ij} that satisfy

$$\Omega = p_{12}p_{34} + p_{13}p_{42} + p_{14}p_{23} = 0,$$

and each $T \in \mathfrak{G}$ defines by its action on lines a projective transformation T^*

of $\mathbf{PGL}(5, \mathbb{C})$ that takes the quadratic hypersurface $\mathcal{M}^*$ defined by $\Omega = 0$ into itself. From Klein's viewpoint the geometry $(\mathfrak{G}, \mathcal{M})$ is equivalent to the geometry $(\mathfrak{G}^*, \mathcal{M}^*)$, where $\mathfrak{G}^*$ denotes the group of transformations T^*, even though $\mathcal{M}^*$ is 4-dimensional. That is, projective geometry and line geometry are equivalent from the Kleinian viewpoint.

Although Klein had made it clear that two geometries $(\mathfrak{G}, \mathcal{M})$ and $(\mathfrak{H}, \mathcal{N})$ could be equivalent even when the underlying manifolds $\mathcal{M}$ and $\mathcal{N}$ have different dimensions, so that there is no possibility of a Hesse–Klein transfer mapping $\Phi : \mathcal{M} \to \mathcal{N}$, he did not set forth a general criterion for determining when two given geometries are to be considered equivalent. However, he concluded Section 5 with a remark that suggested to others such a criterion. According to Klein, the remark was added to explain the type of configurations that could serve as the fundamental elements in a geometry, so as to prepare the reader for the next section, where the idea of a geometry defined by Lie's contact transformations was introduced—with Lie's surface elements (x, y, z, p, q) constituting the fundamental elements. (See Section 1.5 above.) To form the fundamental elements of a geometry, geometrical configurations had to form what Klein called a "field," a term he borrowed from Dedekind, who had used it in his writings on algebraic numbers.[18] Klein explained what he meant as follows:

> Let some particular group ... be given. A single spatial configuration is then chosen, perhaps a point or a line or an ellipsoid, and all the transformations of the group are applied to it. One then obtains a multiply infinite manifold with a number of dimensions, which generally equals the number of arbitrary parameters contained in the group but in special cases can be less, namely when the originally chosen configuration possesses the property that it is taken into itself by infinitely many transformations of the group. Every manifold so generated is called, relative to the generating group, a *field*.

In the subsequent terminology of Lie's theory, Klein's remarks take the following form. Let $\mathcal{F}$ denote a configuration in the space upon which a group $\mathfrak{G}$ of transformations

$$x'_k = f_k(x_1, \ldots, x_n; a_1, \ldots, a_r), \quad k = 1, 2, \ldots, n,$$

acts. Denote the above transformation by T_a, where $a = (a_1, \ldots, a_r)$. Then Klein's field consists of all the configurations $\mathcal{F}_a = T_a[\mathcal{F}]$—the orbit under $\mathfrak{G}$ of the configuration $\mathcal{F}$ as we would now say. In the "general" case to which Klein referred, the $\mathcal{F}_a$ can be coordinatized by $a = (a_1, \ldots, a_r)$ and so form an r-dimensional manifold $\mathcal{M}^*$. The action of each $T_a \in \mathfrak{G}$ on the $\mathcal{F}_b$ defines a new group $\mathfrak{G}^*$ of transformations T^*_a, where $T^*_a : b \mapsto c$ means

$$T^*_a[\mathcal{F}_b] = T_a[T_b[\mathcal{F}]] = (T_a \circ T_b)[\mathcal{F}] = T_c[\mathcal{F}] = \mathcal{F}_c.$$

[18] The German word is *Körper*, meaning literally "body."

If $\mathfrak{M}$ denotes the manifold of elements on which $\mathfrak{G}$ acts, which are coordinatized by $(x_1, \ldots, x_n)$, then Klein's remarks indicate, in effect, how to associate to the geometry $(\mathfrak{G}, \mathfrak{M})$ an equivalent geometry $(\mathfrak{G}^*, \mathfrak{M}^*)$.

In his encyclopedia article on groups and geometry, Fano developed the ideas in Klein's remarks into a general criterion for when two geometries $(\mathfrak{G}, \mathfrak{M})$ and $(\mathfrak{H}, \mathfrak{N})$ are to be regarded as equivalent [1907b:359]. Given geometries $(\mathfrak{G}, \mathfrak{M})$ and $(\mathfrak{H}, \mathfrak{N})$, consider the associated geometries described above, viz., $(\mathfrak{G}^*, \mathfrak{M}^*)$ and $(\mathfrak{H}^*, \mathfrak{N}^*)$. Then $(\mathfrak{G}, \mathfrak{M})$ and $(\mathfrak{H}, \mathfrak{N})$ are equivalent if $\mathfrak{G}^*$ and $\mathfrak{H}^*$ are similar in Lie's sense. The assumed similarity means that the geometries $(\mathfrak{G}^*, \mathfrak{M}^*)$ and $(\mathfrak{H}^*, \mathfrak{N}^*)$ are equivalent by virtue of the mapping $\Phi : \mathfrak{M}^* \to \mathfrak{N}^*$ that establishes the similarity. Whether or not Klein had such a precise notion of the equivalence of geometries in the back of his mind when he wrote his program is not known, but certainly Fano's definition is in agreement with what Klein actually wrote there. That is, Klein evidently regarded the pairs $(\mathfrak{G}, \mathfrak{M})$, $(\mathfrak{G}^*, \mathfrak{M}^*)$ and $(\mathfrak{H}, \mathfrak{N})$, $(\mathfrak{H}^*, \mathfrak{N}^*)$ as equivalent, and he certainly accepted the transitivity of the relation of equivalence, so if $(\mathfrak{G}^*, \mathfrak{M}^*)$ and $(\mathfrak{H}^*, \mathfrak{N}^*)$ are equivalent, then the equivalence of $(\mathfrak{G}, \mathfrak{M})$ and $(\mathfrak{H}, \mathfrak{N})$ would follow.

In the French version of Fano's article, Cartan pushed Fano's remarks further and concluded that the Klein–Fano notion of equivalent geometries was an aspect of the theory of structure [1915:91–92]. According to Cartan a configuration $\mathcal{F}$ always exists such that $T_a[\mathcal{F}] \neq \mathcal{F}$ for all $T_a \neq I$. If e.g., $\mathfrak{G}$ is projective, we may take for $\mathcal{F}$ an orthonormal basis—with the order of elements counting as part of the definition of $\mathcal{F}$. It seems likely that, in the general situation, Cartan was thinking in terms of an n-dimensional analogue of Darboux's moving trihedron. If such configurations are used in the above-described construction of both $(\mathfrak{G}^*, \mathfrak{M}^*)$ and $(\mathfrak{H}^*, \mathfrak{N}^*)$, then the groups $\mathfrak{G}^*$ and $\mathfrak{H}^*$ can be identified with the respective parameter groups of $\mathfrak{G}$ and $\mathfrak{H}$ (Section 3.2). Now, Lie had "proved" [1888:414] that two groups have the same structure (i.e., isomorphic Lie algebras) if and only if their parameter groups are similar. As noted in Section 3.2, he had also "proved" that two groups have the same structure if and only if they are isomorphic, although this result, too, actually holds only in a "local" sense. Cartan, whose sensitivity to the global aspects of Lie's theory was only later aroused by Weyl's work (Section 12.5), apparently accepted these theorems of Lie's, and so he could conclude, "In order that two geometries be equivalent it is necessary and sufficient that their ... groups be isomorphic ... that is have the same structure. Thus one sees the important role played by the structure of groups in geometry" [1915:91–2].

In transforming Fano's article into the French version, Cartan sought to show in particular how the new results on structure—the Killing–Cartan theory — could be applied to geometrical questions. In this way, he was brought back to Kowalewski's problem, which by virtue of his Theorem XVIII was recognized as equivalent to the representation problem. In the sixth section of his article Fano discussed the problem of determining continuous subgroups of the general projective group, what were usually simply referred to as "projective groups." He pointed out that Lie had determined all projective groups in the plane and that in dealing with projective groups in space he had distinguished between the cases in which the group does or does not leave invariant a "planar point figure"—a point, a straight line or a plane. Fano specified the three groups that leave no plane figure invariant and then added that "for $n > 3$ general theorems of this sort are unknown" [1907b:302]. Fano was apparently unaware that Kowalewski had solved the problem for $n = 4$ in 1899 and that Apfelstedt had done the same in 1906 for $n = 5$ (Section 7.5), but of course, the general problem implicit in his remarks remained unsolved. Cartan was certainly aware of the problem as well. It was Kowalewski's problem, and he presented the solution to it in the first of his trilogy of memoirs.

Kowalewski's problem was particularly important to Cartan because he conjectured that every geometry was equivalent to a projective one. Thus he wrote, "Since among finite[-dimensional] groups, projective groups play a preponderant role, the examples of equivalence that I am going to indicate all relate to projective geometries. By the way, it is probable that every geometry with a finite[-dimensional] group is equivalent to a projective geometry" [1915:92]. As he explained in a footnote: "That would be certain if one proved that a projective group exists that admits a given (finite) structure." On the Lie algebra level on which Cartan tended to think, the above theorem says that given any Lie algebra $\mathfrak{g}$, there is a linear Lie algebra isomorphic to it. The truth of this theorem had been suggested tentatively by Cartan in [1908 :449, n.249], and, as we saw in Section 3.3, it had been conjectured by Lie as well. A demonstration was finally provided by Ado in [1934].

Because Cartan believed in the truth of his conjecture, he deemed the problem of determining all projective groups of importance for geometrical reasons. Since Kowalewski's problem is a subproblem, it, too, acquired in Cartan's mind a geometrical significance. These attitudes are expressed clearly in his popular essay "The Theory of Groups" [1914a:451–2]:

> Every geometry with finite [continuous] group is probably equivalent to a *projective* geometry in a space of sufficiently many dimensions; this is not absolutely certain, because a theorem on the structure of finite [continuous] groups is required that has not yet been demonstrated; but there is no example showing this theorem to be defective. This fact is very important, since it makes it possible to make all geometries with finite [continuous] groups subordinate to projective geometry .... The determination of all continuous projective groups is an important problem; M. Cartan has been able to obtain all those that leave no planar variety (point, line, plane, etc.) invariant.

By solving Kowalewski's problem in [1913] Cartan had in effect proved a characterization theorem—in the sense in which this term was used to discuss the work of Study in Section 7.2—for every semisimple structure. Had he been able to prove a corresponding complete reducibility theorem as well, he would have made an even greater contribution to the problem of determining all projective groups, a problem to which he had attached so much geometrical importance in [1914a]. Cartan certainly realized this, for he was quite familiar with Lie's discussion of Study's work in *Transformationsgruppen*. In fact, shortly after Weyl published his complete reducibility theorem in 1925, Cartan wrote to him as follows: "For some time now I have regarded what you call the theorem of complete reducibility as very likely for all simple and semisimple groups: I even think I wrote it down somewhere. But I lacked the necessary tools to prove it."[19] Actually, when he composed his thesis, Cartan realized that he had developed the means to prove Study's claims about $\mathbf{sl}(2, \mathbb{C})$ [1894b:134], which include the complete reducibility theorem for $\mathbf{sl}(2, \mathbb{C})$.[20] To my knowledge, however, Cartan never wrote down, in a publication, the statement of the general complete reducibility theorem for semisimple Lie algebras, although it would have been appropriate to have mentioned this conjectured theorem in his essay "The Theory of Groups" [1914a], along with the one proved by Ado.

Thus Cartan, like Study, had come to suspect the veracity of a complete reducibility theorem for all semisimple Lie algebras, but even he did not see how to provide it with a demonstration. On the other hand, he did succeed in proving the corresponding characterization theorem, a major feat in itself and one that had proved too difficult for Killing, Study, and Kowalewski.

8.4. The Memoir of 1913

Cartan's "trilogy" of memoirs [1913; 1914b; 1914c] contains 175 tersely written pages of substantial mathematics. Using anachronistic terminology and a modern perspective, their content can be summed up as follows. In [1913] he solved the problem of determining, over the complex field, all finite-dimensional irreducible representations of semisimple algebras. In [1914c] he solved the problem of classifying all real semisimple Lie algebras, and in [1914b] he showed how to determine —first over the complex field and then over the real field—all finite dimensional irreducible representations of such algebras. A detailed discussion of these papers is consequently far beyond

[19] "[J]e regardais depuis quelque temps comme très vraisemblable le théorème que vous appelez de réducibilité complète pour tous les groupes simples ou semisimples: je crois même l'avoir écrit quelque part. Mais il me manquait le levier nécessaire pour le démontrer." Letter dated 12 March 1925 (ETH Zürich, Hs 91:501).

[20] Borel [1986:74] has confirmed Cartan's claim by locating the pages (in Cartan's proof of Engel's theorem on solvability) containing the requisite notions; a proof of complete reducibility for $\mathbf{sl}(2, \mathbb{C})$ based on Cartan's ideas is given in [Borel 1998:§12].

the scope of this chapter, which is concerned primarily with the historical background of the trilogy. The purpose of this and the following section is simply to serve as a capstone to the preceding ones by highlighting aspects of these memoirs that are especially germane to the historical developments and background presented earlier. Emphasis will be placed on the manner in which Cartan drew upon existing mathematics, including his own earlier work, for the means and the motivation to solve the various problems posed in these memoirs.

The most fundamental of the three memoirs is the first, in which Cartan solved the problems posed by Killing and Kowalewski and implied by the work of Study. Expressed in the language of the period, the problem was to determine all projective groups that leave nothing planar invariant. By Theorem XVIII of [1909]—see Section 2—these groups are semisimple. In his memoir Cartan regarded the lie algebra $\mathfrak{l}$ of such a group as consisting of differential operators $X = \Sigma a_{ij}x_j\partial/\partial x_i$ defining linear transformations which act upon $x = (x_1, \ldots, x_n)$ by $X(x) = (X(x_1), \ldots, X(x_n))$. In terms of the matrix $A = (a_{ij})$, X sends x into Ax if we regard x as a column matrix. As indicated in Section 6.4, associated with $\mathfrak{l}$ is an indecomposable Lie algebra $\mathfrak{g} = \mathfrak{l} \oplus \mathcal{V}$, $\mathcal{V} = \mathbb{C}^n$, and consequently a system of principal roots $\alpha, \beta, \gamma, \ldots$ and secondary roots $\lambda, \mu, \nu, \ldots$. Whereas in [1909] Cartan had referred to all these roots as weights, that term is reserved for λ, μ, ν in [1913]. Following the notation of Section 6.4 and Section 2 of this chapter, the saturated set of weights "generated" by λ (in the sense of Section 6.3) will be denoted by $S(\lambda)$. Recall that Killing and Cartan realized that $S(\lambda)$ consists of two sorts of weights: those that are, in effect, conjugate to λ under the action of the Weyl group—Cartan said they were homologous to λ in [1913]—and those that are not.

In the examples of systems $S(\lambda)$ calculated in his thesis, Cartan could see that the weights conjugate to λ all have multiplicity one, although he never proved there that this must be the case. Since he did know that conjugate roots necessarily have the same multiplicity, it sufficed to prove that one of them had multiplicity one.[21] With this in mind, he introduced the concept of a dominant weight in the following manner. Any weight λ is expressible as a rational linear combination of the l "fundamental roots" $\alpha_1, \ldots, \alpha_l$, viz., $\lambda = \Sigma m_i\alpha_i$, $m_i \in \mathbb{Q}$. Among all weights consider those such that m_1 is maximal. Among these consider those such that m_2 is maximal, and so on. This is the lexicographical ordering that goes back to Gauss, and in this way a unique weight λ_0 is defined that Cartan called the dominant weight with respect to the given choice of the set of l fundamental roots. When the fundamental roots form a base in the modern sense, then Cartan's definition of a dominant weight coincides with the usual one.[22]

[21] Cartan [1913:58] referred to this property of weights in a portion of his paper summarizing results on secondary roots from pages 127–53 of his thesis, although to my knowledge it is not explicitly stated there.

[22] For the modern version of the theory I have followed the exposition of

Cartan indicated how all the weights composing $S(\lambda)$ could be generated from the dominant weight by a systematic procedure [1913:63–4], so that the system can be denoted by $S(\lambda_0)$.

Although the above considerations pushed the analysis of the saturated weight system $S(\lambda)$ further than had been done previously, they do not in themselves convey the new outlook that Cartan presented in [1913]. In his thesis he had reasoned within the framework of the structure theory of the indecomposable Lie algebra $\mathfrak{g}$. The alternative perspective of the study of linear Lie algebras $\mathfrak{l}$ was significant in his thesis simply as the means for applying the structure theory of $\mathfrak{g}$ to a question of interest in differential Galois theory—the determination of the $\mathfrak{l}$ of a given simple structure in the minimum number of variables. By contrast, in [1913] Cartan used the perspective of linear Lie groups and algebras in a fundamental way. The following considerations, which led first of all to a proof that the dominant weight has multiplicity one, are characteristic of the new outlook.

Let u be a vector with dominant weight λ_0, and let $\mathcal{W}$ denote the linear subspace of $\mathcal{V} = \mathbb{C}^n$ spanned by vectors of the form

$$z = X_{\beta_m} \cdots X_{\beta_2} X_{\beta_1} u, \tag{4.1}$$

where each β_i is a nonzero root and each X_{β_i} is from the corresponding root space $\mathfrak{l}_{\beta_i}$. The number m of factors X_{β_i} can be 0 ($z = u$), $1, 2, \ldots$. (In the modern theory $\mathcal{W}$ would be identified with the standard cyclic module of weight λ_0.) It is easy to see that $\mathfrak{l}$ leaves $\mathcal{W}$ invariant. For if $z \neq 0$ is of the form (4.1) and if $H \in \mathfrak{l}_0$ (the "Cartan subalgebra" of $\mathfrak{l}$), then $Hz = (\beta_1 + \cdots + \beta_m + \lambda_0)u$, which belongs to $\mathcal{W}$. Likewise, for any $X \in \mathfrak{l}_\beta$, $X(z) = z'$, another expression (of length $m+1$ of the form (4.1). Since $\mathfrak{l} = \mathfrak{l}_0 + \sum_\beta \mathfrak{l}_\beta$, we see that $\mathcal{W}$ is left invariant by the transformations of $\mathfrak{l}$. Since $\mathfrak{l}$ leaves "nothing planar invariant," this means that $\mathcal{W} = \mathcal{V}$. The action of $\mathfrak{l}$ on $\mathcal{V}$ can therefore be calculated by applying the X_β and H of $\mathfrak{l}$ to vectors z of the form (4.1). Using this approach, and the fact that λ_0 is dominant, Cartan showed that if z is any vector of the form (4.1) with weight λ_0, then z is a scalar multiple of u.[23] Hence λ_0 has multiplicity one.

Cartan also used the characterization of $\mathcal{V}$ as the span of vectors (4.1) to show that if $\mathfrak{l} \subset \mathbf{gl}\,(\mathcal{V})$ and $\mathfrak{l}^* \subset \mathbf{gl}\,(\mathcal{V}^*)$ are isomorphic and have the same dominant weight λ_0, then they are similar in the sense of Lie. This means, in modern terms, that $\mathfrak{l}$ and $\mathfrak{l}^*$ are equivalent irreducible representations. As in the modern theory, this result is fundamental to the solution of the representation problem. Suppose now that $\mathfrak{l} \subset \mathbf{gl}\,(\mathcal{V})$ and $\mathfrak{l}^* \subset \mathbf{gl}\,(\mathcal{V}^*)$ have the same structure but with possibly different dominant weights, λ_0 and λ_0^*. Thus $\mathfrak{l} \cong \mathfrak{l}^* \cong \mathfrak{g}$, where $\mathfrak{g}$ is semisimple. In modern terms, $\mathcal{V}$ and $\mathcal{V}^*$ are irreducible $\mathfrak{g}$-modules with dominant weights λ_0 and λ_0^*, respectively.

[Humphreys 1972].

[23] The idea is to show that any z of length m with weight λ_0 can be expressed as a sum of z's of length less than m and weight λ_0 [1913:7–9].

Cartan showed in effect how to define the $\mathfrak{g}$-module $\mathcal{V} \otimes \mathcal{V}^*$ and then within this $\mathfrak{g}$-module how to construct an irreducible $\mathfrak{g}$-module with dominant weight $\lambda_0 + \lambda_0^*$. That is, suppose $\mathfrak{l}$ acts on the variables $x_1, \ldots, x_p$ and $\mathfrak{l}^*$ acts on the variables $y_1, \ldots, y_q$, where the x_i, y_i are weight vectors and x_1 and y_1 denote the respective dominant weight vectors. Then the pq variables $x_i y_k$ transform linearly in a natural manner: If $X \in \mathfrak{l}$ is such that $X(x_i) = \sum_j t_{ij} x_j$ and $X^* \in \mathfrak{l}^*$ is such that $X^*(y_p) = \sum_q t^*_{pq} y_q$, let $x_i y_p \to X(x_i) X^*(y_p) = \Sigma_{j,q} t_{ij} t^*_{pp} x_j y_q$. The resulting linear group is thus in pq variables.

As noted in discussing Kowalewski's use of a similar construction in Section 7.5, the notion of a tensor product was still not a standard part of mathematics in 1913. Incidentally, this was the year that Einstein and Grossmann introduced the calculus of tensors into relativity theory. As will be seen in Chapter 11, Weyl's interest in relativity led him to introduce the viewpoint and terminology of tensor algebra into mathematics—and ultimately led to his interest in representation theory. In terms of tensor algebra, we may say that Cartan has constructed the $\mathfrak{g}$-module $\mathcal{V} \otimes \mathcal{V}^*$.

Within the framework of this group in pq variables, Cartan proceeded as follows to define what he called the product of the two given groups. This corresponds to what is nowadays called the *Cartan component* of $\mathcal{V} \otimes \mathcal{V}^*$. As Cartan put the matter:

> We start from the product $x_1 y_1$ and apply to it successively and repeatedly the different transformations of the group. We will obtain a certain number of linear combinations $z_1, z_2, \ldots, z_r$ of the products $x_i y_k$ each of which is formed of products all having the same weight. The group that indicates how the r quantities $z_1, z_2, \ldots, z_r$ are transformed is a linear group that evidently leaves no planar multiplicity invariant. We will call it *the product* of the two given groups. Its dominant weight is equal to the sum of the dominant weights of the two groups. *The preceding thus establishes the existence of a linear group admitting as dominant weight the sum of the dominant weights of the two given groups and gives a simple procedure for constructing this group.* [1913:64]

Except for the terminology, Cartan's construction of an irreducible module of dominant weight $\lambda_0 + \lambda_0^*$ is the now familiar one [Humphreys 1972:117]. However, as Weyl first observed [1925:556–7], the irreducibility of this module is not as evident as Cartan believed. Weyl could not see how to prove it without invoking the complete reducibility theorem for semisimple structures. Cartan presumably did not consciously apply this theorem, since, as indicated in Section 3, although he believed the theorem plausible he was unable to produce a proof.[24]

[24] Borel has pointed out [1986:73] that Cartan could have given a slightly different construction, based on a consideration that is now a familiar part of the theory, which would have avoided any need of the complete reducibility theorem. That

Construction of the Cartan components was fundamental to Cartan's solution of his representation problem by virtue of the following result [1913:66]: For each simple structure of rank l there exist l linearly independent dominant weights $\lambda_1, \ldots, \lambda_l$ such that any dominant weight λ is expressible as $\lambda = p_1\lambda_1 + \cdots + p_l\lambda_l$, where the p_i denote nonnegative integers. Thus if an $\mathfrak{l}_i$ leaving nothing planar invariant and with dominant weight λ_i can be produced for each $i = 1, \ldots, l$, then by the formation of Cartan components an $\mathfrak{l}$ leaving nothing planar invariant can be constructed corresponding to any dominant weight. Cartan's theorem on the similarity of such $\mathfrak{l}$'s with the same dominant weight guaranteed that his procedure exhausted the possibilities for simple projective groups leaving nothing planar invariant. In fact, for the same reason, all semisimple $\mathfrak{l}$ leaving nothing planar invariant can then be constructed: "If, for example, a semisimple group is formed from two simple subgroups, all the linear groups ... [$\mathfrak{l}$] that have its structure are generated by taking any linear group [$\mathfrak{l}_1$] of the first simple structure in variables $x_1, x_2, \ldots, x_m$ and any linear group [$\mathfrak{l}_1$] with the second simple structure in variables $y_1, \ldots, y_n$. The most general group sought is that which indicates how the mn quantities $x_i y_k$ are transformed linearly among themselves by the transformations of ... [$\mathfrak{l}_1$] and ... [$\mathfrak{l}_2$]" [1913:54]. In other words, if $\mathfrak{g} = \mathfrak{g}_1 \oplus \mathfrak{g}_2$, where $\mathfrak{g}_1$ and $\mathfrak{g}_2$ are simple, then every irreducible $\mathfrak{g}$-module is of the form $\mathcal{W}_1 \otimes \mathcal{W}_2$, where $\mathcal{W}_i$ is an irreducible $\mathfrak{g}_i$-module, i.e., $\mathfrak{l}_i \subset \mathbf{gl}\ (\mathcal{W}_i)$.

The problem of determining all projective groups that leave nothing planar invariant was in this way reduced by Cartan to constructing, for each simple structure type, the fundamental dominant weights $\lambda_1, \ldots, \lambda_l$ and the corresponding "fundamental groups" $\mathfrak{l}_1, \ldots, \mathfrak{l}_l$. The remainder of his memoir [1913:67–96] is devoted to these nontrivial constructions. Following the practice adhered to in his thesis and in [1909], he simply described, for each simple structure type in turn, the desired fundamental dominant weights and corresponding fundamental groups and suppressed all calculations involved as well as the ideas that enabled him to obtain these nontrivial results. Given his definition, even the construction of the fundamental dominant weights required some thought; their existence was not established a priori but on a case-by-case basis. Here I will consider only the case of type A_l, projective groups with the structure of $\mathbf{PGL}(l+1, \mathbb{C})$, or, equivalently, that of $\mathbf{SL}(l+1, \mathbb{C})$.

In this case Cartan's task was facilitated greatly by Clebsch's generalized Plücker coordinates, which, as we saw in Section 7.3, Fano had presented in his encyclopedia article and so were quite familiar to Cartan. Recall from Section 7.3 that in $\mathbb{P}^l(\mathbb{C})$, with points expressed in coordinates $x_1, \ldots, x_{l+1}, y_1, \ldots, y_{l+1}, \ldots,$ there are l types of point, line, plane, etc., coordinates $X_i = x_i$, $X_{ij} = x_i y_j - x_j y_i$, $\ldots$, $X_{i_1 i_2 \ldots i_l}$. The kth fundamental

is, if $\mathcal{W}$ is the Cartan component of $\mathcal{V} \otimes \mathcal{V}^*$, then it is not difficult to see that $\mathcal{W}$ contains a maximal proper submodule $\mathcal{U}$ and $\mathcal{W}/\mathcal{U}$ is an irreducible $\mathfrak{g}$-module with $\lambda_0 + \lambda_0^*$ as dominant weight.

group $\mathfrak{l}_k$ is the group induced by the action of $\mathbf{sl}(l+1, \mathbb{C})$ on the kth type of coordinate. Thus $\mathfrak{l}_1 = \mathbf{sl}(l+1, \mathbb{C})$, and so on. In the terminology that Weyl was to bring into the theory from general relativity, $\mathfrak{l}_k$ is the linear Lie algebra induced by $\mathbf{sl}(l+1, \mathbb{C})$ on the skew symmetric tensors of degree k.

In his French version of Fano's encyclopedia article, after dutifully presenting Fano's generalizations of the Hesse–Klein principle of transfer (described in Section 7.3), Cartan added a new section in which he showed how the above ideas could be utilized to systematically create new projective groups of a given structure from known ones. As he explained, referring to his memoir [1913], "From the preceding principle it is possible to deduce all the projective groups that leave nothing planar invariant" [1915:100]. Indeed, Cartan had used this principle repeatedly in [1913]. Thus to construct the Cartan component of $\mathcal{V} \otimes \mathcal{V}^*$, as described in the above quotation, he had introduced the quantities $z_1, \ldots, z_r$, which are linear homogeneous functions of the x_i and y_j, viz., $z_k = f_k(x, y)$. The r forms $f_k(x, y)$ form a linear system in the sense of Fano, and the group that is generated by Fano's transfer principle is Cartan's product of $\mathfrak{l}$ and $\mathfrak{l}^*$. Likewise, it is Fano's version of the transfer principle that underlies Cartan's construction of all irreducible representations of the semisimple $\mathfrak{g} = \mathfrak{g}_1 \oplus \mathfrak{g}_2$; in this case the linear system consists of the mn forms $f_k(x, y) = x_i y_j$, $i = 1, \ldots, m$, $j = 1, \ldots, n$. Finally, the above construction of the l fundamental groups $\mathfrak{l}_k$ involves Fano's transfer principle with the linear system defined by the $\binom{l+1}{k}$ coordinates $X_{i_1 \ldots i_k}$, which are linear homogeneous forms in k sets of variables $x_1, \ldots, x_{l+1}, y_1, \ldots, y_{l+1}, \ldots$.

That Cartan was already aware in 1913 that he was utilizing a generalization of the Hesse–Klein principle of transfer is clear from his introductory description of the contents of his memoir:

> In each fundamental group $[\mathfrak{l}_i]$ a particular variable called a *dominant* variable is taken. Let x_i be the dominant variable of $[\mathfrak{l}_i]$. Then, one takes l arbitrary positive or zero integers $p_1, p_2, \ldots, p_l$ and considers the expression
>
> $$x_1^{p_1} x_2^{p_2} \cdots x_l^{p_l}.$$
>
> When the transformations of the group $[\mathfrak{l}]$ with equations obtained by the juxtaposition of the equations of $[\mathfrak{l}_1, \mathfrak{l}_2, \ldots, \mathfrak{l}_l]$ are applied to this expression, it is transformed into a certain number of other linearly independent ones. *The most general group of the type sought indicates how the group $[\mathfrak{l}]$ linearly transforms all these expressions* .... The results thus obtained include in particular those already known relative to the generation of projective groups with three parameters that leave nothing planar invariant: Each of them indicates in effect how the projective group of the line transforms among themselves the expressions $x_1^p, x_1^{p-1} x_2, \ldots, x_1 x_2^{p-1}, x_2^p$ [1913:54].

The latter groups are the groups of normal curves (Section 7.1) that arise from the original Hesse–Klein principle of transfer.

Cartan thus developed the ideas expounded in Fano's article [1907b] for the "tensor product" constructs required for his solution of the problems of Killing, Study, and Kowalewski, problems that had become important to him for geometrical reasons. Of course, he also drew upon the theory of secondary roots and, in addition, introduced new ideas such as dominant weights and associated weight vectors. Thus, for example, his characterization of the projective groups with the structure of $\mathbf{sl}(n, \mathbb{C})$ is quite different from Study's. A comparison is instructive. Recall from Section 7.2 that Study had claimed, but never proved in print, that all such projective groups are described by the action of this group on the "linear system" $\mathcal{V}(m,n)$ of forms $f = (p,x)^m(q,u)^n$ that satisfy the normality condition $Df = 0$, for some value of m and n. In Cartan's approach, the starting point is the dominant weight $\lambda = m\lambda_1 + n\lambda_2$, which has dominant vector $f_0 = x^m[x_1, x_2]^n = x_1^m(X_{12})^n$ and the linear system $\mathcal{W}$ of forms generated from f_0. Of course, f_0 is normal, as are the forms into which it is taken by $\mathbf{SL}(3, \mathbb{C})$. Thus $\mathcal{W}$ is contained in $\mathcal{V}(m,n)$. Since Study had proved, by invariant-theoretic means, that $\mathcal{V}(m,n)$ is irreducible, it follows from his results that $\mathcal{W} = \mathcal{V}(m,n)$ and hence that $\mathcal{W}$ is irreducible. Recall that Cartan's proof that $\mathcal{W}$ is irreducible had a sizable gap in it that was eventually filled by the complete reducibility theorem that Study, motivated by consideration of series expansions, had conjectured but could not prove. Both Study's and Cartan's treatments of Study's characterization problem for $\mathbf{sl}(3, \mathbb{C})$ were thus not completely satisfactory. A more satisfactory treatment was achieved through the synthesis of ideas from the theory of invariants with Cartan's theory of weights that was provided by Weyl slightly over a decade later (Chapter 12).

8.5. The Memoirs of 1914

Cartan's memoirs [1914b; 1914c] on the structure and representation of real Lie algebras followed on the heels of his 1913 memoir. They, too, seem to have been motivated to a large extent by the prospect of geometrical applications. Indeed, it was in the realm of geometry that the importance and abundance of real groups was perhaps greatest. For example, the space problem posed by Helmholtz's well-known essay of 1868 and studied group-theoretically by Lie in the 1880s and 1890s required consideration of real groups, whereas the rest of Lie's work on groups concentrated on the complex case. Killing also admitted the need in the foundations of geometry for considering real structures [1892:124], although he never attempted to relate his discoveries on complex structures to such matters.

Geometry was also a fertile source of examples of real continuous groups, especially projective ones. Real projective geometry itself of course involved $\mathbf{PGL}(n, \mathbb{R})$. Many geometries also involved real orthogonal groups, that is, groups of real projective transformations leaving $f = 0$ invariant, where f is some nondegenerate quadratic form.[25] It was well known [Meyer 1889:327] that f could be transformed by real linear transformations into a unique canonical type

$$f = x_1^2 + \cdots + x_r^2 - x_{r+1}^2 - \cdots - x_n^2. \tag{5.1}$$

Consider the geometry corresponding to the group of all real projective transformations leaving the variety $f = 0$ invariant. If $n = 4$ and $r = 3$, the result is Klein's projective model of hyperbolic non-Euclidean geometry in three-dimensional space, while $n = 4$, $r = 4$ yields non-Euclidean elliptic geometry. Lie's sphere geometry corresponded to $n = 6$ and $r = 4$, whereas $n = 6$, $r = 3$ yielded real projective line geometry. All these geometries were also considered to some extent as generalized to n homogeneous coordinates. These and other geometries were considered in the complex case as well, although some, such as affine and conformal geometry, were traditionally restricted to the real case [Fano 1907b:298, 313]. Furthermore, it was recognized that many interesting geometrical distinctions evaporate in the transition from the real to the complex case. The projective geometries associated with a nondegenerate quadratic form f, such as the ones given above, afforded a good illustration of this fact. As Fano remarked [1907b:308,310], since there are $n + 1$ canonical types (5.1) for f and since f and $-f$ determine the same group, there are $[(n + 2)/2]$ real projective geometries of a nondegenerate quadratic variety, whereas these distinctions vanish in the complex case.

Another important source of real groups was provided by the "hermitian geometries" introduced independently by Fubini [1904] and Study [1905], who investigated them more thoroughly. These geometries are based upon consideration of hermitian forms $f = \Sigma a_{jk} x_j \overline{x}_k$, $a_{kj} = \overline{a}_{jk}$. Such forms had been introduced into the theory of numbers by Hermite in the early 1850s.[26] Interest in them seems to have been belatedly stimulated by the work of Picard (Hermite's son-in-law) on hyper-Fuchsian functions [1883b; 1884] that involved discontinuous groups of linear transformations that leave invariant a hermitian form with Gaussian integer coefficients. Citing Picard's work as indicative of possible applications, Alfred Loewy systematically studied the purely algebraic properties of hermitian forms (as he called them) in [1898]. Earlier, Segre [1890] had also pointed to Picard's work to justify the potential applicability of a new domain of geometrical investigation in which "hyperalgebraic" varieties played a fundamental role. More will be said about Segre's work further on. Here it suffices to point out that a prime example of a hyperalgebraic variety was obtained by setting a

---

[25] For details and references, see [Cartan 1915:25–39; Fano 1907a:308–18].

[26] See [Hawkins 1977b:134–5]. As noted there, Clebsch independently introduced hermitian symmetric linear systems of equations in his work in mathematical physics, but his work seems to have been forgotten.

nondegenerate hermitian form equal to zero. As Segre had observed, since a hermitian form $f(x)$ is real-valued, the variety $f(x) = 0$ separates $\mathbb{C}^n$ into two parts, $f(x) < 0$ and $f(x) > 0$.

Such properties of a real quadratic form were fundamental to the projective construction of real non-Euclidean geometries, and Study went on to observe that hermitian forms make it possible to define a real metric by an analogous method. For example, one can consider the elliptic hermitian geometry defined by the group of all projective transformations in n homogeneous coordinates which leave $f = \Sigma x_j \overline{x}_j = 0$ invariant. This group is real and of course is now known (in nonprojective form) as the special unitary group $\mathbf{SU}(n)$. Hermite had proved in [1856] that any nondegenerate hermitian form can be transformed by real linear transformations into a canonical form

$$f = x_1\overline{x}_1 + \cdots + x_r\overline{x}_r - x_{r+1}\overline{x}_{r+1} - \cdots - x_n\overline{x}_n, \tag{5.2}$$

so that one could investigate a real group and corresponding geometry for each canonical type (5.2), although no one did so at the time. Study concentrated his attention on the cases $n = 3$ with $r = 3$ or $r = 1$ and the case $r = n$, but he did show in general that the projective group leaving (5.2) invariant is a real simple group of dimension $n^2 - 1$ [1905:363–4]. Study's geometry of radial projectivities, his projective geometry of "somas," and his pseudoconformal geometry of somas also provided further examples of real groups, since he considered these geometries in both the real and complex cases.[27]

The abundance and importance of real projective groups in geometry also suggested the more algebraic problem of their structural classification. As was emphasized in the previous section, the resolution of this problem, in whole or in part, had an immediate bearing on the geometrical problem of systematically establishing equivalences among geometries, a problem that appealed to Cartan because it was such a direct application of the theory of structure he had done so much to develop and promote. A first and important step would have been the resolution of the real analogue of Kowalewski's problem. In Section 3 we noted that in his encyclopedia article Fano had reported some of the results along these lines for complex projective groups. As Fano wrote, "The relevant results mostly assume that all transformations and parameters that occur will be considered to be complex. Not much has been worked out for real groups" [1907b:300]. Fano's note 17 referred to Chapter 19 of the third volume of Lie's *Theorie der Transformationsgruppen*, the sole chapter in the three volumes aside from the treatment of the Helmholtz space problem itself (Vol. 3, Abteilung V) to deal with real groups. In Chapter 19 Lie dealt with the problem of determining all real projective groups of the line and plane.

---

[27] For details and references, consult [Cartan 1915:47–62].

Since the real projective groups that leave nothing planar invariant are also simple or semisimple, the real analogue of Kowalewski's problem required a preliminary classification of real simple structures. This Cartan undertook in [1914c]. His approach was based on the observation that real simple structures arise naturally in two manners from complex ones. It is interesting that these two methods of generating real structures were well known from the group-theoretic study of geometry. The first method, when stated abstractly, was as follows:

Method (i). *Let $\mathfrak{g}$ be a complex Lie algebra of dimension r. Regard $\mathfrak{g}$ as a real Lie algebra $\mathfrak{g}^{(r)}$ of dimension $2r$; thus if X_k, $k = 1, \ldots, r$, is a basis for $\mathfrak{g}$, then X_k and iX_k, $k - 1, \ldots, r$, form a basis for $\mathfrak{g}^{(r)}$.*

A well-known geometrical manifestation of this method was provided by the stereographic projection of the Riemann sphere onto the plane. Since the points of the plane can be identified with points of the complex line, stereographic projection sets up a correspondence between the group of real projective transformations leaving the sphere invariant, viz., the real projective group leaving $f = x_1^2 + x_2^2 + x_3^2 - x_4^2 = 0$ invariant, and the projective group of the line $\mathbb{P}^1(\mathbb{C})$ considered as a real 6-dimensional group. In other words, if $\mathfrak{g} = \mathbf{sl}(2, \mathbb{C})$ (the Lie algebra of the projective group of the line), then $\mathfrak{g}^{(r)}$ can be identified with the real projective group of the Riemann sphere. As Klein put it in his *Erlanger Programm*, "The theory of binary forms of complex variables [viz., projective geometry of the complex line] finds its representation in the projective geometry of the real spherical surface."[28]

A second method for generating real structures from complex ones can be stated abstractly as follows:

Method (ii). *If $\mathfrak{g}$ is a complex r-dimensional Lie algebra with real structure constants c_{ijk}, then the c_{ijk} define a real r-dimensional Lie algebra $\mathfrak{g}_r$.*

According to the Killing–Cartan theory the c_{ijk} can always be taken to be real when $\mathfrak{g}$ is semisimple. Geometrical examples suggested that a single complex simple structure could give rise in this manner to many different real structures. Thus all the real projective groups of an n-variable nondegenerate quadratic form (5.1) arise in this way from $\mathfrak{g} = \mathbf{o}(n, \mathbb{C})$. A less obvious example was afforded by the real groups associated with nondegenerate hermitian forms (5.2). Study showed that in effect, they all arise from $\mathfrak{g} = \mathbf{sl}(n, \mathbb{C})$ [1905:365]. Cartan was in a position to see that these two ways of constructing real Lie algebras from complex ones sufficed to obtain all real simple Lie algebras from the simple complex ones. This insight

[28] [Klein 1872b:§6]. See also [Klein 1875], especially p. 185. These considerations were also referred to by Fano [1907b:364] and Cartan [1915:109] in their discussion of equivalent geometries.

was based upon that aspect of the Killing–Cartan structure theory that originated with Cartan, namely the theory of the so-called Killing form.

Let $\mathfrak{g}$ be a real r-dimensional simple Lie algebra. Then $\mathfrak{g}$ determines a complex Lie algebra $\mathfrak{g}_c$ as follows (following Cartan's presentation). If X_k, $k = 1, \ldots, r$, is a basis for $\mathfrak{g}$, then $\mathfrak{g}_c$ consists of all complex linear combinations of the X_k, such as $Z = \sum_k e_k X_k$. (In modern notation, $\mathfrak{g}_c = \mathfrak{g} \otimes_{\mathbb{R}} \mathbb{C}$.) If we define $\overline{Z} = \sum_k \overline{e}_k X_K$, $P = \text{Re}\,(Z) = \sum_k \text{Re}\,(e_k)X_k$, and $Q = \text{Im}\,(Z) = \sum_k \text{Im}\,(e_k)X_k$, then $Z = P + iQ$, $\overline{Z} = P - iQ$, and $Z \in \mathfrak{g}_c$ is in $\mathfrak{g}$ if and only if $\overline{Z} = Z$. Cartan observed that since $\mathfrak{g}$ is assumed simple, $\mathfrak{g}_c$ must be semisimple.[29] In his thesis Cartan had proved that $\mathfrak{g}_c$ must therefore be a direct sum of simple ideals. By using the simplicity of $\mathfrak{g}$, it can be shown that there are only two possibilities:

> (I) $\mathfrak{g}_c = \mathfrak{s} + \overline{\mathfrak{s}}$, *where* $\mathfrak{s}$ *is a simple ideal and* $\overline{\mathfrak{s}} = \{\overline{Z} : Z \in \mathfrak{s}\}$ *is isomorphic as a complex algebra to* $\mathfrak{s}$; (II) $\mathfrak{g}_c$ *is simple.*

Consider the case (I). If $Z_k = P_k + iQ_k$, $k = 1, \ldots, t$, is a basis for $\mathfrak{s}$, then the $2t$ transformations P_k, Q_k (or Z_k and iZ_k) form a basis for $\mathfrak{g}$. Thus from the complex simple Lie algebra $\mathfrak{s}$ of dimension t we create the real simple Lie algebra $\mathfrak{g}$ of dimension $r = 2t$. This possibility corresponds to the method (i) of constructing real Lie algebras from complex ones discussed above. In the notation used there, $\mathfrak{g} = \mathfrak{s}^{(r)}$. Likewise, case (II) corresponds to method (ii).

As noted in the above discussion of real orthogonal groups, it was geometrically manifest that many distinct real simple groups could be derived from a single complex type, viz., $\mathfrak{o}(n, \mathbb{C})$. It should be noted, however, that in his exposition of these geometries, Fano based this conclusion on the algebraic fact that quadratic forms (5.1) with different values of r could not be transformed into one another by real linear transformations of the variables on which the groups acted [1907b: 308–9]. This did not automatically imply that two groups corresponding to distinct forms (5.1) have different real structures but only that they are not similar (in Lie's sense) by means of a linear transformation of the variables. Fano did not consider the relevance of structure to the notion of equivalent geometries. As indicated in Section 3, Cartan took this step. The question of structural identity depended for its answer upon the possibility or impossibility of real linear transformations of the basis elements of the underlying abstract Lie algebras, not the variables of its representation as an orthogonal group. The forms (5.1) related to these variables and not to the basis elements or, equivalently, to the adjoint representation.

Cartan was in a position to realize that the adjoint representation also comes equipped with its own quadratic form. That is, let $\mathfrak{g}$ be a complex

---

[29] This followed from consideration of the quadratic form $\psi_2(e)$ associated with $\mathfrak{g}$, which showed that if the radical of $\mathfrak{g}_c$ were nonzero, then the same would be true for $\mathfrak{g}$ [Cartan 1914c:266].

simple Lie algebra from which, by method (ii), many real Lie algebras $\mathfrak{g}_r$— real forms of $\mathfrak{g}$ as Cartan himself called them— are to be derived. In [1914c], instead of considering $\psi_2(e)$, Cartan considered the quadratic form $\psi(e)$, which is the sum of the squares of the roots of $\mathfrak{g}$. For simple $\mathfrak{g}$ $\psi = -2\psi_2$ and ψ is the quadratic Killing form, as already noted. Now suppose that $\mathfrak{g}$ and $\mathfrak{g}^*$ are isomorphic complex Lie algebras and that X_k and X_k^* are any bases of $\mathfrak{g}$ and $\mathfrak{g}^*$, respectively, such that the structure constants are all real. Let δ denote the signature of $\psi(e)$ and δ^* the signature of $\psi^*(e)$. If ψ has canonical form (5.1), then $\delta = 2r - n$, i.e., the number of positive squares minus the number of negative ones. If $\delta \neq \delta^*$, it is impossible to transform ψ^* into ψ by a real linear transformation. This means that the corresponding real Lie algebras $\mathfrak{g}_r$ and $\mathfrak{g}_r^*$ defined by method (ii) cannot be isomorphic. They are distinct real forms of the associated complex structure type. For example, if $\mathfrak{g} = \mathbf{sl}(2, \mathbb{C})$ with the usual basis H, P, N of Section 7.1 and $\mathfrak{g}^*$ is $\mathbf{sl}(2, \mathbb{C})$ with the basis $X = iH$, $Y = P - N$, $Z = i(P + N)$, then ψ has signature 1, whereas ψ^* has signature -3, so that the corresponding real Lie algebras $\mathfrak{g}_r = \mathbf{sl}(2, \mathbb{R})$ and $\mathfrak{g}_r^* = \mathbf{su}(2)$ cannot be isomorphic.

In seeking to classify all real Lie algebras that may arise by method (ii) Cartan made systematic use of the number δ, which he called the character of the associated real Lie algebra, as a principle of classification. His solution to this nontrivial classification problem does not make for easy reading. As in his papers [1909] and [1913], he considered each simple complex structure type A through G in turn and presented his conclusions with many of the long and tedious calculations omitted. Fifteen years later, in the light of Weyl's paper on the representation of semisimple Lie groups [1925] and by means of his theory of symmetric spaces, Cartan discovered a theorem [1929:27] that made a unified and greatly simplified derivation of his results possible. Thus, "The calculations, often tedious, which I had to make can be greatly reduced; what is more, the long memoir [1914c] in which I determined all the real forms of simple groups could now be reduced from 90 to about 20 pages." In [1939] Gantmacher gave a purely algebraic proof of Cartan's theorem and used it to obtain the real forms of simple complex structures with the relative ease that Cartan had foreseen.

Although in 1914 Cartan could find no such alternative to a lengthy case-by-case approach to the classification problem, he realized that some of his results were independent of any particular structure type. Thus he wrote, "In every case there is a real type with character equal to the rank [l] .... *There is also a real type for which the character is equal to the negative of the order [= dimension] of the group.* For the latter type the quadratic form $\psi(e)$ is a negative definite form, and the roots of the characteristic equation are thus pure imaginaries" [1914c:265]. Stated in modern terms, Cartan's observation is that a *real compact form* exists for each complex simple (or semisimple) structure type, a property of simple $\mathfrak{g}$—and therefore also semisimple $\mathfrak{g}$— rediscovered by Weyl and employed in a key role in Weyl's above-mentioned paper. Although by italicizing it Cartan evidently thought it was especially notable, as he later explained to Weyl, it was not something for which he was looking and he had no inkling at the time of its importance.[30]

In [1914b] Cartan considered the real analogue of the problem he had solved in [1913]: Determine all real projective groups of real projective space that leave nothing planar invariant. As we shall see, Cartan saw the solution of this problem as having geometrical consequences. In what follows it will be helpful to use notation, as Cartan did, to distinguish three kinds of linear groups considered by Cartan in solving the above problem. Thus $\mathfrak{G}$ will denote a real Lie group such that $\mathfrak{G} \subset \mathbf{GL}(n, \mathbb{R})$ for some n. Likewise, let $\mathcal{G}$ denote a real Lie group such that $\mathcal{G} \subset \mathbf{GL}(n, \mathbb{C})$ for some n, and let Γ denote a complex Lie group such that $\Gamma \subset \mathbf{GL}(n, \mathbb{C})$ for some n. The groups $\mathfrak{G}$ can be interpreted as real projective groups expressed in homogeneous coordinates and acting on $\mathbb{P}^{n-1}(\mathbb{R})$. Likewise, the groups $\mathcal{G}$ can be interpreted as real projective groups on $\mathbb{P}^{n-1}(\mathbb{C})$ and the Γ as complex projective groups on $\mathbb{P}^{n-1}(\mathbb{C})$.

Cartan's problem was to determine all the $\mathfrak{G}$ that leave nothing planar invariant, and to solve it he first solved the intermediary problem of determining all $\mathcal{G}$ that leave nothing planar invariant. As he showed, this problem could be reduced to the two problems he had solved, respectively, in the first two parts of the trilogy [1913; 1914c]. To have an inkling of why this is so, let $\mathcal{G}_0$ denote the Lie algebra of $\mathcal{G}$ and let $\Gamma_0 = \mathcal{G}_0 \otimes_{\mathbb{R}} \mathbb{C}$, so that $\Gamma_0 = (\mathcal{G}_0)_c$ in the notation used above. Then Γ_0 is a complex linear Lie algebra, and if Γ is the corresponding connected Lie group, then if $\mathcal{G}$ leaves nothing planar invariant, the same is true of Γ. Thus Γ is one of the groups determined in Cartan's memoir [1913]. In fact, if the complex dimension of Γ equals the real dimension of $\mathcal{G}$, then $\mathcal{G}$ is a real form of Γ. In general this is not the case, but as Cartan showed [1914b:150–54], the problem of finding all $\mathcal{G}$ leaving nothing planar invariant does reduce to the problems solved in [1913] and [1914c].

Having shown how to determine all $\mathcal{G}$ leaving nothing planar invariant, Cartan proceeded to show that each such $\mathcal{G}$ corresponds to a $\mathfrak{G}$ leaving nothing planar invariant and that all possible $\mathfrak{G}$ are obtained in this manner. Although I shall not enter into all the technical details of Cartan's proof of these results, certain aspects of the proof are historically interesting because they show how deeply he had immersed himself in the geometrical literature of the period. To this end, let us consider how each $\mathfrak{G}$ gives rise to a $\mathcal{G}$. Thus let $\mathfrak{G}$ leave nothing planar invariant. If $x' = Ax$, $x \in \mathbb{R}^n$, defines an element of $\mathfrak{G}$, let $z = x + iy \in \mathbb{C}^n$, where $x, y \in \mathbb{R}^n$, and consider the linear transformation $A^*z = Ax + iAy$. Then $A^* \in \mathbf{GL}(n, \mathbb{C})$, and the totality of A^*'s form a real group $\mathcal{G}$. Now, $\mathcal{G}$ may or may not have the property that it leaves nothing planar invariant. If it has this property, $\mathfrak{G}$ is said to be of the *first class* and $\mathcal{G}$ is the *associate* of $\mathfrak{G}$. When $\mathfrak{G}$ is of the second class, Cartan

[30] Letter from Cartan to Weyl dated 28 March 1925 (ETH Zürich, Hs 91:502).

showed, n must be even and $\mathfrak{G}$ can be used to define a $\mathcal{G} \subset \mathbf{GL}(\frac{n}{2}, \mathbb{C})$ that leaves nothing planar invariant and that is by definition the associate of $\mathfrak{G}$ [1914b:155–6]. Thus for any $\mathfrak{G}$ that leaves nothing planar invariant there is associated to it a $\mathcal{G}$ that leaves nothing planar invariant. Of course, Cartan had to show how to reverse the process and proceed from a $\mathcal{G}$ that leaves nothing planar invariant to the $\mathfrak{G}$ for which $\mathcal{G}$ is the associate. In the case of $\mathfrak{G}$'s of the first class he made use of some of the geometrical ideas of Segre's theory of hyperalgebraic varieties (mentioned in passing at the beginning of this section).

Segre had linked his study of hyperalgebraic varieties to the real representation of points in complex projective space $\mathbb{P}^{n-1}(\mathbb{C})$. By a real representation of these points he meant a one to one mapping ϕ from the points of $\mathbb{P}^{n-1}(\mathbb{C})$ to the points or other figures in some real projective space. The simplest examples of such mappings were afforded by the representation of complex numbers as points in the plane or as points on the Riemann sphere. The first example generalizes readily to a real representation ϕ^* of points in $\mathbb{P}^{n-1}(\mathbb{C})$ for any n, as Segre noted [1890: 239]. Expressed in terms of homogeneous coordinates, ϕ^* takes $(x_1 + iy_1, \ldots, x_n + iy_n)$ into $(x_1, y_1, \ldots, x_n, y_n)$. In [1892] Segre also considered other representations ϕ that were preferable for his geometrical purposes. The special relevance of real representations ϕ to the study of hyperalgebraic varieties was that these varieties were, by definition, the preimages under Segre's ϕ's of real algebraic varieties [1892:365]. Segre systematically studied hyperalgebraic varieties from the standpoint of their real representation.

The geometry of hyperalgebraic varieties also led Segre to consider transformations more general than projective ones, which he called *antiprojective.* These are transformations $A : \mathbb{P}^{n-1}(\mathbb{C}) \to \mathbb{P}^{n-1}(\mathbb{C})$ of the form $A = \Gamma \circ P$, where $P \in \mathbf{PGL}(n, \mathbb{C})$ and Γ is the conjugation map defined in homogeneous coordinates $z = (z_1, \ldots, z_n)$ by $\Gamma(z) = (\overline{z}_1, \ldots, \overline{z}_n)$. An especially important type of antiprojectivity is an *anti-involution.* By analogy with involutions in the projective case, they are defined to be antiprojective transformations A such that $A^{-1} = A$. Anti-involutions gave rise to the simplest sort of hyperalgebraic varieties, which Segre called chains, since in the case of $\mathbb{P}^1(\mathbb{C})$ they coincided with the chains of von Staudt. Since Segre's theory of chains and anti-involutions was employed by Cartan to obtain groups $\mathfrak{G}$ from groups $\mathcal{G}$, a brief summary of its content will facilitate discussion of Cartan's work.[31]

A chain is the set $\mathcal{K}_A \subset \mathbb{P}^n(\mathbb{C})$ of fixed points (or "double points") of an anti-involution A. The only anti-involutions possessing fixed points, and thus giving rise to a chain $\mathcal{K}_A$, are those that are *positive*, i.e., those A for

[31] Segre's ideas were expounded in a more algebraic form by Sforza in [1892]. The proofs of the algebraic theorems implicit in Segre's account are found there. Cartan does not seem to have been aware of Sforza's paper, which I found helpful for an understanding of the algebraic underpinnings of Segre's memoirs.

which a $P \in \mathbf{PGL}(n, \mathbb{C})$ exists such that $PAP^{-1} = \Gamma$. In this case P maps $\mathcal{K}_A$ onto $\mathcal{K}_\Gamma$. Since $\mathcal{K}_\Gamma$ consists of points that are their own conjugates, we see that it can be identified with $\mathbb{P}^{n-1}(\mathbb{R})$, and hence $P : \mathcal{K}_A \to \mathcal{K}_\Gamma$ defines a real representation of $\mathcal{K}_A$ in Segre's sense.

To see how Cartan utilized these considerations, let $\mathfrak{G} \subset \mathbf{GL}(n, \mathbb{R})$ leave nothing planar invariant—the $\mathfrak{G}$'s he is seeking to determine. Assume that $\mathfrak{G}$ is of the first class and let $\mathcal{G} \subset \mathbf{GL}(n, \mathbb{C})$ denote the associate of $\mathfrak{G}$. Recall that the transformations of $\mathcal{G}$ are obtained by applying those of $\mathfrak{G}$ to complex instead of real variables: $T^*(x + iy) = Tx + iTy$, $x, y \in \mathbb{R}^n$. Also, $\mathcal{G}$ leaves nothing planar invariant. It is thus clear how to go from $\mathfrak{G}$ to $\mathcal{G}$, but of course, since Cartan has determined all $\mathcal{G}$ leaving nothing planar invariant, the idea is to recover $\mathfrak{G}$ from its associate $\mathcal{G}$—without knowing a priori whether $\mathfrak{G}$ is of the first or second class. To see how to do this Cartan observed that the associate $\mathcal{G}$ of a first class $\mathfrak{G}$ has the property that $T^*(\overline{z}) = T^*(x - iy) = Tx - iTy = \overline{Tx + iTy} = \overline{Tz}$. That is, $T^* \circ \Gamma = \Gamma \circ T^*$, or the transformations of $\mathcal{G}$ commute with the anti-involution Γ. Now, the property of being the associate of $\mathfrak{G}$ is determined only up to equivalence, i.e., any $\mathcal{G}' = S^{-1}(\mathcal{G})S$ is also understood to be the associate of $\mathfrak{G}$. The groups $\mathcal{G}'$ are obtained by variable changes $z = Sw$ that transform $z' = \Gamma z$ into $w' = \Gamma P w$, where $P = \overline{S}^{-1} S$, and the transformations of $\mathcal{G}'$ commute with $A = \Gamma P$. Now, since $\overline{P}P = I$, it follows that $A = \Gamma P$ defines a positive anti-involution of $\mathbb{P}^{n-1}(\mathbb{C})$. Thus all the $\mathcal{G}'$ equivalent to $\mathcal{G}$ and thereby associated to a fixed $\mathfrak{G}$ of the first class commute with some positive anti-involution A.

Suppose, conversely, that $\mathcal{G}$ leaves nothing planar invariant and commutes with some positive anti-involution $A = \Gamma P$. Then there is a variable change $z = Sw$ such that $S^{-1}(\Gamma A)S = \Gamma$ and the transformations of $\mathcal{G}' = S^{-1}\mathcal{G}S$ commute with Γ. This means that the $T' \in \mathcal{G}'$ leave the chain $\mathcal{K}_I = \{w : \overline{w} = w\}$ invariant. That is, $T'w$ is real for all real w, and therefore $\mathcal{G}'$ restricted to the real points $K_I = \mathbb{R}^n$ of $\mathbb{C}^n$ defines a $\mathfrak{G} \subset \mathbf{GL}(n, \mathbb{R})$ that evidently also leaves nothing planar invariant. This means that $\mathfrak{G}$ is of the first class and $\mathcal{G}$ is an associate. In this way every $\mathcal{G}$ that leaves nothing planar invariant *and* commutes with a positive anti-involution is the associate of a first class $\mathfrak{G}$, and the transformations of $\mathfrak{G}$ are the restrictions of those of $\mathcal{G}$ to the fixed points of the anti-involution. As Cartan put it: "the group [$\mathfrak{G}$] indicates how ... $\mathcal{G}$ transforms the double points of the positive anti-involution" [1914b:159]. Finally, $\mathfrak{G}$ is unique (up to equivalence), since as Cartan showed, $\mathcal{G}$ cannot commute with two distinct anti-involutions [1914b:160].

In this manner Cartan utilized the geometrical researches of the nineteenth century to help solve the algebraic problem posed in [1914b]. The stimulation to solve it also arose from geometrical considerations. To a certain extent these are the same as motivated the first two memoirs of the trilogy, [1913] and [1914c]. That is, given that many real geometries in the sense of Poincaré and Klein had been introduced, it was of interest not only to classify them by their structure but also to generate equivalent geometries in the sense described in Section 3. As in the complex case, the problem posed in [1914b] had a direct bearing on this matter, since the groups $\mathfrak{G}$ correspond to specific realizations of a geometrical "structure" in the setting of real projective space. Cartan, however, perceived a further geometrical application of the results in [1914b], an application that was also inspired by Segre's work on the geometry of hyperalgebraic varieties.

Indeed, Cartan considered this geometrical application sufficiently important to emphasize it in his introductory remarks. As he wrote there with a reference to Segre's papers [1914b:149–50]:

> The results obtained also have an application to the question of the representation of imaginary points by real figures. Stated more precisely, the problem the results obtained enable us to solve is the following: *Given a continuous projective group ... that leaves nothing planar invariant in a space with complex points, to represent in another real space the complex points of the given space as real figures in such a way that by virtue of this representation the group ... becomes, in the new space, a real projective group that leaves no planar multiplicity invariant.*

In other words, if the real group G acts on the points of $\mathbb{P}^{n-1}(\mathbb{C})$, it defines a geometry in Klein's sense. If ϕ takes the points of $\mathbb{P}^{n-1}(\mathbb{C})$ onto certain configurations in some real projective space $\mathbb{P}^m(\mathbb{R})$, then this geometry is transferred to an equivalent geometry in $\mathbb{P}^m(\mathbb{R})$ defined by a group $\mathfrak{G}$ that is one of the groups determined by Cartan in his memoir. As Cartan observed, there are many $\mathfrak{G}$'s that have the same structure as G (in the sense of isomorphic Lie algebras). He therefore proposed to pick out the $\mathfrak{G}$'s in the minimal number of variables and to use them and their associates $\mathcal{G}$ to create a real representation ϕ in the sense of Segre. To illustrate his method, he applied it to three general types of G [1914b: §§XIX–XX]. In what follows I will consider, with Cartan, some of the special cases of this procedure, not only to make it clear but also to convey the evident satisfaction he took in relating the abstract theory he had created to concrete, familiar geometrical relations.

When $G = \mathbf{PGL}(2, \mathbb{C})$, so that G has the structure of $\mathbf{sl}(2, \mathbb{C})$ (as a real six-dimensional Lie algebra), there are two $\mathfrak{G}$'s that leave nothing planar invariant in the minimal number of 4 real variables. The first of these, $\mathfrak{G}_1$, is of the second class, so that the connection $\mathfrak{G}_1 \leftrightarrow \mathcal{G}_1$ was not described above. Suffice it to say that if we take $\mathcal{G}_1 = G$, then $\mathfrak{G}_1$ is obtained from $\mathcal{G}_1$ as follows. Express the transformations of $\mathcal{G}_1$ in terms of (complex) homogeneous coordinates:

$$x_1' = ax_1 + bx_2, \quad x_2' = cx_1 + dx_2. \tag{5.3}$$

To obtain $\mathfrak{G}_1$ from $\mathcal{G}_1$, one sets $x_1 = u_1 + iu_2$, $x_2 = u_3 + iu_4$, $x_1' = u_1' + iu_2'$, $x_2' = u_3' + iu_4'$, $a = \alpha + i\alpha'$, $b = \beta + i\beta'$, $c = \gamma + i\gamma'$, $d = \delta + i\delta'$ in

(5.3) and separates real and imaginary parts to obtain the corresponding linear transformation $u' = Tu$ of $\mathfrak{G}_1$. (Cartan wrote it down explicitly [1914b:186].) If $p = x + iy = x_1/x_2$ represents a point of the complex line $\mathbb{P}^1(\mathbb{C})$, then $x + iy = (u_1 + iu_2)/(u_3 + iu_4)$, which is equivalent to the real equations

$$u_1 = xu_3 - yu_4, \quad u_2 = yu_3 + xu_4.$$

For $p = x+iy$ fixed, the locus of these equations is a straight line $\ell \subset \mathbb{P}^3(\mathbb{R})$. Thus we have a real representation $\phi : p \to \ell$ of the points of $\mathbb{P}^1(\mathbb{C})$ as lines in $\mathbb{P}^3(\mathbb{R})$. These lines, by virtue of their defining equations, have (real) Plücker line coordinates p_{ij} satisfying $p_{13} - p_{24} = 0$ and $p_{14} + p_{23} = 0$, which is equivalent to $(p_{13} - p_{24})^2 + (p_{14} + p_{23})^2 = 0$. Thus the lines ℓ form a line complex $\mathcal{C}$ in $\mathbb{P}^3(\mathbb{R})$, the sort of object (in the complex case) that had played such an important role in Lie's early group-related geometrical studies (Sections 1.1–1.4). The real representation ϕ maps $\mathbb{P}^1(\mathbb{C})$ onto $\mathcal{C}$, and $\mathfrak{G}_1$ takes $\mathcal{C}$ into itself. The geometry defined by $G = \mathbf{PGL}(1, \mathbb{C})$ (regarded as a real group) transfers to the geometry in $\mathbb{P}^3(\mathbb{R})$ defined by the action of the group $\mathfrak{G}_1$ on the line complex $\mathcal{C}$.

The second example in four variables, $\mathfrak{G}_2$, is of the first class and is defined as follows. Let $u_1, \ldots, u_4$ be real variables and consider the hermitian form

$$f = u_1 x_1 \overline{x}_1 + (u_2 + iu_3) x_1 \overline{x}_2 + (u_2 - iu_3) \overline{x}_1 x_2 + u_4 x_2 \overline{x}_2. \tag{5.4}$$

Then $\mathfrak{G}_2$ is the group of real linear transformations of the u_i that result from applying the substitutions (5.3) to x_1, x_2 and setting $a = \alpha + i\alpha'$, etc. (Again, Cartan actually wrote down the transformations of $\mathfrak{G}_2$ explicitly [1914b:182].) Since the discriminant of f is an invariant of $\mathfrak{G}_2$, setting it equal to zero we obtain a quadric surface in $\mathbb{P}^3(\mathbb{R})$,

$$(u_2 + iu_3)(u_2 - iu_3) - u_1 u_4 = u_2^2 + u_3^2 - u_1 u_4 = 0,$$

which is taken into itself by $\mathfrak{G}_2$, and $\mathfrak{G}_2$ consists of the real projective transformations that take this surface into itself. A real representation of the points $p = x + iy \in \mathbb{P}^1(\mathbb{C})$ is obtained by observing that p defines the hermitian form $f_p = (x_1 - px_2)(\overline{x}_1 - \overline{p}\overline{x}_2)$, which has vanishing discriminant. Since f_p corresponds to setting $u_1 = 1$, $u_2 = -x$, $u_3 = y$, $u_4 = x^2 + y^2$ in (5.4), this defines a point P on the quadric surface, and $\phi : p \mapsto P$ is a real representation of the points $\mathbb{P}^1(\mathbb{C})$ as points on a quadric surface in $\mathbb{P}^3(\mathbb{R})$. "This representation is in essence identical to that by which Riemann represents the complex numbers as points on a sphere. Every homographic transformation of the complex line is interpreted in space as a projective transformation leaving the sphere invariant. There is thus an equivalence between the projective geometry of the complex line and the hyperbolic non-Euclidean geometry of real space" [1914b:182–3].

As this quotation indicates, Cartan enjoyed showing how known geometrical results flowed from his theory. One further example is worth giving.

To illustrate his methods Cartan had showed how to determine the groups $\mathfrak{G}$ with the real structure of familiar, low-dimensional groups. One such group was $\mathbf{PGL}(2, \mathbb{C})$, and it was in this way that the above groups $\mathfrak{G}_1$ and $\mathfrak{G}_2$ arose—in effect as the two real representations of $\mathbf{PGL}(2, \mathbb{C})$ of minimal degree. Another group he considered was the projective special unitary group $\mathbf{PSU}(2)$. The transformations of this group in homogeneous coordinates are obtained from those of $\mathbf{PGL}(2, \mathbb{C})$ by the specialization

$$c = -\bar{b}, \quad d = \bar{a} \tag{5.5}$$

in (5.3). The associate groups $\mathcal{G}$ are obtained by the same specialization from the irreducible representations of $\mathbf{PGL}(2, \mathbb{C})$ as a complex group. Thus for any integer $m \geq 2$ and $a_0, a_1, \ldots, a_m$ complex, consider the binary form

$$\phi(a, x) = \sum_{i=0}^{m} \binom{m}{i} a_i x_1^{m-i} x_2^i.$$

Recall from Section 7.1 that the group $\mathfrak{N}_m$ of a normal curve arises as the linear transformations of the a_i induced by transforming x_1, x_2 by the transformations (5.3) of $\mathbf{PGL}(2, \mathbb{C})$. The specialization (5.5) yields the associate groups $\mathcal{G}_m \subset \mathfrak{N}_m$.

When m is even, $\mathcal{G}_m$ is the associate of a group $\mathfrak{G}_m$ of the first class in $m+1$ variables. It is obtained from $\mathcal{G}_m$ by restricting the coefficients a_i of $\phi(a, x)$ by the conditions

$$a_0 = \bar{a}_m, \quad a_1 = -\bar{a}_{m-1}, \quad a_2 = \bar{a}_{m-2}, \quad \ldots \quad a_{m/2} = (-1)^{m/2}\bar{a}_{m/2}$$

and taking the real and imaginary parts of the a_i and a_i' to obtain linear transformations in $m+1$ real variables. In particular, when $m = 2$ the group $\mathfrak{G}_2$ is obtained from $\mathcal{G}_2$ by setting $a_0 = v_1 + iv_2$, $a_2 = v_1 - iv_2$, $a_1 - iv_3$. By applying the transformations of $\mathbf{PSU}(2)$ to x_1, x_2 the real variables v_i are transformed according to $\mathfrak{G}_2$—thereby yielding the real representation of minimal degree of $\mathbf{PSU}(2)$. Cartan's group $\mathfrak{G}_2$, regarded as a group of linear transformations, consists of all positive scalar multiples of orthogonal matrices, the scalar being $|a|^2 + |b|^2$ for the representation of the transformation in (5.3) with the specialization (5.5). If we restrict to the transformations satisfying in addition $|a|^2 + |b|^2 = 1$, as was usually done in discussing these transformations as homographies, then Cartan's representation is a homomorphism of $\mathbf{SU}(2)$ onto $\mathbf{SO}(3, \mathbb{R})$ with kernel $\pm I$. Hence Cartan could conclude, "From another point of view a well known relation between the group of rotations of space and the homographic group $x' = (ax+b)/(-\bar{b}x+\bar{a})$ is recovered" [1914b:180]. The "other point of view" was of course that of Cartan's theory. It yielded the representation customarily obtained via the representation of complex numbers by points on the Riemann sphere—a representation made prominent by Klein's lectures on the icosahedron [1884].[32]

[32] See pp 34–36 of [1884], where Klein also presents the historical background to the connection between homographies and rotations.

In creating his trilogy of memoirs, Cartan drew his methods and motivation from a diverse spectrum of sources: the Killing–Cartan theory of structure, especially the theory of secondary roots (Sections 6.3–6.4); Killing's problem (Section 6.3), an offshoot of his program to classify all Lie algebra structures; Kowalewski's problem (Section 7.5), which derived from Kowalewski's interest in determining primitive transformation groups; Study's characterization problem (Section 7.2), inspired by Study's geometrical-conceptual approach to the theory of invariants; the prospect of a far-reaching differential Galois theory for both finite- and infinite-dimensional groups, a *leitmotiv* for the research of the Paris branch of Lie's school; and last, but not least, nineteenth-century geometry. That geometrical considerations and a geometrical perspective had come to permeate Cartan's thinking when he composed his trilogy should be evident from the presentation in this section and in the concluding portion of the previous section. Cartan's extensive treatment of equivalent geometries in his encyclopedia article also shows how deeply engaged in the subject he had become.[33]

Compared with the later applications of the results of the trilogy, e.g., in the theory of symmetric spaces,[34] the geometrical applications found in the trilogy and in Cartan's *Encyclopédie* article [1915] may seem a relatively minor aspect of Cartan's geometrically oriented mathematical output. But they seem to be of far greater historical significance. That is, Cartan's growing interest in the application of Lie's theory of groups to geometry, particularly the structural aspects of the theory, and the relevance of the subject matter of the trilogy seem to have provided the "last straw" necessary to fix his resolve to continue what he had begun in his doctoral dissertation two decades earlier. In any case, elements of nineteenth-century geometry such as geometry in the sense of Poincaré and Klein and Segre's theory of hyperalgebraic entities form a part of the context of these memoirs and thereby of the prehistory of the representation of Lie groups and algebras.

---

[33] See [Akivis and Rosenfeld 1993:§3.6] for a detailed discussion of this aspect of Cartan's work. In [Rosenfeld 1997] the interested reader will find a systematic treatment of geometrical interpretations of Lie groups in the spirit of Cartan's work.

[34] See [Akivis and Rosenfeld 1993:Ch.6] for a discussion of Cartan's theory of symmetric spaces.

Chapter 9
The Göttingen School of Hilbert

The fourth and final part of this book culminates in the the ground-breaking papers of Hermann Weyl on the representation of semisimple Lie groups (Chapter 12). For a full historical appreciation of Weyl's work and the events leading up to it (Chapter 11), however, it is necessary to be mindful of the historical context in which Weyl developed as a mathematician. That context was provided by the school of mathematics that grew up around David Hilbert at the University of Göttingen. Weyl was there from 1904 until 1913, first as a student, writing his doctoral dissertation under Hilbert's supervision, and then as a lecturer (*Privatdozent*). Weyl himself has made it clear that Hilbert exerted a great influence upon him, and to make this point one can do no better than to quote the eloquent Weyl himself: "I came to Göttingen as a country lad of eighteen .... In the fullness of my innocence and ignorance I made bold to take the course Hilbert had announced for that term, on the notion of number and the quadrature of the circle. Most of it went straight over my head. But the doors of a new world swung open for me, and I had not sat long at Hilbert's feet before the resolution formed itself in my young heart that I must by all means read and study whatever this man had written" [1944b:132]. That Weyl was deeply affected by what he read is also clear. On the occasion of Hilbert's seventieth birthday, he wrote of "the magic that his scientific personality exerts on the following generation," and speaking for that generation, he added, "the mathematical way of thinking conveyed by our works, the special style of our mathematical thought, goes back to him more so than to any other living person" [1944a:128].

The first section is devoted to Hilbert's work on the theory of invariants. Although it was done before he came to Göttingen and consequently was not an area into which Weyl delved while at Göttingen, it is important for an appreciation of what Weyl termed "the peculiarly Hilbertian brand of mathematical thinking" that I discuss in Section 6. It is also relevant to Hurwitz's work on invariants (Section 10.2), which, through the intermediary of I. Schur, inspired Weyl's work on the representation of Lie groups. In Section 2 I survey Hilbert's mathematical activity at Göttingen during Weyl's years there, especially his work on integral equations, which was Weyl's primary research interest at the time. The next section is devoted to a discussion of one of the most remarkable and unique traits of the Göttingen school: the conviction, fostered by developments in relativity theory, that mathematicians could and should be engaged with theoretical physics. As we shall see in Chapter 11, acting on this conviction in response to Einstein's general theory of relativity, Weyl was led to the mathematical considerations that drew him to the study of group representations.

The next two sections are devoted to Weyl's own mathematical activity at Göttingen: in integral equations (Section 4) and on Riemann surfaces (Section 5). Among other things, these sections illustrate Weyl's observation about Göttingen "in the halcyon days before 1914" that, "Once a band of disciples had gathered around Hilbert, intent upon research and little worried by the chore of teaching, it was but natural that in joint competitive aspiration of related aims each should stimulate the other; there was no need that everything come from the master." [1944a:128]. Indeed, from Hilbert's student Erhardt Schmidt, Weyl learned much about integral equations that later proved useful in application to group representations. Likewise, Paul Koebe, who had come to Göttingen as a lecturer, made quite a stir among the young mathematicians there by virtue of his work on the uniformization of Riemann surfaces. Koebe's work was an important motivating factor behind Weyl's Göttingen lectures on Riemann surfaces, which formed the basis for his innovative and influential book *Die Idee einer Riemannschen Fläche* (1913). As we shall see in Chapter 12, this aspect of Weyl's experiences at Göttingen provided him with a point of view that he put to brilliant use in his work on group representations.

In the final section (Section 6) I consider some of the general characteristics of Hilbert's style of doing mathematics that seem especially relevant to Weyl's work on the representation of Lie groups.

9.1. Hilbert and the Theory of Invariants

David Hilbert (1862–1943) was born and raised in the East Prussian city of Königsberg, where he also received most of his university education.[1]

[1] For further biographical information on Hilbert, the reader may consult [Blumenthal 1935] and [Reid 1970].

Figure 26. Hermann Weyl as a young man (top) and Hilbert during his years at Göttingen .

Among his teachers at Königsberg, the one who influenced him the most deeply was Adolf Hurwitz (1859–1919). Hurwitz, who was only three years older than Hilbert, had been Felix Klein's student at Leipzig during the period in which Klein was intensely interested in aspects of complex function theory. When Hurwitz arrived at Königsberg in 1884 as an assistant professor (*ausserordentlicher Professor*) he had done research in complex function theory and in algebraic geometry and had begun to investigate arithmetical questions as well. Hurwitz became Hilbert's guide through the world of mathematics, much of his education taking place on the walks they regularly took together.

The senior mathematician at Königsberg at this time was another former student of Klein's, Ferdinand Lindemann, who is remembered mostly for his proof, in 1882, of the transcendence of π. As sole full professor (*ordentlicher Professor*) of mathematics at the University, Lindemann was Hilbert's "Doctor-Father" and in that capacity suggested to Hilbert a problem in the theory of invariants as the subject of his doctoral dissertation. Hilbert solved it, thereby obtaining his doctorate in 1885 and initiating a phase of research on the theory of invariants. After a semester with Klein in Leipzig and a semester in Paris (in the company of Study), Hilbert became a lecturer (*Privatdozent*) at Königsberg in 1886 with a *Habilitationsschrift* on the theory of invariants. Klein later recalled that when Hilbert lectured in his seminar in Leipzig in 1885 he perceived that "he was the coming man in mathematics" [Reid 1970:19], but it was probably not until 1888 that the rest of the mathematical world began to have such an inkling. In that year Hilbert published a paper in the Göttingen *Nachrichten* that claimed to have solved an important problem that had arisen in connection with a paper published by Paul Gordan in 1868. eb

Before stating the problem, it will be helpful to recall same basic definitions (introduced for binary forms in Section 7.2). Recall that invariants were always defined with respect to one or more base forms (*Grundformen*). Suppose, for example, we take as base form the general homogeneous polynomial of degree d:

$$\phi(a,x) = \sum_{e_1+\cdots+e_n=d} a_{e_1,\ldots,e_n} x_1^{e_1} x_2^{e_2} \cdots x_n^{e_n}, \tag{1.1}$$

where the summation is over all $N = \binom{d+n-1}{d}$ distinct terms $x_1^{e_1} x_2^{e_2} \cdots x_n^{e_n}$ of degree d, and the coefficients $a_{e_1,\ldots,e_n}$ are regarded as variable parameters. If $x = Sx'$ denotes any nonsingular linear change of variables $x = (x_1,\ldots,x_n) \to (x_1',\ldots,x_n')$ of this base form, then

$$\phi(a,x) = \phi(a,Sx') = \phi(a',x'). \tag{1.2}$$

The relation between the N coefficients $a_{e_1,\ldots,e_n}$ and the N $a'_{e_1,\ldots,e_n}$ is given by a nonsingular linear transformation $\Gamma(S)$,

$$a' = \Gamma(S) \cdot a, \tag{1.3}$$

Figure 27. Gordan in academic regalia. Tradition has it that he characterized Hilbert's solution as theology rather than mathematics.

and a homogeneous polynomial $I(a)$ in the N variables $a_{e_1,\ldots,e_n}$ is an invariant (or absolute invariant) if $I(a') = I(a)$, i.e.,

$$I(\Gamma(S) \cdot a) = I(a) \quad \text{for all } S \in \mathbf{GL}(n). \tag{1.4}$$

Here the notation $\mathbf{GL}(n)$ is used to denote either $\mathbf{GL}(n, \mathbb{C})$ or $\mathbf{GL}(n, \mathbb{R})$; both possibilities were considered in the theory of invariants and usually were not carefully distinguished. More generally, $I(a)$ was considered an invariant (or relative invariant) if $I(a') = |S|^w I(a)$, where w is the weight of the invariant. However, from the more general view of invariants introduced by Hilbert, invariants in this broader sense are simply the absolute invariants relative to the group $\mathbf{SL}(n)$. As we shall see in Section 10.2, Hurwitz in particular adopted this point of view.

In 1868 Gordan had showed that when $n = 2$, *i.e.*, for the invariants of a binary base form ϕ, a finite set of invariants $I_1, \ldots, I_k$ can be constructed such that every invariant is expressible as some polynomial in $I_1, \ldots, I_k$. Thus, in effect, all invariants of ϕ are known when the "basis" invariants $I_1, \ldots, I_k$ are known. Gordan's finiteness theorem thus raised the question as to whether the same sort of finiteness result was true for the invariants of ternary forms and, more generally, for the invariants of n-ary forms. Gordan was unable to establish such results by his constructive methods. Thus Hilbert's announcement in 1888 that he could prove the finiteness theorem for the invariants of an n-ary form attracted considerable attention. Hilbert announced his results in a trilogy of notes in the *Göttinger Nachrichten* entitled "Zur Theorie der algebraischen Gebilde"[2] and then presented more developed versions in two papers published in *Mathematische Annalen* [1890; 1893].

As Hilbert himself stressed [1890:246], his proof of the n-ary finiteness theorem involves two steps. The first step is contained in the following theorem, which represents an equivalent formulation of the theorem [1890:199 (Theorem I)] that has evolved into what is now known as Hilbert's basis theorem and so will be designated as such:

(1.5) **Hilbert's Basis Theorem.** *If $\mathfrak{I}$ is any set of homogeneous polynomials in N variables, then there exists an integer m and polynomials $I_1, \ldots, I_m$ from $\mathfrak{I}$ such that any $I \in \mathfrak{I}$ is expressible in the form*

$$I = B_1 I_1 + \cdots + B_m I_m, \tag{1.6}$$

where $B_1, \ldots, B_m$ are homogeneous polynomials in the same N variables.

In particular, if $\mathfrak{I}$ is the set of invariants of an n-ary base form f, then (1.6) holds.

The second step of Hilbert's proof consists in deducing from (1.6) by means of various differential operators that

$$I = J_1 I_1 + \cdots + J_m I_m, \tag{1.7}$$

where each J_k is an invariant or a constant. Thus each B_k is replaced by an invariant J_k. Then Theorem 1.5 can be applied to each nonconstant J_k and then the analogue of (1.7) deduced from the analogue of (1.6), and so on. Since in (1.6) the degrees of the B_i may be taken less than the degree of I, if the operation sending $B_i \to J_i$ does not increase degrees, the above-described process must come to a halt after a finite number of steps: All the J_k at some stage will be constants, and I will be expressed as a polynomial in the I_k. Thus the finiteness question is resolved affirmatively, albeit not constructively, for the general n-ary base form ϕ.

---

[2] See Hilbert, *Gesammelte Abhandlungen* **2**, 176–198.

For n-ary forms with $n > 2$, Hilbert used a differential operator, introduced by Cayley, that became known as the omega process to go from (1.6) to (1.7). This was somewhat complicated, and for binary forms he showed that another differential operator that he had utilized in an earlier paper on invariant theory could be used to the same end. Indeed, his observations had the following implication. Let $\mathfrak{J}$ be any set of homogeneous polynomials that might be invariants in some generalized sense, and suppose an operator $\mathcal{E}$ can be defined on homogeneous polynomials B in the same variables as the $I \in \mathfrak{J}$ that has the following properties:

$$(1.8)\qquad \begin{array}{llll} \text{(i)} & \mathcal{E}[B] \in \mathfrak{J} \quad \text{for all } B; & \text{(ii)} & \mathcal{E}[I] = I \quad \text{for all } I \in \mathfrak{J}; \\ \text{(iii)} & \mathcal{E}[B \cdot I] = \mathcal{E}[B] \cdot I; & \text{(iv)} & \mathcal{E}[B_1 + B_2] = \mathcal{E}[B_1] + \mathcal{E}[B_2]. \end{array}$$

Then Theorem 1.5 applies to $\mathfrak{J}$, and if $\mathcal{E}$ is applied to (1.6), by virtue of the above properties the result is (1.7) with $J_i = \mathcal{E}[B_i]$. Thus the finiteness problem would be solved affirmatively for the "invariants" compromising $\mathfrak{J}$, provided that $\mathcal{E}$ does not increase degrees.

These implications of Hilbert's work stood out because, urged on by Klein, he endorsed the idea of cultivating a theory of invariants based upon a broader conception of an invariant suggested, he said, by Klein's *Erlanger Programm* and by Lie's theory of transformation groups. Traditionally, invariants of an n-ary form were defined by considering *all* transformations of $S \in \mathbf{GL}(n)$ applied to the variables of the form, as in (1.3)–(1.4) above. Hilbert proposed considering the invariants that arise when only the transformations of some subgroup $\mathfrak{G}$ of $\mathbf{GL}(n)$ are applied to the variables of the n-ary form. Thus, in the notation of (1.4), $I(a)$ is an invariant relative to $\mathfrak{G}$ when $I(\Gamma(S) \cdot a) = |S|^w I(a)$ for all $S \in \mathfrak{G}$.[3] In conjunction with the generalized notion of an invariant, Hilbert focused on the associated finiteness problem. He pointed out that his own method, using an analogue of the omega operator, could be extended to deal with certain subgroups $\mathfrak{G}$ connected with linear associative algebras; but these subgroups are very special. For example, Hilbert's method extends, as he showed, to the subgroup $\mathfrak{G}$ of $\mathbf{GL}(3)$ consisting of all real orthogonal transformations with determinant $+1$, i.e., the rotation group in 3-space, which may be defined using quaternions (see Section 12.1). But his method did not extend to the rotation group in n-space for $n > 3$. As we shall see in Section 10.2, Hurwitz took up the challenge of the rotation group inspired by the idea that there were other ways to create an appropriate operator $\mathcal{E}$, using integration instead of differentiation. Hurwitz's contributions to the theory of invariants,

[3] See Hilbert's papers [1889:196–7; 1890:255–6; 1893:288]. Judging by the correspondence between Klein and Hilbert when the former was refereeing Hilbert's notes for the *Nachrichten*, Hilbert had considered a few examples of invariants relative to groups $\mathfrak{G} \neq \mathbf{GL}(n)$, but it was Klein who urged him to formulate, with reference to the *Erlanger Programm* and Lie's theory, the general notion of invariants relative to any $\mathfrak{G} \subset \mathbf{GL}(n)$ and to consider the finiteness problem in this generality [Frei 1985:44, 54–55].

made in 1894 and 1897, are discussed in Chapter 10, which is devoted to the work of the Berlin algebraists—Georg Frobenius, who created the character and representation theory of finite groups, and his student Issai Schur. It was through Schur's reaction to Hurwitz's work that it came to play a vital role in our story.

9.2. Hilbert at Göttingen

In 1886 Klein left Leipzig for Göttingen, having taken care that Lie would be his successor in Leipzig. As we saw in Chapters 6–7, during the decade he spent at Leipzig, Lie, who undoubtedly benefited from Klein's earlier successes in building up the mathematics program at Leipzig,[4] managed to build up about him a lively school, with an outpost in Paris, devoted to the development and application of his theory of transformation groups. Klein, whose creative mathematical productivity had declined sharply after the breakdown that followed his intense competition with Poincaré, did nothing comparable at Göttingen. Instead, he channeled his considerable energy into administrative, pedagogical, and publishing activities that, combined with his skill at political maneuvering within academia, were to lay the foundations for one of the most influential schools of mathematics.[5] In this regard, his single most important act occurred in 1895 when he finally succeeded in bringing Hilbert—whom he had recognized years earlier as a future mathematical star— to Göttingen as a professor.

By the time Hilbert arrived in Göttingen, the theory of invariants was no longer a field of research that interested him. Now all of his attention was focused on the theory of numbers. This "arithmetical phase" of his career culminated in his "*Zahlbericht*" [1897] and his papers on relative quadratic and abelian number fields (1898–9).[6] Ostensibly a report on the theory of algebraic numbers for the Deutsche Mathematiker-Vereinigung, the *Zahlbericht* was far more than the typical report. As Freudenthal has written, "It is one of the classics, a masterpiece of mathematical literature. For half a century it was the bible of all who learned algebraic number theory .... In it Hilbert collected all relevant knowledge ..., reorganized it under striking new unifying viewpoints, reshaped formulations and proofs, and laid the ground work for the still growing edifice of class field theory .... Starting with the quadratic field, Hilbert step by step increases the generality, with a view to a complete theory of relative Abelian fields; but from the beginning he chooses those methods which foreshadow the general principles" [1972:390].

---

4 On Klein's activity at Leipzig see Rowe's account [1989b:191–94].

5 See in this connection the detailed carefully documented account by Rowe [1989b] as well as his essay [1986].

6 See Hilbert's *Gesammelte Abhandlungen* **1**, 370–509.

Figure 28. Klein as he looked during his Göttingen years. To Weyl and other contemporaneous students at Göttingen he seemed a distant god overseeing the heroic activities of Hilbert and Minkowski [Reid 1970: 95].

Hilbert's arithmetical work made a great impression upon Weyl as a student. As noted at the beginning of the chapter, Hilbert's lectures during Weyl's first year at Göttingen had opened up a new world to him, and he resolved to study whatever Hilbert had written: "And after the first year I went home with Hilbert's *Zahlbericht* under my arm, and during the summer vacation I worked my way through it—without any previous knowledge of elementary number theory or Galois theory. These were the happiest months of my life, whose shine, across years burdened with our common share of doubt and failure, still comforts my soul" [1944b:132]. I mention this because Weyl saw in Hilbert's arithmetical work the most striking examples of what he perceived as a characteristic feature of Hilbert's creative power, namely "his sensitivity in registering hints which revealed to him general relations while solving special problems. This is most magnificently exemplified by the way along which, during his number-theoretical period, he was led to the enunciation of his general theorems on class fields and the general law of reciprocity" [1944a:124]. "It is a great pleasure to watch how, step by step, advancing from the special to the general, Hilbert evolves the adequate concepts and methods, and the essential conclusions emerge" [1944b:145]. As we shall see in Chapter 12, this Hilbertian trait can be seen as well in Weyl's own work on the representation of Lie groups, where, beginning with the special linear group and then proceeding to the symplectic and orthogonal groups, he developed the general concepts and methods that enabled him to create a theory valid for any semisimple group.

Having successively investigated an algebraic theory (invariants) and an arithmetical one, Hilbert turned next to geometry, where the subject of non-Euclidean geometry focused his attention of the foundations of geometry. His lectures on this subject culminated in the publication in 1899 of his book *Grundlagen der Geometrie*, in which, following the lead of Pasch, the axiomatic approach to geometry was taken to new heights.[7] The following year Hilbert gave what turned out to be one of the most influential talks on mathematics ever given. Addressing the International Congress of Mathematicians in Paris on the subject "Mathematical Problems," he stressed the importance of concrete problems for the continuing vitalization of mathematics [1900]. By way of conclusion he listed twenty-three unsolved problems, which greatly affected the history of mathematics during the following half-century.[8] In particular, the solutions given by Koebe and Poincaré to the twenty-second problem will be seen to play a role in our story (Section 5). The fifth problem is also of interest because it involved the Helmholtz space problem—the problem of deducing the geometry of space by means of the properties of motions of rigid bodies.

In Chapter 4 we saw the important role played by Helmholtz's problem in leading Killing to the notion of a Lie algebra. From Hilbert's correspondence with Engel, whom Hilbert had known through Klein since his student days, we learn that Hilbert was planning on giving a report on "infinitely small transformations" in the Königsberg Colloquium.[9] Whether he actually did is not clear, but after Killing lectured in Königsberg in 1891 on "the elements of the theory of transformation groups," Hilbert wrote to Engel in the spring of 1892, "Sometime I would like very much to have a thorough discussion of this theory."[10] Hilbert hoped to have such a discussion with Engel at a fall meeting of the *Naturforschersammlung* in Halle, and in September he wrote, "Hopefully, I then will be somewhat more oriented toward infinitesimal transformations than at present, since I am thinking of lecturing on them this winter."[11] Hilbert's interest in infinitesimal trans-

[7] A careful study of the origins of *Grundlagen der Geometrie* has been given by Toepell [1986].
[8] See in this connection [Alexandrov 1971; Bieberbach 1930].
[9] Letter to Engel dated 9 July 1886 (Engel Archive Giessen).
[10] Letter to Engel dated 25 April 1891 (Engel Archive Giessen).
[11] Letter to Engel dated 18 September 1892 (Engel Archive Giessen).

formations (or Lie algebras) was probably due in part to Killing's apparent discoveries regarding simple Lie algebras but seems to have been connected with his growing interest in the foundations of geometry. For in a letter in January of 1894 he informed Engel that he had announced a course of lectures on non-Euclidean geometry and in that connection wanted to make a thorough study of Lie's approach.[12] Lie's approach was contained in his treatment of the Helmholz space problem [1890b; 1890c], which he tackled by means of Lie-algebraic techniques.[13]

Although it is unclear to what extent Hilbert familiarized himself with Lie's theory of groups, he eventually decided that the "infinitesimal transformation" approach to the Helmholtz problem was inappropriate, since it is based upon the presumed differentiability of the functions involved in the definition of the transformation group. Such an assumption seemed out of place in Helmholtz's problem, where only the continuity of the group of motions seemed natural, and Hilbert posed as his fifth problem to investigate whether in fact Lie's differentiability hypothesis could be proved [1900:305]. Hilbert himself showed this to be the case in his study of Helmholtz's problem in the plane [1902; 1903]. His treatment is notable for the pioneering use he made of set-theoretic, essentially topological, reasoning. In particular, in Hilbert's attempt to define a plane, we can see the seeds of the conception of a two-dimensional manifold set forth by Weyl, who consulted Hilbert's treatment when developing his lectures on Riemann surfaces (Section 5).[14]

By the time Hilbert's lectures on the foundations of geometry appeared in 1899 he had begun to focus on aspects of the one basic branch of mathematics he had not yet touched in his research, namely analysis. In 1899 he began lecturing on the calculus of variations and found it so stimulating that three of the twenty-three problems posed in his Paris lecture were drawn from this area. His most well known accomplishment was his reinstatement of Dirichlet's principle as a part of rigorous mathematics.[15]

A new direction was introduced into the analytical phase of Hilbert's research activity by a paper on integral equations published in 1903 by Ivar Fredholm. There is an obvious analogy between a finite system of equations $\sum_{j=1}^{n} a_{ij}x_j = b_i$, $i = 1, 2, \ldots, n$, and an integral equation $\int_a^b K(s,t)\varphi(t)\,dt = f(s)$, but Fredholm was the first to show that it is possible to establish significant theorems about integral equations that are direct analogues of theorems involving the simpler and better-understood theory of linear algebraic equations.[16] He considered the integral equation

[12] Letter to Engel dated 14 January 1894 (Engel Archive Giessen).

[13] Slightly earlier Poincaré [1887] had dealt with Helmholtz's problem for the plane and also used Lie-algebraic techniques.

[14] See Weyl's recollections in [Weyl 1944b:156].

[15] On the history of the principle and its relations with the development of analysis see the book by Monna [1975].

[16] For a more detailed discussion of Fredholm's work and other work on integral equations done before Hilbert entered the picture, see [Bernkopf 1966].

$$\varphi(s) + \int_0^1 K(s,t)\varphi(t)\,dt = f(s), \tag{2.1}$$

where K, φ, and f are all assumed to be continuous and φ is the unknown function. The function K will be referred to as the kernel of the integral equation (or operator). In terms of the linear operator A defined by $(A\varphi)(s) = \int_0^1 K(s,t)\varphi(t)\,dt$, (2.1) is $(I + A)\varphi = f$. Fredholm himself made some use of such operator notation, although Hilbert did not. Nonetheless, it will be used occasionally in what follows. Fredholm introduced an analogue of the the determinant of $I + A$, denoted here by D_A, and he showed that when $D_A \neq 0$, (2.1) has a unique solution φ. He also considered the case $D_A = 0$. In this case, the transposed homogeneous integral equation $\int_0^1 K(t,s)\psi(t)\,dt = 0$, or $A^T\psi = 0$, has a finite number of linearly independent solutions $\varphi_1, \ldots, \varphi_n$, and (2.1) will have a solution if and only if f is such that $\int_0^1 f(s)\psi_k(s)\,ds = 0$, $k = 1, \ldots, n$.[17]

Fredholm's results were direct analogues of known facts about linear systems of equations and they made a great impression upon Hilbert, who learned of them when a student lectured on Fredholm's paper in the Göttingen mathematics seminar. Between 1904 and 1910 Hilbert published six memoirs on the theory of integral equations in the pages of the *Göttinger Nachrichten*, which were then republished as a book [1912]. In these papers he succeeded in pushing the analogy with finite systems of equations further than had Fredholm. With the theory of quadratic forms in mind, he focused considerable attention on integral equations of the form

$$\varphi(s) + \lambda \int_a^b K(s,t)\varphi(t)\,dt = f(s), \quad K(t,s) = K(s,t), \tag{2.2}$$

or $(I + \lambda A)\varphi = f$, with A symmetric. As with Fredolm, all functions K, φ, f are assumed to be continuous.

To begin with, Hilbert obtained the analogues of Fredholm's results. Thus there is a function $\delta(\lambda)$, defined by a power series in λ, that plays the role of the determinant of $I + \lambda A$. By virtue of the symmetry hypothesis all the zeros of $\delta(\lambda)$ are real, and since $\delta(\lambda)$ is a power series, these zeros have no finite limit point. If $\delta(\lambda) \neq 0$, then (2.2) has a unique solution, and for $\delta(\lambda) = 0$ the homogeneous equation $(I + \lambda A)\varphi = 0$ has a finite number of linearly independent solutions $\varphi_1, \ldots, \varphi_n$ such that any solution is a linear combination of them. Also, (2.2) has solutions if and only if f satisfies $\int_a^b \varphi_k(s)f(s)\,ds = 0$, $k = 1, \ldots, n$.

The zeros of $\delta(\lambda)$, which must be countable, Hilbert called eigenvalues, and the corresponding functions $\varphi \neq 0$ such that $(I + \lambda A)\varphi = 0$ he called eigenfunctions. Thus λ is an eigenvalue in Hilbert's sense if $1/\lambda$ is what is now called an eigenvalue of the operator A. (Since $\lambda = \infty$ is not

[17] The operator A is an example of what is now called a *compact* symmetric operator.

an eigenvalue in Hilbert's sense, Hilbert's theory has no counterpart to 0 as an eigenvalue in the modern sense.) For eigenvalues $\lambda_1 \neq \lambda_2$ and corresponding eigenfunctions φ_1, φ_2, he showed that $\int_a^b \varphi_1(s)\varphi_2(s)\,ds = 0$ and that if $\lambda_1, \lambda_2, \lambda_3, \ldots$ denote the eigenvalues each counted according to its multiplicity, then the corresponding eigenfunctions can be chosen to satisfy the orthogonality relations (as he called them)

$$\int_a^b \varphi_i(s)\varphi_j(s)\,ds = \delta_{ij}. \tag{2.3}$$

Hilbert established the following analogue of the principal axes theorem for quadratic forms.

(2.4) Principal Axes Theorem. *If $K(s,t)$ is symmetric, then for any two continuous functions x and y,*

$$\int_a^b \int_a^b K(s,t)x(s)y(t)\,ds\,dt = \sum_k \frac{1}{\lambda_k} c_k(x)c_k(y),$$

where $c_k(x)$, $c_k(y)$ are the Fourier coefficients of x, y with respect to the normalized eigenfunctions φ_k, e.g., $c_k(x) = \int_a^b x(s)\varphi_k(s)\,ds$.

Another generalization of linear algebra obtained by Hilbert that, like his principal axes theorem, was not an analogue of Fredholm's results had to do with the representation of continuous functions as linear combinations of the eigenfunctions of a symmetric kernel. Since the number of eigenfunctions is generally infinite, these linear combinations were of course infinite series, which Hilbert likened to Fourier series. That is, let $\varphi_1, \varphi_2, \ldots$ be the eigenfunctions associated to the integral equation (2.2) with symmetric kernel $K(s,t)$ and assume that they are chosen to satisfy the orthogonality relations (2.3). Corresponding to any continuous function $f(s)$ one may consider a Fourier-type series $\sum_n c_n\varphi_n(s)$ associated to $f(s)$, where the "Fourier coefficients" c_n are given by $c_n = \int_a^b f(s)\varphi_n(s)\,ds$. Hilbert's main theorem about such Fourier-type series was that if $f(s)$ is any continuous function such that $f(s) = \int_a^b K(s,t)g(t)\,dt$ for some continuous function $g(s)$ (in other words, f is in the range of the operator A), then its Fourier-type series converges to it uniformly and absolutely. This result is usually known as the Hilbert–Schmidt theorem. Although Hilbert stated it for any continuous symmetric kernel, he proved it only for certain kernels. His student E. Schmidt published the first general proof in his dissertation [1905]. Schmidt's contributions are discussed further in Section 4.

All of Hilbert's above-mentioned results are contained in his first memoir on integral equations [1904a]. In his fourth memoir [1906a] he then developed what he saw as the appropriate mathematical context for developing and extending his results. This context would now be described as the real Hilbert space ℓ^2 of square summable sequences. Within that context Hilbert developed what would now be described as the theory of compact linear operators and the theory of bounded symmetric operators. Following in the tradition of Weierstrass and his student Frobenius, Hilbert expressed his ideas in terms of bilinear and quadratic forms in infinitely many variables rather than in terms of linear operators acting on ℓ^2. In particular, compact operators correspond to what Hilbert called completely continuous (*vollstetig*) forms. Within this more general context, Hilbert was led to notions that do not arise in the above-described theory of integral equations, such as that of the continuous spectrum of a noncompact operator.

The connection of the more general ℓ^2 theory with that of integral equations was provided in Hilbert's fifth memoir [1906b] by the concept of a *complete orthogonal system of functions*, i.e., a sequence of continuous functions $\varphi_1, \varphi_2, \ldots$ on $[a,b]$ that satisfy the orthogonality relations (2.3) and in addition the following completeness relation: If $x(s)$ and $y(s)$ are any two continuous functions with respective generalized Fourier coefficients $a_n = \int_a^b x(s)\varphi_n(s)\,ds$, $b_n = \int_a^b y(s)\varphi_n(s)\,ds$, then

$$\int_a^b x(s)y(s)\,ds = \sum_{n=1}^{\infty} a_n b_n \quad \text{for all} \quad x, y \in C[a,b]. \tag{2.5}$$

Thus completeness implies that

$$\int_a^b x(s)^2 ds = \sum_{n=1}^{\infty} a_n^2 \quad \text{for all } x \in C[a,b], \tag{2.6}$$

which, as Hilbert noted, in turn implies (2.5), so (2.6) is equivalent to the completeness property.

Hilbert's mentor Hurwitz had just called attention to the fact that when a_n, b_n are the ordinary Fourier coefficients of two Riemann-integrable functions x and y, then the Parseval equality (2.5) always holds, even though the Fourier series may not converge [1901; 1902; 1903; 1904]. As Hurwitz explained in the last two papers cited above, which were published in *Mathematische Annalen*, after announcing his proof he had discovered that de la Vallée-Poussin (1893) and Liapunov (1896) had already arrived at the same result. Thus just prior to Hilbert's work on integral equations considerable attention had been focused on the general validity of Parseval's equality, i.e., upon the fact that the sine and cosine functions of classical Fourier analysis have the completeness property that Hilbert now introduced within the context of his generalized Fourier analysis.[18]

Utilizing a complete system of orthogonal functions $\varphi_1, \varphi_2, \varphi_3, \ldots$ on $[a,b]$, Hilbert showed how the study of integral equations can be deduced from what would now be regarded as the theory of linear operators on the Hilbert space ℓ^2. Consider, for example, Fredholm's equation

[18] Parseval's equality was soon extended to bounded measurable and then arbitrary Lebesgue-integrable functions by Lebesgue (1904) and Fatou (1906), respectively. See [Hawkins 1970:167–171].

$\varphi(s) + \int_a^b K(s,t)\varphi(t)\,dt = f(t)$ with nonsymmetric kernel. Let x_n and b_n denote the respective Fourier coefficients of $\varphi(s)$ and $f(s)$ with respect $\varphi_1, \varphi_2, \ldots$ and set $a_{pq} = \int_a^b \int_a^b K(s,t)\varphi_p(s)\varphi_q(t)\,ds\,dt$. In place of the integral equation one considers the analogous infinite system of equations $x_i + \sum_{j=1}^{\infty} a_{ij}x_j = b_i$, $i = 1, 2, 3, \ldots$, in which the unknowns x_i are the Fourier coefficients of the unknown function $\varphi(s)$. Since the associated form $A(x,y) = \sum_{i,j=1}^{\infty} a_{ij}x_j y_j$ is completely continuous, Hilbert was able to analyze the possibilities for the solution of the infinite system and, by means of the completeness of $\varphi_1, \varphi_2, \ldots$, to translate that into a complete analysis of the solutions to the integral equation. Likewise, the analysis of the integral equations with symmetric kernel (2.2) was reduced to the theory of completely continuous quadratic forms. Thus the notion of a complete system of orthogonal functions was fundamental to Hilbert's theory of integral equations.

It should be pointed out that the notion of a complete system has another interpretation, which Hilbert realized but apparently did not think significant. That is, if $\varphi_1, \varphi_2, \varphi_3, \ldots$ is any system of functions satisfying the orthogonality relations (2.3), then these relations imply the following identity, sometimes referred to as Bessel's identity. For any $f \in C[a,b]$ if $c_n = \int_a^b f(s)\varphi_n(s)\,ds$ denote the Fourier coefficients of f with respect to the system $\varphi_1, \varphi_2, \varphi_3, \ldots$, then

$$\int_a^b \left|f(s) - \sum_{k=1}^{n} c_k\varphi_k(s)\right|^2 ds = \int_a^b \left|f(s)\right|^2 ds - \sum_{k=1}^{n} |c_k|^2. \tag{2.7}$$

Since the left-hand side of (2.7) is nonnegative, it follows that one always has the analogue of Bessel's inequality

$$\sum_{n=1}^{\infty} |c_n|^2 \leq \int_a^b \left|f(s)\right|^2 ds \quad \text{for any} \quad f \in C[a,b]. \tag{2.8}$$

Since the system $\varphi_1, \varphi_2, \varphi_3, \ldots$ is complete precisely when Bessel's inequality (2.8) is an equality for any $f \in C[a,b]$, we see from (2.7) that completeness means that

$$\lim_{n \to \infty} \int_a^b |f(s) - \sum_{k=1}^{n} c_k\varphi_k(s)|^2 ds = 0. \tag{2.9}$$

That Hilbert realized (2.9) is clear, since (2.7) and (2.8) are contained in his paper. To us today (2.9) means that when the system of eigenfunctions $\varphi_1, \varphi_2, \varphi_3, \ldots$ is complete, every $f \in L^2[a,b]$ is represented, in the sense of L^2 convergence, by its Fourier series with respect to $\varphi_1, \varphi_2, \varphi_3, \ldots$. But the fact of convergence in the sense of (2.9) does not seem to have appeared significant to Hilbert, even after F. Riesz and Ernst Fischer independently introduced the space $L^2[a,b]$ of Lebesgue measurable functions f such that $\int_a^b |f(s)|\,ds < \infty$ shortly after Hilbert's paper [1906b] appeared and showed in effect that $L^2[a,b]$ is isomorphic to ℓ^2. Hilbert was primarily interested in the question of pointwise convergence. Of course, it follows from (2.7) that if for all $f \in C[a,b]$ one has $f(s) = \sum_{n=1}^{\infty} c_n\varphi_n(s)$ in the sense of uniform convergence, then the $\varphi_1, \varphi_2, \varphi_3, \ldots$ satisfy (2.6) and so form a complete system. The completeness of $\varphi_1, \varphi_2, \varphi_3, \ldots$ is thus a necessary condition that the Fourier series of any $f \in C[a,b]$ converge uniformly to it, and this may have been one factor behind Hilbert's interest in the question of when a system of eigenfunctions $\varphi_1, \varphi_2, \varphi_3, \ldots$ satisfying the orthogonality relations (2.3) is complete. Hilbert regarded this as an "important question" [1906b:193] and proved in passing that $\varphi_1, \varphi_2, \varphi_3, \ldots$ is complete if the associated kernel $K(s,t)$ is what he called *general*: given any $g \in C[a,b]$ and any $\varepsilon > 0$ there is an $h \in C[a,b]$ such that

$$\int_a^b \left[g(s) - \int_a^b K(s,t)h(t)\,dt\right]^2 ds < \varepsilon.$$

In other words, the range of the operator A defined by $K(s,t)$ is dense in $C[a,b]$ in the sense of the L^2 norm. Indeed, since each of the eigenfunctions φ_k is in the range A, the system of eigenfunctions will be complete if and only if the kernel is general. Earlier, in his second paper on integral equations [1904b:47], Hilbert had showed that the symmetric kernels that arise from Sturm–Liouville equations $p(t)u'' + p'(t)u' + q(t) = 0$ with boundary conditions are always general, so that the corresponding system of eigenfunctions is always complete.

Hilbert's interest in the question of when eigenfunctions $\varphi_1, \varphi_2, \varphi_3, \ldots$ form a complete set is of special relevance to our story because as we shall see in Section 12.7 Weyl observed that the coefficient functions in the irreducible representations of a semisimple Lie group may be taken as eigenfunctions. Together with F. Peter he proved that they form a complete set and went on to obtain a striking result on the question of uniform pointwise approximation as well.

Except for one further application of integral equations that Hilbert published in 1912 (see Section 4), his work on integral equations was completed by 1910. By that time he had made profound contributions to all the major branches of mathematics—algebra, arithmetic, geometry, and analysis. It was thus appropriate that in 1910 he was awarded the second Bolyai prize by the Hungarian Academy of Science. The lucrative prize was awarded to the mathematician whose contributions over the past twenty-five years had most greatly contributed to the progress of mathematics. Poincaré, who had received the first prize in 1905, was a member of the prize committee and wrote the report on Hilbert's work. As Reid has written, "Poincaré's report for the Bolyai Prize appeared in *Acta Mathematica* in 1911. Unsuspected by anyone at the time, he summed up what was to be the totality of Hilbert's contribution to constructive mathematics. The following year Hilbert—now fifty years old—became, as far as fellow mathematicians were concerned, a physicist" [1970:126].

Although, as we shall see, 1912 did mark the beginning of a period of heightened engagement with the physics of his day, Hilbert's active involvement with current physics goes back to 1905 and is connected with the activities of his close friend Hermann Minkowski (1864–1909).[19] What happened in 1905 and thereafter at Göttingen is remarkable from the standpoint of the history of mathematics. The extent to which Göttingen mathematicians participated in, and influenced, current developments in theoretical physics was without precedent, and it greatly influenced both the course of physics and that of mathematics. In this book I will be concerned with one highly important example of how the course of mathematics was affected. As we shall see in Chapter 11, by following in the footsteps of his Göttingen teachers and seeking to understand and contribute to the theory of relativity, Weyl was led to the mathematical considerations that eventuated in his work on the representation of Lie groups. The following section is devoted to depicting the manner in which theoretical physics, and especially the theory of relativity, became a subject of serious interest to the mathematicians of Göttingen. As we shall see, Minkowski was to play a key role.

9.3. The Mathematization of Physics at Göttingen

Hilbert and Minkowski had both grown up in Königsberg, although it was not until Hilbert was a university student there in 1883 that they became friends.[20] Their friendship developed into a close one on both the personal and mathematical levels. At the beginning of their friendship, Minkowski, although two years younger than Hilbert, was ahead of him academically and had already earned fame as a mathematician by winning at age 18 (jointly with H. J. S. Smith) a prize offered by the Paris Academy of Science for work on a problem in the theory of numbers. After obtaining his doctorate from Königsberg in 1885, Minkowski moved to the University of Bonn, where he remained until 1894. It was while in Bonn that he became seriously interested in physics.

At least initially the turn to physics seems to have been due to the combined effects of inadequate mathematical colleagues at Bonn and the presence of Heinrich Hertz as director of the physics institute. Hertz was a gifted experimentalist and theoretician to whom Minkowski was drawn.[21] He had gained fame in the 1880s by experimentally confirming Maxwell's electromagnetic theory of light, and in 1890 he produced a brilliant reformulation of the theory itself. Hertz's influence on Minkowski can be seen in

---

19 As Corry has shown [1997a], Hilbert's interest in the foundations of mechanics goes back even further and seems to be contemporaneous with his interest in the foundations of geometry.

20 For further details, see [Reid 1970:Ch I].

21 This according to Hilbert [1909a:355].

his letter to Hilbert of 22 December 1890. After indicating that he would not be returning as usual to Königsberg for the vacation period, he wrote:[22]

> I do not know if I need to console you though, since I think that this time you would have found me thoroughly infected with physics. Perhaps I even would have had to pass through a ten-day quarantine period before you and Hurwitz would have admitted me again, mathematically pure and unapplied, to your joint walks .... In order to have points in common with other mortals, I have surrendered myself to magic—that is to say, physics. I have my laboratory periods at the Physics Institute; at home I study Thompson, Helmholtz, and their consorts. And, starting at the end of next week, I will even work several days a week in a blue smock in an institute for the production of physical instruments, thus a practical training as shameful as you can imagine.

Despite the jesting nature of these remarks, they reflect the fact that by 1890 mathematics had been professionalized to the point where university mathematicians rarely ventured outside the ever expanding domain of pure mathematics. By then most mathematicians were no longer active participants in the science of their day as Euler, Lagrange, and Gauss had been in their time—Poincaré being a notable exception to the general trend. Klein at Göttingen was an advocate of close bonds between mathematicians and science, but he himself had spent his career as a pure mathematician. So Minkowski's scientific activities were rather remarkable for the period. Later, he told Hilbert that they might have led to more than they did at the time had not Hertz died of blood poisoning at the beginning of 1894, when he was 36. Minkowski's interest in physics survived Hertz's death, however, as his letters to Hilbert and his lectures show.[23]

While Minkowski was surrendering himself to the magic of physics, Hilbert was, as we have seen, deeply involved with pure mathematics—the theory of invariants, arithmetic, and the foundations of geometry. But Hilbert envisioned doing for some other domains of science what he had done for geometry. In his Paris lecture on mathematical problems [1900], he posed as the sixth problem: "The mathematical treatment of the axioms of physics." The recent work on the axiomatic foundations of geometry, he explained, suggested doing the same for "those physical sciences in which mathematics plays an important part," and he saw mechanics as one of the prime candidates for such a treatment, along with probability theory and the kinetic theory of gases. Noting that physicists such as Mach, Hertz, and Boltzmann had already devoted treatises to the foundations of mechanics, he felt it "very desirable that the discussion of the foundations of mechanics be taken up by mathematicians also." In keeping with this interest in

---

[22] [Rüdenberg and Zassenhaus 1973:39–40]. Except for a few changes, I have followed Reid's translation [1970:35].

[23] See [Pyenson 1979:65].

Figure 29. Minkowski as he looked during his years with Hilbert at Göttingen.

mechanics, during 1898–1907, while publishing research in mathematics, Hilbert presented courses of lectures and seminars on various topics in mechanics (elementary mechanics, potential theory, continuum mechanics); and in 1905, in a course on the logical principles of mathematical thinking, he followed up on his Paris lecture by discussing the axiomatic foundations of mechanics, thermodynamics, the calculus of probabilities, the kinetic theory of gases, electrodynamics, and even psychophysics.[24] The inclusion of electrodynamics in the 1905 lectures was probably a consequence of Minkowski's presence in Göttingen. In any case, Minkowski introduced a new dimension into the mathematization of physics at Göttingen.

Through the efforts of Hilbert and Klein, Minkowski had come to Göttingen in the fall of 1902 as the first occupant of a newly created third

[24] Corry [1997a] discusses these lectures in considerable detail; see also his broader study [1996], which contains a discussion (and list) of Hilbert's lectures on physical topics. In [1997b] he argues that Hilbert's program of axiomatizing physics influenced Minkowski's thinking about special relativity.

professorship in mathematics. Now, instead of running seminars jointly with Klein, Hilbert ran them with his friend Minkowski. In 1904 Minkowski had the idea of a seminar on the theory of the electron, which was run during the summer term of 1905. The choice of topic was a good one because it was a subject currently under considerable theoretical and experimental investigation and because many contributions had been made at Göttingen. By 1900 many physicists were toying with the possibility of a unified physics based upon an "electromagnetic world picture," according to which all the phenomena of physics, including those customarily treated by Newtonian mechanics, were to be accounted for by means of electrodynamics. This was a rival to the "mechanical world picture" advocated by Boltzmann, and in his lectures while a professor at the ETH in Zürich, Minkowski had already expressed an interest in whether, in fact, electrical phenomena could be explained mechanically in accordance with Bolzmann's view.[25]

Central to the electrodynamical world picture was a theory of the electron, whose existence and properties were the subject of experimental investigation (some of which was carried out by W. Kaufmann while at Göttingen). The Dutch theoretical physicist H. A. Lorentz had devised the most ambitious theory of the electron, although aspects of it had been criticized by Poincaré and a rival theory had been put forth by Max Abraham, a student of Planck's who was then (in 1905) at Göttingen. The syllabus for the seminar involved heavy doses of material by Lorentz as well as papers by Poincaré and Abraham. The leaders of the seminar were, in addition to Hilbert and Minkowski, Emil Wiechert, director of the geophysical institute at Göttingen, and Gustav Herglotz, an instructor (*Privatdozent*) at Göttingen who was writing a *Habilitationsschrift* on the three-body problem under Klein's direction. Both Wiechert and Herglotz had written papers on electron theory, which were also studied by the seminar participants. Among the participants in the seminar were Max Born and Max von Laue, who went on to distinguished careers as physicists.[26]

Minkowski approached physics with an optimistic belief in the power of mathematics to resolve the riddles of physics. In particular, he maintained the view, first set forth by Leibniz, of a preestablished harmony between mathematics and the physical world. As he put it in his introductory lecture on differential equations a year before the electron theory seminar:[27]

> Mathematics has the task of developing the tools necessary to grasp the logical coherence of external appearances. Its basic concepts, the axioms of physical quantities and of geometry, have arisen from experience. Mathematics constantly derives the most beautiful problems in applications from the natural sciences. And through a peculiar, preestablished harmony, it has been shown

[25] See [Pyenson 1979:65].
[26] Pyenson's paper [1979] contains considerable information on the participants in the seminar and on the syllabus it followed.
[27] Quoted by Pyenson [1979:66].

> that, by trying logically to elaborate the existing edifice of mathematics, one is directed on exactly the same path as though by questions arising from the facts of physics and astronomy.

In his Paris lecture [1900:440] Hilbert had also expressed a belief in the appearance of preestablished harmony, a belief that he reaffirmed throughout his career. In the lecture he saw such a harmony as resulting from the constant interplay between thought and experience:

> Surely the first and oldest problems in every branch of mathematics spring from experience and are suggested by the world of external phenomena .... But in the further development of a branch of mathematics, the human mind, encouraged by the success of its solutions, becomes conscious of its independence. It evolves from itself alone, often without appreciable influence from without ... indeed almost all nicer questions of modern arithmetic and function theory arise in this way. In the meantime, while the creative power of pure reason is at work, the outer world again comes into play, forces upon us new questions from actual experience, opens up new branches of mathematics, and while we seek to conquer these new fields of knowledge for the realm of pure thought, we often find the answers to old unsolved problems and thus at the same time advance most successfully the old theories. And it seems to me that the numerous and surprising analogies and that apparently prearranged harmony that the mathematician so often perceives in the questions, methods, and ideas of the various branches of this science have their origin in this ever-recurring interplay between thought and experience.

In this passage Hilbert discussed the phenomenon of preestablished harmony within various branches of mathematics and without explicit inclusion of theoretical physics, although the same justification could be given for expecting it to surface there as well, namely the "ever-recurring interplay between thought and experience." In any case, a few years after the electron theory seminar, Minkowski was to point out a spectacular instance of preestablished harmony between pure mathematics and a part of physics—the theory of the electron. Two papers on electron theory that appeared after the 1905 seminar played an important role in stimulating Minkowski's vision.

One was Poincaré's paper on the dynamics of the electron [1906] (summarized in [1905]). From Lorentz's work Poincaré seized upon and perfected the "Lorentz transformation" (as he called it)

$$(3.1) \qquad x' = \frac{x - vt}{\sqrt{1 - v^2}}, \quad y' = y, \quad z' = z, \quad t' = \frac{t - v}{\sqrt{1 - v^2}}, \quad |v| < 1,$$

which Lorentz had introduced as a purely mathematical variable change of no physical significance. The position and time variables x, y, z, t referred to a system at rest in the ether (the luminiferous medium); the units are chosen so that the speed of light c is equal to 1. The transformations $T_v : (x, y, z, t) \to (x', y', z', t')$ are linear, and as Poincaré was the first to observe, they form a group, since $T_u T_v = T_w$ where $w = (u + v)/(1 + uv)$. Of course, one can consider the analogous transformations that arise when the y- or z-axis plays the role of the x-axis in (3.1). Each of these separately defines a one-parameter transformation group in the sense of Lie, and when composed with each other they generate a larger transformation group. By considering the infinitesimal transformations X_x, X_y, X_z corresponding to these possibilities, he showed that X_x, X_y, X_z together with the infinitesimal transformations $[X_x, X_y], [X_x, X_z], [X_y, X_z]$ generate a six-parameter Lie group of linear transformations that leave the quadratic form

$$x^2 + y^2 + z^s - t^2 \tag{3.2}$$

invariant. Poincaré called it the Lorentz group.[28] As we have seen, Poincaré had always been a believer in the importance of groups in mathematics and especially the continuous groups of Lie, and so it must have pleased him to see a group hidden away in Lorentz's calculations. Following the lead of Lorentz, Poincaré saw the transformations of the Lorentz group as important because the Maxwell–Lorentz field equations are covariant with respect to them, i.e., the form of the equations is the same in the primed and unprimed coordinate systems.

In the final section of his memoir Poincaré turned to the question of gravitational forces. Lorentz had hypothesized in effect that all force fields, including in particular the gravitational field, transform under the Lorentz transformations as does the electromagnetic force field, and Poincaré set out to explore mathematically the implications of this hypothesis. To find a suitable replacement for the Newtonian inverse square law, Poincaré proposed to look among the invariants of the Lorentz group, and so he had need to determine them.[29] To do so he observed that if x, y, z, it are regarded as the coordinates of a point in four-dimensional space, then transformations of the Lorentz group are simply rotations in this space, since they leave the sum of the squares of the coordinates, viz.,

$$x^2 + y^2 + z^2 + (it)^2 = x^2 + y^2 + z^2 - t^2,$$

[28] Today the Lorentz group is usually defined to consist of all linear transformations leaving (3.2) invariant. This group is disconnected and consists of four components; the component containing the identity (usually called the proper Lorentz group) is what Lie would have regarded as a transformation group (Section 3.2), and this may have been what Poincaré also had in mind. We know that this is what Klein had in mind by the Lorentz group [1910:548]. (The full Lorentz group eventually became of interest to physicists by virtue of time reversal $(x, y, z, t) \to (x, y, z, -t)$ and parity $(x, y, z, t) \to (-x, -y, -z, t)$.) Actually, Lorentz and Poincaré initially considered the more general transformations lT_v, where l is some constant, but then argued that l must equal 1. Poincaré's argument that $l = 1$ involved group-related considerations [1906:513-515].

[29] On this point see Miller's discussion [1981:80–82].

invariant. Thus invariants of the Lorentz group may be regarded as rotational invariants and so determined. Poincaré's observation that the Lorentz group could be considered as the group of rotations of a four-dimensional space was simply a convenient momentary viewpoint for facilitating computation of the desired invariants, whereas for Minkowski, as we shall see, it was the key to a radically different view of space and time, which seemed to confirm the preestablished harmony of mathematics and physics.

Undoubtedly, Einstein's paper [1905] on what became known as the special theory of relativity helped encourage Minkowski to develop the ideas for which Poincaré's observation was the key. Starting from a profoundly simple analysis of the physical meaning of time, Einstein deduced the Lorentz transformations (3.1) and the associated group property as a consequence of two principles. The first, called the principle of relativity, boldly declared that "[t]he laws by which the states of physical systems undergo change are not affected, whether these changes of state be referred to the one or the other of two systems of coordinates in uniform translatory motion," and the second declared the constancy of the speed of light. Since (as Einstein showed) the Lorentz transformations give the coordinate relations between two reference frames in uniform translatory motion, it is a simple mathematical theorem that the electromagnetic field equations are covariant with respect to the Lorentz group and so satisfy the principle of relativity. Minkowski called this the "theorem of relativity" to distinguish it from the the bold physical hypothesis that covariance is required of all physical theories, past and future, which is Einstein's principle of relativity.

Minkowski first publicly set forth his views in a talk on "The Principle of Relativity" presented to the Göttingen Mathematical Society on 5 November 5 1907. The opening passage is worth quoting since, it conveys the self-confidence with which Minkowski and then Hilbert approached, as mathematicians, the physics of their day as a result of the above-described developments in physics:[30]

> Out of the electromagnetic theory of light there recently seems to have come a complete transformation of our representations of space and time which must be of exceptional interest to the mathematician. The mathematician is also especially well prepared to pick up the new views because it is a question of acclimatization to conceptual schemes with which he has long been familiar. The physicist meanwhile must rediscover these concepts and must painfully cut his way through a primeval forest of obscurities. However, close by the old excellently laid out path of the mathematician comfortably leads forward. Above all, the new formulation would be, if in fact it mirrors the right phenomena, practically the greatest triumph applied mathematics has ever shown. Expressed

[30] [Minkowski 1907a:927–928]. I have used the translation by Galison [1976:83–84].

> as briefly as possible, it is this—the world in space and time is in a certain sense a four-dimensional, non-Euclidean manifold .... It would be revealed, to the fame of the mathematician and to the boundless astonishment of mankind, that the mathematician, purely in his imagination, had created a large field to which one day the fullest real existence should be ascribed.

To see what Minkowski had in mind, consider (as he did) a mass moving with a speed v with respect to some coordinate system x, y, z. Let v_x, v_y, v_z denote the components of the velocity with respect to these coordinate axes. In ordinary mechanics one considers the 3-dimensional velocity vector $\mathbf{v} = (v_x, v_y, v_z)$, but in the new relativistic mechanics the 4-dimensional vector $\mathbf{w} = (w_1, \ldots, w_4)$ needs to be considered, where $w_1 = v_x, w_2 = v_y, w_3 = v_z$ and $w_4 = i\sqrt{1-v^2}$. (Recall that the speed of light is taken as $c = 1$ and so $|v| < 1$ in accordance with Einstein's theory.) Thus one always has

$$(3.3) \qquad w_1^2 + w_2^2 + w_3^2 + w_4^2 = -1,$$

and the velocity vectors $\mathbf{w}$ always lie on the surface (3.3), which Minkowski called a four-dimensional hyperboloid, since, expressed in terms of the purely real coordinates $(y_1, \ldots, y_4) = (w_1, \ldots, w_3, \sqrt{1-v^2})$, its equation is $y_1^2 + y_2^2 + y_3^2 - y_4^2 = -1$, analogous to the hyperboloid $y_1^2 + y_2^2 - y_3^2 = -1$ of three-dimensional space.

After introducing the velocity vector $\mathbf{w} = (w_1, \ldots, w_4)$, Minkowski explained, "The non-Euclidean geometry of which I have already vaguely spoken is developed now for these velocity vectors. The transformations about which I have already spoken become the real transformations of this four-dimensional hyperboloid ..." [1907a:932]. It is tempting to add to Minkowski's words the remark that in other words, the geometry of these vectors is the geometry in the sense of Klein's *Erlanger Programm* (Section 1.5) defined on the manifold (3.3) by the group of Lorentz transformations. But Minkowski nowhere refers explicitly to Klein's work, so it is unclear whether he had it specifically in mind. By 1908, the idea of geometries, including especially non-Euclidean geometries, as the study of groups of "motions" was fairly commonplace, going back to the Helmholtz space problem. (Section 4.5) In fact, Poincaré had contributed to the solution of this problem by utilizing Lie's theory in [1887] and had done much to popularize the idea of geometry as the study of groups of motions.[31] Here, then, was the "old, excellently laid out path of the mathematician," which would lead, by means of the machinery of groups and their invariants as perceived within the conceptual framework of higher-dimensional non-Euclidean geometry, to the development of the new theory of relativity. Minkowski ended his talk with the "hope that my report has made one point clear: that the investigations dealt with here will be highly interesting to the mathematician."

[31] See in this connection my study [1984].

Minkowski's point was that the new relativistic physics was so enmeshed with nineteenth-century mathematics that it was the mathematician who was perhaps best equipped by training to develop it! At the least, the mathematician could contribute mathematical insights that might go unnoticed by physicists. This attitude is brought out in the talk by Minkowski's discussion of Planck's work (1907) applying Einstein's relativity principle to heat radiation. "The result at which Planck arrived by a detour through an approach using electrodynamic considerations," Minkowski explained [1907a:937], "would have been very easy to predict. Namely, simply set [the kinetic potential] $Hdt = K(dt^2 - dy^2 - dz^2)^{\frac{1}{2}}$. The second factor is invariant under the Lorentz group, and Planck's result is simply that the first factor should also be invariant." What Planck achieved by a "detour" through physics, Minkowski seems to be saying, could have been anticipated much more easily by a mathematician.

Minkowski followed up his talk with a more technical paper on "The Basic Equations for Electromagnetic Processes in Moving Bodies" [1907b], which contains the first relativistically correct presentation of Maxwell's equations in a ponderable medium.[32] Then in September of 1908, at a scientific congress in Cologne he delivered his now famous lecture on "Space and Time" [1908], which contains the oft-quoted lines, "The views on space and time that I wish to lay before you ... are radical. Henceforth space by itself, and time by itself, are doomed to fade away into mere shadows, and only a kind of union of the two will preserve an independent reality .... Three-dimensional geometry becomes a chapter in four-dimensional physics" [1908:75,80]. Of course, for Minkowski, the new four-dimensional physics was essentially mathematics. In his lectures of 1907 on complex function theory he spoke of the new developments in relativistic physics as a "triumph of mathematics,"[33] and he concluded his "Space and Time" lecture by confidently suggesting that future experiments suggested by the new four-dimensional physics would uphold "the idea of a preestablished harmony between pure mathematics and physics."

Minkowski did not live to further develop his ideas on space and time. A few months after the "Space and Time" lecture, on 12 January 12 1909, he died, a victim of appendicitis.[34] But his belief in the preestablished harmony between mathematics and physics, his confidence that mathematicians were capable of, and well trained for, contributing to the theoretical physics of their day, lived on after him at Göttingen.[35] In particular, Klein stressed the connections between Minkowski's conception of space-time and his own

---

[32] [Galison 1979:90]. Minkowski's attempts to deal with gravitation were less successful [Pyenson 1977:80–84].

[33] Pyenson [1977:90–95] has published a transcription of Minkowski's lecture; see p. 8 of the lecture for the above quotation.

[34] See [Reid 1970:114–115] for further details.

[35] Pyenson [1982] has studied the appeal to a preestablished harmony within the early development of relativity theory.

characterization, in the early 1870s, of non-Euclidean geometries in terms of Cayley's projective metric (Section 1.5). Thus he began a talk before the Göttingen Mathematical Society in May 1910 by saying, "In some form or other you have all heard that the theory of the Lorentz group, or, what is the same thing, the modern principle of relativity of the physicists, falls within the *general theory of the projective metric* as it developed in connection with Cayley's groundbreaking work of 1859" [1910:533]. Much in the spirit of Minkowski's opening remarks in his talk "The Principle of Relativity," Klein continued, "The theory of the projective metric ... gains here a new and surprising application, while on the other hand the modern developments of the physicists, which so easily strike the novice as paradoxical, prove to be, so to speak, a corollary to a general line of thought that was long ago set in order. This confluence of two entirely separated spheres of thought ... cannot fail to be mutually stimulating to a high degree" [1910:533]. Although Klein did not expressly refer to any preestablished harmony, he did express the conviction that "the history of science shows that ... everything that is mathematically healthy, sooner or later will achieve a significance reaching far beyond its own narrow domain" [1910:541].

Acknowledging the presence in the audience of many physicists and young mathematicians who were not familiar with the projective way of thinking, Klein proceeded to give a lucid and elegant sketch of the basic ideas involved. From the projective point of view, Klein showed, plane Euclidean geometry and the spherical non-Euclidean geometries of Riemann and Lobachevsky all correspond to subgroups of the eight-dimensional general projective group, $\mathbf{PGL}(3,\mathbb{C})$. The fact that these subgroups "have little to do with one another in terms of their inner structure" reflected the fact that "it is so difficult to understand non-Euclidean geometry from the standpoint of Euclidean geometry .... But all difficulties vanish as soon as I familiarize myself with the general projective line of thought" [1910:542]. For then all these geometries can be understood within the all-encompassing framework of projective geometry, the geometry of the group $\mathbf{PGL}(3,\mathbb{C})$. Klein likened the projective viewpoint to that attained by a hiker standing atop a mountain; he can look into different valleys at the same time, whereas earlier, standing in a particular valley, it was difficult to imagine the characteristics of other valleys. He felt the same to be true regarding the new mechanics of the theory of relativity. That is, both the metric geometry of the Lorentz group and that of the group of "Galiliean transformations" of Newtonian mechanics could, as he showed, be derived in a similar fashion from four-dimensional projective geometry and so could be viewed simultaneously from the higher perspective of projective geometry. In fact, both the old and new mechanics could, as he showed, be considered within the somewhat more limited framework of four-dimensional affine geometry, the geometry corresponding to the group of all affine transformations $\mathbf{y} = A\mathbf{x} + \mathbf{b}$, where $A \in \mathbf{GL}(4,\mathbb{R})$ and $\mathbf{x}, \mathbf{y}, \mathbf{b} \in \mathbb{R}^4$. Thus what was needed was "a systematic theory of invariants of the affine "world," for which, incidentally, all elements lay already at hand in the multidimensional investigations of the mathematician ..." [Klein 1910:550]. Evidently, Klein, like Minkowski, believed that a new era was dawning in which the mathematical developments of the past century, and perhaps mathematicians themselves, could make significant contributions to the new physics that had emerged—and which they embraced more readily than many physicists.

As for Hilbert, the loss of Minkowski seems to have intensified and broadened his involvement with physics. In his memorial essay on his friend, Hilbert had written that "Minkowski had always striven to master, not just the methods of pure mathematics, but also the content of all scientific domains in which mathematics is called upon to play a decisive role ..." [1909a:362]. These words apply equally well to Hilbert in the years following Minkowski's death. Between 1911 and 1915 he lectured on statistical mechanics, the theory of radiation, the molecular theory of matter, electron theory, electromagnetic oscillations, and the structure of matter. By the time of his lectures on electromagnetic oscillations (1913–14), encouraged by Minkowski's space-time electrodynamics, Hilbert was anticipating the imminent mathematization of theoretical physics in the spirit of what he had proposed in his Paris lecture of 1900:[36]

> Since antiquity geometry has been a discipline within mathematics; the necessary experimental grounds are so close at hand and generally recognized that from the outset it immediately appeared as a theoretical science. Now I believe that the highest distinction for every science is to be assimilated by mathematics and that theoretical physics is now also on the verge of achieving this distinction. This is so above all for relativistic mechanics, or four-dimensional electrodynamics; I have been convinced for some time now that it belongs to mathematics.

In 1912 Hilbert also published his first work dealing with physics.[37] The occasion was the publication of a book [1912] on integral equations in which he reprinted, with minor changes and an overview, his papers of 1904–10. What was new to the book was the final chapter on the mathematical foundations of the kinetic theory of gases. In most applications, the theory of integral equations enters by means of a differential equation, but in Hilbert's formulation of the kinetic theory an integral equation with symmetric kernel enters directly, and the attendant theory of such equations becomes an indispensable part of the kinetic theory. For this reason he regarded this application of his theory of integral equations as "the most splendid" of all the applications he had given [1912:iii].

Hilbert's growing commitment to devote himself to the mathemati-

[36] The following passage is translated from the German text as presented in [Corry 1996:47, n.104].

[37] Hilbert's increasing involvement with physics is depicted in [Corry 1999].

cal study of physics is also reflected in the fact that in 1912 he recruited his first physics assistant, Paul Ewald, who went on to become a distinguished physicist. Hilbert thus had an assistant for each of mathematics and physics to help him assimilate whatever mathematics or physics he was interested in at the moment. At the moment, however, he was mainly interested in physics, primarily the elementary theory of radiation, and so even his mathematics assistant at the time, Erich Hecke, who went on to distinguish himself in the theory of numbers, was involved with the task of teaching Hilbert physics. During 1912–14 Hilbert directed four doctoral dissertations on subjects from physics[38] and himself contributed three papers on the foundations of the theory of radiation.[39]

Hilbert also saw to it that talks on the latest developments in theoretical physics by distinguished scientists were given at Göttingen. A few months after Minkowski's death, he became chairman of the Wolfskehl Prize Committee. A mathematics professor in Darmstadt named Paul Wolfskehl left in his will a prize of 100,000 Marks for the first complete proof of Fermat's last theorem, and until the prize was claimed the accrued interest was to be used at the discretion of a committee of the Göttingen Scientific Society.[40] One of Hilbert's first acts as chairman was to invite Poincaré to Göttingen, where he lectured on integral equations and relativity theory. Later, in 1910, Lorentz came, and in 1913 Hilbert organized a conference on the kinetic theory of gases, which brought together scientists such as Planck, Debye, Nernst, Sommerfeld, and Lorentz.

There were also many talks on theoretical physics within the auspices of the Göttingen Mathematical Society.[41] Hilbert, not surprisingly, gave several, but so did others. In particular, Max Born lectured twice—November 1912 and December 1913—on the unified theory of gravitation and electrodynamics that Gustav Mie had proposed within the framework of special relativity, and F. Böhm spoke about Einstein's joint paper with the mathematician Marcel Grossmann on a general theory of relativity created to deal with gravitation [1913]. Their work utilized the mathematical apparatus (called tensor calculus by Einstein and Grossmann) that had been introduced by mathematicians such as Christoffel and Ricci to develop the invariant theory of the n-dimensional differential geometry hinted at in Riemann's essay of 1854 on the foundations of geometry (Section 4.5). Finally, in June 1915, Einstein himself lectured on gravitation. More will be said about Einstein's work in Section 11.1. Here it will suffice to point out that Einstein had still not taken the final step in the development of his theory of general relativity; he had not accepted the principle that the laws governing gravitation should be covariant with respect to all coordinate

---

[38] See the list of dissertations in Hilbert, *Gesammelte Abhandlungen* **3**, 433.

[39] See Hilbert, *Gesammelte Abhandlungen* **3**, 238–257.

[40] The following is based on Reid's account [1970:120].

[41] In an appendix Corry [1996] has compiled a list of all such talks as announced in the *Jahresbericht der Deutschen Mathematiker-Vereinigung*.

transformations and so had not yet posited the now familiar gravitational field equations.

Just as Minkowski had earlier responded to Einstein's paper on special relativity, so now Einstein's work on general relativity proved inspirational to Hilbert, who in a paper entitled simply "The Foundations of Physics" and dated 20 November 1915 set forth axiomatically a unified theory of gravitation and electromagnetism based on Mie's ideas. Five days later, without knowing of Hilbert's work, Einstein presented his derivation of the now familiar gravitational field equations based upon the assumption of general covariance. By the time Hilbert's paper was published in a revised form at the end of March 1916, he had been able to modify his field equations so that they were in agreement with Einstein's.[42] Even before these modifications had been made, however, he had become convinced that he had done more than provide the foundations for a theory "through which our conceptions of space, time, and motion are transformed in a fundamental way in the sense stipulated by Einstein." That is, he was convinced that "by means of the fundamental equations set forth here the most intimate, hitherto hidden processes within the atom will be clarified" and that "the reduction of all physical constants to mathematical ones must be possible," so that out of physics "a science like geometry in kind" would emerge.[43]

In his second communication on the foundations of physics, presented in December 1916, Hilbert expressed himself less tentatively about the envisioned newly arising geometrical physics:

> According to my formulation, physics is a four-dimensional pseudogeometry, whose metric $g_{\mu\nu}$ is linked to electromagnetic quantities, i.e., to matter, through the fundamental equations (4) and (5) of my first communication. Through this realization an old geometrical question becomes ready for resolution, namely, the question of whether, and in what sense, Euclidean geometry—about which we know from mathematics only that it is a logically consistent structure—also has validity in reality.
>
> The old physics with its notion of absolute time assumed the axioms of Euclidean geometry and made them the basis of every physical theory ....
>
> The new physics of Einstein's general principle of relativity takes an entirely different position regarding geometry. Neither Eu-

[42] The generally accepted view that Hilbert and Einstein independently discovered essentially the same gravitational field equations has recently been shown to be untenable [Corry, et al. 1997]. The revised view, on which I have based the above portrayal of events, is the result of the authors' careful analysis of the initial page proofs of Hilbert's paper.

[43] Although Hilbert's statement—on p. 405 of [1916]—that his field equations are in agreement with Einstein's is not contained in the initial page proofs, his above-quoted speculations about a physical science like geometry—on p. 407—are already there. I am grateful to Leo Corry for this information.

clidean nor any other specific geometry is posited in advance in order to deduce therefrom properly physical laws; instead, as I have shown in my first communication, the new theory of physics supplies with one blow, through one and the same Hamiltonian principle, both geometrical and physical laws ....

Euclidean geometry is *is as foreign to modern physics as action at a distance*: Moreover, by rejecting Euclidean geometry as a general assumption, the theory of relativity teaches us that physics and geometry are similar in nature and, as a single science, rest upon a common foundation.

Hilbert's words and the example he set by pursuing research on the mathematical foundations of physics thus continued the tradition initiated by Minkowski. They encouraged the view that mathematicians were admirably qualified to contribute in a fundamental way to theoretical physics. Indeed, the above passages suggest that Hilbert saw the roles of geometers and physicists as merging into one of a mutual exploration of a single science. In his Göttingen lectures on the foundations of physics during 1916–17, Hilbert developed this point by declaring that "*the physicist must become a geometer*, for otherwise he runs the risk of ceasing to be a physicist and vice versa. The separation of the sciences into professions and faculties is an anthropological one and foreign to reality. For a natural phenomenon does not ask whether it is the business of a physicist or of a mathematician."[44] The advent of general relativity was perhaps the most striking instance of the preestablished harmony between mathematics and reality to which Minkowski and Hilbert were fond of referring; it lent great weight to the conviction, already a part of the Göttingen atmosphere since Minkowski's encounter with special relativity, that (to quote Hilbert[45]) "nature is so arranged that for understanding her the deepest mathematical speculation is required."

Referring to the period surrounding the publication of Hilbert's "Foundations of Physics," Weyl later wrote, "Hopes in the Hilbert circle ran high at that time; the dream of a universal law accounting both for the structure of the cosmos as a whole, and of all the atomic nuclei, seemed near fulfillment" [1944b:172]. Weyl was very much a part of that circle and very much under the optimistic spell of Hilbert's views. As we shall see in Chapter 11, in 1918 he developed his own unified field theory, concerning which he wrote, very much in the spirit of Hilbert's above-quoted lines:[46]

---

44 Quoted by Corry [1996:49, n.115]. I have diverged slightly from his translation.

45 From a talk given in Copenhagen in March 1921, as translated by Sigurdsson [1991:70]

46 [Weyl 1918b:2]. I have followed the translation of Sigurdsson [1994:355]. As Sigurdsson points out (pp. 363-5), later Weyl became far less optimistic about the ability of mathematicians to contribute to theoretical physics on the grand scale envisioned by himself, Hilbert, and Minkowski. Ironically, Hilbert's initial reaction to Weyl's theory, as expressed in lectures [Hilbert 1920:99–101], was

> According to this theory *everything real that is in the world is a manifestation of the world metric*; the physical concepts are no different from the mathematical ones. The only difference that exists between geometry and physics is that geometry establishes in general what is contained in the nature of the metrical concepts, whereas it is the task for physics to determine the law and explore its consequences, according to which the real world is characterized among all the geometrically possible four-dimensional metric spaces.

And a year later, to emphasize that in his theory physical reality had become a pure form, Weyl put a new twist on Hilbert's vision of the fusion of geometry and physics by declaring that "geometry has not become physics; rather physics has become geometry."[47]

Although Weyl's intense involvement with general relativity during 1917–24 was certainly inspired by the appeal of Einstein's revolutionary ideas—as had been the case with Hilbert as well—it was clearly encouraged by the precedent set by Hilbert and, more generally, by the Göttingen tradition of mathematizing physics. This episode in Weyl's career is not just of interest to historians of physics or differential geometry; it is relevant as well to our story because it was within the context of his work related to relativity that Weyl was led to formulate questions that served to convince him of the importance of group representation theory, a conviction which, in conjunction with the appearance of a suggestive paper by I. Schur (Section 10.5), led to his remarkable papers on the representation of Lie groups.

9.4. Weyl's Göttingen Years: Integral Equations

As we have seen in Section 2, when Weyl arrived in Göttingen in 1903, Hilbert was immersed in the theory of integral equations. Thus although the study of Hilbert's *Zahlbericht* had made a tremendous impression upon him at the end of his first year there, it was to Hilbert's theory of integral equations that he turned to write a dissertation under Hilbert's direction. Even much later in his career, Weyl felt that Hilbert's theory, with its many applications, represented possibly the most outstanding mathematical event of his lifetime,[48] so that working in this area then must have seemed a wonderful opportunity. Weyl's dissertation, which was submitted in February 1908 and was followed by papers on related matters, was very

---

negative, although the basis for his criticisms seem to have had more to do with Weyl's views on the foundations of mathematics than with his views on physics. See in this connection [Scholz 1995:1602].

47 My translation of a passage from the third edition of Weyl's *Raum, Zeit, Materie* as quoted in German (and more extensively) by Scholz [1996:28, n.47].

48 Thus wrote Weyl in a draft of his lecture at the Princeton Bicentennial Conference (1946) [Sigurdsson 1991:36].

much in line with Hilbert's idea (discussed below in Section 6) of viewing a problem from a higher point of view. In this case the problem involved the analysis of integral equations, and the higher point of view was that provided by Hilbert's theory of bounded quadratic forms on ℓ^2. Hilbert himself had applied his ℓ^2 theory to integral equations with continuous symmetric kernels $K(s,t)$ defined for s and t in a finite closed interval (in which case the associated form is "compact"), but it remained to attempt to do the same for various types of kernels not covered by Hilbert (called singular kernels) because the interval involved is infinite or K has discontinuities. In his dissertation and early papers Weyl showed how a large class of singular kernels could be dealt with in terms of the ℓ^2 theory of Hilbert, including some results of Hellinger.[49]

Weyl's early achievements in integral equations, which in the work leading up to his habilitation in 1910 he applied to the study of ordinary differential equations,[50] established his reputation as a first-rate mathematician. He himself was more impressed by the discoveries he made in mathematical physics that were inspired by a problem posed by Lorentz in his "Wolfskehl" lectures at Göttingen in 1910 (Section 3); to him they represented his first completely independent mathematical discoveries.[51] To us, they also indicate how keen Weyl was even then to apply his mathematical expertise to solve problems of theoretical physics. In this case the problem involved an assumption at the basis of Jean's theory of radiation, which was deemed important for the theory of blackbody radiation. In a perfectly reflecting cavity standing electromagnetic waves can form, and Jean's theory was based upon the assumption that between frequencies ν and $\nu + d\nu$ the number of sufficiently high overtones is independent of the shape of the cavity and depends only on its volume. For specific simple shapes this was shown to be the case by direct consideration of the asymptotic distribution of the eigenvalues of the wave equation. In 1911, Weyl showed how to use the theory of integral equations to obtain the general result. It is interesting to note that Weyl saw this work, in true Hilbertian fashion, as evidence of the importance, in physical applications, of the great generality of the methods of Hilbert's theory. He was led to conclude that in physical applications of the theory one should not rest content with being able to apply the methods to solve specific problems down to the smallest computational detail. A far more important task was to take advantage of the generality of the methods "to find common features of a large aggregate

---

49 See papers 1 and 3 of volume 1 of Weyl's *Gesammelte Abhandlungen* and for commentary [Hellinger and Toeplitz 1927:1592] and [Dieudonné 1981:160–67; 1988].

50 See papers 6–8 of volume 1 of Weyl's *Gesammelte Abhandlungen*. In this work Weyl obtained the first results on unbounded linear operators. See [Dieudonné 1988].

51 The following draws upon the informative discussion of Weyl's work and its place in the physics of the time given by Sigurdsson [1991:42–52].

of phenomena."[52]

Almost all of Weyl's many publications during his decade at Göttingen (1904–1913) were concerned with the theory of integral equations and its applications to physical problems. Although these publications contain much of considerable mathematical merit, for the purposes at hand it will not be necessary to discuss their contents. What is important is the fact that Weyl was a master of Hilbert's theory and subsequent contributions to it by others. As we shall see in Section 12.7, his mastery of this area of mathematics was to prove useful in his work on the theory of semisimple Lie groups. To facilitate the appreciation and understanding of this aspect of Weyl's work with Lie groups, it will be helpful to amplify somewhat the discussion of integral equations in Section 2 by briefly mentioning here some of the post-Hilbert work on integral equations that is relevant to Weyl's subsequent application to group theory.[53]

Hilbert's theory was enhanced in several ways by his student Erhard Schmidt (1876–1959). Schmidt had studied with H. A. Schwarz and A. Kneser at Berlin during 1899–1901 and then moved on to Göttingen, where under Hilbert's direction he wrote a dissertation on integral equations [1905] that became something of a classic during the ensuing years. As was mentioned in Section 2, one of Schmidt's achievements in his dissertation was to give a completely general proof of the Hilbert–Schmidt theorem that — to state it in modern terms—any continuous function f in the range of a symmetric integral operator A, i.e., $f = Ag$ for some continuous function g, can be developed in a uniformly convergent Fourier-type series with respect to the normalized orthogonal eigenfunctions of A. Recall that Hilbert's proof had assumed, unnecessarily, that the kernel is general. In addition, his result had depended on utilizing the theory of the Hilbert space ℓ^2. Schmidt's proof was not only completely general but also applied directly to the integral operator.

Schmidt based his proof on another result of his, which was also accorded considerable subsequent attention and, in particular, is relevant to Weyl's later work. Hilbert had observed that his principal axes theorem for integral equations with symmetric kernels (2.4) implies, when the number of eigenvalues λ_k is finite (say n), that the kernel has the special form

$$K(s,t) = \sum_{k=1}^{n} \frac{1}{\lambda_k} \varphi_k(s)\varphi_k(t).$$

As a generalization Schmidt showed that the kernel is given by

[52] "[V]ielmehr sollen diese Methoden ... vor allem das leisten ...: *die einem großen Komplex von Erscheinungen gemeinsamen Züge ausfindig zu machen*," *Gesammelte Abhandlungen* **1**, p. 393. Weyl's results are contained in papers 13 and 16 of this volume.

[53] For contemporary accounts of what was known in this period, see [Hahn 1911; Hellinger and Toeplitz 1927; Hilb and Szász 1922; Kneser 1922]; see also the historical accounts by Bernkopf [1966] and Dieudonné [1981].

$$K(s,t) = \sum_k \frac{1}{\lambda_k}\varphi_k(s)\varphi_k(t) \tag{4.1}$$

whenever the right-hand side converges uniformly.[54] Although in some cases this series does not converge [Hellinger and Toeplitz 1927:1521], Schmidt showed that it always does for the kernel $K^{(4)}(s,t)$, which corresponds to the fourfold compositie A^4, A being the operator corresponding to K. Using the above series expansion for $K^{(4)}(s,t)$, he deduced not only the Hilbert–Schmidt theorem but Hilbert's principal axes theorem (2.4) as well. In 1909 J. Mercer showed that Schmidt's kernel expansion is actually valid for $K^{(2)}(s,t)$ and for any kernel $K(s,t)$ that is positive definite.[55] Mercer's theorem for positive definite kernels became a basic part of the theory; various alternative proofs were given in 1914 by A. Kneser, by I. Schur, and by E. W. Hobson.[56] For later reference I summarize below the salient results.

(4.2) **Kernel Expansion Theorem.** *Any symmetric kernel $K(s,t)$ is given by the expansion (4.1) whenever it converges uniformly. In particular, this always happens when $K(s,t)$ is positive definite.*

Another feature of Schmidt's dissertation was his proof of the existence of eigenfunctions for symmetric kernels. Hilbert's treatment had utilized the function $\delta(\lambda)$, which as noted in Section 2 is a power series in λ representing the analogue of the characteristic polynomial of a matrix. The approach was complicated by the possible existence of multiple roots of $\delta(\lambda) = 0$, which required a separate treatment.[57] Utilizing a technique that his former teacher, H. A. Schwarz, had used (in conjunction with the Schwarz integral inequality) in the theory of minimal surfaces, Schmidt showed how to establish the existence of eigenvalues and eigenfunctions of a symmetric kernel $K(s,t)$ without dealing with $\delta(\lambda)$.[58] Schmidt's proof was admired by Weyl, who saw how to use it in his work on the theory of groups (Section 12.7).

Having thus taken care of the above points in the theory of symmetrical kernels, Schmidt proceeded to tackle the nonsymmetric case [1905:22ff.]. The key idea of his approach, which Weyl was also to utilize in his work on groups, was to consider, corresponding to the nonsymmetric kernel $K(s,t)$,

---

[54] See Section 8 of [1905] or of [1907], which contains a reprinting of his dissertation with minor changes, as well as new material.

[55] Here "positive definite" is understood in the weak sense that $(Ax,x) = \int_a^b K(s,t)x(s)x(t)\,ds\,dt \geq 0$ for all x, which is tantamount to all the Hilbert eigenvalues being positive.

[56] See [Hellinger and Toeplitz 1927:1524] for references to the papers by Mercer et al.

[57] See pp. 14–22 and 35–8, of [1904a] for Hilbert's treatment of the cases of simple and multiple roots, respectively.

[58] For a historical and mathematical appreciation of Schwarz's paper see [Dieudonné 1981:47ff.].

Figure 30. Erhard Schmidt eventually became a professor at the University of Berlin.

the symmetric kernel $L(s,t) = \int_a^b K(s,r)K(t,r)\,dr$. In operator notation (not used by either Hilbert or Schmidt), if $Af(s) = \int_a^b K(s,t)f(t)\,dt$ and we set $A^T f(s) = \int_a^b K(t,s)f(t)\,dt$, then $L(s,t)$ is the kernel of AA^T, which is positive definite. Schmidt also introduced the symmetric kernel corresponding to $A^T A$, which has the same eigenvalues as AA^T. If λ^2 is an eigenvalue of AA^T and $A^T A$ with respective eigenfunctions φ and ψ, then λ is said to be an eigenvalue of A by virtue of the relations $\varphi = \lambda A\psi$ and $\psi = \lambda A^T\varphi$; the functions φ, ψ are called a pair of adjoined eigenfunctions for this eigenvalue. Thus Schmidt did not speak of eigenfunctions of A in the same sense as in the symmetric case, but he was able to establish results analogous to the symmetric case such as the Hilbert–Schmidt theorem: If λ_k are the eigenvalues of A with corresponding eigenfunction pairs φ_k, ψ_k and if $f = Ag$ for some continuous g, then f can be expanded in its uniformly converging Fourier series with respect to the φ_k; likewise, if $f = A^T g$, then

f can be expanded in its uniformly converging Fourier series with respect to the ψ_k.

Although in the development of Hilbert's theory of integral equations the functions involved were usually assumed to be real-valued, it was realized, for example, that the theory of symmetric kernels could be imitated for hermitian symmetric kernels, i.e., complex-valued kernels $K(s,t)$ satisfying $K(t,s) = \overline{K(s,t)}$ with orthogonality of complex-valued functions f, g defined by $\int_a^b f(s)\overline{g(s)}\,ds = 0$.[59] In fact, in Schmidt's paper [1908], which is usually cited for making the geometrical concepts of a Hilbert space underlying Hilbert's own work explicit,[60] it is the geometry of the complex Hilbert space $\ell^2(\mathbb{C})$ that is described. In this connection it should be remarked that the "Hilbert space" at the theoretical heart of Hilbert's theory of integral equations remained ℓ^2 even after the theorems of F. Riesz and Ernst Fischer showed that elements of ℓ^2, i.e., square summable sequences, correspond to Lebesgue measurable functions f on $[a,b]$ such that $\int_a^b |f(s)|^2 ds < \infty$, with the elements of the sequence being the Fourier coefficients of f with respect to a prescribed complete set of orthonormal functions. Certainly, it was realized that what is now denoted by $L^2[a,b]$ could be viewed geometrically as a vector space just as well as ℓ^2, but the study of various sorts of operators seemed more congenial in the context of ℓ^2. Of course, a desideratum of the Hilbert school was to translate results of the ℓ^2 theory into results about integral equations, as, e.g., Weyl had done in his doctoral dissertation.

9.5. Weyl's Göttingen Years: Riemann Surfaces

In 1910 Weyl became a *Privatdozent* at Göttingen,[61] which meant that he could give lectures there. During the winter semester of 1911–12 he decided to lecture on the theory of Riemann surfaces. His lectures became his classic book *Die Idee der Riemannschen Fläche* [1913], in which the modern approach to Riemann surfaces was developed for the first time. Although several factors contributed to Weyl's decision to lecture on Riemann surfaces, by far the most compelling and essential motivation came from the

[59] See, e.g., [Hilbert 1906a:162] and [Hellinger and Toeplitz 1927:1535].

[60] I note in passing that Schmidt credited Lie's student Gerhard Kowaleski (Section 7.5) with suggesting to him a geometrically informed presentation of Hilbert's theory. There is, in fact, a letter from Kowalewski to Engel dated 1 February 1907 (Engel Archive Giessen) in which Kowalewski discusses at length such a presentation of integral equations in the geometrical context of an inner product space of functions, essentially as is done today. From an earlier letter to Engel written on a visit to Göttingen we learn that Kowalewski had spent considerable time in the company of Schmidt, who was then finishing his doctoral dissertation [1905] on integral equations (letter dated 15 June 1905, Engel Archive Giessen).

[61] According to [Sigurdsson 1991:56].

significant advances that were being made within complex function theory dealing with uniformization. These advances were of special interest in Göttingen, for they resolved one of the problems posed in Hilbert's Paris lecture and provided the first adequate proof of what Klein termed "the fundamental theorem" of his and Poincaré's theory of automorphic functions. Furthermore, one of Weyl's fellow *Privatdozenten* at Göttingen, Paul Koebe, figured prominently in these advances. Weyl's own involvement with these developments not only led to his influential book, it also familiarized him with the concept of a universal covering surface, which later enabled him to see more deeply into connections between the representation theory of Lie algebras and Lie groups than any of his predecessors. As we shall see in Chapter 12, aided by these insights Weyl was able to prove the complete reducibility theorem that (unbeknownst to him) had been conjectured by Study (Section 7.2) and Cartan (Section 8.3).

The story of uniformization begins with the work of Poincaré. I have mentioned in passing on several occasions that at the beginning of his career, Poincaré had become involved in a competition with Klein involving what is now known as the theory of automorphic functions.[62] One type of automorphic function studied by Poincaré he had called "Fuchsian functions," and in a note [1881a:31] he claimed he could prove that "the coordinates of the points of any algebraic curve can be expressed by Fuchsian functions of an auxiliary variable." That is, as Poincaré explained in [1882a:101], if $f(x, y)$ is a polynomial in the complex variables x and y, then Fuchsian functions F, G defined in an open circular disk D exist such that the nonsingular points on the algebraic curve $f(x, y) = 0$ are parametrized by the equations $x = F(\zeta)$, $y = G(\zeta)$, so that $f(F(\zeta), G(\zeta)) = 0$ for all $\zeta \in D$.[63] Poincaré's assertion can also be expressed as follows. Imagine that the equation $f(x, y) = 0$ defines, say, y as a multiple-valued analytic function x, what was called at the time a *nonuniform* function of x. For example, $f(x, y) = y^2 - x = 0$ defines the two-valued square root function y. Then since $x = F(\zeta)$ and $y = G(\zeta)$ are single-valued, or uniform, functions of ζ, they may be regarded as uniformizing the original multiple-valued function. Thus $x = \zeta^2$, $y = \zeta$ uniformizes the two-valued square root function.

Klein was impressed by Poincaré's uniformization theorem (as we shall call it), and stimulated by Poincaré's achievements, he discovered a remarkable theorem of his own, which he called the "fundamental theorem" of the theory of automorphic functions [1883:698–699] and which had the uniformization theorem as a corollary.[64] Poincaré and Klein discovered that

[62] An engaging account of this competition can be found in [Gray 1986].

[63] Poincaré described $f(x, y) = 0$ as "any algebraic relation," which was usually taken to mean that $f(x, y)$ is a polynomial (frequently assumed to be irreducible when dealing with algebraic curves).

[64] For a discussion of Klein's fundamental theorem and the closely related limit circle theorem (*Grenzkreistheorem*) see [Gray 1986:297–316].

the proofs of their respective theorems involved a similar type of argument, which Poincaré called "the method of continuity" [1884:329]. The method involved reasoning of an essentially topological nature at a time when topology (*analysis situs*) was still in its infancy. Consequently, the method was fraught with difficulties. Klein made no claims of a proof but merely sketched out his intuitive ideas for such a proof. Poincaré, while suggesting that Klein's approach involved a difficulty "that cannot be overcome in a few lines" [1884:332], sought to develop his own continuity arguments more carefully by establishing lemmas that "permit us to apply the method of continuity with all rigor" [1884:368]. Eventually, in the light of advances in set theory and topology, the arguments of both Klein and Poincaré were seen to be inadequate.[65]

An important factor in triggering renewed interest in these matters was Hilbert's Paris lecture of 1900, in which he focused attention on uniformization theorems. Although Hilbert may have privately doubted the adequacy of the proofs of Poincaré's uniformization theorem and Klein's fundamental theorem, he focused his critical attention upon a far more general uniformization theorem that Poincaré had set forth in a paper of 1883. There, without any explicit reference to his uniformization theorem for algebraic curves, he proposed to demonstrate the following theorem: "Let y be any nonuniform analytic function of x. A variable z can always be found such that x and y are uniform functions of z" [1883c:57]. To followers of Riemann it was clear that by considering the Riemann surface of the nonuniform function, x and y could be regarded as uniform functions of a variable tracing out this surface, but Poincaré's variable z traces out a portion of the complex plane and thus makes x and y uniform functions of a complex variable.

Poincaré was not very well versed in Riemannian principles and had developed his own ways of looking at Riemann surfaces [Gray 1986:299]. In particular, the demonstration of his theorem is based upon his own construction of a simply connected Riemann surface for several multivalued analytic functions. The construction involved the now familiar idea behind the construction of a universal covering space and is worth quoting in its entirety [1883c:58–59]:[66]

> We consider m analytic functions of x,
>
> (1) $$y_1, y_2, \ldots, y_m,$$
>
> which are in general not uniform. These functions will be completely defined when not only the value of x is known but also the path, starting from an initial point O, by which the variable has attained this value.

---

[65] A detailed comparative analysis of Klein's and Poincaré's use of the method of continuity is given by Scholz [1980:205–216].

[66] For a lucid elementary account of topological covering spaces and the construction of the universal covering space, see [Singer and Thorpe 1967:Ch.3].

> We do not consider the variable x as moving on a plane but on a Riemann surface S. This surface will be formed of superimposed plane sheets as in the Riemann surfaces by means of which algebraic functions are studied: Only here the number of sheets will be infinite.
>
> We trace in the plane an arbitrary closed contour C that begins and ends at the same point x. The surface will be completely defined if we state the conditions under which the initial and final point of this contour must be regarded as belonging to the same sheet or to different sheets.
>
> Now there are two sorts of contours C:
>
> 1° Those that are such that at least one of the m functions y does not return to its initial value when the variable describes the contour C;
>
> 2° Those that are such that the m functions y return to their initial values when the variable x describes the contour C.
>
> Among the contours of the second sort, I will distinguish two species:
>
> 1° C will be of the first species if by deforming this contour in a continuous manner one can pass to an infinitesimal contour such that the contour never ceases to be of the second sort.
>
> 2° C will be of the second species in the contrary case.
>
> The initial and the final point of C will belong to different sheets if this contour is of the first sort or of the second species of the second sort. They belong to the same sheet if C is of the first species of the second sort ....
>
> The Riemann surface is then defined completely. It is simply connected and does not differ, from the viewpoint of the geometry of position, from the surface of a circle, from a spherical cap, or from one sheet of a hyperboloid of two sheets.

Although the above description of S is rather vague by present-day standards, in view of the first paragraph of the quotation it seems safe to say that Poincaré's remarks suggest the idea of conceiving of the points $\tilde{x}$ of the surface S that lie over the point x in the complex plane as corresponding to paths from O to x that lie in the domain D of analyticity of the functions $y_1, \ldots, y_m$, where two such paths α and β are regarded as determining the same point $\tilde{x}$ of S if the closed curve $-\alpha + \beta$ (α traced in reverse direction from x to O followed by β from O to x) is of the first species of the second sort. Furthermore, if the functions $y_1, \ldots, y_m$ are all assumed to be uniform and so irrelevant to the construction of S, then two paths α and β from

O to x would determine the same point $\tilde{x}$ over x precisely when $-\alpha + \beta$ can be continuously deformed within D to a point. In this case S would be the universal covering space of D. As we shall see, these ideas were set forth explicitly by Poincaré in 1907 when he returned to his general uniformization theorem, spurred on by Hilbert's Paris lecture, which called attention to an unsatisfactory quirk in his demonstration and resultant theorem.

Before describing the quirk, it is necessary to say something about Poincaré's overall proof idea, which was a good one: to establish the existence of what is now called a Green function on the surface S, and then using the simple connectivity of S, to deduce the existence of a one-to-one conformal mapping from S into the unit disk $|z| < 1$. This idea is still central to modern proofs of uniformization.[67] If $\tilde{x} = \Phi(z)$ denotes the inverse of this mapping and $\pi(\tilde{x}) = x$ is the projection mapping from S to D, then, since each y_i is a single-valued function on S, say $y_i = f_i(\tilde{x})$, the desired uniformization of $x, y_1, \ldots, y_m$ is given by $x = \pi(\Phi(z))$, $y_i = f_i(\Phi(z))$.

The quirk arises in the following manner. In order to construct the Green function, Poincaré used a composite of linear fractional transformations and an elliptic modular function to define a function $\psi(\zeta)$ satisfying $|\psi(\zeta)| < 1$ that is analytic, except at three points a, b, c corresponding to the points $0, 1, \infty$, where the elliptic modular function is not analytic. Then by taking $y_m = \psi$ in the construction of the Riemann surface S, he could assume that ψ was defined on S and consider the function $t = \log|1/\psi|$ on S, which is then used to construct a Green function on S. However, because ψ enters into the construction of S, the points a, b, c are excluded from the common domain of analyticity of $y_1, \ldots, y_{m-1}, \psi$. Thus, for example, to uniformize a single function y of x, as is asserted in Poincaré's theorem, one would consider the Riemann surface S for the two functions y, ψ that excludes the points covering a, b, c even though a, b, c have nothing to do with the function y and may well be points at which y is analytic. Thus in the uniformization $x = \pi(\Phi(z))$, $y = f(\Phi(z))$ some perfectly good values of x and y may be excluded due to the introduction of the function ψ. Poincaré himself called attention to this artificial restriction by remarking in a note at the end of his paper that the points a, b, c "being singular points, are outside the Riemann surface."

Although Poincaré's paper [1883c] appears to have been fairly well known,[68] it was not until the turn of the century that attention was focused on its defects and on the desirability of remedying them. It may have been the American W. F. Osgood (1864–1943) who initiated the critical

[67] See in this connection the expository account by Abikoff [1981], especially Section 5. A more detailed description of Poincaré's own construction of the Green function and the associated conformal mapping is given by Gray [1994:§2].

[68] For example, Harnack referred to it in his well-known book on potential theory [1887:121n], and Hurwitz singled it out in his talk on recent developments in complex function theory at the first International Congress of Mathematicians (Zürich, 1897) [1898:469].

interest in Poincaré's theorem. In a series of lectures on complex function theory held at a colloquium in Cambridge, Massachusetts, and published in the *Bulletin* of the American Mathematical Society in November 1898, Osgood devoted two of his six lectures to the subject of Poincaré's general uniformization theorem [1898:69–74]. Noting Poincaré's "novel manner" of constructing a Riemann surface, Osgood sketched the proof and emphasized the quirk involving the points a, b, c, noting that it could be avoided if the Green function could be constructed without the aid of the function ψ. But he was more critical of the fact that more generally, Poincaré's theorem said nothing about the mapping of the boundary points of S such as singular points of the given nonuniform function. An "ideal theorem," he said, would deal with these matters. Osgood also pointed out that "nothing in Poincaré's analysis throws any light" on whether the mapping from S to $|z| < 1$ "fills the whole circle" [1898:73], but in a paper written in June 1900 while crossing the Atlantic [1900] he showed how S could be mapped onto $|z| < 1$.[69]

Whether or not Hilbert was aware of Osgood's 1898 lectures on Poincaré's general uniformization theorem is uncertain. In any case, on 8 August 1900 in his lecture on mathematical problems, Hilbert posed, as one of the ten problems included in his talk, the problem of uniformization, noting that Poincaré's 1883 theorem on the matter was subject to limitations.[70] In the published version of his talk the problem is the twenty-second and reads as follows:[71]

> As Poincaré was the first to prove, it is always possible to reduce any *algebraic* relation between two variables to uniformity by the use of automorphic functions of one variable .... The generalization of this fundamental theorem to arbitrary nonalgebraic analytic relations between two variables has likewise been successfully taken on by Poincaré, though by a way entirely different from that which served him in the special problem first mentioned. From Poincaré's proof of the possibility of reducing to uniformity any arbitrary analytic relation between two variables, however, it does not follow that it is possible to choose the singlevalued functions of the new variable so that, as this variable traverses the *regular* domain of those functions, the totality of *all* regular points of the given analytic configuration are actually represented. On the contrary, in Poincaré's investigation, it seems to be the case that, besides the branch points, there are certain other discrete points, in general infinite in number, that can be reached only by making

69 In [1900] Osgood gave a rigorous proof of the Riemann mapping theorem for the plane, which implies that if T is the image of S under Poincaré's map, then T may be mapped conformally onto $|z| < 1$.

70 According to the account of the Paris congress by Scott [1901:68].

71 [Hilbert 1900:323]. I have followed with some modifications the translation by Newson listed in the bibliography.

> the new variable approach certain limiting points of the functions. *In view of the fundamental importance of Poincaré's formulation of the question, it seems to me that an elucidation and resolution of this difficulty is extremely desirable.*

The "certain other" exceptional points mentioned by Hilbert were the points that would be on S covering the points a, b, c if S were constructed simply for the given nonuniform function y and not for this function and Poincaré's function ψ. The number of points covering a, b, c will generally be infinite, because the number of leaves of S will in general be infinite, since y is not assumed to be algebraic.

Hilbert thus reiterated one of the criticisms Osgood had made in 1898. In his article on complex function theory for Klein's *Enzyklopädie*, Osgood devoted a section to uniformization and noted again the limitations of Poincaré's theorem [1901:75]. In 1905–06 some attempts were made to deal with aspects of the uniformization problem,[72] probably spurred on by its status as one of Hilbert's problems, but it was Poincaré himself and Paul Koebe, a *Privatdozent* at Göttingen, who in 1907 independently provided what is generally regarded as the first satisfactory resolution of the issues raised by Hilbert. Poincaré's paper [1907] is of particular interest because in it he decided to separate the construction of the Riemann surface of a nonuniform function from that of its universal covering surface. To construct the Riemann surface of such a function, Poincaré followed the lead of Weierstrass[73] and conceived of the surface as composed of "function elements" (x, y) corresponding to pairs of series in powers of $\zeta - \zeta_0$, but unlike Weierstrass, he added elements to correspond to poles and branch points as well [1907:73–77]. Weyl was to develop this approach systematically in his book [1913].

Having indicated how to describe in this fashion a Riemann surface for nonuniform functions $y_1, \ldots, y_m$, which he called the "principal domain" D of $y_1, \ldots, y_m$, Poincaré pointed out that other constructions were possible that instead of leading to D would lead to a "multiple domain" Δ in which every point of D corresponds to several, possibly infinitely many, points of Δ. He came closest to the modern notion of a covering space with his definition of a "regular multiple" of D [1907:90]:

> We will say that Δ is a regular multiple of D if it satisfies the following condition. Let M be a point of D; let $M_1, M_2, \ldots$ be the corresponding points of Δ. Let M' be another point of D, infinitely close to M; we suppose that among the points of Δ that correspond to M' there is one that is infinitely close to M_1, one that is infinitely close to M_2, ... If this condition is fulfilled, that

---

[72] By the Finnish mathematician Severin Johansson [1905; 1906] and the Swedish mathematician Torsten Brodén [1905].

[73] See Hettner and Knoblauch's edition of Weierstrass' lectures on abelian functions [Weierstrass 1902:13–19] to which Poincaré referred.

> is to say if Δ is a regular multiple of D, it is clear that if to a certain point M of D a finite number n of points of Δ correspond, then there will correspond to any other point of D the *same* finite number n of points of Δ.

With these preliminaries out of the way, Poincaré returned to the ideas he had presented in his paper [1883c], now presenting them without the added complications of a Riemann surface construction. The result is essentially the modern construction of a universal covering space, which he gave to justify his claim that there is a regular multiple Δ of D that is simply connected. If M_0, M are points of D:

> One can go from M_0 to M on D by many paths. Consider two of these paths. They could be *equivalent*, that is, they could bound a continuous area situated on D; but they might not be, at least if D is not simply connected. That given, let us define the domain Δ. A point of this domain will be characterized by the point M of D to which it corresponds and by the path by which one proceeds from M_0. In order that two points so characterized be identical it is necessary and sufficient that one have come from M_0 by equivalent paths. It is clear that Δ is simply connected [1907:90].

Poincaré then sketched an argument to the effect that Δ is a regular multiple of D.

Poincaré's new construction of more ample Riemann surfaces was then combined by him with his new notion of a covering space (regular multiple) to establish his uniformization theorem without the limitations caused by the points a, b, c [1907:87–92].

Several months before Poincaré's paper [1907] appeared, Klein presented to the Göttingen Academy a paper by Paul Koebe [1907] that also established the general uniformization theorem without any restrictions. Koebe (1882–1945) had been a student of H. A. Schwarz at Berlin, where he received his doctorate in 1905. He became *Privatdozent* in Göttingen in 1907. Following in the footsteps of Schwarz, Koebe was interested from the ouset of his career in conformal mapping problems.[74] His proof of the uniformization theorem was based upon his theorem that any simply connected Riemann surface can be mapped conformally onto one of the following three regions of the Riemann sphere: a spherical cap, the sphere minus a point, or the entire sphere [1907:198]. (In terms of the complex plane $\mathbb{C}$, the three regions may be taken as $|z| < 1$, $\mathbb{C}$, and $\mathbb{C}$ plus the point at infinity.) To establish the uniformization theorem for nonuniform analytic functions $y_1, \ldots, y_m$ of x, Koebe considered the Riemann surface for $y_1, \ldots, y_m$. Then, citing Poincaré's original paper [1883c], he concluded that if this surface is not simply connected, "it is then transformed into a simply connected surface B by means of a definite covering process. The construction of this surface offers no major difficulties" [1907]. If $\Phi : B \to D$ denotes

[74] Regarding Koebe's mathematical career, see [Bieberbach 1968].

the inverse of the conformal mapping into the Riemann sphere posited by Koebe's theorem, then on B, $x, y_1, \ldots, y_m$ are all uniform analytic functions of $b \in B$, say $y_i = F_i(b)$, $i = 0, 1, \ldots, m$, where $y_0 = x$, and so $y_i = (F_i \circ \Phi)(\zeta)$ gives the uniformization. Koebe pointed out that the case in which D is the entire Riemann sphere occurs only when the Riemann surface is algebraic of genus zero, a possibility perhaps tacitly excluded by Poincaré, but Poincaré's proof also did not lead to the possibility in which D is the punctured sphere, *i.e.*, the entire complex plane [1907:193n]. Thus Koebe's results supplemented those of Poincaré, while at the same time making critical use of Poincaré's construction of a simply connected covering surface.

During 1907–1911 Koebe published a total of nineteen papers on various aspects of uniformization, including the algebraic case, where he provided the first proof of Klein's "fundamental theorem" and many detailed studies of the various cases that can arise (due to the nature of the group of linear fractional transformations defining the automorphic functions). Although he developed other methods as well, Koebe also stressed the importance of the "method of covering surfaces."[75] Thanks to Koebe there was no danger that the subject of uniformization or covering surfaces would soon be forgotten around Göttingen. Indeed, he appears to have generated considerable interest in uniformization and conformal mappings among the young mathematicians at Göttingen circa 1909 [Reid 1976:31], which is not surprising given the notoriety he had achieved for sharing with a mathematician of the stature of Poincaré the honor of solving one of Hilbert's problems. Hilbert himself helped generate this interest by showing that the rigorous method he had developed for solving boundary value problems by means of Dirichlet's principle (see Section 2) could be extended to obtain a very general theorem establishing the existence of Green functions and conformal mappings into the complex plane for Riemann surfaces that were not necessarily simply connected [1909c:77–78]. A corollary to his theorem was Koebe's conformal mapping theorem for simply connected Riemann surfaces.

According to Weyl's later recollections, Koebe's work on the fundamen-

[75] See, e.g., [Koebe 1909] as well as subsequent papers. Koebe claimed that the idea of the method of covering surfaces went back to Schwarz, although Poincaré was the first to publish something on it [1909:26]. He had in mind a conversation Klein had with Schwarz in April 1882, which was then reported on in a letter Klein sent to Poincaré the following month. To deal with some of the difficulties encountered regarding the uniformization of algebraic curves by automorphic functions, Schwarz had suggested constructing over a certain "dissected" Riemann surface one with infinitely many leaves, which in some cases would be simply connected and mappable conformally onto $|z| < 1$. (For Klein's letter, see his *Gesammelte Abhandlungen* **3**, p. 616; for the historical background, see [Gray 1986:297–304].) The fact that Schwarz's surface might not always be simply connected suggests that he was not thinking of Poincaré's construction, which is contrived to guarantee simple connectivity. It is of course possible that Poincaré was stimulated by hearing about Schwarz's infinitely leaved surface to conceive of the construction we find in [1883c].

tal uniformization theorems and "Hilbert's establishment of the foundation on which Riemann had built his structure and which was now available for uniformization theory, the Dirichlet principle," were two of the "three events" that "had a decisive influence" on the form of his lectures on Riemann surfaces (presented at Göttingen in the winter semester 1911–12 and published as [1913]). The other "event" was "the fundamental papers of Brouwer on topology from 1909 onwards."[76] By the time Weyl took on the task of lectures on Riemann surfaces, the Dutch mathematician L. E. J. Brouwer (1881–1966) was well known in Göttingen circles. In the spring of 1909 Brouwer began to correspond with Hilbert, whom he met personally that summer at the seaside resort of Scheveningen [van Dalen 1990:18n]. In his letters to Hilbert, Brouwer communicated his latest topological discoveries, which were then published in the journal of the Göttingen school, *Mathematische Annalen*. Brouwer's initial point of contact with Hilbert had been the foundations of geometry. As we noted in Section 2, Hilbert's work on foundations of geometry led him in [1903] to deal with the Helmholtz space problem for the plane and thereby to resolve for the special case of Lie groups of planar motions the fifth problem of his Paris lecture. Brouwer's first *Annalen* paper was a contribution toward the goal of a general resolution of the fifth problem, which he continued in another *Annalen* paper.[77] These papers marked the beginning of the topological study of continuous groups,[78] but Brouwer was diverted from his ultimate goal by other, more pressing, questions raised by these studies.

We noted that when Klein and Poincaré were grappling with their method of continuity, topology was in its infancy. By 1909 some progress had been made. Poincaré had begun to develop the homological aspects of algebraic topology in his papers on "*analysis situs*" (1895-1905),[79] and Cantor's papers on infinite sets had led to some developments within point set topology, as can be seen from A. Schoenflies's writings on the subject during 1904–08. Still, the discipline was in its early stages and, especially as regards point sets, was still somewhat naive in its reasoning about their properties. Brouwer pointed this out impressively with an *Annalen* paper of 1910 containing, among other things, an example of a curve that divides the plane into three open connected sets but is the *complete* boundary of each.[80] This example and the others Brouwer constructed undermined point set topology as developed by Schoenflies and revealed the need for a more rigorous treatment [Johnson 1987:67–71].

Brouwer then went on to show in a succession of papers how point set

[76] Quoted from the preface of the English translation of the third edition of Weyl's book [Weyl 1955]. The importance of these three events is already implicit in the preface to the 1913 edition.
[77] Papers 1909C and 1910H in [Brouwer 1976.]
[78] See on this subject my essay [1999b].
[79] See in this connection the historical studies [Bollinger 1972] [Dieudonné 1989:Ch.I] and [Pont 1974].
[80] Paper 1910C in [Brouwer 1976].

Figure 31. Brouwer (right) with Otto Blumenthal, who became Hilbert's biographer.

topology could be developed rigorously. In particular, he solved a problem of considerable interest in Göttingen, that of invariance of dimension under one-to-one continuous mappings of Euclidean space $\mathbb{R}^n$. The problem first arose when in 1878 Cantor established the existence of one-to-one mappings from $\mathbb{R}^n$ to $\mathbb{R}^1$.[81] It was then assumed by Cantor and others that such a mapping would be impossible if it were continuous, but the efforts to prove this either failed or dealt only with very special cases. Peano's example (1890) of a space-filling curve—a continuous mapping from $[0,1]$ onto the unit square—illustrated just how nonintuitive continuous mapping could be, as did Brouwer's above-mentioned example. At Göttingen a solution of the problem of invariance of dimension and the related problem of invariance of domain[82] was of special interest because the method of continuity utilized by Klein and Poincaré in the uniformization of algebraic curves by automorphic functions had taken such invariance properties for

[81] For a detailed history of the invariance of dimension problem from its origins up to Brouwer's work see [Johnson 1979; 1981].

[82] The problem is to show that the homeomorphic image of an open subset of $\mathbb{R}^n$ is also open in $\mathbb{R}^n$.

granted. Brouwer's proof of these invariance properties appeared in the *Annalen* in 1911, and in 1912 he applied his results to vindicate the method of continuity.[83]

It is clear from the preface to Weyl's lectures on Riemann surfaces that the developments in point set topology had convinced him that the theory of Riemann surfaces should be developed in a form that "completely satisfies all modern requirements of *rigor*," and that Brouwer's work, with its many innovative ideas of an algebraic-topological nature, had convinced him that it could be so developed: "To a far greater extent than follows from the citations, I have been encouraged by the fundamental topological investigations of Brouwer ... [1913:iii]. Although Weyl shared with Klein a conviction that the essence of mathematics is intuitive and that there is a real danger in pushing the rigor and abstraction too far,[84] he also saw the theory of Riemann surfaces as being by its nature especially in need of a rigorous formulation: "A rigorous set-theoretical founding of the topological concepts and theorems that come into question in Riemannian function theory is all the more necessary, since the 'points' of the basic configurations (curves and surfaces) in this case are not points in space in the usual sense but can be arbitrary mathematical entities of a different sort (e.g., function elements)" [1913:iv]. Such a development of the theory was in harmony with Hilbert's own views, but Weyl's stronger beliefs in the centrality of intuition to mathematical truth seem to have made him more defensive about devoting himself to working out such an abstract and rigorous formulation of the theory, which "may not be esteemed particularly highly." His defense of what he had done, however, has a decidedly Hilbertian ring to it: "Still, I think I can say that I have earnestly and conscientiously sought out the *simplest* and *most appropriate* methods that achieve the prescribed goal" [1913:iii].

The influence of both Klein and Hilbert can be seen also in Weyl's definition of a Riemann surface. Following the lead of Klein, he accepted the idea that the theory should be developed within the conceptual framework of the theory of surfaces rather than in terms of a wrapping of the complex plane over itself so as to make an analytic function single-valued. And following the lead of Hilbert, who in [1903] had attempted to define the plane by means of a system of neighborhoods (Section 2), Weyl did the same for surfaces. Thus he began, sounding very much like Hilbert in [1903]: "Let a totality of things, which are called 'the points of the manifold $\mathfrak{F}$,' be given. To every point $\mathfrak{p}$ of the manifold $\mathfrak{F}$ let certain sets of points be defined as 'neighborhoods of $\mathfrak{p}$ on $\mathfrak{F}$.'" The neighborhoods thus define what would now be called a basis for the topology on $\mathfrak{F}$, and Weyl then

[83] See, respectively, papers 1911C-E and 1912C,D,G,H, as well as the informative commentary on the papers by H. Freudenthal. Koebe also published papers vindicating the continuity method.

[84] These convictions are set forth with eloquence and passion on p. iv of the Preface.

went on to stipulate further properties of these neighborhoods to make $\mathfrak{F}$ a surface [1913:17–18]. Thus associated with each neighborhood is a bijective mapping onto an open disk in the plane satisfying certain conditions. Weyl's definitions contained in spirit if not in detail the modern formulation of a 2-dimensional manifold. In addition, $\mathfrak{F}$ is assumed to be arcwise connected. Such an abstract surface $\mathfrak{F}$ then becomes a Riemann surface if it possesses additional structure that makes it possible to speak of an analytic function of points $\mathfrak{p} \in \mathfrak{F}$ [1913:35–36].[85]

From the viewpoint of the history of representation theory what is of interest in Weyl's presentation is that the theory of Riemann surfaces is developed within the broader context of a theory of real two-dimensional manifolds, a theory that readily generalizes to r-dimensional manifolds and in part to the more general topological spaces defined by Hausdorff [1914], who also used systems of neighborhoods. This is true in particular of Weyl's extensive discussion of covering surfaces [1913:47–51], which marks an important step toward the development of a purely topological theory of covering spaces. Likewise, his construction of a "universal covering surface" $\tilde{\mathfrak{F}}$ (as he called it) associated to a given surface $\mathfrak{F}$ is capable of broad generalization. As with Poincaré, the points of $\tilde{\mathfrak{F}}$ consist of "curves" γ—*i.e.*, continuous mappings $\gamma : [0,1] \to \mathfrak{F}$ — emanating from a fixed point $\gamma(0) = \mathfrak{p}_0 \in \mathfrak{F}$, where two such curves γ, γ' define the same point of $\tilde{\mathfrak{F}}$ if $\gamma(1) = \gamma'(1)$ and if they satisfy a condition that nowadays would be described by saying they are homotopic.[86] Thus in effect each point $\tilde{\mathfrak{p}} \in \tilde{\mathfrak{F}}$ can be identified with an equivalence class $[\gamma]$ of homotopic curves γ that start at $\mathfrak{p}_0$ and end at the same point $\mathfrak{p} = \gamma(1)$ and is thereby covered by $\tilde{\mathfrak{p}}$. The neighborhoods $\tilde{\mathfrak{U}}$ of $\tilde{\mathfrak{F}}$ are then defined as follows: "Let γ_0 be a curve in $\mathfrak{F}$ from $\mathfrak{p}_0$ to $\mathfrak{p}$ and $\mathfrak{U}$ a neighborhood of $\mathfrak{p}$. Attach to γ_0 all possible curves γ that start from $\mathfrak{p}$ and lie within $\mathfrak{U}$ and say that the points of $\tilde{\mathfrak{F}}$ defined by all such curves $\gamma_0 + \gamma$ form a 'neighborhood $\tilde{\mathfrak{U}}$ of $\tilde{\mathfrak{p}}$'" [1913:51].[87] With these words Weyl thus gave what is now a customary description of a basis for the topology of the universal covering space.

If we consider the points $\tilde{\mathfrak{p}}_0$ that cover the point $\mathfrak{p}_0$ defined above, each $\tilde{\mathfrak{p}}_0$ corresponds to an equivalence class of curves that begin and end at $\mathfrak{p}_0$ and thus correspond to the elements of the fundamental group of $\mathfrak{F}$ at $\mathfrak{p}_0$, $\pi_1(\mathfrak{F}, \mathfrak{p}_0)$ —a notion Poincaré had introduced in his memoir on *analysis situs* [1895]. Weyl never mentioned $\pi_1(\mathfrak{F}, \mathfrak{p}_0)$ but spoke instead of the group Γ of covering transformations T (*Decktransformationen*), which he defined to be homeomorphisms of $\tilde{\mathfrak{F}}$ with the property that for any $\tilde{\mathfrak{p}} \in \tilde{\mathfrak{F}}$, $T(\tilde{\mathfrak{p}})$ covers the same point as $\tilde{\mathfrak{p}}$, i.e., as is still done today. Considered abstractly, the

---

[85] For a detailed, critical discussion of Weyl's definitions and their relation to ideas of Hilbert and Klein, see [Scholz 1980:193–98].

[86] Weyl's definitions of simple connectivity and of the universal covering surface avoided using homotopic equivalence and were apparently guided by convenience of application to function theory [1913:47, n.2].

[87] Here $\gamma_0 + \gamma$ denotes a curve that first follows γ_0 and then γ.

group Γ, Weyl declared, expresses completely the relation between $\tilde{\mathfrak{F}}$ and $\mathfrak{F}$, "in so far as it possesses an analysis situs character."[88] It is not difficult to see that Γ is isomorphic to $\pi_1(\mathfrak{F}, \mathfrak{p}_0)$. That is, given an equivalence class $[\alpha]$ of $\pi_1(\mathfrak{F}, \mathfrak{p}_0)$, $T_\alpha : [\gamma] \to [\alpha + \gamma]$ defines a covering transformation T_α and the correspondence $[\alpha] \to T_\alpha$ gives an isomorphism onto Γ.[89] No doubt Weyl realized all this but preferred not to introduce fundamental groups per se.

In view of the use Weyl later made of universal covering spaces in his work on group representations (Chapter 12), it is helpful to make a few observations at this point because they follow readily from what has been said. Weyl made them a decade later when he had occasion to think of Lie groups as manifolds. To this end, let us now suppose that $\mathfrak{F}$ denotes the r-dimensional generalization of Weyl's notion of a surface.

The first observation involves the situation in which $\mathfrak{F}$ is compact in the sense employed by Weyl and his contemporaries, namely that every infinite subset of $\mathfrak{F}$ has an accumulation point.[90] Suppose that $\mathfrak{F}$ has finite connectivity, i.e., that $\mathfrak{F}$ has the property that there are N nonhomotopic curves from $\mathfrak{p}_0$ to $\mathfrak{p}$, so that each point $\mathfrak{p} \in \mathfrak{F}$ has N points of $\tilde{\mathfrak{F}}$ lying over it, which is sometimes expressed by saying that $\tilde{\mathfrak{F}}$ consists of N sheets. Then if $\tilde{\mathfrak{K}} \subset \tilde{\mathfrak{F}}$ is infinite, it follows that the set $\mathfrak{K} \subset \mathfrak{F}$ of points covered by points in $\tilde{\mathfrak{K}}$ must also be infinite. Thus $\mathfrak{K}$ has an accumulation point $\mathfrak{q}$. If $\tilde{\mathfrak{q}}_1, \ldots, \tilde{\mathfrak{q}}_N$ are the points that cover $\mathfrak{q}$, then it follows readily from Weyl's definition of the topology of $\tilde{\mathfrak{F}}$ that at least one of the points $\tilde{\mathfrak{q}}_1, \ldots, \tilde{\mathfrak{q}}_N$ must be an accumulation point of $\tilde{\mathfrak{K}}$, so that $\tilde{\mathfrak{F}}$ is also compact.

The second observation is that if the product $\mathfrak{p}\mathfrak{p}'$ of elements of $\mathfrak{F}$ is defined so that $\mathfrak{F}$ is a group, then it easy to extend the multiplication from $\mathfrak{F}$ to the universal cover $\tilde{\mathfrak{F}}$ so that the projection mapping π, which sends $\tilde{\mathfrak{p}}$ into the point $\mathfrak{p}$ it covers, is a group homomorphism from $\tilde{\mathfrak{F}}$ onto $\mathfrak{F}$. Indeed, this can be done in more than one way. To see this, choose as the point $\mathfrak{p}_0$ in the above construction of $\tilde{\mathfrak{F}}$ the identity element e of $\mathfrak{F}$. Then if γ, γ' denote two curves with initial point $\mathfrak{e}$, let the product curve $\gamma \cdot \gamma'$ be defined by $(\gamma \cdot \gamma')(t) = \gamma(t)\gamma'(t)$ for all $t \in [0,1]$. Then $\mathfrak{e}$ is also the initial point of $\gamma \cdot \gamma'$, and if $\tilde{\mathfrak{p}} = [\gamma]$ and $\tilde{\mathfrak{p}}' = [\gamma']$ are two elements of $\tilde{\mathfrak{F}}$, we may define the product $\tilde{\mathfrak{p}}\tilde{\mathfrak{p}}'$ to be $[\gamma \cdot \gamma']$. With multiplication so defined, $\tilde{\mathfrak{F}}$ becomes a group as well, and $\pi : \tilde{\mathfrak{F}} \to \mathfrak{F}$ is a homomorphism.

[88] [Weyl 1913:50]. Here Weyl defined Γ for covering surfaces more general than the above-discussed universal covering surface.

[89] See, e.g., [Singer and Thorpe:67–8].

[90] Weyl spoke of $\mathfrak{F}$ being "closed" [1913:24], but in his classic text on topology [1914:230] Hausdorff followed Fréchet and used the term "compact" for this property, reserving "closed" for the meaning it still has today. Hausdorff also showed that a topological space is compact in this sense if and only if every countable open cover has a finite subcover. Today the term "countably compact" would be used to distinguish it from compactness in the now customary sense. Of course, the two notions are equivalent for an extensive class of topological spaces (Hausdorff spaces with a countable basis) including the Lie groups Weyl later considered.

If $\tilde{\mathfrak{K}} \subset \tilde{\mathfrak{F}}$ denotes the kernel of this homomorphism, then as noted above, since $\tilde{\mathfrak{K}}$ consists of the points covering $\mathfrak{p}_0 = \mathfrak{e}$, its elements can be identified with those of the fundamental group $\pi_1(\mathfrak{F}, \mathfrak{e})$, although the multiplication defined in $\tilde{\mathfrak{K}}$ need not coincide with the multiplication defined in $\pi_1(\mathfrak{F}, \mathfrak{e})$. This can be made to happen, however, by defining multiplication in $\tilde{\mathfrak{F}}$ as follows. Given $\tilde{\mathfrak{p}} = [\gamma]$, $\tilde{\mathfrak{p}}' = [\gamma'] \in \tilde{\mathfrak{F}}$, let $\tilde{\mathfrak{p}} \times \tilde{\mathfrak{p}}' = [\gamma + \gamma(1) \cdot \gamma']$, where $\gamma + \gamma(1) \cdot \gamma'$ denotes the curve that follows γ from e to its endpoint $\gamma(1)$ and then follows the curve $t \to \gamma(1)\gamma'(t)$ from its starting point at $\gamma(1)e = \gamma(1)$ to its endpoint at $\gamma(1)\gamma'(1)$. (Thus $\pi(\tilde{\mathfrak{p}}\tilde{\mathfrak{p}}') = \pi(\tilde{\mathfrak{p}} \times \tilde{\mathfrak{p}}')$.) Then for $\tilde{\mathfrak{p}} = [\gamma]$, $\tilde{\mathfrak{p}}' = [\gamma'] \in \tilde{\mathfrak{K}}$, it follows that $\tilde{\mathfrak{p}} \times \tilde{\mathfrak{p}}' = [\gamma + \gamma']$, in accordance with the definition of multiplication in $\pi_1(\mathfrak{F}, \mathfrak{e})$. As we shall see in Chapter 12, Weyl simply observed that $\tilde{\mathfrak{F}}$ could be made into a group without pausing to explicitly define the multiplication (Section 12.1), whereas Cartan then introduced the definition $\tilde{\mathfrak{p}} \times \tilde{\mathfrak{p}}'$ (Section 12.6).

9.6. Hilbert's Brand of Mathematical Thinking

The purpose of this chapter has been to provide the Göttingen background necessary for an appreciation of Weyl's contributions to the representation of Lie groups, since, as we shall see, what Weyl learned at Göttingen was put to good use in his study of Lie groups and their representations. This included more than mathematical facts and theories. As we have seen from the quotations at the beginning of the chapter, Weyl acknowledged Hilbert's great influence upon him: What Hilbert did was considered important, and the way he did it, what Weyl later termed "the peculiarly Hilbertian brand of mathematical thinking," left its imprint upon him. Having now described Hilbert's activities at Göttingen, let us consider against that backdrop the mathematical way of thinking and the broader disciplinary ideals that were encouraged by it. As I will attempt to indicate here in general terms, these turn out to be quite relevant to Weyl's work on representations and to the intellectual journey that led him to undertake it. In the course of reading Chapters 11 and 12, the reader can decide to what extent the following remarks are relevant to, or enlightening for, an understanding of Weyl's achievement.[91]

In his memorial essay on Hilbert, Weyl singled out five traits of "the peculiarly Hilbertian brand of mathematical thinking" [1944b:134–5]: (1) A lucid literary style; (2) rigor and simplicity going hand in hand; (3) the importance of fruitful concrete problems; (4) his faith in "the methodological unity of mathematics"; (5) a direct approach to problems "unfettered

[91] See also in this connection my discussion in Section 2 of a characteristic of Hilbert's work on algebraic number fields (also singled out by Weyl in his memorial essays), which bears a striking resemblance to Weyl's approach to the representation theory of semisimple groups.

by algorithms" and going back to questions in their original simplicity. The fourth and fifth traits are particularly relevant to Weyl's work on Lie groups.

Regarding (4) Weyl explained in another memorial essay on Hilbert [1944a:123], "The methodological unity of mathematics was for him a matter of belief and experience. It appeared to him essential that—in the face of the manifold interrelations and for the sake of fertility of research—the productive mathematician should make himself at home in all fields." As we have seen in the previous two sections, Hilbert did this to a remarkable degree, embracing not only fields of mathematics but also of theoretical physics. Of course, not every student at Göttingen had the intellectual capacity to emulate Hilbert in this way, but Weyl did. Indeed, he seemed to have a faculty not shared by Hilbert of being able to "feel at home" in a field of mathematics or physics with incredible rapidity (and without the assistants who predigested the substance of a new field for Hilbert). The fact that he did emulate Hilbert in this way enabled him, as we shall see, to bring to bear on the representation problem a host of mathematical ideas and theories from seemingly (at the time) disparate fields such as Frobenius's theory of group characters, tensor calculus, Riemann surfaces, and integral equations.

Hilbert's mastery of many fields was of course a reflection of his conviction of an underlying unity to mathematics (including mathematical physics). In Hilbert's work this conviction was first embodied in his work on the finiteness problem in the theory of invariants (Section 1), where that problem came to be seen within the broad framework of polynomial rings and associated ideals and fields. It should be mentioned that Klein seized upon and encouraged Hilbert in this direction. Thus after Klein saw Hilbert's first announcement of his finiteness theorem in 1889, he criticized Hilbert for pushing the invariant-theoretic question too much into the foreground, since, so it seemed to Klein, what was really involved was "a far more general, and hence far more important, finiteness theorem, from which among other things the much discussed invariant-theoretic theorem proceeds as a corollary."[92] Hilbert agreed with Klein's opinions and in his last major paper on invariants expressed the view that, thanks to his methods, the theory of invariants had become just a special case of the theory of algebraic function fields, just as in arithmetic the theory of cyclotomic fields is merely an especially noteworthy example where the most important theorems of general number fields were first recognized and proved [1893:287]. As we noted already in Section 1, it was also Klein who encouraged Hilbert to propose the consideration of invariants with respect to other groups and in relation to his *Erlanger Programm* and Lie's theory.

Hilbert's belief in the unity of mathematics, exemplified in a limited way by the above-described work on invariants, was proclaimed in his Paris

[92] Letter of 24 February 1889 [Frei 1985:41].

lecture on mathematical problems [1900:478–9] in vivid, general terms:[93]

> Mathematical science is in my opinion an indivisible whole, an organism whose vitality is conditioned upon the connection of its parts. For with all the variety of mathematical knowledge, we are still clearly conscious of the similarity of the logical devices, the *relationship* of the *ideas* in mathematics as a whole and the numerous analogies in its different fields, we also notice that the farther a mathematical theory is developed, the more harmoniously and uniformly does its construction proceed, and unsuspected relations are disclosed between hitherto separate branches .... So it happens that with the extension of mathematics, its organic character is not lost but only manifests itself the more clearly .... [E]very real advance goes hand in hand with the invention of sharper tools and simpler methods that at the same time assist in understanding earlier theories and cast aside older more complicated developments. It is therefore possible for the individual investigator, when he makes these sharper tools and simpler methods his own, to find his way more easily in the various branches of mathematics than is possible in any other science. The organic unity of mathematics is inherent in the nature of this science.

Such a belief brings with it a proclivity to look for unexpected mathematical connections, and Hilbert came to be seen, and admired, for his ability to see such connections. Thus Otto Blumenthal, Hilbert's student and a long-term resident at Göttingen, wrote in his biography of Hilbert, "In the analysis of mathematical talent one has to differentiate between the ability to create new concepts that generate new types of thought structures and the gift for sensing deeper connections and underlying unity. In Hilbert's case, his greatness lies in an immensely powerful insight that penetrates into the depths of a question. All of his works contain examples from far-flung fields in which only he was able to discern an interrelatedness and connection with the problem at hand. From these the synthesis, his work of art, was ultimately created."[94]

Hilbert's success often depended upon viewing matters in a more general fashion. As he put it in his Paris talk, "If we do not succeed in solving a mathematical problem, the reason frequently consists in our failure to recognize the more general standpoint from which the problem before us appears only as a single link in a chain of related problems. After finding this standpoint, not only is this problem frequently more accessible to our investigation, but at the same time we come into possession of a method that is applicable also to related problems."[95] That is why Hilbert created his theory of bilinear forms on ℓ^2. In [1909b] he even envisioned a far more

---

[93] This is a slightly modified version of Newsom's translation.

[94] [Blumenthal 1935:429]. I have followed the translation by Rowe [1989b:200].

[95] [Hilbert 1900:443–4]. I have modified slightly Newson's translation.

general theory of analytic functions of infinitely many variables that would afford the prize of a "methodologically unified organization of algebra and analysis" (p. 57) and would have manifold applications to differential and integral equations, to the calculus of variations, and to function theory (p. 72).

The imprint of Hilbert's views on the unity of mathematics can be seen in Weyl's work relating to group representations. I already mentioned that Weyl discovered deep-seated connections between Lie's theory and seemingly disparate theories. Such insights are, of course, a tribute to his extraordinary genius, and it would be ridiculous to think that any competent disciple of Hilbert's would have had them. But it does seem reasonable to see Weyl's disposition to believe in the possibility and desirability of such connections as a factor in his success at finding them. On a more general level, there seems to be a connection between Hilbert's views on the unity of mathematics and Weyl's habit of formulating a certain type of question. As we shall see in Chapters 11 and 12, many key discoveries on Weyl's part were characterized as revealing the "true" or "proper" or "natural" basis for understanding a particular mathematical or physical phenomenon (e.g., a mathematical theorem or an observed relation). In seeking understanding Weyl, like Hilbert, looked for unexpected connections with other sorts of mathematics than that involved in the customary view of the phenomenon; he thereby sought what seems like Hilbert's "more general standpoint" from which to view the relation as a "link" in a more far-reaching theoretical "chain," thereby obtaining the "true" (or "proper" or "natural") basis for understanding and dealing with it.

In the above quotation from Hilbert's Paris lecture he spoke of "the numerous analogies" in the "different fields" of mathematics as a reflection of the underlying unity of mathematics. Hilbert's work on integral equations (including the concomitant ℓ^2 theory) was based upon the perception of an analogy with finite systems of linear equations. Hilbert was not the first to see this analogy nor the first to attempt to develop some results that could be seen as continuous analogues of results about systems of linear equations. As we have seen in Section 2, Fredholm was the first to take some steps in this direction. But it was Hilbert who showed that a far-reaching, substantial continuous analogue could be created, of which Section 2 contains a mere fragment. It was a marvelous confirmation of his belief in the unity of mathematics. Weyl's work on group representations can be viewed in similar fashion, for it was driven to a remarkable degree by the desire to create a substantial continuous analogue of the representation theory created by Frobenius for finite groups (Section 10.1) This can be seen in the considerations leading up to the complete reducibility theorem for $\mathbf{SL}(n, \mathbb{C})$ (Sections 11.3–11.5) and in the Peter–Weyl paper of Section 12.6 (which grew out of Weyl's quest for a viable continuous analogue of Frobenius's regular representation theorem (Section 10.1). As with Hilbert and his work on integral equations, Weyl was not the first to perceive an analogy between

Frobenius's theory and an analogous theory for some continuous groups, for this was done by Frobenius's gifted student Issai Schur (Sections 10.3–5). But compared to what Weyl did, Schur appears as playing a role more like the one Fredholm played for Hilbert. It is tempting to wonder whether the paradigm of Hilbert's work on integral equations encouraged Weyl to believe in the possibility of a substantial analogue of Frobenius's theory. Weyl certainly knew Hilbert's theory intimately, because, as we have seen in Section 4, most of Weyl's own research at Göttingen was on this subject. He later looked back on the creation of Hilbert's theory as probably the outstanding mathematical event of his lifetime.[96]

Closely connected with Hilbert's belief in the methodological unity of mathematics was what Weyl described above as the fifth trait of Hilbert's brand of mathematical thought: a direct approach to problems "unfettered by algorithms" and going back to questions in their original simplicity. Again one sees this first of all in Hilbert's work on the theory of invariants. As Weyl put it [1932:346], in dealing with the finiteness problem Hilbert freed himself from the entire technical apparatus (*Technik*) that Gordan had devised — the complicated algebraic algorithms by which via difficult constructions invariants were determined. "He saw the problem in its original intellectual simplicity and he solved it with intellectually simple, pure 'existential' means. This type of approach has remained typical for Hilbert." Indeed, it can be seen in his work on arithmetic as well. Thus in his *Zahlbericht*, in presenting Kummer's results on higher reciprocity laws, Hilbert declared his intention "to avoid the great computational apparatus of Kummer, so that here also Riemann's principle would be realized, according to which proofs should be driven not by calculation but solely by ideas [1897:67].[97] The close link between Hilbert's ideal of conceptual rather than computational reasoning and his belief in methodological unity can be seen reflected in a minor paper of Hilbert's devoted to a new proof of a theorem of Dedekind's on commutative algebras. "The following lines," he wrote, "should show how Dedekind's theorem flows without computational aids from a general theorem that I have used in my investigations on the theory of algebraic invariants" [1896:372]. The theorem was a version of what is now called Hilbert's *Nullstellensatz*, and was one of the theorems of what Hilbert termed the "critical" phase in the development of invariant theory, wherein it was subsumed by the general theory of "algebraic function fields." By placing Dedekind's theorem within the more general framework of the theory of algebraic function fields, he was able to prove it without the computational devices that had accompanied the original proof.

[96] See the quotation circa 1946 given by Sigurdsson [1991:36].

[97] "Ich habe versucht, den grossen rechnerischen Apparat von Kummer zu vermeiden, damit auch hier der Grundsatz von Riemann verwirklicht würde, demzufolge man die Beweise nicht durch Rechnung, sondern lediglich durch Gedanken zwingen soll."

Weyl in effect brought Hilbert's "Riemann principle" to bear on two points of Cartan's theory where Cartan resorted to extensive, case-by-case computations, i.e., computations for each simple structure type: (1) Cartan's observation that every complex semisimple Lie algebra has what is now termed a real compact form (Section 8.5) and (2) Cartan's theorem that every dominant weight corresponds to an irreducible representation (Section 8.4). With regard to (1), Weyl succeeded in giving a proof that was valid for any semisimple Lie algebra and did not require the extensive, tedious calculations of Cartan's approach (Section 12.3). As for (2), Weyl's dissatisfaction with Cartan's computational, case-by-case approach led him to seek a continuous analogue of Frobenius's regular representation theorem (Section 10.1), which in turn led him to see connections between his representation problem and Hilbert's theory of integral equations, connections that ultimately led to the Peter–Weyl theorem and the idea of a harmonic analysis on groups (Section 12.7).

Chapter 10
The Berlin Algebraists: Frobenius and I. Schur

In the introductory remarks to Chapter 7 I pointed out that the work of Lie, Study, Kowalewski, and Cartan discussed in Chapters 7–8 belonged to the prehistory of the representation theory of Lie groups in the sense that these mathematicians did not operate within the conceptual framework that we customarily associate with group representation theory today. That framework, which included a complete reducibility theorem and a theory of characters, was provided by the work of Frobenius, not long after Cartan published his doctoral dissertation. Frobenius's theory was developed for finite groups and was motivated by arithmetical considerations, so that it was in a sense far removed from the interests of the above-mentioned mathematicians.[1]

In this chapter the salient features of Frobenius's theory are summarized (Section 1) and extensions of the theory to specific continuous groups by his foremost student, Issai Schur, are discussed (Sections 3–5). These extensions were motivated by the theory of invariants in general and by two seminal papers by Adolph Hurwitz (Section 2) in particular. As we shall see in Chapter 12, the paradigm of Frobenius's theory for finite groups and the ideas of Hurwitz and Schur played a vitally important role in inspiring Weyl to create for semisimple groups a continuous analogue of Frobenius's theory.

---

[1] For an appreciation of the arithmetical background to Frobenius's theory, see my paper [1971] and essay [1999a].

10.1. Frobenius's Theory of Group Characters and Representations

Georg Frobenius (1849–1917) grew up in Berlin, and except for one semester at Göttingen, his entire university education was at the University of Berlin.[2] As we saw in Section 4.1, during his student days, which overlapped with those of Killing, Berlin was the most vibrant center for mathematics in Germany. Frobenius had first come to the attention of his teachers because of his unusual skill at calculations, but he was soon perceived to possess as well creative mathematical ability to a high degree. Judging by his subsequent mathematical research, the work of both Kronecker and Weierstrass made a lasting impression upon him. He wrote his doctoral dissertation under the direction of Weierstrass on a topic in Weierstrassian analysis, the representation of analytic functions by certain types of series of functions (1870).[3] He was graduated *summa cum laude* in the summer of 1870. Weierstrass described his dissertation as "a work of scientific value containing many new ideas and results" [Biermann 1973:191].

Almost immediately after receiving his doctorate, Frobenius was offered a position at the University of Freiburg, but he declined the offer for personal reasons connected with his family. Instead, he taught for a while at one of the local high schools. At the University, he was highly regarded, both as an independent, creative mathematician and as an excellent teacher. In 1872 the mathematics faculty sought to appoint him as an assistant professor (*ausserordentlicher Professor*). Although the appointment finally went through in 1874, he remained in the position less than a year because he was offered, and accepted, a full professorship at the Polytechnicum in Zürich, which subsequently became the Federal Technical Institute—*Eidgenössische Technische Hochschule*, or ETH for short.

Frobenius remained in Zürich until 1892, when he was called back to Berlin. Kronecker had died suddenly in December of 1891, and Weierstrass, who was in poor health, wished to retire now that his chosen successor, H. A. Schwarz, whom Kronecker had opposed, would in all likelihood be approved. Thus two professorships were open; Frobenius became Kronecker's successor and H. A. Schwarz became Weierstrass's. The committee charged with the task of making these appointments stressed, in the case of Frobenius, "the universality of his mathematical talent. There is scarcely a branch of pure mathematics that he has not made the subject of his research" [Biermann 1973:207]. Frobenius's papers had ranged over aspects of the theory

---

[2] The following biographical details are drawn from Biermann's study of the Berlin school of mathematics [1973], which, in particular, contains invaluable documentary material pertaining to Frobenius (who was never the subject of an extensive biography). See especially pp. 95–96, 109–110, 122–123, 189–192, 204–209.

[3] Frobenius's dissertation is in Latin, but he published a paper in German based on it in Crelle's journal. See his *Gesammelte Abhandlungen* **1**, 1–64.

Figure 32. Frobenius as he may have looked in the late 1890s when he created his theory of group characters and representations.

of linear ordinary differential equations, abelian functions (including many papers on theta functions), linear algebra (determinants and the theory of bilinear and quadratic forms), the theory of Pfaffian equations (where Frobenius's theorem on completely integrable vector fields originated), the theory of numbers, geometry, the theory of equations, and the theory of finite groups. Although the problems he considered were thus remarkably diverse, it was always what would now be described as the algebraic aspects of the mathematical theories he considered that engaged his interest. Frobenius was at heart an algebraist.

The year 1892, when Frobenius and Schwarz arrived in Berlin, marked a new era in the history of mathematics at Berlin.[4] Whereas Berlin had been the mecca of German mathematics at the time when Frobenius and Killing were students there and Weierstrass was a charismatic presence,

[4] See [Biermann 1973:Ch. 6] and [Rowe 1998b] for details.

after Hilbert's arrival at Göttingen in 1895 that university soon became the new mecca, as we have seen in Chapter 9. For many years Frobenius was the leading mathematician at Berlin. Schwarz had accomplished much of great significance as a mathematician, but it was all done prior to the call to Berlin; and the third full professor of mathematics at Berlin in 1892 was Lazarus Fuchs, whose best work was also behind him. When Fuchs died in 1902, an attempt was made to bring Hilbert to Berlin, but it failed, and instead, Friedrich Schottky, Frobenius's friend and former collaborator, became the third full professor. He was a good mathematician noted mainly for work in complex function theory (Schottky's theorem) but not on the level that Schwarz had achieved.

As leader of the Berlin mathematicians, Frobenius lacked the personal characteristics and inclinations needed to maintain Berlin as the main mathematical attraction among German universities. For example, during his twenty-five years as Berlin professor he had only five doctoral students. Two of them, however, were exceptional mathematicians, Edmund Landau and Issai Schur, who was clearly Frobenius's favorite. It was also a few years after his arrival in Berlin that some observations by Dedekind led Frobenius to develop his remarkable theory of group characters and representations, which Schur then went on to extend in several important directions. Berlin may not have been any longer in the limelight as the leading center for mathematics in Germany, but mathematics of lasting importance was still being created there. Furthermore, the algebraic style of mathematics that Frobenius cultivated was continued by Schur, who spent most of his career at Berlin, and by Schur's students, who were more numerous than those of Frobenius.

Frobenius's work relating to arithmetical problems had brought him into contact with Richard Dedekind (1831–1916), the founder of the arithmetical theory of ideals of algebraic numbers, and it was in response to a problem posed by Dedekind in 1895 that Frobenius created his theory of group characters and representations.[5] Much earlier, in 1879, Dedekind had introduced a general notion of a group character of a finite abelian group as a generalization of the special cases of this notion that entered into the arithmetical work of Gauss (who introduced the term "character") and Dirichlet. If $\mathfrak{G}$ denotes such a group, then a nonzero, complex-valued function χ on $\mathfrak{G}$ is a character if for all $A, B \in \mathfrak{G}$,

$$\chi(AB) = \chi(A)\chi(B). \tag{1.1}$$

The arithmetical nature of characters is revealed by the theorem, which also first arose within the context of particular groups occurring in arithmetical investigations, that every finite abelian group $\mathfrak{G}$ can be represented as a direct product of cyclic subgroups. If $\mathfrak{G} = (R_1) \times (R_2) \times \cdots \times (R_k)$ is

[5] The following draws upon my fuller accounts of the history of the representation of finite groups in [1999a; 1971; 1972; 1974].

such a direct product representation, where (R_i) denotes a cyclic group with generator R_i and order n_i, then the function χ defined on $R = R_1^{m_1} \times R_2^{m_2} \times \cdots \times R_k^{m_r} \in \mathfrak{G}$ by $\chi(R) = \epsilon_1^{m_1}\epsilon_2^{m_2}\cdots\epsilon_k^{m_r}$, where ϵ_i denotes an n_ith root of unity, is a character on $\mathfrak{G}$. Conversely, the multiplicative property (1.1) of characters implies that all characters must be defined in this manner, so that there are exactly $n_1 n_2 \cdots n_r = g$ distinct characters on $\mathfrak{G}$, where g denotes the order of $\mathfrak{G}$.

Sometime in 1880 Dedekind discovered a connection between group characters and a certain determinant associated to the group, which came to be known as the group determinant. The group determinant was suggested to him by his study of the discriminant of an algebraic number field. Let $\mathbb{K}$ denote such a field, i.e., a normal extension of the field of rational numbers, and let $\mathfrak{G} = \{\pi_1, \pi_2, \ldots, \pi_g\}$ denote its Galois group. Given a system of linearly independent elements $\omega_1, \omega_2, \ldots, \omega_g$ from $\mathbb{K}$, the discriminant Δ of this system is defined as $\Delta = D^2$, where

$$D = \begin{vmatrix} \omega_1\pi_1 & \omega_2\pi_1 & \cdots & \omega_g\pi_1 \\ \omega_1\pi_2 & \omega_2\pi_2 & \cdots & \omega_g\pi_2 \\ \vdots & \vdots & \ddots & \vdots \\ \omega_1\pi_g & \omega_2\pi_g & \cdots & \omega_g\pi_g \end{vmatrix}.$$

An important special case arises when the elements ω_i are the conjugates of a fixed element $\omega \in \mathbb{K}$, namely $\omega\pi_1, \ldots, \omega\pi_g$. Then the discriminant takes the form

$$(1.2)\qquad D = \begin{vmatrix} \omega\pi_1\pi_1 & \omega\pi_2\pi_1 & \cdots & \omega\pi_g\pi_1 \\ \omega\pi_1\pi_2 & \omega\pi_2\pi_2 & \cdots & \omega\pi_h\pi_2 \\ \vdots & \vdots & \ddots & \vdots \\ \omega\pi_1\pi_g & \omega\pi_2\pi_g & \cdots & \omega\pi_h\pi_g \end{vmatrix}.$$

The form of (1.2) led Dedekind to consider the following determinant. Corresponding to each $\pi_i \in \mathfrak{G}$, let x_{π_i} denote a complex variable and consider $\det(x_{\pi_j\pi_i})$. In other words, in (1.2) the row j, column i entry, $\omega\pi_j\pi_i$, has been replaced by the variable $x_{\pi_j\pi_i}$. Thus instead of getting an element of $\mathbb{K}$ one gets a polynomial in g complex variables. Eventually, Dedekind switched to the determinant $\Theta = \det(x_{\pi_i\pi_j^{-1}})$, which makes all the diagonal entries correspond to the variable of the identity element of $\mathfrak{G}$.

The determinant Θ came to be known as the group determinant, and it became the source of Frobenius's theory of group characters and representations. The mathematical connection between group determinants and representations is easily seen. Let $\mathcal{H}$ denote the group algebra of $\mathfrak{G}$, i.e., the g-dimensional linear associative algebra over $\mathbb{C}$ consisting of all expressions $\mathbf{x} = \sum_{i=1}^{g} x_i\pi_i$, $x_i \in \mathbb{C}$, with addition and multiplication defined in the obvious way. The right regular representation of $\mathcal{H}$ is the isomorphism ρ_R that associates to $\mathbf{y} \in \mathcal{H}$ the linear transformation $\rho_R(\mathbf{y})$ on $\mathcal{H}$ sending $\mathbf{x} \to \mathbf{xy}$ for all $\mathbf{x} \in \mathcal{H}$. The restriction of ρ_R to $\mathfrak{G}$ yields the right regular representation of $\mathfrak{G}$. It is easy to check that $(x_{\pi_i \pi_j^{-1}}) = \sum_{\pi \in \mathfrak{G}} \rho_R(\pi) x_\pi = \rho_R(\sum_{\pi \in \mathfrak{G}} x_\pi \pi)$, so that $\Theta(x_1, \ldots, x_g) = \det \rho_R(\mathbf{x})$, where $\mathbf{x} = \sum_{i=1}^g x_i \pi_i$. One final point to keep in mind is that if ρ_R decomposes into the sum of two representations $\mu \oplus \nu$ of $\mathcal{H}$ in the sense that a $g \times g$ invertible matrix M exists for which

$$M \rho_R(\mathbf{x}) M^{-1} = \begin{pmatrix} \mu(\mathbf{x}) & 0 \\ 0 & \nu(\mathbf{x}) \end{pmatrix}$$

for all $\mathbf{x} \in \mathcal{H}$, then this corresponds to the factorization of the group determinant $\Theta(x) = \Phi(x)\Psi(x)$, $x = (x_1, \ldots, x_g)$, where $\Phi(x) = \det \mu(\mathbf{x})$ and $\Psi(x) = \det \nu(\mathbf{x})$. As we shall see, it was the question of how the polynomial Θ factors into irreducible factors, and how this factorization reflects the structure of the underlying group, that ultimately led Frobenius to his theory of matrix representations of finite groups.

The first steps in this direction were taken by Dedekind. In the case that $\mathfrak{G}$ is abelian, he discovered a remarkable connection between the factorization of Θ and the characters of $\mathfrak{G}$: Θ factors into linear factors with the g characters of $\mathfrak{G}$ giving the coefficients of the factors. That is, if $\chi^{(1)}, \ldots, \chi^{(g)}$ denote the characters of $\mathfrak{G}$, then

$$\Theta = \prod_{\lambda=1}^{g} \left(\sum_{r \in \mathfrak{G}} \chi^{(\lambda)}(r) x_r \right). \tag{1.3}$$

Later, in 1886, Dedekind considered the question of the factorization of Θ when $\mathfrak{G}$ is nonabelian. By computing and factoring group determinants of low-order nonabelian groups such as the symmetric group $\mathfrak{G} = S_3$ and the quaternion group $\mathfrak{G} = \{\pm 1, \pm i, \pm j, \pm k\}$, he saw that Θ does not generally factor into linear factors when $\mathfrak{G}$ is not abelian. His response to this fact was to pose the following question: For any finite group $\mathfrak{G}$ is it possible to obtain a factorization into linear factors

$$\Theta(x) = \prod_{\lambda=1}^{g} \left(\sum_{r \in \mathfrak{G}} \mathbf{a}_r^{(\lambda)} x_r \right)$$

by allowing the coefficients $\mathbf{a}_i^{(\lambda)}$ to belong to a linear associative algebra $\mathcal{K}$ over $\mathbb{C}$ that is somehow obtained from $\mathfrak{G}$?

At this time—the mid-1880s—linear associative algebras over the real or complex field were generally called hypercomplex numbers. As the name suggests, they were regarded as generalizations of complex numbers, among which the quaternions of Hamilton stood out as and important an early example. Incidentally, Frobenius himself had contributed to the literature on hypercomplex numbers by proving in 1878 that the only linear associative algebras over $\mathbb{R}$ possessing an identity element and such that all nonzero numbers have inverses are $\mathbb{R}$, $\mathbb{C}$, and the real quaternions. In 1885 Dedekind had published a paper on commuting hypercomplex systems, and this may have been the stimulus for the question he posed about Θ. In any case, for S_3, the quaternion group, and a semidirect product of cyclic groups of orders 2 and 5 he managed to come up with a system $\mathcal{K}$. The most striking example, however, was the quaternion group. By making a linear variable change $(x_1, \ldots, x_8) \to (u_1, \ldots, u_4, v_1, \ldots, v_4)$, he was able to factor Θ into four linear factors and a repeated irreducible quadratic factor $\Phi = v_1^2 + v_2^2 + v_3^2 + v_4^2$, which he recognized as the norm of the quaternion $v_1\mathbf{1} + v_2\mathbf{i} + v_3\mathbf{j} + v_4\mathbf{k}$. Hence using quaternions as coefficients

$$\Phi = (\mathbf{1}v_1 + \mathbf{i}v_2 + \ jv_3 + \mathbf{k}v_4)(\mathbf{1}v_1 - \mathbf{i}v_2 - \mathbf{j}v_3 - \mathbf{k}v_4),$$

and so Θ factors into linear factors.

In 1896 Dedekind communicated the above results and questions regarding the group determinant to Frobenius, together with the half-proved conjecture that in general the number of linear factors in Θ is equal to the order of the abelian group $\mathfrak{G}/\mathfrak{G}'$, where $\mathfrak{G}'$ denotes the commutator subgroup of $\mathfrak{G}$. Frobenius was an expert on what we would now call linear algebra and had also published a number of papers dealing with finite groups. He was therefore ideally suited to study the group determinant, and he did so with great relish. From the outset, however, he did not find Dedekind's "hypercomplex" approach to the factorization of Θ appealing. For him the problem was the following. Let $\Theta(x) = \prod_{\lambda=1}^{l} \Phi_\lambda(x)^{e_\lambda}$ denote the factorization of the homogeneous polynomial Θ into distinct irreducible homogeneous factors Φ_λ of degree f_λ. How is this factorization related to the nature of the underlying group $\mathfrak{G}$? In particular, is the number of linear factors what Dedekind conjectures, and how are l, e_λ, and f_λ related to $\mathfrak{G}$?

Six months after Dedekind had informed him about the group determinant, Frobenius was able to answer all these questions [1896]. Yes, Dedekind's conjecture was correct. Furthermore, the total number of distinct irreducible factors l was equal to the number k of conjugate classes of $\mathfrak{G}$; and $e_\lambda = f_\lambda$, which means that each Φ_λ occurs as often as its degree. In the course of establishing these results he inadvertently hit upon a generalization of the notion of a group character in the following manner. From the definition of Θ it follows that $\Theta = x_e^g + \cdots$, where e denotes the identity element of $\mathfrak{G}$. Thus we may completely specify the factors Φ_λ by requiring that the coefficient of x^{f_λ} also be 1. Thus

$$\Phi_\lambda(x) = x_e^{f_\lambda} + \left(\sum_{r \neq e} \psi^{(\lambda)}(r) x_r\right) x_e^{f_\lambda - 1} + \cdots. \tag{1.4}$$

Initially, $\psi^{(\lambda)}(r)$ was just a notation for the coefficient of $x_r x_e^{f_\lambda - 1}$, $r \neq 1$, although early in his investigations Frobenius could see that $\psi^{(\lambda)}$ with $\psi^{(\lambda)}(e) = f_\lambda$ could be regarded as a generalization of the traditional notion of a character. That is, (1.4) with $f = 1$ says that the linear factors of Θ are of the form $\sum_{r \in \mathfrak{G}} \psi^{(\lambda)}(r) x_r$, whereas, in reaction to Dedekind's conjecture, Frobenius had already proved that the linear factors are all of the form

$\sum_{r\in\mathfrak{G}} \chi(\mathfrak{G}'r)x_r$, where χ denotes one of the (traditional) group characters of the abelian group $\mathfrak{G}/\mathfrak{G}'$ and $\mathfrak{G}'$ denotes the commutator subgroup.

Eventually, Frobenius discovered that Φ_λ is completely determined by $\psi^{(\lambda)}$, and that although $\psi^{(\lambda)}$ does not have the multiplicative property (1.1), it has the property that $\psi^{(\lambda)}(rs) = \psi^{(\lambda)}(sr)$ or, equivalently, $\psi^{(\lambda)}(pqp^{-1})) = \psi^{(\lambda)}(q)$, which means that the functions $\psi^{(\lambda)}$ are constant on the conjugate classes of $\mathfrak{G}$. This fact led him to regard the $\psi^{(\lambda)}$ as functions on these classes and to introduce notation expressive of this fact. Another fundamental property of the $\psi^{(\lambda)}$ discovered by Frobenius is that they satisfy the orthogonality relations

$$\sum_{r\in\mathfrak{G}} \psi^{(\lambda)}(r)\overline{\psi^{(\mu)}(r)} = g \cdot \delta_{\lambda\mu}. \tag{1.5}$$

By virtue of the important role they play in the theory of the group determinant, Frobenius called his functions characters.

In two key papers [1897; 1899], Frobenius showed that his investigations on the group determinant could be recast and extended within the now familiar framework of a theory of the representation of finite groups by matrices, which he deemed to be far more important. In these papers he introduced all the basic notions of the theory, including the term "representation" (*Darstellung*) itself, and, in particular, observed that his generalized characters $\psi^{(\lambda)}$ are simply the trace functions of the irreducible representations of $\mathfrak{G}$.

Some of Frobenius's key results are summarized here for future reference. The first is what I shall refer to as the complete reducibility theorem:

(1.6) **Complete Reducibility Theorem.** *If $\phi : \mathfrak{G} \to \mathbf{GL}(n, \mathbb{C})$ is any representation of $\mathfrak{G}$, there exists a nonsingular matrix M such that*

$$M\phi(r)M^{-1} = \begin{pmatrix} \mu_1(r) & 0 & \cdots & 0 \\ 0 & \mu_2(r) & \cdots & 0 \\ \vdots & \vdots & \ddots & \vdots \\ 0 & 0 & \cdots & \mu_m(r) \end{pmatrix},$$

where the μ_λ denote irreducible representations of $\mathfrak{G}$.

This theorem was also discovered independently (and slightly earlier) by the Estonian mathematician Theodore Molien, whose work is discussed further in Sections 4–5 below. Also Heinrich Maschke discovered a theorem—Maschke's theorem of Section 4—which is equivalent to Theorem 1.6. However, it was as a part of Frobenius's theory and its extension by Schur to compact continuous groups that the complete reducibility theorem influenced Weyl, as will be seen in Chapters 11 and 12.

It is convenient to express the result of Theorem 1.6 in the notation $\phi \sim \mu_1 \oplus \mu_2 \oplus \cdots \oplus \mu_m$. Using this notation, we may state a second key result as follows:

(1.7) **Regular Representation Theorem.** *The number of inequivalent irreducible representations of* $\mathfrak{G}$ *equals* k, *the number of its conjugacy classes. If* $\mu_1, \ldots, \mu_k$ *denote the irreducible representations and* f_λ *denotes the degree of* μ_λ, *then each* μ_λ *occurs* f_λ *times in the complete reduction of the regular representation* ρ_R, i.e.,

$$\rho_R \sim \overbrace{\mu_1 \oplus \cdots \oplus \mu_1}^{f_1} \oplus \quad \cdots \quad \oplus \overbrace{\mu_k \oplus \cdots \oplus \mu_k}^{f_k} .$$

Of course, $\Phi_\lambda(x) = \det\left[\sum_{r\in\mathfrak{G}} \mu_\lambda(r)x_r\right]$ and $\psi^{(\lambda)}(r) = \operatorname{tr}\,\mu_\lambda(r)$, so that this theorem contains all of Frobenius's discoveries about the factorization of the group determinant.

Even in its matrix representation form, the theory of the generalized characters continued to play a central role in Frobenius's theory. This is because two representations of $\mathfrak{G}$, ν and ν', are equivalent if and only if their characters, $\chi = \operatorname{tr}\,\nu$ and $\chi' = \operatorname{tr}\,\nu'$, are equal. This is a consequence of the orthogonality relation (1.5) and the complete reducibility theorem. To see this, the notation

$$\langle \chi, \chi' \rangle = \frac{1}{g} \sum_{r\in\mathfrak{G}} \chi(r)\overline{\chi'(r)},$$

for any two characters, will be used. What needs showing is that if $\chi = \chi'$ then $\nu \sim \nu'$. By the complete reducibility theorem we know that $\nu \sim \sum_\lambda m_\lambda \mu_\lambda$, and $\nu' \sim \sum_\lambda m'_\lambda \mu_\lambda$, so it is necessary to show that $m_\lambda = m'_\lambda$ for all $\lambda = 1, \ldots, k$. This follows from the orthogonality relation, since, e.g., $\chi = \sum_\lambda m_\lambda \psi_\lambda$, and so for any fixed λ_0,

$$\langle \chi, \psi^{(\lambda_0)} \rangle = \sum_\lambda m_\lambda \langle \psi^{(\lambda)}, \psi^{(\lambda_0)} \rangle = m_{\lambda_0},$$

and similarly $\langle \chi', \psi^{(\lambda_0)} \rangle = m'_{\lambda_0}$. Thus $\chi = \chi'$ implies that $m_{\lambda_0} = m'_{\lambda_0}$ for all λ_0, and so $\nu \sim \nu'$.

For a number of examples of nonabelian finite groups Frobenius had computed the k^2 matrix or table of values $\psi_\alpha^{(\lambda)}$ using the orthogonality relations and other devices. In particular, $\psi_1^{(\lambda)} = f_\lambda$, the degree (or dimension) of the λth irreducible representation, since $\psi_1^{(\lambda)} = \operatorname{tr}\,\mu_\lambda(e)$, the trace of the $f_\lambda \times f_\lambda$ identity matrix. His most ambitious undertaking along these lines was the formidable problem of determining formulas for the characters of the symmetric group $\mathfrak{S}_n$, for any n, which he did in [1900] using his theory of induced characters [1898]. He characterized the values of the irreducible characters $\psi^{(\lambda)}$ for elements in a fixed conjugate class as certain coefficients in a polynomial expansion, and using this characterization obtained an explicit formula for the degree $f^{(\lambda)}$.

Although a representation ϕ of $\mathfrak{G}$ is uniquely determined by its character $\chi = \operatorname{tr}\phi$, Frobenius realized that knowing χ was not sufficient for the complete reduction of ϕ by means of "rational operations." Consider, for example, the regular representation ρ_R. Then an equivalence of the form $\rho_R \sim \nu_1 \oplus \cdots \oplus \nu_k$, with $\Phi_\lambda(x)^{f_\lambda} = \det[\sum_{r\in\mathfrak{G}} \nu_\lambda(r)\, x_r]$, can be obtained by means of a matrix M with entries in the field $\mathbb{F}$ obtained by adjoining to the field $\mathbb{Q}$ of rational numbers the k^2 values of the character table of $\mathfrak{G}$.[6] But to further decompose ρ_R into the irreducible representations requires a matrix with coefficients involving "higher irrationalities." Likewise, over $\mathbb{F}$ the irreducible representations cannot be constructed.

The requisite "higher irrationalities" were supplied by what Frobenius called the primitive characteristic units of the group. These were conceived by him in the same way he conceived of characters, namely as complex-valued functions on $\mathfrak{G}$. Thus associated to each irreducible character $\psi^{(\lambda)}(r)$ was a primitive characteristic unit $\varepsilon^{(\lambda)}(r)$. As the term "unit" suggests, Frobenius realized that $\varepsilon^{(\lambda)}(r)$ corresponds to an idempotent element in the group algebra $\mathcal{H}$ of $\mathfrak{G}$, namely $\mathbf{e}^{(\lambda)} = \sum_{r\in\mathfrak{G}} \varepsilon^{(\lambda)}(r)r$, although he preferred not to develop his ideas within the framework of the group algebra. Nowadays, the $\mathbf{e}^{(\lambda)}$ would be called primitive idempotents of $\mathcal{H}$. Not only did the primitive characteristic units determine the irreducible characters by the simple relation $\psi^{(\lambda)}(r) = \sum_{s\in\mathfrak{G}} \varepsilon^{(\lambda)}(s^{-1}r^{-1}s)$, but, as Frobenius showed, the complete reduction of the regular representation ρ_R as given in Theorem 1.7 could be achieved with a matrix M with coefficients from the field $\mathbb{K}$ obtained by adjoining to $\mathbb{Q}$ the kg values $\varepsilon^{(\lambda)}(r)$ of the primitive units. Furthermore, over $\mathbb{K}$ Frobenius derived simple formulas for both the irreducible factors $\Phi_\lambda(x)$ of the group determinant and the corresponding irreducible representation $\mu^{(\lambda)}$.[7]

As we have seen, Frobenius had devoted considerable effort to the formidable problem of determining the irreducible characters of the symmetric group $\mathfrak{S}_n$. Not long after having done this work he discovered that the British mathematician Alfred Young (1873–1940) had written two papers [1901; 1902] that, properly interpreted, yielded a remarkably simple formula for the primitive characteristic units of $\mathfrak{S}_n$. Entitled "On Quantitative Substitutional Analysis," Young's papers were motivated by his work on the theory of invariants. It had led him to the problem of determining polynomials $P = P(x_1, \ldots, x_n)$ having what might be called in a broad sense symmetry properties. Young expressed these properties in terms of what he called "substitutional equations" such as

$$(1.8) \qquad (\lambda_1 + \lambda_2 s_2 + \lambda_3 s_3 + \cdots)P = 0,$$

where $s_2, s_3, \ldots$ denote the substitutions (or permutations) on n objects (other than the "identical" substitution) and the left-hand side of (1.8)

---

6 Frobenius did not use the language of fields but spoke of expressing the coefficients of M rationally in terms of the k^2 character values.

7 These results are summarized in Theorem 15.4 of [Hawkins 1999a].

means $P+\lambda_2 s_2 P+\lambda_3 s_3 P+\cdots$, where $s_m P$ is P with its variables permuted according to s_m.

Equation (1.8) is expressed in Young's notation. In Frobenius own notation, (1.8) may be written as $(\sum_{S\in\mathfrak{S}_n} a_S S)P = 0$. Young showed that all P satisfying (1.8) may be written in the form $P = \sum_{i=1}^m P_i$, where corresponding to each P_i is a "substitutional expression" $\sum_{S\in\mathfrak{S}_n} c_S^{(i)} S$ such that $P_i = (\sum_{S\in\mathfrak{S}_n} c_S^{(i)} S)F_i$, where the polynomial F_i is an arbitrary polynomial in $x_1,\ldots,x_n$. Moreover, each such substitutional expression satisfies the relation

$$\left(\sum_{S\in\mathfrak{S}_n} a_S S\right)\left(\sum_{S\in\mathfrak{S}_n} c_S^{(i)} S\right) = 0,$$

where the multiplication on the left-hand side of this equation was to be carried out "by virtue of the multiplication table of the group" [1901:103]. These equations could then be used to find the expressions $(\sum_{S\in\mathfrak{S}_n} c_S^{(i)} S)$ and thence to find all P satisfying (1.8).

Young's papers were motivated by the theory of invariants, a theory that Frobenius held in low esteem—see Section 2 below—and much of their contents was not of direct interest to Frobenius. But he could see that in effect, Young was investigating the group algebra of the symmetric group; and taking the trouble to go through the entire contents of the papers, Frobenius was rewarded by his discovery that Young had, in effect, obtained a remarkable formula for the primitive idempotents of $\mathfrak{S}_n$. Frobenius wrote up this discovery in a paper "On the characteristic units of the symmetric group" [1903], which was to have a decisive influence upon Weyl, as will be seen in Chapter 11.

In order to state Frobenius's main theorem, it is necessary to introduce some notions from Young's paper relating to what is now called a Young tableau—an arrangement $\mathfrak{T}$ of n objects $a_1,\ldots,a_n$ into rows of decreasing length $\alpha_1 \geq \alpha_2 \geq \cdots \geq \alpha_\mu$:

$$\begin{array}{ccccccccc} a_{1,1} & a_{1,2} & \cdots & \cdot & \cdot & \cdot & \cdot & a_{1,\alpha_1} \\ a_{2,1} & a_{2,2} & \cdots & \cdot & \cdot & a_{2,\alpha_2} & & \\ \cdot & \cdot & \cdots & \cdot & & & & \\ a_{\mu,1} & a_{\mu,2} & \cdots & a_{\mu,\alpha_\mu} & & & & \end{array} \tag{1.9}$$

Let $\mathfrak{R}$ denote the subgroup of $\mathfrak{S}_n$ consisting of all permutations R that permute among themselves the objects in each row of $\mathfrak{T}$, and let $\mathfrak{C}$ denote the subgroup of all those permutations C that permute among themselves the objects in each column of $\mathfrak{T}$. Then Frobenius's version of what he found in Young may be stated in the following manner:

(1.10) **Young–Frobenius Theorem.** *Given $\mathfrak{T}$ and the associated groups $\mathfrak{C}$ and $\mathfrak{R}$, the equation $RC = C'R'$ with $C, C' \in \mathfrak{C}$ and $R, R' \in \mathfrak{R}$, has more solutions such that $C'C$ is even than solutions with $C'C$ odd. Let the excess of the even solutions over the odd solutions be denoted by $n!/f$,* *and let $\zeta(S)$ be defined for $S \in \mathfrak{S}_n$ by $\zeta(S) = \operatorname{sgn} C$ if $S = RC$ with $C \in \mathfrak{C}$ and $R \in \mathfrak{R}$ and $\zeta(S) = 0$ otherwise. Then $\varepsilon(S) = (f/n!)\zeta(S)$ is a primitive characteristic unit of $\mathfrak{S}_n$.*

Frobenius wrote [1903:265] that this theorem "probably contains some of the most noteworthy properties of the symmetric group and its characters."

Since our primary concern is with the remarkable analogue of Frobenius's theory that Weyl constructed for all semisimple Lie groups, it is of interest to ask whether Frobenius ever contemplated an extension of his theory to a broad class of continuous groups. Such an extension, to be at all general, would have to be based upon a theory of continuous groups—Lie's theory. But as we have seen, Lie and his theory were held in low esteem in Berlin (Section 6.1). Lie was regarded as a poorly educated, narrow-minded mathematician, and his theory of transformation groups was regarded as in need of a sound mathematical foundation. These views, which are not without some justification, seem to have been held—perhaps in a less extreme form—by Weierstrass as well as Frobenius.[8] In addition, Frobenius was critical of Lie's theory because although the problem of developing a theory of continuous groups was an original one, Lie's attempt at a solution was deemed totally lacking in originality: What Lie had done, so Frobenius felt, was simply to mechanically transfer to continuous groups "the elegant concepts that the most gifted algebraists had created in the theory of finite groups" evidently in the process defiling them, since the resultant continuous theory was deemed "trivial" and its applications "insignificant."[9]

Frobenius's harsh estimate was not entirely off the mark, for Lie's theory was primarily presented on a rather naive formal level, and although it contained many fertile ideas, it lacked the mathematical coherence, rigor, and depth of, say, Frobenius's theory of group representations. Of course, the Killing–Cartan theory of semisimple groups, as presented by Cartan, was another matter by virtue of its profound theoretical substance. But it is doubtful that Frobenius ever attempted to study Cartan's difficult papers. Frobenius tended to dismiss work coming out of Lie's school as ipso facto containing nothing of mathematical significance.[10] Had Frobenius lived to see Weyl's theory, I have no doubt he would have been favorably impressed, because what Weyl did was far from a mindless extension of Frobenius's ideas. The point here is that Frobenius himself was not disposed to envision a significant extension of his theory within the context of Lie's theory.

All of Frobenius's major contributions to the theory of group charac-

---

[8] Judging by the 1902 evaluation of Lie for a position at Berlin [Biermann 1973:215], which was written by Frobenius but refers to Weierstrass's views.

[9] [Biermann 1973:215]; my translation.

[10] This point is documented in Section 4 in the discussion of Molien. Frobenius did at least eventually read Cartan's paper [1898] on linear associative algebras (hypercomplex numbers), but that was after he had discovered (with Molien) that important mathematics could come out of Lie's school. His respectful but critical reaction to Cartan's paper is described in my essay [1999a].

ters and representations had been made by 1903, and his last paper having anything to do with the theory was published in 1907. By that time the mantle had passed on to his favorite student, Issai Schur (1875–1941), who developed the theory in several directions, one of which is particularly relevant to the present story, since it involved extensions of Frobenius's ideas to specific continuous groups. These extensions were inspired by Hurwitz's work on the theory of invariants, to which I now turn.

10.2. Hurwitz and the Theory of Invariants

When Frobenius left Zürich for a Berlin professorship in 1892, his position at the ETH was given to Hurwitz, who remained there for the remainder of his career. In 1894 and 1897 he published his only papers devoted to the theory of invariants. Hilbert's enthusiasm for the theory and his remarkable achievements in this domain during 1888–93 did not go unnoticed by his friend and resulted in two papers, both of which influenced and inspired Schur, as we shall see in Sections 3 and 5.

In the traditional approach to the study of invariants, they were defined by means of an underlying base form (*Grundform*) or system of such forms. For example, as we saw in Section 9.1, the invariants with respect to the homogeneous polynomial $\phi(a, x)$ of degree d as base form are invariants with respect to the linear transformations $a' = \Gamma(A) \cdot a$, $A \in \mathbf{GL}(n)$, introduced at the beginning of Section 9.1. We saw that at Klein's urging, Hilbert had proposed considering invariants with respect to subgroups of $\mathbf{GL}(n)$ such as the rotation group. In his paper [1894] Hurwitz suggested a generalization in a different direction. He pointed out that the concept of an invariant was only indirectly related to the base forms; what was essential was (in the notation of Section 9.1) that an invariant be a homogeneous polynomial $I(a)$ in the variables a_i such that $I(a) = I(a')$ with respect to a certain system of linear transformations $\Gamma(A) : a \to a'$. Thus Hurwitz proposed thinking of invariants as homogeneous polynomials satisfying $I(a) = I(a')$ with respect to a system of linear transformations $a = Ta'$. Furthermore, he pointed out that if T_1, T_2 are two transformations of this system, so that $I(T_j a') = I(a')$, $j = 1, 2$, and if $T = T_1 T_2$, then $I(Ta') = I(a')$ also; and so the system may be regarded as forming a group $\mathfrak{G}$ of linear transformations.

In effect, in discussing the traditional theory, Hurwitz proposed to emphasize the more general (and now commonplace) view of the theory as the study of homogeneous polynomials left invariant by a group of linear transformations. As we noted in Section 9.1, for finite groups $\mathfrak{G} \subset \mathbf{GL}(n)$ Hurwitz's teacher Klein had already introduced such a notion of invariant, which may have influenced Hurwitz to bring it to bear on the more classical situation in which $\mathfrak{G}$ is a continuous group, such as the group of transformations $T = \Gamma^{-1}(A)$.

Since the goal of Hurwitz's paper was to consider the classical theory within the above conceptual framework, and so without consideration of the intermediary base forms, he showed how to define directly the sort of groups $\mathfrak{G}$ that occur in the classical theory. To this end he introduced several basic types of transformations, which turn out to be important in what is to follow. The first of these he called *product transformations.* If $A \in \mathbf{GL}(m, \mathbb{C})$ and $B \in \mathbf{GL}(n, \mathbb{C})$ transform respectively the variables $x = (x_1, \ldots, x_m)$ and $y = (y_1, \ldots, y_n)$, then $A \times B$ transforms the mn variables $x_i y_j$ in the obvious way. Nowadays, $A \times B$ is frequently called a tensor or Kronecker product of A and B. As we have seen in Section 7.5 and Section 8.4, both Kowalewski and Cartan made use of such transformations. Hurwitz pointed out that $(A, B) \to A \times B$ defines a homomorphism of the group of pairs (A, B).

Next Hurwitz considered *power transformations.* Given the variables $x_1, \ldots, x_n$ and a fixed positive integer d, consider the $N = \binom{n+d-1}{d}$ products $x_1^{e_1} x_2^{e_2} \cdots x_n^{e_n}$, where $e_1 + e_2 + \cdots + e_n = d$. Order them somehow into a fixed sequence $X_1, \ldots, X_N$.[11] Let $X = P_d(A)X'$ be the transformation of the X_i induced by setting $x = Ax'$. It is called the dth power transformation. As he observed, the matrix coefficients of $P_d(A)$ are homogeneous functions of degree d in the coefficients a_{ij} of A. He also showed that $P_d(A)$ has the properties that make $A \to P_d(A)$ a group homomorphism.

Although the product and the power transformations were the basic ones and would be the only ones occurring in the remainder of his paper, Hurwitz nonetheless introduced a third type of transformation, which he called *determinant transformations.* For $m \le n$ let $x^{(k)} = (x_1^{(k)}, \ldots, x_n^{(k)})$, $k = 1, \ldots, m$, denote m series of variables. Then set

$$X_{i_1 \ldots i_m} = \begin{vmatrix} x_{i_1}^{(1)} & \cdots & x_{i_1}^{(m)} \\ x_{i_2}^{(1)} & \cdots & x_{i_2}^{(m)} \\ & \ddots & \\ x_{i_m}^{(1)} & \cdots & x_{i_m}^{(m)} \end{vmatrix}.$$

As Hurwitz acknowledged and we have seen (Section 7.3), Clebsch first introduced these generalized Plücker coordinates into the theory of invariants in 1872 with an eye toward geometrical applications, and Cartan later (1913) utilized them in his determination of the irreducible representations of $\mathbf{sl}(n, \mathbb{C})$ as indicated in Section 8.4. Let these $N = \binom{n}{m}$ variables be ordered in a fixed sequence $X_1^{(m)}, \ldots, X_N^{(m)}$. Then the transformation $X = C_m(A) \cdot X'$ of these variables induced by setting $x^{(k)} = A[x'^{(k)}]$ is called the nth determinant transformation associated to A. Once again Hurwitz established the properties of determinant transformations that show that

[11] Actually, Hurwitz first "prepared" his terms so that his X_α are the above-described divided by $\prod_i \sqrt{e_i!}$, e_i being the degree of x_i in the given term [1894:518].

Figure 33. Adolf Hurwitz (1859–1919). His two papers on invariants played a seminal role in the creation of a representation theory for continuous groups.

$A \to C_m(A)$ is a homomorphism of $\mathbf{GL}(n)$. And once again the coefficients of $C_m(A)$ are homogeneous polynomials of a fixed degree in the coefficients a_{ij} of A.[12]

These three types of transformation actually suffice to create the groups $\mathfrak{G}$ that are usually considered in the theory of invariants. In the case of the invariants of the form $\phi(a;x)$ of Section 9.1, it turns out that the transformations $a = Ta'$ induced by $x = Ax'$ are $T = T(A) =$

[12] The representation $A \to C_m(A)$ is the one with representation module consisting of the skew symmetric tensors of degree m, i.e., the skew symmetric tensors in $\mathcal{V} \otimes \cdots \otimes \mathcal{V}$ (m factors), where $n = \dim \mathcal{V}$.

$[P_d(A)]^C = P_d(A^C)$, where $A^C = [A^{-1}]^T$, the "contragredient" of A.[13] Of course, the *set* of transformations $P_d(A^C)$, $A \in \mathbf{GL}(n)$, is precisely the same as the set $P_d(A)$, $A \in \mathbf{GL}(n)$, so one can just as well study the invariants with respect to the group $T(A) = P_d(A)$. For a second example, consider the invariants with respect to the base form $\phi(a; x, y)$, where ϕ denotes the general polynomial in $x_1, \ldots, x_m, y_1, \ldots, y_n$ that is homogeneous of degree d in the x_i and of degree e in the y_j. Then the group $\mathfrak{G}$ induced by setting $x = Ax'$, $y = By'$ consists of the linear transformations $[P_d(A) \times P_e(B)]^C = P_d(A^C) \times P_e(B^C)$. As in the previous case, the study of the invariants with respect to $\phi(a; x, y)$ may be based upon using the group of transformations $T(A, B) = P_d(A) \times P_d(B)$, or when $m = n$ and $A = B$, upon the group $T(A) = P_d(A) \times P_d(A)$. Finally, in geometrical applications one might consider invariants with respect to, e.g., $\phi(a; x, X^{(m)})$, the general form homogeneous of degree d in point coordinates $x_1, \ldots, x_n$ and of degree e in the generalized Plücker coordinates $X_1^{(m)}, \ldots, X_N^{(m)}$. This would be the theory of invariants with respect to the group of transformations $T(A) = P_d(A) \times P_e(C_m(A))$.

In all these examples where the group $\mathfrak{G}$ of transformations T depends on a single $A \in \mathbf{GL}(n)$, so that $T = T(A)$, we see that $A \to T(A)$ is a group homomorphism and the coefficients of $T(A)$ are homogeneous polynomials of a fixed degree in the coefficients a_{ij} of A. Thanks to Hurwitz's paper, it was easy for Schur to see that $A \to T(A)$ was a special type of representation of $\mathbf{GL}(n)$, and that these are the type of representations underlying the various sorts of invariants traditionally studied. As we shall see in Section 3, in his dissertation [1901] Schur proposed to determine all of them.

In 1896 Minkowski had joined his old friend Hurwitz at the ETH in Zürich, where he remained until he received the call to Göttingen in 1902. In a letter to Hilbert in 1897 he wrote, "I am together with Hurwitz a lot. In recent days he has found really interesting new theorems about invariants and has had the good fortune to prove new finiteness theorems for which other methods did not suffice; he is very pleased."[14] Hurwitz's discoveries were published in the Göttingen *Nachrichten* [1897]. A few weeks thereafter, he explained the gist of his paper to Hilbert as follows:[15]

> I believe the contents of this note will interest you, since it is also related to your finiteness theories. Thus, e.g., I have succeeded in proving in general the finiteness of the orthogonal invariants, and I believe the same methods will suffice for the proof of finiteness of invariants of an arbitrary algebraic group of linear transformations. The fundamental idea of my note is this: If

[13] This can be seen by a line of reasoning used by Hurwitz: In terms of the X_α, the form $\phi(a; x)$ becomes bilinear, $\phi(a; x) = \sum_\alpha a_\alpha X_\alpha = \langle a, X \rangle$, so that setting $x = Ax'$ and $X = P_d(A) \cdot X'$, $\langle a, X \rangle = \langle a, P_d(A) \cdot X' \rangle = \langle P_d(A)^T \cdot a, X' \rangle = \langle a', X' \rangle$. Thus $a = P_d(A)^C a'$.

[14] Letter dated 11 March 1897 [Rüdenberg and Zassenhaus 1973:96].

[15] Letter dated 24 March 1897 (NSUB Göttingen, Nachlass Hilbert).

$$x_i' = f_i^{(\lambda)}(x_1, x_2, \ldots, x_n) \quad (i = 1, 2, \ldots, n),$$

where λ runs through the indices $1, 2, \ldots, r$, form a finite group of r substitutions and if the function $\mathcal{F}(x_1, x_2, \ldots, x_n)$ goes over to $\mathcal{F}_1, \mathcal{F}_2, \ldots, \mathcal{F}_r$ by these substitutions, then the sum

$$\mathcal{F}_1 + \mathcal{F}_2 + \cdots + \mathcal{F}_r$$

represents an invariant and indeed the most general. A corresponding procedure for manufacturing invariants can (always?) be followed for continuous groups of transformations, with a definite integral replacing the sum $\mathcal{F}_1 + \mathcal{F}_2 + \cdots + \mathcal{F}_r$.

In order to appreciate Hurwitz's remarks, it is necessary to recall from Section 9.1 that Hilbert had shown that by virtue of his finite basis theorem (1.5), the finiteness property could be established for the system $\mathfrak{J}$ of invariants defined by any group $\mathfrak{G} \subset \mathbf{GL}(N)$, provided that an operator $\mathcal{E}$ could be defined on all homogeneous polynomials F in N variables such that $\mathcal{E}[F]$ is another such polynomial with degree equal to or less than that of F satisfying the following relations for all F and all $I \in \mathfrak{J}$:

$$\text{(2.1)} \qquad \begin{array}{ll} \text{(i) } \mathcal{E}[F] \in \mathfrak{J}, & \text{(ii) } \mathcal{E}[I] = I, \\ \text{(iii) } \mathcal{E}[F \cdot I] = \mathcal{E}[F] \cdot I, & \text{(iv) } \mathcal{E}[F_1 + F_2] = \mathcal{E}[F_1] + \mathcal{E}[F_2]. \end{array}$$

Furthermore, Hilbert had extended his finiteness theorem to cover the case in which $\mathfrak{G}$ is the group of transformations $\Gamma(S)$ of the parameters $a_1, \ldots, a_N$ of the base form $\phi(a, x)$ generated by restricting S to the rotation group $\mathbf{SO}(3, \mathbb{R})$, but his method, which involved the use of quaternions, did not generalize to cover the general rotation group $\mathfrak{D} = \mathbf{SO}(n, \mathbb{R})$, $n > 3$. In his paper Hurwitz showed how to define a suitable operator $\mathcal{E}$ by integration (rather than the more usual differentiation) so as to establish the finiteness theorem for the invariants relative to the general rotation group.

As his letter to Hilbert indicates, Hurwitz's idea to use integration was inspired by the case of a finite group. Once one sees it, the idea of generating invariants of a finite group by the above summation process seems so obvious that it is perhaps worth noting that several good mathematicians failed to see it. That is, there was some interest in finding for a given finite subgroup of $\mathbf{GL}(n, \mathbb{C})$ a positive definite hermitian form that is an invariant of the group. This problem was posed by Picard in the case $n = 3$ in [1887a]. Noting its truth for $n = 2$, Picard deemed it "entirely natural" to ask whether the same is true for groups in 3 variables. This he sought to prove by utilizing Jordan's classification of all finite subgroups of $\mathbf{GL}(3, \mathbb{C})$. Thus it is a case-by-case construction. However, for one type he was unable to construct the desired hermitian form [1887a:154]. Similar case-by-case methods had been used by Klein's students. Finally, in 1896, G. E. Moore and A. Loewy independently showed how to use summation over the group to obtain the desired hermitian form.[16] In terms of Hurwitz's notation, if $\mathfrak{F}(x_1, x_2, \ldots, x_n)$ is any positive definite hermitian form, then $\mathfrak{F}_1 + \mathfrak{F}_2 + \cdots + \mathfrak{F}_r$ is the desired invariant positive definite hermitian form. Klein was so impressed by the Loewy–Moore result that he lectured on it as an important new discovery [1897].

There is no indication on Hurwitz's part that he was aware of the work of Loewy and Moore or of Picard's problem. It seems that the idea of generating invariants by summation over the group was suggested by similar considerations in the theory of automorphic functions, for after explaining the idea for finite groups Hurwitz remarked, "As is known, in a similar way the invariants of infinite discontinuous groups can be formed" [1897:546]. It would seem that this was an allusion to a technique used in the theory of elliptic functions and, more generally, in the theory of automorphic functions of Klein and Poincaré.[17] In his doctoral dissertation, "Foundations for an Independent Theory of Elliptic Modular Functions ..." [1881:19], written under Klein's direction, Hurwitz isolated the general underlying idea as it applies to modular forms, i.e., homogeneous functions $F(\omega)$, $\omega = (\omega_1, \omega_2)$, of two complex variables ω_1, ω_2 with the property that $F(T\omega) = F(\omega)$ for all T belonging to some subgroup of $\mathbf{SL}(2, \mathbb{Z})$. He stated it as follows. Suppose that $\varphi(\omega)$ is any homogeneous function and that $\Delta \subset \mathbb{C}^2$ is a region with the property that $T[\Delta] = \Delta$ for all T in a subgroup $\mathfrak{G} \subset \mathbf{SL}(2, \mathbb{Z})$ for which the series

$$\sum_{T \in \mathfrak{G}} \varphi(T\omega) \tag{2.2}$$

coverges absolutely on Δ (so the order of summation is irrelevant) and so as to define an analytic function $F(\omega)$ there. Then evidently F is a modular form, since for any $S \subset \mathfrak{G}$ it follows from (2.2) that $F(S\omega) = F(\omega)$, since the map $T \to ST$ is bijective and the order of summation irrelevant by hypothesis. The underlying principle here is essentially the one Frobenius later introduced for finite groups in the 1890s—what I have termed his principle of invariance under group summation in Section 10.4 below. Of course, because the sum is infinite, the choice of φ, Δ, and $\mathfrak{G}$ is not a trivial matter. Shortly after Hurwitz's dissertation was written, the technique figured prominently in Poincaré's theory of "theta Fuchsian functions" [1882a:97; 1882b:184].

For a *finite* subgroup $\mathfrak{G} \subset \mathbf{GL}(n)$, however, the above technique simplifies considerably. Since the summation over the group is finite, convergence is not a concern. Consequently, if $P(x)$ is any function of $x = (x_1, \ldots, x_n)$, then for the same reason as above,

$$I(x) = \sum_{T \in \mathfrak{G}} P(T(x)) \stackrel{\text{def}}{=} \mathfrak{I}[P](x)$$

---

[16] See in this connection Section 14 of my essay [Hawkins 1999a].

[17] Cf. in this connection [Slodowy 1998:§2]. On the contributions of Poincaré and Klein to the theory of automorphic functions, see the book by Gray [1986].

satisfies $I(Sx) = I(x)$ for all $S \in \mathfrak{G}$ and so is an invariant relative to $\mathfrak{G}$. Furthermore every invariant $I(x)$ is generated by this process, since if we start with $P(x) = I(x)/n$, then $\mathfrak{J}[P] = I$. In particular, when P is a positive definite hermitian form, $\mathfrak{J}[P]$ is the desired invariant hermitian form. As obvious an application of the above technique from complex function theory as this may seem to be, it had escaped the attention of two mathematicians who knew the technique and were also keenly interested in the matter of the existence of an invariant form for a given finite group, namely Klein and Picard, who had developed a higher-dimensional analogue of Poincaré's theory utilizing the same technique.

It follows immediately that if g denotes the order of $\mathfrak{G}$, then the operator $\mathcal{E} = (1/g)\mathfrak{J}$ satisfies all the properties in (2.1) and leaves degrees unaffected, so that Hilbert's proof idea may be used to prove the finiteness theorem for invariants relative to finite subgroups of $\mathbf{GL}(n)$, but Hurwitz did not pause to note this fact. His interest was focused upon continuous groups.

Hurwitz was primarily interested in a detailed study of the invariants of $\mathfrak{D}$ and of the special linear group, but he realized that some of his observations had a broader significance. And so, after first showing in considerable detail how integration could be used to deal with the invariant theory of the above-mentioned specific groups, he concluded his paper with a section in which he pointed out that the principle of generating invariants by integration could be extended to "every finite continuous group" in the sense of Lie [1897:560]. To facilitate the exposition of the paper as a whole it is helpful to begin by discussing this historically important section.

In invoking as he did Lie's theory—he referred to the first volume [1888] of the Lie–Engel three-volume *Transformationsgruppen*—Hurwitz was drawing upon a vaguely formulated theory, especially in terms of global considerations, but it was the only theory there was. Recall from Section 3.1 that for Lie a real r-dimensional transformation group $\mathfrak{G}$ consists of invertible transformations T_α from some N-dimensional "piece" of $\mathbb{R}^N$ or $\mathbb{C}^N$ into the same, where the parameters $\alpha = (\alpha_1, \ldots, \alpha_r)$ run through some r-dimensional piece D of $\mathbb{R}^r$ and the correspondence $\alpha \to T_\alpha$ is bijective. Of course, such a global parametrization is not generally possible, but Lie and his students glossed over such matters. Recall also from Section 3.1 that associated with $\mathfrak{G}$ is what Lie called its first parameter group [1888:401ff.], which is as close as he came to the notion of a Lie group as r-dimensional manifold. For $\alpha, \beta \in D$ define $\alpha\beta = \gamma$ if $T_\alpha \circ T_\beta = T_\gamma$. For Lie the parameter group consisted of the transformations $U_\alpha : \beta \to \alpha\beta$ of $\mathbb{R}^r$. They form a simply transitive group isomorphic to the original. In what follows U_α will be identified with α and no notational distinction will be made between $\mathfrak{G}$ and its parameter group.[18]

[18] Hurwitz in effect explained how to define an invariant-producing integral over the parameter group, although he did not make this clear and simply assumed from the outset that the group under consideration is an r-dimensional simply

To see the motivation behind Hurwitz's definition consider a homogeneous polynomial $F(a)$ in the variables $a = (a_1, \ldots, a_N)$. Corresponding to Hurwitz's finite sum $\mathfrak{J} = \mathfrak{F}_1 + \cdots + \mathfrak{F}_r$ is the integral

$$I(a) = \int_{\mathfrak{G}} F(T_\alpha \cdot a)\, d\alpha_1 d\alpha_2 \cdots d\alpha_r = \int_{\mathfrak{G}} f(\alpha)\, d\alpha. \tag{2.3}$$

In order for $I(a)$ to be an invariant one must have $I(a') = I(a)$, where $I(a') = I(T_\beta \cdot a)$. Now for any β,

$$I(a') = \int_{\mathfrak{G}} F(T_\alpha \circ T_\beta \cdot a)\, d\alpha = \int_{\mathfrak{G}} F(T_{\alpha\beta} \cdot a)\, d\alpha = \int_{\mathfrak{G}} f(\alpha\beta)\, d\alpha. \tag{2.4}$$

Thus $I(a') = I(a)$, and $I(a)$ is an invariant, provided that for continuous functions f one has

$$\int_{\mathfrak{G}} f(\alpha\beta)\, d\alpha = \int_{\mathfrak{G}} f(\alpha)\, d\alpha. \tag{2.5}$$

Hurwitz realized that in general one could not expect (2.5) to hold. For example, if $f = \chi_{\mathfrak{K}}$ denotes the characteristic function of an r-dimensional piece of $\mathfrak{G}$, called by Hurwitz a "body" (*Körper*), then we would have $\int_{\mathfrak{K}\beta^{-1}} d\alpha = \int_{\mathfrak{K}} d\alpha$, which says that volume in $\mathfrak{G}$ is preserved under right translations: The volumes of $\mathfrak{K}\beta^{-1}$ and $\mathfrak{K}$ are equal for all $\beta \in \mathfrak{G}$. If this were the case, then (2.5) would follow. Since this is not generally the case, he introduced a "density function" $\psi(\alpha)$ on $\mathfrak{G}$ so that the "mass" of $\mathfrak{K}$ is $m(\mathfrak{K}) = \int_{\mathfrak{K}} \psi(\alpha)\, d\alpha$. Then instead of (2.3) and (2.4) we have

$$I(a) = \int_{\mathfrak{G}} f(\alpha)\psi(\alpha)\, d\alpha, \quad I(a') = \int_{\mathfrak{G}} f(\alpha\beta)\psi(\alpha)\, d\alpha,$$

and $I(a') = I(a)$ will follow, provided that mass is invariant under right translations, so that $m(\mathfrak{K}\beta) = m(\mathfrak{K})$ for all $\beta \in \mathfrak{G}$. Since by a change of variables

$$m(\mathfrak{K}\beta) = \int_{\mathfrak{K}\beta} \psi(\alpha)\, d\alpha \overset{\alpha' = \alpha\beta^{-1}}{=} \int_{\mathfrak{K}} \psi(\alpha'\beta) J_{\beta^{-1}}(\alpha')\, d\alpha',$$

where $J_{\beta^{-1}} = \partial(\alpha_1, \ldots, \alpha_r)/\partial(\alpha'_1, \ldots, \alpha'_r)$ is the Jacobian of the transformation $\alpha \to \alpha' = \alpha\beta^{-1}$, we see that $m(\mathfrak{K}) = m(\mathfrak{K}\beta)$, provided that

$$\int_{\mathfrak{K}} \psi(\alpha)\, d\alpha = \int_{\mathfrak{K}} \psi(\alpha\beta) J_{\beta^{-1}}(\alpha)\, d\alpha.$$

Since this must hold for all $\mathfrak{K}$, the equality of the integrals is equivalent to the equality of the integrands, and therefore

$$\psi(\alpha\beta) = \psi(\alpha) J_\beta(\alpha) \quad \text{for all } \alpha,\ \beta, \tag{2.6}$$

---

transitive group on $\mathbb{R}^r$.

since $[J_{\beta^{-1}}]^{-1} = J_\beta$. Setting $\alpha = \epsilon$ (the identity element of $\mathfrak{G}$) we see that

$$\psi(\beta) = J_\beta(\epsilon)\psi(\epsilon), \tag{2.7}$$

so that ψ is uniquely determined up to a constant $\psi(\epsilon)$.

If ψ is defined by means of (2.7), it is easy to see from the properties of Jacobians under composition of transformations that (2.7) implies (2.6) and so defines a mass on $\mathfrak{G}$ that is invariant under right translations. This was the gist of Hurwitz's reasoning leading up to his theorem that "density can always be determined, and in essentially only one way, so that an arbitrary body of the space of r dimensions conserves its mass under all transformations of the given group ..." [1897:561]. Of course, the reasoning is not rigorous by modern standards because Hurwitz had to base it upon Lie's theory, which lacked a rigorous formulation, as Weierstrass and other Berliners had privately observed (Section 6.1, Section 10.1). A rigorous formulation of Hurwitz's theorem (independent of Lie's theory) was later given by Hilbert's student and Weyl's contemporary at Göttingen Alfred Haar [1933], after the work of Weyl (Chapter 12) focused attention on the desirability of such a formulation. In the following chapters, which cover the period before Haar's paper, these translation-invariant integrals will be referred to as *Hurwitz integrals* and indicated by notation such as $\int_{\mathfrak{G}} f(\alpha)dm$, where $dm = \psi(\alpha)d\alpha$ gives Hurwitz's density function.

By virtue of the above theorem, Hurwitz was able to conclude that $\mathfrak{I}[F] = \int_{\mathfrak{G}} F(T_\alpha \cdot a)\psi(\alpha)\, d\alpha$ transforms an arbitrary homogeneous polynomial $F(a)$ into an invariant function $I = \mathfrak{I}[F]$, provided that the integral converges. The proviso was necessary, since $\mathfrak{G}$ need not be closed and bounded, i.e., compact.[19] In fact, of the two groups that particularly interested him, the rotation group and the special linear group, the latter is unbounded.

In the case of his first example, the rotation group $\mathfrak{D} = \mathbf{SO}(n,\mathbb{R})$, there is no problem, since as he observed, the matrix coefficients of $S = (s_{ij}) \in \mathfrak{D}$ satisfy orthogonality relations, which imply that $-1 \leq s_{ij} \leq 1$ and hence that $\mathfrak{D}$ is bounded.[20] Thus for the rotation group $\mathfrak{D} \subset \mathbf{GL}(n,\mathbb{R})$ the operator

$$\mathcal{E}[F] = I(a) = \frac{1}{m}\int_{\mathfrak{D}} F(\Gamma(S)\cdot a)\, dm, \quad m = \int_{\mathfrak{D}} dm, \tag{2.8}$$

satisfies all the conditions of (2.1) and, since it does not raise degrees, can be used to prove the finiteness theorem for $\mathfrak{D}$. Hurwitz had therefore improved considerably on what Hilbert had been able to achieve for rotation groups. What primarily interested him, however, was the fact that using the n-dimensional version of Euler's representation of rotations, which provides a

[19] Hurwitz stressed the possible unboundedness of $\mathfrak{G}$; for the groups of interest to him closure was not a problem.

[20] [Hurwitz 1897:547]. Of course, $\mathfrak{D}$ is also closed and therefore compact, but Hurwitz did not mention this fact.

unique parametrization of $\mathfrak{D}$ in terms of $N = n(n-1)/2$ (suitably limited) rotation angles $\varphi_1, \ldots, \varphi_N$, he was able to obtain an explicit formula for the invariant $I(a) = \mathfrak{I}[F]$ as an integral with respect to $\phi_1, \ldots, \phi_N$ [1897:eqns. (24)–(26)]. Using this formula in the case of a binary base form $\phi(a, x)$, he was able to describe an algorithm for producing the invariants.

Although in his paper Hurwitz presented the above results on the rotation group first, he explained in his introductory remarks that they were actually a byproduct of his original research program, which was to find an expression for the classical invariants in terms of an integral. For Hurwitz the classical invariants were the absolute invariants $I(a)$ generated by the transformations of the special linear group $\mathbf{SL}(n)$, i.e., $I(\Gamma(S) \cdot a) = I(a)$ for all $S \in \mathbf{SL}(n)$. Since this group is unbounded, however, the method of integration used for the rotation group $\mathfrak{D}$ is not directly applicable, because the type of integrals to be considered will not converge. To overcome this obstacle Hurwitz introduced the following considerations, which gave birth to an extremely fertile idea eventually dubbed "the unitarian trick" by Weyl [1939a; 1939b].

Now it became necessary to replace the ambiguous $\mathbf{SL}(n)$ with the complex group $\mathbf{SL}(n, \mathbb{C})$. Within this group Hurwitz focused attention on the unitary transformations. They form a real $(n^2 - 1)$-dimensional group, now called the special unitary group and denoted by $\mathbf{SU}(n)$. At this time, however, this group was not well known. Although Frobenius and others had studied unitary matrices as analogues of orthogonal matrices, Study's introduction of $\mathbf{SU}(n)$ into the study of hermitian geometries (Section 8.5) had not yet occurred. Hurwitz therefore showed that it was a real group. Since $\mathbf{SU}(n)$ is compact, he could imitate what he had done for the rotation group $\mathfrak{D}$—represent all invariants of $\mathbf{SU}(n)$ in the form analogous to (2.8). Moreover, since the transformations $U \in \mathbf{SU}(n)$ are the complex analogues of rotations in $\mathbb{R}^n$, he was able to show that they have a representation analogous to the Euler representation for rotations, so that each U is uniquely determined by $N = n^2 - 1$ (suitably restricted) real parameters $\psi_1, \ldots, \psi_N$ [1897:556–7]. For example, the elements of $\mathbf{SU}(2)$ are represented in matrix form by

$$U(\psi_1, \psi_2, \psi_3) = \begin{pmatrix} e^{i\psi_2} \cos \psi_1 & e^{i\psi_3} \sin \psi_1 \\ -e^{-i\psi_3} \sin \psi_1 & e^{-i\psi_2} \cos \psi_1 \end{pmatrix},$$

with ψ_1 varying between 0 and $\pi/2$ and the other ψ_j between 0 and 2π. Using this representation he was able to put the integral formula for invariants in the form

$$(2.9) \qquad I(a) = \int_0^{\omega_1} \cdots \int_0^{\omega_N} F(\Gamma(U(\psi)) \cdot a) \Delta(\psi)\, d\psi_i \cdots d\psi_N,$$

where the limits ω_j are $\pi/2$ or 2π, depending on the variable, and Δ is an explicit function of $\psi_1, \ldots, \psi_N$ [1897:558]. This was analogous to what he had done for the rotation group, and, as in that case, was the starting point for a further description of how all the invariants $I(a)$ relative to $\mathbf{SU}(n)$ are constructed.

Hurwitz's real interest, however, was in the invariants relative to $\mathbf{SL}(n, \mathbb{C})$—the classical invariants. His version of "the unitarian trick" is the observation that the formula (2.9) actually gives all the invariants relative to $\mathbf{SL}(n, \mathbb{C})$. The reason for this is that in his parametric representation $U = U(\psi_1, \ldots, \psi_N)$ of the elements of $\mathbf{SU}(n)$, if the parameters ψ are allowed to take on complex values, then the corresponding expressions $U = U(\psi_1, \ldots, \psi_N)$ run through all transformations in $\mathbf{SL}(n, \mathbb{C})$. From this he concluded without further ado that the invariants of $\mathbf{SL}(n, \mathbb{C})$ are given by (2.9). This may be seen as follows. What needs to be shown is that if $I(a)$ is an invariant relative to $\mathbf{SU}(n)$, then it is actually an invariant relative to $\mathbf{SL}(n, \mathbb{C})$, since the converse is obviously true. So suppose that $I(a)$ is an invariant relative to $\mathbf{SU}(n)$. Then for any $\psi = (\psi_1, \ldots, \psi_N) \in \mathbb{C}^N$, let $h(\psi) = I(\Gamma(U(\psi)) \cdot a) - I(a) = I(a') - I(a)$. From the formula for $U(\psi)$ and (2.9) it follows that $h(\psi)$ is a polynomial in expressions like $\sin \psi_p$, $\cos \psi_q$, $e^{i\psi_r}$. Since $I(a)$ is an invariant relative to $\mathbf{SU}(n)$, $h(\psi)$ vanishes identically when the ψ_k are real. To Hurwitz it was then obvious that $h(\psi)$ must vanish for all complex values of ψ as well. Since h is evidently an analytic function of the complex variables $\psi_1, \ldots, \psi_N$, perhaps he had in mind the "identity theorem" from complex function theory, from which the conclusion follows readily.

As we shall now see, it was through the work of Frobenius's student Issai Schur (1875–1941) that the fertile ideas in Hurwitz's papers of 1894 and 1897 were brought within the framework of the theory of group representations.

10.3. Schur's Doctoral Dissertation

Schur, who was born in Russia, had begun attending the University of Berlin in 1894, initially as an undergraduate. He was thus in attendance during the years that Frobenius was creating his theory of group characters. In November of 1901, seven years after commencing his education in Berlin, he received his doctorate. According to Frobenius, Schur had posed his dissertation problem completely on his own. It was inspired by Hurwitz's paper [1894], which, as we saw in Section 2, made it clear that the classical theory of invariants was concerned with homogeneous polynomials left invariant by the group formed by a representation $A \to T(A)$ of $\mathbf{GL}(n, \mathbb{C})$ with the property that all matrix entries of $T(A)$ are polynomials in the coefficients a_{ij} of A. For a student of Frobenius, a natural question to ask would be, What are the possibilities for such representations? Can they be described? Does the problem reduce to finding irreducible ones? If so, can one find formulas for the irreducible characters and the corresponding dimensions? These are natural enough questions to ask, but the task of answering them is formidable: One has only to think how nontrivial the analogous problem was, even in Frobenius's hands, for the symmetric group; and the group $\mathbf{GL}(n, \mathbb{C})$ is continuous, not finite, so Frobenius's theory of characters and representations does not directly apply.

Nonetheless, in his dissertation Schur proposed to determine all such representations. In fact, he found he could solve the more general problem of determining all mappings $A \to T(A)$, $A \in \mathbf{GL}(n, \mathbb{C})$, $T(A)$ an $N \times N$ matrix, satisfying (1) $T(AB) = T(A)T(B)$ for all $A, B \in \mathbf{GL}(n, \mathbb{C})$; (2) the matrix coefficients of $T(A)$ are polynomials in the coefficients a_{ij} of A. He called $T(A)$ "an invariant form or matrix built out of A" or, for brevity, "an invariant form of A." Of course, in the examples $A \to T(A)$ occurring in the theory of invariants, one has $\det T(A) \neq 0$, and the polynomial coefficients of $T(A)$ are homogeneous of fixed degree, but these assumptions were not added because he could prove that if $T(A)$ satisfies (1) and (2), then, in the notation of Section 1,

$$T \sim T_1 \oplus \cdots \oplus T_k \oplus U, \tag{3.1}$$

where (i) $\det T_i(A) \neq 0$; (ii) the coefficients of $T_i(A)$ are all homogeneous polynomials of the same degree; (iii) the singular part, $U(A)$, is the zero matrix for all A [1901:10]. Thus the real problem of the dissertation was to determine all nonsingular invariant forms $T(A)$ whose coefficients are homogeneous polynomials of a fixed degree m, which Schur called the *order* of $T(A)$.[21]

Schur's problem, as he pursued it, was to imitate how all representations are determined in Frobenius's theory. This meant (1) establishing a complete reducibility theorem for invariant forms of order m so as to reduce the problem to that of determining irreducible invariant forms of order m; (2) determining the trace functions of these irreducible invariant forms and therefrom the degrees of the matrices $T(A)$. Schur's success in carrying out (1) and (2) was based upon his discovery of a remarkable connection, discussed below, between homogeneous polynomial representations of $\mathbf{GL}(n, \mathbb{C})$ of homogeneity degree m and representations of the symmetric group $\mathfrak{S}_m$. In my discussion I will adhere to Schur's terminology and speak of "invariant forms of order m" rather than homogeneous degree-m polynomial representations of $\mathbf{GL}(n, \mathbb{C})$.

Schur deduced the above-mentioned connection with the symmetric groups from the following sort of considerations [1901:20ff.]. Suppose $T(A)$ is an $N \times N$ invariant form of order m, where $A \in \mathbf{GL}(n, \mathbb{C})$. Given two m-tuples of integers between 1 and n, $k = (k_1, \ldots, k_m)$ and $l = (l_1, \ldots, l_m)$, let

[21] As Schur pointed out [1901:7, n.1], Hurwitz had, in effect, solved this problem in the special case that $N = 1$, so $T(A)$ is a number $\varphi(A)$, by proving, as a lemma [1894:§5], that if $\varphi(A)$ is a homogeneous polynomial of degree m in the a_{ij} and $\varphi(A)\varphi(A^{-1}) = 1$ for every $A \in \mathbf{GL}(n, \mathbb{C})$, then $\varphi(A) = \pm[\det A]^{\frac{m}{n}}$.

Figure 34. Issai Schur. Frobenius had nothing but praise for Schur, who continued the Berlin algebraic tradition that Frobenius did so much to establish.

$$\begin{bmatrix} k \\ l \end{bmatrix} = \begin{bmatrix} k_1 & k_2 & \cdots & k_m \\ l_1 & l_2 & \cdots & l_m \end{bmatrix} \tag{3.2}$$

denote the $N \times N$ constant matrix whose (i, j) entry is the coefficient of the term $a_{k_1 l_1} \cdots a_{k_m l_m}$ in the (i, j) entry of $T(A)$. Thus we may write

$$T(A) = \sum_{k,l} \begin{bmatrix} k \\ l \end{bmatrix} a_{k_1 l_1} \cdots a_{k_n l_n}, \tag{3.3}$$

where the summation is over all possible m-tuples k, l.

Formula (3.3), which expresses the special polynomial nature of the entries of $T(A)$, was the starting point of Schur's entire investigation. The fact that $T(C) = T(A)T(B)$ when $C = AB$ means that the matrices (3.2) must have special properties, which Schur deduced [1901:§9]—with an impressive display of combinatorial algebraic skill—by comparing

$$T(A)T(B) = \sum_{k,l,p,q} \begin{bmatrix} k \\ p \end{bmatrix} \begin{bmatrix} q \\ l \end{bmatrix} a_{k_1p_1} \cdots a_{k_np_n} b_{q_1l_1} \cdots b_{q_nl_n}$$

with

$$T(C) = \sum_{k,l} c_{k_1l_1} \cdots c_{k_nl_n}, \quad c_{k_i,l_i} = \sum_{p_i=1}^{m} a_{k_ip_i} b_{p_il_i}.$$

As a consequence of these properties, he was able to show that if $m \leq n$ and if the m-tuples k, l represent *permutations* of $1, \ldots, m$, then

$$\begin{bmatrix} k_1 & \cdots & k_m \\ 1 & \cdots & m \end{bmatrix} \begin{bmatrix} 1 & \cdots & m \\ l_1 & \cdots & l_m \end{bmatrix} = \begin{bmatrix} k_1 & \cdots & k_m \\ l_1 & \cdots & l_m \end{bmatrix}. \tag{3.4}$$

This means that if k, l are the above sort of permutation m-tuples, then we can let

$$\pi = \begin{pmatrix} k_1 & \cdots & k_m \\ l_1 & \cdots & l_m \end{pmatrix} \in \mathfrak{S}_m$$

denote the permutation such that $\pi(k_i) = l_i$, and set

$$t(\pi) = \begin{bmatrix} k_1 & \cdots & k_m \\ l_1 & \cdots & l_m \end{bmatrix}.$$

This is well-defined, and by virtue of (3.4) it follows that $t(\pi_1\pi_2) = t(\pi_1)t(\pi_2)$. Thus $\pi \to t(\pi)$ is almost a representation of $\mathfrak{S}_m$. Since $\det t(\pi) = 0$ is possible, however, Schur had to work harder to associate a representation of $\mathfrak{S}_m$ to $T(A)$. He had to show that a subset $\{\nu_1, \ldots, \nu_s\}$ of $\{1, \ldots, N\}$ with $s < N$ exists such that if in the $N \times N$ matrix $T(A)$ the ν_ith row and the ν_ith column are struck out for each $i = 1, \ldots, s$ so as to obtain an $(N - s) \times (N - s)$ submatrix

$$T'(A) = \sum_{k,l} \begin{bmatrix} k \\ l \end{bmatrix}' a_{k_1l_1} \cdots a_{k_ml_m},$$

then

$$\tau(\pi) = \begin{bmatrix} k_1 & \cdots & k_m \\ l_1 & \cdots & l_m \end{bmatrix}'$$

defines a bona fide representation of $\mathfrak{S}_m$ of degree $N - s$.

Having shown that every invariant form T of order $m \leq n$ determines a representation τ of $\mathfrak{S}_m$, Schur showed that, conversely, if $m \leq n$ and τ is a representation of $\mathfrak{S}_m$ of degree f then it is possible to define $f \times f$ matrices $\begin{bmatrix} k \\ l \end{bmatrix}$ such that

$$\begin{bmatrix} 1 & \cdots & m \\ r_1 & \cdots & r_m \end{bmatrix} = \tau \begin{pmatrix} 1 & \cdots & m \\ r_1 & \cdots & r_m \end{pmatrix},$$

and

$$T(A) = \sum_{k,l} \begin{bmatrix} k \\ l \end{bmatrix} a_{k_1,l_1} \cdots a_{k_m l_m}$$

defines an invariant form of order m on $\mathbf{GL}(n, \mathbb{C})$. In general, $\det[T(A)] = 0$; but by (3.1), $T \sim T_1 \oplus \cdots \oplus T_q \oplus U(A)$, and $\overline{T}(A) = T_1(A) \oplus \cdots \oplus T_q(A)$ is a nonsingular invariant form of order m. Schur showed that the representation $\overline{\tau}(\pi)$ of $\mathfrak{S}_m$ associated to $\overline{T}(A)$ in the above-described manner is equivalent to the given representation τ. Thus, with m and n fixed and satisfying $m \leq n$, every representation of $\mathfrak{S}_m$ is realized in Schur's construction process $T \to \tau$.

To capitalize on the connection $T \to \tau$, Schur turned to the analogue of Frobenius's character, the trace function $\Phi(A) = \text{tr } T(A)$, which he called the *characteristic* of $T(A)$. Early in his dissertation [1901:12] he had proved the following result

(3.5) *Let $\omega_1, \ldots, \omega_n$ denote the eigenvalues of $A \in \mathbf{GL}(n, \mathbb{C})$, which as functions of the a_{ij} are distinct. Then the eigenvalues of $T(A)$ are all of the form $\omega_1^{m_1} \omega_2^{m_2} \cdots \omega_m^{m_n}$, where $m_1 + m_2 + \cdots + m_n = m$, the order of $T(A)$, and*

$$\Phi = \text{tr } T(A) = \sum r_{m_1 m_2 \ldots m_n} \omega_1^{m_1} \omega_2^{m_2} \cdots \omega_n^{m_n}$$

is a symmetric function of the ω_i.

When $T(A)$ is irreducible, these symmetric functions are now known as Schur functions, or S-functions [Green 1980:1,47]. Of course, since $\Phi(A)$ is the sum of the eigenvalues of A, the coefficients $r_{m_1 m_2 \ldots m_n}$ are all nonnegative integers. Using Frobenius's characterization of the irreducible characters of the symmetric group in connection with the correspondence $T \to \tau$, Schur was able to show that the characteristic Φ of T is related to the character χ of τ by the remarkable formula

$$\Phi = \sum_{\alpha} \frac{\chi_\alpha}{\alpha_1! \alpha_2! \cdots} \left(\frac{s_1}{1}\right)^{\alpha_1} \left(\frac{s_2}{2}\right)^{\alpha_2} \cdots, \tag{3.6}$$

where χ_α denotes the value of χ on the conjugate class of $\mathfrak{S}_m$ composed of products of α_1 1-cycles, α_2 2-cycles, etc.; summation is over all $\alpha = (\alpha_1, \alpha_2, \ldots)$ such that $\alpha_1 + 2\alpha_2 + 3\alpha_3 + \cdots = m$ and $s_k = \omega_1^k + \cdots + \omega_n^k$ [1901:31].

Using (3.6) Schur was able to establish the equivalence of the following three statements: (i) $T_1 \sim T_2$; (ii) $\Phi_1 = \Phi_2$; (iii) $\tau_1 \sim \tau_2$ [1901:36]. Of course, the resultant equivalence of (i) and (ii) mirrored the above mentioned fundamental theorem of Frobenius's theory. Schur was also able to transfer the complete reducibility theorem for $\mathfrak{S}_m$ to invariant forms of order $m \leq n$ by showing that $T \sim T_1 \oplus T_2$ if and only if $\tau \sim \tau_1 \oplus \tau_2$ [1901:39]. This also shows that T is irreducible precisely when τ is. By capitalizing on the polynomial nature of his invariant forms and limiting his attention to invariant forms of order $m \leq n$, Schur was thus able to connect these special representations of $\mathbf{GL}(n, \mathbb{C})$ to representations of $\mathfrak{S}_m$ in such a way that basic theorems about the representations of $\mathfrak{S}_m$, such as the complete reducibility theorem, can be transferred to these representations. Thus the problem of determining all invariant forms of order $m \leq n$ was reduced to determining the irreducible ones. Schur solved this problem next and then, with all these results for the case $m \leq n$ in hand, he turned, as we shall see, to the case $m > n$. First, however, let us consider how he set about determining the irreducible invariant forms satisfying $m \leq n$ [1901:40ff.].

By virtue of the above results, the number of inequivalent irreducible invariant forms of a fixed order $m \leq n$ is equal to the number of inequivalent irreducible representations of $\mathfrak{S}_m$, which according to Frobenius's theory equals the number k of its conjugacy classes, i.e., the number of partitions of m. Given such a partition $\kappa_1 + \kappa_2 + \cdots + \kappa_\rho = m$ with $\kappa_1 \geq \kappa_2 \geq \cdots \geq \kappa_\rho > 0$, Schur introduced an invariant form of order m composed out of Hurwitz's product and determinant transformations: $C(A) = C_{\kappa_1}(A) \times \cdots \times C_{\kappa_\rho}(A)$. If, following Schur, we set p_i equal to the number of κ's equal to i, then

$$C(A) = C_1(A)^{p_1} \times C_2(A)^{p_2} \times \cdots \times C_\sigma(A)^{p_\sigma},$$

where σ is the largest i for which $p_i > 0$ and $C_k(A)^{p_k} = C_k(A) \times \cdots \times C_k(A)$ (p_k factors). Thus $C(A)$ is the representation that corresponds to the module $\Gamma_1^{p_1} \otimes \Gamma_2^{p_2} \otimes \cdots \otimes \Gamma_\sigma^{p_\sigma}$, where Γ_k denotes the vector space of skew symmetric tensors of degree k.

Since $C(A)$ is an invariant form of order $m = \kappa_1 + \kappa_2 + \cdots + \kappa_\rho$, if we take $n \geq \kappa_1 + \kappa_2 + \cdots + \kappa_\rho$, then Schur's complete reducibility theorem applies to $C(A)$, and it is equivalent to a sum of irreducible invariant forms of order m. Schur singled out one of these by characterizing it in terms of the sequence $m_1, \ldots, m_\sigma$, where

$$(3.7) \qquad m_1 = p_1 + p_2 + p_3 + \cdots, \qquad m_2 = p_2 + p_3 + \cdots, \qquad m_\sigma = p_\sigma.$$

Since $\text{tr } C(A) = \prod_{i=1}^{\rho} [\text{tr } C_k(A)]^{p_k}$, and $\text{tr } C_\kappa(A)$ equals the κth elementary symmetric function of $\omega_1, \ldots, \omega_n$, it is not difficult to see that in the lexicographic ordering of the expressions $\omega_1^{\mu_1} \cdots \omega_m^{\mu_m}$ occurring in $\Phi = \text{tr } C(A)$, the highest term is $\omega_1^{m_1} \cdots \omega_\sigma^{m_\sigma}$ and its coefficient is 1.[22] Now by Schur's complete reducibility theorem, $\Phi = \omega_1^{m_1} \omega_2^{m_2} \cdots \omega_\sigma^{m_\sigma} + \cdots = \sum_i \Phi_i$, where the Φ_i are the characteristics of the irreducible components $T_i(A)$ of $C(A)$. It follows that one and only one of the Φ_i has the term $\omega_1^{m_1} \omega_2^{m_2} \cdots \omega_\sigma^{m_\sigma}$ in it and it occurs there with coefficient 1. This is because by (3.5), Φ and the Φ_i are equal to sums of eigenvalues $\omega_1^{n_1} \cdots \omega_m^{n_m}$, and so have, as symmetric polynomials in $\omega_1, \ldots, \omega_m$, nonnegative integral coefficients.

[22] To see this, recall that $\omega_1^{\mu_1} \omega_2^{\mu_2} \cdots > \omega_1^{\nu_1} \omega_2^{\nu_2} \cdots$ if the first nonvanishing difference $\mu_i - \nu_i$ is positive. Thus the highest term we can get from $[\text{tr } C_1(A)]^{p_1} = (\omega_1 + \cdots + \omega_m)^{p_1}$ is $\omega_1^{p_1}$, and the highest term we can get from the $[\text{tr } C_2(A)]^{p_2} = (\omega_1\omega_2 + \cdots)^{p_2}$ is $(\omega_1\omega_2)^{p_2} = \omega_1^{p_2}\omega_2^{p_2}$, and so on. Thus the highest term in $\prod_{i=1}^{\rho} [\text{tr } C_k(A)]^{p_k}$ is $(\omega_1^{p_1})(\omega_1^{p_2}\omega_2^{p_2})(\omega_1^{p_3}\omega_2^{p_3}\omega_3^{p_3}) \cdots = \omega_1^{m_1}\omega_2^{m_2} \cdots \omega_\sigma^{m_\sigma}$, and it occurs with coefficient 1, since it can occur in only one way.

The irreducible component of $C(A)$ singled out in this manner will be denoted by $T_{m_1,\ldots,m_\sigma}(A)$. The reader may have noticed remarkable similarities between Schur's identification of $T_{m_1,\ldots,m_\sigma}$ as an irreducible component of $C(A)$ and the construction of irreducible representations of $\mathbf{SL}(n, \mathbb{C})$ of given highest weight presented twelve years later by Cartan (Section 8.4), who did not know of Schur's work. Of course, every representation of $\mathbf{GL}(n, \mathbb{C})$ is a representation of $\mathbf{SL}(n, \mathbb{C})$ as well, although the 1-dimensional representation $C_m(A) = \det A$ of the former becomes the trivial representation of the latter and is excluded from consideration in Cartan's Lie-algebraic approach. With the exclusion of the representation $C_m(A)$, every irreducible representation of $\mathbf{GL}(n, \mathbb{C})$ is likewise an irreducible representation of $\mathbf{SL}(n, \mathbb{C})$. Furthermore, Cartan's irreducible representation of highest weight $\lambda = p_1\pi_1 + \cdots + p_l\pi_l = p_1\pi_1 + \cdots + p_{n-1}\pi_{n-1}$, namely the representation whose module is the Cartan component of $\Gamma_1^{p_1} \otimes \cdots \otimes \Gamma_{n-1}^{p_{n-1}}$ (Section 8.4), is precisely Schur's $T_{m_1,\ldots,m_\sigma}(A)$ restricted to $\mathbf{SL}(n, \mathbb{C})$. This follows from the work of Weyl. In particular, as we shall see in Chapter 11, Weyl discovered how to generate the Cartan component using the primitive characteristic unit (or idempotent) associated by Frobenius (Section 10.1) with the Young tableau $\mathfrak{T}$ whose rows have as their lengths Schur's integers $m_1, \ldots, m_\sigma$.

Making use of the key formula (3.6), Schur proceeded to find a formula for the irreducible characteristic $\Phi_{m_1,\ldots,m_\sigma}(A) = \operatorname{tr} T_{m_1,\ldots,m_\sigma}(A)$. It involves the characteristics $\Psi_\nu(A) = \operatorname{tr} P_\nu(A)$ of Hurwitz's power transformations, which are expressible in the form

$$\Psi_\nu = \sum_{p_1+2p_2+\cdots=\nu} \frac{1}{p_1!p_2!p_3!\cdots} \left(\frac{s_1}{1}\right)^{p_1} \left(\frac{s_2}{2}\right)^{p_2} \left(\frac{s_3}{3}\right)^{p_3} \cdots,$$

where $s_k = \omega_1^k + \omega_2^k + \cdots + \omega_\nu^k$. The characteristic of $\Phi_{m_1,m_2,\ldots}$ is expressible as the $\sigma \times \sigma$ determinant

$$\Phi_{m_1,m_2,\ldots} = |\Psi_{\mu,\nu}|, \tag{3.8}$$

where $\Psi_{\mu,\nu} = \Psi_{(m_\mu-\mu+\nu)}$ if $m_\mu - \mu + \nu \geq 0$, and $\Psi_{\mu,\nu} = 0$ otherwise ($\Psi_0 = 1$), for $\mu, \nu = 1, \ldots, \sigma$ [1901:47].

As a simple illustration, consider the irreducible representation Schur's theory associates to $C(A) = C_2(A)^2 \times C_4(A)$. Here $p_1 = 0$, $p_2 = 2$, $p_3 = 0$ and $p_4 = 1$, so $\sigma = 4$ and $m_1 = m_2 = 3$, $m_3 = m_4 = 1$. Since $\sum_i i \cdot p_i = \sum_j \kappa_j = m$, here $m = 8$, and so $T_{3,3,1,1}(A)$ has order 8 (and $n \geq 8$). Then (3.8) states that

$$\Phi_{3,3,1,1} = \begin{vmatrix} \Psi_3 & \Psi_4 & \Psi_5 & \Psi_6 \\ \Psi_2 & \Psi_3 & \Psi_4 & \Psi_5 \\ 0 & 1 & \Psi_1 & \Psi_2 \\ 0 & 0 & 1 & \Psi_1 \end{vmatrix}.$$

The formula (3.8) is more satisfying than what Frobenius had obtained in [1900] for the symmetric group, since (3.8) is explicit, whereas Frobenius had simply characterized the values of the irreducible characters as certain coefficients in a polynomial expansion.

Of course, Frobenius had been able to give an explicit formula for the degrees of the irreducible representations, and so did Schur. To get the degree $d_{m_1,\ldots,m_\sigma}$ of the representation $T_{m_1,\ldots,m_\sigma}(A)$ Schur set all $\omega_i = 1$ in (3.8), which since the ω_i are the eigenvalues of A is equivalent to setting $A = I$. By skillfully manipulating the resultant formula he obtained

$$(3.9) \qquad d_{m_1,\ldots,m_\sigma} = \frac{\Delta(l_1,\ldots,l_m)}{\prod_{\mu=1}^{m}(m-\mu)!} = \frac{\Delta(l_1,\ldots,l_m)}{\Delta(1,\ldots,m)},$$

where $\Delta(x_1,\ldots,x_m) = \prod_{\mu<\nu}(x_\mu - x_\nu)$ and $l_1 = m_1 + m - 1$, $l_2 = m_2 + m - 2, \ldots, l_\sigma = m_\sigma + m - \sigma$, $l_{\sigma+1} = m - \sigma - 1, \ldots, l_{m-1} = 1$, $l_m = 0$ [1901:51].

All the above described results were established by Schur under the assumption that the order m of $T(A)$ satisfies $m \leq n$, where T is an invariant form on $\mathbf{GL}(n, \mathbb{C})$ in the sense that $A \in \mathbf{GL}(n, \mathbb{C})$. In other words T is a homogeneous polynomial representation of $\mathbf{GL}(n, \mathbb{C})$ in which the degree of the polynomials cannot exceed n. We saw that this hypothesis was essential to Schur's method of establishing the correspondence $T \to \tau$ with representations of the symmetric group $\mathfrak{S}_m$ and was thereby the means of transferring many basic results of Frobenius's theory to the theory of invariant forms. Reading the dissertation, it is easy to forget the restriction $m \leq n$, for Schur did not mention it in stating his theorems. The reason was that the theorems also hold for $m > n$, although, in terms of the logical progression of the theory, Schur was not in a position to prove this until the end of the dissertation [1901:§32] and then in a rather curious and dissatisfying way, which involved coaxing the results for the case $m > n$ out of the case $m = n$.

The resulting theory, although a brilliant algebraic tour de force, is nonetheless somewhat unsatisfying. As Schur himself later observed [1927:60], not only is the emphasis on the case $m \leq n$ disconcerting, given that the final results can be expressed in a uniform manner, i.e., without the need to distinguish between $m \leq n$ and $m > n$, but in addition, the construction of the correspondence $T \to \tau$ when $m \leq n$ is formal, rather than conceptual, and rather complicated. As we shall see in Chapter 11, Weyl discovered an entirely different, far more conceptual, way to link representations of $\mathbf{SL}(n, \mathbb{C})$ with those of the symmetric groups.[23]

[23] In [1927], which was prompted by Weyl's discoveries, Schur returned to the subject of his dissertation and derived his earlier results in a manner that avoided the unsatisfactory characteristics of his original approach while at the same time differing from Weyl's. He also pointed out [1927:68, n.1] that the study of rational representations of $\mathbf{GL}(n, \mathbb{C})$ is readily reduced to that of polynomial representations. For a modern exposition of Schur's theory see [Green 1980]. In [1928] Schur determined all continuous representations of $\mathbf{GL}(n, \mathbb{C})$ in the sense of determining all continuous indecomposable representations that are reducible.

10.4. Schur's Career 1901–1923

Understandably, Frobenius was quite impressed by Schur's dissertation. In his evaluation of it, after mentioning some of the main results, he explained that they "are only the most important of the superabundant contents of the work. Between them there are in addition a host of interesting special cases. Some of these less important results would already suffice for the content of a good dissertation."[24] Frobenius went on to say that if Schur had simply showed how his original "beautiful problem"—to determine all (possibly singular) invariant forms—reduced to the problem he eventually studied—that of determining all (nonsingular) invariant forms of order m—the dissertation would have called for the designation "outstanding" (*hervorragend*). "But in addition, he has completely solved the difficult problem and with the simplest means and, so to speak, conformally mapped the theory of invariant forms onto the seemingly totally different theory of the representations of the symmetric group—something that always lends a particular attraction to a mathematical development." With this one work, he concluded, Schur had proved himself to be "a master of algebraic research."

These are strong words coming from Frobenius, especially since his views toward the theory of invariants were generally negative. This comes out in his report, dated 7 June 1902, recommending Hilbert for the professorship at Berlin vacated by Fuchs.[25] As we saw in Chapter 9, Hilbert began his mathematical career by working on the theory of invariants, which Frobenius described as "a subject of minor significance" and a source of much "hack work" (*Kärrnerarbeit*), including thereby Hilbert's own early work. Hilbert's true, path-breaking talent finally shone through, Frobenius continued, with his work on the finiteness theorems in which he "transferred the concepts of higher algebra to this field of research, and hence Hilbert is to be designated as the scientific founder and terminator (*Vollender*) of the theory of invariants." By these remarks Frobenius seems to have meant that first of all, by treating the finiteness problem within the framework of higher algebra — function fields—rather than by the sort of extensive formal calculations that had characterized Gordan's solution for binary forms, Hilbert had made the subject a part of legitimate, mainstream mathematics; he had provided a genuinely "scientific" (*wissenschaftlich*) foundation and context for the theory. On the other hand, by solving its central problem and making it nothing more than a special case of a more general algebraic theory he had in effect terminated invariant theory as a viable independent mathematical discipline.

---

[24] This and the following quotations are my translation of portions of the conclusion of Frobenius's evaluation as transcribed by Biermann [1973:127].

[25] See [Biermann 1973:209–11] for a transcription of this document. Hilbert declined the offer, and Frobenius's friend Friedrich Schottky was subsequently appointed [Biermann 1973:133–4, 211–16].

Schur's dissertation, although inspired by the old theory of invariants, dealt with a difficult general problem that was formulated and worked out within the realm of "higher algebra"—linear algebra in the form of Weierstrass's theory of elementary divisors (Section 4.1) and the attendant matrix algebra developed by Frobenius, the study of groups and their homomorphisms and, of course, Frobenius's theory of group representations and characters.[26] Clearly, in Schur, Frobenius perceived a kindred mathematical spirit of demonstrably great talent. Schur's dissertation was published, as required, as a separate pamphlet. Given Frobenius's great esteem for Schur's results, however, it is rather surprising to find that they were never published by Schur in a journal article, which was a common practice. Perhaps Schur himself was already dissatisfied with the undue emphasis upon the case $m \leq n$ and hoped quickly to find a more satisfactory approach. As a consequence his dissertation was probably not as well known as it deserved to be. In particular, it seems that Weyl did not know of it in 1921 and, possibly, not until 1924, when Schur referred to it in a paper (discussed in the following section) dealing with questions about the invariant theory of the rotation group $\mathbf{SO}(n, \mathbb{R})$.

It would be misleading to regard Schur's dissertation as the inaugural step in a research program aimed at extending Frobenius's theory to continuous groups, for it does not seem that either Schur or Frobenius thought of it in this way. Indeed, in his dissertation Schur did not attempt to deal with all representations of $\mathbf{GL}(n, \mathbb{C})$. Possibly he recognized the failure of the complete reducibility theorem at this level of generality, but in any case, he was primarily interested in the representations that arise in the theory of invariants. And although within this more limited framework he was able to obtain analogues of many of Frobenius's main theorems, he could not do this for all of them. In particular, there was no analogue of the character orthogonality relations for Schur's characteristics. The analogues that were established were all obtained thanks to the special connection between invariant forms of order m and representations of $\mathfrak{S}_m$, so that the work resembled more a brilliant application of the representation theory of the symmetric groups than an independently founded theory of the representations of $\mathbf{GL}(n, \mathbb{C})$ that could be extended to other continuous groups.

Schur's work on invariant forms was a masterful demonstration of how an interesting problem—the determination of invariant forms — could be resolved with the aid of Frobenius's theory. His next research project, inaugurated with his *Habilitationsschrift*, which he completed slightly over a year after his dissertation and published in Crelle's journal [1904], was similar in aim. After leaving continuous groups to Lie and turning to finite groups, Klein had sought to develop a generalization of the Galois theory of equations, which involved the following sorts of problems (pursued mainly

[26] Regarding Frobenius's role in the development of matrix algebra, see my paper [1977a].

by Klein's students): Determine all finite groups of projective transformations in n dimensions that are isomorphic to a given group; determine the minimal n for which a given group can be realized as a group of projective transformations. These problems were solved only in very special cases by Klein and his students. Schur saw that all such problems were special cases of the following general problem: Given a finite group $\mathfrak{G}$, consider mappings $\sigma : \mathfrak{G} \to \mathbf{GL}(n, \mathbb{C})$ such that $\sigma(s)\sigma(t) = C_{s,t}\sigma(st)$ for all $s, t \in \mathfrak{G}$, where the $C_{s,t}$ denote nonzero complex numbers satisfying $C_{p,q}C_{pq,r} = C_{p,qr}C_{q,r}$ for all $p, q, r \in \mathfrak{G}$. Find all such mappings. Schur solved this problem by showing that it reduces to (1) the determination of a certain abelian "multiplier group" $\mathfrak{M}$ and (2) the determination of the representations (in the sense of Frobenius) of the factor group $\mathfrak{G}/\mathfrak{M}$.[27]

Frobenius had nothing but praise for Schur's *Habilitationsschrift*, which "shows, as previously in his dissertation, his extraordinary talent for positing, transforming, breaking down, and solving substantial algebraic problems."[28] Later, in a memorandum supporting Schur for a professorship he wrote, "F. Klein and his school had computationally investigated the representation of groups by linear fractional substitutions, and the general resolution of this problem appeared as a hopeless undertaking. The difficulties that stood in the way of mastering the problem were first completely overcome through the penetrating acumen and persistent reflection of Schur."[29]

As we have seen, what Frobenius wrote about Klein and his students applied equally well to Lie and the majority of his students in their efforts to classify various types of transformation groups. They seemed to lack the combination of talent and sufficiently powerful mathematical tools to do more than solve the admittedly formidable problem computationally for very special cases. This was the type of "hack work" that Frobenius also saw as characteristic of the theory of invariants in the days before Hilbert's finiteness theorems. By contrast, the Berlin school of algebra, which was founded by Weierstrass with his theory of elementary divisors and continued by Frobenius and Schur, was characterized by rigorously established general theories and their application to the solution of general algebraic problems. Thus Frobenius had applied the theory of elementary divisors to solve the problem of the group determinant and then to create his theory of group characters and representations, and Schur had applied the theories of Weierstrass and Frobenius to solve the formidable problems posed in his dissertation and *Habilitationsschrift*.

I mention this by way of characterization of the type of mathematics to which the Berlin school of algebraists aspired and to suggest that as a result, they did not tend to look to the schools of Klein or Lie for significant mathematics. Indeed, when Frobenius learned that Theodor Molien was

[27] Further details are contained in Section 16 of my essay [1999a].
[28] Quoted by Biermann [1973:135]; the translation is mine.
[29] See [Biermann 1973:224]; my translation.

associated with Lie's school he was surprised, because he had not expected something as significant as Molien's work on hypercomplex numbers (linear associative algebras) [1893] to issue from that source.[30]

In addition to the above mentioned development of the theory of projective representations, which also included treating the symmetric and alternating groups in a fashion analogous to what Frobenius had done, Schur made many fundamental contributions to the arithmetical aspects of Frobenius's theory, i.e., those aspects in which attention is paid to the algebraic number field in which the coefficients of a matrix representation lie. These important contributions, which involved what is now known as the Schur index, must have pleased Frobenius, who regarded the theory of finite groups as a part of arithmetic.[31]

More relevant to our story, however, is Schur's paper [1905] presenting a reworking of Frobenius's theory. As we have seen in Section 1, Frobenius's theory had emerged from his study of the group determinant. As a consequence, the main theorems about group characters and representations were not proved in the most direct manner. As William Burnside (1852–1927) pointed out in [1904b:117] with Frobenius's theory specifically in mind, "the historical development of a subject rarely follows the lines of the logical presentment that can be given when a sufficient number of results and their connexion with each other have become familiar." Burnside, who came close, via connections with Lie's theory, to anticipating Frobenius's discoveries and who went on to apply the theory of characters and representations to problems involving finite groups, was the first to publish such a "logical presentment" of the theory.[32] In [1904a] the basic theorems about representations were proved, and then in [1904b] these results were applied to develop the theory of characters.

Burnside based his presentation upon the following theorem already mentioned in Section 9.1:

(4.1) **Loewy–Moore Theorem.** *Given any finite subgroup $\mathfrak{G} \subset \mathbf{GL}(n, \mathbb{C})$, there exists a positive definite hermitian form $\varphi(x)$ in the variables $x = (x_1, \ldots, x_n)$ on which $\mathfrak{G}$ acts that is an invariant of $\mathfrak{G}$, i.e., $\varphi(Tx) = \varphi(x)$ for all $T \in \mathfrak{G}$.*

This theorem says that any representation $t \to R(t)$ of a finite group admits an invariant positive definite hermitian form $\varphi(x)$. If φ' is another such form, linearly independent of the first, then for any real number λ, $\varphi + \lambda\varphi'$ is invariant, hermitian, and definite, but not necessarily positive definite

[30] This is found in a letter by Frobenius dated 6 May 1898 and presented in Russian translation in [Kanounov 1983:97–100]. More will be said about Molien in the next section.

[31] The arithmetical aspects of the work of Frobenius and Schur are discussed in Section 4 and Section 16, respectively, of my essay [1999a].

[32] On Burnside's near discovery of Frobenius's theory see pp. 273–278 of my paper [1972]; a briefer discussion is given in Section 13 of my essay [1999a]; Burnside's applications of representation theory are discussed in Section 15 of the essay.

since λ can always be chosen such that the determinant (or discriminant) of $\varphi + \lambda\varphi'$ vanishes. Burnside's real starting point was thus that every representation admits an invariant definite hermitian form. He showed that if the form is singular, then the representation is reducible. The representation is irreducible if and only if it admits an invariant positive definite hermitian form which is unique up to constant multiples; and so on.

In [1905] Schur proposed to give a different, thoroughly elementary, logical presentation of the theory on the grounds that Burnside's involved a concept "far removed" from the subject matter, namely the concept of a hermitian form. Schur's starting point was the proposition now commonly known as Schur's lemma:

(4.2) **Schur's Lemma**. *Let $s \to A(s)$ and $s \to B(s)$ be two irreducible representations of the finite group $\mathfrak{G}$ of degrees f and f', respectively, and suppose there is an $f \times f'$ constant matrix P such that $A(s)P = PB(s)$ for all $s \in \mathfrak{G}$. Then either $P = 0$, or $f = f'$ and P is invertible, so that the representations are equivalent. In the latter case, when in addition $A(s) = B(s)$, P must be a scalar matrix $P = \lambda I$.*[33]

Actually, this result, in the case $B(s) = A(s)$, had been used—and proved — by Frobenius in the course of reestablishing properties of irreducible characters on the basis of characterizing them as trace functions, although Frobenius's proof was not elementary, since it depended on some of the main results of his theory.[34] As Schur realized, Burnside had also used (without any proof) the same special case at two points in his presentation [Burnside 1904a:384; 1904b:121]. Not only did Schur give a simple, elementary proof of his lemma, but, what is far more remarkable, he showed how it could be used to quickly develop by means of elementary linear algebra all the basic theorems about representations and characters.

In addition to elementary linear algebra, Schur used a simple but extremely useful fact that goes back to Frobenius, who had used it repeatedly throughout his study of group determinants and matrix representations. For purposes of later reference, I will term it the principle of invariance of group summation.

Invariance of Group Summation. *If $M(s)$ is any complex- or matrix-valued function of the elements s in the finite group $\mathfrak{G}$, then for any fixed $t \in \mathfrak{G}$,*

$$\sum_{s \in \mathfrak{G}} M(st) = \sum_{s \in \mathfrak{G}} M(s). \tag{4.3}$$

This, of course, follows because in a finite group, as s runs through all the elements of $\mathfrak{G}$ so does $u = st$, and so the left side of (4.3) equals $\sum_{u \in \mathfrak{G}} M(u)$, which is identical to the right side.

---

[33] The contents of Schur's lemma are given in his Theorems I and II of Section 2 of [1905].

[34] See the end of Section 10 of my essay [1999a].

As an illustration of how Schur used this invariance in conjunction with his lemma, let us consider his proof [1905:411] of the orthogonality relations for irreducible characters:

$$\sum_{r\in\mathfrak{G}} \chi^{(\lambda)}(r)\overline{\chi^{(\mu)}(r)} = g \cdot \delta_{\lambda\mu}. \tag{4.4}$$

In the case $\lambda = \mu$, so that $\chi = \chi^{(\lambda)} = \chi^{(\mu)}$ is the character of an irreducible representation $s \to A(s)$, let U denote an arbitrary matrix of the same size as $A(s)$ and set

$$V = \sum_{s\in\mathfrak{G}} A(s^{-1})UA(s). \tag{4.5}$$

Then for any t, an easy calculation using the fact that $s \to A(s)$ is a homomorphism shows that $A(t)^{-1}VA(t) = \sum_{s\in\mathfrak{G}} A([st]^{-1})UA(st)$, and by the invariance of group summation (4.3) this last expression equals V, which means that $VA(t) = A(t)V$ for all t, and so by Schur's lemma, $V = \lambda(U)I$. It is easily seen that $\lambda(U)$ is a linear function of U, and this fact, together with the fact that the coefficients of U are entirely arbitrary and can thus be chosen conveniently, then leads to the desired orthogonality relation. The relation for distinct characters corresponding to irreducible representations $A(s)$ and $B(s)$ follows in the same manner with $V = \sum_{s\in\mathfrak{G}} A(s^{-1})UB(s)$.

After developing the properties of characters, Schur turned to those of representations (thus proceeding in the reverse order of Burnside). He took as his starting point a theorem due to Heinrich Maschke (1853–1908). Maschke, who had been Klein's student, was interested in the properties of finite subgroups $\mathfrak{G} \subset \mathbf{GL}(n, \mathbb{C})$. In particular, he sought to prove that by a suitable linear change of variables the elements in $\mathfrak{G}$ are given by matrices with entries that are cyclotomic numbers, by which he meant elements of a finite extension of $\mathbb{Q}$ by roots of unity. This turns out to be true but very difficult to prove; after Maschke, Burnside and Schur gave proofs in special cases, and Schur's student Richard Brauer finally gave a general proof in 1945.[35] In the course of his attempt at a proof, however, Maschke established the following theorem, which I state here for representations of an abstract finite group:

(4.6) **Maschke's Theorem.** *If $t \to A(t)$ is a representation of a finite group $\mathfrak{G}$ that is reducible, so that a nonsingular matrix M exists for which*

$$MA(t)M^{-1} = \begin{pmatrix} B(t) & 0 \\ D(t) & C(t) \end{pmatrix}$$

for all $t \in \mathfrak{G}$, then M may be chosen such that $D(t) = 0$ for all $t \in \mathfrak{G}$.

[35] On Maschke see my paper [1972] and Section 14 of my essay [1999a]; the contributions of Burnside, Schur, and Brauer are discussed in Sections 16–17 of the essay.

Maschke proved this theorem for groups of linear transformations in Section 3 of his paper [1899], where it functions as a lemma en route to a theorem about the intransitivity of finite groups $\mathfrak{G} \subset \mathbf{GL}(n, \mathbb{C})$ with an identically vanishing matrix coefficient. He was not familiar with Frobenius's theory, but to those who were, its relevance was clear. Maschke had based his proof on the Loewy–Moore theorem—he was Moore's colleague at the University of Chicago—and this seems to have inspired Burnside to make the existence of invariant hermitian forms fundamental to his exposition of Frobenius's theory. Schur actually proved Maschke's theorem for the more general type of representations considered already in his dissertation, namely mappings $s \to A(s)$ from $\mathfrak{G}$ to possibly singular $n \times n$ matrices satisfying $A(st) = A(s)A(t)$. And of course, hermitian forms are not involved in the proof. However, in a footnote [1905:416], "for the all-important case" in which the $A(s)$ are nonsingular, he gave a simpler proof, due to Frobenius, that used the existence of an invariant positive definite hermitian form.

The proofs of both Maschke and Frobenius were by means of formal algebraic considerations. It is perhaps worth noting in passing that Weyl, who in his book *Raum, Zeit, Materie* (1918) had developed the now familiar conceptual-geometrical approach to linear algebra (Section 11.3), proved Maschke's theorem in line with what Hilbert had called Riemann's principle, according to which proofs should be driven not by calculation but solely by ideas (Section 9.6). That is, the Loewy–Moore theorem says that the $A(s)$ may be regarded as *unitary* transformations on a complex inner product space $\mathcal{V}$, and the hypothesis of Maschke's theorem is that a nontrivial subspace $\mathcal{W} \subset \mathcal{V}$ is left invariant by all the $A(s)$. The conclusion of the theorem follows by observing that $\mathcal{W}^{\perp}$, the orthogonal complement of $\mathcal{W}$, is also left invariant by the $A(s)$.[36]

Although Hurwitz had independently discovered the proof idea behind the Loewy–Moore theorem (4.1), neither Burnside in his papers of 1904 nor Schur in his 1905 paper cites Hurwitz, which suggests that they were not then familiar with Hurwitz's paper [1897]. Once Schur became familiar with the contents of this second and final contribution by Hurwitz to the theory of invariants, he was in a position to see that the theorems of Frobenius's theory, as derived in his 1905 paper, remain valid for any continuous group with an integral in the sense of Hurwitz defined on it and convergent for continuous integrands. That is, the only way the finiteness of the group $\mathfrak{G}$ enters the picture in Schur's 1905 presentation is through the application of the invariance of group summation (4.3), but (4.3) may be replaced by $\int_{\mathfrak{G}} M(st)\, dm = \int_{\mathfrak{G}} M(s)\, dm$, where $dm = \psi(s)ds$ defines the Hurwitz integral as in Section 9.1. Exactly when Schur became acquainted with Hurwitz's paper [1897] and made this observation is unclear. My guess is that it happened when, in the early 1920s, his interest turned back to the theory of invariants and, in particular, focused on Hilbert's finiteness

[36] See [Weyl 1925:559]. Weyl does not mention Riemann's principle.

theorem and its extensions, Hurwitz's extension to the rotation group in [1897] being the prime example.

The first sign of Schur's renewed interest in the theory of invariants was a joint paper with Alexander Ostrowski (1893–1986). Ostrowski had come to Göttingen from the Ukraine in 1918. He was already a published mathematician when he arrived, and he had made a thorough study of the writings of both Hilbert and Klein. One aspect of their work that interested him was the theory of invariants, and he became one of the few mathematicians to contribute to the theory of invariants in the sense of Hilbert—the function-field study of the question of a finite basis for systems of homogeneous polynomial functions—during the period 1893–1923, others being Emil Fischer and Emmy Noether.[37] One major reason for the small number of mathematicians writing about Hilbertian invariant theory was that Hilbert had abandoned research in this area in 1893, and since his students tended to work on what currently interested him, most of them did not work on the theory of invariants.[38] In this respect Weyl was typical: he worked on problems in integral equations because that was what interested Hilbert. Ostrowski's work on invariants is of interest both because of his connection with Schur and also because of the views he expressed about the relevance of Lie's theory to the study of invariant-theoretic problems.

It was Ostrowski's second paper after arriving in Göttingen [1919] that attracted Schur's attention. It was concerned with an aspect of the most classical part of the theory of invariants: the invariants associated to the binary form

$$f(x,y) = a_0x^n + \binom{n}{1} a_1x^{n-1}y + \binom{n}{2} a_2x^{n-2}y^2 + \cdots + a_ny^n \tag{4.7}$$

as base form. The paper was dedicated to Klein on the occasion of the fiftieth anniversary of his doctorate and viewed invariants in accordance with Klein's view of invariants as defined directly by a given group of linear transformations. This was, of course, also the viewpoint of Hurwitz's paper [1894] discussed above in Section 2, and indeed for the binary form f the group $\mathfrak{G}_n$ consists of the power transformations $P_n(s)$, $s \in \mathbf{GL}(2,\mathbb{C})$. Ostrowski does not seem to have known of Hurwitz's paper at this time, but what he did in his paper was similar in spirit to what Hurwitz had done: He gave a direct characterization of the group $\mathfrak{G}_n$ by proving the following theorem. *Let $\Delta(a_0,\ldots,a_n)$ denote the discriminant of the form $f(x,y)$;*

[37] On the "death" of invariant theory in the years following Hilbert's finite basis theorem and its recent resurrection, see, respectively, [Fisher 1966; 1967] and [Dieudonné and Carrell 1970; Kung and Rota 1984].

[38] This can be seen from the list of dissertations directed by Hilbert (*Abhandlungen* **2**, 431–3); only two dealt wth the theory of invariants: S. Marxsen's in 1900 and Andreas Speiser's in 1909. See Section 1 of Ostrowski's paper [1918] for an idea of what had been done by Fischer and Noether and what Ostrowski felt could still be done.

Figure 35. Alexander Ostrowski as he looked somewhat later in his lengthy mathematical career.

then $\mathfrak{G}_n$ consists precisely of all linear transformations $T \in \mathbf{GL}(n+1,\mathbb{C})$ that leave Δ invariant.[39]

A few months after Ostrowski submitted his paper, he realized that Gerhard Kowalewski, in his paper [1902], had obtained results with implications for the type of a problem he had just solved. As we saw in Section 7.5, Kowalewski's paper was part of his research program aimed at determining all projective groups that leave nothing planar invariant—the problem Cartan completely solved in 1913 (Section 8.4). As a step in that program, it had as its goal the determination of all Lie subgroups of $\mathbf{PGL}(n+1,\mathbb{C})$ that contain the projective group of a normal curve, $\mathfrak{P}_3$, which is the projective analogue of Ostrowski's group $\mathfrak{G}_4$. Of course, Kowalewski worked exclu-

[39] Here "leave invariant" means that $\Delta(Ta) = \gamma\Delta(a)$, where γ is a nonzero number that is independent of the a_i, i.e., the *equation* $\Delta(a) = 0$ is left invariant.

sively on the Lie algebra level and utilized results due to Lie as well as his own theory of weights. His results were described in Section 7.5. What was not mentioned there was the geometrical application Kowalewski deduced from them on the final page:

(1) Suppose that $\mathcal{M} \subset \mathbb{CP}^n$ *is a manifold such that* $T[\mathcal{M}] \subset \mathcal{M}$ *for all* $T \in \mathfrak{P}_3$. *(2) If* $T \in \mathbf{PGL}(n+1, \mathbb{C})$ *is such that* $T[\mathcal{M}] \subset \mathcal{M}$, *then* $T \in \mathfrak{P}_3$, *except when* $n = 2\nu > 2$ *is even and* $\mathcal{M}$ *is a nondegenerate* $(n-1)$*-dimensional second-degree manifold, defined, e.g., by the equation* $J = 0$, *where*

$$J = a_0 a_n - \binom{n}{1} a_1 a_{n-1} + \binom{n}{2} a_2 a_{n-2} - \cdots \pm \frac{1}{2}\binom{n}{\nu} a_\nu^2. \tag{4.8}$$

By considering as $\mathcal{M}$ the manifold defined in homogeneous coordinates by $F(a) = 0$, where F is a homogeneous polynomial in $a_0, \ldots, a_n$, Kowalewski's geometrical application translates into the following generalization of Ostrowski's theorem on the discriminant:

(4.9) **Theorem.** *(1) Suppose that* $F(a)$ *is any nonconstant invariant of the binary form* $f(x, y)$. *(2) If* $T \in \mathbf{GL}(n+1, \mathbb{C})$ *leaves* F *invariant, then* $T \in \mathfrak{G}_4$, *except when* $n = 2\nu > 4$ *is even and* F *is a power of* J.

Ostrowski, however, did not ascribe this theorem to Kowaleski, because he realized that what Kowalewski had actually proved under the hypothesis (1) was that, barring the exceptional case, the biggest projective group *in Lie's sense* that leaves $F(a)$ invariant[40] is $\mathfrak{P}_3$. Although admitting that this was a "beautiful theorem," Ostrowski pointed out that it did not imply his theorem on the discriminant (nor, of course, the more general Theorem 4.9[41]) because Lie's groups are assumed "continuous" (in the sense of connected—see Section 3.1), whereas the maximal group need not be continuous in this sense. That is, suppose $\mathfrak{G}_F$ is the group of all $T \in \mathbf{GL}(n+1, \mathbb{C})$ for which a homogeneous polynomial $F(a)$ is a relative invariant, meaning $F(Ta) = \gamma F(a)$, where $\gamma \neq 0$. Then $\mathfrak{G}_F$ contains all scalar transformations $T = \lambda I$, but neither $\mathfrak{G}_F$ nor its projective counterpart $\mathfrak{P}_F = \mathfrak{G}_F/(\lambda I)$ is necessarily connected.[42] In the brief addition to [1919] prompted by the discovery of Kowalewski's paper, Ostrowski gave no example. A simple one is afforded by $F(a) = a_0^2 + a_1^2 + \cdots + a_n^2$. For this choice, $\mathfrak{G}_F \subset \mathbf{GL}(n+1, \mathbb{C})$ consists of all scalar multiples of orthogonal transformations, and when n is odd neither $\mathfrak{G}_F$ nor the corresponding projective group $\mathfrak{P}_F$ is connected.[43]

---

[40] When its transformations are expressed in homogeneous coordinates $a_0, \ldots, a_n$.

[41] This more general theorem is not explicitly mentioned by Ostrowski.

[42] In his paper [1892], discussed in Section 3.1, Engel realized this fact.

[43] This is because if $T \in \mathfrak{G}_F$ is expressible as $T = \lambda_1 U_1 = \lambda_2 U_2$, where the U_i are orthogonal and $\lambda_i \neq 0$, then $U_1 = \pm U_2$, and so $\det U_1 = (-1)^{n+1} \det U_2 = \det U_2$ for $n+1$ even. In this case, the mapping $\varphi(T) = \det U$ for $T = \lambda U$ is well-defined, maps $\mathfrak{G}_F$ onto $\{-1, +1\}$, and is continuous, so that neither $\mathfrak{G}_F$ nor $\mathfrak{P}_F$ is connected.

Thus Lie's theory cannot be assumed to apply to establish Ostrowski's theorem or the more general Theorem 4.9.[44]

Schur's collaboration with Ostrowski came about because Schur saw how to extricate the relevant results of Kowalewski from their Lie group context and use them to establish results such as Ostrowski's. In their joint paper [1922], written in 1921 while Ostrowski was visiting Berlin,[45] they provided a proof of Theorem 4.9, which, they stressed, used "thoroughly elementary means" and thus completely avoided the terminology and results of Lie's theory, although they noted that there was a natural connection between their reasoning and Lie's theory. "It can even be said that we are following an older approach of Lie according to which infinitesimal operations appear merely as the left sides of certain differential equations" (p. 82). What they meant can be seen in their formulation of the theorem they attribute to Kowalewski.

Recall from Chapters 2–3 that in Lie's theory infinitesimal transformations dT were identified with differential operators XF so as to facilitate the calculus of infinitesimal transformations. For linear groups in variables $a_0, \ldots, a_n$ these operators take the form

$$XF = \sum_{k,l=0}^{n} c_{kl} a_l \frac{\partial F}{\partial a_k}. \tag{4.10}$$

As we saw in Section 7.3, already in the 1850s, and therefore independently of Lie's theory, Cayley had proved that a homogeneous polynomial $F = F(a_0, \ldots, a_n)$ is an invariant of the binary form f of (4.7) if and only if F satisfies the differential equations $\mathcal{X}F = 0$ and $\mathcal{Y}F = 0$, where

$$\mathcal{X} = \sum_{j=1}^{n} j a_{j-1} \frac{\partial}{\partial a_j}, \quad \mathcal{Y}F = \sum_{j=0}^{n-1} (d-j) a_{j+1} \frac{\partial}{\partial a_j}\,.$$

It follows, as Cayley had observed, that an invariant F of f will also satisfy the additional equation $\mathcal{Z}F = 0$, where

$$\mathcal{Z} = \mathcal{X}\mathcal{Y} - \mathcal{Y}\mathcal{X} = \sum_{j=0}^{n} (n-2j) a_j \frac{\partial}{\partial a_j}\,. \tag{4.11}$$

Ostrowski and Schur translated the results of Kowalewski's paper into the following theorem, attributed to Kowalewski, but for which they provided their own, purely algebraic, proof:

---

[44] In [1924:183–4] Ostrowski mentioned another example.

[45] According to Ostrowski (in a letter to me dated 22 March 1978). Ostrowski also wrote that it was Schur who saw how to generalize his 1919 result into the main theorem of their paper by making use of the differential equations for the invariants given below.

Kowalewski's Theorem. *Let F be any invariant of the binary form f with the exception of a power of the quadratic invariant J of (4.8) when n is even. Then if $XF = 0$, where XF is of the form (4.10), XF is a linear combination of $\mathcal{X}F$, $\mathcal{Y}F$, and $\mathcal{Z}F$.*

The attribution of this theorem to Kowalewski makes sense, because within the conceptual framework of Lie's theory, what it states is that if an infinitesimal transformation XF leaves F invariant (i.e., $XF = 0$), then XF belongs to the group of infinitesimal transformations (Lie algebra) $\mathfrak{p}_3$ of the projective group $\mathfrak{P}_3$. This is what followed from the results of Kowalewski, who worked on the Lie algebra level. Since the projective group $\mathfrak{P}_F$ of all transformations leaving F invariant need not be connected, however, Kowalewski's geometrical application—and Theorem 4.9—did not follow immediately from Lie's theory as he had implied. After pointing this out, the authors showed that by means of purely elementary algebraic reasoning it is possible to go from Kowalewski's theorem to Theorem 4.9.[46]

If this paper is any indication of their views, Ostrowski and Schur did not see Lie's theory as a suitable tool for dealing with such essentially algebraic topics as represented by Theorem 4.9. This point was developed further by Ostrowski in his *Habilitationsvortrag*, the talk that was part of the formalities of becoming a lecturer (*Privatdozent*) at Göttingen [1924:175]:

> It can probably be said that Lie's theory was completely insignificant for the development of the algebraic theory of invariants. There was too much brute force in Lie's ingenious geometrical intuition to be able to master such fine things as the algebraic properties of invariants. It belongs far more in the domain of the algebraist ....

Among the finer points Ostrowski included the reasoning leading from Kowalewski's theorem to Theorem 4.9.[47] He saw it as indicative of the "group-theoretic asymmetry" of the problem of characterizing the invariants of F (as in Theorem 4.9). "The problem is of an essentially arithmetical character, which is hence far removed from the direct analytical propositions of Lie's theory of groups and requires for its treatment methods that reach further and penetrate more deeply" (p. 184). Schur himself never expressed such views in print, but he also never utilized Lie's theory in his own work.

Ostrowski's views on the incompatibility of Lie's theory and the algebraic theory of invariants now seem dated in view of such developments as the theory of algebraic groups, which so well fuses the "arithmetical" nature of Hilbertian algebra with the structural considerations underlying Lie's theory. Of course, the heart of these structural considerations is the Killing–Cartan theory. It is doubtful that Ostrowski knew this theory well if at all—certainly Schur did not, as will be seen in Section 12.2.

---

[46] This is done in Section 2 of their paper utilizing a result of Section 1 (Satz II).

[47] Ostrowski [1924:183] was referring expressly to his new proof that Kowalewski's theorem implies Theorem 4.9.

Ostrowski's understanding of Lie's theory was probably restricted to the work of Lie and his students, such as Kowalewski, who, as we have seen, had not mastered the Killing–Cartan theory. Indeed, Weyl may have been the first mathematician who did so. As we shall see in Section 12.4, Weyl unwittingly provided a counterexample to Ostrowski's pessimistic view of the applicability of Lie's theory to the theory of invariants by showing how the Killing–Cartan theory, as extended by him, could be used to establish the existence of a finite basis for the invariants of any representation of any semisimple group. This achievement on Weyl's part was inspired by Schur's suggestive work on the Cayley counting problem, the subject of the next section.

10.5. Cayley's Counting Problem Revisited

Although Schur had spent almost his entire career at Berlin—all except the years 1913–16—it was not until 1921 that he finally became a chaired full professor there. Frobenius, who died in 1917, had taken every opportunity to advance Schur's career at Berlin, but his efforts proved less successful than they should have, given Schur's extraordinary talent.[48] The year 1921 was when Schur collaborated with Ostrowski, which may have served to reawaken his interest in the theory of invariants. In any case, during the winter semester of 1923–24 he gave what appears to be the first lectures ever at Berlin on the theory of invariants. A record of these lectures exists, since the notes taken by his foremost student, Richard Brauer, have been preserved.[49] Aside from the requisite preliminary introductory material on the subject, Schur focused on two topics: (1) the work of Cayley on the counting problem discussed in Section 7.3; (2) the finite basis theorems of Hilbert and Hurwitz. As we will see in the next section, these two topics provided the motivation for Schur's paper [1924a], which, in turn, was to inspire Weyl.

Hurwitz's idea of applying the integral calculus to the study of invariants suggested to Schur two ways to go beyond what he had found in the literature on which he based his lectures. He presented a paper on these matters at the 10 January 1924 sitting of the Berlin Academy. It bore the title "New Applications of the Integral Calculus to Problems in the Theory of Invariants," and as we shall see, when Weyl read it he found just what he needed to answer questions about the foundations of tensor calculus that he had been considering.

---

[48] See Biermann's comments and speculations on this matter [1973:139].

[49] I am grateful to the late Mrs. Richard Brauer and to Walter Feit and Jonathan Alperin for kindly making these notes available to me. H. Grunsky also published a version of the lectures [Schur 1968] based primarily on the form in which Schur presented them in 1928 but with consultation of earlier lecture notes, which, according to Grunsky, varied little over time.

Recall from Section 2 that Hurwitz had shown how every classical invariant is given by an integral as in (2.9). The reasoning was complicated by the parametric representation $U = U(\psi)$ of the elements of $\mathbf{SU}(n)$ that Hurwitz had devised. The first part of Schur's paper was devoted to showing that a different integral representation of invariants could be used to the same ends as those for which Hurwitz had devised his integral formula; it was simpler because it did not require Hurwitz's parametric representation. Also, rather than considering only the "classical" invariants of the power transformation representation $H(S) = P_d(S)$ of $\mathbf{GL}(n, \mathbb{C})$ (in the notation of Section 2), he allowed $S \to H(S)$ to denote any representation of $\mathbf{GL}(n, \mathbb{C})$ with the property that the matrix coefficients of $H(S)$ are homogeneous polynomials of fixed degree in the coefficients s_{ij} of S. As we saw in Section 2, all the types of invariants traditionally considered are covered by such a choice of $H(S)$, and Schur naturally cited his dissertation [1901] with the remark that he had there determined all such representations.

Thus in the first part of "New Applications" Schur called attention to his dissertation [1901] and to Hurwitz's paper [1897]; in addition he called attention to the unitarian trick by stating and proving his own, more general, version:

(5.1) *If $F(s_{\alpha\beta})$ is any homogeneous function of the n^2 complex variables $s_{\alpha\beta}$ that vanishes for all $s_{\alpha\beta}$ defining an element of $\mathbf{SU}(n)$, then F vanishes identically.*[50]

In the second part of "New Applications," which deals with Cayley's counting problem for the rotation group, he in effect showed how Frobenius's theory of group characters and representations, including the complete reducibility theorem, could be extended to any continuous group admitting a Hurwitz integral that converges for continuous integrands.[51]

The possibility of dealing with Cayley's counting problem for the invariants relative to the rotation group $\mathfrak{D} = \mathbf{SO}(n, \mathbb{R})$ was suggested by a paper by Theodor Molien [1898]. In 1897 Molien had applied his results on associative algebras to group algebras and independently discovered many of the basic theorems of Frobenius's theory of group characters and representations.[52] Frobenius quickly became aware of Molien's work and praised it. He also submitted Molien's paper [1898] for publication in the proceedings of the Berlin Academy. In this paper Molien showed how Cayley's problem could be solved for the invariants of a representation of a finite group.

---

[50] Schur called the elements of $\mathbf{SU}(n)$ unitary and used the term "semiunitary" for what would now be called unitary [1924a:190].

[51] Slodowy [1998] has recently discussed in present-day mathematical terms the contents of Schur's "New Applications" paper and its two sequels as a prelude to his discussion of his correspondence with Weyl (discussed here in Section 12.2) and Weyl's character formula (discussed here in Section 12.5).

[52] On Molien's work see my [1972] and Section 12 of my essay [1999a].

Actually, there is no indication in Molien's paper that he was aware of Cayley's problem, but the interest in finite groups of linear transformations that led him to rediscover the basics of Frobenius's theory was connected from the outset with an interest in the invariants of such groups. He had studied for a year with Klein, whose program of developing a generalized Galois theory of equations involved the consideration of such invariants.[53] In [1898] Molien considered the following problem. Given a finite group $\mathfrak{H}$ of linear transformations, let m_d denote the number of linearly independent invariants of this group that are of degree d as homogeneous polynomials. The problem was to develop a formula by which to compute m_d for specific groups. His solution was based upon his realization that (1) m_d can be interpreted as the number of times the trivial or 1-representation ($t \to 1$ for all $t \in \mathfrak{H}$) occurs in the complete reduction of a certain representation of $\mathfrak{H}$ (described below); (2) the number of times any irreducible representation occurs in the complete reduction of a given representation can be expressed by a character formula.

Consider first (2). Let $t \longrightarrow R(t)$ denote a representation of $\mathfrak{H}$ and $\chi(t) = \operatorname{tr} R(t)$. Then by the complete reducibility theorem (1.6), $R(t)$ is equivalent to a direct sum of irreducible representations of $\mathfrak{H}$. If $m^{(\lambda)}$ denotes the number of times the λth irreducible representation occurs, then, as we saw following the regular representation theorem (1.7), it follows from the orthogonality relations for irreducible characters that

$$m^{(\lambda)} = \frac{1}{h} \sum_{t \in \mathfrak{H}} \chi(t) \overline{\psi^{(\lambda)}(t)},$$

where h is the order of $\mathfrak{H}$. In particular, if $\lambda = 1$ corresponds to the trivial or 1-representation, then since $\psi^{(1)}(t) \equiv 1$,

$$m^{(1)} = \frac{1}{h} \sum_{t \in \mathfrak{H}} \chi(t). \tag{5.2}$$

To apply (5.2) Molien observed (without going into any details) that the linear transformations $t \in \mathfrak{H}$ act naturally on the homogeneous polynomials of degree d in the variables $x = (x_1, \ldots, x_n)$ so as to define a representation $t \to R(t)$. In the notational style of Hurwitz's paper of 1894 (Section 2) this can be explained as follows. Let $F(x) = F(x_1, \ldots, x_n)$ denote a homogeneous polynomial of degree d. Then $F(x) = F(tx') = F'(x')$ and $R(t) : F \to F'$. There is no evidence that Molien knew Hurwitz's paper, but of course, Schur did; and he could see that $R(t) = P_d(t)$, Hurwitz's power transformation. Now, F is an invariant of $\mathfrak{H}$ precisely when $F' = F$, i.e., when $R(t) \cdot F = F$. From this it follows that the number m_d of linearly independent invariants is precisely the number of times the 1-representation occurs in the complete reduction of $R(t) = P_d(t)$; or in view of (5.2),

[53] See [Hawkins 1972:268–9].

$$m_d = \frac{1}{h}\sum_{t\in\mathfrak{H}} \zeta^{(d)}(t), \quad \zeta^{(d)}(t) = \text{tr } P_d(t). \tag{5.3}$$

In order to compute m_d for specific groups using (5.3) Molien showed that $\zeta^{(d)}(t)$ is the coefficient of ω^d in the expansion of $f(\omega) = [\det(I - \omega t)]^{-1}$ in ascending powers of ω.

Except within Klein's school, however, the invariants of finite groups was not a primary concern of those interested in the theory of invariants, which continued to focus on the invariants of continuous groups of linear transformations. What interested Schur in Molien's paper was that he realized that its approach could be used to solve Cayley's problem for invariants relative to the rotation group $\mathfrak{D} = \mathbf{SO}(n, \mathbb{R})$. As we have seen, Molien's method was based on the fundamentals of Frobenius's theory: the complete reducibility theorem and the orthogonality relations for the irreducible characters, which led to (5.3). By virtue of his reformulation of Frobenius's theory in [1905], Schur could see that what he had done there could be applied just as well to continuous representations $s \to R(s)$ of $\mathfrak{D}$, the only difference being that the technique of summing over the finite group is replaced by integrating over $\mathfrak{D}$ in the sense of Hurwitz.

In particular, he pointed out that the complete reducibility theorem in the form of Maschke's theorem (4.6) follows at once, since if $\varphi(z) = \sum_{i=1}^{N} z_i \bar{z}_i$ is the standard positive definite hermitian form in the variables $z = (z_1, \ldots, z_N)$ on which $H(s)$ acts, then the Hurwitz integral $\Phi(z) = \int_{\mathfrak{D}} \varphi(H(s) \cdot z)\, ds$ defines a positive definite hermitian form that is invariant under the transformations $H(t)$. Thus, unlike the situation he had encountered in his dissertation [1901] when dealing with the group $\mathbf{GL}(n, \mathbb{C})$, for the group $\mathfrak{D}$ orthogonality relations have a meaning, namely

$$\int_{\mathfrak{D}} \chi^{(\lambda)}(s)\overline{\chi^{(\mu)}(s)}\, ds = M\delta_{\lambda\mu}, \quad M = \int_{\mathfrak{D}} ds \tag{5.4}$$

and are proved, as he noted, by reasoning entirely analogous to what had been used for finite groups in [1905]. Thus if (as in Section 3) U is an arbitrary $N \times N$ matrix and the matrix V of (4.5) is now defined via a Hurwitz integral[54] as $V = \int_{\mathfrak{D}} R(s^{-1})UR(s)\, ds$, then, as before, $VR(t) = R(t)V$ for all $t \in \mathfrak{D}$, so by Schur's lemma (4.2), $V = \omega(U) \cdot I$, and (5.4) with $\lambda = \mu$ follows with $\lambda = \mu$ by exactly the same reasoning as before.

Letting the continuous group of linear transformations $R(s)$, $s \in \mathfrak{D}$, play the role of Molien's finite linear group $\mathfrak{H}$, Schur was able to derive the analogue of (5.3), namely that the number m_d of linearly independent invariants of $H(s)$ of degree d is given by

$$m_d = \frac{1}{M}\int_{\mathfrak{D}} \zeta^{(d)}(s)\, ds, \quad \zeta^{(d)}(s) = \text{tr } P_d(H(s)). \tag{5.5}$$

[54] As extended to matrix-valued functions in the obvious way.

In order to calculate from this equation, Schur used Molien's characterization of $\zeta^{(d)}(s)$ but pushed it further. Referring to his dissertation he also assumed that $H(s)$ was a polynomial representation, since "in all problems to which one is led by the algebraic theory of invariants" it is always only a question of this type of representation [1924a:202]. The end result was that (5.5) was turned into an iterated integral with respect to the $n(n-1)/2$ Euler rotation angles φ_i, which provided, in Schur's opinion, a method of calculating the integers m_d that was, from a practical standpoint, thoroughly useful [1924a:203]. To support his claim he considered the classical case of the rotational invariants of the general homogeneous form $\phi(a;x)$ of degree k, so that $R(s) = P_k(s)$. As he pointed out, in the ternary case $k = 3$ his method makes it possible to "completely carry out" the numerical computation of the numbers m_d by means of well-defined computational algorithms.[55]

The above-described treatment of Cayley's problem for the rotation group $\mathfrak{D}$ suggested several avenues of further investigation to Schur. Given that for the group $\mathfrak{D}$ he had developed the basics of a theory "which exhibits far-reaching analogies with Frobenius's beautiful theory of group characters" [1924a:190], it was only natural for him to consider the analogue of the problem Frobenius had solved for the symmetric and alternating groups and that he himself had solved for the polynomial representations of $\mathbf{GL}(n,\mathbb{C})$: Determine formulas for the irreducible characters and obtain therefrom formulas for the degrees of the irreducible representations. Although when he was writing up the above-described solution to the counting problem, the problem seemed very difficult, by the time he had the page proofs he had solved it [1924a:197, n.1]. Other natural problems to consider were the counting problem for the full orthogonal group $\mathfrak{D}' = \mathbf{O}(n,\mathbb{R})$ and the problem of determining for $\mathfrak{D}'$ analogous irreducible character and dimension formulas. Schur showed that the counting problem was easily solved using the fact that

$$\oint_{\mathfrak{D}'} f(t)\,dt = \int_{\mathfrak{D}} f(s)\,ds + \int_{\mathfrak{D}} f(su_0)\,ds, \; u_0 \in \mathfrak{D}', \; \det u_0 = -1,$$

defines a translation-invariant integral on $\mathfrak{D}'$ [1924a:§9]. His results on irreducible character and dimension formulas for both $\mathfrak{D}$ and $\mathfrak{D}'$ were presented to the Academy on 13 November 1924 and published as [1924b].

In the final sentences of his paper [1924a:190], referring to the part of Hurwitz's paper containing the general definition of a translation-invariant integral (as indicated here in Section 2), Schur wrote:

> The results obtained by Hurwitz ... make it possible to extend the theory developed in the present work to more general classes of groups. The groups $\mathfrak{D}$ and $\mathfrak{D}'$ are not simply distinguished by

[55] "Auf Grund dieser Regel oder mit Hilfe der Formeln (27) lassen sich insbesondere die im vorigen Paragraphen erwähnten Abzählungsproblem der Invariantentheorie rechnerisch vollständig durchführen" [1924a].

> the important role they play in applications but rather also by the fact that here the integral calculus provides a solution to the counting problem that is actually practically useful.

In other words, given any linear group admitting a Hurwitz integral that converges for continuous integrands, one could imitate what Schur had done for $\mathfrak{D}$ and develop the essentials of Frobenius's theory, including the complete reducibility theorem and the orthogonality relations for irreducible characters, and obtain a formula for m_d analogous to (5.5). This was worth noting in passing, but what was of real importance to Schur—and worthy of a detailed publication—was what could be done with the formula for m_d by utilizing the specific properties of the groups $\mathfrak{D}$ and $\mathfrak{D}'$. For these groups he had succeeded in doing what Cayley had done, namely in obtaining formulas for the numbers m_d amenable to calculation, and this is what intrigued him about Hurwitz's "integral" approach to the theory of invariants.

As we shall see in Chapters 11–12, Weyl saw the matter differently. By virtue of his different background and interests, it was the passing remark in the first sentence of the above quotation that was of paramount importance. It provided him with the key to answering a question raised by his involvement with the tensor calculus of general relativity and then suggested the possibility of extending Frobenius's theory to all semisimple groups, compact or not.

Chapter 11
From Relativity to Representations

In an essay entitled "Relativity Theory as a Stimulus in Mathematical Research" [1949:400] Weyl wrote:

> For myself I can say that the wish to understand what really is the mathematical substance behind the formal apparatus of relativity theory led me to the study of representations and invariants of groups ....

The aim of this chapter is to attempt to explain what Weyl meant by this remark by attempting to reconstruct his intellectual journey from an intense involvement with Einstein's general theory of relativity and its possible extensions to the equally intense involvement with the representation of Lie groups that produced his great papers of 1925 and 1926, which form the subject of Chapter 12.[1]

11.1. Einstein's General Theory of Relativity

In Chapter 9 we considered Weyl's activities at Göttingen as student (1904–08) and then lecturer (*Privatdozent*) (1908–13). In 1913 a professorship opened up at the ETH in Zürich due to the retirement of the geometer K. F. Geiser.[2] At Geiser's suggestion, the president of the Swiss school board, R. Gnehm, who was professor of chemistry at the ETH, sought the

[1] In writing this chapter I have drawn upon my essay [1998].

[2] The following discussion of Weyl's appointment to and early years at the ETH is based on the documents contained in [Frei and Stammbach 1992:12–15].

opinion of Frobenius on Geiser's replacement. As we saw in Section 10.1, Frobenius had spent 1876–92 as professor at the ETH, where Geiser was his colleague, before moving on to a professorship at Berlin. Hurwitz had in fact been Frobenius's successor at the ETH, where he remained until his death in 1919. Also at the ETH since 1912 was Albert Einstein, who was on friendly terms with Hurwitz as can be seen from the photo in Figure 36. The work of these three men—Frobenius, Hurwitz, and Einstein—was to play an important role in Weyl's mathematical future.

The list of candidates presented by Gnehm to Frobenius included Weyl and many other exceptional mathematicians such as M. Dehn, A. Haar, O. Perron, F. Riesz, and O. Toeplitz. Weyl, however, was their first choice. Given that a decade later Weyl was to brilliantly extend Frobenius's theory to semisimple continuous groups, it is especially interesting to see what Frobenius had to say.

Frobenius's reply was certainly not an unequivocal endorsement of Weyl for the position. He pointed out that although the ETH preferred someone with geometrical talent, only two of the listed candidates were geometers, and Weyl was of course not one of them. Among nongeometers he also mentioned his own student I. Schur, but then added that he was "much too good for Zürich" and would someday be his own successor at Berlin. Weyl was one of five listed candidates from the Göttingen school, a fact that clearly annoyed Frobenius. He pointed out that they all worked on matters related to integral equations, all of which work seemed to Frobenius to have a certain sameness to it. With evident sarcasm he said that their work was "very general, very deep, so abysmally deep that a short sighted person like me finds it difficult to recognize the new ideas ...." Of the five from Göttingen, however, Frobenius agreed that Weyl was the best and mentioned in this connection his highly reputed recent book on Riemann surfaces [1913] (discussed in Section 9.5). And so he concluded, "If you nominate Weyl, you will have made an outstanding, unobjectionable choice."

Although hardly a ringing endorsement, this was evidently enough to confirm Gnehm and his colleagues in the correctness of their first choice. And so it came to be that Weyl was offered a professorship at the ETH, where he remained from the fall of 1913 until 1930. It was here that Einstein and Grossmann had collaborated on their paper [1913] introducing tensor analysis into the search for the gravitational field equations, and although Einstein left Zürich for Berlin in 1914, Grossmann remained a professor at the ETH.[3] Neither Frobenius, who died in 1917, nor Hurwitz, who died in

---

[3] Weyl also had as a colleague George Pólya, who later wrote about Weyl, "We had very little mathematical contact, and we were personally very different. He was always interested in far-reaching generalizations and I was interested in special cases. We never quite understood each other" [1987:105]. Cf. Weyl's strong letter in support of Pólya's style of mathematics [Frei and Stammbach 1992:36–9].

Figure 36. Hurwitz conducting his daughter and Einstein in 1913.

1919, lived to see how Weyl, under the impetus provided by his reaction to Einstein's theory, utilized their ideas in his brilliant papers of 1925–26.

When Weyl came to Zürich, he initially worked along the lines he had established while at Göttingen. Then, in 1914, came the disruption of the war. Although at first Weyl was judged physically unfit for military service by the German army, he was finally drafted in the spring of 1915 but was discharged the following spring at the request of the Swiss government.[4] As he later recounted, "My mathematical mind was as blank as any veteran's, and I did not know what to do. I began to study algebraic surfaces; but before I had gotten far, Einstein's memoir came into my hand and set me afire."[5] The memoir in question was Einstein's definitive paper in *Annalen*

[4] This according to [Scholz 1994:205].

[5] Quoted by S. Sigurdsson [1991:62] from Weyl's unpublished "Lecture at the

der Physik, "Grundlage der allgemeinen Relativitätstheorie" [1916]. It will be helpful to begin by considering briefly what went into the making of this paper and how it drew upon the work of mathematicians such as Riemann, Christoffel, and Ricci.

In 1907, when Minkowski was translating the new theory of relativity into a space-time geometry, Einstein began turning his attention to its generalization so as to account for gravity [1907].[6] Between 1911 and 1916 the theory began to unfold in his publications. He began his major paper on the subject [1916] by acknowledging a debt to Minkowski: "The generalization of the relativity theory was facilitated through the form that Minkowski had given to the special relativity theory. He was the first mathematician to clearly perceive the formal equivalence of the space and time coordinates; this made possible the construction of the [general] theory."[7] In order to see what Einstein drew from Minkowski's writings, it is necessary to say a bit more about their content.

In his lecture "Space and Time" Minkowski expressed the view that the term "relativity postulate" for Einstein's first principle was "too feeble," since "the postulate comes to mean that only the four-dimensional world in space and time is given by the phenomena ... I prefer to call it the *postulate of the absolute world* .... The world-postulate permits identical treatment of the four coordinates x, y, z, t. By this means, as I shall now show, the forms in which the laws of physics are displayed gain in intelligibility" [1908:83]. In this connection he introduced the notion of the "proper time" τ of an object moving through space-time along a curve $P(t) = (x(t), y(t), z(t), t)$. Since in the special theory of relativity velocities greater than the velocity $c = 1$ of light are not considered, $dt^2 - dx^2 - dy^2 - dz^2$ is always positive, because it can be rewritten as

$$1 - (\frac{dx}{dt})^2 - (\frac{dy}{dt})^2 - (\frac{dz}{dt})^2 = 1 - v^2,$$

where v is the speed of an object. Minkowski defined $d\tau$ by

$$d\tau^2 = dt^2 - dx^2 - dy^2 - dz^2, \tag{1.1}$$

and the proper time τ by $\tau = \int_{t_0}^{t} d\tau$, where $P_0 = P(t_0)$ is an arbitrarily fixed point.

In effect, (1.1) defines a metric on the four-dimensional space-time manifold with the inner product of two vectors $\mathbf{v} = (x, y, z, t)$, $\mathbf{w} = (x', y', z', t')$ being $\mathbf{v} \cdot \mathbf{w} = tt' - xx' - yy' - zz'$, so that $|\mathbf{v}|^2 = t^2 - x^2 - y^2 - z^2$. For example, the velocity and acceleration vectors of the object at $P(t) =$

Bicentennial Conference" (Princeton, 1946).

[6] For accounts of Einstein's work on general relativity see the *DSB* article by Balzas [1971], the books by Pais [1982] and Mehra [1974], and the the Einstein conference proceedings [Howard and Stachel 1989].

[7] Quoted in the translation by Miller [1981:243].

(x, y, z, t) are the four-dimensional vectors obtained by differentiating with respect to τ:

$$\mathbf{v} = \left(\frac{dx}{d\tau}, \frac{dy}{d\tau}, \frac{dz}{d\tau}, \frac{dt}{d\tau}\right), \quad \mathbf{a} = \left(\frac{d^2x}{d\tau^2}, \frac{d^2y}{d\tau^2}, \frac{d^2z}{d\tau^2}, \frac{d^2t}{d\tau^2}\right).$$

It follows from (1.1) that

$$1 = \left(\frac{dt}{d\tau}\right)^2 - \left(\frac{dx}{d\tau}\right)^2 - \left(\frac{dy}{d\tau}\right)^2 - \left(\frac{dz}{d\tau}\right)^2,$$

which means that $\mathbf{v}$ is a unit vector and implies, upon differentiation with respect to τ, that $\mathbf{v} \cdot \mathbf{a} = 0$, so that the acceleration is always normal to the velocity in the sense of the metric defined by (1.1) [1908:85]. Thus not only did Minkowski emphasize "the formal equivalence of the space and time coordinates" (as Einstein put it) by advocating the development of the physics of relativity within the context of four-dimensional space-time, he also focused on the role of the metric $d\tau$ given by (1.1) in dictating the geometry of space-time. In [1907b] Minkowski also introduced the notation $x_1, \ldots, x_4$ for x, y, z, it to bring out more effectively the formal equality of space and time variables in the fundamental equations of the theory.

The starting point of Einstein's quest for a more general theory of relativity was what became known as his "principle of equivalence." According to this principle (already stated in 1907) the phenomenon of a gravitational field acting with a constant acceleration such as, e.g., $-g\mathbf{k}$ with respect to a reference frame K is indistinguishable from the phenomenon generated by a second frame of reference K' moving with a uniform acceleration g in the direction of the positive z-axis of K. As Einstein explained in a paper applying the principle to predict the bending of light rays by the gravitational field of the sun [1911:100-101], the equivalence principle

> makes it impossible for us to speak of the absolute acceleration of the system of reference, just as the usual theory of relativity forbids us to talk of the absolute velocity of a system; and it makes the equal falling of all bodies in a gravitational field a matter of course .... But this view of ours will not have any deeper significance unless the systems K and K' are equivalent with respect to all physical processes, that is, unless the laws of nature with respect to K are in entire agreement with those with respect to K'.

On a formal mathematical level this view would imply an invariance of the of fundamental physical equations with respect to more general coordinate transformations than those of the Lorentz group, such as those of uniformly accelerated frames. If we introduce the notation $X_1, \ldots, X_4 = x, y, z, t$ and denote Minkowski's metric (1.1) by $ds^2 = -dX_1^2 - dX_2^2 - dX_3^2 + dX_4^2$, then the more general coordinate transformations $X_i = \varphi_i(x_1, \ldots, x_4)$, $i = 1, \ldots, 4$, will take ds^2 into the more general expression

$$ds^2 = \sum_{i,j=1}^{4} g_{ij} dx_i dx_j, \tag{1.2}$$

where the g_{ij} are variable functions of $x_1, \ldots, x_4$ and $g_{ij} = g_{ji}$ for all i, j. With the guidance of his mathematician friend Marcel Grossmann (1878–1936), Einstein came to recognize that the expression (1.2) can be identified with the line element in a Riemannian-type geometry (Section 4.5).

After Riemann's death the fundamentals of an invariant theory of quadratic differential forms were worked out independently by Christoffel and Lipschitz in Germany in the late 1860s. In the 1880s the Italian mathematician Gregorio Ricci-Curbastro utilized their work, especially that of Christoffel, which went further, to create a general, purely analytical method for dealing with quadratic differential forms and their invariants. Ricci termed it the "absolute differential calculus," although it eventually became known by the name "tensor analysis."[8] Ricci gave a summary of his absolute calculus in French in Darboux's *Bulletin* in 1892, and shortly thereafter, Klein urged him to make his methods more widely known. Finally, in 1900, with the assistance of his student Levi-Civita, he published a lengthy (77-page) memoir in French on "Methods of the Absolute Differential Calculus and Its Applications" in Klein's journal, *Mathematische Annalen*. (Two Italians writing in French in a German journal—quite international!) In their opening remarks they summarized the historical roots of their work and its significance:

> The algorithm of the absolute differential calculus ... can be found in a remark due to Christoffel. — But the methods themselves and the advantages they offer have their justification and their source in the intimate relationships that join them to the notion of an n-dimensional manifold, which we owe to the brilliant minds of Gauss and Riemann. — Accordingly, this notion of a manifold V_n is defined intrinsically in its metric properties by n independent variables and by an entire class of quadratic differential forms in these variables any two of which are transformable into one another by point transformations. Consequently, a V_n is invariant vis-à-vis every transformation of its coordinates. The absolute differential calculus is, in its formulas and results, also independent of the choice of independent variables. Being thus associated in an essential way with V_n, it is the natural instrument of all those studies that have as their subject such a variety, or in which one encounters as a characteristic element a positive quadratic form of the differentials of n variables ... [1900:127–128].

Although the authors here speak of ds^2 as positive definite, as did Riemann, much of the theory, including the entire apparatus of the absolute differ-

[8] For a detailed history of tensor analysis see [Reich 1994].

ential calculus, applies to indefinite metrics ds^2 such as the metric (1.1) of special relativity.

The Ricci–Levi-Civita paper became the principal text by which Grossmann taught Einstein the tools he might need to generalize the theory of relativity, to articulate its laws in much more general reference frames. As Einstein explained in a letter to Sommerfeld in 1912, "I am now exclusively occupied with the problem of gravitation theory and hope, with the help of a local mathematician friend, to overcome all the difficulties. One thing is certain, however, that never in life have I been quite so tormented. A great respect for mathematics has been instilled within me, the subtler aspects of which, in my stupidity, I regarded until now as a pure luxury. Against this problem [of gravitation] the original problem of the theory of relativity is child's play."[9]

Let us now consider the basics of Ricci's calculus by means of which Einstein hoped to overcome the formidable mathematical problems posed by his envisioned theory of gravitation.[10] Given the quadratic differential form

$$(1.3)\quad ds^2 = \sum_{ij} g_{ij} dx_i dx_j,\ g_{ij} = g_{ij}(x_1, \ldots, x_n), \quad |g_{ij}| \neq 0, \quad g_{ij} = g_{ji},$$

by the variable change

$$x_i = \varphi_i(\bar{x}_1, \ldots, \bar{x}_n), \quad i = 1, \ldots, n,$$

which means on the differential level that

$$(1.4)\qquad dx_i = \sum_\alpha \frac{\partial x_i}{\partial \bar{x}_\alpha} d\bar{x}_\alpha = \sum_\alpha p^i_\alpha d\bar{x}_\alpha,$$

the form (1.3) is transformed into $ds^2 = \sum_{\alpha,\beta} \bar{g}_{\alpha\beta} d\bar{x}_\alpha d\bar{x}_\beta$, where from (1.4) by direct calculation

$$(1.5)\qquad \bar{g}_{\alpha\beta} = \sum_{i,j} g_{ij} p^i_\alpha p^j_\beta.$$

The transformation rule given in (1.5) makes the system of n^2 functions g_{ij} an example of a covariant tensor of degree two. Likewise, a system of n^3 functions $T_{ijk} = T_{ijk}(x_1, \ldots, x_n)$ forms a covariant tensor of degree three, provided that $\bar{T}_{\alpha\beta\gamma} = T_{\alpha\beta\gamma}(\varphi_1, \ldots, \varphi_n)$ satisfies

$$(1.6)\qquad \bar{T}_{\alpha\beta\gamma} = \sum_{i,j,k} T_{ijk} p^i_\alpha p^j_\beta p^k_\gamma,$$

---

[9] Letter of 12 October 1912 [Hermann 1968:26]. I have used the English translation by Mehra [1974:1–2].

[10] For the sake of clarity I have not adhered to Ricci's notation and terminology. Reich's account [1994] describes the various notations and terminologies employed in the evolution of the theory.

Figure 37. Marcel Grossmann and Einstein, pictured here together in 1899 at the Grossmann home in Thalwil, had been fellow students at the ETH.

and so on for covariant tensors of other degrees. Thus when a covariant tensor is given in one coordinate system it is known in all others in the sense that the rules for expressing it in another, barred, coordinate system have been specified.

There is also a notion of contravariant tensors. If $\bar{x}_i = \psi_i(x_1, \ldots, x_n)$, $i = 1, \ldots, n$, is the inverse variable transform, let

$$d\bar{x}_i = \sum_\alpha \frac{\partial \bar{x}_i}{\partial x_\alpha} dx_\alpha = \sum_\alpha \pi^i_\alpha dx_\alpha.$$

Then, e.g., a system of n^3 functions (which are written with superscripts rather than subscripts) $T^{ijk} = T^{ijk}(x_1, \ldots, x_n)$ forms a contravariant tensor of degree three if they satisfy the transformation rule

$$\bar{T}^{\alpha\beta\gamma} = \sum_{i,j,k} T^{ijk} \pi^\alpha_i \pi^\beta_j \pi^\gamma_k. \tag{1.7}$$

Since the Jacobian matrices (p^i_j) and (π^i_j) are inverses, (1.7) can be rewritten in the form

$$T^{\alpha\beta\gamma} = \sum_{ijk} \bar{T}^{ijk} p^\alpha_i p^\beta_j p^\gamma_k, \tag{1.8}$$

which can be compared to (1.6). An important example of a contravariant tensor of degree two is given by the system g^{ij}, where the matrix (g^{ij}) is the inverse of (g_{ij}).

Tensors of degree 0, or scalars, are simply functions $T = T(x_1, \ldots, x_n)$. If T is a scalar, then its gradient $\nabla T = (\partial T/\partial x_1, \ldots, \partial T/\partial x_n)$ represents a covariant tensor of degree 1, namely the system $T_i = \partial T/\partial x_i$. However, for higher-degree tensors, this does not work. For example, if T_{ij} is a covariant tensor of degree 2, then $T_{ijk} = \partial T_{ij}/\partial x_k$ does not define a covariant tensor of degree 3. Christoffel had discovered how to make a covariant tensor of degree $k+1$ out of the derivative of one of degree k by adding correction terms, and Ricci saw in this procedure, suitably generalized, the key to his absolute differential calculus. For example, if T_i and T_{ij} are covariant tensors of degrees 1 and 2, respectively, then

$$D_{ij} = \frac{\partial T_i}{\partial x_j} - \sum_\alpha \Gamma^\alpha_{ij} T_\alpha, \quad \text{and} \quad D_{ijk} = \frac{\partial T_{ij}}{\partial x_k} - \sum_\alpha \left(\Gamma^\alpha_{ik} T_{\alpha j} - \Gamma^\alpha_{jk} T_{i\alpha}\right)$$

are covariant tensors of degrees 2 and 3, respectively, where the Γ^i_{jk} (now called Christoffel symbols) are defined in terms of the functions g_{ij} by

$$\Gamma^k_{ij} = \sum_\alpha \frac{1}{2} g^{\alpha i} \left(\frac{\partial g_{j\alpha}}{\partial x_k} - \frac{\partial g_{jk}}{\partial x_\alpha} + \frac{\partial g_{\alpha k}}{\partial x_j} \right), \quad (g^{\alpha\beta}) = (g_{ij})^{-1}. \tag{1.9}$$

(Despite the suggestive notation these symbols do not transform as tensors.) Christoffel's discovery led Ricci to develop his absolute differential calculus based on the notion of covariant differentiation it suggests, namely

$$\frac{DT_i}{Dx_j} = D_{ij}, \quad \frac{DT_{ij}}{Dx_k} = D_{ijk},$$

and so on. He also introduced an analogous contravariant differentiation process applicable to contravariant tensors.

Central to the entire theory was a covariant tensor of degree four, which is now called the Riemann curvature tensor and frequently denoted by R_{ijkl}. It was introduced independently by Riemann and Christoffel. For Riemann it was a generalization of the notion of the Gaussian curvature of a surface. For Christoffel it was the key to the problem of determining the differential invariants of the form ds^2. He showed that R_{ijkl} is expressible in terms of his symbols Γ^i_{jk}, since if

$$R^i_{jkl} = \frac{\partial \Gamma^i_{jk}}{\partial x_l} - \frac{\partial \Gamma^i_{jl}}{\partial x_k} + \sum_\alpha \left(\Gamma^\alpha_{jk} \Gamma^i_{\alpha l} - \Gamma^\alpha_{jl} \Gamma^i_{\alpha k}\right), \tag{1.10}$$

then

$$R_{ijkl} = \sum_{\alpha} g_{i\alpha} R^{\alpha}_{jkl}. \tag{1.11}$$

The problem of determining all differential invariants associated to the differential form ds^2 reduced to determining algebraic invariants of the system consisting of g_{ij}, R_{ijkl}, and the successive covariant derivatives of R_{ijkl}.

In his joint paper with Grossmann [1913] Einstein began to apply tensor analysis to the relativistic physics of gravity. This constituted the first part of the paper; in the second part, which was more or less an appendix to the first, Grossmann developed the purely mathematical aspects of the calculus of tensors. (Incidentally, it is in this paper that the term "tensor" of degree k is used in place of Ricci's "system of order k," and so began the terminology of tensors.) In the first part Einstein began by observing that in the Minkowski space-time of special relativity, where one has $ds^2 = -dx^2 - dy^2 - dz^2 + dt^2$, it was known that the equations of motion of a point mass moving free of forces are given by $\delta \int ds = 0$ and that (as he had shown in 1912) this remains true in a static gravitational field according to his equivalence principle. Thus for a general coordinate system in which $ds^2 = \sum_{i,j=1}^{4} g_{ij} dx_i dx_j$, the equations of motion in a gravitational field become $\delta \int ds = 0$, which means that the space-time path of a particle in a gravitational field is a geodesic in the underlying Riemannian geometry. Einstein regarded the functions g_{ij} (of which there are 10 by virtue of the symmetry of the g_{ij}) as generalizations of the gravitational potential function $\varphi(x, y, z)$ of Newton's theory, which gives the gravitational force as $\mathbf{F} = -\text{grad}\,\varphi$. In Newton's theory φ is determined by Poisson's equation $\nabla^2\varphi = 4\pi\rho$, where ρ is the density of matter and

$$\nabla^2\varphi = \frac{\partial^2\varphi}{\partial x^2} + \frac{\partial^2\varphi}{\partial y^2} + \frac{\partial^2\varphi}{\partial z^2}.$$

To complete the basics of his theory of gravitation Einstein needed an analogue of Poisson's equation to determine his potential functions g_{ij}.

Einstein had an intuition as to what these gravitational field equations would look like. What he expected may be written in the covariant form

$$F_{ij} = \gamma T_{ij}, \tag{1.12}$$

where γ is a constant and T_{ij} is the energy–momentum tensor, which was the analogue of ρ in Poisson's equation.[11] As for F_{ij}, he assumed that it would be a degree-2 tensor that arises from the g_{ij} by differentiation. Since the left-hand side of Poisson's equation is the Laplacian $\nabla^2\varphi$, he assumed

[11] Einstein wrote his equation in terms of contravariant rather than covariant tensors [1913:233]. The two forms are equivalent, since corresponding to any covariant tensor T_{ij} of degree 2 is a contravariant tensor T^{ij} defined by $T^{ij} = \sum_{\alpha,\beta} g^{i\alpha} g^{j\beta} T_{\alpha\beta}$. From this it follows that (1.12) is equivalent to $F^{ij} = \gamma T^{ij}$.

that F_{ij} would involve nothing higher than second-order derivatives of the g_{ij}.

In his part of the paper [1913:256–7] Grossmann added some purely mathematical reflections relating to the sought-for tensor F_{ij} of (1.12). He pointed out the fundamental role played by the Riemann curvature tensor R_{ijkl}. This made it "a priori likely" that the differential tensors arising from R_{ijkl} in the general theory "should also be of significance for the problem of the differential equations of a gravitational field." Indeed, from (1.9)–(1.11) we see that both the "mixed tensor"[12] R^i_{jkl} and R_{ijkl} involve derivatives of the second order but no higher. These tensors are, of course, not of degree two, but by the standard contraction technique of Ricci's calculus for generating tensors of lower degree, Grossman obtained the tensor

$$R_{ij} = \sum_{\alpha} R^{\alpha}_{ij\alpha} = \sum_{\alpha,\beta} g^{\alpha\beta} R_{i\alpha\beta j},$$

which is now called the Ricci curvature tensor.[13] He pointed out that R_{ij} involves no more than the second derivatives of the g_{ij}. Might this be a likely candidate for F_{ij} in (1.12)? Clearly, from a purely mathematical point of view the answer was yes, but Grossmann went on to point out, following Einstein, that R_{ij} did not seem to be a generalization of the Laplacian $\nabla^2\varphi$, since in the special case of an "infinitely weak" gravitational field R_{ij} did not reduce to $\nabla^2\varphi$. Actually, this conclusion is based upon a misconception and proves to be incorrect, but at the time it seemed that the Ricci calculus was not suitable for the purpose at hand. As they put it, *if* F_{ij} actually involves second-order derivatives but not higher, then it cannot be covariant with respect to all coordinate transformations. In a remark added just before publication Einstein contributed a physical argument to this effect as well. In a letter to his friend Michele Besso in March 1914 Einstein expressed his disillusionment with "the subtler aspects" of mathematics (as he had put it to Sommerfeld in the above quotation), complaining now that the "general theory of invariants," i.e., the Christoffel–Ricci calculus, had "acted as an obstacle" in his attempts to work out the physical theory.[14]

By November of 1915 Einstein had made discoveries that reinstated his earlier, more favorable, assessment of the value of the tensor calculus. In the first of three papers on his new discoveries presented to the Berlin Academy he wrote that the general theory of relativity, as he was now able to present it, "represents a real triumph of the method of the general differential calculus founded by Gauss, Riemann, Christoffel, Ricci, and Levi-Civita."[15]

---

[12] A notion introduced in this paper.

[13] According to Speziali [1975:407], Ricci had introduced this tensor in 1896. Grossmann, however, did not refer to Ricci in this connection.

[14] Letter dated "Zurich March 1914" in [Speziali 1972]. I have followed Mehra's translation [1974:13]. There is a considerable literature dealing with the intriguing question of why it took Einstein so long to accept R_{jk} and general covariance. See in this regard [Norton 1989; Stachel 1989].

[15] [Einstein 1915:778]. Quoted in the translation by Mehra [1974:14].

He had discovered that he could maintain the general covariance under all transformations of the laws of relativistic physics. In particular, the desired field equations that he had tentatively written as (1.12) but with the caveat that they could not be generally covariant, he was now able to write as a true tensor equation involving the Ricci curvature tensor:

$$R_{ij} = -\gamma \left(T_{ij} - \frac{1}{2} g_{\alpha\beta} T \right),$$

where T is the scalar obtained from T_{ij} by contraction.[16] With this result, Einstein could conclude that "the general theory of relativity is finally completed as a logical structure. In its most general formulation, which turns the space-time coordinates into parameters devoid of physical significance, the principle of relativity leads, with conclusive necessity, to a quite definite theory of gravitation which can explain the perihelion motion of Mercury."[17] Perhaps as a reflection of his new-found appreciation for Riemannian geometry and the attendant tensor analysis, in his definitive exposition of his completed theory [1916] Einstein pointed out that the equations of motion, which follow from the geodesic nature of the path, $\delta \int ds = 0$, may be expressed in the elegant form

$$\frac{d^2 x^\alpha}{ds^2} + \sum_{\rho,\sigma} \Gamma^\alpha_{\rho\sigma} \frac{dx^\rho}{ds} \frac{dx^\sigma}{ds} = 0.$$

Einstein's theory could thus be interpreted as another instance of the preestablished harmony between mathematics and reality that Minkowski and Hilbert were fond of mentioning.[18] This instance, however, was far more impressive. That is, by introducing the idea of a space-time geometry, Minkowski had linked the special theory of relativity to non-Euclidean geometry and the algebraic theory of invariants, and the hope was that these mathematical theories might prove useful in the further development of the theory. Before the advent of general relativity, however, these hopes had not been realized. But now it was no longer a question of promising connections with high-powered mathematical theories; Riemannian geometry and the Ricci calculus had become essential for the discovery and articulation of the fundamental physical laws themselves. As Einstein had discovered, now there was no escaping higher mathematics. Of course, as we saw in Section 9.3, Hilbert went even further and in his "Grundlagen der Physik" papers [1916; 1917] envisioned a new science of physical geometry, capable of explaining not only gravity but electromagnetic phenomena as well. It went without saying that from Hilbert's perspective, mathematicians were eminently qualified to deal with the new physical geometry. Indeed,

---

[16] $T = \sum_{\alpha,\beta} g^{\alpha\beta} T_{\alpha\beta}$, where as usual, $(g^{ij}) = (g_{ij})^{-1}$.

[17] Quoted in translation by Mehra [1974:16].

[18] Indeed, Hilbert pointed this out [Sigurdsson 1991:70].

Hilbert's papers generated a flurry of activity in Göttingen involving Klein and Emmy Noether.[19] And in Zürich, Weyl also began to react to Einstein's theory.

11.2. The Space Problem Reconsidered

By the summer of 1917 Weyl was lecturing on general relativity at the ETH. These lectures formed the basis for his classic book *Raum, Zeit, Materie* (*Space, Time, Matter*), which went through four editions[20] during 1918–23 and spawned many collateral publications by Weyl aimed at further developing the ideas and implications of his lectures.

One of the outcomes of his reflections on general relativity was his introduction of what he called a "purely infinitesimal geometry."[21] Weyl became convinced that Riemannian geometry, including the quasi-Riemannian geometry of an indefinite metric $ds^2 = \sum_{ij} g_{ij} dx_i dx_j$, $g_{ij} = g_{ij}(x_1, \ldots, x_n)$, on which Einstein's theory was based, was not a consistently infinitesimal geometry.[22] That is, thanks to the work of Levi-Civita [1917] Weyl could see that in Riemannian geometry if two parallel vectors located at point P are moved by "parallel transport" to a point Q along two different paths, they are generally no longer parallel. The direction of the transported vector depends upon the path taken in proceeding from P to Q. That is how it should be in a purely local geometry. But Riemannian geometry was not consistently local, Weyl felt, because the underlying quadratic differential form permitted a way to compare the *length* of a vector at P and at Q that is independent of the path taken from P to Q. For if $v = (dx_1, \ldots, dx_n)$ in the tangent plane at point P of the manifold is transported to $w = (dy_1, \ldots, dy_n)$ in the tangent plane at point Q, the lengths of v and w are absolutely comparable in the sense that, independently of path, the ratio $|v|^2 : |w|^2 = \sum_{i,j} g_{ij}(P) dx_i dx_j : \sum_{i,j} g_{ij}(Q) dy_i dy_j$ is determined.

These considerations led Weyl to a generalization of Riemannian geometries in which the lengths of v and w are not absolutely comparable. He achieved this by introducing, along with the usual quadratic differential form, a linear differential form φ that made the length of a transported vector depend on the path. As a consequence, metric relations are determined locally only up to a positive calibration (or gauge) factor λ and so are given

[19] See in this connection the account by Rowe [1998a], which focuses on the role played by Noether.

[20] There were actually five editions, but the second (1919) was simply a reprint of the first [Scholz 1994:205n].

[21] See [Scholz 1994; 1995; 1996] for a detailed account of the historical context and evolution of Weyl's ideas on this theory during 1917–25.

[22] Sigurdsson [1991:154] presents Weyl's account of how this conviction grew out of a student's question.

Figure 38. Participants in the 1917 meeting of the Zürch Mathematical Society included, in the front row, Hilbert (center), and, on Hilbert's left, Geiser, then Weyl's wife, Hella, and Weyl himself. The general theory of relativity of Einstein was no doubt a topic of conversation.

by $ds^2 = \lambda \sum_{ij} g_{ij} dx_i dx_j$. Here λ varies from point to point in such a way that the comparison of the lengths of v at P and w at Q is also in general a path-dependent process.[23]

Although Weyl's geometry was motivated by the above critique of Riemannian geometry, he discovered that he could use its framework to develop a unified field theory, that is, a theory embracing both the gravitational and

[23] For a complete definition of Weyl's geometry see [Scholz 1994:213] and for a contemporary formulation see [Folland 1970]. Weyl's geometry represented the first of a succession of gauge theories that has continued into present-day physics [Vizgin 1989; 1994].

the electromagnetic fields. The coefficients of his linear differential form φ could be thought of as defining an electromagnetic potential with $d\varphi$ giving the corresponding electromagnetic field. As was noted in Section 9.3, Hilbert had been the first to devise a unified theory within the framework of general relativity in 1915. Certainly, Weyl, as Hilbert before him, was inspired by the great intellectual achievement of Einstein's theory, as well as its requirement of sophisticated mathematics. But he also must have been encouraged by the precedent set by Minkowski and Hilbert in seeking to blur the boundaries between mathematics and physics, and between mathematicians and physicists. Indeed, as already noted at the end of Section 9.3, Weyl expressed views about such a blurring of lines that are quite similar to Hilbert's.

Weyl's theory was presented in several papers during 1918–19 and in the third edition (1919) of *Raum, Zeit, Materie*. Einstein admired it for its mathematical brilliance, but he flatly rejected it as physically impossible.[24] Although Weyl respected Einstein's profound physical intuition and was accordingly disappointed by the negative reaction to his unified theory, Einstein's arguments did not convince him that his own approach was wrong. As he wrote in his rejoinder to Einstein's objections, "It must be emphasized that the geometry that has been developed here is, from the mathematical standpoint, the true local geometry. It would be strange if, instead of this true [geometry], a partial and inconsistent local geometry with the electromagnetic field glued to it were realized in Nature"[1918a:42]. A few days later in a letter to Einstein, Weyl presented his position more colorfully: "If you are right with regard to the real world, then I regret having to point out a mathematical inconsistency to the dear Lord."[25] Weyl admitted that his whole approach was highly speculative and would have to be confirmed by experience, although how, experimentally this could be undertaken remained unclear.

Weyl's interest in the Helmholtz–Lie space problem, which can be traced back to 1919 when he edited for republication Riemann's *Habilitationsvortrag* [1868],[26] may have been encouraged by his desire to find other, more philosophical, types of confirmation of the correctness of his theory. This may have seemed all the more necessary, since the initially favorable attitude toward his theory by theoretical physicists other than Einstein (Planck, Sommerfeld, Pauli, Eddington) had turned critical starting about 1920 [Scholz 1994:220–21]. It was in connection with this problem that Weyl first began to appreciate the value of group theory for investigating questions of interest to him involving the mathematical foundations of physical theories.

---

[24] Einstein's objections, initially expressed in letters to Weyl, were appended to Weyl's presentation of his theory in the *Sitzungsberichte* of the Berlin Academy; see [Weyl 1918a:40]. For details see [Sigurdsson 1991:162–70].

[25] Letter dated 19 May 1918, as quoted (and translated) by Sigurdsson [1991:170].

[26] See [Scholz 1996:33–4].

Recall from Section 4.5 that in 1866 Helmholtz had sought to deduce the geometrical properties of space from facts about the existence and motion of rigid bodies. He concluded that the distance between points (x, y, z) and $(x + dx, y + dy, z + dz)$ is $\sqrt{dx^2 + dy^2 + dz^2}$ and that space is indeed Euclidean. He returned to the matter in 1868, however, after learning from the work of Riemann and Beltrami about geometries of constant curvature. Using the properties of rigid bodies he had singled out earlier, he argued that Riemann's hypothesis that metric relations are given locally by a quadratic differential form is actually a mathematical consequence of these facts. Later, in [1887], Poincaré obtained Helmholtz's results for two-dimensional space by applying Lie's theory of groups and utilizing, in particular, the consideration of Lie algebras. Lie himself considered the problem in n dimensions by means of the consideration of Lie groups and algebras in [1890b; 1890c]. The Helmholtz–Lie treatment of the space problem, however, was rendered obsolete by the advent of general relativity, since, as Weyl put it, "Now we are ... dealing with a four-dimensional [continuum] with a metric based not on a positive definite quadratic form but rather on one that is indefinite. What is more, we no longer believe in the metric homogeneity of this medium—the very foundation of the Helmholtzian metric—since the metric field is not something fixed but rather stands in causal dependency on matter" [1921b:263].

These words were written by Weyl in one of the several essays on the space problem he published during 1921–23, starting with Section 18 of the fourth edition of *Raum, Zeit, Materie*. Here is a brief summary of his approach.[27] Following the Helmholtz–Lie tradition, Weyl conceived of space (including therewith the possibility of space-time) as an n-dimensional differentiable manifold $\mathcal{M}$ with metric relations determined by the properties of congruences that are conceived in terms of groups. Thus at each point $P \in \mathcal{M}$ the rotations at P are assumed to form a continuous group of linear transformations $\mathfrak{G}_P$, and since the volume of parallelepipeds is assumed to be preserved by rotations, the $\mathfrak{G}_P$ are taken as subgroups of the special linear group $\mathbf{SL}(T_P(\mathcal{M}))$. Metrical relations in a neighborhood $\mathcal{U}$ of P are then based on the assumption that all rotations at $P' \in \mathcal{U}$ can be obtained from a single linear congruence transformation A taking P to P' by composition with the rotations at P; that is, every $T' \in \mathfrak{G}_{P'}$ is of the form $T' = ATA^{-1}$, so that $\mathfrak{G}_{P'} = A\mathfrak{G}_P A^{-1}$. By "passing continuously" from P to any point Q of the manifold $\mathcal{M}$, Weyl was led to the assumption that all the groups $\mathfrak{G}_P$ are congruent to a group $\mathfrak{G} \subset \mathbf{SL}(n, \mathbb{R})$ with Lie algebra $\mathfrak{g} \subset \mathbf{sl}(n, \mathbb{R})$. Thus, whereas in the Helmholtz–Lie treatment of the space problem the homogeneity of space entails the identity of the rotation groups at diverse points, in Weyl's formulation the rotation groups have differing "orientations," although they are similar in Lie's sense and so share the

[27] For a more detailed and careful analysis of Weyl's somewhat vaguely formulated ideas and the underlying manifold concept he had in mind, see [Scholz 1994; 1996].

same abstract Lie algebra.

Within this mathematical context Weyl stipulated two postulates: (1) The nature of space imposes no restriction on the metrical relationship; (2) the affine connection is uniquely determined by the metrical relationship. His interesting mathematical interpretation of these two postulates led to the conclusion that the Lie algebra $\mathfrak{g}$ must possess the following properties:

(a) For all $X \in \mathfrak{g}$, tr $X = 0$ (i.e., $\mathfrak{g} \subset \mathbf{sl}(n, \mathbb{R})$),

(b) dim $\mathfrak{g} = \frac{1}{2}n(n-1)$,

(c) For any $X_1, \ldots, X_n \in \mathfrak{g}$ with matrix form $X_k = (a_{ij}^{(k)})$ with regard to some basis, if Col i of X_j = Col j of X_i for all $i, j = 1, \ldots, n$, then $X_i = 0$ for all $i = 1, \ldots, n$.

In the fourth edition of *Raum, Zeit, Materie*, where Weyl first presented his analysis of the space problem [1921a:§18], he pointed out that the Lie algebras $\mathfrak{g}_Q$ of all orthogonal groups with respect to a nonsingular quadratic form Q satisfy (a)–(c), and he conjectured as "highly probable" the following theorem, which he had confirmed for $n = 2, 3$:

(2.1) **Theorem**. *The only Lie algebras satisfying (a)–(c) are the orthogonal Lie algebras* $\mathfrak{g}_Q$ *corresponding to a nondegenerate quadratic form* Q.

Weyl's conjectured theorem thus implied the locally Pythagorean nature of space. Weyl pointed out that when $\mathfrak{g}$ does correspond to an orthogonal Lie algebra, the quadratic form Q is determined only up to a constant of proportionality [1921a:146]. Although he did not say it explicitly at this point, the truth of Theorem 2.1 would thus imply that his generalization of quasi-Riemannian geometry, his purely infinitesimal geometry, was also compatible with the conclusions of his analysis of the space problem.

Within a few months of completing the fourth edition of *Raum, Zeit, Materie*, Weyl had obtained a proof of Theorem 2.1, which he submitted for publication in April 1921 [1921b] and announced in a general talk in September 1921 [1922]. With the proof of Theorem 2.1 his analysis of the space problem was complete. Weyl saw it as confirmation of the legitimacy of his geometrical approach to relativity theory—his purely infinitesimal geometry with its concomitant unified field theory. He was also mindful of the fact that it had been achieved by utilizing the theory of groups: "The establishment by group theory is hence a new support for my conviction that this geometry, as geometry of the world, is the basis for the interpretation of physical field phenomena, rather than, as with Einstein, the more restrictive Riemannian [geometry]" [1922:344]. Indeed, Weyl was so taken up with Theorem 2.1 that he even likened the "confirmation by logic" of his approach to the space problem afforded by Theorem 2.1 to the factual confirmation of the correctness of Einstein's relativistic approach to gravitation afforded by the observed advance of the perihelion of Mercury [1921b:269].

During the spring of 1922 Weyl lectured on the space problem in Spain, and a German version of his lectures was then published as a monograph [1923a], which he regarded as a supplement to *Raum, Zeit, Materie*, since "the deeper conception of the space problem using group theory" was only sketched there. In the eighth lecture, which sketches a proof of Theorem 2.1, Weyl wrote (p. 61):

> While almost all deeper mathematical theories—such as, e.g., the wonderful theory of algebraic number fields—have little to signify within the great philosophical continuum of knowledge and while, on the other hand, what mathematics can contribute to enlighten the general problem of knowledge stems mostly from the surface of mathematics, here we have the rare case that a problem that is fundamental to all knowledge of reality, as is the space problem, gives rise to deeply penetrating mathematical questions.

Within the context of the space problem Weyl had discovered group theory as a powerful tool for dealing with fundamental questions inspired by general relativity and leading to "deeply penetrating mathematical questions." Although he described it as a rare occurrence, as we shall see, this was not the last time that his involvement with the fundamentals of general relativity led to important mathematical questions of a group-theoretic nature and, ultimately, to his papers on the representation of Lie groups.

Before proceeding to consider these further occurrences, however, there is one additional, historically important, consequence of Weyl's involvement with the space problem that needs to be mentioned. In 1922 the fourth edition of *Raum, Zeit, Materie* was translated into French and read by Élie Cartan, who, since 1921, had been interested in Einstein's theory. Unaware that Weyl had already proved the conjectured Theorem 2.1, Cartan provided a proof of his own [1922]. Strictly speaking, Cartan did not prove Theorem 2.1. Instead, he reformulated Weyl's somewhat vaguely articulated version of the space problem in terms of his own approach to geometry based on moving frames and differential forms. Cartan's approach evolved into the modern theory of G-structures.[28] Within that framework, however, Cartan's formulation of the space problem ultimately reduced to the problem of determining all linear Lie algebras $\mathfrak{g}$ satisfying Weyl's conditions (a) and (b) and, in lieu of the rather mysterious condition (c), the condition that $\mathfrak{g}$ leaves no proper subspaces invariant—the condition that as a geometrical property Kowalewski had focused on in his research (Sec-

[28] See in this connection [Scheibe 1988:66] and [Scholz 1994:225]. Scheibe argues that if what Weyl had in mind is made more precise in accord with what his writings seem to suggest, then it is not equivalent to Cartan's formulation, but the theorem Cartan proved implies the theorem Scheibe reconstructs from Weyl's vague statements (*op. cit.*, pp. 68–69).

tion 7.5).[29] As we saw in Section 8.2, Cartan had proved in [1909:148] that if $\mathfrak{g}$ satisfies Kowalewski's condition, it is either simple or semisimple or the direct sum of such an algebra with the one-dimensional algebra of scalar matrices. In view of the trace condition in (a), it followed that $\mathfrak{g}$ must be simple or semisimple. Since Cartan had already determined in [1913; 1914b] all such linear algebras that leave no vector spaces invariant, it was, as he noted, just a matter of checking which of them satisfy the dimension condition (b) to arrive at Weyl's conclusion that $\mathfrak{g}$ must be an orthogonal Lie algebra.[30]

Expressed in modern terms, what Cartan had done in [1913] was to determine all irreducible representations of a complex semisimple Lie algebra, and in [1914b] he did the same for real Lie algebras. However, as I have remarked in Section 7.1 and Section 8.1, Cartan did not conceive of his work within the conceptual framework of Frobenius's theory of finite group representations, a theory that may not have been familiar to him despite Poincaré's enthusiasm for it. Certainly, he did not know of either Schur's extensions of Frobenius's theory to continuous groups or of Hurwitz's integral technique.[31] He conceived of his work as solving the problem of determining all groups of projective transformations that "leave nothing planar" invariant—a problem of importance from the Kleinian view of geometry as the study and classification of groups acting on manifolds. As we shall see, it was Weyl who first brought Cartan's work within the conceptual framework of Frobenius's theory in his papers of 1925–26.

Weyl learned of Cartan's work when the latter sent him the announcement [1922] of his solution to the space problem. In Weyl's reply, dated 5 October 1922,[32] after pointing out that he had already given a proof of Theorem 2.1, he wrote:

> Untraveled on the paved roads of the general theory of continuous groups, which have been laid out and constructed thanks to your masterly skill, I have on my own beat a steep inconvenient footpath through much underbrush to my goal. I have no doubt that your method corresponds better to the nature of the matter; still, I see that you also cannot arrive at the goal without distinguishing many cases.

---

[29] This property of the $\mathfrak{g}$ satisfying (a)–(c) of Weyl's Theorem 2.1 actually follows readily from propositions Weyl deduced from (a)–(c) [1923a; 1923c], although he did not expressly take note of this fact.

[30] Both Cartan and Weyl realized that it suffices to consider the problem for complex Lie algebras. In his announcement Cartan indicated that a detailed solution of the problem in a generalized form was contained in [1923].

[31] According to Cartan's own testimony as contained in his letter to Weyl of 12 March 1925 (ETH Zürich, Hs 91:501).

[32] I am grateful to B. L. van der Waerden, who called these letters to my attention many years ago and sent me copies after obtaining consent of the holders—Henri Cartan in the case of Weyl's letters and the ETH in the case of Cartan's letters (ETH Zürich, Hs. 91:499,501–2).

The general consensus seems to be that Weyl, impressed by Cartan's papers on Lie algebras, studied them carefully and that this study, combined with an interest in the theory of invariants piqued by some critical remarks by the mathematician E. Study (discussed below), led, through the inspiration provided by Schur's paper [1924a] (Section 10.5), to his celebrated papers of 1925–26 on the representation of Lie groups.[33] While there is much truth in such a portrayal of events, it does overlook Weyl's deep-seated, philosophically inclined interest in the mathematical foundations of theoretical physics; in particular, it fails to account fully for Weyl's own statement, quoted at the beginning of this chapter, that "the wish to understand what really is the mathematical substance behind the formal apparatus of relativity theory led me to the study of representations and invariants of groups ...." Weyl's involvement with the space problem was certainly an instance of his interest in the mathematical substance underlying the theory of relativity, and it led him to Cartan's work. But the space problem was not the only focal point of this interest. In what follows I will attempt to give a clearer notion of how other manifestations of his interest in finding the proper mathematical basis for the mathematical apparatus of general relativity increased his involvement with the theory of groups and, in particular, with the theory of their representations, and how this in turn made Cartan's work all the more relevant.

In this connection it should be kept in mind that Weyl's above-quoted reply to Cartan was written when he had only the original proof of his theorem, which he disliked because it was complicated and lacked an overall unifying idea [1922:344]. He compared it deprecatingly to tightrope dancing [1921b:269], and in his popular lectures on the space problem, including those in Spain in the spring of 1922, he declined for this reason to sketch the proof. By the spring of 1923, however, when he published a German version of the lectures in Spain, he included a proof (as Appendix 12) because he had obtained what he felt were "far-reaching simplifications" to his original proof [1923a:v], so that although still complicated in detail due to the need to distinguish many cases, it was now guided by a single idea, which in fact Weyl pushed further in [1923c]. There he wrote in conclusion, "Our game on the chessboard of matrix schemes has been played to its end. As complicated in details as it may be, it—including the first part, which was already laid out in ... [1923a] ... Append. 12—rests ... upon a single construction idea that determined each step and was tenaciously carried out to the end." It is interesting to observe that in Weyl's presentation of his new proof he used another "roadway" analogy in comparing his and Cartan's proofs, but now with a different slant: "By contrast with Cartan's proof mine does not take the detour of the investigation of abstract groups. It is based on the classical theory of the individual linear mapping going back to Weierstrass" [1923a:88]. So now Cartan's solution involves a "detour" because it draws

[33] See [Chevalley and Weil 1957:675–6], [Borel 1986:54], and [Scholz 1994:225].

upon the theory of semisimple Lie's algebras, whereas Weyl's approach is more direct and elementary, depending only on "the classical theory" of the Jordan canonical form of a matrix implicit in Weierstrass's theory of elementary divisors.

That is not to say that Weyl did not appreciate by this time—early 1923—the impressive results and deep theory developed by Killing and Cartan. Indeed, immediately prior to the above quotation, Weyl characterized Cartan's solution to the space problem by writing, "An entirely different proof has been given by Cartan ... based on [his] ... earlier comprehensive and deep investigations ... on the theory of continuous groups, in which he achieved a far-reaching solution to the problem of determining all abstract groups and their realization through linear operations .... Now he needed only to seek out among the groups determined by him those that satisfy my stipulations." These words indicate that Weyl certainly understood the gist of what Cartan had done in his papers and appreciated the profundity of the mathematics. But as far as the space problem was concerned, the extensive detour required by Cartan's approach was not deemed appropriate by Weyl, who was still fascinated by his own approach. I would suggest that Weyl had not yet found sufficient reason to take on the nontrivial task of mastering the details of Cartan's papers so as to put them to his own use. Eventually he did—and he was perhaps the first mathematician to do so—but the motivation to do so seems to have come not from the space problem but from the calculus of tensors. In fact, judging by the context of Weyl's above-quoted remark about his interest in "the mathematical substance behind the formal apparatus of relativity theory," the "apparatus" he seems to have had specifically in mind was that of the tensor calculus.[34] As we shall see, such an interpretation makes good sense when one looks at Weyl's treatment of tensors and the mathematical questions it suggested.

11.3. Tensor Algebra and Tensor Symmetries

In *Raum, Zeit, Materie*, Weyl devoted considerable attention to the calculus of tensors. As he explained in the preface to the first edition of 1918, "The setting up of the tensor calculus—by means of which, alone, it is possible to express adequately the physical knowledge under discussion—occupies a relatively large amount of space. It is therefore hoped that it will be found suitable for making physicists better acquainted with this mathematical instrument ...."[35] Although he credited the Ricci–Levi-Civita paper [1900]

[34] I base this on the fact that earlier on the same page Weyl wrote, "If one tries to understand what is behind the formal apparatus of tensor calculus that is used in general relativity, one arrives with necessity at the general notion of *covariant quantity*" [1949:400]. (Italics in original.)

[35] I have followed the translation as found in [1921a:ix–x].

for the systematic development of tensor calculus,[36] Weyl's own systematic presentation was quite different. Drawing upon his Göttingen background, he recast tensor calculus in a form that became the paradigm for present-day formulations. For one thing, Weyl treated tensor algebra—tensor analysis in a fixed tangent plane—independently as a preliminary to tensor analysis, and in developing tensor algebra he did so within the geometrically flavored context of vector spaces, which had grown out of Hilbert's work on integral equations. In the process, he was confronted with questions about the essence of tensors and the scope of the tensor calculus, which, as will be seen in Sections 4–5, ultimately led to his interest in investigating the representation theory of semisimple groups. To make the questions intelligible I begin with a sketch of the basics of Weyl's approach to tensor algebra.

Underlying tensor algebra for Weyl was the concept of an n-dimensional space $\mathcal{V}$ of vectors $\mathbf{v}, \mathbf{w}, \ldots$, which he defined axiomatically, Hilbert-style [1921a:Ch.I,§2]. In effect, Weyl gave the now familiar definition of a real n-dimensional vector space and within that framework introduced the attendant notions of linear dependence and independence, bases, etc. Metrical relations were introduced by specifying a nondegenerate (but not necessarily positive definite) quadratic form $Q(\mathbf{v}, \mathbf{w})$, $\mathbf{v}, \mathbf{w} \in \mathcal{V}$, defining a "scalar product." Then if $\mathbf{e}_1, \ldots, \mathbf{e}_n$ is a basis for $\mathcal{V}$, we may express $\mathbf{v} \in \mathcal{V}$ in the form

$$\mathbf{v} = \sum_{i=1}^{n} x^i \mathbf{e}_i. \tag{3.1}$$

The x^i are called *contravariant* coordinates of $\mathbf{v}$, since if $\bar{\mathbf{e}}_1, \ldots, \bar{\mathbf{e}}_n$ is another basis related to the first by

$$\bar{\mathbf{e}}_i = \sum_{k=1}^{n} p_i^k \mathbf{e}_k, \tag{3.2}$$

then $\mathbf{v} = \sum_{i=1}^{n} \bar{x}^i \bar{\mathbf{e}}_i$, where

$$x^i = \sum_{k=1}^{n} p_k^i \bar{x}^k. \tag{3.3}$$

Comparision of (3.3) with the definition of a contravariant tensor given in (1.8) shows that this terminology is justified.

To obtain covariant coordinates for $\mathbf{v}$, Weyl observed that $\mathbf{v}$ is also uniquely determined by the n values $y_i = Q(\mathbf{v}, \mathbf{e}_i)$. Given the basis change (3.2), direct calculation of $\bar{y}_i = Q(\mathbf{v}, \bar{\mathbf{e}}_i)$ shows that

$$\bar{y}_i = \sum_{k=1}^{n} p_i^k y_k, \tag{3.4}$$

[36] See note 4 to p. 53 of the fourth edition [1921a].

which, in view of the definition of a covariant tensor (1.6), justifies calling the y_i *covariant coordinates* of $\mathbf{v}$. Nowadays, the y_i would be regarded as coordinates of the element $\mathbf{v}^*$ in the dual space $\mathcal{V}^*$ defined by $\mathbf{v}^*(\mathbf{w}) = Q(\mathbf{v}, \mathbf{w})$. That is, the y_i are the coordinates of $\mathbf{v}^*$ with respect to the basis $\mathbf{e}^1, \ldots, \mathbf{e}^n$ of $\mathcal{V}^*$ dual to $\mathbf{e}_1, \ldots, \mathbf{e}_n$.

The key to tensor algebra for Weyl consisted in identifying tensors with multilinear forms [1921a:37]. Consider, for example, the mixed tensor of degree 3, T^i_{jk}, which transforms contravariantly in i and covariantly in j, k, so that in accordance with the rules laid down in (1.6) and (1.7),

$$\bar{T}^{\alpha}_{\beta\gamma} = \sum_{i,j,k} T^i_{jk} \pi^{\alpha}_i p^j_{\beta} p^k_{\gamma}, \quad (\pi^i_j) = (p^i_j)^{-1}. \tag{3.5}$$

This tensor is determined by a multilinear form $f = f(\mathbf{u}, \mathbf{v}, \mathbf{w})$, where if x_i are the covariant coordinates of $\mathbf{u}$ and y^j, z^k are the contravariant coordinates of $\mathbf{v}$ and $\mathbf{w}$, then

$$f = \sum_{i,j,k} T^i_{jk} x_i y^j z^k. \tag{3.6}$$

By identifying the tensor T^i_{jk} with the form f in (3.6), the transformation rule (3.5) then follows directly by setting

$$x_i = \sum_{\alpha=1}^{n} \pi^{\alpha}_i x_{\alpha}, \quad y^j = \sum_{\beta=1}^{n} p^j_{\beta} y^{\beta}, \quad z^k = \sum_{\gamma=1}^{n} p^k_{\gamma} z^{\gamma} \tag{3.7}$$

in (3.6).

Since, as noted above, the covariant coordinates x_i may be thought of as coordinates of $\mathbf{u}^* \in \mathcal{V}^*$, f may be regarded as a multilinear form on $\mathcal{V}^* \times \mathcal{V} \times \mathcal{V}$ with $u^* = \sum_i x_k \mathbf{e}^k$, from which (3.6) follows with $T^i_{jk} = f(\mathbf{e}^i, \mathbf{e}_j, \mathbf{e}_k)$. Thus Weyl in effect identified tensors of the above type with the vector space $\mathcal{L}$ of all such multilinear forms f, which agrees with the present-day formulation, according to which

$$\mathcal{L} \cong (\mathcal{V}^* \otimes \mathcal{V} \otimes \mathcal{V})^* \cong \mathcal{V} \otimes \mathcal{V}^* \otimes \mathcal{V}^*.$$

In view of these isomorphisms, the reader may wish to identify the tensor f defined by (3.6) with the element

$$\mathbf{f} = \sum_{i,j,k} T^i_{jk}\, \mathbf{e}_i \otimes \mathbf{e}^j \otimes \mathbf{e}^k \in \mathcal{V} \otimes \mathcal{V}^* \otimes \mathcal{V}^*. \tag{3.8}$$

Weyl also introduced a new notion into his discussion of tensor algebra, namely a *tensor density*.[37] It was motivated by the consideration of

[37] This occurred in his paper [1918b] and in *Raum, Zeit, Materie*. In the fourth edition [1921a], see Section 13. Pauli [1921:32n.16] credits Weyl with this notion and cites Weyl's paper [1918b]—see Section 5—and the third edition of *Raum, Zeit, Materie*; I am grateful to John Stachel for calling this to my attention.

an invariant integral $I = \int W(\mathbf{x})d\mathbf{x}$, $\bar{x} = (x^1, \ldots, x^n)$. Given a variable change $\mathbf{x} = \varphi(\bar{\mathbf{x}})$, a scalar function $W = W(\mathbf{x})$ transforms from W to $\bar{W}$, where $\bar{W}(\bar{x}) = W(\varphi(\bar{\mathbf{x}}))$. In the integral I, however, where $W(\mathbf{x})$ can be regarded as giving the density of the manifold at $\mathbf{x}$, so that I represents its mass, we have $I = \int W(\varphi(\bar{\mathbf{x}}))|\det P|d\bar{\mathbf{x}}$, where, in accordance with the notation in (1.4), P denotes the Jacobian matrix with row i, column k entry $p^i_k = \partial x_i/\partial \bar{x}_k$. Hence the function W, as a scalar density function, transforms by the rule $\bar{W} = W|\det P|$. For tensors Weyl introduced an analogous notion as follows. Tensor densities are also identified with multilinear forms, such as the form f given in (3.6), but the rules of transformation are different. Consider, for example, the mixed tensor density $\mathbf{T}^i_{jk}$. (Weyl used boldface for tensor densities to distinguish them from tensors.) Then if $f = \sum_{ijk} \mathbf{T}^i_{jk} x_i y^j z^k$, to obtain the transformation rule for $\mathbf{T}^i_{jk}$ one multiplies f by $|\det P|$ and makes the variable substitutions (3.7). In this way, in lieu of (3.5), the transformation rule

$$\bar{\mathbf{T}}^{\alpha}_{\beta\gamma} = |\det P| \sum_{i,j,k} \mathbf{T}^i_{jk} \pi^{\alpha}_i p^j_{\beta} p^k_{\gamma} \tag{3.9}$$

is obtained. “By contrasting tensors and tensor-densities,” he wrote in *Raum, Zeit, Materie*, “it seems to me that we have rigorously grasped the difference between **quantity** and **intensity**, so far as the difference has a physical meaning ...” [1921a:109]. Weyl's notion of tensor densities is still a part of general relativity today.[38]

The introduction of the concept of a tensor density seems to have prompted the following mathematical question on the part of Weyl's student at the ETH Alexander Weinstein.[39] The answer Weinstein gave became his doctoral dissertation at the ETH and was also published in *Mathematische Zeitschrift* [1922]. Weinstein, who had proofread the third edition (1919) of *Raum, Zeit, Materie*, where the notion of tensor density was introduced, observed that the transformation rules for the various types of tensors and tensor densities corresponded to different rules of variable substitution to be made in a multilinear form. That is (as can be seen from (3.7)), covariant tensors transformed according to the (contravariant) variable change $y^j = \sum_{\beta=1}^n p^j_{\beta} \bar{y}^{\beta}$, or, in matrix notation, $\mathbf{y} = P\bar{\mathbf{y}}$; and contravariant tensors transformed by setting $\mathbf{x} = P'\bar{\mathbf{x}}$, where $P' = (P^T)^{-1}$. Likewise, Weyl's transformation rule for tensor densities could be obtained by setting, in the associated multilinear form, $\mathbf{y} = P'\bar{\mathbf{y}}$ with $P' = |\det P|^a P$ for covariant densities and $\mathbf{x} = P'\bar{\mathbf{x}}$ with $P' = |\det(P^T)^{-1}|^a (P^T)^{-1}$ for contravariant densities. For example, to obtain the transformation rule (3.9)

---

[38] See, e.g., [Misner, et al. 1973:501; Møller 1972:310].

[39] I owe my awareness of Weinstein's paper to Borel [1986:54]. Weinstein was one of Weyl's few students, and one he regarded highly. He went on to distinguish himself as an analyst. See in this connection the biographical sketch by Diaz in [Weinstein 1978].

for $\mathbf{T}^i_{jk}$ take $a = 1$, since $|\det(P^T)^{-1}| = |\det P|^{-1}$; to obtain the transformation rule for the covariant degree-two tensor $\mathbf{T}_{ij}$, take $a = \frac{1}{2}$, and so on.

Thus each type of tensor corresponds to a type of variable substitution $\mathbf{x} = P'\bar{\mathbf{x}}$, where in each case above, the mapping (*Abbildung*) $P \to P'$ preserves the law of matrix composition

$$(P_1P_2)' = P_1'P_2'. \tag{3.10}$$

Suppose now that $P \to P'$ is some mapping of the nonsingular $n \times n$ matrices P into itself that satisfies (3.10). Then one could regard a new sort of "tensor" to be defined by virtue of the transformation rule that follows by setting $\mathbf{x} = P'\bar{\mathbf{x}}$ in the associated multilinear form. The question therefore arises as to whether in this manner one obtains any essentially new sorts of tensors. In other words, what are the possibilities for the mappings $P \to P'$?

According to Weinstein (p. 80), Weyl advised him to deal with the question by working on the more convenient Lie algebra level, which he did. In fact, he used a linear-algebraic technique (p. 86) that Weyl had introduced in his own Lie algebra calculations on the space problem in [1921b]. To transfer the Lie algebra results back to the noninfinitesimal matrices, Weinstein limited his attention to matrices P with positive determinant, the corresponding matrix P' being assumed simply to be nonsingular. The matrices P thus make up $\mathbf{GL}(n, \mathbb{R})^+$, the component of $\mathbf{GL}(n, \mathbb{R})$ containing the identity element. His conclusion was that there are no essentially new types of tensors obtained from other mappings $P \to P'$ in the sense that the ones given above (including $P' = P$, which gives rise to covariant tensors) are the only "elementary ones," all others being composed out of these.[40] As he put it, "for transformations with positive determinant there exist no further possibilities for introducing new types of tensors, i.e., the processes of the tensor calculus are exhaustive. This proposition could be designated as the fundamental theorem of the tensor calculus" (p. 79).

From the vantage point of the present, we can see that Weinstein's mappings $P \to P'$ are special representations of the group $\mathbf{GL}(n, \mathbb{R})^+$, namely those of degree n; but Weinstein (and presumably Weyl) did not point out any connection with Frobenius's theory (Section 10.1) or with Schur's dissertation of 1901 (Section 10.3). Since the mappings $P \to P'$ corresponding to the different types of tensors and tensor densities were the sort of representation Schur had studied in his dissertation, the absence of any reference to Schur suggests that it was unknown to Weinstein and

[40] The mapping $P' = A^{-1}PA$ was dismissed as not leading to a new tensor concept, since it "merely signifies a simultaneous coordinate transformation, applied to all coordinate systems" [1922:79]. Also dismissed as a "degenerate case of no use for the tensor calculus" (p. 91) was the mapping that on the Lie algebra level is $X \to (\mathrm{tr}\, X)\Gamma$, where Γ is a fixed but otherwise arbitrary matrix (p. 89).

Weyl in February 1922, when Weinstein submitted his dissertation for publication in *Mathematische Zeitschrift.* This is not surprising, since Schur's dissertation, unlike Weinstein's, had not been published in a journal of mathematics. There was just the obligatory publication as a separate work. My guess is that Weyl first learned of Schur's dissertation in 1924, when he read about it in Schur's paper on the counting problem [1924a] (Section 10.5). In the meantime, he learned of Cartan's work in October of 1922, as we have seen in Section 2. He also learned of Frobenius's theory, or, more precisely, he learned how useful it could be for answering other questions about tensor algebra suggested to him by its applications. Indeed, as will be seen in Sections 4–5, his involvement with these questions and with Frobenius's theory, combined with the appearance of Schur's paper [1924a], naturally led him to the mathematical considerations that inspired his great papers of 1925–26.

The aspect of tensor algebra that proved significant in this connection had to do with the symmetry properties of tensors. In relativistic physics and geometry the tensors that arose were not totally general; they came with specific symmetry properties, which were essential to their mathematical treatment. For example, in the pages of *Raum, Zeit, Materie* [1921a], the stress tensor S_{ik} is seen to be a symmetric tensor of degree 2 (Section 8), and the four-dimensional relativistic electromagnetic intensity vector F_{ik} of Section 20 is a skew symmetric tensor of degree 2. The symmetry properties of the Riemann curvature tensor $R_{\alpha\beta ik}$ (Section 17) are more complex, being given by the three conditions (a) $R_{\alpha\beta ki} = -R_{\alpha\beta ik}$, (b) $R_{\beta\alpha ik} = -R_{\alpha\beta ik}$, (c) $R_{\alpha\beta ik} + R_{\alpha ik\beta} + R_{\alpha k\beta i} = 0$, from which further conditions may be deduced, such as (d) $R_{ik\alpha\beta} = R_{\alpha\beta ik}$. With such examples in mind, Weyl wrote emphatically at the beginning of Section 7, which is entitled "Symmetrical Properties of Tensors," that "the character of a quantity is not in general described fully, if it is stated to be a tensor of such and such an order [i.e., degree], but **symmetrical characteristics** have to be added" [1921a:54]. In other words, the really important objects of study are not tensors per se but rather tensors of a specific symmetry type, since these are the objects that occur in physical and geometrical applications. Symmetry characteristics are, from this point of view, a basic attribute of tensors comparable to that of degree, a point Weyl made repeatedly during 1921–24, as we shall see.

Weyl realized that permutations could be used to characterize symmetry properties in general. Consider, for example, a covariant tensor of degree 3, T_{ijk}, which, following Weyl, we regard as a multilinear form

$$f = f(x, y, z) = \sum_{i,j,k} T_{ijk} x^i y^j z^k. \tag{3.11}$$

If S is some permutation of the three variable sets x, y, and z, let f_S denote the form that arises from f by permuting the variable series according to S, so $f_S(x, y, z) = f(Sx, Sy, Sz)$. (The form f_S also arises from f by permuting the three indices i, j, and k, by the inverse permutation S^{-1}.) Then f is symmetric if $f_S = f$ for all S and skew symmetric if $f_S = (\operatorname{sgn} S)f$, where as usual the sign of S is ± 1 according to whether S is an even or odd permutation. For the Riemann curvature tensor $f(x, y, z, w) = \sum_{ijkl} R_{\alpha\beta ik} x^\alpha y^\beta z^i w^k$, the S are the permutations of the symmetric group on four objects, $\mathfrak{S}_4$, and, for example, conditions (c) and (d) can be expressed as

$$\begin{aligned} &\text{(c)} \quad f(x,y,z,w) + f(x,w,y,z) + f(x,z,w,y) = 0, \\ &\text{(d)} \quad f(x,y,z,w) - f(z,w,x,y) = 0. \end{aligned}$$

Weyl concluded his discussion of tensor symmetry by observing that the most general form of a symmetry condition is expressible by one or more equations of the form

$$\sum_S e_S f_S = 0, \tag{3.12}$$

where the e_S are numbers and S runs over all permissible permutations of the variables. For example, for the Riemann curvature tensor, the permissible permutations S constitute the symmetric group on four objects, $\mathfrak{S}_4$, and conditions (c) and (d) expressed in the notation of (3.12) are (c) $f_E + f_S + f_{S^2} = 0$, where $S = (234)$, and (d) $f_E - f_S = 0$, with $S = (13)(24)$.[41]

It may be helpful to express the above considerations and some immediate implications in more familiar terms, which are not far removed from Weyl's own mode of presentation. It follows from the discussion surrounding (3.8) that a covariant tensor of degree p, $T_{i_1 \ldots i_p}$, may be identified with an element $\mathbf{f}$ of the p-fold tensor product $\bigotimes^p \mathcal{V}^* = \mathcal{V}^* \otimes \cdots \otimes \mathcal{V}^*$, namely

$$\mathbf{f} = \sum_{i_1, \ldots, i_p = 1}^{n} T_{i_1 \ldots i_p} \mathbf{e}^{i_1} \otimes \cdots \otimes \mathbf{e}^{i_p}. \tag{3.13}$$

From this viewpoint, any permutation $S \in \mathfrak{S}_p$ defines a linear transformation $S : \mathbf{f} \to S \cdot \mathbf{f} = \mathbf{f}_S$ of $\bigotimes^p \mathcal{V}^*$ that sends $\mathbf{v}_1^* \otimes \cdots \otimes \mathbf{v}_p^*$ to $\mathbf{v}_{S(1)}^* \otimes \cdots \otimes \mathbf{v}_{S(p)}^*$. Thus any element $\sum_{S \in \mathfrak{S}_p} e_S S$ of the group algebra of $\mathfrak{S}_p$ defines a linear transformation of $\bigotimes^p \mathcal{V}^*$, and the tensors satisfying the symmetry condition (3.12) form its null space. Also, since each $S \in \mathfrak{S}_p$, regarded as a linear transformation of $\bigotimes^p \mathcal{V}^*$, evidently commutes with the natural action of the general or special linear group on $\bigotimes^p \mathcal{V}^*$, it follows that these null spaces are invariant under the action of these groups and so define representation modules. As we shall see, this property of the symmetry classes of tensors of a given degree figured prominently in Weyl's subsequent deliberations on the foundations of the tensor calculus.

---

[41] Here, e.g., $S = (234)$ means $Sy = z$, $Sz = w$, $Sw = y$.

Weyl laid a special emphasis upon skew symmetric tensors in his discussion of symmetry. This was because if, e.g., $\mathbf{v} = \sum_i \xi^i \mathbf{e}_i$ and $\mathbf{w} = \sum_i \eta^i \mathbf{e}_i$ are two linearly independent vectors, then they determine a plane and thus a "surface element," or two-dimensional element, with coordinates

$$\xi^{ik} = \begin{vmatrix} \xi^i & \xi^k \\ \eta^i & \eta^k \end{vmatrix}.$$

In three-dimensional space this is the plane with normal $\mathbf{v} \times \mathbf{w}$. Likewise, there are three-dimensional elements with coordinates ξ^{ikl} determined by three independent vectors, and so on. These are, of course, precisely the generalized Plücker coordinates that Weyl was eventually to encounter in Cartan's 1913 characterization of all projective groups with the structure of the special linear group that leave nothing planar invariant (Section 8.4). For Weyl in 1921 they were important for geometrical reasons related to the algebraic fact that any skew symmetric covariant tensor of degree ν can be expressed in terms of the coordinates of a ν-dimensional space element.[42] This follows from the observation that if f is skew symmetric, so that $(\operatorname{sgn} S) f_S = f$ for all $S \in \mathfrak{S}_\nu$, then

$$f = \frac{1}{\nu!} \sum_S (\operatorname{sgn} S) f_S. \tag{3.14}$$

Applying (3.14) to, say, the skew symmetric tensor $f = \sum_{i,k} a_{ik} \xi^i \eta^k$, one sees that $f = \frac{1}{2} \sum_{i,k} a_{ik} \xi^{ik}$.

Applied to the Riemann curvature tensor $f = \sum_{\alpha,\beta,i,k} R_{\alpha\beta ik} \xi^\alpha \eta^\beta \xi'^i \eta'^k$, which by (a) and (b) above is skew symmetric in the α, β and i, k index pairs, one has (setting the primed coordinates equal to the unprimed) $f = \frac{1}{4} \sum_{\alpha,\beta,i,k} R_{\alpha\beta ik} \xi^{\alpha\beta} \xi^{ik}$, which, by virtue of symmetry relation (d), is a quadratic form in the surface element variables ξ^{ik}. When the vectors with components ξ^i and η^i are regarded as lying in the tangent plane at some point P of the Riemannian manifold with metric given by the tensor g_{ij}, then the above quadratic form divided by the square of the magnitude of the surface element[43] gives what Riemann called the curvature of the manifold at the point P in the surface direction given by the element ξ^{ik}. As Weyl stressed, the symmetry conditions (a)–(d) of the Riemann curvature tensor "enable us to characterize the curvature tensor completely by means of a quadratic form that is dependent upon an arbitrary element of surface"

---

[42] Translated into modern terms this is the fact that the skew symmetric tensors in the ν-fold tensor product $\mathcal{V} \otimes \cdots \otimes \mathcal{V}$ are expressible in terms of the $\binom{n}{\nu}$ standard basis elements $[e_{i_1}, \ldots, e_{i_\nu}] = \sum_{S \in \mathfrak{S}_\nu} (\operatorname{sgn} S) e_{Si_1} \otimes \cdots \otimes e_{Si_\nu}$.

[43] Namely, $Q(\mathbf{v}, \mathbf{v}) \cdot Q(\mathbf{w}, \mathbf{w}) - Q(\mathbf{v}, \mathbf{w})$, where $\mathbf{v}$, $\mathbf{w}$ are the vectors with coordinates ξ^i, η^k and Q is the quadratic form defined by ds^2 at P. When Q is the usual Euclidean inner product in 3-space, this expression is just the square of the magnitude of the cross product $\mathbf{v} \times \mathbf{w}$.

[1921a:133], since for every quadratic form in ξ^{ik} there is one and only one quadrilinear form satisfying the symmetry relations (a)–(d) [1921a:57].

The algebra of tensors and, in particular, their classification into symmetry types was thus for Weyl an important part of the new physical geometry envisioned by him and by Hilbert. In view of the importance of tensor symmetries, Weyl eventually became interested in obtaining, mathematically, an overview of possible symmetry types. In true Hilbertian fashion he looked for a higher vantage point from which to regard these symmetries, and once again he found it in the theory of groups. This time it was not the theory of continuous groups to which he turned but rather Frobenius's theory of finite group representations. Notice in this connection that the general symmetry relation (3.12) suggests writing $(\sum_{S\in\mathfrak{G}} e_S S)f = \sum_{S\in\mathfrak{G}} e_S f_S = 0$, where $\mathfrak{G}$ is the appropriate group of permutations and $\sum_{S\in\mathfrak{G}} e_S S$ is an element of what is now called the group algebra of $\mathfrak{G}$. As we saw in Section 10.1, these hypercomplex number systems are intimately related, through their primitive idempotents, to the characters and representations of $\mathfrak{G}$; and in particular, the work of Young had prompted Frobenius to deal with the symmetric group in this fashion. The results of Weyl's application of Frobenius's theory to the study and classification of symmetry types were presented in a paper [1924a], which he submitted in January of 1924. Although he may have had the idea of dealing with tensor symmetries by applying Frobenius's theory to the group algebra of the symmetric group well before that time, it would seem that his paper was precipitated by some critical remarks that Study aimed at him in 1923.

11.4. Weyl's Response to Study

As we saw in Section 7.2, Study was an idiosyncratic mathematician whose primary mathematical research interest was in the theory of invariants and its geometrical applications. For a while, in the late 1880s and early 1890s, he became an active participant in Lie's school, being charged by Lie with the task of relating his theory of transformation groups to the theory of invariants. During this period his work on ternary invariants even led him to conjecture what amounts to the complete reducibility theorem for semisimple Lie groups—the theorem Weyl first succeeded in proving in 1925. But as we saw, Study finally gave up trying to advance Lie's theory because the problems posed were too abstract for him; instead, he turned to more concrete, geometrically flavored problems involving the theory of invariants and specific groups.[44] For example, in [1897] he classified the invariants of

[44] See Engel's biographical essay [1931] for an idea of the full range of Study's mathematics.

the orthogonal group using the symbolical method of Clebsch and Aronhold, thereby contributing to the projective geometry of manifolds defined by the vanishing of a nonsingular quadratic form; and in [1905] he studied the real projective geometries defined by hermitian forms, thereby calling attention to the group $\mathbf{SU}(n)$, which, as we saw in Section 10.2, had already figured in Hurwitz's work on the theory of invariants in 1897.

Study also had a sharp tongue and used it to criticize the work of other mathematicians, particularly in the realm of geometry. Like Weierstrass and other Berliners, he rejected the legitimacy of generic reasoning, which abounded in geometrical work. As we have seen (Section 3.2), Lie tended to express his mathematical ideas on the generic level, and so it is not entirely surprising to find that even Lie, despite Study's open admiration for his genius, was the object of his criticism. We saw in Section 3.2 in connection with Lie's exponential map theorem that already in 1890 Study was privately skeptical of Lie's general mathematical claims. In an article [1908] in the journal of the German Association of Mathematicians[45] and bearing as caption a line from the opening soliloquy of *Richard III*—"I am determined to prove a villain"[46]—Study went public and at length with his criticism of Lie, using the latter's theory of equivalent manifolds to illustrate what he saw as the "basic error"[47] behind Lie's methods.

Lie's above-mentioned theory was included in his *Lectures on Continuous Groups* [1893b]. Given a group $\mathfrak{G}$ of transformations of $\mathbb{C}^n$, two m-dimensional submanifolds are said to be equivalent if there is an element of $\mathfrak{G}$ taking the one into the other. This notion is important in the study of the geometry defined by $\mathfrak{G}$. Using Lie's own example of the group of all rigid motions in $\mathbb{C}^3$ and the one-dimensional manifolds formed by space curves, Study illustrated in great detail Lie's flawed reasoning and conclusions. He summarized his findings as follows [1908:136]:

> The fundamental theorem on certain differential equations is partly untrue, partly incorrectly founded. As a consequence one large family of curves has been entirely overlooked, the criteria for a second are all incorrect—even in the real domain—and also the treatment of the third family is not free from objections. These unsatisfying results have their source in defects in the method applied, which are not insignificant .... In addition there are formal defects that we have not discussed here, some of which put the patience of the reader to a rather hard test.

According to Study, the defects of Lie's treatment had their source in his "basic error," which was to employ a special form of generic reasoning: using a thoroughly *local* theory to deal with matters that are really *global* in nature. How could a theory that treats differential invariants as given by

[45] *Deutscher Mathematiker-Vereinigung.*
[46] Study quoted Shakespeare's line as, "Ich bin gewillt ein Bösewicht zu werden."
[47] *Grundirrtum*; see [1908:137].

power series be expected to deal with actual differential invariants that are given globally and are multivalued? In its present form, Study concluded, it could not.

Study thus put his finger on the fundamental weakness of Lie's theory—its thoroughly local orientation. His criticisms are especially interesting in the context of the present discussion of Weyl, because as we shall see, Weyl successfully remedied problems caused by the local nature of Lie's theory by infusing it with the global constructs he had learned and developed in lecturing on Riemann surfaces. Whether Weyl ever read Study's essay [1908] is unknown, although it is not unlikely, since it was published in the journal of the Association of German Mathematicians. In any case, he did read another critical essay by Study, which had as its principal target Weyl's book *Raum, Zeit, Materie.* Weyl's responses to Study's harsh words form an important part of the developments leading up to his initial foray into the representation of continuous groups, where he established the need for global considerations within Lie's theory.

At the beginning of 1923 Study published a book on the theory of invariants [1923], and in the lengthy introduction he chastised those working on the theory of relativity for their neglect of the tools of the algebraic theory of invariants in favor of tensor calculus. He pointed out that for over fifty years a highly developed theory of invariants with respect to the general linear group had been in existence, and citing his own work on orthogonal invariants, he noted that an invariant theory of other groups had also been indicated. But "with the majority of authors there is nothing to indicate that they live in an age in which the theory of groups is in full bloom" (p. 3). "In short," Study continued (p. 4), "they are behind the times, and not just a little. Even with an otherwise knowledgeable writer one can read, for example, the following: 'Many will be appalled at the deluge of formulas and indices with which the leading ideas are inundated. It is certainly regrettable that we have to enter into the purely formal aspect in such detail and to give it so much space, but nevertheless, it cannot be avoided.'" That quotation, although not identified as such, was drawn from Weyl's book *Raum, Zeit, Materie.*[48]

Study went on to criticize Weyl for accepting the formalism of the tensor calculus as an unavoidable, necessary evil. That is not to say that Study was against formalism. He believed that the formal aspects of mathematics were important, but the formalism must be of the right kind: "Mathematics is neither the art of calculation nor the art of avoiding calculations. To mathematics, however, belongs the art of avoiding superfluous calculations and carrying out the necessary ones adroitly. In this regard, one can learn

[48] I am grateful to Erhard Scholz for informing me that Study was using the first edition of 1918 or its 1919 reprint as second edition. The quotation is from p. 111. In subsequent editions published before 1923 the passage was changed and is not as vulnerable to Study's criticism. See p. 123 of the third edition and p. 137 of the fourth edition [Weyl 1921a].

from the older authors" (p. 4). What Study had principally in mind was the symbolical method of the theory of invariants that went back to Aronhold and Clebsch. This method reduced the problem of determining invariants to the far easier problem of determining symbolical or vector invariants. In sum (to use Study's own metaphor), mathematicians had thought that in the tensor calculus they were borrowing from the garden of their neighbor the physicist the seeds of the golden apples of the Hesperides but were actually contenting themselves with a harvest of potatoes! The neglected theory of invariants and in particular the symbolical method, Study implied, would prove to be far more valuable.

Weyl responded to Study's criticism in two papers. The first response was contained in a paper [1923b] submitted at the end of October 1923. This paper was intended as the first of what Weyl hoped would be a series of papers dealing with mathematical topics of interest to all mathematicians and with an emphasis on the clarification of known results rather than on the presentation of new ones. Weyl dealt with two topics in this fashion: I The theory of invariants and II The fundamental theorem of algebra in relation to foundational issues. In I the emphasis was upon the symbolical method, which Weyl applied to the problem of determining the vector invariants of "the most important linear groups," namely the orthogonal group and Lie's projective group of a linear line complex—the symplectic group (Section 3.5), which Weyl referred to simply as the "complex group" and made here for the first time an object of study within the theory of invariants. That Part I was prompted "in part" by Study's criticisms was expressly acknowledged by Weyl in a lengthy footnote, which reveals that he had been stung and annoyed by Study's remarks, particularly by the fact that he alone had been singled out by virtue of the quotation. Weyl admitted that his knowledge and experience in the field of algebraic invariant theory were limited, and in a sense Part I was a gesture to Study; it acknowledged the theory of invariants and particularly Study's beloved symbolical method as something of general interest to mathematicians. But Weyl was adamant about the superiority of tensor calculus for the physical geometry of space-time. Even if he possessed Study's great command of the the theory of invariants, he declared, "I would not apply the symbolical method in my book 'Space, Time, Matter' and not a single word would have been said about the completeness theorems of the theory of invariants. Everything in its proper place!"

Weyl's paper [1923b] is sometimes seen as revealing an awakening interest in the theory of invariants, which in turn encouraged his work on the representation of Lie groups. Although there may be some truth in such an interpretation, it fails to make much sense out of Weyl's remark, quoted at the beginning of the chapter, that it was his "wish to understand what really is the mathematical substance behind the formal apparatus of relativity theory" that led to his study of the representations and invariants of groups. More enlightening in this regard is what I consider to be his second response to Study, namely the above-mentioned paper [1924a] on tensor symmetries. Submitted in January 1924 and so a few months after [1923b], it bore the title "On the symmetry of tensors and the scope of the symbolical method in the theory of invariants."[49] As the title suggests, the paper has two parts. The first analyzes symmetry classes of tensors along the lines indicated above, i.e., by applying Frobenius's theory to the group algebra of the symmetric group, and in the second the results are applied to deal with a question that may well have been prompted by his encounter with Study, even though he is never mentioned by name in the paper.

Recall from Section 9.1 that during this period, invariants were linked to one or more underlying base forms (*Grundformen*), which generated the transformations with respect to which invariants were considered. Translated into modern terms, the base forms constituted a representation module for the general linear group, and one studied the polynomial invariants of the associated representation. Since with Weyl tensors are identified with multilinear forms, a tensor such as, e.g., $f = \sum_{i,j,k=1}^{n} T_{ijk}x^i y^j z^k$ could be chosen as base form. Under the linear variable change $x \to \bar{x}$, $y \to \bar{y}$, $z \to \bar{z}$ determined by the linear transformation with coefficients p_j^i as in (3.3), $T_{ijk} \to \bar{T}_{ijk}$ (by the rules of the tensor calculus) and, for example, absolute invariants are homogeneous polynomials I in n^3 variables T_{ijk} such that $I(\ldots, T_{ijk}, \ldots) = I(\ldots, \bar{T}_{ijk}, \ldots)$. The symbolical method utilized the Aronhold "polarization process" to transform each invariant I into a symbolical or vector invariant $i(\alpha, \beta, \ldots)$, i.e., a homogeneous polynomial in vectors $\alpha = (\alpha_1, \ldots, \alpha_n)$, $\beta = (\beta_1, \ldots, \beta_n)$, $\ldots$ such that $i(\bar{\alpha}, \bar{\beta}, \ldots) = i(\alpha, \beta, \ldots)$.[50] Since the original invariant can be recaptured from the vector invariant, the problem of determining the invariants associated to f as base form reduces to the simpler problem of finding vector invariants.

There was, however, one problem with this method from Weyl's standpoint. In the traditional theory of invariants the base forms are always "general," which means that their coefficients are unrestricted. In the case of the trilinear base form f the coefficients T_{ijk} must be completely arbitrary and not subject to symmetry relations. (In modern terms, as indicated following (3.13), the representation module is all of $\mathcal{V}^* \otimes \mathcal{V}^* \otimes \mathcal{V}^*$.) But as we have seen, for Weyl tensors, as they arise within the physical–geometrical context of relativity theory, possess symmetry properties. Echoing this sentiment, Weyl wrote in the first part of [1924a:472], "For every tensor that arises, a category characterized by symmetry relations must be specified a priori, inside of which the tensor is to be thought of as freely variable." The

49 In Weyl's *Gesammelte Abhandlungen II* this paper is misleadingly placed (with the date of submission omitted) after Weyl's two notes of November 1924 [1924b; 1924c] announcing his principal results on the representation of semisimple Lie groups.

50 For a clear, elementary exposition of the polarization process and its role in the symbolical method see Weyl's book on the classical groups [1946:5–6, 243–5].

type of invariants of interest to a relativist would correspond, for example, to a trilinear form of a specific symmetry type defined by equations of the form (3.12). (Hence the representation module is a *proper* submodule of $\mathcal{V}^* \otimes \mathcal{V}^* \otimes \mathcal{V}^*$.)

Now it turns out, as Weyl observed, that the symbolical method breaks down when applied to a tensor f as base form that is not completely general but is restricted to some symmetry class. For example, because the trilinear form f is not the "general" degree-3 covariant tensor but is restricted by symmetry conditions (3.12), the link between its invariants and the symbolical ones that is fundamental to the symbolical method is severed. As a simple illustration of this fact, consider the general skew symmetric degree-two tensor $f = a_{12}x^1y^2 - a_{12}x^2y^1$, which has the linear invariant $I(a) = a_{12}$. The symbolical method associates with I the expression $i(\alpha, \beta) = \alpha^1\beta^2$, where $\alpha = (\alpha^1, \alpha^2)$, $\beta = (\beta^1, \beta^2)$. But i is not a vector invariant. Thus an ordinary invariant need not give rise in the usual manner to a symbolical one, and so direct application of the method fails. A reasonable problem from the standpoint of a relativist—to determine the algebraic invariants with respect to a tensor f of a given symmetry type—cannot be treated directly by the symbolical method.

Weyl, however, perceived a way to salvage the symbolical method for dealing with such a problem. Suppose, for example, that f belongs to the class of skew symmetric tensors. Then (3.14) holds, so $f = \frac{1}{\nu!}\sum_S (\operatorname{sgn} S) f_S$, and an application of Frobenius's technique of "invariant summation" (Section 10.4) shows that (3.14) can be strengthened as follows:

(4.1) *A covariant tensor f of degree ν is skew symmetric if and only if f is expressible in the form $f = \frac{1}{\nu!}\sum_{S\in\mathfrak{S}_\nu} (\operatorname{sgn} S) f^*_S$, where f^* is a covariant tensor of degree ν.*

If in (4.1) we think of f^* as the general covariant tensor of degree ν, then we see that the general skew symmetric tensor arises from f^* in the manner described.

To illustrate this point let us consider again the above skew symmetric tensor f of degree 2. By (4.1), f can be obtained from the completely general degree-two tensor $f^* = a_{11}x^1y^1 + a_{12}x^1y^2 + a_{21}x^2y^1 + a_{22}x^2y^2$:

$$f = \frac{1}{2}f^* - \frac{1}{2}f^*_{(12)} = \frac{1}{2}(a_{12} - a_{21})x^1y^2 - \frac{1}{2}(a_{12} - a_{21})x^2y^1 \equiv \bar{a}_{12}x^1y^2 - \bar{a}_{12}x^2y^1.$$

Thus f is expressible in terms of the coefficients of the completely general f^*. As a consequence the linear invariant $I = \bar{a}_{12} = \frac{1}{2}(a_{12} - a_{21})$ is expressible in terms of the coefficients of f^* and is an invariant with respect to f^* as base form. The symbolical method, which requires that the coefficients of the base form be completely unconstrained, may now be applied to $I = \frac{1}{2}(a_{12} - a_{21})$ to obtain the (skew symmetric) vector invariant

$$i(\alpha, \beta) = \frac{1}{2}\alpha_1\beta_2 - \alpha_2\beta_1 = \frac{1}{2}\begin{vmatrix} \alpha_1 & \alpha_2 \\ \beta_1 & \beta_2 \end{vmatrix}.$$

It is easily seen that an analogue of (4.1) holds for symmetric tensors as well: f is symmetric if and only if $f = \frac{1}{\nu!}\sum_{S\in\mathfrak{S}_\nu} f_S^*$ for some tensor f^*. Weyl went beyond these simple observations and discovered how to generalize (4.1) so as to apply to any symmetry class of tensors. This constitutes the first part of his paper. There he proved the following result:

(4.2) **Theorem.** *Let $f(x,y,z,\ldots) = \sum_{ijk\ldots} T_{ijk\ldots}x^iy^jz^k\cdots$ denote a covariant tensor of degree ν. Then:*

(a) If $\mathcal{W}$ is the class of such tensors f that satisfy several symmetry relations of the form (3.12), there is a single such relation that characterizes $\mathcal{W}$.

(b) Given any symmetry class $\mathcal{W}$, there exist constants c_S, $S \in \mathfrak{S}_\nu$, such that if f^* is arbitrary, then $f = \sum_{S\in\mathfrak{S}_\nu} c_S f_S^* \in \mathcal{W}$. Moreover, every $f \in \mathcal{W}$ is obtained in this manner, since if $f \in \mathcal{W}$, then $f = \sum_{S\in\mathfrak{S}_\nu} c_S f_S \in \mathcal{W}$.

How did Weyl obtain these results? As he tells us, "By applying the representation theory of Frobenius to the group of all permutations one easily obtains a complete insight into the possible symmetry characteristics of tensors" [1924a:468–9]. In [1924a] Weyl wrote (3.12) in the equivalent form

$$\sum_{S\in\mathfrak{S}_\nu} k_S f_{S^{-1}} = 0 \tag{4.3}$$

and observed that one can associate to the form f an element $\mathbf{f}$ in the group algebra $\mathcal{H}$ of the symmetric group $\mathfrak{S}_\nu$, namely

$$\mathbf{f} = \sum_{S\in\mathfrak{S}_\nu} f_S S. \tag{4.4}$$

Since the $\mathbf{f}$ given in (4.4) depends upon the values of the variables $x, y, z, \ldots$ defining f, (4.4) actually defines a family of elements in $\mathcal{H}$. Weyl glossed over this point, but his results go through nonetheless.[51] Direct calculation then shows that (4.3) is equivalent to $\mathbf{kf} = \mathbf{0}$, where $\mathbf{k} = \sum_{S\in\mathfrak{S}_\nu} k_S S$, and, for example, part (b) of Theorem 4.2 can be deduced from the following result about the group algebra $\mathcal{H}$:

(4.5) **Theorem.** *Given $\mathbf{k} = \sum_{S\in\mathfrak{S}_\nu} k_S S \in \mathcal{H}$, there is an element $\mathbf{c} = \sum_{S\in\mathfrak{S}_\nu} c_S S \in \mathcal{H}$ such that $\mathbf{kf} = \mathbf{0}$ if and only if $\mathbf{f} = \mathbf{cf}^*$ for some $\mathbf{f}^*$ in $\mathcal{H}$. Moreover, $\mathbf{cf} = \mathbf{f}$ for all $\mathbf{f}$ satisfying $\mathbf{kf} = \mathbf{0}$.*

Theorem 4.5 was proved using the fact that the group algebra of the symmetric group decomposes into a direct sum of complete matrix algebras.

---

[51] Weyl later touched on this point in his exposition of tensor symmetries and the group algebra $\mathcal{H}$ in his book on group theory and quantum mechanics [1931:283].

This is the group algebra version of Frobenius's complete reducibility theorem (Section 10.1). By virtue of this decomposition into complete matrix algebras the proof of Theorem 4.5 is reduced to basic linear algebra, as Weyl showed. The element $\mathbf{c} = \sum_S c_S S \in \mathcal{H}$ produced in Theorem 4.5 produces the constants c_S of part (b) of Theorem 4.2. The same technique was used to prove a theorem about $\mathcal{H}$ that yields part (a) of Theorem 4.2.

So Weyl had discovered the value of Frobenius's theory of group characters and representations for answering questions from tensor algebra as well as questions about the scope of the symbolical method. If this paper is viewed as a response to Study, then the message would seem to be, I have discovered in the representation theory of groups something of far greater importance than the essentially formal symbolical method. Not only does it enable me to gain insight into questions about the foundations of tensor algebra that interest me, it also provides insight into the scope of the symbolical method itself. Such a reading of Weyl's state of mind when writing [1924a] is, of course, purely speculative. But it is in keeping with the following later opinion, expressed in his essay "Relativity Theory as a Stimulus in Mathematical Research" in the paragraph immediately preceding the statement quoted at the beginning of this chapter: "In the last decades a quite elaborate theory of representations of continuous groups has developed .... Here those problems which according to Study's complaint the relativists had let go by the board are attacked on a much deeper level than the formalistically minded Study had ever dreamt of" [Weyl 1949].

Once again, Weyl had discovered in the theory of groups the means to answer questions motivated by relativity theory. In the case of the space problem, however, he had utilized only the basic elements of Lie's theory—primarily the fact that one could deal with certain group theoretic-questions more easily by working with the infinitesimal group or Lie algebra and applying much linear algebra. The same can be said for Weinstein's work on his "fundamental theorem of tensor calculus." Now for the first time, Weyl had gone beyond the elements of group theory to achieve his goal. He had discovered the power of the theory of representations, particularly the complete reducibility theorem, for answering the questions that intrigued him. Furthermore, as we shall now see, this work on tensor symmetries leads naturally to his discovery of what he termed "the group-theoretic foundation of tensor calculus"—the starting point of his intense involvement with the theory of group representations.

11.5. The Group-Theoretic Foundation of Tensor Calculus

On 25 January 1924, the month he submitted his tensor symmetry paper, Weyl wrote a letter to the president of the Swiss school board reporting on his work over the past year.[52] It documents the importance to Weyl at this time of his involvement with the representation theory of finite groups: "The theory of finite groups, to which I was originally drawn by the theory of relativity, is more and more becoming my real area of work; as a result of the abundance of questions that arise there, I am again strongly inclined to work on individual, purely mathematical problems. Recently, I have been considering the invariant theories associated to the most important linear groups." This passage leaves no doubt as to the importance of Weyl's work on tensor symmetries in reorienting the direction of his research away from relativity and towards purely mathematical aspects of the theory of groups. Weyl's next report to the president, on 12 January 1925, shows how far he had already come by then: "In the past year my scientific activity has been directed above all toward *group theory*—faithful to the program I developed in my last letter. After lengthy, unsuccessful efforts, in the last few months I succeeded in overcoming the decisive difficulties, based in part on researches of Prof. E. Cartan, Paris, and partly in collaboration with Prof. Schur, Berlin. The results combine into one of the most wonderful theories to be found in mathematics."[53]

Weyl's interest in relativity theory had thus drawn him to the theory of finite groups, which, as his tensor symmetry paper makes clear, meant primarily the representation theory of finite groups. And the abundance of research questions that this theory raised in his mind had drawn him back to pure mathematics and to questions about the invariant theory (or representation theory) of the most important linear groups, presumably the groups already discussed in his paper [1923b], namely the special linear, orthogonal, and symplectic groups. Whether he had already discovered Schur's paper on Cayley's counting problem [1924a] by the time of the first report is unclear, but in any case, it was not long before he discovered it, since about nine months later, on 21 November 1924, Runge presented a paper by Weyl to the Göttingen Scientific Society for publication in its *Nachrichten*. It bore the title "The Group-Theoretic Foundation of the Tensor Calculus" [1924b] and made critical use of ideas from Schur's paper. The details of how Weyl utilized those ideas, and the sense in which he collaborated with Schur, will be discussed in the next chapter. In the present section I will concentrate on how his fascination with the representation theory of finite groups inspired the abundance of questions related to linear continuous groups.

The opening paragraph of Weyl's *Nachrichten* note conveys well the connections, in Weyl's mind, between the deliberations in his tensor symmetry paper and his latest discoveries. I will quote it in full and then provide commentary.

In the coordinate space of the x_i $(i = 1, 2, \ldots, n)$ we consider the

[52] A transcript of the letter is given in [Frei and Stammbach 1992:62–3]; the following translation is mine.

[53] Translation based on the transcript presented in [Frei and Stammbach 1992:64].

group $\mathfrak{G}$ of homogeneous linear transformations of determinant 1. The tensors of degree ν in that space that satisfy prescribed symmetry conditions form on their part, if they possess N independent components, a linear manifold of N dimensions; under the influence of the group $\mathfrak{G}$ they undergo a group Γ of homogeneous linear transformations that is homomorphic to $\mathfrak{G}$. The true mathematical foundation of the tensor calculus seems to me to be the theorem that *in this way every linear homogeneous group Γ homomorphic to $\mathfrak{G}$, every "representation of $\mathfrak{G}$," is obtained.* In general, in addition to degree, symmetry conditions are part of the characterization of a definite type of quantity in coordinate space. An overview of the possible symmetry characteristics of tensors of degree ν is easily attained on the basis of the representation theory of the symmetric permutation group $\mathfrak{S}_\nu$ of ν objects developed by Frobenius, as I have recently shown. I call a symmetry characteristic *irreducible*, the associated type of quantity *simple*, if every further symmetry condition imposed on the quantity leaves open no possibility except 0. The tensors of each symmetry class can be additively composed out of independent components that are simple in this sense. To the irreducible symmetry classes of tensors correspond the irreducible representations Γ of $\mathfrak{G}$.

To begin with, it should be clear from the above that Weyl is thinking of various types of tensors as representation modules for the special linear group $\mathfrak{G}$—linear manifolds on which $\mathfrak{G}$ acts to produce the representation Γ. Therefore, in what follows I will speak in terms of modules. To understand the theorem that Weyl sees as the foundation of the tensor calculus as well as its connections with the tensor symmetry paper, consider a symmetry class $\mathcal{W}$ of degree-ν tensors. As noted following (3.13), it is easily seen that the action of $\mathfrak{G}$ on $\mathcal{W}$ takes it into itself so that $\mathcal{W}$ is a representation module. It was the invariants with respect to the representation Γ associated to such a module that concerned Weyl in his tensor symmetry paper. Recall that $\mathcal{W}$ is completely determined by an element $\mathbf{c}$ in the group algebra of $\mathfrak{S}_\nu$ in accordance with Theorems 4.2 and 4.5. Now, $\mathbf{c}$ is an idempotent, i.e., $\mathbf{c}^2 = \mathbf{c}$, as Weyl realized.[54] As we saw in Section 10.1, Frobenius had shown that the primitive idempotents determine the irreducible representations of the symmetric group $\mathfrak{S}_\nu$. It would be natural to wonder whether therefore a primitive idempotent $\mathbf{c}$ defines a symmetry class $\mathcal{W}$ that is an irreducible representation module for $\mathfrak{G}$ and thus identifiable with one of the irreducible representation modules of $\mathfrak{G}$ described by Cartan in [1913] in terms of its highest dominant weight. Indeed, although Cartan did not couch his results in terms of tensors, to anyone with a background in tensor algebra it would have been clear that the representation modules he constructed

[54] This fact is not mentioned explicitly in Part I of Weyl's tensor symmetry paper, but in Part II it is taken for granted [1924a:474]. It follows immediately from Weyl's proof of Theorem 4.5.

could be regarded as consisting of tensors. As we saw in Section 8.4, the irreducible module he described (the Cartan component) was a submodule of "direct products" of generalized Plücker coordinates, which for Weyl meant that this ambient module consisted of tensors with an assortment of skew symmetry properties.

Weyl discovered that Cartan's irreducible modules do correspond to a symmetry class $\mathcal{W}$ of tensors with primitive idempotent defined by a Young tableau that can be read off from the highest weight of the representation module. To illustrate his discovery, let us consider $\mathfrak{G} = \mathbf{SL}(4, \mathbb{C})$ and the irreducible representation module of highest weight $\lambda = \sum_{i=1}^{3} p_i \pi_i$, where the π_i are the fundamental dominant weights and $p_i \geq 0$ in accordance with Cartan's theory. If we set $m_1 = p_1 + p_2 + p_3$, $m_2 = p_2 + p_3$, $m_3 = p_3$, then $m_1 \geq m_2 \geq m_3$, and the m_i define the shape of a Young tableau, with first row of length m_1, second row of length m_2, and third of length m_3, which corresponds to the symmetric group $\mathfrak{S}_\nu$ with $\nu = \sum_{i=1}^{3} m_i$. For example, if $\lambda = 2\pi_1 + \pi_2 + 3\pi_3$, so $m_1 = 6, m_2 = 4, m_3 = 3$, and $\nu = 13$, then

$$\mathfrak{T} = \begin{array}{cccccc} 1 & 2 & 3 & 4 & 5 & 6 \\ 7 & 8 & 9 & 10 & & \\ 11 & 12 & 13 & & & \end{array}$$

is a Young tableau of the given shape. If $\mathbf{c}$ is the primitive idempotent defined by $\mathfrak{T}$ (as indicated in the Young–Frobenius theorem of Section 10.1), the Cartan component of highest weight $\lambda = 2\pi_1 + \pi_2 + 3\pi_3$ is precisely the symmetry class of degree-13 tensors generated by $\mathbf{c}$, i.e., all degree-13 tensors f satisfying the symmetry relation $\sum_S k_S f_{S^{-1}} = 0$, where $\mathbf{k} = \mathbf{1} - \mathbf{c}$.

The general theorem characterizing in this manner the Cartan components of $\mathbf{SL}(n, \mathbb{C})$ as symmetry classes of tensors associated to primitive idempotents will be referred to in what follows as *Weyl's characterization theorem.* Weyl's discovery of this theorem is not surprising given what he had learned in writing the tensor symmetry paper. That is, as already noted, the results of that paper suggest that the symmetry class of a primitive idempotent $\mathbf{c}$ might be an irreducible representation module for $\mathfrak{G}$; once one is looking for a characterization of a Cartan component in terms of an appropriate tableau $\mathfrak{T}$ and associated $\mathbf{c}$, it is relatively easy to see how Cartan's description of the Cartan component of highest weight λ leads to the appropriate choice of $\mathbf{c}$ and $\mathfrak{T}$.[55]

In the *Nachrichten* paper under discussion, noting that Cartan had not indicated this connection between symmetry classes of tensors and irreducible representations, Weyl expressed the conviction that it was through this connection "that the entire matter is first placed in the right light" [1924b:462]. Indeed, as a consequence of his characterization theorem, Weyl

[55] Weyl sketched the procedure by way of a simple example in [1925:557–58]. One finds it explained more fully, within the context of $\mathbf{GL}(n, \mathbb{C})$, in his lectures [1934–5:26–29].

could see that the irreducible representation modules are actually classes of tensors with what he termed (in the above quotation) irreducible symmetry characteristics, an observation that must have pleased him given his emphasis on symmetry characteristics. This observation makes especially compelling what would seem to be the next obvious question: Does the complete reducibility theorem of Frobenius's theory also hold for $\mathfrak{G} = \mathbf{SL}(n, \mathbb{C})$? As the above passage indicates, Weyl had discovered that this is the case, and so every finite-dimensional representation module with representation Γ can be decomposed into a sum of simple tensor types characterized by irreducible symmetry characteristics. This is the theorem that formed for him the foundation of the tensor calculus. It implied, in particular (as he states), that every symmetry class of tensors can be broken down into a sum of independent simple tensor types. It also showed, more generally, that the general tensor of a given degree can be thought of as a sum of tensors with irreducible symmetry characteristics.

The title of Weyl's *Nachrichten* note, "The Group-Theoretic Foundation of the Tensor Calculus," is reminiscent of the title of Weinstein's dissertation of 1922, "The Fundamental Theorem of the Tensor Calculus," and I think Weyl had it in mind when he chose his own title. Recall that Weinstein had asked whether there might exist any other essentially different tensor-like quantities besides the ones already known. He had formulated his question in a precise mathematical way and showed that there are no new types. This result was his fundamental theorem of the tensor calculus. From Weyl's viewpoint in 1924, however, Weinstein had formulated his problem in too limited a fashion. He had imagined all possible tensor-like quantities as characterized by a variable transformation $x = W(P)\bar{x}$ satisfying $W(P_1P_2) = W(P_1)W(P_2)$. Now, of course, Weyl could see that this was a very special type of representation of the general linear group, but that observation is relatively unimportant. What is important is that there is another representation associated to $P \to W(P)$. For example, if f denotes the degree-three covariant tensor $f = \sum_{i,j,k} T_{ijk}x^iy^jz^k$, then by virtue of the substitutions $x = W(P)\bar{x}$, $y = W(P)\bar{y}$, $z = W(P)\bar{z}$ we have $f = \sum_{ijk} \bar{T}_{ijk}\bar{x}^i\bar{y}^j\bar{z}^k$, and $R(P) : T_{ijk} \to \bar{T}_{ijk}$ defines a representation of the general or special linear group. The point of all this is that Weyl came to see tensor-like quantities as defined by arbitrary representations or representation modules of these groups, rather than as characterized by Weinstein's transformation rules $x = W(P)\bar{x}$, etc. Thus the question underlying Weinstein's paper now becomes, Are there any representations besides those given by tensors (as representation module)? In view of his characterization theorem it would be natural to consider this question with respect to representations of $\mathbf{SL}(n, \mathbb{C})$, which is what Weyl did in his *Nachrichten* note; and of course it is the complete reducibility theorem for this group that provides the analogue of Weinstein's fundamental theorem to the effect that all tensor-like quantities, i.e., all representations of $\mathbf{SL}(n, \mathbb{C})$, are expressible as sums of tensors with symmetry properties.

In the first chapter of his detailed paper [1925], submitted two months after the *Nachrichten* note and bearing the same title, Weyl expressed himself in ways that bring out more clearly the above-described connection with Weinstein's paper. Let $\mathfrak{G} = \mathbf{SL}(n, \mathbb{C})$ and $t \in \mathfrak{G}$. Then:

> A system of numbers $a_1, \ldots, a_N$ ... is called a *linear quantity* ... if its components a_i in any two coordinate systems that are transformed from one to the other by the transformation t are connected by a homogeneous linear transformation T in N variables associated to t. The correspondence rule $t \to T$ must transform $[\mathfrak{G}]$ ... into a homomorphic group $[\Gamma]$ ... of linear transformations. Γ characterizes the *type* (or *class*) of the quantity under consideration. Hence we can also express ourselves as follows: A *representation* of the group $[\mathfrak{G}]$ ... is given ..., or to every coordinate system in the n-dimensional affine space $\mathfrak{r}$ corresponds one in the N-dimensional space $\mathfrak{R}$. The quantities of the type $([\Gamma])$ are the *vectors* in the image space $\mathfrak{R}$. The problem of obtaining an overview of all possible types of linear quantities in affine space is accordingly nothing else but the *representation problem for the continuous group* $[\mathfrak{G}]$ .... In geometrical and physical applications it always happens that a type of quantity is not characterized solely by tensor degree but in addition by symmetry conditions. The experiences of mathematicians and physicists suggests ... that *there are no other linear quantities besides tensors* (whereby symmetry conditions are to be assumed in the tensor concept). This proposition, in which I perceive the proper group-theoretic justification of the tensor calculus, will in fact be proved in what follows (pp. 544–46).

Weyl's proposition, particularly in the vague form of the final italicized statement, is evidently his version of Weinstein's fundamental theorem, as can be seen by comparing it with the quotation from Weinstein's paper given in Section 3. Weyl called his version of the fundamental theorem "one of the main results of the theory that is to be developed" [1925:545].

Weyl's "fundamental theorem" follows from three results: (1) Cartan's classification of the irreducible modules for $\mathfrak{G} = \mathbf{SL}(n, \mathbb{C})$, (2) Weyl's characterization theorem, and (3) the complete reducibility theorem for $\mathfrak{G}$, i.e., that every finite-dimensional representation module for $\mathfrak{G}$ decomposes into a direct sum of irreducible modules. From the perspective of present-day mathematics, in addition to (1), it is (3) and its generalizations to all semisimple groups that stands out as of the greatest importance, not Weyl's "fundamental theorem." The fact that Weyl turned to the task of extending the complete reducibility theorem to other semisimple groups—he had already succeeded in doing so for the orthogonal and symplectic groups by the time of the *Nachrichten* note—indicates that he, too, came to see the importance of continuing and combining the work of Cartan and Schur so as to create a theory of the representation of semisimple groups comparable to Frobenius's theory for finite groups. The great emphasis that he placed upon tensors in his *Nachrichten* note and in the first chapter of [1925], however, shows that his initial appreciation of the representation theory of groups—both finite and continuous—was based upon the realization that the theory was a powerful instrument for providing understanding of the structure of tensors (in terms of symmetry types) and of the scope of the tensor calculus. This, I believe, is what Weyl meant when he wrote, "for myself I can say that the wish to understand what really is the mathematical substance behind the formal apparatus of relativity theory led me to the study of representations and invariants of groups." The "formal apparatus" was the tensor calculus and "the mathematical substance" behind it was the representation theory of the group $\mathfrak{G}$.

Weyl's interest in tensorial interpretations of group-theoretic results comes out in two other places in Chapter I of [1925] that are worth mentioning as further illustrations of the extent to which he had tensors on his mind as he began to turn to the investigation of group representations. Both are related to Weinstein's fundamental theorem of tensor calculus.

In his paper [1922] Weinstein, as Weyl could now appreciate, was really studying possible representations of degree n of $\mathbf{GL}(n, \mathbb{R})^+$. As we will see in Chapter 12, stimulated by correspondence with Schur, Weyl developed the theory of characters for $\mathbf{SL}(n, \mathbb{C})$ and obtained thereby a formula for the degree of the irreducible representation with given highest weight. He pointed out that his formula made it possible to say various things about low-degree representations, but for him "the most important" such observation was, "The [irreducible] representations with the lowest number of variables except for the identity are ... the group ... itself and the contragredient" [1925:570n.1]. In other words, excluding the trivial representation $P \to I$, the minimal degree of an irreducible representation of $\mathbf{SL}(n, \mathbb{C})$ is n, and the two possibilities are $P \to P$ and $P \to (P^T)^{-1}$. Since the distinction between tensors and tensor densities disappears within the context of $\mathbf{SL}(n, \mathbb{C})$, this is, of course, the analogue for this group of Weinstein's result, and Weyl expressly invited comparison with Weinstein's paper. Within the far broader framework of the theory of characters and representations of $\mathbf{SL}(n, \mathbb{C})$ that he had created, this Weinstein-like result drops out as a simple special-case consequence of a vastly more general formula.

At the end of Chapter I of [1925] Weyl turned to the matter of the representations of the general linear group. In keeping with the previous notation, let $\mathfrak{G} = \mathbf{SL}(n, \mathbb{C})$ and set $\widetilde{\mathfrak{G}} = \mathbf{GL}(n, \mathbb{C})$. A representation $\phi : \widetilde{\mathfrak{G}} \to \widetilde{\Gamma}$ could just as well be regarded as a "linear quantity" in Weyl's sense; indeed, the restriction of this notion to the context of $\mathfrak{G}$ seems a bit contrived. It was due to the fact that Weyl realized that the complete reducibility theorem does not hold for $\widetilde{\mathfrak{G}}$. Making use of Schur's lemma (Section 10.4), he saw that $\widetilde{\Gamma}$ could be completely reduced when the restriction of $\widetilde{\Gamma}$ to the scalar matrices is semisimple (i.e., the representation matrices are diagonalizable), thereby anticipating in this special case a basic theorem about the representations of reductive groups. "But the exceptional cases known from the theory of elementary divisors can also occur; accordingly, for ... $[\widetilde{\mathfrak{G}}]$ the theorem of the full reducibility of representations is no longer valid without exception"[1925:579].

The exceptions can arise, as Weyl showed, when the restriction Γ of $\widetilde{\Gamma}$ to $\mathfrak{G}$ is not irreducible as a representation of $\mathfrak{G}$.[56] However, when Γ is irreducible, Weyl was able to say something more positive, which once again recalls Weinstein's dissertation. That is, by restricting his attention to the special linear group $\mathfrak{G}$ in articulating his version of Weinstein's fundamental theorem, Weyl had lost the distinction between tensors and tensor densities that had figured so prominently in Weinstein's paper. At least for "linear quantities" $\widetilde{\Gamma}$ such that Γ is irreducible he was able to obtain something analogous to Weinstein's results.

To see this, let, as above, $\phi : \widetilde{\mathfrak{G}} \to \widetilde{\Gamma}$. Then since $\tilde{t} \in \widetilde{\mathfrak{G}}$ is expressible as $\tilde{t} = \alpha t$, where $t \in \mathfrak{G}$ and $\alpha^n = \det \tilde{t}$, $\phi(\tilde{t}) = \phi(\alpha I)\phi(t)$, where I denotes the identity matrix. Since Γ is assumed to be irreducible, it is induced by the simple tensors of an irreducible symmetry characteristic corresponding to a symmetric group $\mathfrak{S}_\nu$, from which Weyl could see that the coefficients of the matrices of Γ are homogeneous polynomials of degree ν in the coefficients of $t \in \mathfrak{G}$ [1925:559]. In other words, for all $t \in \mathfrak{G}$, $\phi(t) = P_\nu(t)$, where P_ν denotes a Schur polynomial representation (or invariant form) of homogeneity degree ν (Section 10.3). Also, since for any fixed $\alpha \in \mathbb{C}$ and all $t \in \mathfrak{G}$, $\phi(\alpha I)\phi(t) = \phi(\alpha I \cdot t) = \phi(t \cdot \alpha I) = \phi(t)\phi(\alpha I)$, the matrix $\phi(\alpha I)$ commutes with all $\phi(t)$, and so by Schur's Lemma $\phi(\alpha I)$ is a scalar matrix. When $\alpha^n = \det \tilde{t}$, it turns out that $\phi(\alpha I) = (\det \tilde{t})^{\frac{\gamma}{n}} I$, where γ is some constant. From the homogeneity property of P_ν, it follows that[57]

$$\phi(\tilde{t}) = (\det \tilde{t})^{\frac{\gamma}{n}} P_\nu(t) = (\det \tilde{t})^{\frac{\gamma - \nu}{n}} P_\nu(\tilde{t}).$$

And so Weyl could conclude:

> In place of the concept of a tensor that of a *tensor density* has arisen; however, in the general sense that with the transition to a new coordinate system in [the underlying module] $\mathfrak{r}$ multiplication by an *arbitrary* power of the transformation determinant occurs, the exponent being not necessarily $= 1$ or even integral [1925:579].

Even as Weyl was turning to the study of representations, he could not resist viewing results of his new-found area of research in terms of his earlier interests and experiences involving tensors.

****

[56] More precisely, the exceptions can arise when in the decomposition of Γ into a sum of irreducible representations (in accordance with the complete reducibility theorem for $\mathfrak{G}$) some of the summands are equivalent.

[57] In [1925:579] the exponent of $\det \tilde{t}$ is $\gamma - \frac{\nu}{n}$, which seems to be a misprint.

Before proceeding to Chapter 12 and a discussion of how Weyl was able to prove the complete reducibility theorem for $\mathbf{SL}(n, \mathbb{C})$, two points are worth making by way of conclusion of the present chapter. The first is that Weyl was not the only mathematician interested in tensor algebra who saw a connection with the representation theory of the symmetric group. As early as 1919 the Dutch mathematician J. A. Schouten (1883–1971) studied the problem of expressing a tensor as a sum of "irreducible" tensors with symmetry properties. To this end he utilized the group algebra of the symmetric group and Frobenius's theory of group characters (in a formula for the principal idempotents). He was, however, unaware of Frobenius's paper [1903] or the work of Young that had motivated it; and he independently developed notions akin to that involving Young tableaux. Schouten's work was expounded (with complete references to his earlier publications) in his book *The Ricci Calculus. An Introduction to the Recent Methods and Problems of Multi-dimensional Differential Geometry* [1924:Abschn. VII], which Weyl cited in his *Nachrichten* note [1924b]. There is no evidence that Weyl knew of Schouten's work earlier and was influenced by it. Schouten actually submitted a note [1923] illustrating his method (on an example suggested by Study!) to the *Rendiconti del Circolo Matematico di Palermo* a year before Weyl's paper on tensor symmetries [1924a] was submitted to the same journal. Had Weyl known of this note he most certainly would have cited it in his own.

In this chapter we have seen how Weyl's involvement with relativistic physics led him to an ever deepening appreciation of the theory of groups, particularly the theory of group representations, as a powerful mathematical instrument for dealing with mathematical questions inspired by the physics. Sensing the depth and beauty of the purely mathematical theory of the representation of semisimple groups, he went on to produce the profound and influential papers [1925; 1926], which are the subject of the next chapter. Here I simply wish to point out that in the midst of creating these papers, Weyl kept in touch with contemporary developments in theoretical physics and once again was led to pose mathematical questions for which group representations provided the answer. This time, however, group representations were utilized to deal with questions arising from the new quantum mechanics initiated by the work of Heisenberg and Schrödinger in 1924–25. By 1927 their work had led to further developments by theoretical physicists such as Born, Pauli, Jordan, and Dirac and, from the mathematician's side, by von Neumann.

Weyl seems to have assimilated and mastered these developments as rapidly as he had mastered relativity theory a decade earlier. Thus in a paper "Quantum Mechanics and Group Theory" [1927a] we find him posing the question, "How do I arrive at the matrix, the hermitian form, that represents a given quantity in a physical system of known constitution?" To explain the precise meaning of these words would take us too far afield. But his further words should have a familiar ring to them by now: "Here with the help of group theory I believe I have succeeded in arriving at a deeper insight into the true nature of things" (p. 91). By "group theory" Weyl meant representation theory. This time it was to the study of unitary projective (or ray) representations of the abelian Lie group $\mathbb{R}^{2f}$ that he turned for the deeper insight. Once again the mathematics generated by the question went on to have a fruitful life of its own.[58]

[58] See [Mackey 1988:140ff.] for an interesting account of Weyl's question and its historical background as well as a discussion of Weyl's answer in terms of group representations and the subsequent mathematics it generated.

Chapter 12
Weyl's Great Papers of 1925 and 1926

At the end of the last chapter we noted that Weyl's first two accomplishments in the study of the representations of Lie groups were his characterization of the irreducible representations of the special linear group in terms of symmetry classes of tensors and his establishment of the complete reducibility theorem for this group. These results were important to him because they provided the proper (group-theoretic) foundation for the calculus of tensors. Impressive as these results are, they are but a part of the contents of his first detailed paper on the representation of semisimple groups, published in *Mathematische Zeitschrift* with the title, "Theory of the Representation of Continuous Semisimple Groups by Linear Transformations" [1925]. The following chapter is devoted to a discussion of this remarkable paper (Sections 1–5) and two related matters: Cartan's response to it (Section 6) and, in Section 7, Weyl's paper with F. Peter [1926], in which the details of an idea already presented in the 1925 paper are worked out. The Peter–Weyl paper effectively launched what is now called harmonic analysis on groups. Together these two papers constitute an important landmark in the history of Lie's theory and, more generally, in the history of mathematics.

12.1. The Complete Reducibility Theorem

Let us now consider how Weyl established the complete reducibility theorem for the special linear group—what he thought of as providing the group-theoretic foundation of the tensor calculus. In discussing the proof I will use the following modification of Weyl's notation. Let $\mathfrak{G} = \mathbf{SL}(n, \mathbb{C})$ and $\mathfrak{G}_u = \mathbf{SU}(n)$, and let the corresponding Lie algebras be denoted by $\mathfrak{g}$ and $\mathfrak{g}_u$. For brevity I will say that a Lie algebra or Lie group has the *complete reducibility property* if every finite-dimensional representation decomposes into a direct sum of irreducible representations. As Schur had observed in 1924 and Weyl realized, this property holds precisely when Maschke's theorem (Section 10.4) holds. That is, the complete reducibility property holds precisely when to every representation module $\mathcal{V}$ containing a proper submodule $\mathcal{V}_1$ there is another representation submodule $\mathcal{V}_2 \subset \mathcal{V}$ that is complementary to $\mathcal{V}_1$ in the sense that $\mathcal{V} = \mathcal{V}_1 \oplus \mathcal{V}_2$.

Now, the compact group $\mathfrak{G}_u = \mathbf{SU}(n)$ admits a translation-invariant integral, as Hurwitz had shown in 1897, so that, as Schur had shown in 1924, $\mathfrak{G}_u$ has the complete reducibility property. But what about $\mathfrak{G} = \mathbf{SL}(n, \mathbb{C})$? Weyl had the happy idea to try to use something like the unitarian trick to push this property from $\mathfrak{G}_u$ to $\mathfrak{G}$, as had been done for an invariance property by Hurwitz (Section 10.2) and Schur (Section 10.5). Since in his earlier and more elementary applications of Lie groups Weyl tended to work for convenience on the Lie algebra level, it is not surprising to see him doing the same here. His first discovery may have been the following easily proved proposition, which I call Weyl's unitarian trick.

(1.1) **Weyl's Unitarian Trick.** *If* $\mathfrak{g}_u = \mathbf{su}(n)$ *has the complete reducibility property, then so does* $\mathfrak{g} = \mathbf{sl}(n, \mathbb{C})$.

Here is the gist of the proof (slightly modernized in terminology) [Weyl 1925:560]. Suppose that $\mathcal{V}$ is a finite-dimensional $\mathfrak{g}$-module with a proper submodule $\mathcal{V}_1$. Then $\mathcal{V}$ and $\mathcal{V}_1$ are $\mathfrak{g}_u$-modules, and since $\mathfrak{g}_u$ is assumed to have the complete reducibility property, there is a complementary $\mathfrak{g}_u$-module $\mathcal{V}_2$. Thus with respect to a suitable basis for $\mathcal{V}$ the representation matrices of $X \in \mathfrak{g}$ have the form

$$\varphi(X) = \begin{pmatrix} A(X) & B(X) \\ 0 & C(X) \end{pmatrix},$$

where $B(X) = 0$ for all $X \in \mathfrak{g}_u$. If $X = (\tau_{jk}) \in \mathfrak{g}$, then, since φ is a Lie algebra representation, it is easily seen that every entry of $\varphi(X)$ is a homogeneous polynomial of the first degree in the variables τ_{jk}. Thus a typical entry of $B(X)$ is of the form $f(\tau) = \sum_{jk} a_{jk}\tau_{jk}$. Now, f has the property that it vanishes for all choices of τ_{ik} defining elements in $\mathfrak{g}_u$, which consists of all skew hermitian matrices of trace zero. The τ_{ij} determining elements of $\mathfrak{g}_u$ therefore satisfy the relations $\tau_{jk} + \bar{\tau}_{kj} = 0$, $\sum_k \tau_{kk} = 0$, and $f(\tau) = 0$ for such τ_{jk}. By means of suitable choices of these τ_{jk} it is easy to show that all the coefficients a_{jk} of f must be zero. Hence $f \equiv 0$, and so so $B(X) \equiv 0$ [1925:561]. In other words, $\mathfrak{g}$ will inherit the complete reducibility property from $\mathfrak{g}_u$, provided that it is possessed by $\mathfrak{g}_u$ in the first place.

Figure 39. Weyl with Hilbert in a photo taken in the mid 1920s, hence about the time he was creating the theory here being described.

On the basis of Weyl's unitarian trick (1.1) it would have been tempting for him to jump to the conclusion that he had proved the complete reducibility theorem for the special linear group, that is, to conclude that since $\mathfrak{G}_u$ has the complete reducibility property, so does $\mathfrak{g}_u$ and thus by his unitarian trick so does $\mathfrak{g} = \mathbf{sl}(n, \mathbb{C})$. Indeed, as we have seen, Lie and Cartan usually never made any clear distinction in their mathematical discourse between Lie groups and Lie algebras. But Weyl realized that a representation of a Lie algebra such as $\mathfrak{g}_u$ need not correspond to a representation of the Lie group from which it was derived. Hence even though $\mathfrak{G}_u$ has the complete reducibility property, $\mathfrak{g}_u$ need not have it, which means that the unitarian trick cannot be directly applied.

Weyl seems to have arrived at this insight by perceiving an analogy with analytic continuation and covering surfaces, a subject with which he was well acquainted (Section 9.5). Given a representation $X \to \varphi(X)$ of the infinitesimal group $\mathfrak{g}_u$, Weyl reasoned [1925:561–2], one obtains by integration, following Lie, a matrix $\Phi(x)$ for every $x \in \mathfrak{G}_u$ in a neighborhood $\mathfrak{U}$ of the identity element e. For $x_0 \in \mathfrak{U}$ consider the translated neighborhood $x_0\mathfrak{U}$ of x_0. Since all elements in this neighborhood are of the form $x_0 u$ where $u \in \mathfrak{U}$, the representation can be continued by setting $\Phi(x_0 u) = \Phi(x_0)\Phi(u)$. Repeating this process of continuation could lead to a multivalued representation on $\mathfrak{G}_u$ that first becomes single-valued on a "covering manifold" ("*Überlagerungsgebilde*") over $\mathfrak{G}_u$. Referring to the universal covering surface of uniformization theory (Section 9.5), Weyl declared, "This universal covering manifold $[\widetilde{\mathfrak{G}}_u]$ over $[\mathfrak{G}_u]$ is the true abstract group whose representations are under consideration; $[\mathfrak{G}_u]$ is only *one* of its representations, and indeed it is [unfaithful] ... when the covering manifold is many-sheeted." Weyl did not pause to explain what he meant by this "true abstract group," leaving it to his readers, who were assumed to know about covering manifolds from uniformization theory, to see how to define a group multiplication in $\widetilde{\mathfrak{G}}_u$. As noted in Section 9.5, this is not difficult to do.

It was mentioned in passing in Section 9.1 that Hilbert had extended his finite basis theorem to the rotation group of ordinary space, $\mathfrak{G} = \mathbf{SO}\,(3, \mathbb{R})$, by using the well-known connection between rotations in ordinary space and quaternions. That is, if $\mathbf{a} = a_0 + a_1\mathbf{i} + a_2\mathbf{j} + a_3\mathbf{k}$ is a real quaternion such that $\sum a_i^2 = 1$, then the transformation $(x_1, x_2, x_3) \to (y_1, y_2, y_3)$ of $\mathbb{R}^3$ defined by $\mathbf{y} = \mathbf{a}^{-1}\mathbf{x}\mathbf{a}$, where $\mathbf{x} = x_1\mathbf{i} + x_2\mathbf{j} + x_3\mathbf{k}$, $\mathbf{y} = y_1\mathbf{i} + y_2\mathbf{j} + y_3\mathbf{k}$, is an orthogonal transformation $t(\mathbf{a})$ with determinant $+1$. It is readily seen that quaternions $\mathbf{a}$ of the above type form a group $\widetilde{\mathfrak{G}}$ under quaternion multiplication and that $\Psi : \mathbf{a} \to t(\mathbf{a})$ is a homomorphism from $\widetilde{\mathfrak{G}}$ to $\mathfrak{G}$ with kernel consisting of the quaternions ± 1. Weyl was familiar with this construction as well, and realized that it shows that $\mathfrak{G}$ is not simply connected and that $\widetilde{\mathfrak{G}}$ is a two-sheeted covering [1925:598]. Here, then, was a well-known example illustrating the fact that a group need not coincide with its covering group.

To apply the "integration method" of Hurwitz to establish the complete reducibility property for a Lie algebra $\mathfrak{g}$ of a group $\mathfrak{G}$, it was thus necessary that the covering group $\widetilde{\mathfrak{G}}$ be compact. (In the above example, of course, both $\mathfrak{G}$ and $\widetilde{\mathfrak{G}}$ are compact.) As noted at the conclusion of Section 9.5, by virtue of his work on Riemann surfaces, Weyl could see that the compactness of $\widetilde{\mathfrak{G}}$ would follow from that of $\mathfrak{G}$, provided that $\widetilde{\mathfrak{G}}$ is a covering with finitely many sheets (as occurs in the above example). He also realized that a universal covering manifold of a compact group need not be compact. One has only to consider, as Weyl did, the rotation group of the plane, which in terms of complex variables can be thought of as the group $\mathfrak{G}$ consisting of all transformations $t_\theta : x' = e^{i\theta}x$ of the complex variable x. As a topological object $\mathfrak{G}$ can be identified with a circle, and its covering group $\widetilde{\mathfrak{G}}$ thought of as an infinite spiral. Hence $\widetilde{\mathfrak{G}}$ is not compact.

Using this example, Weyl pointed out some of the phenomena that can occur when the universal covering group is not compact. Since the Lie algebra $\mathfrak{g}$ of $\mathfrak{G}$ is the real 1-dimensional vector space spanned by the linear transformation $T_0 : x' = ix$, if r is any irrational real number, then $\varphi_r : T_0 \to rT_0$ defines an irreducible representation of $\mathfrak{g}$ that integrates to the many-valued representation Φ_r of $\mathfrak{G}$ associating to t_θ the linear transformations $x' = e^{ir(\theta+2\pi k)}x$, $k = 0, \pm 1, \pm 2, \ldots$. This representation first becomes single-valued on $\widetilde{\mathfrak{G}}$. Thus there are irreducible representations of $\mathfrak{g}$ that do not correspond to (single-valued) representations of $\mathfrak{G}$. Furthermore, as Weyl also observed, even though $\mathfrak{G}$ has the complete reducibility property, $\mathfrak{g}$ does not. For if $\varphi : \mathfrak{g} \to \mathbf{gl}(2, \mathbb{R})$ is the two-dimensional representation defined by

$$\varphi(T_0) = \begin{pmatrix} 1 & 1 \\ 0 & 1 \end{pmatrix},$$

then it has an invariant subspace but cannot be completely reduced, since the above matrix cannot be diagonalized.

Weyl could thus see something that had eluded Cartan, who, like Lie, tended to think of groups on the infinitesimal and group germ level: The irreducible representations of the infinitesimal group $\mathfrak{g}$ associated to a given group $\mathfrak{G}$ do not necessarily correspond to representations of $\mathfrak{G}$ but rather to its simply connected universal covering group $\widetilde{\mathfrak{G}}$. This meant that in order to utilize the unitarian trick (1.1), Weyl had to show that the covering group $\widetilde{\mathfrak{G}}_u$ of $\mathfrak{G}_u = \mathbf{SU}(n)$ is compact. He confirmed that this is the case by proving the following proposition:

(1.2) *The group* $\mathfrak{G}_u = \mathbf{SU}(n)$ *is simply connected.*

This meant that in this case $\widetilde{\mathfrak{G}}_u$ can be identified with $\mathfrak{G}_u$ and so is compact. In this manner Weyl established the complete reducibility theorem for the special linear group.

In his first announcement of this theorem in the Göttingen *Nachrichten*, in connection with (1.2) Weyl emphasized that, "Here analysis situs plays a decisive role; the real impediment to the universal application of Hurwitz's method (which would lead to the thoroughly false theorem that every linear group is fully reducible) lies in the realm of topology: the noncompactness [*Ungeschlossenheit*] of most group manifolds" [1924b:464]. In this way, thanks to Weyl, the importance of topological considerations in Lie's theory was first recognized, (1.2) being the first such theorem.[1]

By the time of the *Nachrichten* note Weyl had realized that his proof that $\mathbf{SL}(n, \mathbb{C})$ has the complete reducibility property could be modified to cover the other main classes of simple groups: the symplectic groups $\mathfrak{G} = \mathbf{Sp}(2n, \mathbb{C})$—termed "complex groups" by Weyl as an abbreviation of Lie's name for them (Section 3.5)—and the special orthogonal groups $\mathfrak{G} = \mathbf{SO}(n, \mathbb{C})$. In both cases $\mathfrak{G}_u$ is defined to consist of all $t \in \mathfrak{G}$ that are unitary, and the analogue of the unitarian trick (1.1) is proved by similar

[1] Shortly thereafter, however, and without knowing of Weyl's work, Otto Schreier [1925] introduced universal covering groups of connected, finite-dimensional topological groups and described their properties with Lie's theory in mind. His work is briefly described in Section 6 below.

considerations.[2] For $\mathfrak{G} = \mathbf{Sp}(2n, \mathbb{C})$ Weyl established the analogue of (1.2), namely that $\mathfrak{G}_u$ is simply connected, and for $\mathfrak{G} = \mathbf{SO}(n, \mathbb{C})$ he showed that $\mathfrak{G}_u = \mathbf{SO}(n, \mathbb{R})$, $n > 2$, is doubly connected [1925:588ff., 598ff.]. Thus in both cases $\widetilde{\mathfrak{G}}_u$ is compact, and the complete reducibility theorem follows.

The above outline of Weyl's proof of the complete reducibility theorem for the classical groups does not do it full justice because it glosses over the nontrivial technical considerations needed to turn the outline into an actual proof, e.g., to actually establish the connectivity properties of the groups $\mathfrak{G}_u$. Weyl's masterful skills in the application of linear algebra, which had been honed by his work on general relativity and, especially, on the space problem, stood him in good stead as he grappled with the details of the proof. Had he considered only the classical groups his achievement would have been notable enough, but he did not stop there. As will be seen in Section 2, not long after his *Nachrichten* note was submitted, he was able to tell Schur that he had succeeded in proving the complete reducibility theorem for any semisimple group. In what follows the broad outline of the proof will be sketched. Some of the details, which involved innovative contributions to the structure theory of semisimple groups, will be indicated in Section 3. It seems likely that by the time Weyl's *Nachrichten* note was presented he had already envisioned the following outline and begun to fill in the details.

Let $\mathfrak{G}$ denote a complex semisimple group with Lie algebra $\mathfrak{g}$. Then $\mathfrak{G}$ does not necessarily consist of linear transformations, but as anyone acquainted with Lie's theory would realize, one can consider the adjoint group $\mathfrak{A} \subset \mathbf{GL}(\mathfrak{g})$ (Section 3.2.). To get a real subgroup inside $\mathfrak{A}$ that might play the role of the groups $\mathfrak{G}_u$ utilized in dealing with the classical groups, Weyl considered what are now called real forms of $\mathfrak{g}$. That is, it was clear from the Killing–Cartan classification of semisimple $\mathfrak{g}$ that bases for $\mathfrak{g}$ exist for which the structure constants c_{ijk} are all real. The real span of such a basis defines a real Lie algebra $\mathfrak{g}_r$ with the same structure constants. Let $\mathfrak{a}_r \subset \mathfrak{a}$ denote the image of $\mathfrak{g}_r$ under the adjoint mapping and let $\mathfrak{A}_r \subset \mathbf{GL}(\mathfrak{g})$ denote the corresponding connected real Lie group. Of course, to play the role of $\mathfrak{G}_u$, $\mathfrak{A}_r$ must be compact.

From the Killing–Cartan theory it was known that $\mathfrak{A}$ leaves invariant the nonsingular quadratic form $k(X,X) = \mathrm{tr}\ X \circ \mathrm{tr}\ X$, $X \in \mathfrak{g}$. This is, of course, the now familiar Killing form, which Weyl was the first to introduce explicitly as a trace function, along with the corresponding bilinear form $k(X,Y)$. Thus the linear transformations of $\mathfrak{A}_r$ take the real vector space

[2] In the case of $\mathfrak{G} = \mathbf{SO}(n, \mathbb{C})$, the necessity of distinguishing between the cases of n even and n odd lead Weyl to formulate the unitary restriction somewhat differently [1925:595], but his formulation is equivalent to saying that there is a positive definite hermitian symmetric form $\langle v, w\rangle$ on the complex vector space $\mathcal{V}$ on which the $t \in \mathfrak{G}$ act such that $t \in \mathfrak{G}_u$ if and only if $\langle tv, tw\rangle = \langle v, w\rangle$ for all $v, w \in \mathcal{V}$. Furthermore, there is a basis for $\mathcal{V}$ such that the $t \in \mathfrak{G}_u$ correspond to all real orthogonal matrices with determinant 1, and $\mathfrak{G}_u$ can be identified with the rotation group of real n-dimensional space, $\mathbf{SO}(n, \mathbb{R})$.

$\mathfrak{g}_r$ into itself and leave $\varphi(X) = k(X, X)$ invariant. It is easily seen that the group of all real linear transformations that take a nonsingular quadratic form ψ into itself is bounded, and therefore compact, precisely when ψ is definite, so that the group is the familiar real orthogonal group.[3] To obtain a compact $\mathfrak{A}_r$, Weyl proved the following:

(1.3) **Real Compact Form Theorem.** *Given any complex semisimple $\mathfrak{g}$ there exists a real form of $\mathfrak{g}$, $\mathfrak{g}_u$, such that the Killing form restricted to $\mathfrak{g}_u$ is negative definite.*

We saw in Section 8.5 that in his study of real simple Lie algebras [1914c] Cartan had noted in passing what amounts to this theorem. That is, he had noted that each complex simple type has a real form for which the quadratic form φ is negative definite. Weyl, however, was not aware of this remark, for in his publications of 1924–25 he never referred to Cartan in connection with Theorem 1.3. In fact, he nowhere cited Cartan's paper [1914c] and so may not have studied it.

On the basis of Theorem 1.3 Weyl concluded that the real Lie group $\mathfrak{A}_u$ that corresponds to $\mathfrak{g}_u$ by the above construction is compact: Theorem 1.3 implies that $\mathfrak{A}_u$, being contained in a real orthogonal group, is bounded; it is also closed, although Weyl glossed over this point.[4] Thus $\mathfrak{A}$ and $\mathfrak{A}_u$ are the candidates to fill the roles played by $\mathfrak{G}$ and $\mathfrak{G}_u$ when $\mathfrak{G}$ is one of the classical groups. As in Weyl's proofs for those groups, in order to prove that $\mathfrak{g}$ has the complete reducibility property, two theorems remain to be proved. For the general semisimple case they take the following form:

(1.4) **Finite Connectivity Theorem.** *If $\mathfrak{G}$ is semisimple, then the universal covering group of $\mathfrak{A}_u$ has finitely many sheets.*

(1.5) **General Unitarian Trick.** *If $\mathfrak{g}$ is semisimple and $\mathfrak{g}_u$ has the complete reducibility property, then so does $\mathfrak{g}$.*

From the finite connectivity theorem it follows that since $\mathfrak{A}_u$ is compact, so is the universal covering group $\widetilde{\mathfrak{A}}_u$ (as indicated in Section 9.5), and so $\mathfrak{a}_u$ has the complete reducibility property. Since $\mathfrak{g}_u$ is isomorphic to $\mathfrak{a}_u$, $\mathfrak{g}_u$ must have the property as well. Then by the general unitarian trick $\mathfrak{g}$ also has the property. This is the outline of Weyl's proof of the complete reducibility theorem for any semisimple Lie algebra.

Judging by Weyl's correspondence with Schur in 1924 (Section 2), it would seem that the details behind this outline, given below in Section 3, had been more or less worked out by the time he was in communication with him. These details, especially what I term in Section 3 the conjugation theorem, seem to have enabled him to respond with lightening speed to the

---

[3] If ψ is indefinite, a basis may be chosen such that $\psi = x_1^2 - x_2^2 + \varepsilon_3 x_3^2 + \cdots + \varepsilon_n x_n^2$, $\varepsilon_k = \pm 1$. Then the group contains the unbounded sequence of "hyperbolic rotations" T_n, defined by $x_1' = (\cosh n)x_1 + (\sinh n)x_2$, $x_2' = (\sinh n)x_1 + (\cosh n)x_2$, $x_k' = x_k$, $k > 2$.

[4] See in this connection note 10 to p. 57 of [Borel 1986].

problems posed by Schur and to develop a theory of characters for any semisimple group (Section 5).

12.2. Schur and the Origins of Weyl's 1925 Paper

It seems that by about the end of October 1924 Weyl had devised his proof of the complete reducibility theorem for the classical groups as indicated in Section 1. He wrote these results up for presentation as a note to the Göttingen Academy for publication in the *Nachrichten* and sent a copy, along with an offprint of his paper on tensor symmetry [1924a], to Schur in Berlin, since as we have seen, both Schur's 1901 dissertation and his 1924 paper were related to the content of Weyl's note. Schur responded with a letter dated 10 November 1924, which is of considerable historical interest because it seems to have inspired a significant part of Weyl's paper [1925], namely his theory of characters.[5]

To appreciate Schur's reply and its effect upon Weyl, it must be kept in mind that when Weyl sent Schur his draft *Nachrichten* note, he knew only Schur's paper [1924a] on Cayley's counting problem. He knew nothing of Schur's calculation of the dimensions of the irreducible representations of the rotation group $\mathfrak{D} = \mathbf{SO}(n, \mathbb{R})$ and for the group $\mathfrak{D}'$ of all real orthogonal transformations or of his formulas for the irreducible characters of $\mathfrak{D}'$. These results form the sequel [1924b], which was not presented to the Berlin Academy until 13 November, three days after Schur replied to Weyl's communication.[6] It is possible that by the time he wrote to Schur, Weyl had already discovered Schur's dissertation, had seen his character and dimension formulas for the homogeneous polynomial representations of $\mathbf{GL}(n, \mathbb{C})$ considered there, and, inspired by those results, had already begun to attempt to obtain similar formulas for the irreducible representations of the classical groups. But this seems unlikely. The focus of Weyl's attention seems to have been on the complete reducibility theorem and its extension to all semisimple groups in accordance with the outline of Section 1. Judging by the Schur–Weyl correspondence discussed below, it would seem that Schur's reply to Weyl's communication inspired—and to a certain extent guided—Weyl's successful efforts to develop a theory of characters for semisimple groups that would provide formulas for the irreducible characters and the degrees of the corresponding representations.

---

[5] Schur's letter is located in ETH Zürich (Hs.91:734). The disposition of Weyl's letter is unknown to me, but from Schur's reply (pp. 1 and 13) and later published comment [Schur 1924b: 299n.1], it is clear what Weyl had sent to Schur. Weyl's *Nachrichten* note was presented to the Göttingen Academy on November 21 and published as [Weyl 1924b]. Slodowy [1998] has also discussed the mathematical and historical import of Schur's letter and related papers, as well as Weyl's theory of characters.

[6] For the group $\mathfrak{D}$ itself Schur was still working on the character formulas in the case of even n [1924b:298].

Most of Schur's fourteen-page letter was about his most recent work—that in his sequel [1924b] plus additional results—and its relation to the methods of Cartan, whose work, Schur explained, had not been known to him. Accompanying Schur's letter was an eight-page summary of the main results of the sequel. Referring to the summary, Schur wrote:[7]

> As far as I can see, my results on the representations of the rotation group $\mathfrak{D}$ and its (real) extension $\mathfrak{D}'$ are in essence more precise and provide more of an overview than those of Cartan. The exact specification of the number of variables (I denote them by $N^{(n)}_{\alpha_1\ldots\alpha_\nu}$ in the accompanying note) seems to me to be the main problem. Perhaps even more important is the determination of all the irreducible characters. In this way one first obtains a purely algebraic approach to the ennumerative calculus of the theory of orthogonal invariants .... It would be of considerable interest to me to see how my latest results on the number of variables and characters could be derived on the basis of Cartan's methods. My first impression is that this would still require a separate investigation. My method of proof, which merely uses the theory of characters ... seems simpler to me than the deep methods of Cartan. (pp. 2–3)

To illustrate his point, Schur described the key to his derivation of the dimension numbers $N^{(n)}_{\alpha_1\ldots\alpha_r}$.[8] Here is the general idea as illustrated for the rotation group $\mathfrak{D} = \mathbf{SO}(n, \mathbb{R})$. Each $t \in \mathfrak{D}$ has eigenvalues of the form $e^{\pm i\varphi_1}, \ldots, e^{\pm i\varphi_\nu}$ if n is even and of the form $1, e^{\pm i\varphi_1}, \ldots, e^{\pm i\varphi_\nu}$ if n is odd, where in either case $\nu = [\frac{n}{2}]$. Thus each t has associated to it ν rotation angles $\varphi_1, \ldots, \varphi_\nu$. Since t is conjugate to a diagonal matrix with the above entries, if $\gamma(t)$ is a class function, i.e., takes the same value on all elements in a conjugate class, such as, e.g., a character, then γ may be regarded as a function of the rotation angles: $\gamma(t) = f(\varphi_1, \ldots, \varphi_\nu)$. Schur's computation of the dimension numbers had made critical use of the fact that for such functions γ the Hurwitz integral over $\mathfrak{D}$ could be transformed into an integral over the rotation angles

$$\text{(2.1)} \quad \int_{\mathfrak{D}} \gamma \, dm = \int_0^{2\pi} \cdots \int_0^{2\pi} f(\varphi_1, \ldots, \varphi_\nu) M(\varphi_1, \ldots, \varphi_\nu) \, d\varphi_1 \cdots d\varphi_\nu,$$

where the function M is independent of f. In the case $n = 3$, $\nu = 1$, which Schur first presented, it is easy to show that M exists and equals a constant multiple of $(1 - cos\varphi)$. For $n > 3$ Schur had also obtained a formula for M, namely that M is a certain constant multiple of

---

[7] ETH Zürich, Hs.91:740, pp. 2–3. Schur used the term *Charakteristik*, rather than Frobenius's term *Charakter*, in speaking of continuous groups, and Weyl followed him for a while in this practice, which I have abandoned in the translations. I have also translated *einfache Charakteristik* as irreducible character in keeping with present-day terminology.

[8] ETH Zürich, Hs.91:740, pp. 4–8. Cf. the presentation in [Schur 1924b:299–306].

Figure 40. Issai Schur as Professor.

$$(2.2)\qquad \begin{cases} \Delta^2 & n \text{ even,} \\ \Delta^2 \prod_{k=1}^{r}(1-\cos\varphi_k) & n \text{ odd,} \end{cases}, \quad \Delta = \prod_{k<l}(\cos\varphi_k - \cos\varphi_l),$$

although in this case he had to admit that the proof was not so simple. He particularly disliked the fact that at one point he had to draw upon the theory of orthogonal invariants. A direct proof, he stressed, was much to be preferred (p. 8). Indeed, in the sequel [1924b] Schur noted the key role of (2.1) in going from the case $n = 3$ to $n > 3$ but did not give (2.2), promising it for a future note and taking instead a different approach, which was then used in the derivation of (2.2).[9]

Having thus given Weyl a glimpse into his methods, Schur returned to the comparison with the Lie–Cartan approach:[10]

---

[9] Schur presented his derivation of (2.2) in Anwendungen III [1924c], even though by that time Weyl had admittedly found a simpler, direct proof.

[10] ETH Zürich, Hs.91:740, pp. 8–9. Here I have translated *Kunstgriff* as "trick." Schur seems to have been the first to use this term—it is not in Weyl's

> Now you see that my approach (*Auffassung*) is different from yours. As in the theory of finite groups, in any representation problem I generally ask first of all whether an analogue of the theory of characters is at hand. If that is the case, or if, by the beautiful trick of Hurwitz it can be reduced [to the case of] a subgroup of the group to be studied, then, first of all, I wish to know the system of irreducible characters. The construction of the associated representations is then a *cura posterior*. In general this proceeds much more simply once the characters are known than if the problem is tackled directly. I took this approach already in my dissertation.

To proceed in similar fashion using Cartan's methods, Schur opined, one must proceed type by type and develop for each a theory of characters (pp. 9–10). Noting that this had already been achieved for "some principal types"—presumably a reference to his results on polynomial representations of $\mathbf{GL}(n, \mathbb{C})$ and on all representations of the rotation and real orthogonal groups—Schur expressed optimism that such a research program would be successful. And then "one would . . . have a clearer overview of all irreducible groups than is obtained by Cartan" (pp. 10–11).

It was only toward the end of the letter (p. 11) that Schur mentioned Weyl's own results about the special linear group. Although he declared them to be "very beautiful," he was evidently not overwhelmed by the achievement. And with good reason. After all, in his dissertation he had considered a similar representation problem—the polynomial representations of $\mathbf{GL}(n, \mathbb{C})$—and had gone much further. As we saw in Section 10.3, following the above-described approach, he not only proved a complete reducibility theorem for these representations, but also obtained formulas for the characters, showed how they relate to characters of symmetric groups, and obtained formulas for the dimensions of the irreducible representations, which were themselves fairly well described. These results apply to $\mathbf{SL}(n, \mathbb{C})$ as well, but of course they still come with the restriction to polynomial representations. Schur, however, suggested that the restriction was "not at all so essential." That is, one could obtain analogous results assuming that the coefficients of the representation matrix were analytic functions of the coefficients of $t \in \mathbf{GL}(n, \mathbb{C})$—the sort of assumption (Schur may have been thinking) that seemed implicit in the whole Lie group enterprise anyway. But Schur did not want to press his point, and he ended by allowing that "in any case, you have gone further with this group than I" (p. 13).

A lesser mathematician than Weyl might have found Schur's letter a bit discouraging given the belated praise for his *Nachrichten* note preceded by what amounts to a lecture on the proper, Berlin, way to deal with a representation problem, which implied that Weyl had only scratched the surface

---

Nachrichten note. Weyl then used it in his letter of reply [1924c:453]. When he later began writing in English, Weyl spoke of the "unitarian trick" [1939a; 1939b].

in his study of the representation problems suggested by Cartan's methods. But Weyl was not discouraged. He rose to Schur's challenge, took his advice to heart, and in his reply to Schur, written sometime before 26 November, he was able to announce that he had succeeded in solving the "representation problem" in Schur's sense for semisimple groups in general.[11] As he put it in his "official" letter to Schur of 28 November destined for the Berlin *Sitzungsberichte*, "By a modification of your beautiful method I have succeeded in solving the representation problem, including the calculation of the characters, for all simple and semisimple groups."[12] Weyl backed up this stunning claim with concrete formulas. For each type of classical group and then for an arbitrary semisimple group, he wrote down formulas for (i) the analogue of Schur's equations (2.1)–(2.2); (ii) the irreducible characters; (iii) the degrees of the corresponding representations. In the case of the rotation group itself, Weyl went further than Schur in the sense that he determined the irreducible characters of the universal covering group $\widetilde{\mathfrak{D}}$ of $\mathfrak{D}$, thereby calling attention for the first time to what are now called spin representations—representations of $\widetilde{\mathfrak{D}}$ that are not representations of $\mathfrak{D}$ itself.[13] And of course, by obtaining formulas for irreducible characters and dimension numbers valid for all semisimple groups, Weyl went beyond Schur's initial expectation that it would be necessary to develop a separate theory of characters for each type.

Schur was understandably amazed by what Weyl had accomplished in such a brief amount of time. No matter how speedy the postal service between Berlin and Zürich was in 1924, there were less than two weeks for Weyl to accomplish all that he had![14] "I am filled with the highest admiration (*größte Bewunderung*)," Schur wrote in his reply of 26 November, "for what you have now achieved—and in such a brief time. Your results are among the most beautiful and most important that have been attained in the field of group theory." In addition to soliciting the announcement for the *Sitzungsberichte*, Schur urged Weyl to write up a detailed presentation of his results for publication in *Mathematische Zeitschrift*, which he edited

[11] This letter is known to me only from Schur's reply to it on 26 November (ETH Zürich, Hs.91:735). But presumably it was an abbreviated version of Weyl's "official" letter to Schur of 28 November, which Schur requested on 26 November for publication in the Berlin *Sitzungsberichte*. With minor changes made by Schur, it appeared there as [Weyl 1924c].

[12] In the published form of the letter as [Weyl 1924c] the word "beautiful" is missing, having been struck out by Schur. See p. 2 of Schur's letter to Weyl dated 13 December 1924 (ETH Zürich, Hs.91:736).

[13] See [Weyl 1924c:458; 1925:597–8]. Weyl dealt simultaneously with $\mathfrak{D} = \mathbf{SO}(n, \mathbb{R})$ and $\mathbf{SO}(n, \mathbb{C})$, for which spin representations also exist. Since spin representations of $\mathbf{SO}(n, \mathbb{C})$ occur as representations of its Lie algebra $\mathbf{o}(n, C)$, they were already accounted for in Cartan's classification [1913:33–4,38–9] but without giving them them any special significance. After spin representations were found important in quantum physics, Cartan published a detailed general study [1938].

[14] Regarding Weyl's ability to quickly achieve results in a new direction or area, see also [Borel 1986:54–5].

along with fellow Berliners K. Knopf and E. Schmidt.[15] In his letter of 28 November, Weyl had also defended Cartan's methods as "natural" and "simple" so that "everything that is needed can be set down on a couple of pages,"[16] and so Schur encouraged Weyl to expound Cartan's methods as well. Schur repeated his request in a letter of 13 December, and Weyl complied with his paper "Theorie der Darstellung kontinuierlicher halbeinfacher Gruppen durch lineare Transformationen" [1925], which was published in three installments that were submitted in January, February, and April of 1925.

In the first installment Weyl presented his results on the special linear group, including the theory of characters he had now developed for this group. The second installment is devoted to the extension of these results to the other classical groups, and to the general theory of the structure of semisimple Lie algebras. As will be seen in Section 3, Weyl developed the Killing–Cartan structure theory in new directions that yielded proofs of the real compact form Theorem (1.3) as well as of the general unitarian trick (1.5) and the finite connectivity theorem (1.4). I would conjecture that Weyl may have been thinking deeply about these matters already when he communicated with Schur toward the beginning of November 1924. In any case, as will be seen in Section 5, the extensions of the Killing–Cartan theory Weyl made to establish the complete reducibility theorem, especially what I call in Section 3 the conjugation theorem, proved to be just what was needed in order to develop the theory of characters for an arbitrary semisimple group, replete with Weyl's now familiar formulas for the irreducible characters and the degrees of the associated representations. These results are contained in the third and final installment, which is devoted to representation theory and its application to the theory of invariants, namely the extension of Hurwitz's proof of the finiteness theorem to all semisimple groups, discussed briefly below in Section 4.

12.3. Weyl's Extension of the Killing–Cartan Theory

In the second installment of his paper [1925:606ff] Weyl turned in Chapter III to "the structure of semisimple groups." His guide to the Killing–Cartan theory was Cartan, particularly his doctoral thesis [1894b]. It will be helpful to recall some of the main results of Cartan's version of the theory.

If $\mathfrak{g}$ is a complex semisimple Lie algebra, then as a vector space $\mathfrak{g}$ can be expressed as a direct sum of root spaces, $\mathfrak{g} = \mathfrak{g}_0 \bigoplus_{\alpha\in\Phi} \mathfrak{g}_\alpha$, where Φ

[15] Knopf was at Berlin when the journal was founded in 1918, but by 1925 he had moved to Leipzig. Schmidt had become professor in Berlin in 1917. Incidentally, Weyl was one of the mathematicians (along with Study) listed as a "scientific advisor" (*wissenschaftlicher Beirat*).

[16] See [Weyl 1924c:453–4]. In a footnote he added, "The only difficulty is the construction of all simple groups, which with Cartan seems mixed up with the representation problem."

denotes the set of nonzero roots of $\mathfrak{g}$, $\dim \mathfrak{g}_\alpha = 1$ for all $\alpha \in \Phi$, $\mathfrak{g}_0 = \mathfrak{h}$ is an l-dimensional Cartan subalgebra of $\mathfrak{g}$, and $[\mathfrak{g}_\alpha, \mathfrak{g}_\beta] \subset \mathfrak{g}_{\alpha+\beta}$ for all roots α, β, where $\mathfrak{g}_{\alpha+\beta} = 0$ if $\alpha + \beta$ is not a root. The roots are linear functionals on $\mathfrak{h}$, and if $\alpha \in \Phi$, then $c\alpha \in \Phi$ if and only if $c = \pm 1$. These properties of $\mathfrak{g}$ made it possible to write down a fairly simple multiplication table for $\mathfrak{g}$. That is, let $H_i = H_{\alpha_i}$, $i = 1, \ldots, l$, denote a basis for $\mathfrak{h}$, take $X_\alpha \neq 0$ from $\mathfrak{g}_\alpha$, and let $H_\alpha = [X_\alpha, X_{\alpha'}]$, where $\alpha' = -\alpha$. Then

$$[H_i, H_j] = 0; \tag{3.1}$$

$$[H_i, X_\beta] = \beta(H_i) X_\beta; \tag{3.2}$$

$$[X_\alpha, X_{-\alpha}] = H_\alpha = c_1 H_1 + \cdots + c_l H_l; \tag{3.3}$$

$$\text{For } \beta \neq \alpha,\ [X_\alpha, X_\beta] = \begin{cases} 0 & \text{if } \alpha + \beta \notin \Phi, \\ c_{\alpha\beta} X_{\alpha+\beta} & \text{if } \alpha + \beta \in \Phi, \end{cases} \quad c_{\alpha\beta} \neq 0. \tag{3.4}$$

Of great importance to the Killing–Cartan theory was the fact that the X_α may be chosen such that $a_{\beta\alpha} = \beta(H_\alpha)$ is an integer for all $\alpha, \beta \in \Phi$ and $a_{\alpha\alpha} = -2$.[17] Although Cartan's examples of the simple types indicated that the c_i and $c_{\alpha\beta}$ could be taken as integers as well, such conclusions had not been forthcoming from the general theory. The classification of simple Lie algebras had depended on the many properties of the integers $a_{\alpha\beta}$. Of fundamental significance in this connection was

$$\text{For all } \alpha, \beta \in \Phi,\ \beta + a_{\beta\alpha}\alpha \in \Phi. \tag{3.5}$$

Cartan interpreted (3.5) as defining a permutation S_α of the roots Φ, i.e., $S_\alpha(\beta) = \beta + a_{\beta\alpha}\alpha$.

The results contained in (3.1)–(3.5) were presented by Weyl more or less following Cartan's thesis [1894b], although he did take advantage of Cartan's subsequent theory of weights to obtain (3.5) as a special case of the more general result that if λ is a weight of a representation of $\mathfrak{g}$ and $\alpha \in \Phi$, then

$$\lambda + \frac{2\lambda(H_\alpha)}{\alpha(H_\alpha)}\alpha$$

is also a weight. If this result is applied to the adjoint representation, then (3.5) results, since then $\lambda = \beta \in \Phi$ and

$$a_{\beta\alpha} = \beta(H_\alpha) = \frac{-2\beta(H_\alpha)}{-2} = \frac{-2\beta(H_\alpha)}{\alpha(H_\alpha)}. \tag{3.6}$$

Weyl, however, did not interpret (3.5) in terms of the permutations S_α as had Cartan. As he explained, "Up till now I have followed ... in essence the

[17] As indicated in Section 5.1, these "Cartan integers" are the negatives of what are now (following Weyl) usually called the Cartan integers.

thesis of Cartan. From now on I go my own way" [1925:619]. The starting point of Weyl's own way was a different interpretation of (3.5). In view of (3.6), (3.5) suggests considering the linear transformation σ_α defined on the vector space $\mathfrak{h}^*$ of all linear functionals ξ on $\mathfrak{h}$ by

$$\sigma_\alpha(\xi) = \xi - \frac{2\xi(H_\alpha)}{\alpha(H_\alpha)}\alpha. \tag{3.7}$$

In dealing with the Cartan integers $a_{\alpha\beta}$ and (3.5), Weyl thus turned to the geometrically flavored linear algebra he had fashioned in *Raum, Zeit, Materie* (sketched in Section 11.3 above).

The gist of what Weyl discovered may be described briefly as follows. From Cartan's thesis it followed that the Killing form $k(X,Y)$, restricted to $\mathfrak{h}$, is nonsingular and so establishes a duality between the vector spaces $\mathfrak{h}$ and $\mathfrak{h}^*$.[18] Let (ξ, η) denote the dual of the restricted Killing form, so that (ξ, η) is a symmetric bilinear form on $\mathfrak{h}^*$. Weyl observed that $\xi(H_\alpha) = (\xi, \alpha)$, so that

$$\sigma_\alpha(\xi) = \xi - \frac{2(\xi,\alpha)}{(\alpha,\alpha)}\alpha. \tag{3.8}$$

Moreover, if $\alpha_1, \ldots, \alpha_l \in \Phi$ constitute a basis for $\mathfrak{h}^*$, then since every $\alpha \in \Phi$ is a rational linear combination of the α_i, a result going back to Killing (Section 5.1), Φ can be regarded as a subset of the real vector space $\mathcal{V}$ spanned by the α_i, and (ξ, η) is real-valued and positive definite on $\mathcal{V}$. Thus $\mathcal{V}$ becomes an l-dimensional Euclidean space on which σ_α represents a reflection in the hyperplane with α as normal vector.

Weyl also introduced the finite group of orthogonal transformations generated by the σ_α, which I will denote by $\mathfrak{W}$. In developing the theory of semisimple $\mathfrak{g}$ Cartan had not found it necessary to explicitly introduce the corresponding group generated by the permutations S_α, but Weyl stressed the importance of what has since become known as the Weyl group. For Weyl (3.5) became the proposition that Φ is left invariant under the transformations of $\mathfrak{W}$ [1925:620, Satz 4], and he declared that, "The actual determination of all simple and semisimple continuous groups is based upon the determination of the associated finite groups [$\mathfrak{W}$] ..., i.e., as those finite groups with generating reflections in an ... [l]-dimensional Euclidean space satisfying certain integrality conditions" [1925:621]. Although Weyl did not pause to show how this could be done, the theoretical importance of the group $\mathfrak{W}$ becomes manifest in the theory of characters he devised for semisimple groups (Section 5).

By virtue of Weyl's reformulation of (3.5) he could define the Cartan integers by

[18] As noted in Section 11.3, this duality can be expressed by distinguishing between co- and contravariant coordinates in a fixed vector space, which was Weyl's approach.

$$a_{ij} = \frac{2(\alpha_i, \alpha_j)}{(\alpha_i, \alpha_i)}.$$

He pointed out that as a consequence, many properties of these integers that Killing and Cartan had obtained by "complicated calculations with determinants" could be more easily derived and expressed in a more transparent form. (Riemann's principle of Section 9.6 once again at work!) He was also thereby able to see relations that had eluded his predecessors. For example, if $\alpha, \beta, a\alpha + b\beta \in \Phi$, then since $\xi(H_{a\alpha+b\beta}) = (\xi, a\alpha + b\beta) = a(\xi, \alpha) + b(\xi, \beta) = a\xi(H_\alpha) + b\xi(H_\beta) = \xi(aH_\alpha + bH_\beta)$ for all ξ, it follows that $H_{a\alpha+b\beta} = aH_\alpha + bH_\beta$. As Weyl pointed out, this result was simpler and more consequential than Cartan's relation $H_{\alpha+a_{\alpha\beta}\beta} = H_\alpha + a_{\beta\alpha}H_\beta$ [1894b: 57 (Thm VII)]. For example, applied to (3.3), it shows that since $\alpha = q_1\alpha_1 + \cdots + q_l\alpha_l$ where the q_i are rational, $[X_\alpha, X_{\alpha'}] = q_1H_1 + \cdots + q_lH_l$, i.e., the structure constants c_i of (3.3) are rational. Weyl also showed that the X_α can be chosen such that the constants $c_{\alpha\beta}$ in (3.4) have important properties. In particular, $c_{\alpha\beta}$ is real for all α, β, $c_{\alpha\beta} \neq 0$ when $\alpha + \beta \in \Phi$, and $c_{\alpha'\beta'} = c_{\alpha\beta}$.[19]

Weyl's basis played a fundmental role in his extension of the general theory of structure of semisimple groups. First of all, he was able to establish the existence of a real compact form $\mathfrak{g}_u$ of $\mathfrak{g}$, thereby establishing the real compact form theorem of Section 1. If $\{H_i, X_\alpha\}$ denotes a basis with the X_α so chosen, then $\mathfrak{g}_u$ consists of all elements of the form

$$X = \sum_{k=1}^{l} s_k H_k + \sum_{\alpha \in \Phi} t_\alpha X_\alpha \tag{3.9}$$

such that the coefficients satisfy the conditions

$$s_k \quad \text{is purely imaginary and} \quad t_{\alpha'} = \bar{t}_\alpha. \tag{3.10}$$

In other words, $\mathfrak{g}_u$ is the real linear span of the basis elements iH_k, $X_\alpha + X_{\alpha'}$, $i(X_\alpha - X_{\alpha'})$, where $k = 1, \ldots, l$ and $\alpha \in \Phi$. Using the properties of the $c_{\alpha\beta}$ and the relation $H_{a\alpha+b\beta} = aH_\alpha + bH_\beta$, Weyl showed that $\mathfrak{g}_u$ is a Lie algebra and that the Killing form is negative definite on it.

An easy consequence of the construction of $\mathfrak{g}_u$ was a proof of the general unitarian trick (1.5), which asserts that if $\mathfrak{g}_u$ has the complete reducibility property, then so does $\mathfrak{g}$. The general proof for any semisimple $\mathfrak{g}$ is essentially the same as that for $\mathfrak{g} = \mathrm{sl}(n, \mathbb{C})$ (Section 1). If $X \to \phi(X)$ denotes a matrix representation of $\mathfrak{g}$, then each entry of $\phi(X)$ is a linear form in the variables s_k, t_α, and the proof of (1.5) reduces to the fact that if such a form vanishes for all values of s_k, t_α satisfying the conditions (3.10), then it vanishes identically.

[19] See Satz 5 on p. 626 of [1925] for a complete list of the properties of the $c_{\alpha,\beta}$ in what has become known as a "Weyl basis." The basis introduced by Chevalley in [1955] is a refinement of Weyl's basis.

Up to this point Weyl's extensions of the Killing–Cartan theory have been on the infinitesimal level that had characterized the approach of both Killing and Cartan to the theory of Lie groups. But of course, as indicated in Section 1, Weyl had to deal with global topological properties of the group $\mathfrak{A}_u$, the real connected Lie group in $\mathbf{GL}(\mathfrak{g})$ with Lie algebra $\mathfrak{a}_u$, and its universal covering group $\widetilde{\mathfrak{A}}_u$, in order to establish the complete reducibility theorem. The key tool in his study of these groups is the real connected abelian subgroup $\mathfrak{T}_u \subset \mathfrak{A}_u$ with Lie algebra $\mathfrak{t}_u$ consisting of the real linear span of $i \cdot \operatorname{ad} H_1, \ldots, i \cdot \operatorname{ad} H_l$. Today $\mathfrak{T}_u$ would be called a maximal toral subgroup of $\mathfrak{A}_u$, and the reason for this nomenclature is already evident in Weyl's treatment. In terms of Weyl's basis $\{H_k, X_\alpha\}$, the elements $T = \operatorname{ad} H \in \mathfrak{t}_u$ are represented by diagonal matrices

$$T = \begin{pmatrix} 0_l & & & \\ & \alpha(H) & & \\ & & \beta(H) & \\ & & & \ddots \end{pmatrix},$$

where 0_l denotes the $l \times l$ zero matrix. Since the s_k are purely imaginary, Weyl wrote them in the form $s_k = 2\pi i \varphi_k$, so that[20]

$$\alpha(H) = 2\pi i \alpha\left(\sum_{k=1}^{l} \varphi_k H_k\right) \stackrel{\text{def}}{=} 2\pi i \varphi_\alpha(H). \tag{3.11}$$

Taking for granted, correctly, the surjective nature of exponentiation from $\mathfrak{t}_u$ to $\mathfrak{T}_u$,[21] he was able to represent the elements $t \in \mathfrak{T}_u$ in the diagonal matrix form

$$t = \exp(T) = \begin{pmatrix} I_l & & & \\ & e(\alpha) & & \\ & & e(\beta) & \\ & & & \ddots \end{pmatrix}, \quad e(\alpha) = e^{2\pi i \varphi_\alpha}, \tag{3.12}$$

where the $e(\alpha)$ notation is Weyl's. The real numbers φ_k occurring in (3.11) and so in the $e(\alpha)$ he referred to as the "rotation angles" of t. In Section 5 we will see that they play the role of Schur's rotation angles in Weyl's generalization of Schur's reduction formula (2.1) for the Hurwitz integral. Of course, if $\varphi_\alpha(H_1) \equiv \varphi_\alpha(H_2) \pmod 1$ for all $\alpha \in \Phi$, then $t_1 = t_2$.

Rotation angles had occurred concretely from the eigenvalues of unitary matrices in Weyl's treatment of the classical groups. In this connection he had also made critical use of a theorem that, articulated for semisimple groups in general, I will refer to as the conjugation theorem. It has become

[20] Weyl denoted φ_α by α, thereby changing the meaning of the latter.

[21] The surjectivity of the exponential map is a property of maximal toral subalgebras.

a basic theorem about maximal toral subgroups in the present-day theory of Lie groups.

(3.13) **Conjugation Theorem.** *Any element* $x \in \mathfrak{A}_u$ *is conjugate to an element in* $\mathfrak{T}_u$, *i.e.*, $x = u^{-1}tu$, *where* $u \in \mathfrak{A}_u$ *and* $t \in \mathfrak{T}_u$.

When the underlying semisimple group $\mathfrak{G}$ is one of the classical groups, Weyl explained [1925:629], "this theorem coincides with known algebraic facts." For example, in the case of the special linear group, so that $\mathfrak{A}_u = \mathbf{SU}(n)$, (3.13) follows readily from the known fact of linear algebra that a unitary matrix x can be "unitarily diagonalized," i.e., $x = u^{-1}du$, where u is unitary and d is diagonal (and unitary).[22] For the other classical groups the diagonalization theorems that were analogues of (3.13) may not have been known, but Weyl showed how they followed by similar algebraic considerations. To establish the general conjugation theorem, however, he had to resort to more than linear algebra: "Here [the theorem] will be established generally by means of the method of continuity" [1925:629]. By the "method of continuity" Weyl meant a "topological method" a phrase he also could have used, as can be seen from the quotations from Weyl given in Section 1. His choice here to speak of the "method of continuity," however, imbued his remark with historical resonance, for, as indicated in Section 9.5, that term had evolved out of the work of Klein and Poincaré on the former's fundamental theorem of automorphic functions, where topological reasoning had proved to be indispensable—and in need of further development. In the course of working on his lectures on Riemann surfaces, Weyl had occasion to absorb the newly developing topology (especially as found in Brouwer's papers), and now, a dozen years later, he saw that Lie's theory, like complex function theory earlier, was in need of topological concepts and reasoning to further its development.

In dealing with the classical groups Weyl had already utilized topological notions and reasoning to establish the connectivity properties of these groups. His proof of the general conjugation theorem (3.13), however, was the first topological proof within the framework of the general theory of semisimple groups. What follows is a sketch of his proof.[23] Let $\mathfrak{C}$ denote the totality of all elements $c \in \mathfrak{A}_u$ that are expressible in the form $c = u^{-1}tu$, where $u \in \mathfrak{A}_u$ and $t \in \mathfrak{T}_u$ is *nonsingular* in the sense that in the representation (3.12) of t, $e(\alpha) \neq 1$ for all $\alpha \in \Phi$.[24] Weyl showed that if $c \in \mathfrak{C}$, then $\mathfrak{C}$ contains an entire neighborhood of c, so that, in modern

[22] [Weyl 1925:562].

[23] [Weyl 1925:630]. Essentially the same proof, but with more details added, was given by Weyl in his IAS lectures [1934–5:108–9], which I have used in my sketch to add clarity to the original proof.

[24] This means, in particular, that $\alpha(H) \neq 0$ for all $\alpha \in \Phi$, so that H is a regular point, i.e., within a Weyl chamber, but despite their name, such chambers play no role in Weyl's paper; they were introduced into the theory by Cartan as indicated in Section 6 below.

terms, $\mathfrak{C}$ is open.[25] Now consider the totality of the boundary elements of $\mathfrak{C}$, which I will denote by $\partial\mathfrak{C}$. Weyl observed that the compactness of $\mathfrak{A}_u$ implies that every element of $\partial\mathfrak{C}$ is also expressible in the form $u^{-1}tu$, albeit with $t \in \mathfrak{T}_u$ necessarily singular in the above sense.[26] The conjugation theorem will then follow if $\mathfrak{C} \cup \partial\mathfrak{C} = \mathfrak{A}_u$. To show that this must be the case, Weyl argued that $\partial\mathfrak{C}$, which is easily seen to consist precisely of all "singular elements" of $\mathfrak{A}_u$, namely all $x \in \mathfrak{A}_u$ of the form $x = u^{-1}tu$ with t singular, has codimension 3, from which he concluded that "$[\mathfrak{A}_u]$ is not decomposed by $[\partial\mathfrak{C}]$," i.e., $\mathfrak{X} = \mathfrak{A}_u \sim \partial\mathfrak{C}$ ($\mathfrak{A}_u$ with $\partial\mathfrak{C}$ removed) is connected. This means that $\mathfrak{C}$ must coincide with $\mathfrak{X}$, since otherwise $\mathfrak{X}$ is divided into disjoint open sets, $\mathfrak{C}$ and $\mathfrak{A}_u \sim (\mathfrak{C} \cup \partial\mathfrak{C})$.

Weyl's first application of the conjugation theorem was to prove what was referred to in Section 1 as the finite connectivity theorem.

Finite Connectivity Theorem. *If $\mathfrak{G}$ is semisimple, then the universal covering group of $\mathfrak{A}_u$ has finitely many sheets.*

By virtue of this theorem, the covering group $\widetilde{\mathfrak{A}}_u$ inherits the compactness of $\mathfrak{A}_u$ (as explained in Section 9.5). With the proof of this theorem, Weyl's proof of the complete reducibility theorem, as outlined in Section 1, is complete.

Here is the gist of Weyl's proof of the finite connectivity theorem. Fix a point $t \in \mathfrak{T}_u$ that is nonsingular—no $e(\alpha) = 1$ in the representation (3.12)—and has the additional property that all the $e(\alpha)$ are distinct. Let γ denote a closed curve that begins and ends at t. Weyl assumed that γ contains no points of $\partial\mathfrak{C}$, presumably because being of codimension 3, $\partial\mathfrak{C}$ can be avoided by a closed curve homotopic to γ.[27] This means that if $\gamma(\tau)$, $\tau \in [0, 1]$, denotes the points of γ, then by the conjugation theorem,

$$\gamma(\tau) = u(\tau)^{-1}t(\tau)u(\tau) \tag{3.14}$$

and $t(\tau)$ is nonsingular. It is assumed that $u(\tau)$ and $t(\tau)$ vary continuously with τ and that $u(0) = e$, the identity element of $\mathfrak{A}_u$. In effect, Weyl thought of $t(\tau)$ as

(3.15)
$$t(\tau) = \exp \operatorname{ad} H(\tau) = \begin{pmatrix} I_l & & & \\ & e(\alpha) & & \\ & & e(\beta) & \\ & & & \ddots \end{pmatrix}, \quad e(\alpha) = e^{2\pi i \varphi_\alpha(\tau)},$$

with $\tau \to H(\tau)$ presumed continuous.

---

[25] Weyl refers to $\mathfrak{C}$ as a domain (*Gebiet*).

[26] If $c \in \partial\mathfrak{C}$, then $c = \lim_n c_n$ with $c_n \in \mathfrak{C}$ and is therefore expressible as $c_n = u_n^{-1}t_n u$. By compactness of $\mathfrak{A}_u$ a subsequence n_k exists such that $u_{n_k} \to u \in \mathfrak{A}_u$ and $t_{n_k} \to t \in \mathfrak{T}_u$. Thus $c = u^{-1}tu$.

[27] Cf. [Borel 1986:75].

Given this setup, the idea is to consider the behavior of the eigenvalues $e^{2\pi i\varphi_\alpha(\tau)}$ of $t(\tau)$ as $\gamma(\tau)$ moves around γ. Since by (3.14) $t = \gamma(0) = \gamma(1) = u(1)^{-1}t(1)u(1)$, $t(1)$ is similar to t and so has the same eigenvalues. In other words, if r is the dimension of $\mathfrak{g}$, then, in the notation of (3.15), the $r - l$ distinct numbers $e^{2\pi i\varphi_\alpha(1)}, e^{2\pi i\varphi_\beta(1)}, \ldots$ are some *permutation* of $e^{2\pi i\varphi_\alpha(0)}, e^{2\pi i\varphi_\beta(0)}, \ldots$.

Consider first the case in which it is the identical permutation. Then $e^{2\pi i\varphi_\alpha(1)} = e^{2\pi i\varphi_\alpha(0)}$ for all α, and since $\varphi_\alpha \not\equiv 0 \pmod 1$, $\varphi_\alpha(0) = \varphi_\alpha(1)$ for all α. In this case, Weyl showed that γ is homotopic to a point. For each $\sigma \in [0,1]$ he defined the curve γ_σ by $\gamma_\sigma(\tau) = u(\tau)^{-1}t_\sigma(\tau)u(\tau)$, where $u(\tau)$ is as in (3.14) and

$$t_\sigma(\tau) = \exp[\sigma \operatorname{ad} H(\tau)] = \begin{pmatrix} I_l & & & \\ & e^{2\pi i\sigma\varphi_\alpha(\tau)} & & \\ & & e^{2\pi i\sigma\varphi_\beta(\tau)} & \\ & & & \ddots \end{pmatrix}.$$

Using the fact that the eigenvalues $e(\alpha)$ of t are all distinct, he showed that each γ_σ is a closed curve. Furthermore, for $\sigma = 1$ it is the curve γ, whereas for $\sigma = 0$ it is the point e, the identity element of $\mathfrak{A}_u$. Thus γ is homotopic to a point.

Having dealt with this special case, Weyl concluded his proof with the observation that in the general situation the $r - l$ eigenvalues of $t(1)$, i.e., $e^{2\pi i\varphi_\alpha(1)}, e^{2\pi i\varphi_\beta(1)}, \ldots$, are some permutation of $e^{2\pi i\varphi_\alpha(0)}, e^{2\pi i\varphi_\beta(0)}, \ldots$, so that the universal covering group "is at most $(r - l)!$-sheeted" [1925:632]. That is, if γ_1 is another closed curve that begins and ends at t and if, in the above sense, the $r - l$ eigenvalues of $t(\tau)$ undergo the *same* permutation in going from $\tau = 0$ to $\tau = 1$ along γ_1, then $\gamma - \gamma_1$, i.e., the curve that first follows γ from t to t and then follows γ_1 in reverse from t to t, is a closed curve which begins and ends at t and the eigenvalues at $\tau = 1$ are identical to those at $\tau = 0$. Therefore, by Weyl's above argument, $\gamma - \gamma_1$ is homotopic to a point, which means that γ_1 is homotopic to γ. In other words, for γ_1 to be nonhomotopic to γ, the $r - l$ eigenvalues along γ_1 must undergo a permutation different from the one along γ. Thus there are at most as many homotopy classes as there are permutations of $r - l$ objects, which means $\widetilde{\mathfrak{A}}_u$ has at most $(r - l)!$ sheets.[28]

As we have seen, Weyl's extensions of the Killing–Cartan theory of structure were motivated by the goal of proving the complete reducibility theorem for all semisimple groups. However, he quickly realized that the new ideas and results he had obtained could be applied to other matters as well. Two such applications from his paper [1925] are considered in the following sections.

---

[28] This explanation of what Weyl meant is confirmed by his lectures [1934–5:110–11], where essentially the same proof is given but with somewhat more detail.

12.4. Weyl's Finite Basis Theorem

Hurwitz's proof of the finite basis theorem for the classical invariants of $\mathfrak{O} = \mathbf{SO}\,(n,\mathbb{R})$ and $\mathbf{SL}(n,\mathbb{C})$ had provided much of the inspiration behind Weyl's successful proof of his complete reducibility theorem. At the end of [1925], Weyl pointed out that by generalizing Hurwitz's ideas he had also obtained, as an easy consequence, a generalized finite basis theorem. That is, just as Hurwitz had proved that the homogeneous polynomial invariants of the power representations $g \to P_d(g)$ of $\mathbf{SL}(n,\mathbb{C})$ have a finite basis, so now Weyl was in a position to show that the invariants of any representation $g \to R(g)$ of a semisimple group $\mathfrak{G}$ have a finite basis. Let me explain.[29]

First of all, since every representation of $\mathfrak{G}$ is also a representation of its universal covering group $\tilde{\mathfrak{G}}$ (but not conversely), we may assume without loss of generality that we are concerned with a representation of $\tilde{\mathfrak{G}}$. Now, since the Lie algebra $\mathfrak{g}$ of $\mathfrak{G}$ is isomorphic to $\mathfrak{a} = \operatorname{ad}\mathfrak{g}$, the covering groups of $\mathfrak{G}$ and $\mathfrak{A} = \operatorname{Ad}\mathfrak{G}$ are isomorphic, and so we may assume that we are concerned with a representation of $\tilde{\mathfrak{A}}$. Let $\mathfrak{J}$ denote the set of all invariants with respect to some representation $R : \tilde{\mathfrak{A}} \to \mathbf{GL}(N,\mathbb{C})$. Hence a homogeneous polynomial $F(a) = F(a_1,\ldots,a_N)$ is in $\mathfrak{J}$ if and only if $F(R(g)a) = F(a)$ identically in a for all $g \in \tilde{\mathfrak{A}}$.

As Weyl realized, it is convenient to characterize the invariance property on the infinitesimal, or Lie algebra, level. Thus for g sufficiently close to the identity element we have $g = \exp tX$, $X \in \mathfrak{a}$, and so for $F \in \mathfrak{J}$ the function $\varphi(t) = F(R(\exp tX \cdot a))$ is equal to $F(a)$ for all sufficiently small t. Differentiating with respect to t and setting $t = 0$ one sees that F belongs to $\mathfrak{J}$ if and only if the identity in a

$$dF(a) \cdot r(X)a = 0 \tag{4.1}$$

holds for all $X \in \mathfrak{a}$, where $r = dR$ is the representation of $\mathfrak{a}$ corresponding to the given representation R.

Now consider the unitary restriction $\mathfrak{a}_u \subset \mathfrak{a}$. Evidently, $r(X)$ restricted to $X \in \mathfrak{a}_u$ is a representation of $\mathfrak{a}_u$ and so corresponds to a representation of $\tilde{\mathfrak{A}}_u$. Since $\tilde{\mathfrak{A}}_u$ is compact, Hurwitz's integral method goes through and shows that the totality $\mathfrak{J}_u$ of invariants of this representation has a finite basis. Now, $F \in \mathfrak{J}_u$ precisely when the identity (4.1) holds for all $X \in \mathfrak{a}_u$. In other words, as in Hurwitz's original proof, $\mathfrak{J} \subset \mathfrak{J}_u$, and it remains to show that $\mathfrak{J} = \mathfrak{J}_u$. Hurwitz had established the equality in his proof by utilizing his version of the unitarian trick (Section 10.2), and the ideas behind Weyl's infinitesimal version of the trick can now be used to the same end. Recall from Section 3 that the existence of a real compact form of a semisimple Lie algebra followed from the existence of a special basis (3.9), which, applied to the semisimple algebra $\mathfrak{a}$ (rather than to $\mathfrak{g}$), shows that

[29] Weyl's remarks are extremely brief; what follows is my exposition of those remarks.

$X = \sum_k s_k H_k + \sum_\alpha t_\alpha X_\alpha \in \mathfrak{a}$ belongs to $\mathfrak{a}_u$ if and only if the coefficients s_k, t_α satisfy the conditions (3.10). By setting $X = \sum_k s_k H_k + \sum_\alpha t_\alpha X_\alpha$ in (4.1) the left-hand side of (4.1) becomes a linear form in the variables s_k, t_α. Thus for $F \in \mathfrak{J}_u$ this form vanishes when these variables satisfy the conditions (3.10). As in Weyl's proof of the real compact form theorem, it then follows readily that this form must vanish for all values of the variables, which means that (4.1) holds for all $X \in \mathfrak{a}$, and so $F \in \mathfrak{J}$.

In this way Weyl had in effect discovered that Hurwitz's integral method is viable in a much more general context—the context of semisimple groups. And so he wrote enthusiastically, "Now we recognize the import of this method: It *leads to the goal for all semisimple groups*, not just for the projective and orthogonal groups to which Hurwitz had applied it. It exhibits therewith its great superiority to the purely algebraic methods, which draw upon differentiation processes modeled after Cayley's Ω process. *For the first time a group-theoretic domain of validity is thereby marked off in a natural way.*"[30]

Weyl's result was the first progress in the area of finiteness theorems since Hurwitz's paper of 1897. Around 1900 it had seemed that Ludwig Maurer had extended the finiteness theorem to the invariants of any continuous subgroup of $\mathbf{GL}(n, \mathbb{C})$, "continuous" being used here in the sense of Lie, and this together with Hurwitz's result motivated the fourteenth problem of Hilbert's Paris lecture. But Maurer was never able to complete the proof he had begun publishing, with Hilbert's encouragement, in [1903].[31] Weyl's finiteness theorem remained for many years the most general. It was not until the development of the theory of linear algebraic groups in the 1950s that more general results were forthcoming such as M. Nagata's finiteness theorem for reductive linear algebraic groups (1964).

In this connection, a word is in order about Weyl's original proof, which represents a generalization of Hurwitz's. Nowadays, thanks to the theory of linear algebraic groups, we are accustomed to seeing finiteness theorems for invariants follow from the complete reducibility property for polynomial representations implicit in the concept of reductivity.[32] In 1925, however, it was the other way around. That is, as we have seen, it was the perceived viability of Hurwitz's ideas on the finiteness theorem for invariants that led Weyl to develop them to establish his complete reducibility theorem. In 1925 he did not realize that complete reducibility, once established, could lead more directly to finiteness theorems. This insight was later provided to him through an unpublished manuscript by Menahem Schiffer, which contains the theorem that if a group $\mathfrak{G}$ of linear transformations T is such

30 [Weyl 1925:642]. Later he became more interested in purely algebraic methods [Borel 1986:64ff].

31 However, as Borel has shown in [2000], Maurer's papers contained many of the basic notions and propositions of the subsequent theory of linear algebraic groups.

32 See, e.g., [Springer 1977:21–25].

that the representations $\pi_d : T \to P_d(T)$, where $P_d(T)$ denotes Hurwitz's power transformation (Section 10.1), are completely reducible for all d, then the finiteness theorem holds for the corresponding invariants. Weyl expounded Schiffer's theorem in his treatise on the classical groups and their invariants [1946:300–303], deeming it "of considerable interest" to see that the finiteness theorem holds "wherever full reducibility prevails." In calling attention to Schiffer's discovery Weyl set the stage for the finiteness theorems of the theory of linear algebraic groups.[33]

As another, far less straightforward, application of his extension of the Killing–Cartan structure theory, Weyl applied it to solve what Schur thought of as the heart of the representation problem for a group (Section 2), namely to determine formulas for the characters of the irreducible representations and to obtain therefrom formulas for the degrees of the irreducible representations. How he did this is the subject of the next section.

12.5. Weyl's Theory of Characters

The beauty of Weyl's theory of characters is due to the way it brings together the Frobenius–Schur approach to the study of representations via characters with Cartan's approach via his theory of weights (Section 8.4). I will begin by summarizing the main points of Cartan's theory in the notation of the previous section. If $r : \mathfrak{g} \to \mathbf{gl}(\mathcal{V})$ is an irreducible representation of the semisimple Lie algebra $\mathfrak{g}$, then $\mathcal{V}$ decomposes into weight spaces, $\mathcal{V} = \bigoplus_{\mu \in \Pi(\lambda)} \mathcal{V}_\mu$, where $\mu \in \mathfrak{h}^*$ and

$$r(H)v = \mu(H)v \text{ for all } v \in \mathcal{V}_\mu \text{ and all } H \in \mathfrak{h}. \tag{5.1}$$

The set of weights $\Pi(\lambda)$ form what is now called a saturated set with λ as highest weight. For both Cartan and Weyl, the ordering of the weights is lexicographic. That is, if $H = \sum_{k=1}^{l} \lambda_k H_k$, then $\mu(H) = \sum_{k=1}^{l} m_k \lambda_k$, where the $m_k = \mu(H_k)$ are integers. Then $\mu > 0$ if the first nonzero coefficient m_k is positive and $\mu > \nu$ if $\mu - \nu > 0$. In this sense, $\lambda \geq \mu$ for all $\mu \in \Pi(\lambda)$. Cartan realized that $\Pi(\lambda)$ is closed under the "conjugation" $\mu \to S_\alpha(\mu)$ and that all conjugates have the same multiplicity. Stated in terms of the Weyl group $\mathfrak{W}$, Cartan's results imply that each $\sigma \in \mathfrak{W}$ takes $\Pi(\lambda)$ onto itself and that μ and $\sigma\mu$ have the same multiplicity for all $\sigma \in \mathfrak{W}$. Weyl also introduced the totality of integral forms $\mu = \sum_{k=1}^{l} m_k \lambda_k$, $m_k \in \mathbb{Z}$, which will be denoted here by Λ, and realized that $\mu \in \Lambda$ precisely when $2(\mu, \alpha)/(\alpha, \alpha) \in \mathbb{Z}$ for all $\alpha \in \Phi$. Today the elements of Λ would

33 See the historical comments by Springer [1977:36–38], who concludes with what seems to be an updated version of the quotation given above from Weyl's paper [1925]: "The result of the developments sketched is that reductive groups give the natural range of validity of the finiteness results of the theory of invariants." In this connection it should be noted that in 1959 Nagata had given an example of a linear group for which the finiteness theorem fails.

be called weights, since every $\mu \in \Lambda$ occurs as a weight in an irreducible representation of $\mathfrak{g}$. Cartan's main theorem was that every $\lambda \in \Lambda$ that is a nonnegative integral linear combination of certain fundamental weights $\pi_1, \ldots, \pi_l$ is the highest weight of an irreducible representation of $\mathfrak{g}$. These λ are now called dominant weights because $\lambda \geq \sigma(\lambda)$ for all $\sigma \in \mathfrak{W}$.

Because a representation r of $\mathfrak{g}$ corresponds to a representation R of the universal covering group $\widetilde{\mathfrak{G}}$ (with $r = dR$ in modern notation) the character associated to the irreducible representation r_λ of highest weight λ is $\chi_\lambda(g) =$ tr R_λ, $g \in \widetilde{\mathfrak{G}}$. The key to the Frobenius–Schur theory of characters is the fact that the irreducible characters satisfy orthogonality relations. We saw in Section 10.5 that Schur had extended this property of Frobenius's characters to those of the rotation group by using the invariant integral of Hurwitz in such a way that the same could be done for any compact Lie group, as he himself realized. Since complex semisimple groups $\mathfrak{G}$ are not generally compact, however, Weyl naturally turned to the compact group $\widetilde{\mathfrak{A}}_u$, the universal covering group of the compact connected Lie group $\mathfrak{A}_u$ with Lie algebra $\mathfrak{a}_u$ as defined in (3.9)–(3.10). That is, as in Section 4, since $\mathfrak{g}$ is isomorphic to $\mathfrak{a} = \operatorname{ad} \mathfrak{g}$, one can just as well consider the representations of $\mathfrak{a}$ and the associated characters on $\widetilde{\mathfrak{A}}$. Now, by the logic behind Weyl's unitarian trick—Theorems 1.1 and 1.5 from Section 1—it can be seen that the irreducible representations of $\mathfrak{a}_u$ are precisely the restrictions of the irreducible representations of $\mathfrak{a}$. This means that the restrictions of the irreducible characters χ_λ of $\widetilde{\mathfrak{A}}$ to $\widetilde{\mathfrak{A}}_u$ are the irreducible characters of this real compact group. As a consequence one has the orthogonality relations

$$(5.2) \qquad \int_{\widetilde{\mathfrak{A}}_u} \chi_\lambda(x)\overline{\chi_{\lambda'}(x)}\, dm(x) = \delta_{\lambda,\lambda'}\Omega,$$

where $\Omega = \int_{\widetilde{\mathfrak{A}}_u} dm(x)$ represents the volume of $\widetilde{\mathfrak{A}}_u$ as measured by the Hurwitz integral.

In discussing how Weyl utilized (5.2), I will use boldface to simplify the notation by setting $\mathbf{\mathfrak{G}} = \widetilde{\mathfrak{A}}_u$ and letting $\mathbf{\mathfrak{T}}$ denote the connected subgroup of $\mathbf{\mathfrak{G}}$ with Lie algebra $\mathfrak{t}_u$ consisting of all $T = \operatorname{ad} H$ such that $H \in \mathfrak{a}_u$. Hence $\mathbf{\mathfrak{T}}$ is another example of what is now called a maximal toral subgroup of $\mathbf{\mathfrak{G}}$. Weyl realized that his conjugation theorem (3.13) extends to $\mathbf{\mathfrak{G}}$, so that every element in $\mathbf{\mathfrak{G}}$ is conjugate to an element in $\mathbf{\mathfrak{T}}$.[34] Since characters take the same value on conjugate elements, it was necessary to evaluate χ_λ only on $\mathbf{\mathfrak{T}}$. If $r_\lambda = dR_\lambda$ is the representation of $\mathfrak{a}$ corresponding to R_λ, then since every $t \in \mathbf{\mathfrak{T}}$ is expressible as $t = \exp T$ with $T \in \mathfrak{t}_u$,[35] $\chi_\lambda(t) =$ tr $R_\lambda(t) =$ tr $e^{r_\lambda(T)}$. In view of (5.1), $\exp[r_\lambda(T)]v = e^{\mu(T)}v$ for $v \in \mathcal{V}_\mu$, and so $R_\lambda(t)$ is a diagonal matrix with trace

[34] Weyl justified this extension of the conjugation theorem with the remark that $\widetilde{\mathfrak{A}}_u$, like $\mathfrak{A}_u$, is compact. In his lectures [1934–5:111–2] he gave a proof.

[35] This property of $\mathbf{\mathfrak{T}}$, which holds generally for maximal toral subgroups, was taken for granted by Weyl.

(5.3) $$\chi_\lambda(t) = \sum_{\mu\in\Pi(\lambda)} m(\mu)e(\mu),$$

where $m(\mu)$ denotes the multiplicity of μ and $e(\mu) = e^{\mu(T)}$. It should be noted that (5.3) is well-defined for all $t \in \mathfrak{T}$. That is, even though $e(\mu)$ depends upon T and there are many $T \in \mathfrak{t}_u$ such that $t = \exp T$, for all such T's the value of $e(\mu)$ is the same for *any* $\mu \in \Lambda$.[36]

Since in the sort of notation employed in (3.11), $e(\mu) = e^{2\pi i\varphi_\mu(T)}$, where $\varphi_\mu(T) = \mu(\sum_k \varphi_k T_k)$, $T_k = \operatorname{ad} H_k$, Weyl described the expression in (5.3) as a *finite Fourier series*. More generally, any function on $\mathfrak{T}$ of the form $\xi(t) = \sum_{\mu\in\Lambda} c_\mu e(\mu)$ such that only a finite number of the c_μ are nonzero is called a finite Fourier series. (Note that ξ is well-defined on $\mathfrak{T}$ for the same reason (5.3) makes sense.) All the finite Fourier series considered by Weyl have integer coefficients and so today can be identified with elements in the character ring of the group. Weyl took advantage of the fact that since $e(\mu)e(\nu) = e(\mu+\nu)$, finite Fourier series may be multiplied (as well as added) to yield new ones. He also made use of the natural action of $\mathfrak{W}$ on finite Fourier series defined by

$$\sigma : \sum_{\mu\in\Lambda} c_\mu e(\mu) \to \sum_{\mu\in\Lambda} c_\mu e(\sigma\mu).$$

Since $m(\sigma\mu) = m(\mu)$, it follows that $\sigma\chi_\lambda = \chi_\lambda$ for all $\sigma \in \mathfrak{W}$, or as Weyl put it, χ_λ is a *symmetric* finite Fourier series. Likewise, a finite Fourier series ξ is said to be *alternating* if $\sigma_\alpha\xi = -\xi$ for all reflections σ_α. Hence alternating finite Fourier series ξ have the property that $\sigma\xi = (\operatorname{sn}\sigma)\xi$ for any $\sigma \in \mathfrak{W}$, where $\operatorname{sn}\sigma$ is $+1$ or -1 depending on whether σ is a product of an even or an odd number of reflections, respectively.

In order to study the nature of the characters as finite Fourier series, one obviously needs to utilize their fundamental property, namely that they satisfy orthogonality relations. These relations as stated in (5.2), however, are not directly applicable to χ_λ as a finite Fourier series, because such series are defined only on $\mathfrak{T}$. Weyl saw that what was needed was an appropriate generalization of Schur's key equations (2.1)–(2.2). That is, analogous to what was done regarding the adjoint group $\mathfrak{A}_u$—see (3.11)–(3.12)—to each $t \in \mathfrak{T}$ may be associated the rotation angles $\varphi_1, \ldots, \varphi_l$, where $t = \exp \operatorname{ad} H$ and $H = 2\pi i\big(\sum_{k=1}^{l} \varphi_k H_k\big) \in \mathfrak{a}_u$. It will be helpful to indicate the dependence on the φ_k by writing $t = t(\varphi_1, \ldots, \varphi_l)$. As in the case of the adjoint group, t has more than one set of rotation angles. Here

[36] This is because every $\mu \in \Lambda$ occurs as a weight in some irreducible representation r_λ of $\mathfrak{a}$. Thus if $t = \exp T_1 = \exp T_2$, it follows from (5.1) that $R_\lambda(t)v = e^{\mu(T_1)}v = e^{\mu(T_2)}v$, so $e^{\mu(T_1)} = e^{\mu(T_2)}$. Weyl probably realized this in 1925; it is explicit in his lectures [1934–5:120].

$t(\varphi_1, \ldots, \varphi_l) = t(\psi_1, \ldots, \psi_l)$ precisely when $\varphi_k \equiv \psi_k (\text{mod } 1)$ for all k.[37] Thus if we restrict the rotation angles φ_k by requiring that $\varphi_k \in [0, 1)$ for all k, we may think of the characters of $\mathfrak{G}$ restricted to $\mathfrak{T}$ as functions of the rotation angles: $\chi(t) = \chi(\varphi_1, \ldots, \varphi_l)$. This is the way that Weyl thought of them in the following presentation of his theory of characters.

To reduce integrals over $\mathfrak{G}$ to multiple integrals in $\varphi_1, \ldots, \varphi_l$ as Schur did in (2.1)–(2.2), Weyl invoked the conjugation theorem for $\mathfrak{G}$ to express any $x \in \mathfrak{G}$ in the form $x = u^{-1} t(\varphi_1, \ldots, \varphi_l) u$, with $t(\varphi_1, \ldots, \varphi_l) \in \mathfrak{T}$. Then $\varphi_1, \ldots, \varphi_l$ may be thought of as a set of rotation angles for x. For $\varphi_1, \ldots, \varphi_l$ fixed with $\varphi_k \in [0, 1)$, Weyl considered the infinitesimal portion $\delta\mathfrak{G}$ of $\mathfrak{G}$ consisting of all elements x that possess a set of rotation angles ψ_k lying between φ_k and $\varphi_k + d\varphi_k$. Let $d\Omega$ denote the volume of $\delta\mathfrak{G}$. By means of a nontrivial line of geometrical reasoning that he had already applied in the more concrete situations of the classical groups, he obtained the remarkable formula

$$d\Omega = \prod_{\alpha \in \Phi} [e(\alpha) - 1]\, d\varphi_1 \cdots d\varphi_l,$$

where, since $e(\alpha) = e^{2\pi i \alpha(\varphi_1 H_1 + \cdots + \varphi_l H_l)}$, $\prod_{\alpha \in \Phi} [e(\alpha) - 1]$ is actually a function $M(\varphi_1, \ldots, \varphi_l)$. Schur had singled out for special praise this "exceedingly beautiful" (*wunderschön*) result.[38] As Weyl had noted in deriving an analogous expression $d\Omega = M(\varphi_1, \ldots, \varphi_l)\, d\varphi_1 \cdots d\varphi_l$ for the special linear group [1925:564], the function M gives the density with which the elements $x \in \mathfrak{G}$ are distributed into conjugate classes. That is, each l-tuple $(\psi_1, \ldots, \psi_l)$ with ψ_k between φ_k and $\varphi_k + d\varphi_k$ determines the class $\mathfrak{C}(\psi_1, \ldots, \psi_l)$ of all elements $x \in \mathfrak{G}$ conjugate to $t(\psi_1, \ldots, \psi_l)$. Since all such x evidently belong to $\delta\mathfrak{G}$, $\delta\mathfrak{G}$ consists precisely of all the conjugate classes $\mathfrak{C}(\psi_1, \ldots, \psi_l)$ with ψ_k between φ_k and $\varphi_k + d\varphi_k$.

Weyl realized that his formula for $d\Omega$ could be factored as $d\Omega = \Delta\overline{\Delta}\, d\varphi_1 \cdots d\varphi_l$, where

$$\Delta = \prod_{\alpha > 0} \left[e\left(\frac{\alpha}{2}\right) - e\left(\frac{-\alpha}{2}\right) \right]. \tag{5.4}$$

Thus as a generalization of Schur's (2.1)–(2.2) he obtained for $\gamma(x) = f(\varphi_1, \ldots, \varphi_l)$

$$\int_{\mathfrak{G}} \gamma(x) dm(x) = \int_0^1 \cdots \int_0^1 f(\varphi_1, \ldots, \varphi_l) \Delta\overline{\Delta}\, d\varphi_1 \cdots d\varphi_l. \tag{5.5}$$

[37] This fact seems tacit in Weyl's terse treatment, although he asserts only that $t(\varphi_1, \ldots, \varphi_l) = t(\psi_1, \ldots, \psi_l)$ when $\varphi_k \equiv \psi_k (\text{mod } 1)$ [1925:641]; it is stated and proved in his lectures [1934–5:113], along with the fact that $t(\varphi_1, \ldots, \varphi_l)$ and $t(\psi_1, \ldots, \psi_l)$ correspond to the same element of the adjoint group precisely when $\alpha(\sum_{k=1}^{l} \varphi_k H_k) \equiv \alpha(\sum_{k=1}^{l} \psi_k H_k)$ (mod 1) for all roots $\alpha \in \Phi$ (p. 111).

[38] Letter to Weyl dated 26 November 1924 (ETH Zürich, Hs.91:735).

The expression Δ given in (5.4) is sometimes now called Weyl's q-function or denominator function (for reasons to be seen below).

The importance of (5.5) to Weyl was that it enabled him to write the orthogonality relations (5.2) in the form

$$\int_0^1 \cdots \int_0^1 \chi_\lambda \Delta \cdot \overline{\chi_{\lambda'} \Delta}\, d\varphi_1 \cdots d\varphi_l = \delta_{\lambda\lambda'} \Omega. \tag{5.6}$$

Judging by his remarks [1925:634], (5.6) suggested consideration of the expressions $\xi_\lambda = \chi_\lambda \Delta$. Since χ_λ is a finite Fourier series, ξ_λ will be one as well, provided that Δ is. Using the relation $e(\mu)e(\nu) = e(\mu+\nu)$ it is easily seen from the definition of Δ that

$$\Delta = \left[\prod_{\alpha>0} e\left(\frac{\alpha}{2}\right)\right]\left[\prod_{\alpha>0} [e(0) - e(-\alpha)]\right] = e(\delta)\left[\prod_{\alpha>0} [e(0) - e(-\alpha)]\right],$$

where $\delta = \frac{1}{2}\sum_{\alpha>0} \alpha$. Since the second factor is a product of finite Fourier series and hence a finite Fourier series, Δ will be a finite Fourier series, provided that $e(\delta)$ is, i.e., provided that $\delta \in \Lambda$. Weyl showed this to be the case. Since, as he also showed, $\sigma_\alpha \Delta = -\Delta$, we see that both Δ and $\xi_\lambda = \chi_\lambda \Delta$ are alternating finite Fourier series.

The ξ_λ are thus (i) alternating finite Fourier series that are (ii) divisible by Δ (in the sense that ξ can be factored as a product of finite Fourier series with Δ as a factor) and (iii) satisfy the orthogonality relations

$$\int_0^1 \cdots \int_0^1 \xi_\lambda \cdot \overline{\xi_{\lambda'}}\, d\varphi_1 \cdots d\varphi_l = \delta_{\lambda\lambda'} \Omega. \tag{5.7}$$

Weyl's strategy was to define "from scratch," for each dominant weight $\lambda \in \Lambda$, a finite Fourier series ξ_λ with the properties (i)–(iii). Then the finite Fourier series $\psi_\lambda = \xi_\lambda / \Delta$ will be symmetric and will satisfy the same orthogonality relations (5.6) as the characters χ_λ, and it can be argued that as a consequence one must have $\psi_\lambda = \chi_\lambda$. In this manner a formula for χ_λ restricted to $\mathfrak{T}$ was obtained.

To construct the ξ_λ, Weyl introduced special finite Fourier series, called "elementary sums," which correspond to dominant weights $\mu \in \Lambda$:

$$\omega_\mu = \sum_{\sigma \in \mathfrak{W}} (\operatorname{sn} \sigma) e(\sigma\mu).$$

These elementary sums are easily seen to be alternating. Weyl realized that any alternating finite Fourier series can be expressed as a linear combination of elementary sums, and he showed that in particular,

$$\Delta = \omega_\delta.$$

For any dominant weight λ, ξ_λ is then defined by

$$\xi_\lambda = \omega_{\lambda+\delta}$$

and (5.7) is deduced, but with Ω equal to the order of $\mathfrak{W}$. Taking $\xi_\lambda = \xi_{\lambda'} = \xi_0 = \omega_\delta$ in (5.7) shows that the volume Ω of $\mathfrak{G}$ is equal to the order of $\mathfrak{W}$. Next, Weyl showed that ξ is divisible by Δ, i.e., there is a finite Fourier series ψ_λ such that $\xi_\lambda = \psi_\lambda \Delta$. Since both ξ_λ and Δ are alternating, the ψ_λ are symmetric finite Fourier series, which by virtue of (5.7) satisfy the same orthogonality relations (5.6) as the irreducible characters χ_λ, whence the identity of χ_λ and ψ_λ is deduced. In this way Weyl obtained his beautiful character formula

$$\chi_\lambda = \frac{\omega_{\lambda+\delta}}{\omega_\delta} = \frac{\sum_{\sigma\in\mathfrak{W}} \operatorname{sn}\sigma \cdot e(\sigma(\lambda+\delta))}{\sum_{\sigma\in\mathfrak{W}} \operatorname{sn}\sigma \cdot e(\sigma\delta)}, \tag{5.8}$$

thereby relating Cartan's weights with the characters of Frobenius and Schur.

If N_λ denotes the dimension of the irreducible representation corresponding to χ_λ, then $N_\lambda = \chi_\lambda(e)$. Since the identity element e corresponds to setting all rotation angles $\varphi_k = 0$, to use (5.8) to determine N_λ, one must set all $\varphi_k = 0$ in this formula. Since for any $\mu \in \Lambda$,

$$e(\mu) = \exp\Big[2\pi i \mu(\sum_k \varphi_k T_k)\Big] = \exp\Big[2\pi i \sum_k \mu(T_k)\varphi_k\Big],\ T_k = \operatorname{ad} H_k, \tag{5.9}$$

every expression $e(\mu)$ equals 1 when all $\varphi_k = 0$, so that (5.8) takes on the indeterminate form $\frac{0}{0}$. Weyl circumvented this difficulty in the following manner. From (5.9) we see that $m = (m_1, \ldots, m_l)$, with $m_k = \mu(T_k)$, may be thought of as coordinates of μ with respect to the basis dual to the T_k. If $n = (n_1, \ldots, n_l)$ gives the coordinates of ν, then the Killing form is expressible as $(\mu, \nu) = mAn^T$, with A symmetric. A simple calculation shows that if $m'_1, \ldots, m'_l$ and $n'_1, \ldots, n'_l$ are defined by $m' = Am$ and $n' = An$, respectively, then

$$e(\lambda+\delta)(n'_1\varphi, \ldots, n'_l\varphi) = \exp[2\pi i(\lambda+\delta, \delta)] = e(\delta)(m'_1\varphi, \ldots, m'_l\varphi).$$

Since the $\sigma \in \mathfrak{W}$ are orthogonal transformations with respect to the Killing form, $(\sigma(\lambda+\delta), \delta) = (\lambda+\delta, \sigma^{-1}\delta)$, and the above equality implies that

$$\omega_{\lambda+\delta}(n'_1\varphi, \ldots, n'_l\varphi) = \omega_\delta(m'_1\varphi, \ldots, m'_l\varphi) = \Delta(m'_1\varphi, \ldots, m'_l\varphi).$$

Direct calculation of Δ at the given values using its definition (5.4) then shows that

$$\omega_{\lambda+\delta}(n'_1\varphi, \ldots, n'_l\varphi) = (2\pi i\varphi)^{\frac{r-l}{2}} \prod_{\alpha>0} (\lambda+\delta, \alpha) \ +\ o(\varphi^{\frac{r-l}{2}}), \tag{5.10}$$

where $r = \dim \mathfrak{g}$. Since (5.10) with $\lambda = 0$ gives a similar expression for ω_δ, Weyl could use these expressions in his character formula (5.8) with φ infinitely small (as he put it) to obtain his remarkable dimension formula

$$N_\lambda = \frac{\prod_{\alpha>0}(\lambda+\delta,\alpha)}{\prod_{\alpha>0}(\delta,\alpha)}.$$

12.6. Cartan's Response

In Section 1 we saw that although Cartan had already observed in 1914 what amounts to Weyl's real compact form theorem (1.3), Weyl was unaware of this fact when at the end of February 1925 he submitted the installment of [1925] containing it. As for Cartan, once he saw Weyl's 1924 announcements of the complete reducibility theorem—but before he had any idea of how Weyl proved it for semisimple algebras in general—he realized that the existence of a real form with associated quadratic form ψ_2 negative definite is the key to complete reducibility and devised a proof based on this fact. He sketched his proof in a letter to Weyl in March 1925, where he correctly conjectured that Weyl's general proof must also depend on the existence of such a real form. His proof was published in the May issue of Darboux's *Bulletin* and has been described by Borel [1986:76 (n.16)]. It drew on many ideas that were already in Weyl's 1924 announcements such as the idea of utilizing the Hurwitz integral in conjunction with the unitarian trick, but Cartan avoided Weyl's idea of introducing the universal covering group and proving its compactness. The advantage of his approach, Cartan explained in the above-mentioned letter, was that complete reducibility could be proved "without being obligated to devote oneself to studies of analysis situs, which are always delicate."[39]

In his reply (dated 22 March 1925) Weyl defended his use of analysis situs:[40]

> In the general theory I proceed as you from the adjoint group, respectively, its above-mentioned "unitary restriction." I am not

---

[39] "sans être obligé de se livrer à des études d'analysis situs toujours délicates" (ETH Zürich, Hs 91: 501). As Borel has pointed out, however, Cartan's proof takes for granted results that seem to require the sort of topological reasoning he wished to avoid.

[40] "Bei der allgemeinen Theorie gehe ich ebenso wie Sie aus von der adjungierten Gruppe, bezw. ihrer eben erwähnten "unitären Beschränkung". Ich bin in der Lie'schen Theorie nicht so beschlagen, dass ich es wagen konnte, aus ihr mehr zu entnehmen, als dass, sich aus der infinitesimalen Gruppe durch Integration die Umgebung des Einheitselements der kontinuierlichen Gruppe konstruieren lasse; die ganze Gruppe gewinne ich daraus erst durch einen Prozess der "Fortsetzung" und orientiere mich durch eine topologische Betrachtung über ihre Zusammenhangsverhältnisse. Diese Analysis-situs-Betrachtung ist übrigens sehr einfach und betrifft ohne Fallunterscheidungen die sämtlichen halb-einfachen Gruppen. Diese Vorgehen liegt meiner ganzen Denkweise näher als Ihre mehr algebraische Methode, die ich vorläufig nur halb und halb verstehe." (Original in possession of H. Cartan.)

so well-versed in Lie theory as to dare to take more from it than the [fact] that from the infinitesimal group, by integration, the neighborhood of the identity element of the continuous group can be constructed; the entire group I obtain first by a process of "continuation" and I orient myself about its connectivity relations by means of a topological consideration. Incidentally, this consideration of analysis situs is very simple and applies to all semisimple groups without distinguishing cases. This approach lies closer to my whole way of thinking than your more algebraic method, which at the moment I only half understand.

In his reply of 28 March 1925, Cartan praised Weyl for establishing the connections between the "infinitesimal" and the "integral" approach to the study of Lie groups.[41] It was needed because of "the difficulty—I dare not say the impossibility—of finding a direct proof that does not depart from the strictly infinitesimal domain."[42] But Cartan also envisioned the dual infinitesimal–integral approach as bearing more fruit in the future. Here "integral" of course referred to the Hurwitz integral but also included thereby the global consideration of Lie groups that it presupposes. Before long, Cartan himself championed the global approach to Lie groups and along with it accepted the value of concomitant topological considerations. His book [1930] represented the first treatise on Lie groups from the global, topological, point of view.

Although Cartan's many mathematical accomplishments in the "post-Weyl" period lie outside the scope of this book, I will take this occasion to say a few words about Cartan's rendition of the finite connectivity theorem and its proof [1927:217–8], which appears in one of the earliest papers of his written from the global topological point of view.

The reader may have noticed that in the preceding exposition of Weyl's contributions to the theory of structure of semisimple Lie groups no mention was made of the now familiar Weyl chambers. The reason is that Weyl did not utilize them in [1925]. They first seem to have made their appearance in Cartan's proof, which proceeds as follows.[43] Recall from (3.11) and (3.12) that each element T in the maximal toral subalgebra $\mathfrak{t}_u \subset \mathfrak{a}_u$ is represented by a diagonal matrix with l 0's followed by the eigenvalues $\alpha(T) = 2\pi i \varphi_\alpha(T)$, $\alpha \in \Phi$, $T = \operatorname{ad} H$. Cartan made use of the connection

[41] ETH Zürich, Hs 91:502.

[42] Such a proof was first given by Casimir and van der Waerden [1935]. See [Borel 1986:63; 1998:§8] for an indication of how a construct in Weyl's book on group theory and quantum mechanics, the "moment of momentum" [1928:156; 1931:179], led Casimir to the operator that bears his name and ultimately to his joint paper with van der Waerden. See [Jacobson 1962:79] for a proof valid for any field of characteristic zero, algebraically closed or not.

[43] Usually Weyl is credited with introducing the Weyl chambers in the course of proving the finite connectivity theorem [Akivis and Rosenfeld 1993:62; Borel 1986:75 n.11; Dieudonné 1976:284], although strictly speaking, as we have seen in Section 3, that is not the case.

between conjugation in $\mathfrak{t}_u$ and elements of the Weyl group $\mathfrak{W}$, which Weyl had observed in passing and had expressed on the global level of the maximal toral subgroup $\mathfrak{T}$ of the universal covering group $\mathfrak{G}$ of Section 5 [Weyl 1925:641]. Cartan explained the connection as follows. Let $u \in \mathfrak{A}_u$ be such that $T_1 = u^{-1}Tu \in \mathfrak{t}_u$ for all $T \in \mathfrak{t}_u$. (This means that $u^{-1}\mathfrak{T}_u u \subset \mathfrak{A}_u$, where $\mathfrak{T}_u \subset \mathfrak{A}_u$ is the maximal toral subgroup corresponding to $\mathfrak{t}_u$, so that $u \in N(\mathfrak{T}_u)$, the normalizer of $\mathfrak{T}_u$ in $\mathfrak{A}_u$.) Then since T_1 and T are similar, they have the same eigenvalues. In terms of their representation as diagonal matrices (3.12), this means that the diagonal entries of T_1 must be some permutation of those of T. Thus $\varphi_\alpha(T_1) = \varphi_{\alpha'}(T)$. Cartan observed that the permutations $\alpha \to \alpha'$ are the elements of the group $\mathfrak{W}$.[44] Thus $\mathfrak{W}$ can be regarded as acting on $\mathfrak{t}_u$ by conjugation.

In a footnote, Cartan acknowledged that the interpretation of $\mathfrak{W}$ as a group of rotations and reflections, as well as the interpretation of its generating operations σ_α as reflections in hyperplanes, was due to Weyl [1927: 215 n.2]. He then proceeded to push Weyl's geometrical viewpoint further. The hyperplanes $\varphi_\alpha = 0$ (or equivalently $\alpha = 0$), he observed, partition $\mathfrak{t}_u$, conceived as an l-dimensional Euclidean space with metric relations given by the Killing form, into a certain number of unbounded convex polyhedral regions $D, D', D'', \ldots$. These are the now familiar Weyl chambers. In particular, for a fixed chamber, say D, each T that is nonsingular—$\alpha(T) \neq 0$ for all $\alpha \in \Phi$—is conjugate to a unique T_1 in the interior of D.

Turning next to the connection between $\mathfrak{t}_u$ and $\mathfrak{T}_u$ via exponentiation, Cartan introduced the lattice $\mathcal{L} \subset \mathfrak{t}_u$, which consists of all T such that $\varphi_\alpha(T) \in \mathbb{Z}$ for all $\alpha \in \Phi$. As would now be said, this is the kernel of the exponential mapping, so that $t_1 = \exp T_1$ and $t_2 = \exp T_2$ are equal precisely when $T_1 \equiv T_2 (\text{mod } \mathcal{L})$. Since $t = \exp T$ is also left invariant under the group of translations $\tau(T) = T + L$, $L \in \mathcal{L}$, Cartan considered the group $\mathfrak{W}_1$ generated by $\mathfrak{W}$ and the translations τ. Then if $\sigma(T) = T_1$ for $\sigma \in \mathfrak{W}_1$, $t = \exp T$ and $t_1 = \exp T_1$ are conjugate. The transformations of $\mathfrak{W}_1$ permute the family of hyperplanes $\varphi_\alpha = n$, $\alpha \in \Phi$, $n \in \mathbb{Z}$. These hyperplanes partition the Eucildean space $\mathfrak{t}_u$ into convex polyhedra $E, E', E'', \ldots$, with each E being contained in some D.[45] Fix an E and let γ be a closed curve in the adjoint group $\mathfrak{A}_u$. Without loss of generality it can be assumed that γ begins and ends at the identity element e (being homotopic to such a curve) and (by Weyl's "codimension-three theorem") that $\gamma(t)$ is nonsingular for all $t \in (0,1)$. Then corresponding to γ is a continuous curve Γ such that $\gamma(t) = \exp \Gamma(t)$ for all $t \in [0,1]$, $\Gamma(t)$ is in the interior of E for all $t \in (0,1)$, and $\Gamma(0)$, $\Gamma(1)$ are vertices of E that belong to $\mathcal{L}$ (since $\gamma(0) = \gamma(1) = e$).

---

[44] Each $u \in N(\mathfrak{T}_u)$ determines a $\sigma(u) \in \mathfrak{W}$, but evidently $\sigma(u_1) = \sigma(u_2)$ if u_1 and u_2 are congruent modulo $\mathfrak{T}_u$. However, $\mathfrak{W}$ can be identified with $N(\mathfrak{T}_u)/\mathfrak{T}_u$ and is sometimes defined in this manner. (See e.g., [Bröcker and Dieck 1985:158].) Such observations are not to be found explicitly in Cartan's (or Weyl's) brief discussions.

[45] The E's are sometimes called Stiefel chambers [Bröcker and Dieck 1985:226].

By applying if necessary a translation τ it can be assumed without loss of generality that E is such that $\Gamma(0) = 0$.

"We are now prepared," Cartan continued, "to place ourselves at the general point of view of H. Weyl, to study the topology of the space of the adjoint group $[\mathfrak{A}_u]$ and, in particular, to see whether there exist in this space closed contours that are not reducible to a point by continuous deformation" [1927: 217]. Among the vertices of E there will be a certain number, denoted by $V_1, \ldots, V_m$, that belong to $\mathcal{L}$, where the notation is chosen such that $V_1 = 0$. Thus $\Gamma(0) = V_1$ and $\Gamma(1) = V_k$, $k \geq 1$. If $\Gamma(1) = V_1$ also, then what is essentially Weyl's argument shows that γ can be continuously deformed to a point: Γ is continuously deformed to the point $V_1 = 0$ by $\Gamma_\sigma(\tau) = \sigma\Gamma(\tau)$, and likewise $\gamma_\sigma(\tau) = \exp \Gamma_\sigma(\tau)$ deforms γ to the point e. Thus $\mathfrak{A}_u$ will be simply connected if $m = 1$. When m, which Cartan called the *index of connection*, is greater than 1, $\mathfrak{A}_u$ is not simply connected: There are m homotopy classes of closed curves $\gamma \subset \mathfrak{A}_u$, the kth class being determined by $\gamma_k = \exp \Gamma_k$, where Γ_k is a curve in E from V_1 to V_k. The universal covering group $\widetilde{\mathfrak{A}}_u$ (which Cartan explicitly constructed) thus has m points over each point of $\mathfrak{A}_u$.

In this way Cartan refined Weyl's proof, bringing into play more of the geometry of the Euclidean space $\mathfrak{t}_u$. No doubt nothing in Cartan's proof would have seemed surprising to Weyl, who may well have had similar considerations in the back of his mind. Nonetheless, it was Cartan who made the above geometrical notions an explicit part of the theory. Incidentally, in his construction of the universal covering group $\widetilde{\mathfrak{A}}_u$ Cartan introduced the definition of multiplication in $\widetilde{\mathfrak{A}}_u$ that brings to light the connection with Poincaré's fundamental group, $\pi_1(\mathfrak{A}_u)$ (as explained in Section 9.5). Thus, as he pointed out, the covering map $\pi : \widetilde{\mathfrak{A}}_u \to \mathfrak{A}_u$ is a group homomorphism with kernel $\widetilde{\mathfrak{K}}$ that is immediately identifiable with $\pi_1(\mathfrak{A}_u)$. Cartan, however, called $\widetilde{\mathfrak{K}}$ the *connection group* of $\mathfrak{A}_u$ because "fundamental group" was his term for the group defining a geometry in the sense of Klein's *Erlanger Programm*.[46] He pointed out that the elements of $\widetilde{\mathfrak{K}}$ may be identified with the m vertices $V_1, \ldots, V_m$ of the polyhedron E described above, since each V_k has associated to it a curve Γ_k defining the kth element $[\Gamma_k]$ of $\widetilde{\mathfrak{K}}$. Cartan actually computed $\widetilde{\mathfrak{K}}$ for each simple type. For example, when $\mathfrak{G} = \mathbf{SL}(n, \mathbb{C})$, so $\mathfrak{G}_u = \mathbf{SU}(n)$, the adjoint group is $\mathfrak{A}_u = \mathbf{SU}(n)/\mathfrak{C}$, where $\mathfrak{C}$, the center of $\mathbf{SU}(n)$, consists of the n scalar matrices $\varepsilon_k I$, ε_k being an nth root of unity. Thus $\widetilde{\mathfrak{A}}_u = \mathbf{SU}(n)$, and $\widetilde{\mathfrak{K}} = \mathfrak{C}$ is cyclic of order n. Weyl's paper raised in Cartan's mind the question as to whether $\widetilde{\mathfrak{A}}_u$ could be realized as a group of linear transformations for every simple type, as in the case of

[46] Cartan explained his choice of terminology in [1930:13]. Although in [1927] there is no mention of Poincaré's fundamental group, given Cartan's definition of multiplication in $\widetilde{\mathfrak{A}}_u$ it is likely that he had it in mind. The connection between $\widetilde{\mathfrak{K}}$ and $\pi_1(\mathfrak{A}_u)$ was apparently first made explicit by Cartan in [1928:1818n.2].

$\mathfrak{G} = \mathbf{SL}(n, \mathbb{C})$. From Weyl's work it was evident that this was the case for the types A–D, and Cartan showed that it was true for all simple types.

Cartan also gave a general proof that for any semisimple $\mathfrak{G}$, $\widetilde{\mathfrak{K}}$ is always in the center of $\widetilde{\mathfrak{A}}_u$ and hence always abelian. This can be seen as follows: Since $\widetilde{\mathfrak{K}}$ is normal, for any fixed $k \in \widetilde{\mathfrak{K}}$ the continuous function $f(x) = xkx^{-1}$ maps $\widetilde{\mathfrak{A}}_u$ into $\widetilde{\mathfrak{K}}$; because $f[\widetilde{\mathfrak{A}}_u]$ is connected and $\widetilde{\mathfrak{K}}$ is finite, it must be a single point; and since $f(x) = k$ when x is the identity element, we must have $xkx^{-1} = k$ for all x, and so k is in the center of $\widetilde{\mathfrak{A}}_u$.[47] Of course, the fact that $\widetilde{\mathfrak{K}}$ is finite is what Weyl's finite connectivity theorem states. Thus, as Cartan first noted, an immediate consequence of Weyl's theorem is the fact that the fundamental group of $\mathfrak{A}_u$ is always abelian.

Unbeknownst to Weyl when he introduced the idea of universal covering groups in [1925] or to Cartan when he made the above observations about $\widetilde{\mathfrak{K}}$, a far more general result had been established independently of Weyl's work by Otto Schreier (1901–1929). Schreier's work [1925] was very modern in spirit. He dealt rigorously and abstractly with groups—called continuous groups by him—that would now be described as topological groups that are Hausdorff, connected, and locally Euclidean. His main result was undoubtedly motivated by Lie's misleading pronouncements (discussed in Section 3.2) regarding the global isomorphism of groups with isomorphic Lie algebras. Schreier's theorem may be stated as follows: *Consider the class of all continuous groups $\mathfrak{G}$ that are locally isomorphic. Then there exists within this class a group $\widetilde{\mathfrak{G}}$ with the following properties: If $\mathfrak{G}$ is any group of this class then $\mathfrak{G}$ is isomorphic to a factor group $\widetilde{\mathfrak{G}}/\mathfrak{D}$, where $\mathfrak{D}$ is a discrete subgroup of the center of $\widetilde{\mathfrak{G}}$ and is isomorphic to $\pi_1(\mathfrak{G})$, the Poincaré fundamental group of $\mathfrak{G}$.* The group $\widetilde{\mathfrak{G}}$, which he showed to be simply connected and unique up to global isomorphisms, was called a covering group.[48] By the time Cartan acknowledged the work of Schreier in [1930:1] the latter was dead, the victim of a "general sepsis."[49] Otherwise,

---

[47] In [1927:226] Cartan simply said that since each $k \in \widetilde{\mathfrak{K}}$ "leaves invariant each transformation of $[\mathfrak{A}_u]$ and, by continuity, each transformation of $[\widetilde{\mathfrak{A}}_u]$," the connection group is abelian. The proof given above, which is the proof he gave in [1936:236–7], could be regarded as an amplification of what he had in mind. The proof he gave in [1930:12] is somewhat different.

[48] The name was suggested by Emil Artin, who also gave Schreier the example (described above in Section 1) of the covering group of the rotation group of $\mathbb{R}^3$ [Schreier 1925:24n.6, 32n.9]. Schreier's construction of $\widetilde{\mathfrak{G}}$ is quite different from the Poincaré construction of a universal covering surface (Section 9.5), but in [1927] he adopted this approach. Weyl's papers on group representations are first mentioned by Schreier in [1928]. Schreier also remarked as his paper was going to press [1925:15] that Henri Mineur, in France, in a paper [1925] dealing with Lie groups of birational transformations, had presented results closely related to his own. Mineur's paper, however, was obscurely written and lacked the rigor and generality of Schreier's work. Its contents have been described by Freudenthal [1932].

[49] See "Otto Schreier †," *Abhandlungen aus dem Mathematischen Seminar der Hamburgischen Universität* **7** (1929), p. 107.

he might have had a coworker in the investigation of the topological properties of Lie groups, for in [1928:122] Schreier had indicated an interest in this subject. As it was, for almost a decade Cartan was the sole contributor to the topology of Lie groups.

The above presentation of a bit of Cartan's "post-Weyl" work was included because of its connections with Weyl's proof of the finite connectivity theorem. It fails to convey the beauty, depth, and originality of his mathematical achievements during this period.[50] For example, the above description of Cartan's paper [1927] deals only with the first, introductory, part of what is primarily a study of the connections between Riemannian geometries and Lie groups. Recall that in his study of real simple groups (Section 8.5) Cartan had considered all the real forms $\mathfrak{g}_r$ of a complex simple Lie algebra $\mathfrak{g}$, only one of which is the compact form $\mathfrak{g}_u$. Corresponding to each real form $\mathfrak{g}_r$ one can consider the associated adjoint group $\mathfrak{A}_r$, its universal covering group $\widetilde{\mathfrak{A}}_r$, and the connection group. Cartan showed that these connection groups could be realized as the rotation groups (isotropy groups) of an associated Riemannian geometry and thereby studied.

Cartan's interest in links between Lie groups and Riemannian geometry, including his impressive theory of symmetric spaces, was not inspired by Weyl's work. Nonetheless, Weyl's paper [1925] does seem to have been the primary motivation for Cartan to adopt a global, topological viewpoint in his group theoretic study of Riemannian geometries. It can be seen in the paper under discussion, where Cartan wrote, "The study presented here is no longer local; rather, it concerns the properties of space that depend on *analysis situs* ..." (p. 210).[51]

Having mentioned Weyl's influence upon Cartan, it is fitting that I remind the reader just how much Weyl was influenced by Cartan. Except perhaps for the proof of the complete reducibility theorem for the classical groups, Weyl's contributions as discussed in Sections 2–4—including his development of a Frobenius–Schur theory of characters for semisimple groups—built upon the theoretical foundation laid by Cartan. Although Schur's paper [1924a] on the Cayley counting problem seems to have precipitated Weyl's extensive involvement with the structure and representation of Lie groups and Schur's letter of 10 November 1924 seems to have been the fillip for Weyl's theory of characters, it was Cartan who provided the tools by means of which Weyl was able to deal with continuous groups on a more general level than that represented by a single specific group, such as the rotation group. The existence of an extensive and profound theory of semisimple groups, which is what Weyl discovered in Cartan's work, was a major source of stimulation and inspiration for him. As Weyl put it at the

---

[50] Such an appreciation can be gained from [Borel 2000], which contains a careful, detailed analysis of Cartan's work on groups and geometry in the 1920s and 1930s, including, in particular, his theory of symmetric spaces.

[51] See the discussion by De Rham [1975:11–12] of slightly earlier manifestations of Cartan's new interest in global, topological aspects of his group related work.

Figure 41. The photo of Cartan (top) was taken in 1932, and that of Weyl (below) probably dates from about the same period. They finally met at the International Congress of Mathematicians in Bologna in 1928.

conclusion of his above-quoted letter to Cartan:[52]

> Since I became acquainted with the general theory of relativity, nothing has so stirred me and filled me with inspiration as the study of your works on continuous groups. I do not at all value my present share in this theory very highly; I really seem to be only the fortuitous meeting point for you and Schur.

Of course, Weyl's estimate of his own contributions is far too modest, but it does reflect the extent to which his work in [1925] was dependent upon the work of Schur and Cartan. Slightly over a year after he wrote those words, however, he published with F. Peter a sequel to [1925] that pushed the theory in entirely new directions inspired by the theory of integral equations he had learned at Göttingen and far removed from what Schur or Cartan had done.[53]

12.7. The Peter–Weyl Paper

In developing his theory of characters, Weyl had taken it for granted that to each dominant weight λ corresponds an irreducible representation. He was relying upon Cartan's theorem to this effect, although he was evidently not satisfied with the case-by-case construction underlying the proof: For each simple type of rank l Cartan had to construct l fundamental irreducible modules, which then formed the building blocks for all irreducible representation modules (Section 8.4). Weyl envisioned an analogue of Frobenius's regular representation theorem (Section 10.1) as a means of attaining a more general procedure for constructing all irreducible representations [1925:640–42]. The regular representation of a finite group $\mathfrak{G}$ may be conceived as follows. An element in the group algebra of $\mathfrak{G}$ can be expressed in the form

$$x = \sum_{t \in \mathfrak{G}} x(t)t, \tag{7.1}$$

where $x(t)$ denotes a complex number for each $t \in \mathfrak{G}$. The (right) regular representation $s \to T(s)$ is then given by setting

$$T(s)x = xs = \sum_{t \in \mathfrak{G}} x(t)ts \overset{u=ts}{=} \sum_{u \in \mathfrak{G}} x(us^{-1})u, \tag{7.2}$$

[52] "Seit der Bekanntschaft mit der allgemeinen Relativitätstheorie hat mich nichts so ergriffen und mit Begeisterung erfüllt wie das Studium Ihrer Arbeiten über die kontinuierlichen Gruppen. Meinen gegenwärtigen Anteil an dieser Theorie schätze ich gar nicht besonders hoch ein; ich komme mir eigentlich nur vor wie der zufällige Treffpunkt von Ihnen und Schur." Letter dated 22 March 1925 (in possession of H. Cartan).

[53] A notable exception to this generality is Schur's lemma (Section 10.4), which was used repeatedly by Weyl in his paper with Peter.

where I have used the standard Frobenius–Schur technique of changing the variable of summation. According to the regular representation theorem, $T(s)$ decomposes into a direct sum that contains each irreducible representation of $\mathfrak{G}$ as often as its degree.

None of the above was explicitly stated by Weyl, but it is implicit in what he wrote about his envisioned generalization, in which the finite group $\mathfrak{G}$ is replaced by the real, compact, simply connected universal covering group $\mathfrak{G} = \widetilde{\mathfrak{A}}_u$ underlying his theory of characters (Section 5). The finite-dimensional space of all hypercomplex numbers x in (7.1) is replaced by the space of all continuous complex-valued functions defined on $\mathfrak{G}$, which will be denoted here by $\mathcal{C}(\mathfrak{G})$. This makes good sense, because the x in (7.1) can also be thought of as the *function* $x = x(t)$ defined on $\mathfrak{G}$. This function is then replaced by a complex-valued continuous function on the continuous group $\mathfrak{G}$. Also, if one thinks of x in (7.1) as a function, then from (7.2) we see that $T(s)$ sends $x(t)$ into the function $x'(t) = x(ts^{-1})$.[54] Thus for $x \in \mathcal{C}(\mathfrak{G})$ Weyl's regular representation is given by

$$(7.3) \qquad T(s) : x \to x', \quad s \in \mathfrak{G}, \quad x \in \mathcal{C}(\mathfrak{G}), \quad x'(t) = x(ts^{-1}).$$

Here $T(s_1 s_2) = T(s_1)T(s_2)$, provided that we follow Weyl's practice and read the right-hand side from right to left, i.e., as meaning that first $T(s_1)$ and then $T(s_2)$ is applied.

Weyl had evidently studied the representation (7.3) sufficiently to be able to declare that "by reduction of this regular representation" one arrives at the theorem that the irreducible characters form, relative to the Hurwitz integral on $\mathfrak{G}$, a complete orthogonal system of class functions on $\mathfrak{G}$, i.e., functions that take the same value on elements in the same conjugate class. Here, Weyl was using "complete" in the sense of Hilbert's theory (Section 9.2) to mean that if $x \in \mathcal{C}(\mathfrak{G})$ is any class function, then Parseval's equality holds, i.e., $\int_{\mathfrak{G}} |x(t)|^2 \, dt = \sum_n c_n^2$, where the c_n are the generalized Fourier coefficients of x with respect to the system of normalized, irreducible characters $\varphi_n = \frac{1}{\sqrt{\Omega}} \chi_n$, $\Omega = \int_{\mathfrak{G}} dm$. In [1924a:200] Schur had remarked that the normalized irreducible characters φ_n form an orthonormal system, perhaps thereby piquing Weyl's interest in the question of completeness. Weyl also envisioned this completeness theorem as a way to prove Cartan's above-mentioned theorem, but in April 1925, when he submitted the third and final installment of [1925], he chose not to attempt an exact development of "this entire method," since "at the present stage of research" it seemed to him to require too great an expenditure of effort to achieve a result already established by Cartan (p. 642).

It was only about 14 months later that Weyl presented the details behind the proof of his completeness theorem. This was done in a paper,

---

[54] In [1925:641] Weyl considered the left regular representation, so that $x'(t) = x(s^{-1}t)$; but later in his paper with Peter the right regular representation is used. For consistency of exposition I use the right regular representation throughout.

written jointly with Fritz Peter (1899–1949) and submitted to *Mathematische Annalen* on 24 June 1926 [1926]. Peter, whom Weyl referred to as one of his students [1927b:83], had received his doctorate from Göttingen in 1923 with a work on the indices of refraction and the absorption constants of diamonds [1923]. It would seem that this paper and his joint paper with Weyl were his only scientific publications. He became assistant master (*Studienrat*) and then headmaster (*Oberstudiendirektor*) of a secondary school (*Schule Schloß Salem*). Exactly what his contribution to the paper was is unknown to me, although I would guess that it had to do with working out the details of proving what I call below the construction theorem. Although, as we have seen, initially Weyl was primarily interested in the compact group $\mathfrak{G} = \widetilde{\mathfrak{A}}_u$, he and Peter realized that the reasoning of their joint paper applied, as we would now say, to any compact topological group with a translation-invariant measure defined on it. Since at the time general results about the existence of "Haar measures" were not known—see below—they assumed "that Lie's infinitesimal conceptual apparatus" is applicable to $\mathfrak{G}$ "in order to have an invariant volume measure" [1926:58]. Thus in what follows $\mathfrak{G}$ denotes any such group.

In 1925, when Weyl revealed his idea of studying the regular representation $T(s)$ of (7.3), he did not disclose his method of decomposing it. At the time, the only class of linear operators acting on spaces with an infinite number of dimensions with a coherent theory behind them was that of the integral operators of the Hilbert–Schmidt theory, but $T(s)$ is not such an operator. Given Weyl's background, a link with integral equations would have been foremost in his mind as a means of dealing with the infinite-dimensional linear algebra required by $T(s)$. He seems to have found that link in the observation that if $x = \sum_s x(s)s$ and $y = \sum_t y(t)t$ are two elements in the group algebra of a finite group $\mathfrak{G}$, then their product is

$$xy = \sum_u \Big[\sum_{st=u} x(s)y(t)st\Big] = \sum_u \Big[\sum_t x(ut^{-1})y(t)\Big] u,$$

which means that viewed as a function on $\mathfrak{G}$,

$$(7.4) \qquad (xy)(u) = \sum_t x(ut^{-1})y(t).$$

With the group algebra replaced by $\mathcal{C}(\mathfrak{G})$ the above definition of multiplication becomes what is now known as the convolution $x * y$ of $x, y \in \mathcal{C}(\mathfrak{G})$:

$$(7.5) \qquad (x * y)(s) = \int_{\mathfrak{G}} x(st^{-1})y(t)\, dm(t).$$

It should be noted that in [1926] the product of $x, y \in \mathcal{C}(\mathfrak{G})$ is simply denoted by xy, and the term "convolution" is not used. Before Weyl's paper convolution integrals, that is, integrals of the form $\int f(s-t)g(t)\, dt$, usually involving specific functions f and g, occurred in diverse analytical contexts but had not been singled out as especially important and given a special name or notation. Furthermore, such integrals were not generally regarded as defining a multiplication of f and g. It seems to have been Weyl's paper that prompted the introduction of the notion of the convolution product of functions into analysis.[55] As we have seen, for Weyl this notion was a natural analogue of multiplication in the group algebra of a finite group, and in fact, the functions $x, y \in \mathcal{C}(\mathfrak{G})$ are referred to as a "group numbers" (*Gruppenzahlen*), which had been Weyl's term for the elements of the group algebra of the symmetric group [1924a:469]. In what follows, however, the customary $*$ notation will be used to denote Weyl's product, which will be referred to as a convolution.

Comparison of (7.4) and (7.5) makes it tempting to think of the continuous analogue of the operator $T(s)$ in (7.2) as a convolution $T(s)\cdot x = x * s$ and hence an integral operator. However, the $s \in \mathfrak{G}$ of (7.2), regarded as a function, is defined by $s(t) = 1$ if $t = s$ and $s(t) = 0$ otherwise. In $\mathcal{C}(\mathfrak{G})$ there is no analogue of such a function; the group elements $s \in \mathfrak{G}$ cannot in general be identified with continuous functions on $\mathfrak{G}$. But for any fixed $x \in \mathcal{C}(\mathfrak{G})$ one can consider the integral operator A_x defined by

$$A_x y(s) = x * y(s) = \int_{\mathfrak{G}} x(st^{-1}) y(t)\, dm(t),$$

which has the continuous kernel $K(s,t) = x(st^{-1})$. This operator is not hermitian symmetric, so that Hilbert's strongest results do not apply, but Schmidt's method of dealing with nonsymmetric kernels (Section 9.4) suggests here in the complex case considering the symmetric operator $A_x A_x^H$, where A_x^H denotes the hermitian (or conjugate) transpose of A_x. Weyl did this and was able to take Schmidt's idea one step further because of the special nature of the kernel of the operator A_x. That is,

$$A_x^H y = \int_{\mathfrak{G}} \overline{x(ts^{-1})} y(t)\, dm(t) = \int_{\mathfrak{G}} \tilde{x}(st^{-1}) y(t)\, dm(t) = A_{\tilde{x}} y,$$

where the function $\tilde{x}$ is defined by

$$\tilde{x}(s) = \overline{x(s^{-1})}.$$

A simple calculation then shows that $A_x A_x^H = A_z$, where $z = x * \tilde{x}$.

In other words, if we take any $x \in \mathcal{C}(\mathfrak{G})$ and set $z = x * \tilde{x}$, we obtain a symmetric, positive definite[56] integral operator A_z with kernel $z(st^{-1})$ to which the Hilbert–Schmidt theory applies. Consequently, if λ is any eigenvalue of A_z—by which I mean (following Weyl) the reciprocal of a "Hilbert eigenvalue" and thus a nonzero eigenvalue in the modern sense—then $\lambda > 0$, and the space $\mathcal{E}_\lambda$ of solutions to $(A_z - \lambda I)\varphi = 0$ is a finite-dimensional subspace of $\mathcal{C}(\mathfrak{G})$. The relevance of A_z to the decomposition of

[55] See in this connection [Bourbaki 1974:292–95] and [Dieudonné 1981:194–201].

[56] As in Section 9.5, "positive definite" means that all the Hilbert eigenvalues are positive, i.e., all the nonzero eigenvalues in the modern sense are positive.

the regular representation $T(s)$ of (7.3) stems from the fact that $T(s)$ takes $\mathcal{E}_\lambda$ into itself, thereby determining a finite-dimensional representation of $\mathfrak{G}$. If one thinks heuristically of $T(s)y$ as $y*s$—right multiplication by s—then evidently A_z, being left multiplication by z, commutes with $T(s)$. Of course, $T(s)$ is not literally $y*s$ when the underlying group is continuous, not finite, but it is easy to check by direct calculation (using the translation-invariance of the Hurwitz integral) that $T(s)$ does indeed commute with A_z, which means that $T(s)[\mathcal{E}_\lambda] \subset \mathcal{E}_\lambda$. Perhaps Weyl was led by such considerations to make the operator A_z central to his decomposition of $T(s)$.

In any case, he considered the finite-dimensional representation $R(s)$ of $\mathfrak{G}$ induced by the action of $T(s)$ on $\mathcal{E}_\lambda$. It is unitary, since by the translation-invariance of the Hurwitz integral, $\|T(s) \cdot \varphi\|^2 = \int_{\mathfrak{G}} |\varphi(ts^{-1})|^2 dm = \int_{\mathfrak{G}} |\varphi(t)|^2 dm = \|\varphi\|^2$. The matrix form $R(s) = (r_{ij}(s))$ of $R(s)$ is defined as follows (in order to account for Weyl's right-to-left reading of $T(s_1)T(s_2)$): Let $\varphi_1, \ldots, \varphi_N$ denote an orthonormal basis for $\mathcal{E}_\lambda$, and define the functions $r_{ij}(s)$ by

$$T(s) \cdot \varphi_i = \varphi(ts^{-1}) = \sum_{j=1}^{N} \overline{r_{ij}(s)}\varphi_j(t). \tag{7.6}$$

Then $R(st) = R(s)R(t)$ in the sense of ordinary matrix multiplication. Finally, since the complete reducibility theorem holds for the compact group $\mathfrak{G}$, $R(s)$ can be replaced by one of its irreducible components so as to obtain from $T(s)$ an irreducible unitary representation.

To sum up: By thinking to consider the integral operators $A_z = A_x A_x^H$, Weyl could see how to bring the Hilbert–Schmidt theory to bear on the problem of breaking off finite-dimensional irreducible representations from the infinite-dimensional regular representation $T(s)$. But what does this have to do with the announced completeness theorem for the irreducible characters? Let us consider how Weyl saw the connection.

The first point to observe is that if $R^{(\mu)}(s) = \left(r_{ij}^{(\mu)}(s)\right)$, $\mu = 1, 2, \ldots$, denotes the inequivalent irreducible representations of $\mathfrak{G}$, then the coefficient functions $r_{ij}^{(\mu)}(s)$ satisfy the orthogonality relations

$$\begin{aligned} \int_{\mathfrak{G}} r_{ij}^{(\mu)}(s) r_{kl}^{(\mu)}(s^{-1})\, dm &= \frac{\Omega}{n_\mu}\delta_{ik}\delta_{jl}, \\ \int_{\mathfrak{G}} r_{ij}^{(\mu)}(s) r_{kl}^{(\nu)}(s^{-1})\, dm &= 0 \quad (\mu \neq \nu), \end{aligned}$$

where n_μ is the degree of $R^{(\mu)}(s)$ and, as before, $\Omega = \int_{\mathfrak{G}} dm$. These relations were a special case of relations established by Schur en route to proving the orthogonality relations for the irreducible characters of the rotation group [1924a:199–200]. Since, as Schur had also observed [1924a:198], each $R^{(\mu)}(s)$ can be taken to be unitary, so that $r_{ij}^{(\mu)}(s^{-1}) = \overline{r_{ji}^{(\mu)}(s)}$, it follows that in this case the functions $r_{ij}^{(\mu)}(s)$ form an orthogonal system with respect to the inner product $(x, y) = \int_{\mathfrak{G}} x(s)\overline{y(s)}\, dm$. Of course, these functions are not normalized in Hilbert's sense, since

$$\int_{\mathfrak{G}} \left|r_{ij}^{(\mu)}(s)\right|^2 dm = \frac{\Omega}{n_p}.$$

It should be noted that although Schur pointed out that the irreducible characters, suitably normalized, form a "unitary" orthonormal system [1924a:200], he did not explicitly mention that the same is true for the functions $r_{ij}^{(\mu)}(s)$, although there can be no doubt that he realized it. For Weyl this property of the $r_{ij}^{(\mu)}(s)$ became important because it raised the question of completeness and suggested treating the $r_{ij}^{(\mu)}(s)$ from the viewpoint of Hilbert's generalized Fourier analysis.

Adopting that viewpoint, Weyl introduced for the orthogonal system $r_{ij}^{(\mu)}(s)$ the generalized Fourier coefficients of $x \in \mathcal{C}(\mathfrak{G})$, namely $c_{ij}^{(\mu)}(x) = \int_{\mathfrak{G}} x(s)\overline{r_{ij}^{(\mu)}(s)}\, dm(s)$. He also introduced the Fourier coefficient matrix $C^{(\mu)}(x) = (c_{ij}^{(\mu)}(x))$, which (following Peter and Weyl) may also be expressed by the matrix equation

$$C^{(\mu)}(x) \;=\; \int_{\mathfrak{G}} x(s)\overline{R^{(\mu)}(s)}\, dm(s).$$

Bessel's inequality (Section 9.2) then takes the form

$$\text{(7.7)} \qquad \sum_{\mu,i,j} \frac{n_p}{\Omega} |c_{ij}^{(\mu)}(x)|^2 \;\le\; \int_{\mathfrak{G}} |x(s)|^2 dm(s),$$

and the system of functions $r_{ij}^{(\mu)}(s)$ is complete when for every $x \in \mathcal{C}(\mathfrak{G})$ we have equality in (7.7), i.e., when Bessel's inequality becomes Parseval's equality. Since for any p, $\sum_{i,j} |c_{ij}^{(\mu)}(x)|^2 = \operatorname{tr}\left\{C^{(\mu)}(x)\cdot [C^{(\mu)}(x)]^H\right\}$, Parseval's equality may be written in matrix form as

$$\text{(7.8)} \qquad \sum_{\mu} \frac{n_\mu}{\Omega} \operatorname{tr} C^{(\mu)}(x)[C^{(\mu)}(x)]^H \;=\; \int_{\mathfrak{G}} |x(s)|^2 dm(s).$$

As the fundamental theorem of their paper, Peter and Weyl singled out [1926:72] what is now called the Peter–Weyl theorem.

(7.9) **Peter–Weyl Theorem.** *The orthogonal system of functions $r_{ij}^{(\mu)}(s)$ obtained by letting μ range over all inequivalent irreducible unitary representations of $\mathfrak{G}$ is complete.*

Weyl's earlier announced completeness theorem for the irreducible characters $\chi^{(\mu)}(s) = \operatorname{tr} R^{(\mu)}(s)$ follows as a corollary [1926:62–4]. Presumably, this is how Weyl originally intended to prove it. In any case, it is of interest to see how they deduce it from the Peter–Weyl theorem, because it involves further relations satisfied by the Fourier coefficients and yet another application of Schur's lemma.

What needs to be proved is that for any class function $x \in \mathfrak{C}(\mathfrak{G})$,

$$\sum_{\mu} \frac{1}{\Omega} |c_\mu(x)|^2 = \int_{\mathfrak{G}} |x(s)|^2 dm(s), \tag{7.10}$$

where $c_\mu(x) = \int_{\mathfrak{G}} x(s)\overline{\chi^{(\mu)}(s)}\, dm(s)$ are the Fourier coefficients of x with respect to the orthogonal system of irreducible characters. To show that (7.10) follows from (7.8), they observed that for any unitary representation $R(s)$ one has by by direct calculation

$$R(s)C^T(x) = (R * x)(s), \quad C^T(x)R(s) = (x * R)(s). \tag{7.11}$$

Now, when x is a class function, so that $x(uv) = x(vu)$, it is easily seen from the definition of convolution and the translation-invariance of the integral that $x * y = y * x$ for any y. In view of (7.11) this means that when x is a class function, the matrix $C^T(x)$ commutes with $R(s)$. Thus if $R(s)$ is irreducible (as well as unitary), we may conclude by Schur's lemma that $C(x)^T$, and therefore $C(x)$, is a scalar matrix. Applying this observation to the matrices $C^{(\mu)}(x)$ in (7.8), we see that

$$C^{(\mu)}(x) = \frac{c_\mu(x)}{n_\mu} I,$$

and (7.8) becomes (7.10), which implies the completeness theorem for characters.

Let us now consider in outline how the authors proved the Peter–Weyl theorem itself. The proof utilizes the Hilbert–Schmidt theory as it applies, suitably modified, to the hermitian symmetric operators A_z, $z = x * \tilde{x}$, discussed above. We have already seen how these operators can be used to "break off" from the regular representation $T(s)$ a finite-dimensional irreducible unitary representation $R(s)$. Their idea is to keep doing this, and to do it in a systematic fashion that guarantees that they have obtained all the representations they need to make Bessel's inequality an equality, thereby establishing completeness.

In the course of their reasoning they capitalized on the properties of the Fourier coefficients $C(x) = \int_{\mathfrak{G}} x(s)\overline{R(s)}\, dm$ of a function with respect to a unitary representation. In addition to (7.11) these include the following:

$$C(x * y) = C(x)C(y); \tag{7.12}$$

(7.13) *If $C(x) \neq 0$, then $R(s)$ is equivalent to a unitary representation $R'(s)$ for which the Fourier coefficient matrix $C'(x)$ is diagonal and the coefficient functions $r'_{ij}(s)$ are eigenfunctions for A_z.*

The relation (7.12) follows by straightforward calculation using $R(st) = R(s)R(t)$ and translation-invariance of the integral. As for (7.13), by (7.12) we have $C(z) = C(x)C(\tilde{x}) = C(x)C(x)^H \neq 0$, assuming $C(x) \neq 0$, and so $C(z)$ is a positive definite hermitian symmetric matrix that can be diagonalized by a unitary matrix, i.e., $P^H C(z) P = \Gamma$, where P is unitary and Γ is diagonal. If $R'(s) = \bar{P}^H R(s)\bar{P}$, then it is easily seen that the Fourier coefficients of $R'(s)$ are $C'(z) = \Gamma$. Since $A_z r'_{ij} = z * r'_{ij}$, it follows from (7.11) that $A_z r'_{ij} = \gamma_i r'_{ij}$, where γ_i is the ith diagonal entry of Γ. Thus the entire ith row of $R(s)$ consists of eigenfunctions for γ_i.

The systematic procedure for breaking off unitary representations of $\mathfrak{G}$ from the regular representation $T(s)$ utilized a variant of Schmidt's algorithm for generating an eigenfunction φ_1 for the largest eigenvalue λ_1 of A_z (Section 9.4). Using the fact that $\varphi_1 * \varphi_1 = \varphi_1$, they were able to show that if $R(s)$ is the unitary representation obtained (as explained earlier) by restricting $T(s)$ to the eigenspace $\mathcal{E}_{\lambda_1}$, then $C(x) \neq 0$. If $R(s)$ is not irreducible, it can be replaced by an irreducible unitary component $R^{(1)}(s)$ satisfying $C^{(1)}(x) \neq 0$.[57]

The next step is to repeat the above procedure with z replaced by z minus the finite Fourier series corresponding to $R^{(1)}(s)$, i.e.,

$$z_1 = z - \sum_{i,j} \frac{n_1}{\Omega} c_{ij}^{(1)}(z) r_{ij}^{(1)}(s).$$

Thus from A_{z_1} an irreducible unitary representation $R^{(2)}(s)$ is obtained, which satisfies $C^{(2)}(x) \neq 0$ and is not equivalent to $R^{(1)}(s)$, since the coefficient functions in the two representations turn out to be mutually orthogonal. Continuing this process, a sequence of functions z_p and representations $R^{(p)}$ is obtained, with z_p being z minus the finite Fourier series corresponding to $R^{(1)}, \ldots, R^{(p)}$. It is then proved that $z_p \to 0$ uniformly, which means that the Fourier series of z with respect to the functions $r_{ij}^{(p)}(s)$ converges uniformly to it:

$$z(s) = \sum_{p,i,j} \frac{n_p}{\Omega} c_{ij}^{(p)}(z) r_{ij}^{(p)}(s). \tag{7.14}$$

It is worth mentioning that, as Weyl realized [1926:74], (7.14) is a disguised form of the kernel expansion theorem of Section 9.4, which in the hermitian symmetric case and with Weyl's eigenvalue notation would be

$$K(s,t) = \sum_{\nu} \lambda_\nu \varphi_\nu(s)\overline{\varphi_\nu(t)}.$$

That is, if one replaces s by st^{-1} in (7.14), then the left-hand side is $K(s,t) = z(st^{-1})$ and the right-hand side takes the form $\sum_\nu \lambda_\nu \varphi_\nu(s)\overline{\varphi_\nu(t)}$,

57 Rather than invoke the complete reducibility theorem to do this, it is done constructively by a procedure they liken to Frobenius's method of generating primitive idempotents (called characteristic units) [1926:67–9].

since in view of (7.13), $C^{(p)}(z)$ can be assumed diagonal and the $r_{ij}^{(p)}$ taken as eigenfunctions for A_z with eigenvalues $\gamma_i^{(p)} = c_{ii}^{(p)}(z)$.[58]

Of course, in the kernel expansion theorem for A_z the φ_ν are assumed to be a full set of eigenfunctions for A_z, whereas in the version obtained from (7.14) the φ_ν are the eigenfunctions $r_{ij}^{(p)}(s)$ obtained by the above procedure. However, (7.14) also implies that the procedure yields all of them and, in fact, a complete set. That is, if one sets $s = e$ in (7.14), e the identity element of $\mathfrak{G}$, the left-hand side becomes $(x * \tilde{x})(e) = \int_{\mathfrak{G}} |x(s)|^2 ds$, whereas since $c_{ij}(z) = 0$ for $i \neq j$, the right-hand side is

$$\sum_p \frac{n_p}{\Omega} \operatorname{tr} C^{(p)}(z) = \sum_p \frac{n_p}{\Omega} \operatorname{tr} \left[C^{(p)}(x) C^{(p)}(x)^H \right],$$

so that (7.14) with $s = e$ is precisely (7.8), which, as we have seen, is just another way to express the fact of equality in Bessel's inequality, i.e., Parseval's equality.

The result of these considerations may be summed up as the following theorem:

(7.15) **Construction Theorem.** *For any fixed $x \neq 0$ in $\mathcal{C}(\mathfrak{G})$, it is possible to construct from the regular representation $s \to T(s)$ of (7.3) inequivalent irreducible unitary representations $R^{(p)}(s)$, $p = 1, 2, \ldots$, such that (1) $C^{(p)}(x) \neq 0$ for each p, and (2) Parseval's equality (7.8) holds.*

An immediate consequence of the construction theorem is the Peter–Weyl theorem because the functions $r_{ij}^{(p)}$ may be taken as a subset of the functions $r_{ij}^{(\mu)}$ corresponding to all inequivalent irreducible unitary representations. By (7.15) this subset already gives equality in Bessel's inequality, and so equality must hold as well for the full set (with $C^{(\mu)}(x) = 0$ for any $R^{(\mu)}(s)$ not among the $R^{(p)}(s)$).[59]

Another consequence of the construction theorem is the analogue of Frobenius's regular representation theorem of Section 10.1:

(7.16) *If $s \to R(s)$ is any irreducible representation of $\mathfrak{G}$ then (1) there is an $x \in \mathcal{C}(\mathfrak{G})$ such that a unitary representation $R^{(p)}(s)$ equivalent to $R(s)$ is produced in the construction theorem; (2) $R^{(p)}(s)$ occurs n_p times in $T(s)$.*

I mention this theorem because, as I have emphasized in the above presentation, the proof of the Peter–Weyl theorem consisted in seeking to imitate

---

58 Since then $c_{ij}^{(p)}(z) = 0$ for $i \neq j$ and $r(st^{-1}) = \sum_k r_{ik}^{(p)}(s)\overline{r_{ik}^{(p)}(t)}$, with each $r_{ik}^{(p)}$ an eigenfunction for $\gamma_i^{(p)}$.

59 A simpler proof of the Peter–Weyl theorem, which is based upon ideas in the Peter–Weyl paper but avoids the complications of the Schmidt iteration process and the kernel expansion (7.14), was given by Weil [1940:72–6], who achieved simplicity by working in the context of the Hilbert space $L^2(\mathfrak{G})$.

the regular representation theorem of Frobenius's theory. However, it is the Peter–Weyl theorem that the authors highlight and not (7.16). Nonetheless, they establish (1) in passing [1926:73], and although (2) is not explicitly mentioned, it follows readily from what is at hand.

The proof of (1) is of interest because it involves what are now called approximate identities. If $R(s)$ is the given irreducible representation, then it is equivalent to one that is unitary and so may be assumed unitary. To prove (1), all that is needed is to produce an $x \in \mathcal{C}(\mathfrak{G})$ such that $C(x) \neq 0$, where $C(x)$ is the Fourier coefficient matrix with respect to the given representation $R(s)$. This is because (as already remarked) the construction theorem implies that *all* inequivalent irreducible unitary representations $R(s)$ with Fourier coefficients $C(x) \neq 0$ are constructed by its procedure. The construction of the requisite x was inspired by the analogy with the case of a finite group. In that case, for any $x = \sum_s x(s)s$ in the group algebra, $C(x) = \sum_s x(s)\overline{R(s)}$, and so if $x = e$, the identity element of the group, which means that $x(s) = 0$ for all $s \neq e$ and $x(e) = 1$, $C(e) = \overline{R(e)} = I \neq 0$. Of course, as already noted, in the continuous case there is no $x \in \mathcal{C}(\mathfrak{G})$ corresponding to e, since the function $x_0(s)$, which is 1 for $s = e$ and zero otherwise, is not in $\mathcal{C}(\mathfrak{G})$. Peter and Weyl observed that x_0 can nonetheless be approximated by elements of $\mathcal{C}(\mathfrak{G})$, which is enough to get the same result. That is, we may certainly construct $x_n \in \mathcal{C}(\mathfrak{G})$ such that $x_n(s) \geq 0$ for all $s \in \mathfrak{G}$, $\int_{\mathfrak{G}} x_n(s)\, ds = 1$, and $x_n(s) = 0$ for all s outside of a neighborhood $\mathfrak{U}_n$ of e, where $\cap_{n=1}^{\infty} \mathfrak{U}_n = \{e\}$. Then the Fourier coefficients of the x_n with respect to R satisfy

$$\lim_{n\to\infty} C(x_n) = \lim_{n\to\infty} \int_{\mathfrak{U}_n} x_n(s)\overline{R(s)}\, ds = \overline{R(e)} = I.$$

Thus for n sufficiently large $C(x_n) \neq 0$.

As we saw in Section 9.2, Hilbert was more interested in ordinary uniform convergence than in convergence in the L^2 norm. As Hilbert's student, Weyl naturally considered the question of uniform convergence. Of course, it was clear that one could not expect that for any $x \in \mathcal{C}(\mathfrak{G})$ its Fourier series with respect to the system $r_{ij}^{(\mu)}$ would converge uniformly to x. It was known that even for ordinary Fourier series this is not the case because continuous functions with divergent Fourier series exist. It was in response to this phenomenon that Weierstrass showed that any continuous function on $[0, 2\pi]$ can be uniformly approximated by finite trigonometric sums $a_0 + \sum_{k=1}^{n}(a_k \cos kt + b_k \sin kt)$ [1885:22]. Weierstrass deduced this theorem from his polynomial approximation theorem. In several subsequent proofs of Weierstrass's theorem, including the one in Lebesgue's book on trigonometric series [1906], a book Weyl knew well,[60] the idea was to show that the given continuous function $x(s)$ could be uniformly approximated by a type of function $w(s)$ for which the Fourier series was known to converge

[60] Lebesgue's book is frequently cited in papers 2 and 4 of *Abhandlungen* **1**.

uniformly to w; for then the partial sums of the Fourier series for the w's can be used to uniformly approximate x.[61] Peter and Weyl showed that the same idea, with the role of the w-functions being played by functions that are convolutions, could be used to establish the following analogue of Weierstrass's theorem:

(7.17) **Approximation Theorem.** *Any $x \in \mathfrak{C}(\mathfrak{G})$ can be uniformly approximated by finite linear combinations of the functions $r_{ij}^{(\mu)}(s)$ corresponding to the irreducible unitary representations of $\mathfrak{G}$.*

Their proof of the approximation theorem proceeds as follows. As already noted, the first step is to show that the Fourier series with respect to $r_{ij}^{(\mu)}(s)$ of any function $w = x * y$, $x, y \in \mathfrak{C}(\mathfrak{G})$, converges uniformly to it. To this end let x_N denote the difference between $x(s)$ and the partial sum of its generalized Fourier series up through the representation corresponding to $\mu = N$:

$$x_N(s) = x(s) - \Big\{\sum_{\mu=1}^{N} \frac{n_\mu}{\Omega} \sum_{ij} c_{ij}(x) r_{ij}^{(\mu)}(s)\Big\}. \tag{7.18}$$

Completeness means that

$$\lim_{N\to\infty} \int_{\mathfrak{G}} |x_N(s)|^2 dm(s) = 0. \tag{7.19}$$

Now, if $w = x * y$, then $w_N = (x * y)_N = x_N * y$.[62] Thus by Schwarz's inequality for integrals, a staple in the Hilbertian theory of integral equations, together with translation invariance,

$$|w_N(s)| = |x_N * y| \leq \Big[\int_{\mathfrak{G}} |x_N(t)|^2 dm(t)\Big]^{\frac{1}{2}} \Big[\int_{\mathfrak{G}} |y(t)|^2 dm(t)\Big]^{\frac{1}{2}}.$$

From (7.19), we see that $w_N \to 0$ uniformly, i.e., the generalized Fourier series of functions that are convolutions converge uniformly to them. The second step is to prove that any continuous $x \in \mathfrak{C}(\mathfrak{G})$ can be uniformly approximated by convolutions. To this end they utilized the sequence y_n of approximate identity elements of $\mathfrak{C}(\mathfrak{G})$ defined in the course of proving Theorem 7.16 and denoted there by x_n. It is easily seen that $x * y_n \to x$ uniformly. Since $x * y_n$ is a convolution, it can be uniformly approximated by the partial sums of its generalized Fourier series. These two uniform approximations combine to yield the approximation theorem.

The theory developed by Peter and Weyl, when applied to the compact group $\mathfrak{G} = \widetilde{\mathfrak{A}}_u$, tied up the loose ends left dangling in Weyl's paper [1925]

[61] This idea also underlies A. Kneser's generalization of Weierstrass's theorem to finite linear combinations of the eigenfunctions of a Sturm–Liouville differential equation [1905:§5].

[62] The fact that $(x * y)_N = x_N * y$ follows by direct calculation from (7.18) using the relations in (7.11) and (7.12).

regarding an analogue of Frobenius's regular representation theorem; but of course much more had been achieved. An analogue of Fourier's theory had been created for any compact group $\mathfrak{G}$ equipped with a translation-invariant integration process to play the role of the integral of Hurwitz. Before long, Alfred Haar (1885–1933), whose 1909 Göttingen dissertation had been on systems of orthogonal functions, showed that a translation-invariant integral exists whenever $\mathfrak{G}$ is a separable locally compact topological group [1933].[63] Thus, as Haar observed, "the beautiful theory of F. Peter and H. Weyl" extends to these compact topological groups. This fact was the key to John von Neumann's solution to Hilbert's fifth problem for compact, locally Euclidean groups [1933]. His inspiration came from what Peter and Weyl regarded as the most important implication of their completeness theorem, namely that the irreducible representations of $\mathfrak{G}$ separate the points of $\mathfrak{G}$: If $R^{(\mu)}(g_1) = R^{\mu)}(g_2)$ for all μ, then $g_1 = g_2$ [1926:74].

Weyl did not directly contribute to the above-mentioned offshoots of the Peter–Weyl paper except for some provocative observations about the new theory of almost periodic functions. Harald Bohr, the brother of physicist Niels Bohr and a frequent visitor to Göttingen during Weyl's years there, had given a lecture in Zürich on the theory of almost periodic functions he had developed during 1923–26.[64] Bohr's theory was conceived by him as a generalization of the classical theory of Fourier series for periodic functions. At the heart of this conception was his theorem that any almost periodic function $x(t)$ has a mean value in the sense that

$$M[x] = \lim_{T\to\infty} \frac{1}{T} \int_a^{a+T} x(t)\, dt$$

exists and is independent of a. Bohr showed that if for any real λ, $c_\lambda = M[xe^{-i\lambda t}]$, then only countably many of the c_λ are nonzero and that the c_λ are analogous to the Fourier coefficients of a periodic function. In particular, one of his fundamental theorems is the analogue of Parseval's equality: $M[|x|^2] = \sum_\lambda c_\lambda^2$. Bohr's proof of this theorem ran to almost forty printed pages. Weyl recalled in a letter to Bohr that Bohr's Zürich lecture had rekindled his desire to find "a natural basis" (*natürliche Begründung*) for this fundamental theorem.[65] (As we saw in Chapter 11, Weyl had a Hilbertian penchant for seeking "natural" or "true" or "proper" ways to

[63] Haar assumed that $\mathfrak{G}$ was metrizable, although this was actually a consequence of known topological results. Weil later showed that the separability assumption was also unnecessary. See [Hawkins 1999b:§5].

[64] Bohr's main results were announced in two notes in the *Comptes Rendus* of the Paris Academy [1923a; 1923b]. His detailed papers are in volume 2 of his *Collected Mathematical Works*. See especially the *Acta Mathematica* papers C3, C7, C12 (1924–26). See also [Jessen 1951:172].

[65] The letter was published (without indicating the date of the original letter) as [Weyl 1926a], which was submitted (by L. Bieberbach) on 17 June 1926.

view mathematical phenomena.) In that connection he ventured the prediction that if a new sort of proof of Parseval's equality in the classical periodic case could be found, it would generalize to almost periodic functions.

With the Peter–Weyl theory in hand, Weyl could see that if it is applied to the compact group $\mathfrak{G}$ of real numbers modulo 2π under addition, then a completely different proof of Parseval's equality results. In the case of almost periodic functions the group $\mathfrak{G}$ is the additive group of real numbers, which is not compact. But if instead of considering $\mathcal{C}(\mathfrak{G})$ one restricts oneself to the almost periodic functions on $\mathfrak{G}$, then $\int_{\mathfrak{G}} x(t)dt = M[x]$ defines a translation-invariant "integral" on these functions. In this way, Weyl saw how to give an entirely different proof of Bohr's fundamental theorem [1926a; 1926b], and before long von Neumann was creating a theory of almost periodic functions on groups [1934:445]. An idea of the results achieved in the immediately following years can be obtained from A. Weil's book [1940: Ch. VII]. An interesting upshot of these developments is that the almost periodic functions on a Hausdorff topological group can be identified with the continuous functions on a certain compact group, so that the theory can actually be regarded within the original framework of the Peter–Weyl theory.

Afterword
Suggested Further Reading

With the publication of the Peter–Weyl paper [1926], Weyl had completed what he had begun in his paper [1925]. Within the context of semisimple groups he had fused into a splendid whole the theory of continuous groups of Lie, Killing, and Cartan and the theory of group characters and representations of Frobenius and Schur. Although this paper thus forms a fitting conclusion to the story of the emergence of the theory of Lie groups, the development of "Lie theory" had really just begun.

Thanks in large part to Weyl's impressive results, the mathematical community became more generally aware of the theory, including, of course, the extraordinary work of Killing and Cartan, and research in various directions gradually commenced. Although no detailed, comprehensive history of all these subsequent developments exists, there are a number of works from which the interested reader can gain some idea of these further developments. The purpose of this afterword is to make some suggestions along these lines, most of them being limited to matters touched upon in the present volume.

It should be clear from Section 12.7 that in order to tie up the loose ends of his paper [1925] on the structure and representation of semisimple groups, Weyl in his paper with Peter had ended up laying the foundations for a new theory—the still-burgeoning field of harmonic analysis on groups. On the historical aspects of these developments the writings of Mackey [1992] are very instructive, as is the expository essay by Gross [1978].

In other ways as well, Weyl's papers [1925,1926] provided a significant stimulus to the development of various aspects of the more general theory of topological groups. For example, as noted in Section 12.7, von Neumann's contribution to the solution of Hilbert's fifth problem made critical use of the Peter–Weyl Theory. It represented the first solution to Hilbert's problem for an extensive class of groups and provided inspiration for further work on the problem for general classes of groups. For an historical discussion of Weyl's influence on the development of topological groups see my paper [1999b] and its references.

In Section 12.6 we saw that Weyl's introduction of the universal covering group $\widetilde{\mathfrak{A}}_u$ and his study of its connectivity properties had stimulated Cartan's interest in the topology of Lie groups. Cartan's work along these lines (e.g., on the betti numbers of the simple groups) gradually stimulated the interest of a growing number of mathematicians, and a sizable literature on the topology of Lie groups now exists. For an overview of research in this area, see the survey articles by Samelson [1952] and Borel [1955]. For an idea of Cartan's own results in this area, see his monographs [1930,1936], de Rham's essays [1981:641–64], and the chapter on Cartan in Borel's forthcoming collection of essays [2000].

In Section 12.6 passing mention was made of Cartan's beautiful fusion of Lie theory with differential geometry, and, in particular, his theory of symmetric spaces. Cartan's work along these lines is discussed carefully and at length in Borel's above-mentioned essay on Cartan in [2000]. See also the book on Cartan by Akivis and Rosenfeld [1993].

Weyl, too, went on to contribute more to the theory of Lie groups and related matters than has been discussed here. For an illuminating and careful overview of all this work in this area as well as some of its subsequent ramifications, I recommend Borel's masterful essay "Hermann Weyl and Lie Groups" [1986], which is to be included in [Borel 2000].

In Section 12.4 I hinted at the close connections between Lie theory, the theory of invariants, and the theory of algebraic groups. The reader interested in this line of development should study the several chapters in [Borel 2000] devoted to various aspects of the history of algebraic groups in the nineteenth and twentieth centuries.

Finally, for one mathematician's expository mathematical summation of "A Century of Lie Theory" see the survey essay by Howe [1992], which also includes a fairly extensive bibliography.

References

Published and Unpublished Sources

Principal Manuscript Sources

In referring to archival sources I have used the same abbreviations as in the main body of the text. They are as follows:

Engel Archive Giessen. The papers of F. Engel are located in the Mathematisches Institut, Justus-Liebig Universität Giessen, Germany.

ETH Zürich. Wissenschaftshistorische Sammlungen der Bibliothek der Eidgenössischen Technischen Hochschule, Zürich, Switzerland.

NSUB Göttingen. Handschriftenabteilung, Niedersächsische Staats- und Universitätsbibliothek, Göttingen, Germany.

UB Oslo. Department of Manuscripts, National Library of Norway, Oslo Division (Haandskriftsamlingen, Nasjonalbiblioteket, avdeling Oslo). Formerly known as the University Library of Oslo (Universitetsbiblioteket i Oslo).

List of Figures and Their Sources

Figs. **1**, **2-top** and **7** (Ch. 1), Fig. **10** (Ch. 3), Fig. **19-bottom** (Ch. 7), Fig. 25 (Ch. 8): Department of Mathematics, University of Oslo.

Fig. **2-bottom**: Reproduced from [Plücker 1895].

Fig. **3** (Ch. 1): Reproduced from [Galois 1897].

Fig. **4** (Ch. 1): Reproduced from [Hudson 1905] with the permission of Cambridge University Press.

Fig. **5-top** (Ch. 1), Fig. **8** (Ch. 2), Fig. **17-bottom** (Ch. 6), Fig. **23** (Ch. 7): D. E. Smith Collection of Portraits of Eminent Mathematicians, Rare Book and Manuscript Library, Columbia University, New York.

Fig.**5-bottom** (Ch. 1), Fig. **9** (Ch. 2), Fig. **12** (Ch. 4), Figs. **26-bottom**, **27** & **30** (Ch. 9), Fig. **34** (Ch 10): Archives of Mathematisches Forschungsinstitut, Oberwolfach.

Fig. **6** (Ch. 1): Reproduced from [Lie 1896: 260].

Fig. **11** (Ch. 4): Porträt-Sammlung, Universitäts- und Landesbibliothek Münster.

Fig. **13-top** (Ch. 4), Fig. **28** (Ch. 9): Deutsches Museum, Munich.

Fig. **13-bottom** (Ch. 4): Reproduced from [Helmholtz 1892].

Fig. **14** (Ch. 5): Reproduced from [Oellers 1925].

Fig. **15** (Ch. 6), Fig. **41-top** (Ch. 12): Courtesy of Professor Henri Cartan, Paris.

Fig. **16** (Ch. 6): Reproduced from [Lebon 1913].

Fig. **17-top** (Ch. 6). Reproduced from [Picard 1899].

Fig. **18** (Ch. 6): Reproduced from [Picard 1926].

Fig. **19-top** (Ch. 7): Reproduced from [Lie 1921].

Fig. **20** (Ch. 7): Mathematisches Institut, Universität Bonn.

Fig. **21** (Ch. 7): Archivio dell'Accademia della Scienze di Torino, Membri effettivi. Ritratti, mazzo 103.

Fig. **22** (Ch. 7): Fonti iconografiche, Biblioteca Giuseppe Peano, Dipartimento di Matematica, Torino.

Fig. **24** (Ch. 7): Reproduced from [Kowalewski 1950].

Fig. **26-top** (Ch. 9), Fig. **33** (Ch. 10): Springer-Verlag New York.

Fig. **29** (Ch. 9): Reproduced from [Minkowski 1911].

Fig. **31** (Ch. 9): L. E. J. Brouwer Archive (D. van Dalen, custodian).

Fig. **32** (Ch. 10): Institut Mittag-Leffler, Djursholm, Sweden.

Fig. **35** (Ch. 10): Courtesy of Prof. R. Conzelmann, Basel.

Figs. **36** and **38** (Ch. 11), **40** (Ch. 12): George Pólya Papers (Sc337), Stanford University Archives.

Fig. **37** (Ch. 11): From a print in the archives of the collected papers of Albert Einstein at Boston University with permission of the copyright holder, Anna-Regula Revay-Grossmann, Ittigen, Switzerland.

Fig. **39** (Ch. 12): Reproduced from [Reid 1970: 239].

Fig. **41-bottom** (Ch. 12): Courtesy of Martha Weyl Thompson, Amherst, Mass.

References

Abel, N. H., 1829. "Mémoire sur une classe particulière d'équations résolubles algébriquement, *Jl. für die reine u. angew. Math.* **4** (1829). Reprinted in *Oeuvres* **1**, 478–507.

Abikoff, W., 1981. "The uniformization theorem," *American Mathematical Monthly* **88** (1981), 574–592.

Ado, I. D., 1934. "Über die Darstellung der endlichen kontinuierliche Gruppen durch lineare Substitutionen," *Bull. Soc. Physico-Math. Kazan* **7** (1934–35), 3–43.

Akivis, M. A. and B. A. Rosenfeld, 1993. *Elie Cartan (1869–1951). Translations of Mathematical Monographs Vol. 123*, Providence, American Mathematical Society, 1993.

Alexandrov, P. S. ed., 1971. *Die Hilbertschen Probleme*, Leipzig 1971. German translation of *Problemy Gil'berta* under the editorship of H. Wussing.

Antoine, J.-P. and D. Speiser, 1964. "Characters of irreducible representations of simple groups," *J. Math. Phys.* **5** (1964), 1226–1234; 1560–1572.

Apfelstedt, M., 1906. *Über eine Gattung von projektiven Transformationsgruppen in fünf Veränderlichen*, Leipzig, Inauguraldissertation, Greifswald.

Balazs, N. L., 1971. "Einstein: Theory of Relativity," *Dictionary of Scientific Biography* **4** (1971), 319–332.

Beltrami, E., 1868. "Saggio di interpretazione della geometria non-euclidea," *Giornale di Matematiche* **6** (1868), 284–312. Reprinted in *Opere* **1**, 374–405.

Bernkopf, M., 1966. "The Development of the Theory of Function Spaces with Particular Reference to their Origins in Integral Equation Theory," *Archive for History of Exact Sciences* **3** (1966), 1–96.

Bieberbach, L., 1930. "Über den Einfluss von Hilberts Pariser Vortrag über 'Mathematische Probleme' auf den Entwicklung der Mathematik in den letzten dreissig Jahren," *Naturwissenschaften* **18** (1930), 1101–1111.

Bieberbach, L., 1968. "Das Werk Paul Koebes," *Jahresbericht der Deutschen Mathematiker-Vereinigung* **70** (1968), 148–158.

Biermann, K., 1973. *Die Mathematik und ihre Dozenten an der Berliner Universität 1810–1920*, Berlin, Akademie-Verlag, 1973.

Birkeland, B. ed., 1995. *Ludvig Sylow Forelæsninger over algebraisk Ligninger og Substitutions theorie holdte ved Universitet i Christiania som Prof. O. J. Brochs Vikar 1862. Utskrevet 1995 fra Sylows manuskript av Bent Birkeland*, Olso 1995.

Birkeland, B., 1996. "Ludvig Sylow's Lectures on Algebraic Equations and Substitutions, Christiania (Oslo), 1862: An Introduction and a Summary," *Historia Mathematica* **23** (1996), 182–199.

Bishop, R. L. and R. J. Chrittenden, 1964. *Geometry of Manifolds*, New York and London, Academic Press, 1964.

Blumenthal, O., 1935. "Lebensgeschichte," *David Hilbert Gesammelte Abhandlungen*, Vol. 3, Berlin, Springer, pp. 388–429.

Bohr, H., 1923a. "Sur les fonctions presque périodiques," *Comptes Rendus, Acad. Sci. Paris* **177** (1923), 737–739. Reprinted in *Collected Mathematical Works* **2**, paper C1.

Bohr, H., 1923b. "Sur l'approximation des fonctions presque périodiques par des sommes trigonométriques," *Comptes Rendus, Acad. Sci. Paris* **177** (1923), 1090–1092. Reprinted in *Collected Mathematical Works* **2**, paper C2.

Bollinger, M., 1972. "Geschichtliche Entwicklung des Homologiebegriffs," *Archive for History of Exact Sciences* **9** (1972), 94–166.

Bonola, R., 1912. *Non-Euclidean Geometry*, Chicago, 1912. Reprinted by Dover Publications (New York, 1955).

Borel, A., 1955. "Topology of Lie groups and characteristic classes," *Bulletin of the American Mathematical Society* **61** (1955), 397–432. Reprinted in *Papers* **1**, No. 33.

Borel, A., 1986. "Hermann Weyl and Lie Groups," *Hermann Weyl 1885–1985*, K. Chandrasekharan, ed., Berlin and Heidelberg, Springer-Verlag, pp. 655–685.

Borel, A., 1998. "Full Reducibility and Invariants for $SL_2(\mathbb{C})$," *L'Enseignment Mathématique* **(2) 44** (1998), 71–90.

Borel, A., 2000. *Essays in the History of Lie Groups and Algebraic Groups*, Providence, American Mathematical Society (to appear).

Bourbaki, N., 1974. *Éléments d'histoire des mathématiques. Nouvelle édition augmentée*, Paris, Hermann, 1974.

Bourbaki, N., 1975. *Elements of Mathematics. Lie Groups and Lie Algebras. Part I*, Reading, Mass., Addison-Wesley, 1975.

Brodén, T., 1905. *Bemerkungen über die Uniformisierung analytischer Funktionen*, Lund, 1905.

Brouwer, L. E. J., 1976. *Collected Works 2*, Amsterdam, North-Holland, H. Freudenthal, ed., 1976.

Bröcker, T. and T. t. Dieck, 1985. *Representations of Compact Lie Groups*, New York, Springer-Verlag, 1985.

Burnside, W., 1904a. "On the reduction of a group of homogeneous linear substitutions of finite order," *Acta Mathematica* **28** (1904), 369–387 [December 6, 1902].

Burnside, W., 1904b. "On the representation of a group of finite order as an irreducible group of linear substitutions and the direct establishment of the relations between the group-characteristics," *Proceedings London Math. Soc.* **(2) 1** (1904), 117–123 [February 28, 1903].

Cartan, E., 1893a. "Sur la structure des groupes finis et continus," *Comptes Rendus, Acad. Sci. Paris* **116** (1893), 962–64. Reprinted in *Oeuvres* **1**, 103–5.

Cartan, E., 1893b. "Sur la structure des groupes simples finis et continus," *Comptes Rendus, Acad. Sci. Paris* **116** (1893), 784–786. Reprinted in *Oeuvres* **1**, 99–101.

Cartan, E., 1893c. "Über die einfachen Transformationsgruppen," *Berichte über d. Verh. d. Sächsischen Gesell. der Wiss., math.-phys. Klasse, 1893* **45** (1893), 395–420. Reprinted in *Oeuvres* **1**, 107–132.

Cartan, E., 1894a. "Sur la réduction de la structure d'un groupe à sa forme canonique," *Comptes Rendus, Acad. Sci. Paris* **119** (1894), 640–642. Reprinted in *Oeuvres* **1**, 133–135.

Cartan, E., 1894b. *Sur la structure des groupes finis et continus*, Thèse, Paris, 1894. Reprinted in *Oeuvres* **1**, 137–287.

Cartan, E., 1894c. "Sur un théorème de M. Bertrand," *Comptes Rendus, Acad. Sci. Paris* **119** (1894), p. 902. Reprinted in *Oeuvres* $\mathbf{3}_1$, 1.

Cartan, E., 1895. "Sur certains groupes algébriques," *Comptes Rendus, Acad. Sci. Paris* **120** (1895), 545–548. Reprinted in *Oeuvres* **1**, 289–292.

Cartan, E., 1896a. "Le principe de dualité et certaines intégrales multiples de l'espace tangentiel et de l'espace réglé," *Bull. Soc. Math. France* **24** (1896), 140–177. Reprinted in *Oeuvres* **2**, 265–302.

Cartan, E., 1896b. "Sur la réduction à sa forme canonique de la structure d'un groupe de transformations fini et continu," *American Journal of Mathematics* **18** (1896), 1–61. Reprinted in *Oeuvres* **1**, 293–253.

Cartan, E., 1897a. "Sur les systèmes de nombres complexes," *Comptes Rendus, Acad. Sci. Paris* **124** (1897), 1217–1220. Reprinted in *Oeuvres* **2**, 1–4.

Cartan, E., 1897b. "Sur les sytèmes réels de nombres complexes," *Comptes Rendus, Acad. Sci. Paris* **124** (1897), 1296–97. Reprinted in *Oeuvres* **2**, 5–6.

Cartan, E., 1898. "Les groupes bilinéaires et les systèmes de nombres complexes," *Ann. Fac. Sci. Toulouse* **12** (1898), B1–B99. Reprinted in *Oeuvres* **2**, 7–105.

Cartan, E., 1899. "Sur certaines expressions différentieles et le problème de Pfaff," *Annales scientifiques École Normale Sup. Paris* **(3)16** (1899), 239–332. Reprinted in *Oeuvres* **2**, 303–396.

Cartan, E., 1901a. "Sur quelques quadratures dont l'élément différentiel contient des fonctions arbitraires," *Bulletin des sciences mathématiques* **29** (1901), 118–130. Reprinted in *Oeuvres* **2**, 397–409.

Cartan, E., 1901b. "Sur l'intégration des systèmes d'équations aux différentieles totales," *Annales scientifiques École Normale Sup. Paris* **18** (1901), 241–311. Reprinted in *Oeuvres* **2**, 411–481.

Cartan, E., 1901c. "Sur l'intégration des systèmes de Pfaff de caractère deux," *Bulletin des sciences mathématiques* **29** (1901), 233–302. Reprinted in *Oeuvres* **2**, 483–553.

Cartan, E., 1902. "Sur la structure des groupes infinis," *Comptes Rendus, Acad. Sci. Paris* **135** (1902), 851–853. Reprinted in *Oeuvres* **2**, 567–569.

Cartan, E., 1904. "Sur la structure des groupes infinis de transformations," *Annales scientifiques École Normale Sup. Paris* **(3)21** (1904), 153–206. Reprinted in *Oeuvres* **2**, 571–624.

Cartan, E., 1905. "Sur la structure des groupes infinis de transformations," *Annales scientifiques École Normale Sup. Paris* **(3)22** (1905), 219–308. Reprinted in *Oeuvres* **2**, 625–714.

Cartan, E., 1907. "Les groupes de transformations continus, infinis, simples," *Comptes Rendus, Acad. Sci. Paris* **144** (1907), 1094–1098. Reprinted in *Oeuvres* **2**, 715–718.

Cartan, E., 1908. "Nombres complexes. Exposé d'après l'article allemand de E. Study ...," *Encyclopédie des sciences mathématiques* **I,5** (1908), 331–468. Reprinted in *Oeuvres* **2**, 107–246.

Cartan, E., 1909. "Les groupes de transformations continus, infinis, simples," *Annales scientifiques École Normale Sup. Paris* **(3) 26** (1909), 93–161. Reprinted in *Oeuvres* **2**, 857–925.

Cartan, E., 1910a. "La structure des groupes de transformations continus et la théorie du trièdre mobile," *Bulletin des sciences mathématiques* **(2) 34** (1910), 250–283. Reprinted in *Oeuvres* $\mathbf{3}_1$, 145–178.

Cartan, E., 1910b. "Les développables isotropes et la méthode du trièdre mobile," *Comptes Rendus, Acad. Sci. Paris* **151** (1910), 919–921. Reprinted in *Oeuvres* $\mathbf{3}_1$, 141–143.

Cartan, E., 1913. "Les groupes projectifs qui ne laissent invariante aucune multiplicité plane," *Bulletin des sciences mathématiques* **41** (1913), 53–96. Reprinted in *Oeuvres* **1**, 355–398.

Cartan, E., 1914a. "La théorie des groupes," *Revue du Mois* **17** (1914), 438–468.

Cartan, E., 1914b. "Les groupes projectifs continus réels qui ne laissent invariante aucune multiplicité plane," *Jl. des math. pures et appl.* **10** (1914), 149–186. Reprinted in *Oeuvres* **1**, 493–530.

Cartan, E., 1914c. "Les groupes réels simples finis et continus," *Annales scientifiques École Normale Sup. Paris* **(3)31** (1914), 263–355. Reprinted in *Oeuvres* **1**, 399–492.

Cartan, E., 1915. "La théorie des groupes finis et continus et la géométrie. Exposé d'après l'article allemand de G. Fano," *Encyclopédie des sciences mathématiques* **III, 1** (1915), 332f. Due to the vicissitudes of the First World War, only the first part of the article appeared in 1915. The entire article was first published in Cartan's *Oeuvres* $\mathbf{3}_2$, 1727–1861.

Cartan, E., 1922. "Sur un théorème fondamental de M. H. Weyl dans la théorie de l'espace métrique," *Comptes Rendus, Acad. Sci. Paris* **175** (1922), 82–85. Reprinted in *Oeuvres* $\mathbf{3}_1$, 629–632. [10 July 1922]

Cartan, E., 1923. "Sur un théorème fondamental de M. H. Weyl," *Jl. des math. pures et appl.* **(2)2** (1923), 167–192. Reprinted in *Oeuvres* $\mathbf{3}_1$, 633–658.

Cartan, E., 1927. "La géométrie des groupes simples," *Annali di Matematiche* **4** (1927), 209–256. Reprinted in *Oeuvres* **1**, 793–840.

Cartan, E., 1928. "Sur les espaces de Riemann clos admettant un groupe continu transitif de déplacements," *Comptes Rendus, Acad. Sci. Paris* **186** (1928), 1817–1819. Reprinted in *Oeuvres* **1**, 995–997.

Cartan, E., 1929. "Groupes simples clos et ouverts et la géométrie riemannienne," *Jl. des math. pures et appl.* **(6) 8** (1929), 1–29. Reprinted in *Oeuvres* **1**, 1011–1043.

Cartan, E., 1930. *La théorie des groupes finis et continus et l'anaysis situs*, Paris, Gauthier-Villars, 1930. Reprinted in *Oeuvres* **1**, 1165–1225.

Cartan, E., 1931. *Notice sur les travaux scientifiques*, Paris, 1931. Reprinted (with additional notes) in [1939] and [1974].

Cartan, E., 1936. "La topologie des espaces représentatifs des groupes de Lie," *L'Enseignement mathématique* **35** (1936), 177–200. Also published separately (with same title) as No. 358 in the series *Actualités Scientifiques et Industrielles* (Paris, Hermann, 1936). Reprinted in *Oeuvres* **1**, 1307–1330.

Cartan, E., 1938. *Leçons sur la théorie des spineurs*, 2 vols., Paris, 1938. Translated into English as *The theory of Spinors* (Paris, Hermann, 1966; New York, Dover, 1981).

Cartan, E., 1939. "Allocution à la Sorbonne 18 mai 1939," *Jubilé scientifique de M. Elie Cartan célébré à la Sorbonne 18 mai 1939*, Paris, Gauthier-Villars, 1939, pp. 51–59. Reprinted as Appendix C in [Akivis and Rosenfeld 1993].

Cartan, E., 1947. "L'Oeuvre scientifique de M. Ernest Vessiot," *Bulletin des sciences mathématiques*, **75** (1947), 1–8.

Cartan, E., 1971. "A Centenary: Sophus Lie," *Great Currents of Mathematical Thought*, Vol. 1, F. L. Lionnais, ed., New York, Dover, pp. 262–267.

Casimir, H. L. and B. L. van der Waerden, 1935. "Algebraischer Beweis der vollständigen Reduzibilität der Darstellungen halbeinfacher Liescher Gruppen," *Math. Ann.* **111** (1935), 1–12.

Cayley, A., 1854. "Nouvelles recherches sur les covariantes," *Jl. für die reine u. angew. Math.* **47** (1854), 75–143. Reprinted in *Papers* **2**, 164–178.

Cayley, A., 1856. "A Second Memoir Upon Quantics," *Phil. Trans. R. Soc. London* **146** (1856), 101–126. Reprinted in *Papers* **2**, 250–75.

Cayley, A., 1859. "A sixth memoir upon quantics," *Phil. Trans. R. Soc. London* **149** (1859), 61–90. Reprinted in *Papers* **2**, 561–592.

Cayley, A., 1868. "On the curves which satisfy given Conditions," *Phil. Trans. R. Soc. London* **158** (1868), 75–143. Reprinted in *Papers* **6**, 191–262.

Cecil, T. E., 1992. *Lie sphere geometry with applications to submanifolds*, New York, Springer-Verlag, 1992.

Chern, S. and C. Chevalley, 1952. "Elie Cartan and His Mathematical Work," *Bulletin of the American Mathematical Society* **58** (1952), 217–250.

Chevalley, A. and A. Weil, 1957. "Hermann Weyl," *L'Enseignment Mathématique* **3** (1957), 53–82. Reprinted in H. Weyl, *Abhandlungen* **4**, 655–685.

Chevalley, C., 1948. "Sur la classification des algèbres de Lie simples et de leurs représentations," *Comptes Rendus, Acad. Sci. Paris* **227** (1948), 1136–1138.

Chevalley, C., 1955. "Sur certains groupes simples," *Tôhoku Mathematical Jl.* **(2) 7** (1955), 14–66.

Clebsch, A., 1864. "Über die Anwendung der Abelschen Functionen in der Geometrie," *Jl. für die reine u. angew. Math.* **63** (1864), 189–243.

Clebsch, A., 1866. "Über die simultane Integration linearer partieller Differentialgleichungen," *Jl. für die reine u. angew. Math.* **65** (1866), 257–268.

Clebsch, A., 1872a. *Theorie der binären algebraischen Formen*, Leipzig, 1872.

Clebsch, A., 1872b. "Über eine Fundamentalaufgabe der Invariantentheorie," *Abh. K. Ges. der Wiss. Göttingen, Math. Cl.* **17** (1872), 3–62.

Clifford, W. K., 1870. "On the space-theory of matter," *Proc. Cambridge Philos. Soc.* **2** (1870), 157–158. Reprinted in *Papers*, 21–22.

Coolidge, J. L., 1945. *A History of the Conic Sections and Quadric Surfaces*, Oxford, 1945.

Corry, L., 1996. "Hilbert and Physics (1900–1915)," *Max Planck Institut für Wissenschaftsgeschichte. Preprint No. 43 (1996)*. Published in *The Symbolic Universe. Geometry and Physics 1890–1930*, J. Gray, ed., Oxford, Oxford University Press, 1999, pp. 145-188.

Corry, L., 1997a. "David Hilbert and the Axiomatization of Physics (1894–1905)," *Archive for History of Exact Sciences* **51** (1997), 89–197.

Corry, L., 1997b. "Hermann Minkowski and the Postulate of Relativity," *Archive for History of Exact Sciences* **51** (1997), 273–314.

Corry, L., 1999. "David Hilbert between Mechanical and Electromagnetic Reductionism (1910–1915)," *Archive for History of Exact Sciences* **53** (1999), 489–527.

Corry, L., J. Renn, and J. Stachel, 1997. "Belated Decision in the Hilbert–Einstein Priority Dispute," *Science* **278** (1997), 1270–1273.

Cotton, E.-C., 1900. *Sur les variétés à trois dimensions*, Paris (Thèse), 1900.

Cotton, E.-C., 1905. "Généralisation de la Théorie du trièdre mobile," *Bull. Soc. Math. France, 33: 1–23.*," *Bull. Société Math. de France* **33** (1905), 1–23.

Crilly, A. J., 1986. "The Rise of Cayley's Invariant Theory (1841–1862)," *Historia Mathematica* **13** (1986), 241–254.

Darboux, G., 1882. "Sur le problème de Pfaff," *Bulletin des sciences mathématiques* **(2) 6** (1882), 14–36, 49–68.

Darboux, G., 1887–96. *Leçons sur la théorie générale des surfaces et les applications géométriques*, Paris, 1887–1896. A second edition of Vols. 1 and 2 was published in 1914 and 1915, respectively. Vols. 1 and 2 (2nd edn.) and 3 and 4 (1st edn.) have been reprinted by Chelsea Publishing Co., New York, 1972.

Darboux, G., 1912. *Éloges académiques et discours*, Paris, 1912.

De Rham, G., 1975. "L'œuvre d'Elie Cartan et la topologie," *Hommage à Elie Cartan*, Bucarest, Editura Academici Republicii Socialiste România, 1975, pp. 11–20. Reprinted in *Oeuvres*, pp. 641–650.

De Rham, G., 1981. *Oeuvres mathématiques*, L'Enseignement mathématique, Université de Genève, 1981.

de Tilly, J. M., 1880. "Essai sur les principes fondamentaux de la géométrie et de la mécanique," *Mém. Soc. des Sci. Physiques et Naturelles de Bordeaux* **(2) 3** (1880).

Demoulin, A., 1905. "Sur l'emploi d'un tétraèdre de référence mobile en géométrie Caleyenne," *Comptes Rendus, Acad. Sci. Paris* **139** (1905), 393–395.

Deruyts, J., 1892. "Essai d'une théorie générale des formes algébriques," *Mémoires Soc. Roy. Sci. de Liège* **(2) 17** (1892), 1–156.

Dieudonné, J., 1961. "Notes sur les travaux de C. Jordan relatifs à sa théorie des groupes finis," *Oeuvres de Camille Jordan* **1** (1961), xvii–xlii.

Dieudonné, J., 1976. "Weyl, Hermann," *Dictionary of Scientific Biography* **16** (1976), 280–285.

Dieudonné, J., 1981. *History of Functional Analysis*, Amsterdam, North Holland Publishing Co., 1981.

Dieudonné, J., 1988. "The Early Papers of Hermann Weyl on Spectral Theory," *Exakte Wissenschaften und ihre philosophische Grundlegung. Vorträge des Internationalen Hermann-Weyl Kongress, Kiel 1985*, W. Deppert, ed., Frankfurt am Main, Peter Lang, pp. 41–50.

Dieudonné, J., 1989. *A History of Algebraic and Differential Topology 1900–1960*, Boston, Birkhäuser, 1989.

Dieudonné, J. and J. Carrell, 1970. *Invariant Theory, Old and New*, New York, Academic Press, 1970. Reprinted from *Advances in Mathematics* 4_1 (1970).

Drach, J., 1898. "Essai sur une théorie générale de l'intégration et sur la classification des transcendentes," *Annales scientifiques École Normale Sup. Paris* **(3) 15** (1898), 243–384.

Du Bois-Reymond, P., 1864. *Beiträge zur Interpretation der partiellen Differentialgleichungen mit drei Variablen*, Leipzig, 1864.

Dugac, P., 1973. "Éléments d'analyse de Karl Weierstrass," *Archive for History of Exact Sciences* **10** (1973), 41–176.

Dugac, P. and R. Taton, 1981. "Éléments pour une étude sur Jules Drach," *Cahiers du Séminaire d'Histoire des Mathématiques* **2** (1981), 17–57.

Edwards, H. M., 1980. "The Genesis of Ideal Theory," *Archive for History of Exact Sciences* **23** (1980), 321–378.

Edwards, H. M., 1983. "Dedekind's Invention of Ideals," *Bulletin of the London Mathematical Society* **15** (1983), 8–17.

Edwards, H. M., 1984. *Galois Theory*, New York, Springer-Verlag, 1984.

Einstein, A., 1905. "Zur Elektrodynamik der bewegter Körper," *Annalen der Physik* **17** (1905), 891–921. Reprinted in *Papers* **2**, 275–310. A translation is contained in H. A. Lorentz, A. Einstein, H. Minkowski, H. Weyl, *The Principle of Relativity* (London, 1923), pp. 37–65. Reprinted by Dover Publications (New York, n.d.). Unless otherwise noted, all references are to the English translation.

Einstein, A., 1907. "Über das Relativitätsprinzip und die aus demselben gezogenen Folgerungen," *Jahrbuch der Radioaktivität und Elektronik* **4** (1907), 411–462.

Einstein, A., 1911. "Über den Einfluss der Schwerkraft auf die Ausbreitung des Lichtes," *Annalen der Physik* **35** (1911), 898–908. A partial translation is contained in H. A. Lorentz, A. Einstein, H. Minkowski, H. Weyl, *The Principle of Relativity* (London, 1923), pp. 99–108. Reprinted by Dover Publications (New York, n.d.). Unless otherwise noted, all references are to the English translation.

Einstein, A., 1915. "Zur allgemeinen Relativitätstheorie," *Sitzungsberichte der Akademie der Wiss. zu Berlin 1915* (1915), 778–786.

Einstein, A., 1916. "Die Grundlage der allgemeinen Relativitätstheorie," *Annalen der Physik* **49** (1916), 769–822. A partial translation is contained in H. A. Lorentz, A. Einstein, H. Minkowski, H. Weyl, *The Principle of Relativity* (London, 1923), pp. 111–164. Reprinted by Dover Publications (New York, n.d.). Unless otherwise noted, all references are to the English translation.

Einstein, A. and M. Grossmann, 1913. "Entwurf einer verallgeminerten Relativitätstheorie und die Theorie der Gravitation. I. Physikalischer Teil von A. Einstein. II. Mathematischer Teil von M. Grossmann," *Zeitschr. für Math. und Physik* **62** (1913), 225–261.

Engel, F., 1885. *Über die Definitionsgleichungen der continuirliche Transformationsgruppen*, Habilitationsschrift, Leipzig, 1885. Reprinted in *Math. Ann.*, **27** (1886), 1–57.

Engel, F., 1886. "Zur Theorie der Zusammensetzung der endlichen continuirlichen Transformationsgruppen," *Berichte über d. Verh. d. Sächsischen Gesell. der Wiss., math.-phys. Klasse, 1886* **38** (1886), 83–94.

Engel, F., 1887. "Kleinere Beiträge zur Gruppentheorie," *Berichte über d. Verh. d. Sächsischen Gesell. der Wiss., math.-phys. Klasse, 1887* **39** (1887), 89–99.

Engel, F., 1889–90. "Zur Invariantentheorie der Systeme von Pfaff'schen Gleichungen," *Berichte über d. Verh. d. Sächsischen Gesell. der Wiss., math.-phys. Klasse*, **41** (1889), 157–176; **42** (1890), 192–207.

Engel, F., 1892. "Die Erzeugung der endlichen Transformationen einer projektiven Gruppe durch die infinitesimalen Transformationen der Gruppe. Erste Mitteilung," *Berichte über d. Verh. d. Sächsischen Gesell. der Wiss., math.-phys. Klasse, 1892* **44** (1892), 279–291.

Engel, F., 1893a. "Die Erzeugung der endlichen Transformationen einer projektiven Gruppe durch die infinitesimalen Transformationen der Gruppe. Zweite Mitteilung, mit Beiträgen von E. Study," *Berichte über d. Verh. d. Sächsischen Gesell. der Wiss., math.-phys. Klasse, 1893* **45** (1893), 659–696.

Engel, F., 1893b. "Kleinere Beiträge zur Gruppentheorie. VIII," *Berichte über d. Verh. d. Sächsischen Gesell. der Wiss., math.-phys. Klasse, 1893* **45** (1893), 360–369.

Engel, F., 1893c. "Sur un groupe simple à quatorze paramètres," *Comptes Rendus, Acad. Sci. Paris* **116** (1983), 786–788.

Engel, F., 1899. "Sophus Lie," *Berichte über d. Verh. d. Sächsischen Gesell. der Wiss., math.-phys. Klasse* (1899), xi–lxi.

Engel, F., 1900. "Ein neues, dem linearen Komplexe analoges Gebilde," *Berichte über d. Verh. d. Sächsischen Gesell. der Wiss., math.-phys. Klasse, 1900* **52** (1900), 63–76, 220–239.

Engel, F., 1902. Review of [Slocum 1900a; 1900b; 1901] in *Jahrbuch über die Fortschritte der Mathematik* **31** (Jahrgang 1900) (1902), 148–150.

Engel, F., 1931. "Eduard Study," *Jahresbericht der Deutschen Mathematiker-Vereinigung* (1931), 133–156.

Enriques, F., 1893. "Le superficie con infinite transformazioni proiettive in sè stesse," *Atti R. Ist. Veneto* **(7) 51** (1893), 1590–1635.

Ernst, W., 1933. *Julius Plücker. Eine zusammenfassende Darstellung seines Lebens und Werkens als Mathematiker und Physiker auf Grund unveröffentlicher Briefe und Urkunden, Inaugural Dissertation Bonn*, Bonn, 1933.

Estanave, E., 1901. *Revue décennale des Thèses presentée à la Faculté des Sciences de Paris du 1*^{er} *Janvier 1891 au 31 Décembre 1900*, Paris, 1901.

Estanave, E., 1903. *Nomenclature des Thèses de sciences mathématiques soutenues en France dans le courant du XIX*^{e} *siècle devant les Facultés des sciences de Paris et des départements*, Paris, 1903.

Fano, G., 1896. "Sulle varietà algebriche con un gruppo continuo non integrabile de trasformazioni proiettive in sè," *Mem. Accad. Sci. Torino, Cl. Sci., fiz., mat., nat.* **(2) 46** (1896), 187–218.

Fano, G., 1899. "Sulle equazione differentiali lineari del 5° ordine du cui curve integrali sono contenute in una varietà algebrica," *Rend. R. Ist. Lomb.* **(2) 32, no.4** (1899),

Fano, G., 1900. "Über lineare Differentialgleichungen mit algebraischen Relationen zwischen den Fundamentallösungen," *Math. Ann.* **53** (1900), 493–590.

Fano, G., 1907a. "Gegensatz von synthetischer und analytischer Geometrie in seiner historischen Entwicklung im XIX Jahrhundert," *Enzyklopädie der mathematischen Wissenschaften* **III 4**^{a} (1907), 221–288.

Fano, G., 1907b. "Kontinuierliche geometrische Gruppen. Die Gruppentheorie als geometrisches Einteilungsprinzip," *Enzyklopädie der mathematischen Wissenschaften* **III 4**^{b} (1907), 289–388.

Fisher, C. S., 1966. "The Death of a Mathematical Theory: A Study in the Sociology of Knowledge," *Archive for History of Exact Sciences* 3_2 (1966), 137–159.

Fisher, C. S., 1967. "The Last Invariant Theorists," *Archiv. europ. sociol.* **8** (1967), 216–244.

Folland, G., 1970. "Weyl Manifolds," *J. Differential Geometry* **4** (1970), 145–153.

Frei, G. ed., 1985. *Der Briefwechsel David Hilbert-Felix Klein (1886–1918)*, Göttingen (Vandenhoeck and Ruprecht), 1985.

Frei, G. and U. Stammbach, 1992. *Hermann Weyl und die Mathematik an der ETH Zürich, 1913–1930*, Basel, Birkhäuser Verlag, 1992.

Frenkel, I. B. and V. G. Kac, 1980. "Basic Representations of Affine Lie Algebras and Dual Resonance Models," *Inventiones Math.* **62** (1980), 23–66.

Freudenthal, H., 1932. Review of [Mineur 1925] in *Jahrb. über die Fortschritte der Mathematik* **51** (1932), 318–19.

Freudenthal, H., 1972. "Hilbert, David," *Dictionary of Scientific Biography* **6** (1972), 388–395.

Frischauf, J., 1876. *Elemente der Absoluten Geometrie*, Leipzig, 1876.

Frobenius, G., 1877. "Über das Pfaffische Probleme," *Jl. für die reine u. angew. Math.* **82** (1877), 230–315.

Frobenius, G., 1878. "Über lineare Substitutionen und bilineare Formen," *Jl. für die reine u. angew. Math.* **84** (1878), 1–63. Reprinted in *Abhandlungen* **1**, 343–405.

Frobenius, G., 1896. "Über die Primfaktoren der Gruppendeterminante," *Sitzungsberichte der Akademie der Wiss. zu Berlin 1896* (1896), 1343–1382. Reprinted in *Abhandlungen* **3**, 38–77.

Frobenius, G., 1897. "Über die Darstellung der endlichen Gruppen durch lineare Substitutionen," *Sitzungsberichte der Akademie der Wiss. zu Berlin 1897* (1897), 994–1015. Reprinted in *Abhandlungen* **3**, 82–103.

Frobenius, G., 1898. "Über Relationen zwischen den Charakteren einer Gruppe und denen ihrer Untergruppen," *Sitzungsberichte der Akademie der Wiss. zu Berlin 1898* (1898), 501–515. Reprinted in *Abhandlungen* **3**, 104–118.

Frobenius, G., 1899. "Über die Darstellung der endlichen Gruppen durch lineare Substitutionen II," *Sitzungsberichte der Akademie der Wiss. zu Berlin 1899* (1899), 482–500. Reprinted in *Abhandlungen* **3**, 129–147.

Frobenius, G., 1900. "Über die Charactere der symmetrischen Gruppe," *Sitzungsberichte der Akademie der Wiss. zu Berlin 1900* (1900), 516–534. Reprinted in *Abhandlungen* **3**, 148–166.

Frobenius, G., 1903. "Über die charakteristischen Einheiten der symmetrischen Gruppe," *Sitzungsberichte der Akademie der Wiss. zu Berlin 1903* (1903), 328–358. Reprinted in *Abhandlungen* **3**, 244–274.

Fubini, G., 1904. "Sulle metriche definite da una forma hermitiana," *Atti del real Istitito Veneto di sci., lett. ed arti* **63**$_2$ (1904), 501–513.

Galison, P. L., 1976. *Minkowski's Space–Time: From Visual Thought to the Absolute World*, B.A. thesis, Harvard University 1976.

Galison, P. L., 1979. "Minkowski's Space-Time: From Visual Thinking to the Absolute World," *Historical Studies in the Physical Sciences* **10** (1979), 85–121.

Galois, E., 1846. "Mémoire sur les conditions de résolubilité des équations par radicaux," *Jl. des math. pures et appl.* **11** (1846). [16 January 1831] Reprinted in [Galois 1962:33–50].

Galois, E., 1897. *Oeuvres mathématiques d'Evariste Galois, publiées sous les auspices de la Société mathématique de France, avec une introduction par M. Emile Picard*, Paris, Gauthier-Villars, 1897.

Galois, E., 1962. *Écrits et mémoires mathématiques. Ed. critique intégrale de ses manuscrits et publications, par Robert Bourgne et J.-P. Azra*, Paris, Gauthier-Villars, 1962.

Gantmacher, F., 1939. "On the classification of real simple Lie groups," *Matematicheskii sbornik* **5** (1939), 217–249.

Godeaux, L., 1949. "Notice sur Jacques Deruyts," *Annuaire de l'Académie Royale de Belgique* **115** (1949), 21–43.

González-López, A., N. Kamran, and P. J. Olver, 1992a. "Lie algebras of differential operators in two complex variables," *Amer. Jl. of Math.* **114** (1992), 1163–1185.

González-López, A., N. Kamran, and P. J. Olver, 1992b. "Lie algebras of vector fields in the real plane" *Proceedings London Math. Soc.* **64** (1992), 339–368.

González-López, A., N. Kamran, and P. J. Olver, 1996. "Real Lie algebras of differential operators, and quasi-exactly solvable potentials," *Phil. Trans. R. Soc. London* **A 354** (1996), 1165–1193.

Gordan, P., 1871. "Über die Bildung der Resultante zweier Gleichungen," *Math. Ann.* **3** (1971), 355–414.

Gordan, P., 1872. "Über Combinanten," *Math. Ann.* **5** (1872), 95–122.

Gordan, P., 1887. *Vorlesungen über Invariantentheorie. Herausgegeben von Dr. Georg Kerschensteiner. Zweiter Band: Binäre Formen*, Leipzig, 1887.

Goursat, E., 1891. *Leçons sur l'intégration des équations aux derivées partielles du premier ordre*, Paris, 1891.

Goursat, E., 1896–98. *Leçons sur l'intégration des équations aux derivées partielles du seconde ordre*, 2 vols., Paris, 1896 and 1898.

Goursat, E., 1922. *Leçons sur le problème de Pfaff*, Paris, 1922.

Gray, J., 1979. "Non-Euclidean geometry—A reinterpretation," *Historia Mathematica* **6** (1979), 236–258.

Gray, J., 1981. "Les trois suppléments au Mémoire de Poincaré, écrit en 1880, sur les fonctions fuchsiennes et les équations différentielles," *Comptes Rendus, Acad. Sci. Paris* **293** (1981), 87–90.

Gray, J., 1982. "The Three Supplements to Poincaré's Prize Essay of 1880 on Fuchsian Functions and Differential Equations," *Arch. int. d'histoire des sci.* **32** (1982), 221–235.

Gray, J., 1986. *Linear Differential Equations and Group Theory from Riemann to Poincaré*, Boston, Birkhäuser, 1986.

Gray, J., 1994. "On the History of the Riemann Mapping Theorem," *Supplemento ai Rendiconti Circolo Mat. Palermo* **(2) 34** (1994), 47–94.

Green, J. A., 1980. *Polynomial Representations of* GL_n, Berlin, Springer-Verlag, 1980.

Green, J. A., 1991. "Classical Invariants and the General Linear Group," *Progress in Mathematics* **95** (1991), 247–272.

Gross, K. I., 1978. "On the evolution of noncommutative harmonic analysis," *American Mathematical Monthly* **85** (1978), 359–371.

Haar, A., 1933. "Der Massbegriff in der Theorie der kontinuierlichen Gruppen," *Annals of Math.* **34** (1933), 147–169.

Hahn, H., 1911. "Bericht über die Theorie der linearen Integralgleichungen," *Jahresbericht der Deutschen Mathematiker-Vereinigung* **20** (1911), 69–117.

Halphen, G., 1878. *Sur les invariantes différentielles*, Thèse, Paris, 1878. Reprinted in *Oeuvres* **2**, 197–253.

Halphen, G., 1884. "Sur la réduction des équations différentielles linéaires aux formes intégrables," *Mémoires présentés par divers savants à la Académie des Sciences* **(2) 28** (1884), No. 1 (301pp). Reprinted in *Oeuvres* **3**, 1–260.

Hamilton, W. R., 1834. "On a General Method in Dynamics by which the Study of the Motions of all Free Systems of Attracting or Repelling Points is Reduced to the Search and Differentiation of One Central Relation, or Characteristic Function," *Phil. Trans. R. Soc. London, 1834 (part II)* (1834), 247–308. Reprinted in *Papers* **2**, 103–161.

Hamilton, W. R., 1835. "Second Essay on a General Method in Dynamics," *Phil. Trans. R. Soc. London, 1835 (part I)* (1835), 95–144. Reprinted in *Papers* **2**, 162–211.

Harish-Chandra, 1951. "On Some Applications of the Universal Enveloping Algebra," *Trans. Amer. Math. Soc.* **70** (1951), 28–96.

Harnack, A., 1887. *Die Grundlagen der Theorie des Logarithmischen Potentials und der eindeutigen Potentialfunktion*, Leipzig, 1887.

Hausdorff, F., 1914. *Grundzüge der Mengenlehre*, Berlin and Leipzig, 1914. Reprinted in 1949 and 1965 by Chelsea Publishing Co., New York.

Hawkins, T., 1971. "The Origins of the Theory of Group Characters," *Archive for History of Exact Sciences* **7** (1971), 142–170.

Hawkins, T., 1972. "Hypercomplex Numbers, Lie Groups, and the Creation of Group Representation Theory," *Archive for History of Exact Sciences* **8** (1972), 243–287.

Hawkins, T., 1974. "New Light on Frobenius' Creation of the Theory of Group Characters," *Archive for History of Exact Sciences* **12** (1974), 217–43.

Hawkins, T., 1975a. *Lebesgue's Theory of Integration. Its Origins and Development*, 2nd edn., New York, Chelsea, 1975.

Hawkins, T., 1975b. "The Theory of Matrices in the 19th Century," *Proceedings of the International Congress of Mathematicians, Vancouver 1974* **2** (1975), 561–570.

Hawkins, T., 1977a. "Another Look at Cayley and the Theory of Matrices," *Archives internationales d'histoire des sciences* **26** (1977), 82–112.

Hawkins, T., 1977b. "Weierstrass and the Theory of Matrices," *Archive for History of Exact Sciences* **17** (1977), 119–163.

Hawkins, T., 1980. "Non-Euclidean Geometry and Weierstrassian Mathematics: The Background to Killing's Work on Lie Algebras," *Historia Mathematica* **7** (1980), 289–342.

Hawkins, T., 1982. "Wilhelm Killing and the Structure of Lie Algebras," *Archive for History of Exact Sciences* **26** (1982), 127–192.

Hawkins, T., 1984. "The Erlanger Programm of Felix Klein: Reflections on Its Place in the History of Mathematics," *Historia Mathematica* **11** (1984), 442–470.

Hawkins, T., 1987. "Cayley's Counting Problem and the Representation of Lie Algebras," *Proceedings, International Congress of Mathematicians, Berkeley 1986,* **2** (1987), 1642–1656.

Hawkins, T., 1988. "Hesse's Principle of Transfer and the Representation of Lie Algebras," *Archive for History of Exact Sciences* **39** (1988), 41–73.

Hawkins, T., 1989. "Line Geometry, Differential Equations and the Birth of Lie's Theory of Groups," *The History of Modern Mathematics*, Vol. 1, J. McCleary and D. Rowe, eds., New York, Academic Press, pp. 275–327.

Hawkins, T., 1991. "Jacobi and the Birth of Lie's Theory of Groups," *Archive for History of Exact Sciences* **42** (1991), 187–278.

Hawkins, T., 1992. "Jacobi and the Birth of Lie's Theory of Groups," *Amphora. Festschrift für Hans Wussing zu seinem 65. Geburtstag*, M. F. S. Demidov, D. Rowe, C. Scriba, eds., Basel, Birkhäuser Verlag, pp. 289–313.

Hawkins, T., 1994. "Lie Groups and Geometry: The Italian Connection," *Supplemento ai Rendiconti del Circolo Matemetico di Palermo* **(2) 36** (1994), 185–206.

Hawkins, T., 1998. "From General Relativity to Group Representations. The Background to Weyl's Papers of 1925–26," *Matériaux pour l'histoire des mathématiques au XX*e *Siècle. Actes du colloque à la mémoire de Jean Dieudonné. Société Mathématique de France, Séminaires et Congrès* **No. 3**, 1998, pp. 69–100.

Hawkins, T., 1999a. "Group Characters from Gauss to Brauer," Part I of R. Haubrich and T. Hawkins, *The Frobenius–Dedekind Correspondence in Historical Context*, New York, Springer-Verlag (to appear).

Hawkins, T., 1999b. "Weyl and the Topology of Continuous Groups," *A History of Topology*, I. M. James, ed., Oxford, Oxford University Press, pp. 169–198.

Hein, W. ed., 1997. *Wilhelm Killing. Briefwechsel mit Friedrich Engel zur Theorie der Lie-Algebren. Dokumente zur Geschichte der Mathematik (Deutsche Mathematiker-Vereinigung), Band 9*, Braunschweig/Wiesbaden (F. Vieweg and Sohn), 1997.

Hellinger, E. and O. Toeplitz, 1927. "Integralgleichungen und Gleichungen mit unendlichvielen Unbekannten," *Enzyklopädie der mathematischen Wissenschaften* **II C** (1923–27), 1189–1228.

Helmholtz, H. v., 1868a. "Über die thatsächliche Grundlagen der Geometrie," *Verh. des Naturhistorisch-Medizinischen Vereins Heidelberg* **4** (1868), 197–202. Reprinted in *Abhandlungen* **2**, 610–617.

Helmholtz, H. v., 1868b. "Über die Thatsachen, die der Geometrie zum Grunde liegen," *Göttinger Nachrichten 1868* (1868). Reprinted in *Abhandlungen* **2**, 618–639.

Helmholtz, H. v., 1892. *Wissenschaftliche Abhandlungen, Band 3*, Leipzig, J. A. Barth, 1892.

Hermann, A. ed., 1968. *Einstein–Sommerfeld Briefwechsel*, Basel and Stuttgart (Schwabe and Co.), 1968.

Hermann, R., 1976. *Sophus Lie's 1884 Differential Invariant Paper. Translated by M. Ackerman. Comments and Additional Material by Robert Hermann.* Brookline, Mass., Math Sci Press, 1976.

Hermite, C., 1856. "Sur la nombre des racines d'une équation algébrique comprises entre des limites données," *Jl. für die reine u. angew. Math.* **52** (1856). Reprinted in *Oeuvres* **1**, 397–414.

Hesse, L. O., 1866. "Ein Übertragungsprinzip," *Jl. für die reine u. angew. Math.* **66** (1866). Reprinted in *Werke*, 531–538.

Hilb, E. and O. Szász, 1922. "Allgemeine Reihenentwicklungen," *Enzyklopädie der mathematischen Wissenschaften* **II C** (1923–27), 1229–1226.

Hilbert, D., 1887. "Über eine Darstellungsweise der invarianten Gebilde im binären Formengebiete," *Math. Ann.* **30** (1878), 15–29. Reprinted in *Abhandlungen* **2**, 102–116.

Hilbert, D., 1889. "Zur Theorie der algebraischen Gebilde III," *Göttingen Nachrichten 1889* (1889), 423–430. Reprinted in *Abhandlungen* **2**, 192–198.

Hilbert, D., 1890. "Über die Theorie der algebraischen Formen," *Math. Ann.* **36** (1890), 473–534. Reprinted in *Abhandlungen* **2**, 199-257.

Hilbert, D., 1893. "Über die vollen Invariantensysteme," *Math. Ann.* **42** (1893), 313–373. Reprinted in *Abhandlungen* **2**, 287–344.

Hilbert, D., 1896. "Zur Theorie der aus n Haupteinheiten gebildeten komplexen Größen," *Göttinger Nachrichten 1896* (1896), 179–183. Reprinted in *Abhandlungen* **2**, 371–375.

Hilbert, D., 1897. "Die Theorie der algebraischen Zahlkörper," *Jahresbericht der Deutschen Mathematiker-Vereinigung* **4** (1897), 175–546. Reprinted in *Abhandlungen* **1**, 63–363.

Hilbert, D., 1900. "Mathematische Probleme," *Göttinger Nachrichten 1900* (1900), 253–297. Originally presented as a lecture at the International Congress of Mathematicians (Paris 1900). It was also published, with additions by Hilbert, in *Archiv der Mathematik und Physik* **(3) 1**, 44–63; 213–237. The *Archiv* version is reprinted in *Abhandlungen* **3**, 290–329. An English translation was published by M. Newson in *Bulletin of the American Mathematical Society* **8** (1902), 437–479. Unless otherwise noted, page references and quotations are from the Newson translation.

Hilbert, D., 1902. "Über die Grundlagen der Geometrie," *Göttinger Nachrichten 1902* (1902), 233–241. [Presented 8 November 1901.]

Hilbert, D., 1903. "Über die Grundlagen der Geometrie," *Grundlagen der Geometrie. Zweite durch Zusätze vermehrte und mit funf Anhängen versehene Auflage*, Leipzig, Teubner, 1903, Anhang IV.

Hilbert, D., 1904a. "Grundzüge einer allgemeinen Theorie der linearen Integralgleichungen [erste Mitteilung]," *Göttinger Nachrichten 1904* (1904), 49–91. Reprinted with minor changes as "Erster Abschnitt" of [Hilbert 1912: 1–38]. Unless otherwise noted, all page references are to the 1912 republication.

Hilbert, D., 1904b. "Grundzüge einer allgemeinen Theorie der linearen Integralgleichungen [zweite Mitteilung]," *Göttinger Nachrichten 1904* (1904), 213–259. Reprinted with minor changes as "Zweiter Abschnitt" of [Hilbert 1912: 39–81]. Unless otherwise noted, all page references are to the 1912 republication.

Hilbert, D., 1906a. "Grundzüge einer allgemeinen Theorie der linearen Integralgleichungen [vierte Mitteilung]," *Göttinger Nachrichten 1906* (1906), 157–227. Reprinted with minor changes as "Vierter Abschnitt" of [Hilbert 1912: 109–174]. Unless otherwise noted, all page references are to the 1912 republication.

Hilbert, D., 1906b. "Grundzüge einer allgemeinen Theorie der linearen Integralgleichungen [fünfte Mitteilung]," *Göttinger Nachrichten 1906* (1906), 439–480. Reprinted with minor changes as "Fünfter Abschnitt" of [Hilbert 1912:174–212]. Unless otherwise noted, all page references are to the 1912 republication.

Hilbert, D., 1909a. "Hermann Minkowski," *Göttinger Nachrichten 1909* (1909), 72–101. Reprinted in Math. Ann. **68**, 445–471, in Hilbert's *Abhandlungen* **3**, 339–364 and in Minkowski's *Abhandlungen* **1**, v–xxxi.

Hilbert, D., 1909b. "Wesen und Ziele einer Analysis der unendlichvielen unabhängigen Variablen," *Rendiconti Circolo Mat. Palermo* **27** (1909), 59–74. Reprinted in *Abhandlungen* **3**, 56–72.

Hilbert, D., 1909c. "Zur Theorie der konformen Abbildung," *Göttingen Nachrichten 1909* (1909), 314–323. Reprinted in *Abhandlungen* **3**, 73–80.

Hilbert, D., 1912. *Grundzüge einer allgemeinen Theorie der linearen Integralgleichungen*, Leipzig, B.G. Teubner, 1912.

Hilbert, D., 1916. "Die Grundlagen der Physik. (Erste Mitteilung.)," *Göttinger Nachrichten 1916* (1916), 395–407. [Presented 20 November 1915.] Reprinted with some textual changes and additional comments and notes in D. Hilbert, "Die Grundlagen der Physik," *Math. Ann.* **92** (1924), 1–32. Reprinted in *Abhandlungen* **3**, 258–289.

Hilbert, D., 1917. "Die Grundlagen der Physik. (Zweite Mitteilung.)," *Göttinger Nachrichten 1917* (1917), 53–76. [Presented 23 December 1916.] Reprinted with some textual changes and additional comments and notes in D. Hilbert, "Die Grundlagen der Physik," *Math. Ann.* **92** (1924), 1–32. Reprinted in *Abhandlungen* **3**, 258–289.

Hilbert, D., 1920. *Natur und mathematisches Erkennen. Vorlesungen, gehalten 1919–1920 in Göttingen. Nach der Ausarbeitung von Paul Bernays. Herausgegeben von David Rowe*, Basel, Birkhäuser, 1992.

Hoüel, J., 1867. *Essai critique sur les principes fondamentaux de la géométrie élémentaire ou Commentaire sur les XXXII premières propositions des Éléments d'Euclide*, Paris, 1867.

Howard, D. and J. Stachel eds., 1989. *Einstein and the History of General Relativity. Based on the Proceedings of the 1986 Osgood Hill Conference*, Boston (Birkhäuser), 1989.

Howe, R., 1992. "A Century of Lie Theory," *American Mathematical Society Centennial Proceedings, Vol. 2: Mathematics into the Twenty-first Century*, Providence, R. I., American Mathematical Society, 1992, pp. 101–320.

Hudson, R., 1905. *Kummer's Quartic Surface*, Cambridge, 1905. Reprinted by Cambridge University Press, 1990.

Humphreys, J. E., 1972. *Introduction to Lie Algebras and Representation Theory*, New York, Springer-Verlag, 1972. Third printing (revised), 1980.

Hurwitz, A., 1881. "Grundlagen einer independenten Theorie der elliptischen Modulfunktionen und Theorie der Multiplikator-Gleichungen erste Stufe," *Math. Ann.* **18** (1881), 528–592. Reprinted in *Werke* **1**, 1–67. Published also separately in 1881 as Hurwitz's Leipzig *Inauguraldissertation.*

Hurwitz, A., 1894. "Zur Invariantentheorie," *Math. Ann.* **45** (1894), 381–404. Reprinted in *Werke* **2**, 508–532.

Hurwitz, A., 1897. "Über die Erzeugung der Invarianten durch Integration," *Göttinger Nachrichten 1897* (1897), 71–90. Reprinted in *Werke* **2**, 546–564.

Hurwitz, A., 1898. "Über die Entwickelung der allgemeinen Theorie der analytischen Funktionen in neuerer Zeit," *Vehandlungen des 1. internationalen Mathematiker-Kongresses in Zürich 1897* (1898), 91–112. Reprinted in *Werke* **1**, 461–480.

Hurwitz, A., 1901. "Sur les séries de Fourier," *Comptes Rendus, Acad. Sci. Paris* **132** (1901), 1473–1475. Reprinted in *Werke* **1**, 506–508.

Hurwitz, A., 1902. "Sur quelques applications géométriques des séries de Fourier," *Annales scientifiques École Normale Sup. Paris* **(3) 19** (1902), 357–408. Reprinted in *Werke* **1**, 509–554.

Hurwitz, A., 1903. "Über die Fourierschen Konstanten integrierbarer Funktionen," *Math. Ann.* **57** (1903), 425–446. Reprinted in *Werke* **1**, 555–576.

Hurwitz, A., 1904. "Über die Fourierschen konstanten integrierbaren Funktionen," *Math. Ann.* **59** (1904), 553. Reprinted in *Werke* **1**, 576.

Imschenetsky, V. G., 1869. "Sur l'intégration des équations aux dérivées partielles du premiere ordre," *Archiv für Math. u. Physik* **50** (1869), 278–474.

Jacobi, C. G. J., 1837. "Über die Reduction der Integration der partiellen Differentialgleichungen erster Ordnung zwischen irgend einer Zahl Variablen auf die Integration eines einzigen Systems gewöhnlicher Differentialgleichungen," *Jl. für die reine u. angew. Math.* **17** (1837), 97–162. Reprinted in *Werke* **4**, 59–127.

Jacobi, C. G. J., 1840. "Sur un théorème de Poisson," *Comptes Rendus, Acad. Sci. Paris* **11** (1840), 529–30. Reprinted in *Werke* **4**, 145–146.

Jacobi, C. G. J., 1862. "Nova methodus, aequationes differentiales partiales primi ordinis inter numerum variabilium quemcumque propositas integrandi," *Jl. für die reine u. angew. Math.* **60** (1862), 1–181. Reprinted in *Werke* **4**, 3–189. [A. Clebsch, ed.] Translated into German and annotated by G. Kowalewski as *Ostwald's Klassiker der exakten Wissenschaften, Nr. 156* (Leipzig, 1906).

Jacobi, C. G. J., 1866. *Vorlesungen über Dynamik*, Berlin, A. Clebsch, ed., 1866.

Jacobson, N., 1962. *Lie Algebras*, New York, Dover, 1962.

Jessen, B., 1951. "Harald Bohr," *Acta Mathematica* **86** (1951), i–xiv. Reprinted in H. Bohr's *Collected Mathematical Works* **3**, Supplement 25 (pp. 163–76 of supplementary material).

Johansson, S., 1905. "Über die Uniformisierung Riemannscher Flächen mit endlicher Anzahl Windungspunkte," *Acta societatis scientiarum fennicae* **33** (1905), No. 7.

Johansson, S., 1906. "Ein Satz über die konforme Abbildung einfach zusammenhängender Riemannscher Flächen auf den Einheitskreis," *Math. Ann.* **62** (1906), 177–183.

Johnson, D. M., 1979. "The Problem of Invariance of Dimension in the Growth of Modern Topology, Part I," *Archive for History of Exact Sciences* **20** (1979), 97–188.

Johnson, D. M., 1981. "The Problem of Invariance of Dimension in the Growth of Modern Topology, Part II," *Archive for History of Exact Sciences* **25** (1981), 85–267.

Johnson, D. M., 1987. "L. E. J. Brouwer's Coming of Age as a Topologist," *Studies in the History of Mathematics. MAA Studies in Mathematics, Vol. 26*, E. R. Phillips, ed., pp. 61–97.

Jordan, C., 1869a. "Commentaire sur Galois," *Math. Ann.* **1** (1869), 142–160. Reprinted in *Oeuvres* **1**, 211–230.

Jordan, C., 1869b. "Mémoire sur les groupes de mouvements," *Annali di matematiche* **2** (1868–69), 167–215, 322–345. Reprinted in *Oeuvres* **4**, 231–302.

Jordan, C., 1869c. "Sur les équations de la division des fonctions abéliennes," *Math. Ann.* **1** (1869), 583–591. Reprinted in *Oeuvres* **1**, 231–239.

Jordan, C., 1869d. "Sur les équations de la géométrie," *Comptes Rendus, Acad. Sci. Paris* **68** (1869), 656–659. Reprinted in *Oeuvres* **1**, 199–202.

Jordan, C., 1869e. "Sur une équation du 16e degré," *Jl. für die reine u. angew. Math.* **70** (1869), 182–184. Reprinted in *Oeuvres* **1**, 207–209.

Jordan, C., 1870. *Traité des substitutions et des équations algébriques*, Paris, 1870.

Kanounov, N. F., 1983. *Fedor Eduardovich Molin*, Moscow, 1983.

Katz, V. J., 1981. "The History of Differential Forms from Clairaut to Poincaré," *Historia Mathematica* **8** (1981), 161–188.

Katz, V. J., 1985. "Differential Forms—Cartan to De Rham," *Archive for History of Exact Sciences* **33** (1985), 321–336.

Killing, W., 1872. *Der Flächenbüschel zweiter Ordnung. Inaugural Dissertation*, Berlin, 1872.

Killing, W., 1876. Review of [Frischauf 1876] in *Zeitschrift für mathematischen und naturwissenschaftlichen Unterricht* **7** (1876), 464–469.

Killing, W., 1878. "Über zwei Raumformen mit konstanter positiver Krümmung," *Jl. für die reine u. angew. Math.* **86** (1878), 72–83.

Killing, W., 1880a. "Die Rechnung in den nicht-Euklidische Geometrie," *Jl. für die reine u. angew. Math.* **89** (1880), 265–287.

Killing, W., 1880b. *Grundbegriffe und Grundsätze der Geometrie. Programm des Gymnasiums zu Brilon (Ostern 1880).*

Killing, W., 1883a. *Die Mechanik in den nichteuklidischen Raumformen, Programm des Gymnasiums zu Brilon (Ostern 1883).*

Killing, W., 1883b. *Über die Nicht-Euklidischen Raumformen von n Dimensionen. Festgabe für das Briloner Gymnasium zum 23. Oktober 1883* (Braunsberg).

Killing, W., 1884. *Erweiterung des Raumbegriffes (Programm Braunsberg).* Reprinted in [Hein, 1997].

Killing, W., 1885. *Die nichteuklidische Raumformen in analytischer Behandlung*, Leipzig, 1885.

Killing, W., 1886. *Zur Theorie der Lie'schen Transformationsgruppen, Programm Braunsberg*, 1886.

Killing, W., 1888a. "Die Zusammensetzung der stetigen endlichen Transformationsgruppen," *Math. Ann.* **31** (1888), 252–90.

Killing, W., 1888b. "Die Zusammensetzung der stetigen endlichen Transformationsgruppen. Zweiter Theil," *Math. Ann.* **33** 1–48.

Killing, W., 1889a. "Die Zusammensetzung der stetigen endlichen Transformationsgruppen. Dritter Theil," *Math. Ann.* **34** (1889), 57–122.

Killing, W., 1889b. *Über eine gewisse Determinante*, Programm, Braunschweig, 1889.

Killing, W., 1890a. "Bestimmung der grössten Untergruppen von endlichen Transformationsgruppen," *Math. Ann.* **36** (1890), 239–254.

Killing, W., 1890b. "Die Zusammensetzung der stetigen endlichen Transformationsgruppen. Vierter Theil," *Math. Ann.* **36** (1890), 161–89.

Killing, W., 1890c. "Erweiterung des Begriffes der Invarianten von Transformationsgruppen," *Math. Ann.* **35** (1890), 423–432.

Killing, W., 1891. "Über die Clifford–Kleinschen Raumformen," *Math. Ann.* **31** (1891), 257–278.

Killing, W., 1892. "Über die Grundlagen der Geometrie," *Jl. für die reine u. angew. Math.* **109** (1892), 121–186.

Killing, W., 1893. *Einführung in die Grundlagen der Geometrie, Band I*, Paderborn, 1893.

Killing, W., 1895. *Bemerkungen über die Transformationsgruppen vom Range Null*, Programm, Münster, 1893.

Killing, W., 1897. *Karl Weierstrass. Rede, gehalten beim Antriff des Rektorats an der kgl. Akademie zu Münster am 15 Okt. 1897*, Münster, 1897.

Killing, W., 1898. *Einführung in die Grundlagen der Geometrie, Band II*, Paderborn, 1898.

Killing, W., 1904. "Der Bau einer besonderen Klasse von Transformationsgruppen," *Festschrift für Ludwig Boltzmann*, Leipzig, 1904, pp. 384–401.

Kleiman, S. L., 1980. "Chasles's Ennumerative Theory of Conics: A Historical Introduction," *Studies in Algebraic Geometry. MAA Studies in Mathematics, vol. 20.*, A. Seidenberg, ed., 1980, pp. 117–138.

Klein, F., 1869. "Zur Theorie der Liniencomplexe des ersten und zweiten Grades," *Math. Ann.* **2** (1870). Reprinted in *Abhandlungen* **1**, 53–80 [4 August 1869].

Klein, F., 1871a. "Über die sogenannte Nicht-Euklidische Geometrie," *Göttinger Nachrichten 1871* (1871). Reprinted in *Abhandlungen* **1**, 244–253.

Klein, F., 1871b. "Über die sogenannte Nicht-Euklidische Geometrie," *Math. Ann.* **4** (1871). Reprinted in *Abhandlungen* **1**, 254–301 [19 August 1871].

Klein, F., 1871c. "Über einen Satz aus der Theorie der Linienkomplexe, welcher dem Dupinschen Theorem analog ist," *Göttinger Nachrichten 1871* (1871). Reprinted in *Abhandlungen* **1**, 98–105 [4 March 1871].

Klein, F., 1871d. "Über Liniengeometrie und metrische Geometrie," *Math. Ann.* **5** (1872). Reprinted in *Abhandlungen* **1**, 106–126 [October 1871].

Klein, F., 1871ms-a. "Inhaltsangabe zu: *Über räumliche Gebilde mit mehrfach unendlichen vielen linearen Transformationen in sich selbst*," NSUB Göttingen, Cod. Ms. F. Klein 22 G II. [3 pages. Undated. Probably written in the first half of 1871.]

Klein, F., 1871ms-b. Untitled and undated essay by Klein on W-configurations in space. Papers of Sophus Lie, UB Oslo. [22 pages. Probably written in the first half of 1871. Draws upon material on W-curves and surfaces in space in Cod. Ms. F. Klein 22 G II (NSUB Göttingen).]

Klein, F., 1872a. "Über die sogenannte Nicht-Euklidische Geometrie (Zweiter Aufsatz)," *Math. Ann.* **6** (1873). Reprinted in *Abhandlungen* **1**, 311–343 [June 8 1872].

Klein, F., 1872b. *Vergleichende Betrachtungen über neuere geometrische Forschungen. Programm zum Eintritt in die philosophische Fakultät und den Senat der k. Friedrich-Alexanders-Universität zu Erlangen*, Erlangen, 1872. Reprinted with additional notes in *Math. Ann.* **43** (1893) = *Abhandlungen* **1**, 460–497. Italian translation by G. Fano (*Annali di Mat.* **(2) 17** (1889–90), 307–43). French translation by H. Padé (*Ann. Ec. Norm Sup.* **(3)** (1891), 87–102, 173–99). English translation by M.W. Haskell (*Bull. N.Y. Math. Soc.***2** (1892–3), 215–49).

Klein, F., 1873. "Über die allgemeinen Funktionsbegriff und dessen Darstellung durch die willkürliche Kurve," *Sitzungsber. Physikalisch-Medizinischen Societät in Erlangen* (1873). Reprinted with minor alterations in Math. Ann. **22** (1883) and in *Abhandlungen* **2**, 214–224.

Klein, F., 1875. "Über binäre Formen mit linearen Transformationen in sich selbst," *Math. Ann.* **9** (1875–76), 183–208. Reprinted in *Abhandlungen* **2**, 275–301.

Klein, F., 1883. "Neue Beiträge zur Riemannschen Funktionentheorie," *Math. Ann.* **21** (1883). Reprinted in *Abhandlungen* **3**, 630–710.

Klein, F., 1884. *Vorlesungen über das Ikosaeder*, Leipzig, 1884.

Klein, F., 1890. "Zur Nicht-Euklidische Geometrie," *Math. Ann.* **37** (1890). Reprinted in *Abhandlungen* **1**, 353–382.

Klein, F., 1892ms. "Über unsere Arbeiten aus den Jahren 1870–72, von F. Kl[ein] und S. Lie. Über Lie's und meine ältere geometrische Arbeiten, von F. Klein (Erster, Borkumer, Entwurf von August 1892)," NSUB Göttingen, Cod. Ms. F. Klein 22 G II [24 pages].

Klein, F., 1893a. *Einleitung in der höhere Geometrie, I. Vorlesung gehalten im Wintersemester 1892–3. Ausgearbeitet von Fr. Schilling.* Göttingen, 1893 (lithograph of handwritten copy). Reprinted with some revisions in F. Klein, *Vorlesungen über höhere Geometrie. Dritte Auflage. Bearbeitet und herausgegeben von W. Blaschke*, Berlin, 1926.

Klein, F., 1893b. *Einleitung in der höhere Geometrie, II. Vorlesung gehalten im Sommersemester 1893. Ausgearbeitet von Fr. Schilling.* Göttingen, 1893 (lithograph of handwritten copy).

Klein, F., 1897. "Über einen Satz aus der Theorie der endlichen (discontinuirlichen) Gruppen linearer Substitutionen beliebig vieler Veränderlicher," *Jahresbericht der Deutschen Mathematiker-Vereinigung* **5** (1897), 57.

Klein, F., 1898. "Gutachten, betreffend den dritten Band der Theorie der Transformationsgruppen von S. Lie, anlässlich der ersten Verteilung des Lobatschewsky-Preises," *Bull. Soc. Physico-Mathématique de Kasan* **8** (1898). Reprinted in *Abhandlungen* **1**, 384–401.

Klein, F., 1910. "Über die geometrischen Grundlagen der Lorentzgruppe," *Jahresbericht der Deutschen Mathematiker-Vereinigung* **19** (1910). Reprinted in *Abhandlungen* **1**, 533–552. Lecture delivered to the Göttingen Mathematical Society on 10 May 1910.

Klein, F., 1916ms. Manuscript entitled "Mitteilungen von Felix Klein über seine Beziehungen zu Lie. (Nach Aufzeichnungen, die Engel im Sept. 1916 in Göttingen gemacht hat, wohin zu kommen ihn Klein aufgefordert hatte.)" 25 pages. Manuscript in the hand of Friedrich Engel. In possession of Professor Ernst Hölder, Mainz, Federal Republic of Germany.

Klein, F., 1921. *Gesammelte mathematische Abhandlungen* **1**, Berlin, 1921.

Klein, F., 1927. *Vorlesungen über die Entwicklung der Mathematik im 19. Jahrhundert, Teil II*, Berlin, 1927. Reprinted by Chelsea Pub. Co. (New York, 1950).

Klein, F., 1928. *Vorlesungen über Nicht-Euklidischen Geometrie. Für den Druck neu bearbeitet von W. Rosemann*, Berlin, 1928.

Klein, F. and S. Lie, 1870. "Sur une certaine famille de courbes et de surfaces," *Comptes Rendus, Acad. Sci. Paris* **70** (1870), 1222–1226, 1275–1279. Reprinted in Klein, *Abhandlungen* **1**, 415–423 and Lie *Abhandlungen* **1**, 78–85 [6 June and 13 June 1870].

Klein, F. and S. Lie, 1871. "Über diejenigen ebenen Kurven, welche durch ein geschlossenes System von einfach unendlich vielen vertauschbaren linearen Transformationen in sich übergehen," *Math. Ann.* **4** (1871), 50–84. Reprinted in Klein, *Abhandlungen* **1**, 424–459 and Lie, *Abhandlungen* **1**, 229–285 [March 1871].

Kneser, A., 1905. "Beiträge zur Theorie der Sturm-Liouvilleschen Darstellung willkürlicher Funktionen," *Math. Ann.* **60** (1905), 402–423.

Kneser, A., 1922. *Die Integralgleichungen und ihre Anwendung in der mathematischen Physik. Zweite umgearbeitete Auflage*, Braunschweig, Vieweg, 1922.

Knoblauch, J., 1906. "Die Biegungs-Invarianten und Kovarianten von gegebener Ordnung," *Jl. für die reine u. angew. Math.* **131** (1906), 247–264.

Koebe, P., 1907. "Über Uniformisierung beliebiger analytischer Kurven," *Göttingen Nachrichten 1907* (1907), 191–210.

Koebe, P., 1909. "Über ein allgemeines Uniformizierungsprinzip," *Atti del IV Congresso internazionale dei Matematici (Roma, 6–11 Aprile 1908)*, Vol. 1, Rome, pp. 25–30.

Koenigs, G., 1895. "Applications des invariants intégraux à la réduction au type canonique d'un système quelconque d'équations différentielles," *Comptes Rendus, Acad. Sci. Paris* **121** (1895), 875–878.

Kowalewski, G., 1899. "Die primitiven Transformationsgruppen in fünf Veränderlichen," *Berichte über d. Verh. d. Sächsischen Gesell. der Wiss., math.-phys. Klasse,* **51** (1899), 69–144.

Kowalewski, G., 1902. "Über die projektive Gruppe der Normkurve und eine charakteristische Eigenschaft des sechsdimensionalen Raumes," *Ber. K. Ges. Wiss. Leipzig, Math.-Phys. Kl.* **54** (1902), 371–92.

Kowalewski, G., 1906. "Über die projektive Gruppe einer Mannigfaltigkeit zweiten Grades," *Berichte über d. Verh. d. Sächsischen Gesell. der Wiss., math.-phys. Klasse, 1906* **58** (1906), 394–414.

Kowalewski, G., 1913. "Projektive Transformationen, die nichts Ebenes invariant lassen und zweiteilige Normalgruppen enthalten," *S'ber. Akad. Wiss. Wien* **122, 2a, 2** (1913), 1949–2013.

Kowalewski, G., 1914. "Projektive Transformationen, die nichts Ebenes invariant lassen und zweiteilig Normalgruppen enthalten," *S'ber. Akad. Wiss. Wien* **123, 2a, 2** (1914), 2363–2379.

Kowalewski, G., 1950. *Bestand und Wandel. Meine Lebenserinnerungen zugleich ein Beitrag zur neueren Geschichte der Mathematik*, Munich, 1950.

Kronecker, L., 1853. "Über die algebraisch auflösbaren Gleichungen," *Monatsberichte K. Akademie der Wissenschaften Berlin, 1853* (1853), 365–74. Reprinted in *Werke* **4**, 1–11.

Kronecker, L., 1874. "Über Schaaren von quadratischen und bilinearen Formen," *Monatsberichte K. Akademie der Wissenschaften Berlin, 1874* (1874). Reprinted in *Werke* **1**, 163–174.

Kronecker, L., 1877. "Über Abelsche Gleichungen," *Monatsberichte K. Akademie der Wissenschaften Berlin, 1877* (1877), 845–51. Reprinted in *Werke* **4**, 65–71.

Kummer, A., 1908. *Über eine Gattung von projektiven Transformationsgruppen in sechs Veränderlichen*, (Inaug.-Diss. Bonn), Göttingen, 1908.

Kung, J. P. S. and G.-C. Rota, 1984. "The Invariant Theory of Binary Forms," *Bulletin of the American Mathematical Society* **(N.S.) 10** (1984), 27–85.

Lampe, E., 1897. *Karl Weierstrass*, Leipzig, 1897.

Laplace, P. S., 1835. *Exposition du système du monde*, 6th. edn., Paris, 1835. Reprinted in *Oeuvres* **6**.

Lebesgue, H., 1906. *Leçons sur les series trigonométriques. Professées au Collège de France*, Paris, Gauthier-Villars, 1906.

Lebesgue, H., 1923. "Notice sur la vie et les travaux de Camille Jordan (1838–1922)," *Mém. Acad. Sci.* **(2) 58** (1923), 29–66. Reprinted in C. Jordan, *Oeuvres* **4**, x–xxxiii.

Lebon, E., 1913. *Gaston Darboux. Biographie, bibliographie analytique des écrits. Seconde édition, entièrement refondue*, Paris, Gauthier-Villars, 1913.

Levi, E. E., 1905. "Sulla struttura dei gruppi finiti e continui," *Atti R. Accad. sci. Torino* **40** (1905), 551–565. Reprinted in *Opere* **1**, 101–115.

Levi-Civita, T., 1917. "Nozione di parallelismo in una varietà qualunque e consequente specificacione geometrica della curvatura Riemanniana," *Rendiconti Circolo Mat. Palermo* **42** (1917), 173–205.

Lie, S., 1870. "Über die Reziprocitätsverhältnisse des Reyeschen Komplexes," *Göttinger Nachrichten 1870* (1870), 53–66. Reprinted in *Abhandlungen* **1**, 68–77 [Berlin January 1870].

Lie, S., 1871a. "Over en Classe geometriske Transformationer," *Forhandlinger Christiania 1872* 67–109. Reprinted in *Abhandlungen* **1**, 105–152 [March 25 1871]. This represents Lie's doctoral dissertation, submitted with a letter dated 25 March 1871 (*Abhandlungen* **1**, 674–675). In the edition in Lie's *Abhandlungen* it is translated into German, along with the explanatory remarks Lie had added to the text.

Lie, S., 1871b. "Über eine Klasse geometrischer Transformationen," *Forhandlinger Christiania 1871* (1872), 182–245. Reprinted in *Abhandlungen* **1**, 153–210 [Summer 1871].

Lie, S., 1871c. "Über Komplexe, inbesondere Linien- und Kugelkomplexe, mit Anwendung auf die Theorie der partieller Differentialgleichungen," *Math. Ann.* **5** (1872), 145–208, 209–256. Reprinted in *Abhandlungen* **2**, 1–121 [10 October 1871, 15 November 1871].

Lie, S., 1872a. "Zur Theorie partieller Differentialgleichungen erster Ordnung, inbesondere über eine Klassifikation derselben," *Göttinger Nachrichten 1872* (1872), 473–489. Reprinted in *Abhandlungen* **3**, 16–26 [Erlangen 11 October 1872].

Lie, S., 1872b. "Zur Theorie partieller Differentialgleichungen erster Ordnung, inbesondere über eine Klassifikation derselben," *Göttinger Nachrichten 1872* (1872), 473–89. Reprinted in *Abhandlungen* **3**, 16–26.

Lie, S., 1873a. "Partielle Differentialgleichungen erster Ordnung, in denen die unbekannte Funktion explizite vorkommt," *Forhandlinger Christiania 1873* (1874), 52–85. Reprinted in *Abhandlungen* **3**, 64–95 [21 March 1873].

Lie, S., 1873b. "Über partielle Differentialgleichungen erster Ordnung," *Forhandlinger Christiania 1873* (1874), 16–51. Reprinted in *Abhandlungen* **3**, 32–63 [21 March 1873].

Lie, S., 1873c. "Zur analytischen Theorie der Berührungstransformationen," *Forhandlinger Christiania 1873* (1874), 237–262. Reprinted in *Abhandlungen* **3**, 96–119 [August 1873].

Lie, S., 1873–4. Excerpts from letters received by A. Mayer between 3 December 1873 and summer 1874. Edited with comments by F. Engel as "Lie über die Anfänge seiner Theorie der Transformationsgruppen. Aus Briefen an Adolph Mayer," Lie, *Abhandlungen* **5**, 583–64.

Lie, S., 1874a. "Begründung einer Invariantentheorie der Berührungstransformationen," *Math. Ann.* **8** (1874), 215–303. Reprinted in *Abhandlungen* **4**, 1–96 [Christiania 5 July 1874].

Lie, S., 1874b. "Verallgemeinerung und neue Verwertung der Jacobischen Multiplikatortheorie," *Forhandlinger Christiania 1874* (1875), 255–274. Reprinted in *Abhandlungen* **3**, 188–205 [November 1874].

Lie, S., 1874c. "Über Gruppen von Transformationen," *Göttinger Nachrichten 1874* (1874), 529–542. Reprinted in *Abhandlungen* **5**, 1–8 [3 December 1874].

Lie, S., 1876a. "Theorie der Transformationsgruppen. Abhandlung II," *Archiv for Mathematik* **1** (1876), 152–193. Reprinted in *Abhandlungen* **5**, 42–75 [Cristiania May 1876].

Lie, S., 1876b. "Theorie der Transformationsgruppen. Erster Abhandlung," *Archiv for Mathematik* **1** (1876), 19–57. Reprinted in *Abhandlungen* **5**, 9–41 [Christiania March 1876].

Lie, S., 1878a. "Theorie der Transformationsgruppen III. Bestimmung aller Gruppen einer zweifach ausgedehnten Punktmannigfaltigkeit," *Archiv for Mathematik* **3** (1878), 93–128. Reprinted in *Abhandlungen* **5**, 78–133.

Lie, S., 1878b. "Theorie der Transformationsgruppen. Abhandlung IV," *Archiv for Mathematik* **3** (1878), 375–460. Reprinted in *Abhandlungen* **5**, 136–197 [Christiania 15 October 1878].

Lie, S., 1879. "Theorie der Transformationsgruppen V," *Archiv for Mathematik* **4** (1879), 232–261. Reprinted in *Abhandlungen* **5**, 199–222.

Lie, S., 1880. "Theorie der Transformationsgruppen I," *Math. Ann.* **16** (1880), 441–528. Reprinted in *Abhandlungen* **6**, 1–94 [Christiania, December 1879].

Lie, S., 1882. "Über Flächen, die infinitesimale und lineare Transformationen gestatten," *Archiv for Mathematik* **7** (1882), 179–193. Reprinted in *Abhandlungen* **5**, 224–234 [June 1882].

Lie, S., 1883a. "Über unendliche kontinuierliche Gruppen," *Forhandlinger Christiania 1883* (1883), Nr. 12. Reprinted in *Abhandlungen* **5**, 314–360.

Lie, S., 1883b. "Untersuchungen über Differentialgleichungen III," *Forhandlinger Christiania 1883* (1883), Nr. 10. Reprinted in *Abhandlungen* **5**, 311–313 [4 May 1883].

Lie, S., 1884a. "Über Differentialinvarianten," *Math. Ann.* **24** (1884), 537–578. Reprinted in *Abhandlungen* **6**, 95–138 [Christiania, 29 May 1884].

Lie, S., 1884b. "Zur Theorie der Transformationsgruppen," *Archiv for Mathematik* **9** (1884), 449–451. Reprinted in *Abhandlungen* **5**, 447–448.

Lie, S., 1885a. "Allgemeine Untersuchungen über Differentialgleichungen, die eine kontinuierliche, endliche Gruppe gestatten," *Math. Ann.* **25** (1885), 71–151. Reprinted in *Abhandlungen* **6**, 139–223 [Christiania, 5 July 1884].

Lie, S., 1885b. "Über gewöhnliche lineare Differentialgleichungen," *Forhandlinger Christiania 1885*, Nr. 21. Reprinted in *Abhandlungen* **5**, 503–505.

Lie, S., 1886. "Untersuchungen über Transformationsgruppen. II," *Archiv for Mathematik* **10** (1886), 353–413. Reprinted in *Abhandlungen* **5**, 507–552 [17 December 1885 and March 1886].

Lie, S., 1888. *Theorie der Transformationsgruppen. Erster Abschnitt. Unter Mitwirkung von ... Friedrich Engel*, Leipzig, 1888. Reprinted by Chelsea Publishing Co. (New York 1970).

Lie, S., 1889. "Über irreduzible Berührungstransformationsgruppen," *Berichte über d. Verh. d. Sächsischen Gesell. der Wiss., math.-phys. Klasse, 1889* (1890), 320–327. Reprinted in *Abhandlungen* **6**, 260–266 [1 July 1889].

Lie, S., 1890a. *Theorie der Transformationsgruppen. Zweiter Abschnitt. Unter Mitwirkung von ... Friedrich Engel*, Leipzig, 1890. Reprinted by Chelsea Publishing Co. (New York 1970).

Lie, S., 1890b. "Über die Grundlagen der Geometrie (Erste Abhandlung)," *Berichte über d. Verh. d. Sächsischen Gesell. der Wiss., math.-phys. Klasse, 1890* (1890), 284–321. Reprinted in *Abhandlungen* **2**, 380–413.

Lie, S., 1890c. "Über die Grundlagen der Geometrie (Zweite Abhandlung)," *Berichte über d. Verh. d. Sächsischen Gesell. der Wiss., math.-phys. Klasse, 1890* (1890), 355–418. Reprinted in *Abhandlungen* **2**, 414–468.

Lie, S., 1891a. "Die Grundlagen für die Theorie der unendlichen kontinuierlichen Transformationsgruppen. I. Abhandlung," *Berichte über d. Verh. d. Sächsischen Gesell. der Wiss., math.-phys. Klasse, 1891* (1891), 316–352. Reprinted in *Abhandlungen* **6**, 300–330.

Lie, S., 1891b. "Die Grundlagen für die Theorie der unendlichen kontinuierlichen Transformationsgruppen. II. Abhandlung," *Berichte über d. Verh. d. Sächsischen Gesell. der Wiss., math.-phys. Klasse, 1891* (1891), 353–393. Reprinted in *Abhandlungen* **6**, 331–64.

Lie, S., 1893a. *Theorie der Transformationsgruppen. Dritter Abschnitt. Unter Mitwirkung von ... Friedrich Engel*, Leipzig, 1893. Reprinted by Chelsea Publishing Co. (New York 1970).

Lie, S., 1893b. *Vorlesungen über kontinuierliche Gruppen. Mit geometrischen und anderen Anwendungen. Bearbeitet und herausgegeben von Dr. Georg Scheffers*, Leipzig, 1893. Reprinted by Chelsea Publishing Co. (New York 1971).

Lie, S., 1893ms. "Zur Auseinandersetzung mit Felix Klein." In possession of Professor Ernst Hölder, Mainz, Germany. 41 pages. Written by Friedrich Engel based on documents written by Lie and originally intended for inclusion in volume 7 of Lie's collected works. (See the forward to Lie *Ges. Abh.* **7**.) Contains extensive transcribed excerpts from the Klein–Lie correspondence (1891–93) regarding [Klein, 1892ms] and the revised version prompted by Lie's criticisms (Lie Papers, UB Oslo, Ms. fol. 3839, LXVII: 11). Also contains extensive transcribed excerpts from manuscripts by Lie in which he attempted to present his views on their geometrical work in 1869–72. Many of the excerpted documents are in the form of letters to Klein that Lie apparently never sent. The Lie documents quoted by Engel are undated but were probably written in 1892–93.

Lie, S., 1895a. "Bestimmung aller Flächen, die eine kontinuierliche Schar von projektiven Transformationen gestatten," *Berichte über d. Verh. d. Sächsischen Gesell. der Wiss., math.-phys. Klasse, 1895* (1895), 209–260. Reprinted in *Abhandlungen* **6**, 494–538.

Lie, S., 1895b. "Verwerthung des Gruppenbegriffes für Differentialgleichungen I," *Berichte über d. Verh. d. Sächsischen Gesell. der Wiss., math.-phys. Klasse*, **47** (1895), 261–322. Reprinted in *Abhandlungen* **6**, 539–91.

Lie, S., 1896. *Geometrie der Berührungstransformationen. Dargestellt von Sophus Lie und Georg Scheffers*, Leipzig, 1896.

Lie, S., 1921. *Gesammelte Abhandlungen, Band 1*, Leipzig, Teubner, 1921.

Lilienthal, R. v., 1931. "Karl Weierstrass," *Westfälische Lebensbilder* **2** (1931), 164–179.

Lipschitz, R., 1887. "Beweis des Satzes aus der Theorie der Substitutionen," *Acta math.* **10** (1887), 137–144.

Loewy, A., 1898. "Über bilineare Formen mit conjugirt imäginaren Variabeln," *Math. Ann.* **50** (1898), 557–576.

Loria, G., 1888. *Die hauptsächlichsten Theorien der Geometrie in ihrer früheren und heutigen Entwicklung*, Leipzig, 1888, F. Schütte, transl.

Lüroth, J., 1868. "Über Polartetraeder und die Schnittcurve zweier Flächen zweiter Ornung," *Zeitschr. f. Math. und Physik* **13** (1868), 404–413.

Lützcn, J., 1990. *Joseph Liouville 1809–1882: Master of Pure and Applied Mathematics*, New York, Springer-Verlag, 1990.

Mackey, G. W., 1988. "Hermann Weyl and the Application of Group Theory to Quantuum Mechanics," *Exakte Wissenschaften und ihre philosophische Grundlegung. Vorträge des Internationalen Hermann-Weyl Kongress, Kiel 1985*, W. Deppert, ed., Frankfurt am Main, Peter Lang, pp. 131–159.

Mackey, G. W., 1992. *The Scope and History of Commutative and Noncommutative Harmonic Analysis*, Providence, R.I., American Mathematical Society, 1992.

Marotte, F., 1898. *Les équations differentielles et la théorie des groupes*, Paris (Thèse), 1898. Published also in *Ann. Fac. Sci. Toulouse* **12** (1898), 69 ff., 84 ff.

Maschke, H., 1899. "Beweis des Satzes, dass diejenigen endlichen linearen Substitutionsgruppen, in welchen einige durchgehends verschwindende Coefficienten auftreten, intransitiv sind," *Math. Ann.* **52** (1899), 363–368.

Maurer, L., 1903. "Über die Endlichkeit der Invariantensysteme," *Math. Ann.* **57** (1903), 265–313.

Mayer, A., 1872. "Über unbeschränkt integrable Systeme von linearen totalen Differentialgleichungen und die simultane Integration linearer partieller Differentialgleichungen," *Math. Ann.* **5** (1872), 448–470 [Leipzig, February 1872].

Mehra, J., 1974. *Einstein, Hilbert and the Theory of Gravitation*, Dordrecht, D. Rcidcl, 1974.

Meyer, W. F., 1883. *Apolarität und rationale Curven. Eine systematische Voruntersuchung zu einer allgemeinen Theorie der linearen Räume*, Tübingen,

Meyer, W. F., 1889. "Invariantentheorie," *Encyklopädie der math. Wiss.* **I B**$_2$ (1889), 320–403.

Miller, A. I., 1981. *Albert Einstein's Special Theory of Relativity*, Reading, Mass., Addison-Wesley, 1981.

Mineur, H., 1925. "Sur la théorie analytique des groupes continus," *Jl. des math. pures et appl.* **(9) 4** (1925), 23–108.

Minkowski, H., 1907a. "Das Relativitätsprinzip," *Annalen der Physik* **47** (1915), 927–938. Lecture delivered to the Göttingen Mathematical Society on 5 November 1907. Edited posthumously for publication by A. Sommerfeld.

Minkowski, H., 1907b. "Die Grundgleichungen für die elektromagnetischen Vorgänge in bewegten Körpern," *Göttinger Nachrichten 1908* (1908), 53–111. Reprinted in *Abhandlungen* **2**, 352–430. [Presented 21 December 1907.]

Minkowski, H., 1908. "Raum und Zeit," *Physikalische Zeitschrift* **20** (1909), 104–111. Reprinted in *Abhandlungen* **2**, 431–444. Lecture delivered to the 80th *Naturforscherversammlung* at Cologne on 21 September 1908. An English translation is contained in H. A. Lorentz, A. Einstein, H. Minkowski, H. Weyl, *The Principle of Relativity* (London, 1923), reprinted by Dover Publications (New York, n.d.). All references are to the English translation.

Minkowski, H. *Gesammelte Abhandlungen, Bd. 1*, Leipzig (B. G. Teubner), 1911.

Misner, C., K. Thorne, and J. Wheeler, 1973. *Gravitation*, New York, W. H. Freeman, 1973.

Molien, T., 1893. "Über Systeme höherer complexer Zahlen," *Math. Ann.* **41** (1893), 83–156.
Molien, T., 1898. "Über die Invarianten der linearen Substitutionsgruppen," *Sitzungsberichte der Akademie der Wiss. zu Berlin 1897* (1898), 1152–1156.
Monna, A. F., 1975. *Dirichlet's Principle. A Mathematical Comedy of Errors and Its Influence on the Development of Analysis*, Utrecht, Oosthoek, Scheltema and Holkema, 1975.
Montel, P., 1953. "Notices sur les membres décédés: Vessiot (Ernest)," *Association amicale de secours des anciens éléves de l'Ecole Normale Supérieure*, Paris, 1953, pp. 22–24.
Møller, C., 1972. *The Theory of Relativity*, 2nd. edn., Oxford, Clarendon Press, 1972.
Newcomb, S., 1877. "Elementary theorems relating to the geometry of a space of three dimensions and of uniform positive curvature in the fourth dimension," *Jl. für die reine u. angew. Math.* **83** (1877), 293–299.
Norton, J., 1989. "How Einstein Found His Field Equations, 1912–1915," *Einstein and the History of General Relativity*, D. Howard and J. Stachel, ed., Boston, Birkhäuser, 1989, pp. 101–159.
Oellers, P., 1925. *Wilhelm Killing. Ein Universitätsprofessor im Tertiarenkleide*, Westphalia, 1925.
Oellers, P., 1927. "W. Killings wissenschaftliche Bedeutung," *Franziskanische Studien* **14** (1927), 267–278.
Oellers, P., 1929. *Wihelm Killing. Religiösen Quellenschriften, Heft 54*, Düsseldorf, 1929.
Olver, P. J., 1986. *Applications of Lie Groups to Differential Equations*, New York, Springer-Verlag, 1986.
Osgood, W. F., 1898. "Selected Topics in the General Theory of Functions. Six Lectures Delivered Before the Cambridge Colloquium August 22–27, 1898," *Bulletin of the American Mathematical Society* **5** (1898), 59–87.
Osgood, W. F., 1900. "On the existence of the Green's function for the most generally simply connected plane region," *Trans. American Mathematical Society* **1** (1900), 310–314.
Osgood, W. F., 1901. "Allgemeine Theorie der analytischen Funktionen einer und mehrer komplexen Grössen," *Enzyklopädie der mathematischen Wissenschaften* **II B** (1901), 1–114.
Ostrowski, A., 1918. "Über die Existenz einer endlichen Basis bei gewissen Funktionensysteme," *Math. Ann.* **78** (1918), 94–119. Reprinted in *Papers* **2**, 13–38.
Ostrowski, A., 1919. "Über eine neue Eigenschaft der Diskriminanten und Resultanten binärer Formen (Felix Klein zum fünfzigjährigen Doktorjubiläum)," *Math. Ann.* **79** (1919), 360–87. Reprinted in *Papers* **2**, 39–66 [11 December 1918, with addition dated 17 April 1919].
Ostrowski, A., 1924. "Über eine neue Fragestellung in der algebraischen Invariantentheorie," *Jahresbericht der Deutschen Mathematiker-Vereinigung* **33** (1924), 174–84. Reprinted in *Papers* **2**, 152–162 [10 January 1924].
Ostrowski, A. and I. Schur, 1922. "Über eine fundamentale Eigenschaft der Invarianten einer allgemeinen binären Form," *Mathematische Zeitschrift* **15** (1922), 81–105. Reprinted in Ostrowski, *Papers* **2**, 127–151 [30 July 1921].
Page, J. M., 1888. "On the Primitive Groups of Transformations in Space of Four Dimensions," *Amer. Jl. Math.* **10** (1888), 293–346.
Pais, A., 1982. *'Subtle is the lord...': The Science and the Life of Albert Einstein*, Oxford and New York, Oxford University Press, 1982.
Parshall, K., 1989. "Toward a History of Nineteenth-Century Invariant Theory," *The History of Modern Mathematics*, Vol. 1, J. McCleary and D. Rowe, eds., New York, Academic Press, pp. 157–187.
Pasch, M., 1882a. *Einleitung in die Differential- und Integralrechnung*, Leipzig, 1882.
Pasch, M., 1882b. *Vorlesungen über neuere Geometrie*, Leipzig, 1882.
Pasch, M., 1930. *Eine Selbstschilderung*, Giessen, 1930.

Pauli, W., 1921. *Relativitätstheorie*, Leipzig, Teubner, 1921. Sonderabdruck aus der Enzyklopädie der mathematischen Wissenschaften. English translation by G. Field (New York, Pergamon Press, 1958).

Peter, F., 1923. "Brechungsindizes und Absorptionskonstanten der Diamenten zwischen $\lambda = 644$ und 226 $m\mu$," *Zeitschrift für Physik* **15** (1923), 358–368.

Picard, E., 1883a. "Sur les groupes de transformations des équations différentielles linéaires," *Comptes Rendus, Acad. Sci. Paris* **96** (1883), 1131–1134. Reprinted in *Oeuvres* **2**, 97–100.

Picard, E., 1883b. "Sur une classe de groupes discontinues de substitutions linéaires et sur les fonctions de deux variables independantes restant invariables par ces substitutions," *Acta Mathematica* **1** (1882–83), 297–320. Reprinted in *Oeuvres* **1**, 311–334.

Picard, E., 1884. "Sur les formes quadratiques ternaires indéfinies à indéterminées conjugées et sur les fonctions hyperfuchsiennes correspondantes," *Acta Mathematica* **5** (1884), 121–182. Reprinted in *Oeuvres* **1**, 427–488.

Picard, E., 1887a. "Remarque sur les groupes linéaires d'ordre fini à trois variables," *Bull. Soc. Math. France* **15** (1887), 152–155. Reprinted in *Oeuvres* **1**, 597–600.

Picard, E., 1887b. "Sur les équations différentielles linéaires et es groupes algébriques de transformations," *Ann. de la Fac. d. Sci. Toulouse* **1** (1887), A1–A15. Reprinted in *Oeuvres* **2**, 117–131.

Picard, E., 1889. "Mémoire sur la théorie des fonctions algébriques de deux variables," *Jl. des math. pures et appl.* **(4) 5** (1889), 135–319. Reprinted in *Oeuvres* **3**, 201–385.

Picard, E., 1890. "Revue annuelle d'analyse," *Revue générale des sci. pures et appl.* **1** (1890), 702–708.

Picard, E., 1892a. "A propos de quelques récentes travaux mathématiques," *Revue générale des sci. pures et appl.* **3** (1892), 725–729.

Picard, E., 1892b. "Sur certaines systèmes d'équations aux dérivées partielles généralaisant les équations de la théorie des fonctions d'une variable complexe," *Jl. des math. pures et appl.* **(4) 8** (1892), 217–232. Reprinted in *Oeuvres* **2**, 501–516.

Picard, E., 1897a. Introduction to *Oeuvres mathématiques d'Evariste Galois*, Paris, 1897.

Picard, E., 1897b. "Revue de quelques travaux mathématiques recentes," *Revue générale des sci. pures et appl.* **8** (1897), 957–961.

Picard, E., 1899. *Lectures on Mathematics*, Worcester, Mass., Clark University, 1899.

Picard, E., 1900. "Quelques vues générales sur la théorie des équations différentielles," *Revue générale des sci. pures et appl.* **11** (1900), 229–237.

Picard, E., 1905. *La science moderne et son état actuel*, Paris, 1905.

Picard, E., 1915. "L'histoire des sciences et les prétentions de la science allemande," *Revue de deux-mondes* (1 July 1915). Reprinted in E. Picard, *Discours et Mélanges*, Paris, 1922, pp. 151–175.

Picard, E., 1917. *Les sciences mathématiques en France depuis un demi-siècle*, Paris, 1917.

Picard, E., 1926. *La vie et l'Oeuvre de Jules Tannery*, Paris, Gauthier-Villars, 1926.

Plücker, J., 1866. "Fundamental views regarding mechanics," *Philos. Trans. R. Soc. London* **156** (1866), 361–380. Reprinted in *Abhandlungen* **1**, 546–568.

Plücker, J., 1868. *Neue Geometrie des Raumes gegründet auf die Betrachtung der geraden Linie als Raumelement*, Leipzig, A. Clebsch, ed., 1868.

Plücker, J., 1869. *Neue Geometrie des Raumes gegründet auf die Betrachtung der geraden Linie als Raumelement*, Leipzig, F. Klein, ed., 1869.

Plücker, J., 1895. *Gesammelte wissenschaftliche Abhandlungen. Band 1, Mathematische Abhandlungen*, A. Schoenflies, ed., Leipzig (Teubner), 1895.

Plücker, J. and J. W. Hittorf, 1965. "On the spectra of ignited gases and vapours ...," *Philos. Trans. R. Soc. London* **155** (1965), 1–29. Reprinted in *Abhandlungen* **2**, 665–700.

Poincaré, H., 1881a. "Sur les fonctions fuchsiennes," *Comptes Rendus, Acad. Sci. Paris* **93** (1881), 301–303. Reprinted in *Oeuvres* **2**, 29–31.

Poincaré, H., 1881b. "Sur les équations différentielles linéaires à intégrales algébriques," *Comptes Rendus, Acad. Sci. Paris* **92** (1881), 698–701. Reprinted in *Oeuvres* **3**, 95–97.

Poincaré, H., 1882a. "Sur les fonctions uniformes qui se reproduisent par des substitutions linéaires," *Math. Ann.* **19** (1882), 553–564. Reprinted in *Oeuvres* **2**, 92–105.

Poincaré, H., 1882b. "Sur les fonctions fuchsiennes," *Acta Mathematica* **1** (1882), 193–295. Reprinted in *Oeuvres* **2**, 169–257.

Poincaré, H., 1883a. "Sur l'intégration algébrique des équations linéaires," *Comptes Rendus, Acad. Sci. Paris* **97** (1883), 984–985. Reprinted in *Oeuvres* **3**, 101–102.

Poincaré, H., 1883b. "Sur l'intégration algébrique des équations linéaires," *Comptes Rendus, Acad. Sci. Paris* **97** (1883), 984–985. Reprinted in *Oeuvres* **3**, 101–102.

Poincaré, H., 1883c. "Sur un théorème général des fonctions," *Bulletin de la Société Mathématique de France* **11** (1883), 113–125. Reprinted in *Oeuvres* **4**, 57–69.

Poincaré, H., 1884. "Sur les groupes des équations linéaires," *Acta mathematica* **4** (1884), 201–311. Reprinted in *Oeuvres* **2**, 401.

Poincaré, H., 1886. "Sur les transformations des surfaces en elles-mêmes," *Comptes Rendus, Acad. Sci. Paris* **103** (1886), 732–734. Reprinted in *Oeuvres* **6**, 1–4.

Poincaré, H., 1887. "Sur les hypothèses fondamentales de la Géométrie," *Bull. Soc. Math. de France* **15** (1887), 203–216. Reprinted in *Oeuvres* **11**, 79–91.

Poincaré, H., 1889. "La logique et l'intuition dans la science mathématique et dans l'enseignement," *Enseign. math.* **1** (1889), 157–162. Reprinted in *Oeuvres* **11**, 129–133.

Poincaré, H., 1890. "Sur le problème des trois corps et les équations de la dynamique," *Acta Mathematica* **13** (1890), 1–270. Reprinted in *Oeuvres* **7**, 262–479.

Poincaré, H., 1895. "Analysis situs," *Jl. Ec. Polyt.* **(2) 1** (1895), 1–121. Reprinted in *Oeuvres* **6**, 193–288.

Poincaré, H., 1898. "Sur les rapports de l'analyse et de la physique mathématique," *Verh. des erstern intern. Mathematiker-Kongress in Zürich ...*, Leipzig, 1898, pp. 81–90. A version of this talk appeared as Chapter V of Poincaré's *The Value of Science*.

Poincaré, H., 1899a. "Sur les groupes continus," *Cambridge Philosophical Trans.* **18** (1899), 220–255. Reprinted in *Oeuvres* **3**, 173–212.

Poincaré, H., 1899b. "Sur les groupes continus," *Comptes Rendus, Acad. Sci. Paris* **128** (1899), 1065–1069. Reprinted in *Oeuvres* **3**, 169–172.

Poincaré, H., 1901. "Quelques remarques sur les groupes continus," *Rendiconti Circolo Mat. Palermo* **15** (1901). Reprinted in *Oeuvres* **3**, 213–260.

Poincaré, H., 1902. "Les fondements de la géométrie," *Bulletin des sciences mathématiques* **(2) 26** (1902), 249–272. Reprinted in *Oeuvres* **11**, 92–113.

Poincaré, H., 1903. "'Sur l'intégration algébrique des équations linéaires et les périodes des intégrales abéliennes," *Jl. des math. pures et appl.* **(5) 9** (1903), 139–212. Reprinted in *Oeuvres* **3**, 106–166.

Poincaré, H., 1905. "Sur la dynamique de l'électron," *Comptes Rendus, Acad. Sci. Paris* **140** (1905), 1504–1508. Reprinted in *Oeuvres* **9**, 489–493.

Poincaré, H., 1906. "Sur la dynamique de l'électron," *Rendiconti Circolo Mat. Palermo* **21** (1906), 129–175. Reprinted in *Oeuvres* **9**, 494–550. An English translation of the introduction and §§1–4,9 is given in C.W. Kilmeister, ed., *Special Theory of Relativity* (Pergamon, 1970).

Poincaré, H., 1907. "Sur l'uniformisation des fonctions analytiques," *Acta mathematica* **31** (1907), 1–63. Reprinted in *Oeuvres* **4**, 70–146.

Poincaré, H., 1908a. "L'avenir des mathématiques," *Atti IV Congresso int. dei matematici*, Vol. 1, Rome, pp. 167–182. A translation into English is contained in *Science and Method*, Ch. 2.

Poincaré, H., 1908b. "Nouvelles remarques sur les groupes continus," *Rendiconti Circolo Mat. Palermo* **25** (1908). Reprinted in *Oeuvres* **3**, 261–302.

Poincaré, H., 1912. "Rapport sur les travaux de M. Cartan. Fait à la Faculté des Sciences de l'Université de Paris," *Acta Mathematica* **38** (1921), 137–145. Reprinted in [Akivis and Rosenfeld 1993:263–271].

Poisson, S., 1809. "Mémoire sur la variation des constantes arbitraires dans les questions de mécanique," *Journal de l'Ecole Polytechnique* **15** (1809), 266–344.

Pont, J. C., 1974. *La topologie algebrique: des origines à Poincaré*, Paris, Presses universitaires de France, 1974.

Pyenson, L., 1977. "Hermann Minkowski and Einstein's Special Theory of Relativity," *Archive for History of Exact Sciences* **17** (1977), 71–95.

Pyenson, L., 1979. "Physics in the Shadow of Mathematics: The Göttingen Electron-Theory Seminar of 1905," *Archive for History of Exact Sciences* **21** (1979), 55–89.

Pyenson, L., 1982. "Relativity in Late Wilhelmian Germany: The Appeal to a Preestablished Harmony between Mathematics and Physics," *Archive for History of Exact Sciences* **27** (1982), 137–155. Reprinted as Chapter 6 of L. Pyenson, *The Young Einstein. The Advent of Relativity.*, Bristol, Adam Hilger Ltd., 1985.

Pólya, G., 1987. *The Pólya Picture Album: Encounters of a Mathematician*, Boston, Birkhäuser, G. L. Alexanderson, ed., 1987.

Reich, K., 1994. *Die Entwickelung des Tensorkalküls. Vom absoluten Differentialkalkül zur Relativitätstheorie*, Basel, Birkhäuser, 1994.

Reid, C., 1970. *Hilbert*, New York, Springer-Verlag, 1970.

Reid, C., 1976. *Courant in Göttingen and New York. The Story of an Improbable Mathematician*, New York, Springer-Verlag, 1976.

Ricci, G. and T. Levi-Civita, 1900. "Méthodes de calcul différentiel absolu et leurs applications," *Math. Ann.* **54** (1900–1901), 125–201.

Riemann, B., 1868. "Über die Hypothesen, welche der Geometrie zu Grunde liegen," *Abhandlungen der K. Gesellschaft der Wissenschaften zu Göttingen* **13** (1868). Reprinted in *Werke*, 272–287.

Riemann, B., 1990. *Gesammelte mathematische Werke, wissenschaftlicher Nachlass und Nachträge. Nach der Ausgabe von Heinrich Weber und Richard Dedekind neu herausgegeben von Raghavan Narasimhan*, Berlin and Leipzig, Springer-Verlag and B. G. Teubner, 1990.

Rosanes, J., 1871. *Über den neusten Untersuchungen in Betreff unserer Anschauung von Raume. Ein Vortrag gehalten zur Habilitation an der Universität Breslau am 30. April 1870*, Breslau, 1871.

Rosenfeld, B. A., 1988. *A History of Non-Euclidean Geometry. Evolution of the Concept of a Geometric Space*, New York, Springer-Verlag, 1988. Abe Shenitzer, transl., 1988.

Rosenfeld, B. A., 1997. *Geometry of Lie Groups*, Dordrecht, Kluwer, 1997.

Rota, G.-C., 1982. Introduction to the 1982 reprint of [Study 1889a].

Rowe, D., 1986. "'Jewish Mathematics' at Göttingen in the Era of Felix Klein," *Isis* **77** (1986), 422–449.

Rowe, D., 1989a. "The Early Geometrical Works of Felix Klein and Sophus Lie," *The History of Modern Mathematics*, Vol. 1, J. McCleary and D. Rowe, eds., New York, Academic Press, pp. 209–273.

Rowe, D., 1989b. "Klein, Hilbert and the Göttingen Mathematical Tradition," *Osiris* **(2) 5** (1989), 186–213.

Rowe, D., 1998a. "The Göttingen Response to General Relativity and Emmy Noether's Theorems" (preprint). Published under the same title in *The Symbolic Universe. Geometry and Physics 1890–1930*, J. Gray, ed., Oxford, Oxford University Press, 1999, pp. 189–233.

Rowe, D., 1998b. "Mathematics in Berlin, 1810–1933," *Mathematics in Berlin*, H. Begehr, et al, eds., Basel, Birkhäuser, pp. 9–25.

Rüdenberg, L. and H. Zassenhaus, 1973. *Hermann Minkowski. Briefe an David Hilbert. Mit Beiträgen und herausgegeben von L. Rüdenberg und H. Zassenhaus*, Berlin, Springer-Verlag, 1973.

Salmon, G., 1866. "On Some Points on the Theory of Elimination," *Quart. Jl. Math.* **7** (1866), 27–37.

Salmon, G., 1867. "On the Number of Surfaces of Second Degree which can be described to satisfy nine Conditions," *Quart. Jl. Math.* **8** (1867), 1–7.

Samelson, H., 1952. "Topology of Lie Groups," *Bulletin of the American Mathematical Society* **58** (1952), 2–37.

Samelson, H., 1969. *Notes on Lie Groups and Lie Algebras*, New York, Van Nostrand, 1969. Reprinted by Springer-Verlag, New York, 1990.

Scheibe, E., 1988. "Hermann Weyl and the Nature of Spacetime," *Exakte Wissenschaften und ihre philosophische Grundlegung. Vorträge des Internationalen Hermann-Weyl Kongress, Kiel 1985*, W. Deppert, ed., Frankfurt am Main, Peter Lang, 1988.

Schmid, W., 1982. "Poincaré and Lie Groups," *Bulletin of the American Mathematical Society* **(2) 6** (1982), 175–186.

Schmidt, E., 1905. *Entwickelung willkurlicher Functionen nach Systemen vorgeschriebener*, Göttingen, 1905. Schmidt's Inaugural-Dissertation at Göttingen. Reprinted as part of [Schmidt, 1907].

Schmidt, E., 1907. "Zur Theorie der linearen und nichtlinearen Integralgleichungen. I. Teil," *Math. Ann.* **63** (1907), 433–476.

Schmidt, E., 1908. "Über die Auflösung linearer Gleichungen mit unendlich vielen Unbekannten," *Rendiconti Circolo Mat. Palermo* **25** (1908), 53–77.

Scholz, E., 1980. *Geschichte des Mannigfaltigkeitsbegriffes von Riemann bis Poincaré*, Boston, Birkhäuser Verlag, 1980.

Scholz, E., 1989a. "Crystallographic Symmetry Concepts and Group Theory (1850–1880)," *The History of Modern Mathematics.* , Vol. 2, D. Rowe and J. McCleary, eds., Boston, Academic Press, 1989, pp. 3–27.

Scholz, E., 1989b. *Symmetrie, Gruppe, Dualität: zur Beziehung zwischen theoretischer Mathematik und Anwendungen in Kristallographie und Baustatik des 19. Jahrhunderts*, Basel, Birkhäuser Verlag, 1989.

Scholz, E., 1994. "Hermann Weyl's Contribution to Geometry, 1917–1923," *The Intersection of History and Mathematics*, S. Chiara, S. Mitsuo and J. Dauben, ed., Basel, Birkhäuser Verlag, 1994, pp. 203–230.

Scholz, E., 1995. "Hermann Weyl's 'Purely Infinitesimal Geometry,'" *Proceedings of the International Congress of Mathematicians, Zürich, Switzerland 1994*, Basel, Birkhäuser, 1995, pp. 1592–1603.

Scholz, E., 1996. "Weyls Infinitesimalgeometrie, 1917–1925," *Hermann Weyl's Raum · Zeit · Materie and a General Introduction to His Scientific Work*, E. Scholz, ed., DMV Seminar Proceedings, Basel, Birkhäuser (to appear).

Schouten, J. A., 1923. "Über die Anwendung der allgemeinen Reihenentwickelungen auf eine bestimmte quaternäre Form sechsten Hauptgrades," *Rendiconti Circolo Mat. Palermo* **47** (1923), 409–425 [January 1923].

Schouten, J. A., 1924. *Der Ricci-Kalkül. Eine Einführung in die neueren Methoden und Probleme der mehrdimensionalen Differentialgeometrie*, Berlin, Springer-Verlag, 1924.

Schreier, O., 1925. "Abstrakte kontinuierliche Gruppen," *Abhandlungen aus dem Mathematischen Seminar der Hamburgischen Universität* **4** (1925), 15–32 [January 1925].

Schreier, O., 1927. "Die Verwandschaft stetiger Gruppen im grossen," *Abhandlungen aus dem Mathematischen Seminar der Hamburgischen Universität* **5** (1927), 233–244 [February 1927].

Schreier, O., 1928. "Über neuere Untersuchungen in der Theorie der kontinuierlichen Gruppen," *Jahresbericht der Deutschen Mathematiker-Vereinigung* **37** (1928), 113–122.

Schur, I., 1901. *Über eine Klasse von Matrizen, die sich einer gegebenen Matrix zuordnen lassen*, Berlin, 1901. Schur's doctoral dissertation. Reprinted in *Abhandlungen* **1**, 1–72.

Schur, I., 1904. "Über die Darstellung der endlichen Gruppen durch gebrochene lineare Substitutionen," *Jl. für die reine u. angew. Math.* **127** (1904), 20–50. Reprinted in *Abhandlungen* **1**, 86–127.

Schur, I., 1905. "Neue Begründung der Theorie der Gruppencharaktere," *Sitzungsberichte der Akademie der Wiss. zu Berlin 1905. Physikalisch-Math. Kl.* (1905), 406–432. Reprinted in *Abhandlungen* **1**, 143–169.

Schur, I., 1924a. "Neue Anwendungen der Integralrechnung auf Probleme der Invariantentheorie," *Sitzungsberichte der Akademie der Wiss. zu Berlin 1924* (1924), 189–208. Reprinted in *Abhandlungen* **2**, 440–459 [10 January 1924].

Schur, I., 1924b. "Neue Anwendungen der Integralrechnung auf Probleme der Invariantentheorie. II. Über die Darstellung der Drehungsgruppe durch lineare Transformationen," *Sitzungsberichte der Akademie der Wiss. zu Berlin 1924* (1925), 297–321. Reprinted in *Abhandlungen* **2**, 460–484 [13 November 1924].

Schur, I., 1924c. "Neue Anwendungen der Integralrechnung auf Probleme der Invariantentheorie. III. Vereinfachung des Integralkalküls. Realitätsfragen," *Sitzungsberichte der Akademie der Wiss. zu Berlin 1924* (1925), 346–355. Reprinted in *Abhandlungen* **2**, 485–494 [11 December 1924].

Schur, I., 1927. "Über die rationalen Darstellungen der allgemeinen linearen Gruppe," *Sitzungsberichte der Akademie der Wiss. zu Berlin 1927* (1927), 58–75. Reprinted in *Abhandlungen* **3**, 68–85.

Schur, I., 1928. "Über die stetigen Darstellungen der allgemeinen linearen Gruppe," *Sitzungsberichte der Akademie der Wiss. zu Berlin 1928* (1928), 100–124. Reprinted in *Abhandlungen* **3**, 89–113.

Schur, I., 1968. *Vorlesungen über Invariantentheorie*, Berlin, Springer-Verlag, H. Grunsky, ed., 1968.

Scott, C. A., 1901. "The International Congress of Mathematicians in Paris," *Bulletin of the American Mathematical Society* **7** (1901), 57–79.

Segre, C., 1885. "Considerazioni intorno alla geometria delle coniche di un plano e alla sua rappresentazione sulla geometria dei lineari di retti," *Atti R. Accad. Sci. Torino* **20** (1884–85), 487–504. Reprinted in *Opere* **4**, 1–17.

Segre, C., 1890. "Un nuovo campo di richerche geometriche," *Atti R. Accad. Sci. Torino*, **25** (1889–90), 180–205, 292–317, 376–96; **26** (1890–91), 35–71. Reprinted in *Opere* **2**, 237–337.

Segre, C., 1891. "Sulle varietá che rappresentano le coppie di punti di due piani o spazi," *Rendiconti di Circolo Matematico di Palermo* **5** (1891), 192–204. Reprinted in *Opere* **1**, 173–184.

Segre, C., 1892. "Le rappresentazioni reali delle forme complesse et gli enti iperalgebriche," *Math. Ann.* **40** (1892), 413–67. Reprinted in *Opere* **2**, 338–95.

Serre, J.-P., 1966. *Algèbres de Lie semi-simples complexes*, New York, Benjamin, 1966.

Serre, J.-P., 1987. *Complex Semi-Simple Lie Algebras*, New York, Springer-Verlag, G. A. Jones, transl., 1987.

Serre, J.-P., 1992. *Lie algebras and Lie groups*, New York, Springer-Verlag, 1992.

Sforza, G., 1892. "Contributo alla geometria complessa," *Giornale di matematiche* **30** (1982), 159–187.

Sigurdsson, S., 1991. *Hermann Weyl, Mathematics and Physics, 1900–1927*. Doctoral dissertation, Harvard University, 1991.

Sigurdsson, S., 1994. "Unification, Geometry and Ambivalence: Hilbert, Weyl and the Göttingen Community," *Trends in the Historiography of Science*, K. Gavroglu, ed., Dordrecht, Kluwer, 1994, pp. 355–367.

Singer, I. and S. Sternberg, 1965. "The Infinite Groups of Lie and Cartan, Part I (The Transitive Groups)," *Journal d'analyse mathématique* **15** (1965), 1–114.

Singer, I. M. and J. A. Thorpe, 1967. *Lecture Notes on Elementary Topology and Geometry*, Glenview, Illinois, Scott, Foresman, 1967.

Slocum, S. E., 1900a. "Note on the chief theorem of Lie's theory of continuous groups," *Proc. of the Amer. Acad. of Arts and Sciences* **35**$_2$ (1900), 239–250.

Slocum, S. E., 1900b. "Supplementary note on the chief theorem of Lie's theory of finite continuous groups," *Proc. of the Amer. Acad. of Arts and Sciences* **35**$_2$ (1900), 483–485.

Slocum, S. E., 1901. "On the continuity of groups generated by infinitesimal transformations," *Proc. of the Amer. Acad. of Arts and Sciences* **36**$_1$ (1901), 85–109.

Slodowy, P., 1998. "The early development of the representation theory of semisimple Lie groups: A. Hurwitz, I. Schur, H. Weyl," *Hamburger Beiträge zur Mathematik* (1998). A revised version has been published under the same title in *Jahresbericht der Deutschen Mathematiker-Vereinigung* **101** (1999), 97–115.

Sommerville, D. M. Y., 1910. "Classification of the geometries with projective metric," *Proc. Edinburgh Math. Soc.* **28** (1910), 25–41.

Sommerville, D. M. Y., 1911. *Bibliography of non-Euclidean Geometry*, London, 1911. An essentially identical second edn. by Chelsea Pub. Co. (New York, 1970).

Speiser, A., 1927. "Naturphilosophische Untersuchungen von Euler und Riemann," *Jl. für die reine u. angew. Math.* **156** (1927), 105–114.

Speziali, P. ed., 1972. *Albert Einstein, Michele Besso, Correspondence, 1903–1955*, Paris, Hermann, 1972.

Speziali, P., 1975. "Ricci-Curbastro, Gregorio," *Dictionary of Scientific Biography* **11** (1975), 406–411.

Springer, T. A., 1977. *Invariant Theory. Lecture Notes in Mathematics,* **585**, Springer-Verlag, Berlin and New York, 1977.

Stachel, J., 1989. "Einstein's search for general covariance, 1912–1915," *Einstein and the History of General Relativity*, D. Howard and J. Stachel, eds., Boston, Birkhäuser, pp. 63–100.

Strøm, E., 1994. "Marius Sophus Lie," *The Sophus Lie Memorial Conference. Oslo 1992. Proceedings*, Oslo, Scandinavian University Press, 1994, pp. ix–xxviii.

Study, E., 1889a. *Methoden zur Theorie der ternären Formen*, Leipzig, 1889. Reprinted with an introduction by Gian-Carlo Rota by Springer-Verlag (New York, 1982).

Study, E., 1889b. Review of [Lie 1888] in *Zeitschr. f. Math. u. Phys.* **34** (1889), Hist.-Lit. Abt., 171–191.

Study, E., 1897. "Über die Invarianten der projectiven Gruppe einer quadratischen Mannigfaltigkeit von nicht verschwindender Discriminante," *Berichte über d. Verh. d. Sächsischen Gesell. der Wiss., math.-phys. Klasse, 1897* **49** (1897), 443–461.

Study, E., 1905. "Kürzeste Wege im komplexen Gebiet," *Math. Ann.* **60** (1905), 321–377.

Study, E., 1908. "Kritische Betrachtungen über Lies Invariantentheorie der endlichen kontinuierlichen Gruppen," *Jahresbericht der Deutschen Mathematiker-Vereinigung* **17** (1908), 125–142.

Study, E., 1923. *Einleitung in die Theorie der Invarianten linearer Transformationen auf Grund der Vektorenrechnung*, Branunschweig, Vieweg, 1923.

Stäckel, P. and F. Engel, 1895. *Theorie der Parallellinien von Euclid bis auf Gauss*, Leipzig, 1895.

Stäckel, P. and F. Engel, 1899–1913. *Urkunden zur Geschichte der nichteuklidische Geometrie*, 2 vols., Leipzig, 1899 and 1913.

Sylvester, J. J., 1852. "On the Principles of the Calculus of Forms," *Cambridge and Dublin Math. Jl.* **7** (1852), 52–97. Reprinted in *Papers* **1**, 284–363.

Sylvester, J. J., 1878. "Proof of the Hitherto Undemonstrated Fundamental Theorem of Invariants," *Philosophical Magazine* **5** (1878), 178–88. Reprinted in *Papers* **3**, 117–126.

Terracini, A., 1926. "Corrado Segre (1863–1924)," *Math. Ann.* **35** (1926), 209–250.

Tits, J., 1966. "Sur les constants de structure et le théorème d'existence des algèbres de Lie semi-simples," *Publ. math. Inst. Hautes Études Sci.* **31** (1966), 21–58.

Tobies, R. and D. Rowe, 1990. *Korrespondenz Felix Klein–Adolph Mayer. Auswahl aus den Jahren 1871–1907. Herausgegeben, eingeleitet und kommentiert von R. Tobies und D.E. Rowe*, Leipzig, B.G. Teubner, 1990.

Toepell, M. M., 1986. *Über die Entstehung von David Hilberts "Grundlagen der Geometrie,"* Göttingen, Vandenhoeck and Ruprecht, 1986.

Umlauf, K. A., 1891. *Über die Zusammensetzung der endlichen continuierlichen Transformationsgruppen, inbesondere der Gruppen vom Range Null*, Inaugural-Dissertation, Leipzig, 1891.

van Dalen, D., 1990. "The War of the Frogs and the Mice, or the Crisis of the *Mathematische Annalen*," *The Mathematical Intelligencer* **12**$_4$ (1990), 17–31.

Varadarajan, V. S., 1984. *Lie Groups, Lie Algebras, and Their Representations*, 2nd edn., New York, Springer-Verlag, 1984.

Veronese, G., 1881. "Behandlung der projektivischen Verhältnisse der Räume von verschiedenen Dimensionen durch das Prinzip des Projicirens und Schneidens," *Math. Ann.* **19** (1881), 161–234.

Veronese, G., 1884. "La superficie omaloide normale a due dimensioni e del quarto ordine dello spazio a cinque dimensioni e le sue projezioni nel piano e nello spazio ordinario," *Atti R. Accad. dei Lincei. Mem. cl. sci., fis. mat. e nat.* **19** (1884), 344–371.

Vessiot, E., 1892. "Sur l'intégration des équations différentielles linéaires," *Annales scientifiques École Normale Sup. Paris* **(3) 9** (1892), 196–280.

Vessiot, E., 1903. "Sur la théorie des groupes continus," *Annales scientifiques École Normale Sup. Paris* **(3) 20** (1903), 411–451.

Vessiot, E., 1904. "Sur l'intégration des systèmes différentiels qui admettent des groupes continus de transformations," *Acta Mathematica* **28** (1904), 307–50.

Vessiot, E. and W. de Tannenberg, 1889. Review of [Lie, 1888], *Bulletin des sciences mathématiques* **(2) 13** (1889), 113–148.

Vizgin, V. P., 1989. "Einstein, Hilbert, and Weyl: The Genesis of the Geometrical Unified Field Theory Program," *Einstein and the History of General Relativity*, D. Howard and J. Stachel, eds., Boston, Birkhäuser, 1989, pp. 300–314.

Vizgin, V. P., 1994. *Unified Field Theories in the First Third of the 20th Century*, Science Networks, Vol. 13, Basel, Birkhäuser, 1994.

Viète, F., 1591. *In artem analyticen isagoge . . .*, Tours, 1591. References are to the English translation by J. Klein, *Greek Mathematical Thought and the Origin of Algebra*, Cambridge, Mass., 1968.

von Neumann, J., 1933. "Die Einführung analytischer Parameter in topologischen Gruppen," *Annals of Mathematics* **34** (1933), 170–190. Reprinted in *Works* **2**, 366–386.

von Neumann, J., 1934. "Almost Periodic Functions in a Group I," *Transactions Amer. Math. Soc.* **36** (1934), 445–492. Reprinted in *Works* **2**, 454–501.

Weierstrass, K., 1858. "Über ein die homogenen Functionen zweiten Grades betreffendes Theorem," *Monatsberichte K. Akademie der Wissenschaften Berlin, 1858*, (1858). Reprinted in *Werke* **1**, 233–246.

Weierstrass, K., 1868. "Zur Theorie der bilinearen und quadratischen Formen," *Monatsberichte K. Akademie der Wissenschaften Berlin, 1868*, (1868), 311–338. Reprinted with modifications in *Werke* **2**, 19–44.

Weierstrass, K., 1885. "Über die analytische Darstellbarkeit sogenannter willkürlicher Functionen reeller Argumente," *Sitzungsberichte der Akademie der Wiss. zu Berlin 1885* (1885). Reprinted in *Werke* **3**, 1–37.

Weierstrass, K., 1895. *Mathematische Werke* **2**, Berlin, Weierstrass, K., 1902. *Vorlesungen über die Theorie der Abelschen Transcendenten, bearbeitet von G. Hettner und J. Knoblauch, Werke* **4**, Berlin, 1895.

Weil, A., 1940. *L'intégration dans les groupes topologiques et ses applications*, Paris, Hermann, 1940.

Weiler, J. A., 1863. "Integration der partiellen Differentialgleichungen 1.O. mit $n+1$ Veränderlichen," *Ztschr. für Math. u. Physik* **8** (1863), 264–292.

Weinstein, A., 1922. "Fundamentalsatz der Tensorrechnung," *Mathematische Zeitschrift* **16** (1923), 78–91 [received 14 February 1922]. Published separately as Weinstein's doctoral dissertation (*Inaugural Dissertation*) (Berlin, 1922). Reprinted in [Weinstein, 1978:2–15].

Weinstein, A., 1978. *Alexander Weinstein Selecta*, London and San Francisco, Pitman, J. B. Diaz, ed., 1978.

Weinstein, A., 1983. "Sophus Lie and symplectic geometry," *Expositiones Mathematicae* **1** (1983), 95–96.

Weiss, E. A., 1933. "E. Studys mathematische Schriften," *Jahresbericht der Deutschen Mathematiker-Vereinigung* **43** (1933–34), 108–24, 211–25.

Werner, H., 1889. "Bestimmung der grössten Untergruppen derj. projectiven Gruppe, welche eine Gleichung 2. Grades in n Veranderlichen invariant lässt," *Math. Ann.* **35** (1889), 113–160.

Weyl, H., 1913. *Die Idee der Riemannschen Fläche*, Leipzig, Teubner, 1913. Second edn. 1923; 3rd edn. 1955. English translation of 3rd. edn. by Gerald R. Maclane as *The Concept of a Riemann Surface*, Reading, Mass. (Addison-Wesley), 1955.

Weyl, H., 1918a. "Gravitation und Elektrizität," *Sitzungsberichte der Akademie der Wiss. zu Berlin 1918* (1918), 465–480. Reprinted in *Abhandlungen* **2**, 29–42.

Weyl, H., 1918b. "Reine Infinitesimalgeometrie," *Mathematische Zeitschrift* **2** (1918), 384–411. Reprinted in *Abhandlungen* **2**, 1–28. [Received 8 June 1918.]

Weyl, H., 1921a. *Raum, Zeit, Materie. Vierte Auflage*, 1921. Translated into English by Henry L. Brose as *Space–Time–Matter* (1950). Reprinted by Dover Publications, New York (1952). Translated into French by Gustave Juvet and Robert Leroy as *Temps, espace, matière*, Paris, 1922.

Weyl, H., 1921b. "Dic Einzigartigkcit dcr Pythagoreischen Massbestimmung," *Mathematische Zeitschrift* **12** (1922), 114–146. Reprinted in *Abhandlungen* **2**, 263–295 [24 April 1921].

Weyl, H., 1922. "Das Raumproblem. Vortrag, gehalten auf der Jenenser Tagung, September 1921," *Jahresbericht der Deutschen Mathematiker-Vereinigung* **31** (1922), 205–221. Reprinted in *Abhandlungen* **2**, 328–344 [1 February 1922].

Weyl, H., 1923a. *Das mathematische Analyse des Raumproblems. Vorlesungen gehalten in Barcelona und Madrid*, Berlin, 1923. [Preface dated "Zürich, im April 1923."]

Weyl, H., 1923b. "Randbemerkungen zu Hauptproblemen der Mathematik," *Mathematische Zeitschrift* **20** (1924), 131–150. Reprinted in *Abhandlungen* **2**, 433–452 [25 October 1923].

Weyl, H., 1923c. "Zur Charakterisierung der Drehungsgruppe," *Mathematische Zeitschrift* **17** (1923), 293–320. Reprinted in *Abhandlungen* **2**, 345–372 [17 February 1923]. This paper is stated as received on 17 February 1922, but since it refers to Weyl's lectures in Spain, which occurred in 1922, it would seem to be a typographical error. I therefore assume the correct date of receipt was 17 February 1923.

Weyl, H., 1924a. "Über die Symmetrie der Tensoren und die Tragweite der symbolischen Methode in der Invariantentheorie," *Rendiconti del Circolo Matematico di Palermo* **48** (1924), 29–36. Reprinted in *Abhandlungen* **2**, 468–475 [January 1924].

Weyl, H., 1924b. "Das gruppentheoretische Fundament der Tensorrechnung," *Göttinger Nachrichten 1924* (1924), 218–224. Reprinted in *Abhandlungen* **2**, 461–467. [Presented 21 November 1924.]

Weyl, H., 1924c. "Zur Theorie der Darstellung der einfachen kontinuierlichen Gruppen (Aus einem Schreiben an Herrn I. Schur)," *Sitzungsberichte der Akademie der Wiss. zu Berlin 1924* (1924), 338–345. Reprinted in *Abhandlungen* **2**, 453–460. [Letter dated 28 November 1924.]

Weyl, H., 1925. "Theorie der Darstellung kontinuierlicher halbeinfacher Gruppen durch lineare Transformationen. I–III," *Mathematische Zeitschrift* **23** (1925), 271–301 [21 January 1925], **24** (1926), 328–376 [11 February 1925], **24** (1926), 377–395 [23 April 1925]. Reprinted in *Abhandlungen* **2**, 544–645.

Weyl, H., 1926a. "Beweis des Fundamentalsatzes in der Theorie der fastperiodischen Funktionen. Aus einem Schreiben an Hrn. Harold Bohr," *Sitzungsberichte der Akademie der Wiss. zu Berlin 1926* (1826), 211–214. Reprinted in *Abhandlungen* **3**, 34–37. Presented by L. Bieberbach at the 17 June 1926 session.
Weyl, H., 1926b. "Integralgleichungen und fastperiodische Funktionen," *Math. Ann.* **97** (1927), 338–356. Reprinted in *Abhandlungen* **3**, 38–57 [7 July 1926].
Weyl, H., 1927a. "Quantenmechanik und Gruppentheorie," *Zeitschrift für Physik* **46** (1927), 1–46. Reprinted in *Abhandlungen* **3**, 90–135.
Weyl, H., 1927b. "Sur la représentation des groupes continus," *L'Enseignement mathématique* **26** (1927). Reprinted in *Abhandlungen* **3**, 76–89.
Weyl, H., 1928. *Gruppentheorie und Quantenmechanik*, Leipzig, 1928.
Weyl, H., 1931. *The Theory of Groups and Quantum Mechanics*, London, 1931. Translated from the second edition (1930) of *Gruppentheorie und Quantenmechanik* by H. P. Robertson. Reprinted by Dover Publications (New York, 1950).
Weyl, H., 1932. "Zu David Hilberts siebzigstem Geburtstag," *Die Naturwissenschaften* **20** (1932), 57–58. Reprinted in *Abhandlungen* **3**, 346–347.
Weyl, H., 1934–5. *The Structure and Representation of Continuous Groups*, Lithographed notes by R. Brauer of Weyl's lectures at the Institute of Advanced Study, Princeton, 1934–35.
Weyl, H., 1939a. *The Classical Groups. Their Invariants and Representations*, Princeton, 1939. In 1946, as a second edition, the original edition was reprinted along with a new supplementary chapter.
Weyl, H., 1939b. "Invariants," *Duke Mathematical Journal* **5** (1939), 489–502. Reprinted in *Abhandlungen* **3**, 670–683.
Weyl, H., 1944a. "David Hilbert 1862–1943," *Obituary Notices of Fellows of the Royal Society* **4**$_1$3 (1944), 547–553. Reprinted in *Abhandlungen* **4**, 121–129.
Weyl, H., 1944b. "David Hilbert and his mathematical work," *Bulletin of the American Mathematical Society* **50** (1944), 612–654. Reprinted in *Abhandlungen* **4**, 130–172.
Weyl, H., 1946. *The Classical Groups. Their Invariants and Representations*, 2nd. edn., Princeton, 1946.
Weyl, H., 1949. "Relativity theory as a stimulus in mathematical research," *Proceedings, American Philosophical Society* **93** (1949), 535–541. Reprinted in *Abhandlungen* **4**, 394–400.
Weyl, H., 1955. *Die Idee der Riemannschen Fläche*, 3rd. edn., Leipzig, Teubner, 1955. English translation by Gerald R. Maclane as *The Concept of a Riemann Surface*, Reading, Mass. (Addison-Wesley), 1955.
Weyl, H. and F. Peter, 1926. "Die Vollständigkeit der primitiven Darstellungen einer geschlossenen kontinuierlichen Gruppe," *Math. Ann.* **97** (1927), 737–755. Reprinted in *Abhandlungen* **3**, 58–7 [24 July 1926].
Whitehead, A. N., 1898a. "The geodesic geometry of surfaces in non-Euclidean space," *Proceedings London Math. Soc.* **29** (1898), 275–324.
Whitehead, A. N., 1898b. *A Treatise on Universal Algebra*, Cambridge, 1898.
Witt, E., 1941. "Spiegelungsgruppen und Aufzählung halbeinfacher Liescher Ringe," *Abh. math. Sem. Univ. Hamburg* **14** (1941), 289–322.
Young, A., 1901. "On Quantitative Substitutional Analysis," *Proceedings London Math. Soc.* **33** (1901), 97–146. Reprinted in *Papers*, 42–91.
Young, A., 1902. "On Quantitative Substitutional Analysis (Second Paper)," *Proceedings London Math. Soc.* **34** (1902), 361–397. Reprinted in *Papers*, 92–128.
Zorawski, K., 1892. "Über Biegungsinvarianten. Eine Anwendung der Lie'schen Gruppentheorie," *Acta Mathematica* **16** (1892), 1–64.

Index

Sources and Studies in the History of Mathematics and Physical Sciences

Continued from page ii

E. Kheirandish
The Arabic Version of Euclid's *Optics*, Volumes I and II

J. Lützen
Joseph Liouville 1809–1882: Master of Pure and Applied Mathematics

J. Lützen
The Prehistory of the Theory of Distributions

G.H. Moore
Zermelo's Axiom of Choice

O. Neugebauer
A History of Ancient Mathematical Astronomy

O. Neugebauer
Astronomical Cuneiform Texts

F.J. Ragep
Naṣīr al-Dīn al-Ṭūsī's Memoir on Astronomy
(al-Tadhkira fī ʿilm al-hay'a)

B.A. Rosenfeld
A History of Non-Euclidean Geometry

J. Sesiano
Books IV to VII of Diophantus' *Arithmetica*: In the Arabic Translation Attributed to Qusṭā, ibn Lūqā

B. Stephenson
Kepler's Physical Astronomy

N.M. Swerdlow/O. Neugebauer
Mathematical Astronomy in Copernicus's De Revolutionibus

G.J. Toomer (Ed.)
Appolonius *Conics* Books V to VII: The Arabic Translation of the Lost Greek Original in the Version of the Banū Mūsā, Edited, with English Translation and Commentary by G.J. Toomer

G.J. Toomer (Ed.)
Diocles on Burning Mirrors: The Arabic Translation of the Lost Greek Original, Edited, with English Translation and Commentary by G.J. Toomer

C. Truesdell
The Tragicomical History of Thermodynamics, 1822–1854

K. von Meyenn, A. Hermann, V.F. Weisskopf (Eds.)
Wolfgang Pauli: Scientific Correspondence II: 1930–1939

Sources and Studies in the History of Mathematics and Physical Sciences

Continued from the previous page

K. von Meyenn (Ed.)
Wolfgang Pauli: Scientific Correspondence III: 1940–1949

K. von Meyenn (Ed.)
Wolfgang Pauli: Scientific Correspondence IV, Part I: 1950–1952

K. von Meyenn (Ed.)
Wolfgang Pauli: Scientific Correspondence IV, Part II: 1953–1954